INSECT PHYSIOLOGY AND BIOCHEMISTRY

INSECT PHYSIOLOGY AND BIOCHEMISTRY

James L. Nation
Department of Entomology and Nematology
University of Florida
Gainesville, Florida

CRC PRESS
Boca Raton London New York Washington, D.C.

Cover Art: A transmission electron micrograph of muscle from *Tachinaephagus zealandicus* (Hymenoptera) showing well-defined Z lines (arrow), numerous, large, irregular mitochondria between the myofibrils, and intracellular tracheoles. Inset A shows an enlarged view of one of the intracellular tracheoles and several mitochondria. Inset B shows an enlargement of a mitochondrion. Membranes of the cristae in mitochondria can be seen in the enlargements. The I bands (the light areas on each side of the dark Z line) are very narrow in this muscle, indicating that the myosin filaments extend nearly to the Z line. There is just a hint of a lighter, narrow H zone at the middle of the sarcomeres (see Chapter 9). (Photographs courtesy of Jimmy Becnel and Alexandra Shapiro.)

Library of Congress Cataloging-in-Publication Data

Nation, James L.
Insect physiology and biochemistry / James L. Nation.
p. cm.
Includes bibliographical references (p.).
ISBN 0-8493-1181-0 (alk. paper)
1. Insects--Physiology. I. Title.
QL495 .N37 2001
571.1'57—dc21

2001004460

Direct all inquiries to CRC Press LLC, 2000 N.W. Corporate Blvd., Boca Raton, Florida 33431.

International Standard Book Number 0-8493-1181-0
Library of Congress Card Number 2001004460
Printed in the United States of America 1 2 3 4 5 6 7 8 9 0
Printed on acid-free paper

Preface

This book has grown out of my teaching of insect physiology and biochemistry for most of my career at the University of Florida. Most of the graduate students in entomology and nematology at the University of Florida are not planning to be physiology or biochemistry majors, but are majoring in biocontrol, integrated pest management, systematics, toxicology, or another of the numerous disciplines in entomology and nematology. I think similar situations exist at many other academic institutions.

A textbook is needed that presents the principles of insect physiology and biochemistry and that can provide a foundation for both physiology majors and those whose main interest is in another discipline of entomological or biological science. Another goal for the book is to provide background for, and an entry into, the voluminous literature of insect physiology and biochemistry. Approximately 1500 references to the literature are included. I also hope the book will serve working entomologists and scientists who use insects in their work in molecular biology, genetics, endocrinology, virology, microbiology, and various plant sciences.

Although students with a primary interest in insect physiology and biochemistry must read the original literature, it is unlikely that many whose primary interests lie in other areas of entomology, biology, and plant sciences will find time to read and digest great volumes of literature in physiology and biochemistry. Most students, regardless of interests, are likely to take only one course in insect physiology. They need a book that describes the principles and fundamentals of insect physiology, and explains the terminology and jargon needed to read and understand the primary literature that is most relevant to their particular interests. An appendix is included to help those who have not studied entomology to understand the phylogenetic relationship of insect groups, the relationship of insects to other arthropods, and current understanding of the evolution of insects.

Books in the sciences quickly become dated. In insect biology, even some of the principles are subject to change as new data become available about this great and diverse group of animals. The use of the internet and World Wide Web facilities can help one to keep up with new concepts, ideas, and data. In this regard, it is noteworthy that Dr. Tom Miller, University of California, Riverside, and Dr. Elaine Roberts, Colorado State University, have established a Web site (http://lamar.colostate.edu/~insects/ or http://lut.ucr.edu/sites/insects/index.html) for Insect Physiology Online as a resource for teaching and research. Readers may find it useful to consult this site frequently.

Glenn Hall and Marie Nation Becker read the chapter on embryology and made many helpful suggestions. Jon Harrison reviewed an earlier version of the entire manuscript and provided much helpful advice, but especially good advice on excretion, respiration, and acid-base physiology. Tom Miller read a later version of the entire book and provided very helpful advice on all parts, but especially on circulation, nerves, and endocrines. Three anonymous reviewers read earlier versions and made many helpful suggestions. A number of scientists provided photographs and other illustrations. I thank all these individuals for taking the time to be so helpful. I thank the many students on whom earlier versions were tested; it was a joy for me, even if not always for them. They offered, and I have included, many good suggestions. I thank Kathy Milne, who made many excellent computer drawings and figures. I thank my editors at CRC Press, John Sulzycki and Christine Andreasen, for encouraging me and offering help as needed. Finally, I thank my wife, Dorothy, who often helped in checking references and spelling, and put up with countless hours I worked on the manuscript.

James L. Nation
Department of Entomology and Nematology
University of Florida, Gainesville

About the Author

James L. Nation, Ph.D., is Professor of Entomology at the University of Florida, Gainesville, where he teaches a graduate course on insect physiology and insect biochemistry, and an honors course on global environmental issues. He has a B.S. degree in entomology with a minor in chemistry from Mississippi State University and a Ph.D. in entomology from Cornell University, where he specialized in insect physiology with a minor in insect biochemistry. His research activities have included nitrogen excretion, pheromones, cuticular hydrocarbons, and, currently, butterfly rearing and nutrition. He served as associate editor of *The Florida Entomologist* from 1967 to 1969 and as an editor of the *Journal of Chemical Ecology* from 1994 to 2000.

Contents

1 Embryogenesis

CONTENTS

PREVIEW

Insect eggs have a central yolk surrounded by a layer of cytoplasm. A proteinaceous chorion put on the egg while it is in the ovary provides a protective covering for the egg. Sperm released from the spermatheca of the female pass through the micropyle, a narrow channel through the chorion, as the egg passes down the oviduct on its way to be deposited in the environment. Usually, the egg nucleus is diploid until the entry of the sperm stimulates meiotic division leading to the haploid egg nucleus. Union of a sperm nucleus with the egg nucleus produces the zygote and stimulates the zygote to begin divisions. Complete cleavage of the zygotic yolk and cytoplasm occurs in eggs of some species during the first few divisions, but yolk cleavage ceases after a few divisions. In most species, cleavage of yolk and cytoplasm is incomplete from the beginning. Ultimately, zygotic divisions in all insect eggs produce large numbers of nuclei lacking cell membranes but each surrounded with a small field of cytoplasm. These nuclei and associated cytoplasm are called energids. Energids gradually migrate into a single layer near the periphery of the egg, forming the blastoderm. Cell membranes become complete after blastoderm formation. A few cells, the Pole cells, aggregate at the posterior end of the egg and are the first to become committed to a future developmental track; they will become the gametes of the adult. Cells on the ventral side of the blastoderm enlarge and become committed as the germ band — the cells that will become the embryo. Maternal and zygotic genes control subsequent development of the germ band. Maternal genes are present and active in the nurse cells of the mother during oogenesis. The mother's nurse cells pass maternal gene transcripts (mRNAs) into the developing oocyte in the ovary, and these begin to function in the zygote. The maternal gene transcripts are translated into proteins in the zygote, and one of the earliest actions of these proteins is control of anterior-posterior and dorsal-ventral axes orientation of the embryo. Later-acting zygotic genes include gap genes that divide the embryo into large domains, pair-rule genes that divide the domains into parasegments, and finally segment polarity genes that control formation of true segments. Homeotic genes begin to function during parasegment formation to give each segment its characteristic identity. Organogenesis leads to formation of the organ systems of the embryo. Insects with complete development retain within the larval body small embryonic clusters of cells called imaginal discs that divide, differentiate, and grow into adult structures during pupation.

1.1 INTRODUCTION

The three major divisions in the Insecta — the Apterygota, Hemimetabola, and Holometabola — are not directly ancestral to one another and, consequently, embryological developments in the groups, although similar in some respects, often are divergent. The Apterygota (Protura, Collembola, Diplura, and Thysanura) never evolved wings and lack metamorphosis. The immatures look just like small versions of the adults. The Hemimetabola (Orthoptera, Hemiptera, Heteroptera, and others) evolved wings and have gradual metamorphosis. Immatures have some adult features but lack wings. The Holometabola (Coleoptera, Hymenoptera, Lepidoptera, Diptera, and others) evolved wings and have complete metamorphosis. Immature forms are typically wormlike, and thus look very different from the adults. Wingless adults occur in the Hemimetabola and Holometabola, but the wingless condition evolved secondarily from winged forms.

The goals for this chapter are to describe the morphogenetic events and the action of some genes in formation of the embryo. The work by Johannsen and Butt (1941) is still a very valuable source for understanding variations in morphogenesis, as are more recent reviews by Anderson (1972a, 1972b); Jura (1972); Sander, Gutzeit, and Jäckle (1985); and Campos-Ortega and Hartenstein (1985). A review of the morphology of embryogenesis in the silkworm, *Bombyx mori*, has been provided by Miya (1985), and the early stages of embryogenesis are described for several species of fireflies by Kobayashi and Ando (1985).

More details of genetic control of insect embryogenesis are available for *Drosophila melanogaster* than for any other insect, and a timeline for some of the major morphogenetic events may be helpful (Table 1.1); however, one should keep in mind that *Drosophila* is a fast-developing insect, and many other insects do not develop so rapidly. Good reviews of the development and genetics of *Drosophila* are provided by Gehring and Hiromi (1986), Gehring (1987), French (1988), Nüsslein-Volhard (1991), Lawrence (1992), and Bate and Martinez-Arias (1993). Melton (1991) provides a good comparative review of certain aspects of animal development.

TABLE 1.1
Developmental Stages of *Drosophila* Embryogenesis

Stage	Duration	Morphological Events (25°C)	Hours[a]
Stage 1	25 min	Cleavage divisions 1 and 2	0:25
Stage 2	40 min	Divisions 3–8 occur	1:05
Stage 3	15 min	Pole bud formation, division 9 occurs	1:20
Stage 4	50 min	Final 4 divisions, syncytial blastoderm formed, stage 4 ends at beginning of cellularization	2:10
Stage 5	40 min	Cellularization occurs	2:50
Stage 6	10 min	Early stages of gastrulation	3:00
Stage 7	10 min	Gastrulation complete	3:10
Stage 8	30 min	Formation of amnioproctodeal invagination and rapid germ band elongation	3:40
Stage 9	40 min	Transient segmentation of mesodermal layer, stomodeal invagination	4:20
Stage 10	60 min	Stomodeum invaginates, germ band growth continues	5:20
Stage 11	120 min	Growth stage, no major morphogenetic changes, parasegmental furrows develop	7:20
Stage 12	60 min	Germ band shortens	9:20
Stage 13	60 min	Germ band shortening complete, head involution begins	10:20
Stage 14	60 min	Closure of midgut, dorsal closure	11:20
Stage 15	30 min	Gut forms complete tube and encloses yolk sac	13:00
Stage 16	3 h	Intersegmental grooves evident, shortening of ventral nerve cord	16:00
Stage 17	Stage 17 extends to hatching		

Note: Times and stages will probably be different in other insect species.

[a] The time is elapsed time after the egg has been laid, in hours.

Data from Campos-Ortega and Hartenstein (1985).

1.2 MORPHOGENESIS

1.2.1 THE EGG, FERTILIZATION, AND ZYGOTE FORMATION

Insect eggs are **centrolecithal**, which means that the eggs have a central yolk surrounded by a layer of cytoplasm. The yolk is a nutrient source to be used by the developing embryo. A **vitelline membrane** surrounds the peripheral cytoplasm (sometimes called the **periplasm**), and a proteinaceous **chorion** provides a protective cover for the egg contents (Figure 1.1). The cytoplasm is a layer of variable thickness in eggs of different groups. In some there is so little cytoplasm that it is not visually obvious, as, for example, in eggs of the Apterygota. The egg nucleus may lie at the periphery of the egg, on top of the yolk and surrounding cytoplasm, or it may lie in the cytoplasm. When an egg is laid, the nucleus usually is still in the diploid state. The entry of sperm as the egg passes down the oviduct of the female often initiates maturation divisions. The first maturation division divides the chromosomes equally but the nuclear plasm is divided unequally, resulting in a large egg nucleus and a small **polar body** (Figure 1.2). The egg nucleus divides once more to become the haploid female gamete, with production of another small polar body. The first polar body may or may not divide again. If it does divide, two more polar bodies are produced. In any case, polar bodies eventually are reabsorbed into the yolk. The haploid female nucleus usually migrates toward the center of the egg and unites on the way with the sperm nucleus; the developing organism is then called the **zygote**.

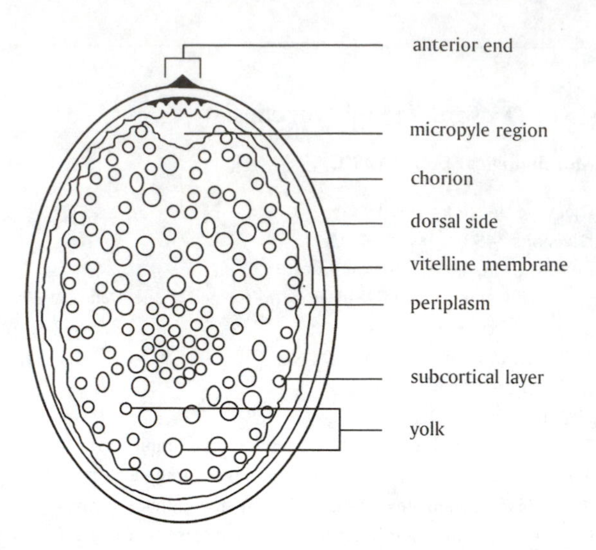

FIGURE 1.1 Diagram of egg structure.

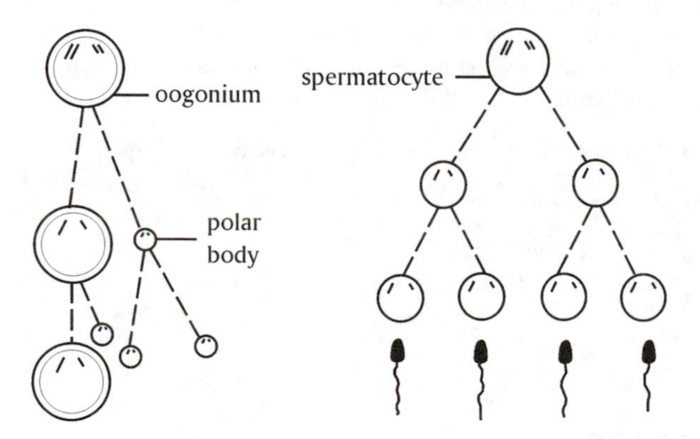

FIGURE 1.2 Maturation divisions of oocyte and sperm. Oogonia in the germarial region of an ovary divide by meiosis to produce an oocyte and a polar body. A second meiotic division, which may not occur until the oocyte is united with the sperm, produces the final oocyte. The polar bodies are reabsorbed as food for the developing oocyte. Spermatocytes in the germarium of the testes give rise to mature spermatozoa by meiotic divisions. Union of a sperm and egg produces the zygote.

1.2.2 Variations in Zygotic Nucleus Cleavage, Formation of Energids, and Blastoderm Formation

Zygotic nucleus divisions are influenced by the quantity of yolk and cytoplasm. Division in eggs with little yolk, such as in the collembolan *Tetrodontophora bielanensis* (Apterygota), partition the yolk in a few early divisions (Figure 1.3), but not after the eight-cell stage. In the great majority of insect groups, the zygotic nuclei divide from the beginning without cleavage of the yolk, and without formation of cell membranes between nuclei. Repeated nuclear divisions produce thousands of nuclei, each surrounded by a small island of cytoplasm. Each nucleus with its island of cytoplasm is called an **energid** (Figure 1.4). Energids migrate toward the periphery, and when a few thousand nuclei have been formed, they distribute themselves in a single layer around the perimeter. Some energids remain in the yolk and become **vitellophages** that digest (liquefy) the yolk and make the nutrients available to the developing embryo. Cytoplasmic strands extend from the **blastomeres**,

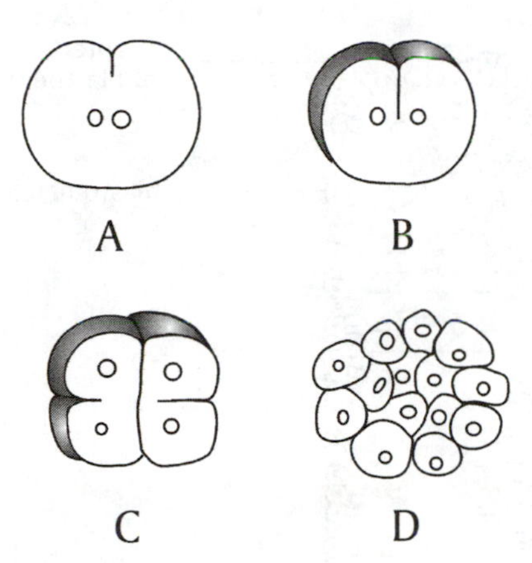

FIGURE 1.3 The first few cleavages of the yolk may be complete, as in some Collembola, but complete cleavage ceases after a few divisions. A: the first division is depicted as beginning after the nucleus has divided by mitosis; B: division into two cells is illustrated; C: cleavage into four cells is under way, and the four may divide into eight cells, after which the yolk usually is not cleaved equally with subsequent nuclear divisions; D: a ball of cells has formed with yolk that has not been partitioned accumulated in the center (not shown) of the mass of cells. (Adapted from Jura, 1972.)

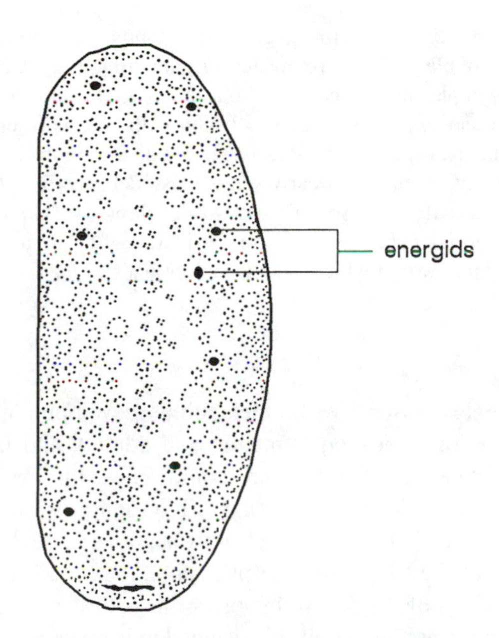

FIGURE 1.4 An example of an egg in which the yolk is not partitioned and cleavage nuclei (energids) are produced and surrounded by a small amount of cytoplasm. The yolk remains in the interior of the egg. (Redrawn and modified from Johannsen and Butt, 1941.)

as the energids are now usually called, into the yolk as a route for nutrient uptake. Eventually, cell membranes become complete, the cytoplasmic strands disappear, and the layer of cells is called the **blastoderm** (Figure 1.5). There are numerous differences in the way the blastoderm forms, and in subsequent morphogenetic movements among the different groups of insects. A brief summary of major differences is given below; the reviews and reference works cited in Section 1.1 should be consulted if more details about specific groups are desired.

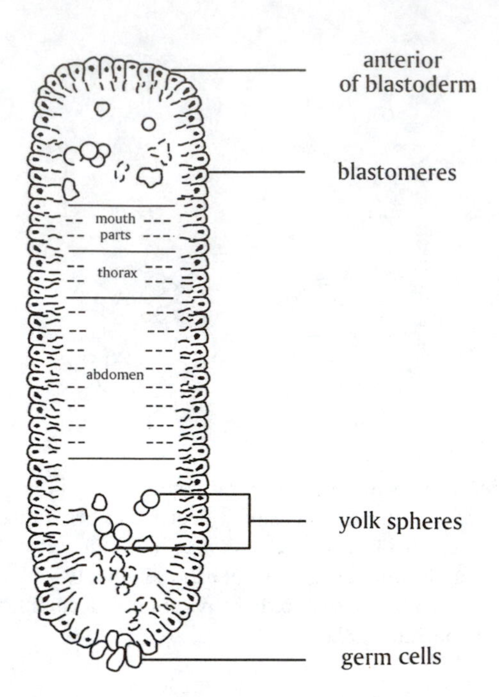

FIGURE 1.5 An illustration of the blastoderm stage in development. Energids gradually migrate to the periphery to form a single layer of blastomeres around the periphery of the egg. Cell membranes are incomplete in an early blastoderm and cytoplasmic strands from blastomeres extend into the yolk, but later the cell membranes become complete and junctions develop between cells to hold them together. The dotted lines across the blastoderm diagrammatically indicate a fate map based on development of *Drosophila melanogaster*, which has a determinate type egg. Even at this early stage, blastoderm cells in *Drosophila* are known (from marking experiments) to be committed to a specific path of development. The broken lines indicate regions developing into mouthparts, thorax, or abdomen. This first evidence of developmental commitment marks the parasegments, developmental units within which genetic action leads to the final segmentation pattern of the larva and adult.

1.2.2.1 Apterygota

Apterygotes are small, wingless insects with ametabolous development (no metamorphosis, and no major changes in morphology between immature and adults), and include the orders Protura (small insects in soil and leaf litter), Collembola (commonly called springtails), Diplura (called bristletails), Archeognatha (also called bristletails), and Thysanura (some bristletails, silverfish, and firebrats) (Romoser and Stoffolano, 1998). Details and variations in development of the Apterygota have been reviewed by Jura (1972). Even in the Apterygota, the processes of division and cleavage are not the same in all members of the group. In some, the yolk is cleaved at each division, but in others, nuclear division occurs without yolk cleavage. Division continues to make many small blastomeres that move toward the periphery of the egg and gradually align themselves in a single layer around the perimeter of the egg to form the blastoderm. At one pole of the blastoderm, blastomeres become increasingly columnar as the **dorsal organ** forms. At the other pole (usually the ventral side), they are smaller but represent the cells that will form the future embryonic rudiment (= germ band) and embryonic membranes (Figure 1.6).

The exact function of the dorsal organ is not clear; it may be secretory. If the cells that form it are damaged or destroyed, the embryo does not develop normally and does not hatch. The dorsal organ cells invaginate into the underlying yolk and take the shape of a mushroom with a stalk, and tendrils grow out and contact the developing germ band (the embryo) after gastrulation has occurred. A dorsal organ does not develop in a recognizable form in Hemi- and Holometabola.

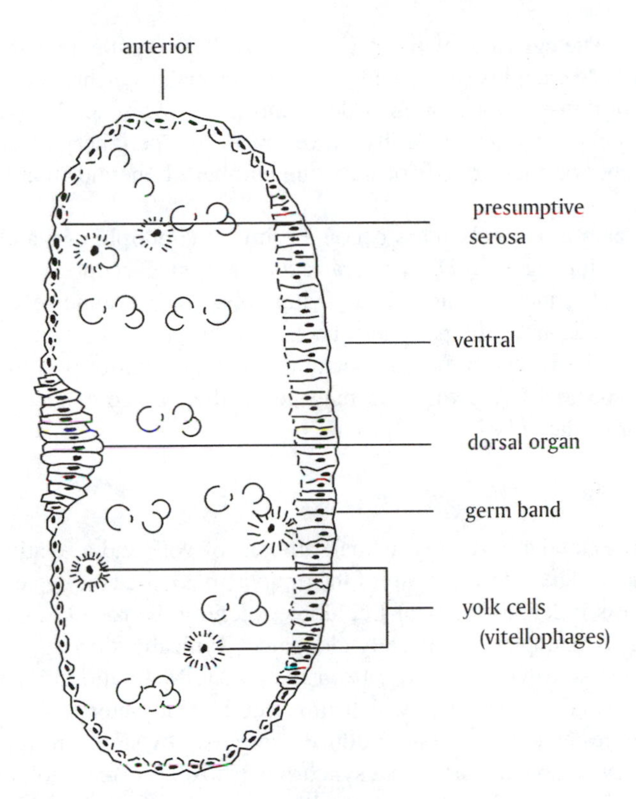

FIGURE 1.6 A late blastoderm stage with germ band formed on the ventral side of egg and dorsal organ on the dorsal side. The germ band will subsequently grow into the embryo. The function of the dorsal organ, not present in the blastoderm of some species, is not known in detail. (Adapted from Johannsen and Butt, 1941.)

In *Japyx solifugus* (Diplura), cleavage is superficial from the start. The blastomeres migrate toward the periphery of the egg at about the 64-cell stage and, after additional divisions, the blastoderm is formed. A dorsal organ is present and behaves much like that in Collembola. Cleavages of the zygote of Thysanura (silverfish) are superficial, and the yolk is not cleaved. Synchrony is lost after a few divisions. Some cleavage nuclei migrate to the periphery while some remain in the yolk, functioning as yolk nuclei, later to become vitellophages. The germ anlage or germ band forms at the posterior pole of the blastoderm. A few blastomeres at the anterior pole of the blastoderm may form a dorsal organ, but some researchers have questioned whether a dorsal organ is present. After gastrulation, a part of the serosal membrane sinks into the yolk to form a secondary dorsal organ, but it soon degenerates and its function is unknown. The yolk cells in Thysanura are true vitellophages that digest the yolk; some yolk cells later disintegrate and contribute to the formation of the midgut epithelium. The blastoderm stage exists only briefly in Apterygota and is followed by gastrulation.

1.2.2.2 Hemimetabola

Embryogenesis of the Hemimetabola has been reviewed by Anderson (1972a). In general, eggs of hemimetabolous insects develop slowly, taking weeks or months to hatch. When the egg is released from the ovary, the oocyte is in metaphase of the first maturation division. The nucleus with a small amount of cytoplasm lies at the periphery of the egg, where it stays as maturation divisions produce three polar nuclei and one haploid female pronucleus. The female pronucleus migrates to the interior while the polar nuclei stay at the periphery. The union with the male pronucleus occurs near the middle of the egg. The three polar nuclei and any unsuccessful male pronuclei are reabsorbed during early cleavage. The eggs are relatively rich in yolk. The zygotic nucleus undergoes division without yolk cleavage and energids are formed. Division is usually synchronous until the blastoderm is formed. The rate of division varies a great deal among the Hemimetabola, but none divides as fast as the Holometabola. Energids gradually move toward the periphery of the egg and form the blastoderm. The number of nuclear divisions and the number of energids that form the blastoderm vary with species.

Energids that remain in the yolk mass become primary **vitellophages**, and continue to divide and produce more vitellophages. In Dictyoptera (cockroaches), Plecoptera (stoneflies), and Gryllotalpidae (mole crickets), there are no primary vitellophages, but some secondary ones develop from energids that migrate from the periphery back into the yolk. The final position of the germ band is ventral and usually posterior, but the position along the anterior/posterior axis is somewhat variable in different species. The germ band may lie on the surface of the yolk mass, or it may grow into the interior of the yolk in some groups.

1.2.2.3 Holometabola

The eggs of the Holometabola have only a small amount of yolk and a relatively large peripheral periplasm (= cytoplasm). The outermost part of the periplasm is called the egg cortex. Eggs typically are small, 1 mm or less in length. Eggs of Lepidoptera tend to be round to ovoid in shape, while those of Diptera and Hymenoptera are usually elongated. Typically, the egg is in the metaphase of the first maturation division when released from the ovary. Maturation division results in three polar nuclei and the female pronucleus, which migrates into the interior of the egg. The yolk is not cleaved and zygotic nuclear divisions produce energids. Divisions are typically synchronous through eight, ten, or even more divisions, but synchrony is lost in various Holometabola at different times after about the eighth division. The rate of division is also variable, with higher Diptera (the Cyclorrhapha) having the fastest division rate. The number of cells in the blastoderm varies in the Holometabola from about 500 to 8000. Some of the blastoderm cells may migrate back into the yolk as secondary vitellophages. Although the nuclei of the vitellophages cease dividing and remain in the central yolk region, their DNA replicates and they become polyploid. The pole buds and syncytial blastoderm nuclei continue to divide independently of each other. The zygote usually lies centrally but may be displaced toward either end. Cell division and growth of the embryo are rapid, and eggs usually hatch in a few days in most cases.

In *D. melanogaster,* the morphogenetic events, as well as their genetic control, have been extensively studied, and there typically are 13 synchronous division cycles before cell boundaries are established between nuclei. After the first seven synchronous divisions, there are 128 nuclei arranged in an ellipsoid shape around the central yolk (Zalokar and Erk, 1976). Most of the nuclei begin to migrate to the periphery of the embryo, but about 26 nuclei stay near the yolk in the center of the egg after the seventh division and become vitellophages. The vitellophages and all the other nuclei undergo the eighth nuclear division together. At this time, the first cells become determined (committed to a particular developmental fate) and a few nuclei are incorporated into the posterior pole plasma to become the polar buds or future germ cells. These will give rise to the gametes (reproductive cells) of the adult insect. The remaining energids are destined to become somatic cells of the embryo.

The vitellophages, polar buds, and somatic nuclei divide in synchrony two more times, making a total of ten divisions for the somatic nuclei. In the eighth, ninth, and tenth divisions, the somatic nuclei progressively move toward the surface of the egg, forming a single layer of nuclei around the perimeter of the egg, the syncytial blastoderm (syncytial denotes the lack of cell boundaries) (Foe and Alberts, 1983).

After 13 divisions of the somatic nuclei during the first 3 hours of embryo life in *D. melanogaster,* there are about 5000 syncytial blastoderm nuclei layered around the periphery of the egg (Chan and Gehring, 1971). Cell membranes begin to form, and **desmosomes** form between cells to hold them together (Mahowald, 1963). The cytoplasmic strands that reach into the yolk gradually disappear as cell membranes are completed. The time at which cells become committed to the formation of specific structures is variable in different insects, but in *D. melanogaster* formerly totipotent energids become determined in the blastoderm stage, after which they can only develop into certain body segments (Simcox and Sang, 1983; Gehring, 1987). The ultimate development of blastoderm cells in *Drosophila* has been ascertained by marking cells to note their future fate, and the representation of the commitment of blastoderm cells as done diagrammatically in Figure 1.5 is called a fate map (Campos-Ortega and Hartenstein, 1985).

1.2.3 Formation of the Germ Band

Initially, the cells of the blastula are uniform in size and shape; but along the ventral side of the blastula, the cells rapidly thicken and enlarge into the **germ band**, the cells destined to give rise to the embryo. In *D. melanogaster* and other Diptera with large amounts of cytoplasm, nearly the entire cell number of the blastoderm becomes the germ band, leaving only a few cells to build the extra-embryonic membranes. In other insects, a variable number of cells along the ventral side of the blastoderm enlarge and become more columnar in shape, while lateral and dorsal to the ventral region the cells become more flattened and squamous and are destined to form the **extra-embryonic membranes** called the **amnion** and **serosa**.

The initial size of the germ band varies in different groups of insects, and three major types occur, characterized as **short germ band**, **long germ band**, and **intermediate germ band** (reviewed by Sander et al., 1985). Short germ band eggs tend to be indeterminate, large eggs from panoistic ovaries, contain a large yolk with relatively little cytoplasm, have a relatively small portion of the blastoderm that becomes the germ band, and develop slowly over days, weeks, or months. Long germ band eggs usually are smaller eggs that come from meroistic ovaries. They tend to have a relatively large amount of cytoplasm and a small amount of yolk. The germ band initially covers a large portion of the blastoderm, and development to hatching is rapid, often hours to days. Long germ band eggs tend to be determinate. **Indeterminate** and **determinate** refer to how soon the blastoderm cells become committed to a specific developmental fate. In determinate eggs, the blastoderm cells become committed very early to a specific developmental pathway. Regardless of the size of the germ band initially, elongation and growth occur as development continues. Short germ eggs tend to be characteristic of the Orthoptera and Odonata (insects with panoistic ovaries), while long germ eggs tend to be produced by Lepidoptera, Coleoptera, Diptera, and Hymenoptera. However, there are some groups that do not fit easily into one category, so the correlation between taxon and egg type is not strong (Sander et al., 1985). Evolutionary selection may have led to long germ eggs as an adaptation to use rapidly decaying vegetable, fruit, or dead animal hosts, as well as reduced exposure of a relatively immobile stage (Sander et al., 1985). Some insects have an intermediate germ band egg in which segmentation in the gnathocephalon and thorax occurs relatively rapidly, but the abdominal portion of the germ band grows slowly and segmentation takes longer to occur. Eggs of the cricket *Acheta domesticus* are of intermediate germ band type (Sander et al., 1985).

1.2.4 Gastrulation

During **gastrulation**, part of the germ band sinks into the ball of blastoderm cells (Figure 1.7A), and germ layers are formed that will give rise to different organs and tissues. Immediately after gastrulation occurs, the structure is sometimes referred to as a **gastrula**. Gastrulation is highly variable, and so different in some insects from the process in other animals that some embryologists have questioned whether the events occurring are really gastrulation in the classical sense. Deep invagination of the germ band, so characteristic of many other types of organisms, does not occur in insects (Johannsen and Butt, 1941). In other organisms, gastrulation results in an outer layer

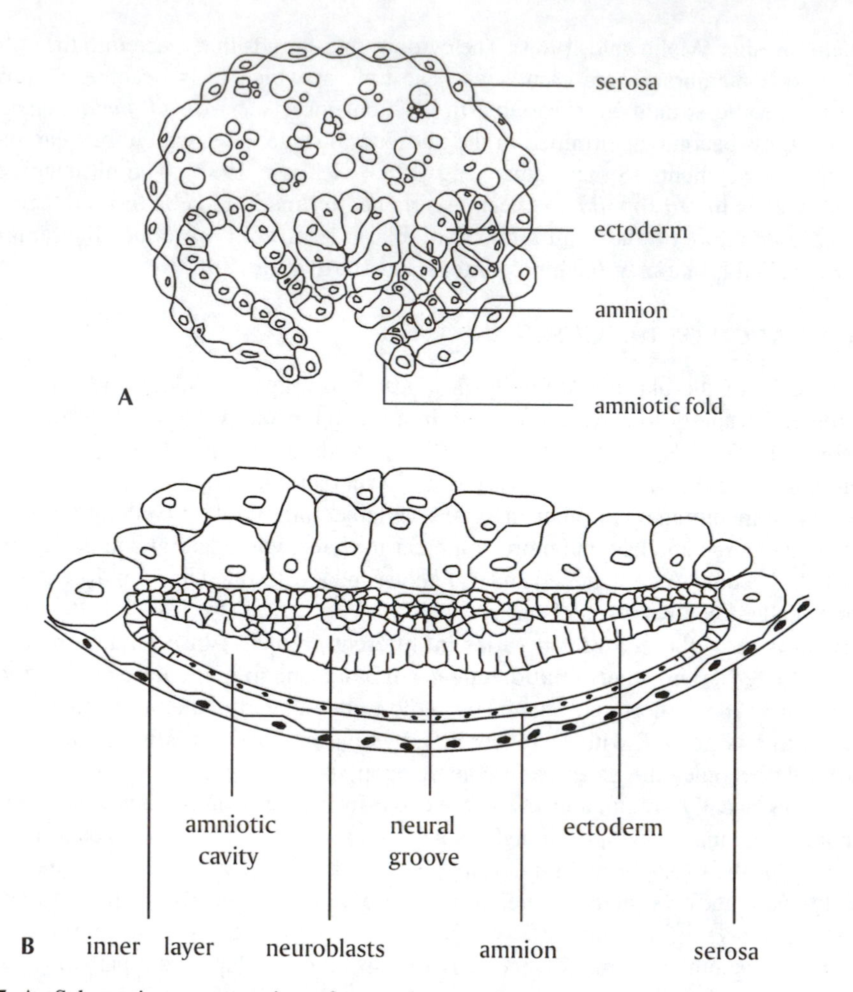

FIGURE 1.7 A: Schematic representation of an early stage in gastrulation during which the germ band invaginates. B: A later stage in which germ layers have formed. (Redrawn and modified from Johannsen and Butt, 1941.)

(the ectoderm), an inner layer (the endoderm), and a middle layer (the mesoderm). Typically, in insects at the end of gastrulation, there is an outer **ectodermal** layer of cells and an inner layer of cells termed the **mesentodermal** layer (Figure 1.7B). Most of the controversy concerning gastrulation has focused on the formation, or some would argue the lack of formation, of a classical **endoderm** (Johannsen and Butt, 1941). The only structure formed from the endoderm, providing it is accepted as a germ layer, is the midgut. The outer layer of **ectoderm** gives rise to the nervous system, the tracheal system, the fore- and hindgut, and the integument. Formation of the mesentoderm layer varies with different insect groups. In Coleoptera, invaginated cells along the ventral midline fold into a tube that subsequently unfolds to becomes an irregular inner layer of cells. In honeybees, *Apis mellifera*, a ventral plate of cells sinks inward and is overgrown by the remaining lateral plates of the germ band, and a somewhat similar formation of an inner layer of cells occurs in Orthoptera. The **mesentoderm** gives rise to muscles, circulatory system, fat body, hemocytes, and the midgut (but see the note regarding the controversy about separate endoderm and origin of midgut above).

During gastrulation, the ectoderm and mesentoderm become overgrown by some of the remaining surface cells, eventually enclosing the developing embryo, an **amnionic cavity**, and remaining yolk. The layer of squamous cells lining the ventral portion of the amnionic cavity is called the **amnion**, and a thin layer of cells on the outside of the **gastrula** becomes the **serosa**, the two extra-embryonic membranes of the embryo (Figure 1.7B). In *Drosophila,* the two membranes fuse into the **amnioserosa**.

1.2.5 Germ Band Elongation

The germ band grows and elongates in all insects regardless of its initial size. The anterior part of the germ band, the **protocephalon**, includes the antennal segments, intercalary segments (which give rise to the tritocerebrum and parts of the head capsule), and three gnathal segments (primordia of the mandible, maxilla, and labium). The protocephalon is bilaterally widened at the anterior end (like a double-headed hammer), with a "finger-like" tail. In *D. melanogaster,* the posterior tail of the germ band as well as the procephalon are fully formed at the blastoderm stage, and segmentation can proceed at once, usually occurring within hours of formation of the blastoderm and while gastrulation is in progress.

The tail portion of the germ band grows at variable rates in different groups. Dorso-ventral furrows rapidly appear behind the protocephalon in *Drosophila* embryos as segmentation is initiated. These first segments are called **parasegments**, and they are slightly out of register with the final segmentation pattern that will develop. Nevertheless, they represent the first evidence of **metamerization** in the embryo.

Six segments fuse into the head. The thorax consists of three segments. The number of abdominal segments is variable in different insects, but 11 (or 12 if the terminal telson is counted) is the primitive number. In rapidly developing Holometabola, body appendages soon appear as bilateral evaginations or small cellular buds from the ectoderm. Buds from the protocephalon form the antennae and mouthparts, and buds from the thoracic segments give rise to the legs and wings. Bilateral outgrowths of buds appear on the abdominal segments, but are later reabsorbed in segments 1 to 7 and 10. In some insects, abdominal buds on segments 8 and 9 continue to develop into the external genitalia, and those on segment 11 form cerci. The exact form that the abdominal limb buds might take, if they were not reabsorbed, is unknown, but some have interpreted them as gill flaps in an ancient insect ancestor (Wigglesworth, 1972).

1.2.6 Blastokinesis

Blastokinesis refers to movements and rotations of the embryo, processes that are variable in different insect groups. Sometimes, blastokinesis is divided into two phases: **anatrepsis** and **katatrepsis**. Anatrepsis movements carry the embryo away from the posterior pole of the egg, and katatrepsis moves the embryo from the ventral region to the dorsal region of the egg. Various degrees of these movements occur in different insects. Only Lepidoptera undergo marked blastokinetic movements among the Holometabola. Coleoptera and Diptera display little or no blastokinetic movements. One consequence of extensive movements is that the embryo reverses its position relative to the yolk, which initially lies outside the embryo, but after blastokinesis, the yolk is enclosed within the embryo. The extra-embryonic membranes, the amnion and serosa, grow over the embryo, but later sink into the yolk and usually are digested. Final dorsal closure occurs as the ectoderm grows over the surface of the embryo.

In those insects in which there is little or no movement of the embryo, the extra-embryonic membranes nevertheless rearrange so that the developing embryo encloses the yolk. In Coleoptera, the amnion breaks ventrally and begins to grow dorsally from the broken edges so that the amnion and the embryo lying on its inner face surround the yolk. In Diptera, the amnionic cavity is extended as both amnion and ectoderm grow around the yolk and embryo, with the amnion forming the outermost cover of the embryo. Similar growth of amnion and ectoderm around the embryo occurs in Lepidoptera and sawflies (Hymenoptera, Tenethredinidae), but in this case a thin layer of yolk is trapped between the serosa and amnion. Upon hatching, Lepidoptera larvae and larvae of sawflies usually feed on the egg shell and the entrapped yolk layer as the first food.

1.3 GENETIC CONTROL OF EMBRYOGENESIS

What causes cells to differentiate and become committed to one pathway as opposed to another? A simple answer to this question is not possible, but two principal mechanisms have been identified for determining cell fate during development. One mechanism is **cell-to-cell interaction**, in which one cell influences or induces its neighbor to follow a certain developmental pathway. A second mechanism is that of **regional localization** of molecules, which provides information to nuclei or cells in contact with the molecule(s) as to a pathway for development. A molecule whose concentration influences a local pattern of determination is called a **morphogen** (Slack, 1987). These two systems influencing development are not mutually exclusive and both seem to work in many systems. Relatively few genes in organisms as diverse as invertebrates and vertebrates may specify cell fates during development (Melton, 1991). Apparent differences in development appear to be much more similar at the molecular and genetic levels than the phenotypic and organismal levels might suggest.

An incredible amount of information is now available regarding genetic control of development in *Drosophila melanogaster* (Bate and Martinez-Arias, 1993). The examples and explanations of genetic control to follow are based on *Drosophila* unless otherwise stated. Although there are important implications for all organisms in the large body of genetic information developed from studies of *Drosophila*, it should not be considered representative of all insects. Many genes expressed in the embryo as mutated genes result in death of the embryo prior to hatching, but it remains possible in many cases to determine how the embryo develops differently from normal ones up to the point of death, and thus identify genes that have specific functions.

More than 70 genes are involved in the embryonic development of *D. melanogaster*, and most of them have been characterized. They are usually classified broadly into (1) **maternal genes** and (2) **zygotic genes.** Maternal genes are present in nurse cells of the maternal ovary, and gene products (tRNA and mRNA) are transferred to the oocyte by nurse cells while it is developing in the ovary. These gene products begin to function in the oocyte during growth in the ovary, and some continue to function in the egg for several hours after the egg is laid. Zygotic genes begin to function in the zygote, and some maternal and some zygotic genes function simultaneously and interactively. The early functioning zygotic genes are divided into **segmentation genes** and **homeotic genes**. Segmentation genes specify the number and polarity of segments (Nüsslein-Volhard and Wieschaus, 1980). Homeotic genes regulate development after segmentation by determining identity and sequence of body segments (Gehring and Hiromi, 1986). In addition to gene transcripts, the mother primes the egg for development with mitochondria, ribosomes, and food (yolk). Entry of a sperm into the egg sets some developmental events in motion, and cascades of genetic actions are initiated.

1.3.1 Development of a Model for Patterning

The earliest genetically controlled events are to order the axes of the egg and thereby determine the axes of the embryo that will develop. Some pre-*Drosophila* work on the formation of the anterior-posterior axis (Sander, 1960, 1976) gave rise to a general model for anterior-posterior development. Sander initially demonstrated a posterior activity center in *Eucelis* (a leafhopper) eggs by ligating the egg into two parts. Neither half could form a perfect embryo after the ligature, but the anterior half of the egg could form a complete, but small embryo, when cytoplasm from the posterior pole was transferred into the anterior half of the egg. Sander inferred from these experiments that both a posterior and an anterior activity center existed in the egg, and that diffusion gradients spread to other parts of the egg from these centers. Similar ligation experiments and genetic analysis with eggs of *Drosophila* verified Sander's analysis. This work led to a model in which it was proposed that anterior (A) and posterior (P) factors (**morphogens**) diffuse through the egg from the initial site of deposition (Nüsslein-Volhard et al., 1987). The A gradient is highest at the anterior end and becomes progressively less concentrated as it diffuses toward the posterior end. The direction of the P gradient is just the opposite: highest near the posterior end and lowest at the anterior end of the embryo. The concentration ratio A:P varies continuously throughout the egg and influences some segmentation genes to respond by initiating a **cascade of gene action**. A cascade of action occurs when one or more genes are activated, and they in turn activate other genes, and those activate still others, etc. The location of a particular cell within the gradient ratio determines which genes respond and what the cell will become, whether part of the head, thorax, or abdomen. The A:P ratio may be a general model for insects, and the major aspects of it have been verified in the development of *Drosophila*.

In *D. melanogaster,* both maternal effect and zygotic genes are involved in establishing the anterior-posterior and dorsal-ventral axes (Nüsslein-Volhard, 1979; Anderson, 1987; and Lehmann, 1988). Patterning along the anterior-to-posterior axis is controlled by three systems of genes: (1) anterior, (2) posterior, and (3) terminal. Each system requires the action of some maternal genes and some zygotic genes. The anterior system is responsible for development in the segmented region of the head and thorax, and the posterior system determines segmentation in the abdomen. The terminal system controls development in the nonsegmented acron at the anterior end (along with the anterior system which has some control over the acron) and the telson at the posterior end.

1.3.1.1 The *bicoid* Gene and Anterior Determination in *Drosophila*

The anterior pole of the egg is determined by a **morphogen** called **Bicoid protein**. **Transcripts** (mRNA) of the maternal gene *bicoid* in nurse cells are passed to the oocyte through cytoplasmic strands called ring canals (Figure 1.8) connecting nurse cells and oocytes (Frigerio et al., 1986; Bopp et al., 1986). Several maternal genes, including *exuperantia*, *staufen,* and *swallow,* help localize *bicoid* transcript, possibly through binding action between their gene products and the *bicoid* transcript. In embryos with mutations of one or more of these genes, *bicoid* transcript is not well localized and development of the anterior region of the embryo is abnormal. The *bicoid*-transcribed mRNA is not translated into protein (the actual morphogen) until shortly after the egg is laid, and then a diffusion gradient of Bicoid protein (Figure 1.9) is established over the anterior half of the egg (Driever and Nüsslein-Volhard, 1988a). Factors promoting the gradient include (1) synthesis at the site of localization of *bicoid* transcripts; (2) diffusion of the Bicoid protein away from the site (cellularization of the blastoderm is not complete yet, so diffusion is not inhibited by cell membranes); and (3) constant rate of proteolytic degradation of Bicoid protein throughout the embryo. The *bicoid* transcripts soon disappear from the egg, but Bicoid protein persists for about 1 hour after the mRNA disappears. Overall Bicoid protein is present for about 4 hours in the early life of the embryo (until gastrulation is under way), and then, its job completed, it disappears.

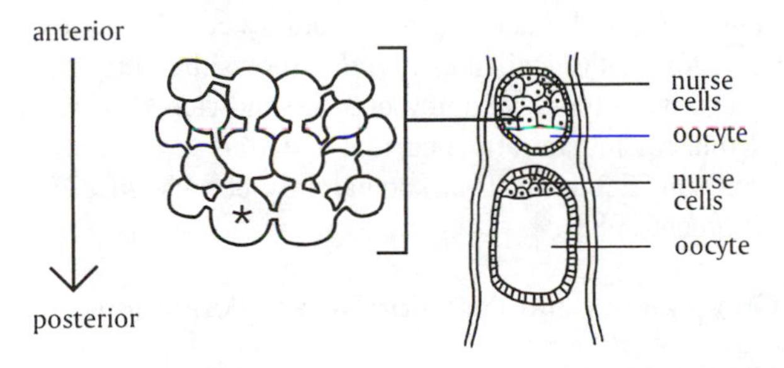

FIGURE 1.8 Left: Nurse cells in the follicle of an insect with meroistic ovarioles (such as *Drosophila*) with interconnecting ring canals. The 16 cells are diploid and represent four mitotic divisions of an oogonium. The cell marked by the asterisk is, for this example, assumed to have become the oocyte in the ovarian follicle diagrammed on the right. The remaining 15 cells serve as nurse cells, supplying nutrients and gene transcripts to the developing oocyte. The oocyte remains diploid, like the nurse cells, while it grows in the ovarian follicle.

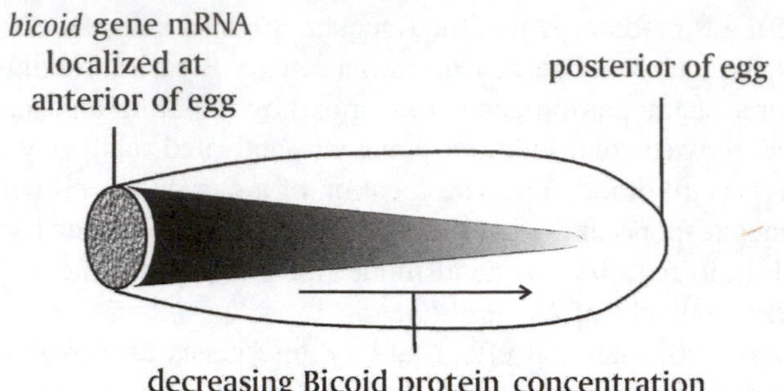

FIGURE 1.9 An illustration of the gradient established by diffusion of Bicoid protein translated from *bicoid* transcript localized at the anterior end of the egg. The concentration of Bicoid protein, necessary for head development, is greatest at the anterior end, and the concentration decreases toward the posterior.

Bicoid protein has a **homeodomain** and thus it is a **transcription factor** that binds to DNA, initiating gene cascades. Local concentrations of Bicoid are believed to be important to its ability to bind to high-affinity and low-affinity binding sites at the promoter region of target genes. At least three groups of zygotic genes are target genes for Bicoid cascade action, including (1) the gap genes *hunchback* and *Krüppel*; (2) several genes involved in formation of the head, including *tailless*, *giant*, and *Deformed* (Driever and Nüsslein-Volhard, 1988b; Melton, 1991); and (3) the pair rule gene *even stripe* 2. Bicoid activates some genes, such as *hunchback*, and sets sharp borders (sometimes only across a few nuclei) where the gene will be expressed. In other cases, it only activates a particular gene, and the borders for that gene action are set by other transcription factors. High concentrations of Bicoid protein are needed at the anterior end to specify head development, where it acts as a positive transcription factor that binds to and activates zygotic *hunchback*. Bicoid protein also activates a cascade of gap genes (segmentation genes, see below) that help specify the parasegmentation pattern (Driever and Nüsslein-Volhard, 1988b; French, 1988). Both positive and negative interactions between genes and their products are common. For example, *Krüppel*, which promotes posterior thoracic and abdomen development, is inhibited from expression in the anterior embryo by high concentrations of *bicoid*, but it can be expressed and influence development in the posterior part of the thorax and posterior embryo because the gradient of Bicoid protein is much lower (Gaul and Jäckle, 1987).

Mothers with a mutant *bicoid* gene may produce an embryo in which the head and thorax are missing, and/or sometimes with anterior parts partially replaced by a second abdomen (because Bicoid protein is not present to inhibit *Krüppel*). An egg containing a mutant *bicoid* gene can be (partially) rescued experimentally by injecting cytoplasm from the anterior pole of a wild-type (normal) egg, resulting in normal thoracic segments and a nearly complete head (Frohnhöfer and Nüsslein-Volhard, 1986). A mutation called *dicephalic* allows the nurse cells in an ovarian follicle to be split into two groups, with the developing oocyte sandwiched between two groups of nurse cells. Some embryos from this mutant are abnormal and start to form an anterior parasegment at each end of the embryo (Lohs-Schardin, 1982), apparently because *bicoid* mRNA is transferred to both ends of the egg (French, 1988).

1.3.1.2 Posterior Group Genes and Posterior Pattern Formation

Posterior development is influenced by the maternal genes *oskar*, *staufen*, *tudor*, *valois*, *vasa*, *nanos*, *pumilio*, *hunchback*, and the zygotic gene *knirps*. The maternal gene ***nanos*** (***nos***) is the principal controlling gene in the posterior group (French, 1988; Nüsslein-Volhard, 1991). The mRNA that *nanos* specifies is localized at the posterior end of the oocyte (Lehmann and Sander, 1988). Research indicates that *nos*-dependent activity spreads over the posterior half of the developing embryo, and that the maternal gene *pumilio* is necessary for the spread. One of the critical functions of *nos* is to clear from the posterior of the embryo the transcript of maternal *hunchback* (Lawrence, 1992). The gene *hunchback* is functional both as a maternal gene (in the nurse cells) and as a zygotic gene in the embryo. Although zygotic *hunchback* transcripts are made only at the anterior end (controlled by Bicoid concentration), maternal *hunchback* transcripts are uniformly distributed in the egg. The maternal *hunchback* transcripts repress the zygotic gene *knirps*, whose activity is essential to the formation of posterior structures. Additional genes that work in conjunction with *nanos* and influence posterior development function in localization, packaging, and deployment of *nanos* mRNA posteriorly (Lawrence, 1992). Mutants in which maternal *hunchback* transcripts are not present also do not require the activity of *nanos*; thus, *nanos* is described as a permissive gene for posterior development, in that it allows posterior development by destroying transcripts of maternal *hunchback*, but *nanos* function itself is not determinative and not a transcription factor (Lawrence, 1992). In the absence of maternal *hunchback* transcripts, other gene(s) promote posterior development. The development of germ cells (pole cells) is also influenced by *nanos*.

1.3.1.3 Genes Required in the Acron and Telson

The terminal group maternal genes are needed to specify normal development of the nonsegmented ends of the embryo. In *D. melanogaster,* the acron at the anterior includes the labrum and dorsal bridge, while the telson at the posterior end includes the anal pads, anal tuft, Fitzkorper, and anal spiracle (Lehmann, 1988). Embryos from mothers that have a mutation in the maternal gene *torso* fail to develop the acron and telson, and sometimes part of the abdominal structures posterior to the 7th abdominal segment, including hindgut and posterior midgut (Schüpbach and Wieschaus, 1986; Denglemann et al., 1986; French, 1988). The gene *torso* appears to be activated by a ligand (a molecule that binds to it), and the ligand is probably only present at the extreme ends of the embryo. There is some evidence to suggest that the ligand may be the product of the maternal gene *torsolike*. When activated, one of the roles for *torso* is to coordinate the activities of two (zygotic) gap genes, *tailless*, involved in development of the telson, and *huckebein*, involved in development of the midgut (Weigel et al., 1990).

1.3.1.4 The Dorsal-Ventral Axis

The dorsal-ventral axis of a *D. melanogaster* egg is established by the concerted action of at least 18 genes (Chasan and Anderson, 1993). One of the principal genes is the maternal gene *dorsal*. The *dorsal* transcript (mRNA) is uniformly distributed in the ooplasm and cytoplasm of the egg, and is detectable for approximately 2 hours after the egg is laid. Translation of the message results in **Dorsal protein**. The protein becomes localized in blastomere nuclei on the ventral side of the egg (Steward, 1987), where it acts as a morphogen and influences the zygotic genes *snail* and *twist*, as well as other genes. Mutations in which Dorsal protein is abnormally localized in blastomere nuclei result in abnormal dorsalization and eventual death of the embryo. The **gene cascade** involved in correct localization of Dorsal protein in blastomeres on the ventral side of the egg is complex. A number of maternal genes are involved. The maternal gene *Toll* encodes a receptor protein that probably acts to bind Dorsal. The maternal genes *easter*, *snake*, and possibly others may be expressed in follicle cells touching the part of the oocyte that is destined to become the ventral side, and encode gene products that promote binding Toll protein to a particular region of the blastoderm.

1.4 SEGMENTATION GENES

Segmentation genes are zygotic genes, and a cascade action by segmentation genes divides the embryo into broad domains and then into smaller regions, resulting finally in parasegments. **Parasegments** are the first evidence of **metamerization** (segmentation) and are the site and boundaries of future gene action (Martinez-Arias and Lawrence, 1985; Lawrence, 1988). A parasegment includes the posterior one-fourth of a segment and the anterior three-fourths of the segment behind, with "segment" used here to mean the final segmentation pattern of the adult insect. Thus, parasegments are a little out-of-register with the final segmentation pattern that will ultimately exist. The parasegments are important because genetic analysis has shown that specific gene control for a body region occurs within a parasegment. Thus, parasegments narrow down the region in which specific genes are responsible for coordinating events.

Drosophila segmentation genes have been divided into three classes (Nüsslein-Volhard and Wieschaus, 1980; Howard, 1988). These are (1) **gap genes** that divide the embryo into a series of major domains, (2) **pair-rule genes** that further divide the major domains into parasegments, and (3) **segment polarity genes** that specify the pattern within each segment (Gehring, 1987; Ingham and Gergen, 1988). Gap genes, which are expressed prior to the pair-rule and segment polarity genes (Jäckle et al., 1986), are required in broad aperiodic regions of the blastoderm to divide the embryo into broad domains; mutations in the gap genes result in embryos that lack segments where particular gap genes should function. Pair-rule genes function with two-segment periodicity; mutants fail to develop structures in alternate segments. Segment polarity genes are required in every segment (Howard, 1988).

Examples of gap genes are zygotic *hunchback*, *Krüppel*, *Knirps,* and *giant*. Pair-rule genes include *hairy*, *runt*, *even skipped*, *fushi tarazu*, *paired*, *odd paired*, and *sloppy-paired*. Segment polarity genes include *engrailed*, *wingless*, *patched*, *hedgehog*, *dishevelled*, and *fused*. In general, the segmentation genes are expressed very early in the development of the embryo. Segmentation genes have been characterized by the defects resulting from mutations in the genes, which typically cause the deletion of segments or some alteration in the polarity of a segment. Mutants of *fushi tarazu* (*ftz*⁻), for example, have only about half the normal number of segments (Wakimoto and Kaufman, 1981). Mutants of *even skipped* (i.e., *eve*⁻) lack all segmentation in the middle region of the embryo, and also show altered patterns of expression of two other segmentation genes (pair-rule *fushi tarazu* and *engrailed*), which contributes to the evidence that the expression of pair-rule segmentation genes probably involves cross-regulatory interactions.

Most of the genes involved in development do not act independently of each other, and many appear to be influenced by the action of other genes (see Howard, 1988, for brief review). Gap genes are regulated by maternal genes and by other gap genes. Pair-rule gene expression is variable in embryos with different types of maternal and gap gene mutations. Expression of some pair-rule genes is dependent on the expression of other pair-rule genes (Harding et al., 1986; Ingham and Gergen, 1988), and segment-polarity genes are regulated by pair-rule genes. Thus, gene interactions and gene hierarchies provide instructions for successively dividing the embryo into smaller units, and regulate development within these small units.

1.5 HOMEOTIC GENES

Homeotic genes give each segment its own identity and control the proper sequence for development of segments so that, for example, the three thoracic segments are in the proper sequence, as opposed to being scattered among abdominal segments (reviewed by Gehring and Hiromi, 1986). The name "homeotic" comes from the observation that mutant alleles of these genes alter the expression of some feature of a segment so that the expressed feature looks like that of another segment. For example, a homeotic gene that should control development of an antenna might mutate (mutant *antennapedia,* for example) to cause development of a leg at the normal site of an antenna. By studying such mutants, many of which can be induced in *Drosophila melanogaster* by various treatments, it has been possible to identify a family of homeotic genes and to specify much of their control. The homeotic genes cluster in two gene complexes: the ***antennapedia complex*** (***ANT-C***) **and *bithorax complex*** (***BX-C***), collectively called the ***homeotic complex*** (***HOM-C***) located on the right arm of chromosome 3. The genes are arranged in a linear sequence on the chromosome correlated with the linear axis of the embryo. In *Tribolium*, Ant-c and Bx-c are adjacent to each other on the same chromosome; but in *Drosophila*, *ANT-C,* which controls parasegments in the head and the first two thoracic segments, is located proximally on chromosome 3; while *BX-C*, which controls the third thoracic segment and abdominal segments, is located distally on chromosome 3 (Lewis, 1978). The linear arrangement may mean that each parasegment requires the cumulative activity of genes anterior to it (Lewis, 1978), but there is also evidence that the homeotic genes have regulatory interactive effects on each other (Struhl, 1982). Examples of homeotic genes are *proboscipedia*, *deformed*, *sex combs reduced*, *Antennapedia*, and *Ultrabithorax*.

The same homeotic genes are expressed in more than one segment (Gehring, 1987), but mutations of a gene seem to be preferentially expressed in particular segments, which suggests that the normal gene also has its principal role in that segment. For example, deletion of *Antennapedia* (*Antp*) affects each of the three thoracic segments, but the main effect is to cause the second thoracic segment to look more like the first thoracic segment. This is interpreted to mean that the main role for *Antp* is to control development of segment 2 of the thorax (Gehring, 1987). Expression of *Antp* where it is not supposed to be expressed causes the embryo to grow, or try to grow, an antenna in the wrong place. Homeotic genes regulate, at least in part, the activities of each other. There is evidence that those genes that act more posteriorly may inhibit the expression of more anterior genes. Gehring (1987) has suggested that one mechanism by which homeotic genes can regulate one another is through competition (by their gene products) for the same binding sites on DNA, which could thus control expression of a particular gene or genes.

1.5.1 The Homeobox

An important functional part of homeotic genes is the **homeobox**, a particular DNA segment about 180 base pairs in length discovered by characterizing the total nucleotide sequence of the *Antp* gene of *D. melanogaster.* The homeobox sequence is highly conserved and has since been found in every homeotic gene examined, and in a few non-homeotic but developmental genes. The homeotic genes are transcription factors and the homeobox codes for the DNA binding sequence. Differences in the homeobox sequence of nucleotides may alter the transcription factor, and thus alter the final product (leg, antenna, or bristles) that the homeotic gene controls. The homeobox sequence has been found in vertebrate (including human) and plant developmental genes. Homeotic genes, and the proteins they encode, may have other highly conserved sequences. The sequence of amino acids encoded by the *engrailed* gene from *Drosophila*, honeybee (*A. mellifera*), and a mouse, organisms representing an evolutionary span covering greater than 500 million years, are conserved near the carboxyl terminus of the protein. The *Deformed* gene encoding protein from *Drosophila*, frogs, and humans contains conserved sequences of amino acids near the amino terminus. These conserved functions suggest that the homeotic genes have fundamental roles in the development of insects and vertebrates (Gehring, 1987).

1.6 ORGANOGENESIS

1.6.1 Neurogenesis

Neurogenesis has been recently reviewed by Campos-Ortega (1994). Development of the nervous system begins in the early germ band stage, and typically is the first tissue to differentiate. In each parasegment, ectodermal cells differentiate into three types of nervous system precursor cells: enlarged **neuroblasts** (NBs), **midline precursor cells** (MPCs), and **nonneuronal cells** (NNCs). The NBs divide repeatedly and produce a chain of **ganglion mother cells** (GMCs). The GMCs and the MPCs each divide only once, producing pairs of progeny cells.

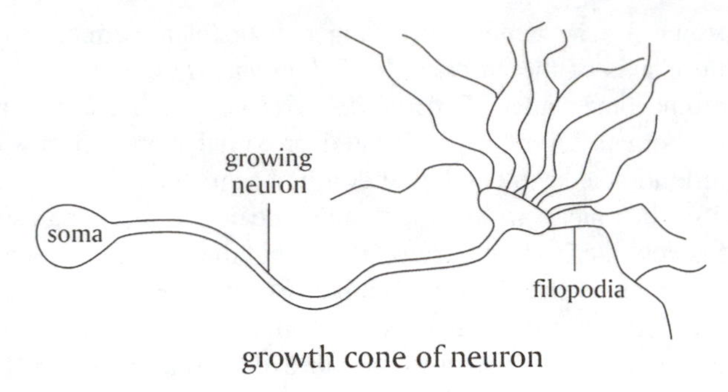

FIGURE 1.10 Diagrammatic representation of a growing neuron with its growth cone and filopodia. (Modified from Goodman and Bastiani, 1984.)

Initially, any cell in the neuroectoderm can become a neuroblast, but once cells differentiate into NBs, they inhibit neighboring cells from also becoming NBs and promote their ultimate differentiation into NNCs. A number of genes are involved in neurogenesis, including many segmentation genes that are expressed in the developing CNS. The *Drosophila* gene *Notch*, expressed in the neuroectoderm, influences whether cells become NBs or NNCs. By 8 to 9 hours after development starts in *D. melanogaster*, there are about 250 neurons in each parasegment, and the segmentation gene *fushi tarazu* (*ftz*) is expressed in a segmentally repeated pattern in about 30 of the neurons in each segment (Doe et al., 1988), although the significance of *ftz* to the developing nervous system is not yet clear. Midline precursor cells briefly express *ftz* (during hours 8 to 9 in different MPCs) in GMCs and in cells that later become glial cells.

Ganglionic masses of cells become differentiated and separated as segmentation occurs in the embryo. Three bilaterally paired groups of neuroblasts in the protocephalon will give rise to the protocerebrum, deutocerebrum, and tritocerebrum. Three ganglionic masses in the gnathal segments fuse into the subesophageal ganglion. Each thoracic segment and the 11 abdominal segments initially have paired ganglionic masses, but fusion of some of the abdominal ganglia always occurs, and in some insects all abdominal ganglia and thoracic ganglia fuse into a single thoracic ganglion, or there may be two thoracic ganglia. The stomatogastric ganglia that send nerves to the foregut are derived from ectoderm associated with the stomodeum. Sensory organs are derived from modifications of ectodermal cells in localized parts of the body where they occur. The optic lobes, a part of the protocerebrum, do not arise from neuroblasts, but develop from ectodermal cells, and like many other aspects of embryogenesis, they arise differently in different groups.

When nerve cells have located and attached to each other to form a ganglion, they send out axonal processes to make contact with neurons in other ganglia, and/or with effector organs such as muscles and glands. Dendritic processes grow toward the central nervous system from peripheral sensory cells. A growing neuron exhibits a **growth cone** (Figure 1.10). The growth cone or leading edge of a growing neuronal process contains "finger-like filopodia" that are constantly extending and retracting, exploring the environment (Goodman and Bastiani, 1984; Harrelson and Goodman, 1988). The filopodia may recognize certain chemical gradients, some of which may attract the growing tip while others repel it, and "guide cells" that provide a pathway to follow. A guide cell may be another axon that already has a connection, and the growing neuronal process partially envelops the guide and grows along it. In this way, multiple neurons going to the same general location would aggregate into larger nerves. The cellular and molecular mechanisms that guide neuronal growth cones in insects and vertebrates appear to be very similar and highly conserved. Two glycoprotein **cell adhesion molecules (CAMs)**, fasciclin II and amalgam, expressed on the surface of certain developing neurons, function in specific adhesion and nerve cell recognition in both vertebrates and insects.

The growth cone regions of developing neurons contain high concentrations of actin, which is involved in the neuron's ability to move and respond to growth gradients from its target. These actin filaments have the same basic molecular structure as those in muscle cells. Myosin is also present in the growth cone lamellae. Smith (1988) has proposed that ATP provides the energy for actin polymerization, which enables the growth cone to send out filopodia, while retraction of the filopodia is also energized by ATP and possibly involves an interaction between actin and myosin, which is also present in growth cone lamellae.

1.6.2 Development of the Gut

Soon after germ band elongation begins, a group of ectoderm cells at the anterior tip of the embryo invaginate to form the stomodeum, hypopharynx, and other parts of the gnathal segments. Similar invagination of ectodermal cells from the posterior ectoderm indicate the beginnings of the proctodeum. The midgut is derived from multiplication of cells at each end of the invaginating tissue. The three segments of the gut at first develop independently, and the complete alimentary canal is formed when plugs of cells at the end of the foregut, each end of the midgut, and at the end of the hindgut die and the three gut segments unite. As the anterior and posterior midgut primordia come together, they enclose the remaining yolk sac within the midgut. In bees and wasps (Hymenoptera), the foregut is open to the midgut before hatching, but the midgut is not open to the hindgut until just before pupation of mature larvae, so undigested food (such as pollen shells) accumulates in the midgut. Prior to pupation, cells plugging the posterior end of the midgut and anterior end of the hindgut die, the open connection is made, and the larva empties the gut.

1.6.3 Malpighian Tubules

The Malpighian tubules develop from evaginations of the anterior proctodeum and mark the junction between the midgut and hindgut. Although they are derived from the proctodeum, which has a cuticular lining, the tubules themselves are not lined by cuticle, and have a cellular morphology more similar to midgut cells than to hindgut cells. In gryllid crickets and mole crickets (Gryllotalpidae), a cuticle-lined excretory tube several millimeters long leads from the gut to a cuticle-lined bladder from which many Malpighian tubules arise. The tubules do not have a cuticular lining.

1.6.4 Tracheal System

The tracheal system develops from ectoderm. Bilaterally paired tracheal pits appear on most segments, and the pits are connected to short tubes shaped like an upside-down "T." Eventually, the pieces fuse into continuous longitudinal tracheae with segmentally arranged spiracular openings.

1.6.5 Oenocytes

Oenocytes are large cells in most larval and adult insects that stain evenly pink with eosin. In the embryo, they are derived from ectoderm in abdominal segments. They usually appear as isolated cells scattered here and there at various places in the body. They often occur between epidermal cells beneath the cuticle, and they also commonly lie between fat body cells. Their function is not clearly defined, but they have smooth endoplasmic reticulum, suggesting lipid synthesis, and they have been implicated as possible sources of ecdysteroids.

1.6.6 Cuticle Secretion in the Embryo

Several cuticles are secreted and shed by the embryos of some Apterygote insects. Embryonic epidermal cells in hemimetabolous insects and some holometabolous ones secrete a cuticle after blastokinesis. Some insects, including some acridids, *Dysdercus* spp. (Hemiptera), and *Hyalophora cecropia* (Lepidoptera), secrete two embryonic cuticles. The first one is shed, and the second one becomes the cuticle of the first instar (Mueller, 1963).

1.6.7 Cell Movements during Embryogenesis

Many cells move in amoeba-like fashion from their site of origin to another point in the embryo where they join with like cells and become functionally active. The processes by which functionally similar cells find each other and form an organ have fascinated embryologists for decades. A current model for cell migration suggests that gradients of diffusing chemicals, cell adhesion molecules (CAMs), and tactile cues guide cells and allow cells of similar function to aggregate into tissues and organs. CAMs are molecules in cell membranes that recognize another CAM of like or unlike molecular structure in another cell membrane. Approximately ten CAMs currently have been identified, but many more are believed to exist. When two cells with homophilic binding CAMs contact each other, the CAMs bind the two cells together. As additional cells bind, a tissue or organ is gradually built. Some heterophilic CAMs may also exist.

1.6.8 Programmed Cell Death: Apoptosis

Cell death is a normal part of embryogenesis in *Drosophila* (Abrams et al., 1993) and in all other multicellular organisms. Programmed cell death is called **apoptosis**. Death of cells depends in some cases on hormonal cues and, in others, on cell-to-cell interactions (Kimura and Truman, 1990; Wolff and Ready, 1991; Campos et al., 1992). A gene, ***reaper*** (***rpr***), plays a major role in control of apoptosis in *Drosophila* and embryos that are homozygous for a small deletion that includes the *reaper* gene exhibit no apoptosis and contain many extra cells that should have died. These embryos fail to survive. Although the mechanisms involved in the functioning of *reaper* are not yet clear, reaper mRNA is expressed in cells that are destined for apoptosis.

1.7 HATCHING

Development to hatching takes days, weeks, or even months in the Hemimetabola, but is much faster in the Holometabola, usually a matter of a few days in most. Faster development may have been an evolutionary process driven by selection for ability to take advantage of rapidly decaying resources (as in decaying fruit or dung breeders) and rapidly changing plant growth resources.

1.8 IMAGINAL DISCS

Imaginal discs are derived from ectoderm and are small groups of embryonic cells that persist in larvae of the Holometabola (Figure 1.11). When the insect pupates, the imaginal discs provide the cells to make the adult structures. Imaginal discs in Diptera include discs for wings, legs, halteres, compound eye, antennae, genital structures, and some mouthparts. Isolated small groups of abdominal histoblasts scattered in the larval abdomen give rise to abdominal structures in the adult, and groups of imaginal cells in the larval salivary glands and proventriculus give rise to the corresponding adult structures. Imaginal discs have been most intensively studied in Diptera, and to some extent in Lepidoptera. Reviews of imaginal disc structure and development (primarily in *Drosophila*) have been provided by Madhavan and Schneiderman (1977), Oberlander (1985), and Larsen-Rapport (1986). Anderson (1963a, 1963b; 1964a, 1964b) described the origin and development of imaginal discs in a tephritid fly, *Bactrocera* (formerly *Dacus*) *tryoni*, and van Ruiten and Sprey (1974) described the development of a leg disc in the blowfly, *Calliphora erythrocephala*.

During larval life, the discs grow by mitotic cell division. Ecdysteroid secretion in the third instar (the last instar) of *Drosophila* signals the imaginal discs to begin rapid differentiation into

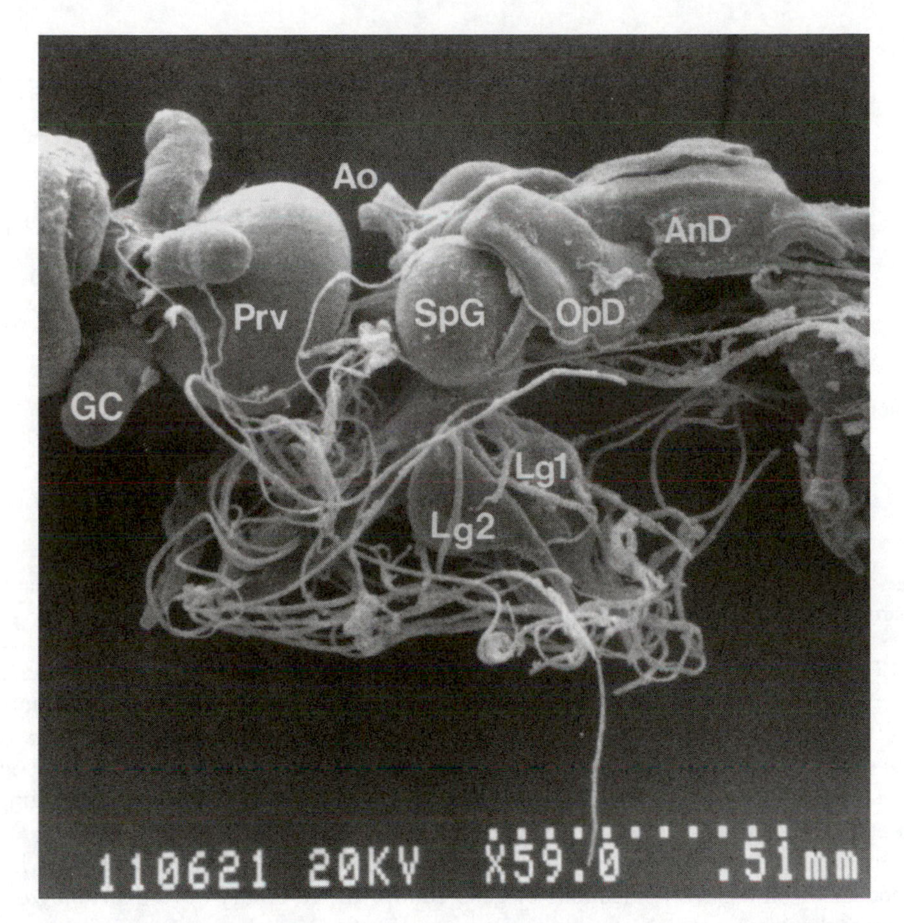

FIGURE 1.11 Some of the imaginal discs in a 3rd instar of the tephritid fruit fly *Anastrepha suspensa*. AnD, antennal disc; Ao, aorta; Gc, gastric caeca; L1 and L2, leg discs for the prothoracic and mesothoracic legs; OpD, optic lobe and compound eye disc; Prv, proventriculus; SpG, supraesophageal or brain imaginal disc. The short, finger-like ventral nerve cord to which L1 and L2 leg discs are attached contains the imaginal disc cells for developing the adult ventral nerve cord and nerves. The leg disc for the metathoracic pair of legs is not attached to the nerve cord, but is located adjacent to the wing discs near the spiracle and not shown in the photograph. The image is from a late 3rd instar that was almost ready to pupariate. (Photograph by the author.)

adult structures during the pupal stage. Discrete discs are first evident in a larva as a thickening of the epidermis in an early instar. The discs separate from the epidermis and migrate to new locations, but they remain in contact with the epidermis by a stalk. The time of appearance and growth rate of different discs in *Drosophila* and other Diptera that have been studied are variable. As a disc grows, it assumes a pocket- or tube-like structure. The pocket or cavity in the disc is called the peripodial cavity. As more cells are produced by mitosis, the tube-like growth folds upon itself to form concentric layers. During metamorphosis, the disc unfolds, elongates, and cells differentiate into an adult structure. In the case of a *Drosophila* leg disc, the innermost part of the disc becomes the most distal part of the leg, the tarsal segments, while the peripheral part becomes the most proximal leg structure, the coxa. The other parts of the leg are sequentially layered within the disc. Similarly, various cells in a wing disc give rise to structures of the thorax as well as wings (Figure 1.12).

Bryant (1993) suggests that many of the same genes, gene products, and pathways functioning in the embryo control imaginal disc development. Patterning of *Drosophila* discs is under the

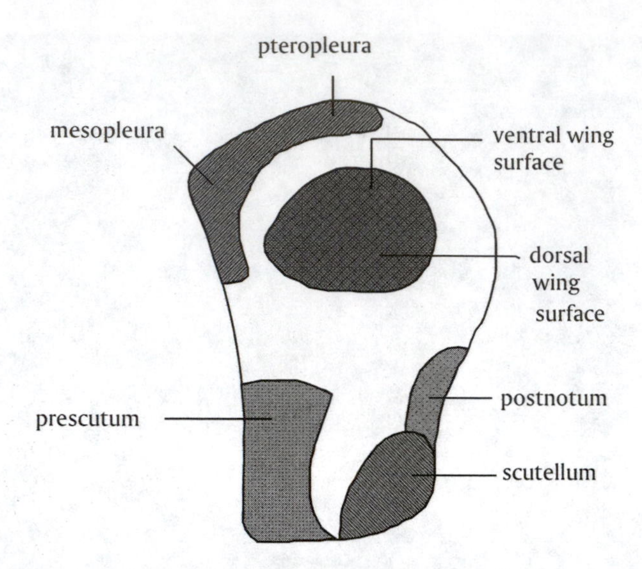

FIGURE 1.12 A wing imaginal disc of *Drosophila melanogaster* showing regions of the disk that will give rise to parts of the thorax and wing at pupation. (Modified from Bryant, 1975.)

control of some segment polarity genes necessary for embryonic development, with at least four of the *wingless*-subclass of segment polarity genes required for development of normal limb pattern (Hatano, 1991; Peifer et al., 1991). Discs do not form in embryos mutant for *wingless* (Simcox et al., 1989). Couso et al. (1993) have shown that *wingless* provides information in a polar coordinate system, a system earlier postulated for development of imaginal discs and regenerating limbs (French et al., 1976).

In conclusion, insect embryogenesis is an active and fertile field of research with important results for all of biology. Investigations of gene control in the embryo of *D. melanogaster* have often pointed the way for vertebrate studies. For example, the homeobox in *Drosophila* now appears to be universal in developmental genes. Although there clearly are major differences between some gene functions in insects and vertebrates, there are also fascinating similarities.

REFERENCES

Abrams, J.M., J. White, L.I. Fessler, and H. Steller. 1993. Programmed cell death during *Drosophila* embryogenesis. *Development,* 117:29-43.

Anderson, D.T. 1963a. The embryology of *Dacus tryoni*. 2. Development of imaginal discs in the embryo. *J. Embryol. Exp. Morphol.,* 11(Part 2):339-351.

Anderson, D.T. 1963b. The larval development of *Dacus tryoni* (Frogg.) (Diptera: Trypetidae). I. Larval instars, imaginal discs, and haemocytes. *Aust. J. Zool.,* 11:202-218.

Anderson, D.T. 1964a. The embryology of *Dacus tryoni* (Diptera). 3. Origins of imaginal rudiments other than principal discs. *J. Embryol. Exp. Morphol.,* 12 (Part 1):65-75.

Anderson, D.T. 1964b. The larval development of *Dacus tryoni* (Frogg.) (Diptera: Trypetidae). II. Development of imaginal rudiments other than the principal discs. *Aust. J. Zool.,* 12:1-8.

Anderson, D.T. 1972a. The development of hemimetabolous insects, pp. 95-163, in J. Counce and C.H. Waddington (Eds.), *Developmental Systems: Insects*, Vol. 1, Academic Press, New York.

Anderson, D.T. 1972b. The development of holometabolous insects, pp. 165-242, in J. Counce and C.H. Waddington (Eds.), *Developmental Systems: Insects*, Vol. 1, Academic Press, New York.

Anderson, K.V. 1987. Dorso-ventral embryonic pattern genes of *Drosophila*. *Trends Genet.,* 3:91-96.

Bate, M., and A. Martínez-Arias (Eds.). 1993. *The Development of Drosophila melanogaster*, Vol. I and II, Cold Spring Harbor Laboratory Press, Cold Spring Harbor, NY.

Bopp, D., M. Burri, S. Baumgartner, G. Frigerio, and M. Noll. 1986. Conservation of a large protein domain in the segmentation gene paired and in functionally related genes of *Drosophila. Cell,* 47:1033-1040.

Bryant, P.J. 1975. Pattern formation in the imaginal wing disc of *Drosophila melanogaster*: fate map, regeneration, and duplication. *J. Exp. Zool.,* 193:49-78.

Bryant, P.J. 1993. The polar coordinate model goes molecular. *Science,* 259:471-472.

Campos, A.R., K.-F. Fischbach, and H. Steller. 1992. Survival of photoreceptor neurons in the compound eye of *Drosophila* depends on connections with the optic ganglia. *Development,* 114:355-366.

Campos-Ortega, J.A., and V. Hartenstein. 1985. *The Embryonic Development of Drosophila melanogaster,* Springer-Verlag, Berlin/New York, 227 pp.

Campos-Ortega, J.A. 1994. Genetic mechanisms of early neurogenesis in *Drosophila melanogaster. Adv. Insect Physiol.,* 25:75-103.

Chan, L.-N., and W. Gerhing. 1971. Determination of blastoderm cells in *Drosophila melanogaster. Proc. Natl. Acad. Sci. U.S.A.,* 68:2217-2221.

Chasan, R., and K.V. Anderson. 1993. Maternal control of dorsal-ventral polarity and pattern in the embryo, pp. 387-424, in M. Bate and A. Martínez-Arias (Eds.), *The Development of Drosophila melanogaster,* Vol. I, Cold Spring Harbor Laboratory Press, Cold Spring Harbor, NY.

Couso, J.P., M. Bate, and A. Martínez-Arias. 1993. A *wingless*-dependent polar coordinate system in *Drosophila* imaginal discs. *Science,* 259:484-489.

Denglemann, A., A. Hardy, N. Perrimon, and A. Mahowald. 1986. Developmental analysis of the torso-like phenotype in *Drosophila* produced by a maternal-effect locus. *Dev. Biol.,* 115:479-489.

Doe, C.Q., Y. Hiromi, W.J. Gehring, and C.S. Goodman. 1988. Expression and function of the segmentation gene *fushi tarazu* during *Drosophila* neurogenesis. *Science,* 239:170-175.

Driever, W., and C. Nüsslein-Volhard. 1988a. A gradient of *bicoid* protein in *Drosophila* embryos. *Cell,* 54:83-93.

Driever, W., and C. Nüsslein-Volhard. 1988b. The *bicoid* protein determines position in the *Drosophila* embryo in a concentration-dependent manner. *Cell,* 54:95-104.

Foe, V.A., and B.M. Alberts. 1983. Studies of nuclear and cytoplasmic behaviour during five mitotic cycles that precede gastrulation in *Drosophila* embryogenesis. *J. Cell Sci.,* 61:31-70.

French, V., P.J. Bryant, and S.V. Bryant. 1976. Pattern regeneration in epimorphic fields. *Science,* 193:969-981.

French, V. 1988. Gradients and insect segmentation. *Development,* 104 (Suppl.):3-16.

Frigerio, G., M. Burri, D. Bopp, S. Baumgartner, and M. Noll. 1986. Structure of the segmentation gene *paired* and the *Drosophila* PRD gene set as part of a gene network. *Cell,* 47:735-746.

Frohnhöfer, H.G., and C. Nüsslein-Volhard. 1986. Organization of anterior pattern in the *Drosophila* embryo by the maternal gene *bicoid. Nature (London),* 324:120-125.

Gaul, U., and H. Jäckle. 1987. Pole region-dependent repression of the *Drosophila* gap gene *Krüppel* by maternal gene products. *Cell,* 51:549-555.

Gehring, W.J., and Y. Hiromi. 1986. Homeotic genes and the homeobox. *Annu. Rev. Genet.,* 20:147-173.

Gehring, W.J. 1987. Homeo boxes in the study of development. *Science,* 236:1245-1252.

Goodman, C.S., and M.J. Bastiani. 1984. How embryonic nerve cells recognize one another. *Sci. Am.,* 251(6):58-66

Harding, K., C. Rushlow, H.J. Doyle, T. Hoey, and M. Levine. 1986. Cross-regulatory interactions among pair-rule genes in *Drosophila. Science,* 233:953-959.

Harrelson, A.L., and C.S. Goodman. 1988. Growth cone guidance in insects: Fasciclin II is a member of the immunoglobulin Superfamily. *Science,* 242:700-708.

Hatano, Y. 1991. Molecular cloning and analysis of *forked* locus in *Drosophila ananassae. Mol. Gen. Genet.,* 226:17-23.

Howard, K. 1988. The generation of periodic pattern during early *Drosophila* embryogenesis. *Development,* 104 (Suppl.):35-50.

Ingham, P., and P. Gergen. 1988. Interactions between the pair-rule genes *runt, hairy, even-skipped* and *fushi tarazu* and the establishment of periodic pattern in the *Drosophila* embryo. *Development,* 104 (Suppl.):51-60.

Jäckle, H., D. Tautz, T. Schuh, E. Seifert, and R. Lehmann. 1986. Cross regulatory interactions among gap genes of *Drosophila. Nature (London),* 324:668-670.

Johannsen, O.A., and F.H. Butt. 1941. *Embryology of Insects and Myriapods,* McGraw-Hill, New York.

Jura, C. 1972. Development of apterygote insects, pp. 49-94, in J. Counce and C.H. Waddington (Eds.), *Developmental Systems: Insects,* Vol. 1, Academic Press, New York.

Kimura, K.-I., and J.W. Truman. 1990. Postmetamorphic cell death in the nervous and muscular systems of *Drosophila melanogaster. J. Neurosci.,* 10:403-411.

Kobayashi, H., and H. Ando. 1985. Early embryogenesis of fireflies, *Luciola cruciata, L. lateralis* and *Hotaria parvula* (Coleoptera, Lampyridae), pp. 157-169, in H. Ando and K. Miya (Eds.), *Recent Advances in Insect Embryology in Japan*, ISEBU Co. Ltd., Tsukuba.

Larsen-Rapport, E.W. 1986. Imaginal disc determination: Molecular and cellular correlates. *Annu. Rev. Entomol.,* 31:145-175.

Lawrence, P.A. 1988. The present status of the parasegment. *Development,* 104 (Suppl.):61-65.

Lawrence, P.A. 1992. *The Making of a Fly: The Genetics of Animal Design.* Blackwell Scientific Publications, Oxford, 228 pp.

Lehmann, R. 1988. Phenotypic comparison between maternal and zygotic genes controlling the segmental pattern of the *Drosophila* embryo. *Development,* 104 (Suppl.):17-27.

Lehmann, R., and K. Sander. 1988. *Drosophila* nurse cells produce a posterior signal required for embryonic segmentation and polarity. *Nature (London),* 335:68-70.

Lewis, E.B. 1978. A gene complex controlling segmentation in *Drosophila. Nature (London),* 276:565-570.

Lohs-Schardin, M. 1982. *Dicephalic* — a *Drosophila* mutant affecting polarity in follicle organization and embryonic patterning. *Roux' Arch. Dev. Biol.,* 191:28-36.

Madhavan, M.M., and H.A. Schneiderman. 1977. Histological analysis of the dynamics of growth of imaginal discs and histoblasts nests during the larval development of *Drosophila melanogaster. Roux' Arch. Dev. Biol.,* 183:269-305.

Mahowald, A.P. 1963. Electron microscopy of the formation of the cellular blastoderm in *Drosophila melanogaster* embryo. *Exp. Cell. Res.,* 32:457-468.

Martínez-Arias, A., and P.A. Lawrence. 1985. Parasegments and compartments in the *Drosophila* embryo. *Nature (London),* 313:639-642.

Melton, D.A. 1991. Pattern formation during animal development. *Science,* 252:234-241.

Miya, K. 1985. Determination and formation of the basic body pattern in embryo of the domesticated silkmoth, *Bombyx mori* (Lepidoptera, Bombycidae), pp. 107-1123, in H. Ando and K. Miya (Eds.), *Recent Advances in Insect Embryology in Japan*, ISEBU Co. Ltd., Tsukuba.

Mueller, N.S. 1963. An experimental analysis of molting in embryos of *Melanoplus differentialis. Dev. Biol.,* 8:222-240.

Nüsslein-Volhard, C. 1979. Maternal effect mutations that alter the spatial coordinates of the embryo of *Drosophila melanogaster*, pp. 185-211, in S. Subtelny and I.R. Königsberg (Eds.), *Determinants of Spatial Organization*, Academic Press, New York.

Nüsslein-Volhard, C. 1991. Determination of the embryonic axes of *Drosophila. Development,* 1(Suppl.):1-10.

Nüsslein-Volhard, C., and E. Wieschaus. 1980. Mutations affecting segment number and polarity in *Drosophila. Nature (London),* 287:795-801.

Nüsslein-Volhard, C., H.G. Frohnhöfer, and R. Lehmann. 1987. Determination of anteroposterior polarity in *Drosophila. Science,* 238:1675-1681.

Oberlander, H. 1985. The imaginal discs, pp. 151-182, in G.A. Kerkut and L.I. Gilbert (Eds.), *Comprehensive Insect Physiology, Biochemistry and Pharmacology, Vol. 1, Embryogenesis and Reproduction*, Pergamon Press, Oxford.

Peifer, M., C. Rauskolb, M. Williams, B. Riggleman, and E. Weischaus. 1991. The segment polarity gene *armadillo* interacts with the *wingless* signaling pathway in both embryonic and adult pattern formation. *Development,* 111:1029-1045.

Romoser, W.S., and J.G. Stoffolano, Jr. 1998. *The Science of Entomology*, 4th ed., WCB/McGraw-Hill, Boston, 605 pp.

Ruiten, Th.M. van, and Th.E. Sprey. 1974. The ultrastructure of the developing leg disk of *Calliphora erythrocephala. Z. Zellforsch.,* 147:373-400.

Sander, K. 1960. Analyse des ooplasmatischen Reakionssystems von Eucelis plebejus Fall. (Circadina) durch Isolieren und Kombinieren von Keimteilen. II. Die Differenzierungsleistungen nach Verlagern von Hinterpolmaterial. *Wilhelm Roux Arch. Entw. Mech. Org.,* 151:660-707.

Sander, K. 1976. Specification of the basic body pattern in insect embryogenesis. *Adv. Insect Physiol.,* 12:125-238.

Sander, K., J.O. Gutzeit, and H. Jäckle. 1985. Insect embryogenesis: Morphology, physiology, genetical and molecular aspects, pp. 319-385, in G.A. Kerkut and L.I. Gilbert (Eds.), *Comprehensive Insect Physiology, Biochemistry and Pharmacology, Vol. 1, Embryogenesis and Reproduction*, Pergamon Press, Oxford.

Schüpbach, T., and E. Wieschaus. 1986. Maternal-effect mutations altering the anterior-posterior pattern of the *Drosophila* embryo. *Roux'Arch. Dev. Biol.,* 195:302-317.

Simcox, A.A., and J.H. Sang. 1983. When does determination occur in *Drosophila* embryos? *Dev. Biol.,* 97:212-215.

Simcox, A.A., I.J.H. Roberts, E. Hersperger, M.C. Gribbin, A. Shearn, and J.R.S. Whittle. 1989. Imaginal discs can be recovered from cultured embryos mutant for the segment-polarity genes *engrailed*, *naked* and *patched* but not from *wingless*. *Development,* 107:715-722.

Slack, J.M.W. 1987. Morphogenetic gradients — past and present. *Trends Biochem. Sci.,* 12:200-204.

Smith, S.J. 1988. Neuronal cytomechanics: The actin-based motility of growth cones. *Science,* 242:708-715.

Steward, R. 1987. *Dorsal*, an embryonic polarity gene in *Drosophila*, is homologous to the vertebrate proto-oncogene, c-*rel*. *Science,* 238:692-694.

Struhl, G. 1982. Genes controlling segmental specification in the *Drosophila* thorax. *Proc. Natl. Acad. Sci. U.S.A.,* 79:7380-7384.

Wakimoto, B.T., and T.C. Kaufman. 1981. Analysis of larval segmentation in lethal genotypes associated with the *Antennapedia* gene complex in *Drosophila melanogaster. Dev. Biol.,* 81:51-64.

Weigel, D., G. Jürgens, M. Klingler, and H. Jäckle. 1990. Two gap genes mediate maternal terminal pattern information in *Drosophila*. *Science,* 248:495-498.

White, K., M.E. Grether, J.M. Abrams, L. Young, K. Farrell, and H. Steller. 1994. Genetic control of programmed cell death in *Drosophila*. *Science,* 264:677-683.

Wigglesworth, V.B. 1972. *The Principles of Insect Physiology,* 7th ed., Chapman & Hall, New York.

Wolff, T., and D.F. Ready. 1991. Cell death in normal and rough eye mutants of *Drosophila*. *Development,* 113:825-839.

Zalokar, M., and I. Erk. 1976. Division and migration of nuclei during early embryogenesis of *Drosophila melanogaster. J. Microsc. Biol. Cell,* 25:97-106.

2 Digestion

CONTENTS

PREVIEW

The structure and function of the alimentary canal, usually called the gut, co-evolved with the food habits of insects. Insect guts are diverse, and modified in special ways for solid vs. liquid food, and animal vs. plant food. Nevertheless, the basic evolutionary plan for three major divisions of the gut — the foregut, midgut, and hindgut — has been retained in all insects. The midgut is the principal site for secretion of digestive enzymes, digestion of food, and absorption of nutrients in most insects, although some insects display significant or major digestion in the foregut or hindgut. The fore- and hindgut are lined with a cuticular intima on the surface of the epithelial cells that is shed with each molt. The midgut does not have an attached cuticular intima, but in many insects the midgut cells secrete a detached peritrophic membrane, an envelope that encloses the food and within which most of the digestion occurs. The peritrophic membrane is not universally present in all insects. Several types of midgut epithelial cells occur in various species; the principal cells have microvilli at the gut lumen surface, a modification providing an extensive surface area for secretion of digestive enzymes and for absorption. Many insects have protein digesting enzymes with the general characteristics of trypsin and chymotrypsin, and some have protein digesting enzymes, the cathepsins, that function at acidic pH. A wide variety of carbohydrate digesting enzymes and general lipases that digest lipids have been identified. In general, insects do not secrete a complete complement of the three principal enzymes needed to digest cellulose, but many insects utilize cellulose digesting enzymes made by their gut symbionts. This chapter describes the principles of gut function and structure, and briefly reviews gut modifications and function in major insect orders.

2.1 INTRODUCTION

The evolutionary success and diversity of insects have been driven by their ability to occupy many ecological niches and utilize many different food sources. Usually, newly hatched insects must obtain food soon or die. Some newly hatched insects eat the egg shell and the small amount of yolk left in the shell after hatching, or begin eating the food on or in which the egg was laid; some have to search for food. With suitable food, a larva may successfully grow, molt, and eventually become an adult. Thus, processing of food and functioning of the alimentary canal (typically called the gut) are activities critical to life. Insects are extraordinarily diverse in food and feeding habits, and have correspondingly high diversity in gut structure and function. Major changes in gut structure and function almost always occur in those insects in which the larval and adult food is different, as in those insects with complete metamorphosis. There is, then, no "typical" insect gut, but the less specialized gut of a honeybee (Figure 2.1) or an orthopteroid (Figure 2.2) can be used to illustrate major gut structures.

2.2 RELATIONSHIPS BETWEEN FOOD HABITS AND GUT STRUCTURE AND FUNCTION

2.2.1 PLANT VS. ANIMAL ORIGIN OF FOOD: SOLID VS. LIQUID DIET

The most likely ancestral-type feeding behavior was probably that of a **general scavenger**, similar to the present-day cockroach, followed by later evolution toward more specialized **phytophagous** or **carnivorous** feeding (Southwood, 1973; Dow, 1986). The gut in such a generalist feeder was probably fairly simple, not much convoluted, and not much longer than the body — conditions that prevail today in generalist feeders. As insects evolved and adapted to new foods, there was concomitant evolution in gut structure and function.

The food that insects consume can be roughly divided into broad categories of **solid food** vs. **liquid food** and of **plant** or **animal** origin. In some insects, solid food is broken up mechanically with the mandibles, and with grinding action by a muscular **proventriculus**. The gut of solid feeders

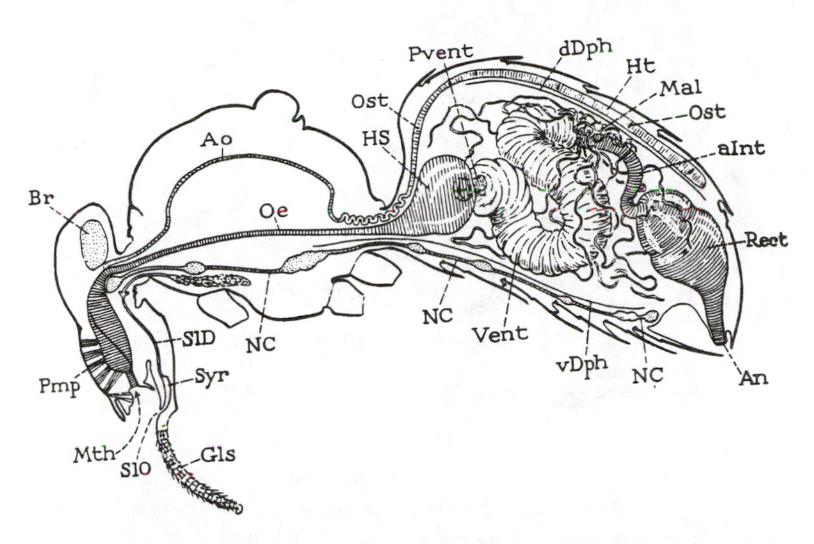

FIGURE 2.1 The body outline of a honeybee showing gut structure, dorsal vessel, and ventral nerve cord with ganglia. AInt, anterior intestine (= ileum); An, anus; Ao, aorta; Br, brain; dDph, dorsal diaphragm; Gls, tongue; HS, honey stomach (= crop); Ht, heart; Mal, Malpighian tubules; Mth, mouth; NC, nerve cord; Oe, esophagus; Ost, ostia in the heart; Pmp, pharyngeal pump; Pvent, proventriculus; Rect, rectum; SlD, salivary duct; Slo, salivary orifice; Syr, salivary syringe; Vent, ventriculus (= midgut); vDph, ventral diaphragm. (From *The Hive and The Honey Bee*, 1975, Dadant & Sons, editors and publisher. Permission granted with acknowledgment.)

tends to be a relatively straight tube, not much if any longer than the body, possibly because solid food does not easily pass through a very convoluted gut. Lepidopterous caterpillars, for example, have a simple, straight-through type of gut. They are, for the most part, phytophagous and often have an abundance of their food source. They tend to feed frequently, and some almost continuously. Cellulose in the plant food is not digested by caterpillars or by most other phytophagous insects, and the incompletely digested food passes rapidly through the gut.

A liquid diet more easily passes through a convoluted gut, which many liquid feeders have, but in some cases the diet may be so dilute that it presents new problems — such as how to get rid of so much water and sometimes other components in excess, most notably sugar. Insects that ingest a dilute liquid food have evolved specialized adaptations to deal with the excess water and other components (e.g., salts or sugars). *Rhodnius prolixus* (Hemiptera: Reduviidae), which takes one large blood meal each instar, and *Dysdercus fasciatus* (Reduviidae), which feeds on the phloem sap of plants, use hormonal controls (diuretic hormones) that regulate rapid water excretion by Malpighian tubules. Homoptera, also plant sap feeders, have morphological modifications of the gut called filter chambers, in which a loop of the hindgut is in close contact with the foregut to allow a large volume of water to bypass the midgut, thus moving fluid directly from foregut to hindgut. This causes some nutrient loss as well, primarily sugars, but sugar is in excess in the phloem sap. Amino acids are present in plant sap in low concentrations, and to extract the quantity of amino acids needed for growth and development from the limited volume of fluid that actually goes through the midgut, homopterans must feed voraciously and excrete a large quantity of honeydew.

The slightly lower nutrient quality of plant food, as compared to animal sources, requires large intake and results in a steady elimination of frass droppings; or if phloem or xylem sap is the food source, elimination of excess fluid. Plants generally supply sufficient carbohydrates and certain lipids, including phytosterols that are important to insect nutrition. Plant tissues usually contain lower levels of amino acids than animal tissues, and some amino acids may be critically low or absent. Some amino acids, as well as vitamins and other important dietary components, often are supplied by symbionts and do not always have to come from the diet.

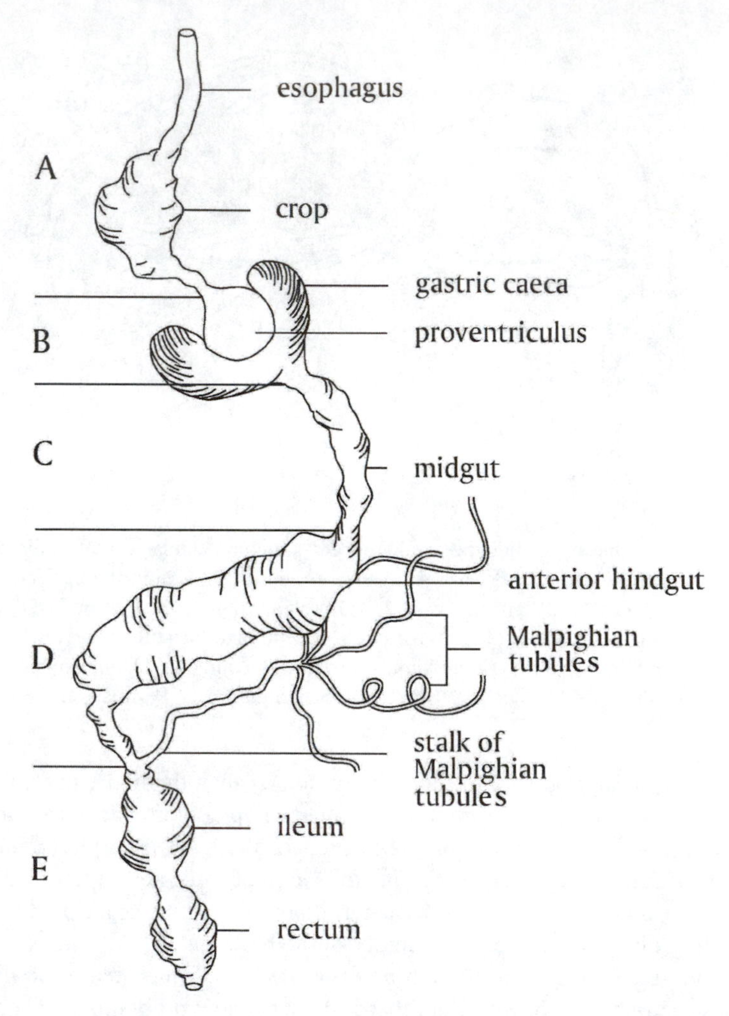

FIGURE 2.2 Gut structure in a generalized feeder such as the cricket *Gryllus rubens*. A: Foregut, including the proventriculus. B: Two large gastric caeca cupped around the proventriculus. The gastric caeca are part of the midgut. C: The short and relatively unspecialized midgut. D: The hindgut is divided into an anterior portion that has a cuticular lining on the surface of the cells (Nation, 1983). The Malpighian tubules do not originate at the junction of the midgut and hindgut in gryllid crickets, but arise from a cuticular lined stalk. The stalk arises near the junction of the anterior and posterior hindgut. E: The posterior hindgut consisting of the ileum and rectum.

Animal feeders feed less frequently and at irregular intervals, as opportunity affords, so they tend to have a gut specialized for storage (large crop, for example, in a praying mantis) that enables them to feast when food is available, and to hold a large meal for digestion over time. Insects that take food of animal origin generally obtain a better balance of amino acids than those that feed on plants. In addition, animal tissues are a rich source of carbohydrates, cholesterol, and other lipids.

2.3 THE MAJOR STRUCTURAL REGIONS OF THE GUT

2.3.1 The Foregut

Despite the diversity mentioned above, three basic divisions — the **foregut**, **midgut**, and **hindgut** — can be recognized on embryological, morphological, and physiological grounds in all insects. During embryonic development, the foregut develops from invaginating ectodermal tissue at the anterior end of the body. The foregut epithelial cells secrete a cuticular lining that is attached to the surface of the cells on the lumen (apical) side. This lining contains both chitin and proteins and is essentially the same as the epicuticle and endocuticle on the body surface. Heavily sclerotized regions of the foregut lining, such as in the proventriculus of some insects, contain hard exocuticle. The foregut can be divided into a buccal cavity (mouth), pharynx, esophagus, crop, proventriculus, and esophageal invagination, and any part may be highly modified. At each molt, the old cuticular lining from the foregut is sloughed off into the gut, and any undigested residue is excreted with the feces. Epithelial cells in the foregut are usually flattened squamous cells that do not secrete digestive enzymes into the lumen of the foregut.

The mouth or buccal cavity is usually just an enlarged opening that receives the slightly chewed food in mandibulate insects or fluid ingested by insects with piercing and sucking mouthparts. Powerful muscles in the wall of the pharynx pump fluid into the buccal cavity and aid swallowing in blood feeders, and xylem and phloem feeders. Salivary glands are diverticula from the anterior part of the foregut that secrete fluid and carbohydrate digesting enzymes (mostly amylases) into the buccal cavity. Salivary secretions contain amylase, lubricate the food, and contribute to digestion in the crop.

The pharynx passes food to the esophagus, which may be a simple tube that continues to the proventriculus at end of the foregut in some insects. Alternatively, the foregut may expand into a much enlarged and dilated **crop**, or the crop may be a diverticulum from the main part of the foregut, as in some Diptera. Substantial digestion occurs in the crop of some insects, for example, in many Orthoptera and some Coleoptera (Dow, 1986), but the enzymes (except for salivary enzymes) come from the midgut and are present in fluid that is regurgitated from the midgut into the crop. The foregut cuticle is impermeable and little or no absorption occurs from the foregut.

The crop periodically releases some of its contents to pass through the proventriculus and into the midgut. In the cockroach, *Leucophaea maderae*, crop emptying is inversely related to the concentration of the crop contents and to hemolymph levels of some nutrients. Thus, the crop releases materials to the midgut at a rate that allows it to digest and absorb nutrients more efficiently (Englemann, 1968).

Extra-oral digestion occurs in many insects. By injecting hydrolytic enzymes into the food source (animal or plant material) and then sucking back the digested products, insects utilize very high percentages of the nutrient value of the food source (see review by Cohen, 1995). Some insects reflux enzyme secretions and partially digested products by repeatedly sucking up and reinjecting the liquefied juices into the food. Refluxing mixes the secretions and fluids and extends the effective life of the digestive enzymes. Refluxing is particularly effective when the food contains a limiting boundary, such as the shell of a seed or the cuticle of an insect, that acts as a container for the liquefying body contents. Larval carabids, which normally feed extra-orally on small arthropods, were allowed to feed on a large portion of meat and then could recover only about 50% of the proteins available, including their own digestive enzymes, apparently because the enzymes and some digested products diffused into the piece of meat (Cheeseman and Gillot, 1987).

The **proventriculus** at the end of the foregut may be very muscular and contain heavily sclerotized teeth, ridges, and spines (Figure 2.3) for further grinding and tearing of the food, or it may be reduced to a simple valve at the entry to the midgut. The proventriculus of worker honeybees, called the **honey stopper** (Figure 2.4), consists of four converging fingers projecting anteriorly into the crop. The fingers, each bearing intermeshing spines, open and close rhythmically to capture pollen grains and sweep them from the crop into a bolus that later enters the midgut for digestion. Nectar is strained through the interlocking spines and retained in the crop for later deposition in the honeycomb. Flap-like or valve-like extensions of the proventriculus sometimes project into the midgut, forming the **esophageal valves** or **cardiac** sphincter as the junction between foregut and midgut. There is a great deal of variability in structure at the junction, and in the depth of its invagination into the midgut. Wigglesworth (1961) suggested that the main function of the invagination is to channel food entering the midgut into the peritrophic membrane. In mole crickets (Gryllotalpidae), four large esophageal valves (Figure 2.5) channel food and sand grains, often

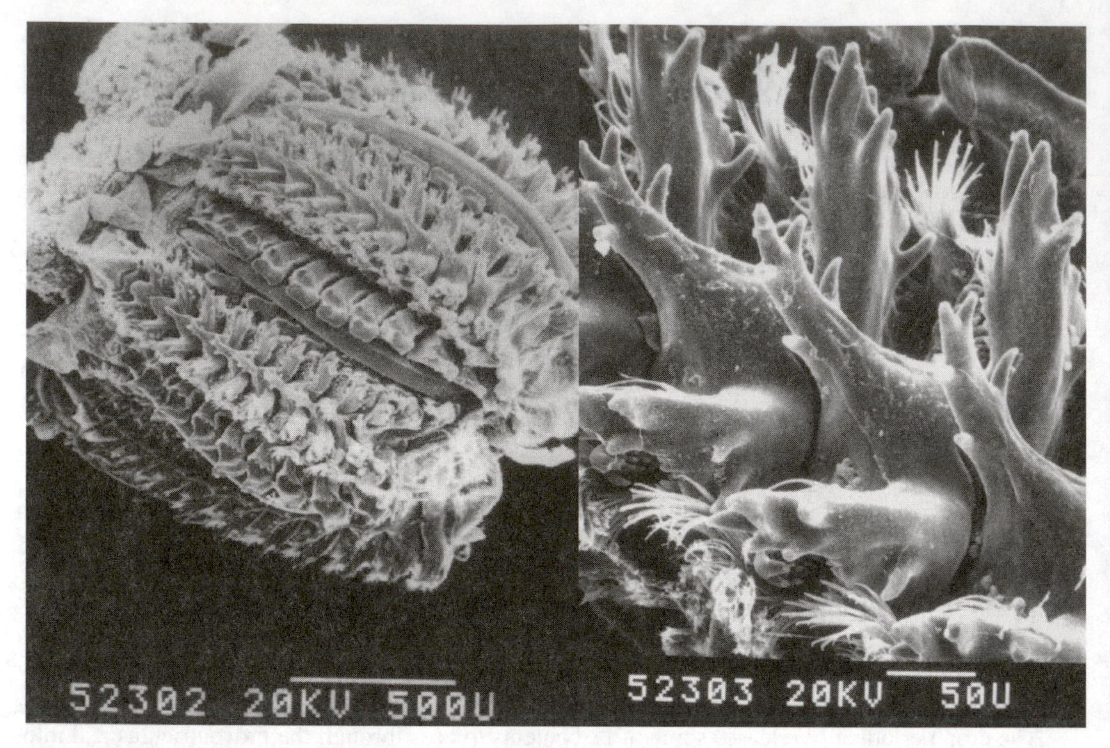

FIGURE 2.3 Proventriculus from a bush cricket. The barrel-shaped proventriculus has been cut longitudinally and turned inside out so that the heavily sclerotized ridges and "teeth" on the internal surface can be observed.

present because of their feeding habits, past two large, cup-shaped gastric caeca, protecting the delicate microvilli on the lumen surface of the gastric caecal cells from abrasive food particles.

2.3.2 The Midgut

The midgut is the principal site for secretion of digestive enzymes and for digestion and absorption in most insects. In many insects, **gastric caeca** arise at or near the origin of the midgut, but they may be located at various points along the midgut. In some insects, the gastric caeca are a major site of absorption of digestion products, and they also produce digestive enzymes. The role of gastric caeca in absorption, especially if the gastric caeca are near the origin of the midgut, depends on a countercurrent flow (see later section) that brings midgut contents back to the gastric caeca.

The origin of the midgut in insects is a controversial topic. Dow (1986) reviewed some of the recent literature regarding whether the midgut develops from endodermal tissue (McFarlane, 1985; Richards and Richards, 1977) or whether it develops from buds of tissue at the invaginated ends of the fore- and hindgut, in which case it would be derived from ectodermal tissue. Dow concluded that in some insects at least, a case can be made that the midgut may be derived from ectodermal tissue, as are the fore- and hindguts. The midgut does not have an attached cuticular lining on the surface of the cells, but midgut cells in the majority of insects can secrete a chitin- and protein-containing membrane, the peritrophic membrane, that surrounds the food and shields the delicate midgut cells from contact with potentially rough and abrasive food particles.

2.3.3 The Hindgut

The hindgut, like the foregut, develops in the embryo from ectodermal tissue and, consequently, hindgut cells have an attached cuticular lining on their surface (Figure 2.6). The Malpighian tubules usually mark the beginning of the hindgut (but see Chapter 13, Excretion, for some exceptions).

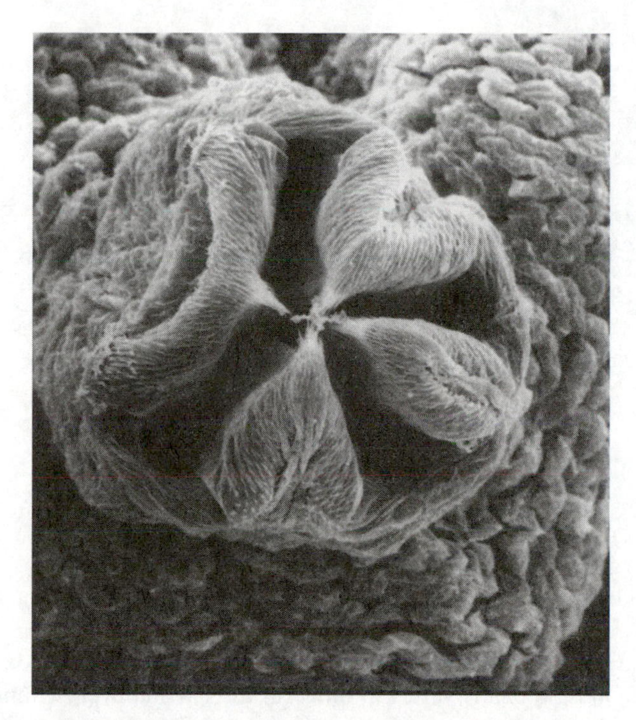

FIGURE 2.4 The proventriculus (also called the honey stopper) of an adult worker honeybee. The view is from the crop looking toward the midgut. The finger-like proventricular flaps containing setae strain pollen grains from the nectar in the crop and pass the pollen into the midgut. Most of the nectar can be left in the crop for honey production in the hive.

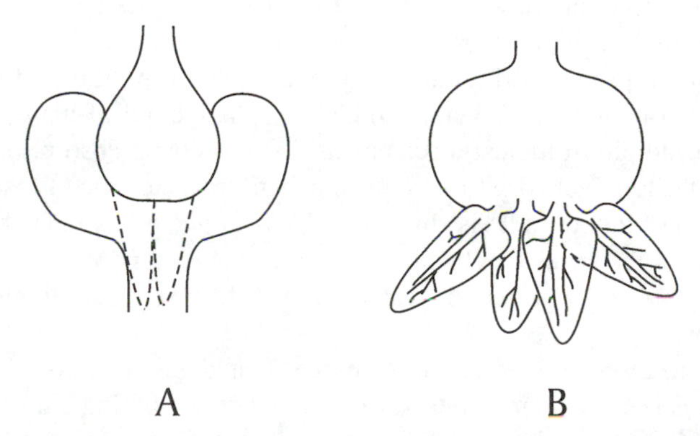

FIGURE 2.5 Proventricular valve flaps in a mole cricket, *Scapteriscus abbreviatus*. The long valve flaps prevent sand (which is common in the gut of mole crickets) and other rough food particles from entering the delicate gastric caeca just posterior to the proventriculus. A: The view shows the position of two of the valves (dotted lines) on the inside of the gut. B: The gut has been dissected and the four valve flaps spread apart; the valve flaps are cusp shaped and project about 1.2 mm past the opening to the gastric caeca. The cells of the gastric caeca protected by the flaps are the only cells in the gut of this mole cricket that have microvilli on the lumenal surface; the remainder of the gut contains a cuticular lining on the surface of the cells. (From Nation, 1983.)

The junction between the mid- and hindgut has been called the pylorus by some authors, and a valvular structure may occur here. The part of the hindgut immediately past the Malpighian tubules has been called the ileum. Sometimes the ileum simply grades into the rectum, but in some insects

FIGURE 2.6 The thick cuticular layer on the lumen surface of cells in the hindgut of a mole cricket, *Scapteriscus borelli.*

there is a distinct middle region called the colon by some authors. The terminal part of the hindgut is the rectum.

Both circular and longitudinal muscles lie on the outer or hemolymph side of the hindgut. The arrangement of these muscle bands varies in different insects and even within the same insect at different points along the hindgut as to which band of muscle is outermost; circular or longitudinal may be outermost (Gupta and Berridge, 1966; Hopkins, 1967). The entire hindgut has a chitinous lining (sometimes referred to as an intima) on the lumen surface of the cells. Typically, the cells of the hindgut wall are arranged in a single layer of irregularly shaped epithelial cells. Numerous septate desmosomes connect the lateral borders of cells and serve to hold cells together and present a barrier to fluid and molecules that might otherwise pass between adjacent cells.

The contents of the hindgut are generally fluid as they pass into the rectum. The rectum plays a critically important role in the reabsorption of water, ions, and dissolved substances (including some nutrients) from the primary urine flushed into the hindgut by the Malpighian tubules. Specialized cells, the rectal papillae and rectal pad cells, in the rectum of many insects have characteristic ultrastructure and physiological mechanisms that promote reabsorption. Water recovery by the rectum results in the relatively dry frass or fecal pellets characteristic of many terrestrial insects. The cuticular lining on hindgut cells is thinner and has larger pores in it than the lining in the foregut, and numerous substances can be absorbed from the lumen. More details on the anatomy of the hindgut, and its role in secretion, reabsorption, and water conservation, are provided in Chapter 13, Excretion.

The hindgut is most specialized in those insects that digest cellulose, such as termites. In termites, the hindgut is usually divided into several chambers harboring either bacteria or protozoa that digest cellulose. Glucose is the principal carbohydrate liberated from cellulose digestion, and it is usually fermented by the resident microorganisms, with the end products being short-chain fatty acids (principally acetic acid) that can be absorbed by the termite and used as an energy source. Additional details are given later in this chapter (Section 2.13.3, Isoptera).

2.4 MIDGUT CELL TYPES

2.4.1 COLUMNAR CELLS

Columnar cells, the most numerous cells in the midgut, conduct most of the absorption of digested products and secretion of enzymes. "Columnar" refers to the tall shape of the cells, but some insects have more than one morphological type. The cells have microvilli on the apical or lumen surface

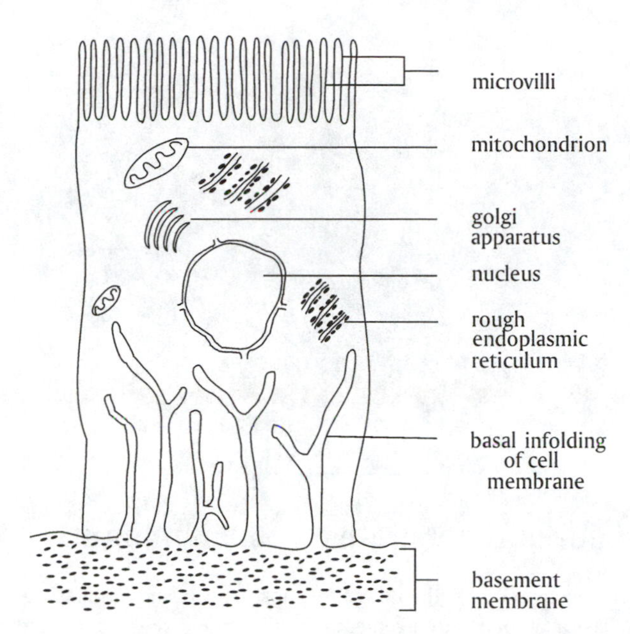

FIGURE 2.7 Diagrammatic drawing of the main ultrastructural features of a midgut cell. Basal infoldings of the cell membrane project into the cell. Microvilli occur on the gut lumen side. Mitochondria are numerous.

and extensive invaginations of the basal cell membrane (Figure 2.7). The extensive membrane infolding at the basal face and the microvilli at the apical face of midgut cells are adaptations to present a large membrane surface for metabolic functions such as absorption and secretion. Midgut cells exhibit other characteristics typical of secretory epithelium, such as a rough endoplasmic reticulum, a large Golgi complex, and secretory vesicles.

2.4.2 Regenerative Cells

Midgut cells wear out rapidly and are replaced by new cells that grow from small **regenerative cells** lying randomly near the base of mature cells in larval Diptera and Lepidoptera, or as small cell clusters called nidi (nests) (Figure 2.8) in Orthoptera and Odonata, and at the apex of crypts or caeca projecting through the gut muscle layers in Coleoptera. Regenerative cells grow into mature cells gradually and replace cells lost through age, wear, and apocrine and holocrine secretion. House (1974) reported that midgut cells in *Periplaneta americana* are replaced about every 40 to 120 hours.

Damage to the midgut regenerative cells after irradiation for inducing sterility in insects for population control (**sterile insect technique, SIT**) has been one of the limiting factors in the use of the technique, particularly in the boll weevil, *Anthonomus grandis*. Irradiated boll weevils soon die from midgut disruption because regenerative cells are unable to successfully divide and replace normal loss of midgut cells after irradiation (Riemann and Flint, 1967).

2.4.3 Goblet Cells

Goblet cells are, indeed, somewhat goblet shaped, with a large central cavity lined with microvilli (Figure 2.9). The sides of a goblet cell curve around to enclose the cavity, and at the apex the apical lips have interdigitating microvilli that control fluid exchange between cavity and midgut lumen. Goblet cells are interspersed among midgut epithelial cells in lepidopterous larvae, and in Ephemeroptera, Plecoptera, and Trichoptera. The apical membrane facing the goblet cavity houses a vacuolar-type H^+-ATPase (**proton ATPase pump**) (Figures 2.10 and 2.11) that establishes a voltage across the apical membrane and pumps H^+ into the goblet cavity (Chao et al., 1991). A similar proton pump occurs in the apical membrane of Malpighian tubule cells and drives the

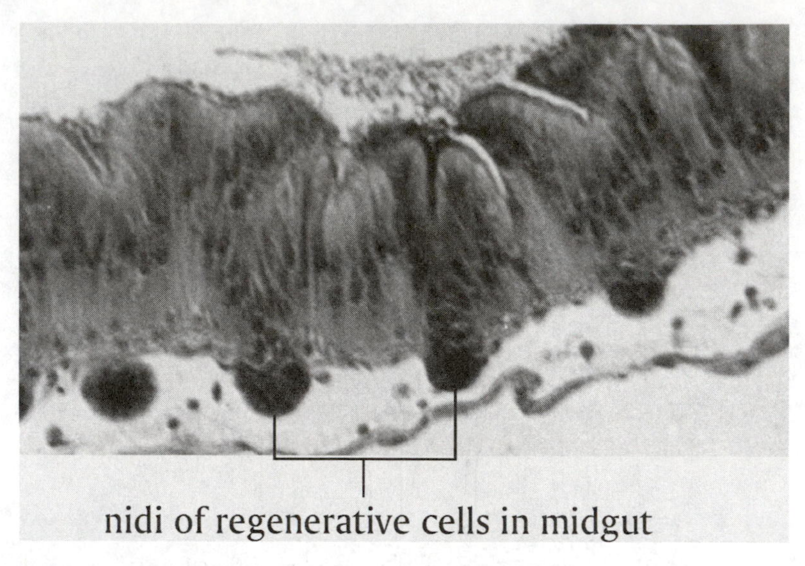

FIGURE 2.8 Regenerative cells occur in the midgut of most insects and gradually grow into mature cells to replace worn-out cells. The anatomical arrangement of regenerative cells varies in different species; nidi or nests of regenerative cells are shown in this illustration from gastric caeca of a mole cricket, *Scapteriscus vicinus*.

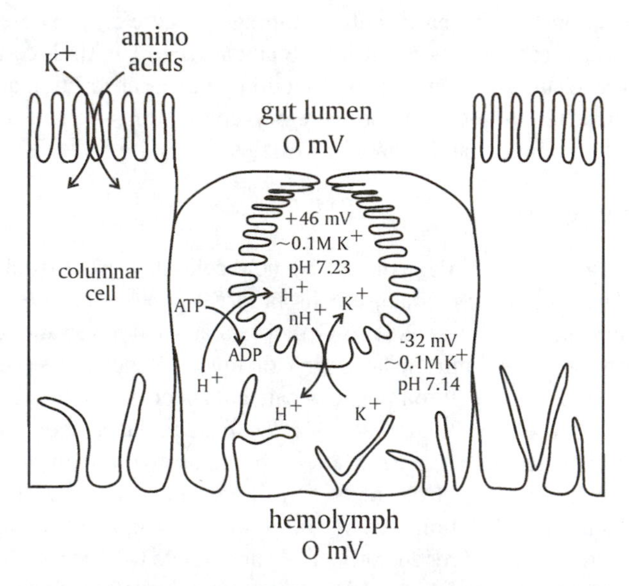

FIGURE 2.9 Structure and function of goblet cells from the midgut of a lepidopteran. A proton pump actively secretes protons (H^+) into the goblet cavity, and an antiporter mechanism in the goblet cell membrane transports K^+ into the goblet cell cavity in exchange for H^+. Goblet cavity contents are eventually emptied into the midgut lumen, creating the strongly alkaline midgut of Lepidoptera.

formation of fluid (urine) in Malpighian tubules (see Chapter 13). In goblet cells, a $K^+\backslash nH^+$ **anti-porter mechanism** (Wieczorek et al., 1989; 1991, 2000) exchanges H^+ for K^+ in the goblet cavity. The pump is strongly electrogenic and creates transmembrane voltages that can exceed 240 mV and transmembrane pH gradients that may exceed 4 pH units (Harvey, 1992). The transmembrane voltage created by the pump enables a co-transporter mechanism in the apical membrane of columnar cells to reabsorb K^+ and amino acids from protein digestion (Giordana et al., 1989). The molecular mechanism of the co-transporter has not been elucidated.

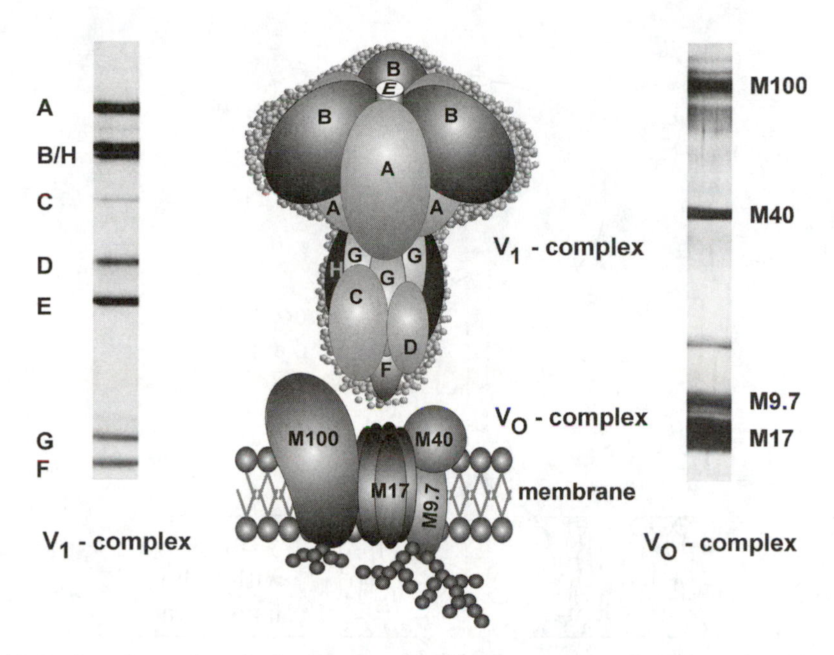

FIGURE 2.10 Projected complex of subunits comprising the proton pump (V-ATPase) that drives H^+ into the lumen of goblet cells in *Manduca sexta*. The head part of the pump, designated V_1, is located within the goblet cell cytoplasm, and the pump base (Vo) forms a transmembrane channel through which protons are pumped into the goblet cell cavity. ATP hydrolysis in the V_1 part of the pump provides the energy for the pumping of protons across the cell membrane. A separate mechanism in the goblet cell membrane that has not been elucidated yet exchanges K^+ for H^+ in the goblet cell cavity. The proton pump is the driving force for the concentration of K^+ against a concentration gradient in the goblet cell cavity. Eventually, the K^+ is released into the gut lumen, creating high pH in the lumen. The relative sizes of the various protein subunits are indicated in the gel bands shown on each side of the pump. (Illustration provided by William Harvey. Reproduced with permission from Wieczorek et al., 2000.)

The pump consists of a complex of protein subunits with the Vo base embedded in the goblet cell apical membrane (the surface facing the goblet cavity) and the V_1 head piece projecting into the goblet cell cytoplasm (Merzendorfer et al., 1997; Wieczorek et al., 2000). The V_1 complex uses energy from ATP breakdown to drive protons into the goblet cavity through the Vo transmembrane complex. The pH of the goblet cavity remains near neutral, at 7.23 ± 0.11, because of the rapid exchange of K^+ for H^+, probably with two or more H^+ per K^+ exchanged (Chao et al., 1991). Potassium ions enter goblet cells from the hemolymph at the basal side through K^+ channels located in the basal membrane (Zeiske et al., 1986; Moffett and Koch, 1988a, 1988b). There may be electrical coupling of the basal and apical membranes involving anion transport at the basolateral membrane (with concomitant cation movement), as has been demonstrated in Malpighian tubules (Beyenbach, 1995; Beyenbach et al., 2000). Electrical coupling, if it occurs in the goblet cells, would permit the electrical driving forces at the basolateral and apical membranes to rise and fall in parallel so that cation entry from the hemolymph matches cation extrusion into the goblet cavity (Beyenback et al., 2000). Ultimately, goblet cavity contents are emptied into the lumen of the midgut. Several different anions, including bicarbonate (HCO_3^-) may be secreted into the lumen with potassium, giving rise to the high pH characteristic of the larval midgut in many Lepidoptera (Dow, 1984; Dow and O'Donnell, 1990). Goblet cells appear to metabolize amino acids preferentially to support pump activity, and L-alanine, L-glutamine, L-glutamate, and L-malate experimentally support pump activity and maintain the transepithelial potential, but glucose is ineffective (Parenti et al., 1985). Only apical (goblet cavity facing) membranes of goblet cells contain significant amounts of V-ATPase proteins, and other plasma membranes and endomembranes of goblet

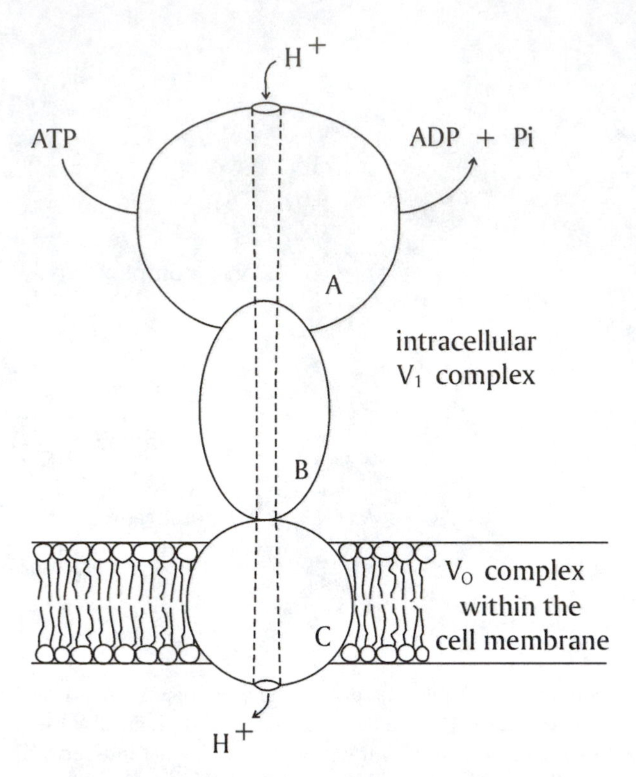

FIGURE 2.11 A schematic diagram of the proton pump. Parts A and B correspond to the V_1 complex of Figure 2.10; These parts of the pump are located within the goblet cell cytoplasm, while part C represents the Vo complex located in the goblet cell membrane. Protons are pumped from the cell into the goblet cell cavity.

and columnar cells contain little or no pump proteins (Klein et al., 1991). Although the detailed mechanism for regulation of pump function in the midgut has yet to be determined, it is likely that part of the regulatory process involves dissociation/reassociation of the V_1 cytoplasmic complex from the Vo transmembrane complex, as demonstrated in yeast proton pumps (Kane and Parra, 2000).

The high midgut pH may help protect against tannins that are common in the plant hosts of Lepidoptera larvae. Tannins can complex with the insect's own enzymes and proteins in the food, and may result in reduced digestion of proteins. Although other factors may also aid in reducing the impact of tannins, there is less formation of nonsoluble protein-tannin complexes at high pH.

2.5 MICROVILLI OR BRUSH BORDER OF MIDGUT CELLS

Microvilli at the apical surface of midgut cells greatly increase the surface area for enzyme secretion and for absorption of digested products (Figure 2.12A, B). Center-to-center spacing (= width) of microvilli in midgut cells of various insects ranges from about 150 to 200 nm (Richards and Richards, 1977), hence the difficulty of resolving them clearly with light microcopy. Microvilli have been described variously as a **brush border** or a **striated border**. The brush border gives the appearance of many very fine, closely spaced, and relatively short hairs, whereas the striated border gives the appearance of less-numerous, "stick-like" extensions from the apical surface. Although the terms "brush border" and "striated border" frequently are used in the literature, both are borders of microvilli. The microvilli on midgut cells of the mosquito *Aedes aegypti* are covered by a network of fine strands called the microvilli-associated network (Zieler et al., 2000). A major function of this dense network may be to protect the midgut microvilli from phagocytes and other cells in the blood meal until digestion begins.

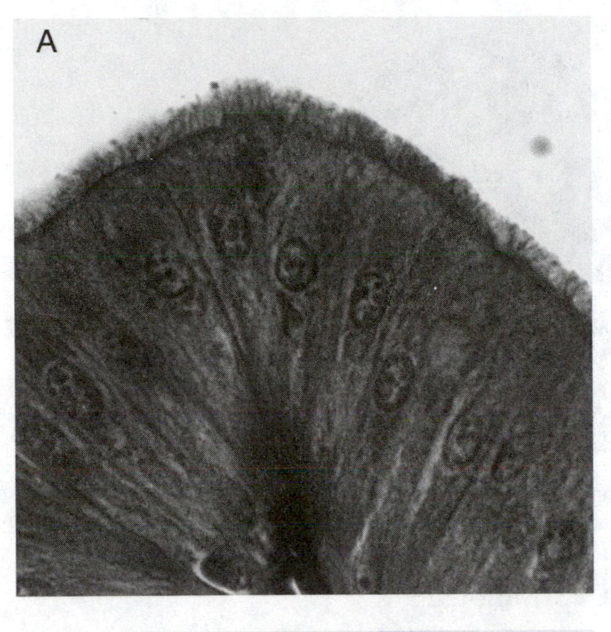

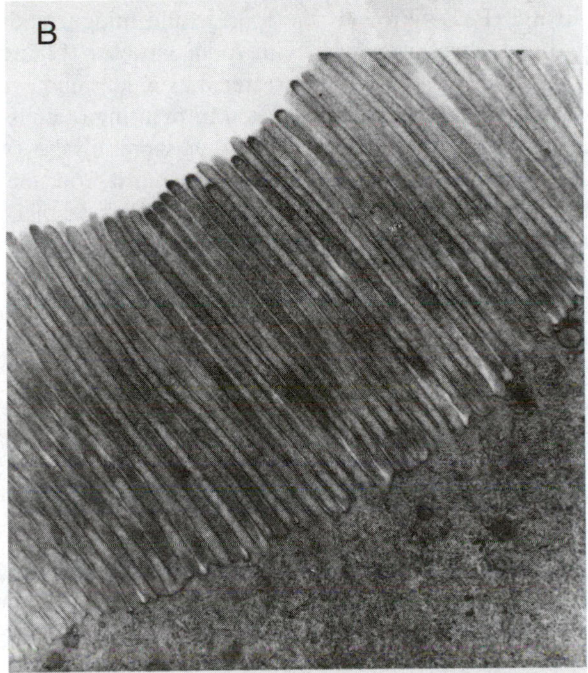

FIGURE 2.12 A: The brush border on gastric caeca cells from a mole cricket, *Scapteriscus vicinus* (oil immersion, light microscope). B: Transmission electron micrograph of midgut microvilli from a chironomid larva.

2.6 THE GLYCOCALYX

Most insects do not have a mucous lining in the midgut that is directly comparable to the mucous lining of various parts of the digestive system of vertebrates, but often there is a viscous secretion consisting of protein and carbohydrate on the surface of, and between, microvilli called the **glycocalyx** (Figure 2.13). The viscous glycocalyx traps and concentrates secreted enzymes and products of digestion (Santos and Terra, 1984; Santos et al., 1986).

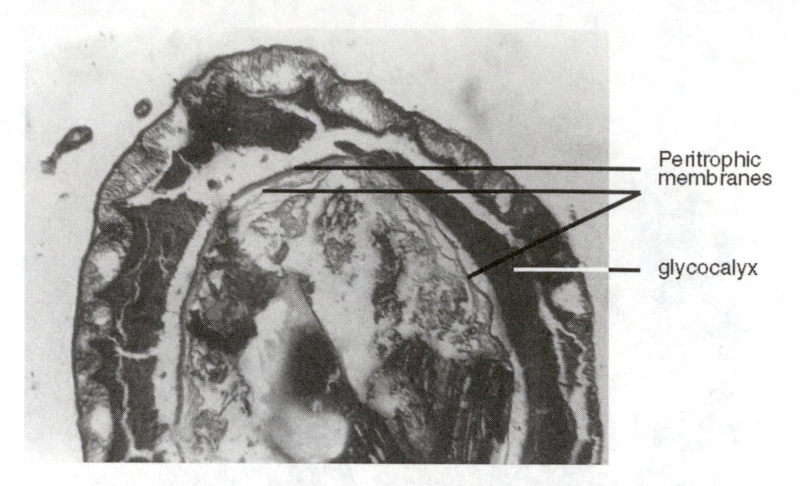

FIGURE 2.13 Photograph of the peritrophic membrane and the dark layer of glycocalyx material between the peritrophic membrane and the surface of the midgut cells in the midgut of a gryllid cricket.

2.7 PERITROPHIC MEMBRANES

The **peritrophic membrane** (**PM**) surrounds the food in the midgut and serves as a shield to protect microvilli from direct physical contact with food particles (Figure 2.13). Wigglesworth (1961) characterized the PM as **Type I** when it is secreted as a continuous delamination all along the length of the midgut, and **Type II** when it is secreted from a ring of cells at the anterior margin of the midgut. No particular cell type has been identified as secreting the Type I PM. The Type II PM is secreted continuously like a stocking as food pushes into it from the foregut. Most insects that produce a PM produce the Type I PM, but Dermaptera and larvae of Diptera produce a Type II PM. Larval mosquitoes form a Type II PM, but adult mosquitoes secrete a Type I PM. Some insects, including adult mosquitoes, secrete a PM only after taking food into the gut. Stretching of the gut, rather than a secretogogue mechanism, appears to be involved in the case of mosquitoes because an enema of saline can induce PM production. Not all insects fit into the Type I/Type II model, and some do not form a PM at all. *Ptinus* spp. beetles secrete a PM starting only some distance along the middle of the third region of the midgut; members of two weevil genera (*Cionus* and *Cleopus*) secrete a PM only toward the posterior of the midgut (Rudall and Kenchington, 1973).

Hemiptera and Homoptera as a group appear not to form a PM; at least a PM has not been unequivocally identified in any of them. In some Hemiptera, a perimicrovillar membrane, a thin membrane over the microvilli, has been described from EM studies. Gryllid crickets have a PM, but several reports indicate that mole crickets (Gryllotalpidae) do not form a PM. Some adult lepidopterans and some adult tabanids have a PM while others do not; differences occur even within the same family (Waterhouse, 1953). *Drosophila* embryos and newly hatched *Aedes aegypti* mosquito larvae have a PM, but newly hatched honeybee larvae do not acquire one until several days after hatching. Although attempts have been made to relate the presence or absence of a PM to diet and to phylogeny, too many exceptions occur to establish any satisfactory relationship.

Peritrophic membranes contain both chitin and protein, with chitin making up from 4 to about 20%, and protein composing up to 40% in various insects. Other components that have been reported are acid mucopolysaccharides, neutral polysaccharides, mucins, hyaluronic acid, hexosamine, glucose, and glucuronic acid. Chitin occurs in its α, β, and γ forms (see Chapter 4, The Integument) in the PM of various insects, but α-chitin is most common. Enzymes must pass through the PM to get at the food, and small molecules resulting from digestion must pass out, so the PM is porous. Reported pore sizes vary, perhaps with mode of estimation and with species tested. Santos and Terra (1986) reported the presence of 7- to 7.5-nm diameter pores in the PM of the sphingid caterpillar, *Erinnyis ello*. Pore size has been estimated at 200 nm in *Locusta* (Baines, 1978) and 150 nm in some cockroaches (Skaer, 1981). Using the permeability of fluorescently labeled dextrans, Edwards and Jacobs-Lorena (2000) determined that the main part of the PM of two mosquito larvae was permeable to 148-kDa or smaller particles, but that part in the gastric caeca was only permeable to 19.5-kDa or smaller particles. Mechanical damage and possible attack by protein digesting enzymes of the gut probably act to shorten the life of the PM, and may cause some breakup of it in the posterior midgut and/or hindgut. Perhaps to counter such destructive action, some insects produce several peritrophic membranes per day, each encasing the one before it. Multiple layers in the PM are often observable with a transmission electron microscope; five layers occur in the dipteran, *Stomoxys calcitrans*, but overall the PM is thin, varying from 0.13 to about 0.4 μm thick (Lehane, 1976).

2.7.1 Functions of the Peritrophic Membrane

A great deal of discussion in the literature (reviewed by Terra, 1990; Lehane, 1997) has been devoted to possible functions of the PM. Suggested functions include:

1. Protection of the delicate microvilli on the surface of midgut cells from contact with rough food particles
2. A barrier against entry of viruses, bacteria, or other parasites that would be too large to pass through the unbroken PM
3. An aid in preventing the rapid excretion of digestive enzymes
4. Compartmentalization of digestion within the midgut
5. Prevention of nonspecific binding of undigested materials or plant allelochemicals to midgut microvillar surfaces and/or binding to transport proteins at the midgut surface

Although protection of the delicate midgut tissue from rough food particles is the most often mentioned function of the peritrophic membrane in the literature, Lehane (1997) believes that protection from pathogens ingested with the food is probably the most important function.

The PM may protect some phytophagous insects from the toxic effects of ingested phenolic compounds, which are common in many plants, by preventing passage of the phenolics through the membrane and/or by complexing the substance within the PM (Bernays and Chamberlain, 1980). The PM of the grasshopper *Melanoplus sanguinipes* (Orthoptera: Acrididae), however, allowed some gallotannins to penetrate and adsorbed less than 1% of the tested tannins (Barbehenn et al., 1996).

A PM is present in *Tomocerus minor* (Humbert, 1979), a collembolan and a generalist scavenger and representative of very early evolution of insects. The PM probably evolved very early in a generalist scavenger feeder, in which protection of midgut microvillar surfaces from food particles, sand, or other hard substances coincidentally ingested was likely to be important. The PM has been conserved over long evolutionary time, even if some of its supposed functions are no longer important in a particular insect. A PM is present in many insects that do not feed on rough or solid food, such as some blood feeders, and in adult lepidopterans that take flower and plant nectars. In these cases, it may be an evolutionary relic, but it may also serve some or all of the other protective functions enumerated previously. Other insects appear to get along just fine without any PM.

2.8 DIGESTIVE ENZYMES

The secretion of digestive enzymes into the gut lumen is characterized as **constitutive secretion** when the enzymes are released from the cells as soon as they are synthesized, and as **regulated secretion** when the enzyme is synthesized and stored, often as a **zymogen** (protein containing a peptide sequence that prevents enzymatic activity until the sequence is removed), until a signal to release it is received (Lehane et al., 1995). Most insects studied to date utilize constitutive secretion rather than regulated secretion. Two well-studied features of vertebrate digestive systems — the storage of enzymes as inactive zymogens and the stimulation of enzyme secretion by the food itself — occur in insects (Moffatt et al., 1995; Blakemore et al., 1995).

Signals to secrete digestive enzymes may come from stimulation of ingested food, in which case it is called **prandial control**; from **hormonal stimulation**; and from **paracrine control** (release of factors from putative endocrine cells in the gut) (Lehane et al., 1995). Clear-cut distinctions between these mechanisms of enzyme control are not always obvious from experimental analyses, and subtle overlap of mechanisms may occur (Lehane et al., 1995). In general, paracrine and prandial control mechanisms of enzyme secretion are most common in insects. Proteins in the food are stimulants for digestive enzyme secretion in many insects (Blakemore et al., 1995, and references therein). Whether these act directly on enzyme secreting cells (prandial mechanism), or act through the putative endocrine cells (paracrine mechanism) present in the gut of many insects, is not well established.

Midgut cells secrete enzymes in three ways. In the most common type of enzyme secretion, called **merocrine secretion** and also called **exocytosis**, enzymes are processed in the Golgi complex of the columnar cells and enclosed in small vesicles. These enzyme-containing vesicles fuse with the cell plasma membrane, and the enzymes are released to the gut lumen. In another, probably more costly and thus less common form of secretion, the entire midgut cell breaks down, and the cytoplasmic contents are discharged into the lumen of the gut. This is called **holocrine secretion**. In a variation of this, called **apocrine secretion**, only parts of the cell, typically just the microvillar membranes, fragment and disintegrate into the gut lumen. A further variation on apocrine secretion is **microapocrine secretion**, in which small single- or double-membrane vesicles are pinched-off from the cell microvilli. Apocrine and microapocrine secretions are typical of the anterior part of the midgut, while exocytosis most often occurs in the posterior midgut. The mechanism of enzyme secretion may be related to the region of the gut and its particular function. For example, the anterior part of the midgut is often involved with absorption of digested products, and apocrine or microapocrine secretion in this region may be an adaptation to promote dispersion of secretory vesicle contents into the midgut lumen in a region undergoing absorption processes (Cristofoletti et al., 2000).

In midgut cells of Lepidoptera, trypsin is incorporated into the membrane of small vesicles within the midgut cells. The vesicles migrate to the microvilli of the columnar cells, where trypsin is processed to become soluble within the vesicles. Through an exocytotic process, the vesicles bud from the microvilli as double-membrane vesicles as they are released into the gut lumen. Trypsin is released into the gut lumen as the inner vesicle membrane fuses with the outer membrane and/or as the vesicles disintegrate due to the high pH in the lumen (Santos and Terra, 1984; Santos et al., 1986). A similar process occurs in *Aedes aegypti* (Graf et al., 1986), in which the vesicles containing soluble trypsin fuse with the membranes of the microvilli, releasing trypsin by exocytosis. Jordao et al. (1996) found that trypsin in larval midgut cells of the housefly, *Musca domestica*, is initially bound to membranes by a small peptide anchor, processed in the Golgi complex, and enclosed in the membrane-bound form in secretory vesicles. These vesicles fuse with the plasma membrane at the gut/lumen interface, and the trypsin, thus exposed to the gut pH (near neutral), is released by a conformation change of the anchoring peptide. Enzyme processing and secretion is very likely a costly process in any case, but holocrine and apocrine secretion cause the loss of all or parts of cells, necessitating extensive repair or replacement. Replacement cells grow in from regenerative cells.

2.8.1 Carbohydrate Digesting Enzymes

Carbohydrate digesting enzymes are secreted by the salivary glands as well as by the midgut epithelium. Dietary starch is the typical nutritive complex carbohydrate (excepting cellulose, which most insects cannot digest) ingested by phytophagous insects, and glycogen is a complex carbohydrate ingested by carnivorous insects. **α-Amylase**, acting upon starch and glycogen, is a common digestive enzyme in insects. It attacks interior glucosidic linkages of starch and glycogen, thus giving rise to a mixture of shorter dextrins. **α-Glucosidase** and oligo-1,6-glucosidase (**isomaltase**) assist in digesting the smaller dextrins, releasing glucose. Many insects also have one or more **α-** or **β-glycosidases** that digest a broad range of small carbohydrates. α-Glucosidase hydrolyzes maltose, sucrose, trehalose, melezitose, raffinose, and stachyose; α-galactosidase hydrolyzes melibiose, raffinose, and stachyose. An **α,α-trehalase** in the gut of some insects digests the trehalose that occurs in the body of other insects preyed upon as food. **β-Glucosidase** attacks cellobiose, gentiobiose, and methyl-β-glycosides. Lactose is hydrolyzed to glucose and galactose by **β-galactosidase**, and **β-fructofuranosidase** acts upon sucrose and raffinose to release simple sugars. An insect usually has only a few of these carbohydrate digesting enzymes, depending on the food it eats. Honeybees, *Apis mellifera,* have several α-glucosidases or sucrases that act rapidly upon sucrose, usually the principal carbohydrate in the nectar taken by honeybees. They utilize the resulting glucose and fructose for an immediate energy source and for making honey.

Termites feed on and digest cellulose, as do some beetles, a few cockroaches, and woodwasps in the family Siricidae. Cellulose cannot be completely digested by one cellulase enzyme; the crystalline structure and the β-linkage of glucose units in cellulose make it difficult to hydrolyze, and complete digestion requires a complement of three enzymes — **C_x-cellulase**, **C_1-cellulase**, and **β-glucosidase** — acting in sequential attacks. Although there are some reports to the contrary, Martin (1983) concluded that no insect is able to secrete the complete complement of enzymes from its own cells. C_x-cellulase and β-glucosidase enzymes are secreted by many insects, but the C_1-cellulase must be obtained from protozoa or bacterial symbionts (some termites, beetles, and cockroaches) or from fungi ingested with the food (some fungus-culturing termites, some beetle larvae, and woodwasp larvae). The C_x-cellulase disrupts the crystalline structure of cellulose, and the C_1-cellulase digests long chains of cellulose into shorter cellobiose chains. β-Glucosidase releases glucose from cellobiose.

2.8.2 Lipid Digesting Enzymes

Most of the fat eaten by an insect consists of **triacylglycerols**. **Lipases** secreted from the midgut, and in some insects probably from symbionts as well, release fatty acids and glycerol from triacylglycerols. In the few insects studied, hydrolysis of triacylglycerols appears to proceed slowly. Slow hydrolysis may be caused by the choice of substrate tested; natural triacylglycerols are often complex mixtures of different chain-length fatty acids esterified with glycerol, whereas test substrates are often triolein or tripalmitin, in which all three fatty acids are oleic or palmitic, respectively. Emulsifying agents that enable hydrophilic enzymes to contact the hydrophobic surface of the triacylglycerol also are likely to be important in the digestion of lipids; natural emulsifying agents are largely unknown in the insect gut, but amino acids, proteins, and fatty acylamino complexes act as emulsifiers in some insects. Components of the glycocalyx layer in the gut may aid in emulsifying fats and in promoting contact between lipases and triacylglycerols.

2.8.3 Protein Digesting Enzymes

Protein digestion in insects is mediated by several classes of protein digesting enzymes, some of which are free in the gut lumen while others are membrane bound. The significance of having different protein digesting enzymes is that some proteases are **endoproteases** that attack large proteins internally at the linkage between certain amino acid, thus breaking the protein into smaller polypeptides, while **exopeptidases** attack the smaller pieces by cutting off the terminal amino acid. In addition, the presence of different types of proteinase inhibitors in the food eaten, especially plant-derived foods, may make it advantageous to have a variety of protein digesting enzymes so that some enzyme molecules will escape inhibition. **Trypsin**, an endoproteinase, is a common component of midgut secretions in many insects. It attacks a protein at peptide bonds in which the carbonyl function comes from lysine or arginine. **Chymotrypsin-like endoproteinases** have been found in several insects, including a cockroach, some beetles, some mosquito larvae, and some wasps and hornets. Chymotrypsin can attack at phenylalanine, tryptophan, and tyrosine residues.

Proteinases are classified as **serine**, **cysteine**, **aspartic acid**, and **metallo-proteinases**, depending on the amino acid or metal at the active site of the enzyme (Barrett and Rawlings, 1991). The enzymes are further characterized by their sensitivity to specific inhibitors that act on the amino acid(s) at the active site, by use of specific substrates that are attacked only by certain types of proteinases, and by the pH for optimum activity. Some proteinases act optimally at alkaline pH, while others have maximum activity at acidic pH.

Serine proteinases include trypsin- and chymotrypsin-like proteinases and elastases that work at alkaline pH. These endoproteinases have been demonstrated in the midgut of many insects. Cysteine and aspartic acid proteinases have mildly acidic pH optima. Proteinases with acidic pH optima also have been called **cathepsins**. Trypsin, chymotrypsin, and aminopeptidases are common in many insects, but there are no reports of secretion in the same insect of proteinases active at both acidic and alkaline pHs. The type of proteinase secreted and gut pH obviously must be coordinated in order for effective digestion to occur. The members of a taxonomic group, however, may not all have the same type of proteinases. Many beetles have cysteine proteinases most active at slightly acidic pH (Thie and Houseman, 1990; Wolfson and Murdock, 1990). Some members of the family Scarabaeidae have serine proteinases that act at the high midgut pH typical of these insects, and they have no detectable cysteine proteinases (McGhie et al., 1995). Lepidoptera typically secrete trypsin-like enzymes (Valaitis, 1995), which are most active at alkaline pH and thus have a generally favorable pH in the midgut due to goblet cell secretion of potassium into the gut lumen.

Evidence for the influence of food on type of proteinase secreted is weak (Wolfson and Murdock, 1990; Thie and Houseman, 1990). Blood-feeding insects may have either serine proteinases or cathepsins. For example, the major proteolytic enzyme in the midgut of female *Aedes aegypti* is trypsin-like (Borovsky & Schlein, 1988), as is also the case in *Stomoxys calcitrans*, two blood-feeding dipterans. *Rhodnius prolixus* (Hemiptera), also a blood feeder, secretes cathepsins active at acidic pH, but has no trypsin-like enzymes. In *S. calcitrans* and *R. prolixus,* the blood meal acts as a stimulus for secretion of the extracellular proteinases, but not for levels of membrane-bound aminopeptidases (Houseman et al., 1985).

The endoproteinases and some exoproteinases must pass through the peritrophic membrane to promote digestion of the large proteins. Smaller proteins and peptides released by the digestive processes may diffuse out the peritrophic membrane through pores of the PM, with final digestion taking place at the surface of the microvilli, where there are both free and membrane-bound **aminopeptidases.** Aminopeptidases (also called **exopeptidases**) are exoenzymes that remove one amino acid after another from the end of a small peptide. Membrane-bound aminopeptidase activity has been found in *Rhodnius prolixus* (Hemiptera), *Stomoxys calcitrans* and various other Diptera, Lepidoptera, and Coleoptera. A carabid beetle, *Pheropsophus aequinoctialis*, has both free and membrane-bound aminopeptidases. When the exopeptidase is bound at the microvillar surface, the amino acids cut from the end of a polypeptide are already at the site, the microvilli, for absorption.

2.8.4 Do Proteinase Inhibitors in the Food Influence Evolution of Proteinase Secreted?

There may be a relationship between the type of proteinase inhibitors in food and the evolutionary selection for type and number of proteinase secreted by insects (Terra, 1988), but only a few insects have been studied. Many plants contain varying levels of naturally occurring proteinase inhibitors, some of which inhibit serine proteinases while others act on cysteine proteinases. Some insects respond to the consumption of a trypsin inhibitor by secreting additional trypsin-like enzyme(s)

TABLE 2.1
The 20 Amino Acids Found in Proteins

Amino Acid	Three-Letter Designation	Single-Letter Designation	Molecular Weight
Alanine	Ala	A	89
Arginine	Arg	R	174
Asparagine	Asn	N	132
Aspartic acid	Asp	D	133
Cysteine	Cys	C	121
Glutamic acid	Glu	E	147
Glutamine	Gln	Q	146
Glycine	Gly	G	75
Histidine	His	H	155
Isoleucine	Ile	I	131
Leucine	Leu	L	131
Lysine	Lys	K	146
Methionine	Met	M	149
Phenylalanine	Phe	F	165
Proline	Pro	P	115
Serine	Ser	S	105
Threonine	Thr	T	119
Tryptophan	Trp	W	204
Tyrosine	Tyr	Y	181
Valine	Val	V	117

that allow differential susceptibility to the inhibitor and/or hyperproducing enzymes to compensate for inhibition (Broadway, 1995; Broadway and Villani, 1995).

Many insects secrete more than one proteinase, including multiple molecular forms of some proteinases (Moffatt et al., 1995, and references therein). At least 20 trypsin isozyme bands were detected by isoelectric focusing in midgut homogenates of *A. aegypti*, with 5 of them representing major bands (Graf and Briegel, 1985). Multiple isozymes and hyperproduction of some enzymes may enable some measure of escape from ingested inhibitors. Transgenic plants containing proteinase inhibitors have been tested and proven to have adverse effects on the growth of some insects (McManus et al., 1994). Nevertheless, it remains to be seen if this technique can be successfully used in insect control, and if so, for how long before insects develop coping mechanisms.

2.9 HORMONAL INFLUENCE ON MIDGUT

Hormonal control of digestive enzyme secretion does not appear to be a widespread mechanism among insect groups, but a well-defined case occurs in *Aedes aegypti* female mosquitoes. The females must have a series of blood meals to mature successive batches of eggs. The terminal oocytes in each of the numerous ovarioles of both ovaries mature together, and this set of mature eggs must be laid to make room for the growth and maturation of a second set of eggs. A decapeptide hormone (**trypsin modulating oostatic factor**, **TMOF**) (amino acid sequence YDPAPPPPPP) (Table 2.1), synthesized by ovarian follicular epithelium cells, is released 24 to 42 h post feeding and transported by the hemolymph to receptors on the hemolymph side (basal side) of midgut cells (Borovsky et al., 1994a, 1994b). The hormone, possibly through a second messenger, signals midgut cells to cease producing late trypsin (see below) that is necessary to finish digesting the blood meal. Younger and secondary oocytes cease growing, apparently from lack of nutrients, until the inhibition of digestion is released after the female lays the mature eggs (Borovsky et al., 1990, 1993). It still remains to be determined how TMOF acts upon midgut cells, and how the inhibition is released so that the next batch of eggs can be nourished and matured. The overall process is highly adaptive for the mosquito because the abdomen of the female is not large enough to hold continuously maturing eggs.

Two forms of trypsin are secreted in *Aedes aegypti* and both forms appear to be critical to the ultimate formation of eggs. An **early trypsin** (Graf and Briegel, 1989) is present in the midgut during the first 4 hours after feeding (Noreiga et al., 1996). Transcription of the early trypsin gene is controlled by juvenile hormone (JH) level, and occurs in the midgut after adult emergence (Noriega et al., 2001). Translation of mRNA occurs in the midgut cells prior to feeding (Felix et al., 1991). Early trypsin begins the process of blood digestion, and its digestive action on the blood meal provides the necessary stimulus (not yet identified, but amino acids fail to stimulate) for inducing gene transcription for late trypsin synthesis and secretion (Barillas-Mury et al., 1995). **Late trypsin** is the major endoproteinase in the midgut and it is necessary to finish the blood meal digestion. Late trypsin synthesis is the form primarily influenced by TMOF. Sufficient amino acids result from early trypsin action to promote maturation of a batch of eggs before TMOF stops late trypsin synthesis. TMOF, which has been synthesized, can survive potential protease degradation and be absorbed in an active form from the midgut when administered to mosquitoes in a blood meal, offering hope that the synthetic hormone may have potential for population control (Borovsky and Mahmood, 1995).

A similar hormonal mechanism involving TMOF has been demonstrated in the grey fleshfly *Neobellieria bullata*, except that the neuropeptide is a hexapeptide (amino acid sequence NPTNLH), called **Neb-TMOF** (Borovsky et al., 1996). Neb-TMOF has also been isolated from larvae of the blue blowfly, *Calliphora vicina*, and it seems to have dual functions, acting in larvae as an **ecdysiostatin** that inhibits the synthesis of ecdysteroid, while inhibiting synthesis of trypsin in the midgut of adult females (Hua et al., 1994; Hua and Koolman, 1995).

The midgut of many insects appears to be a rich source of (putative) endocrine secretions (Sehnal and Zitnan, 1990), but the identity of the secretory products and their functions are largely unknown. The midgut in liver-fed black blowflies, *Phormia regina*, releases a hormone targeted for the brain apporximately 4 to 8 hours after the meal. In the brain, it stimulates median neurosecretory cells to initiate the neuroendocrine cascade (see Chapter 15), leading to ovary and egg development (Yin et al., 1993, 1994). Additional research is needed to determine further details and whether similar hormones occur in other insects. Midgut cells may be involved in secreting neuropeptides that mediate many aspects of digestion and/or enzyme secretion, but few details are currently available.

2.10 COUNTERCURRENT CIRCULATION OF MIDGUT CONTENTS AND ABSORPTION OF DIGESTED PRODUCTS

Berridge (1970) proposed that insects might have a **countercurrent circulation** of fluid contents in the midgut, called the **endo-ectoperitrophic countercurrent flow**. Such a process could (1) serve to increase digestive efficiency, (2) conserve nutrients that might be lost by a rapid passage through a short midgut, (3) conserve and reuse enzymes that would otherwise be excreted rapidly with the bulk of food moving through the gut, and/or (4) allow absorption of digested products along the entire length of the midgut and/or absorption in the gastric caeca by permitting fluid containing digested products to flow forward. Dow (1986) has suggested that the evolutionary driving force for a countercurrent circulation of midgut contents may have been nutrient conservation.

In a countercurrent circulation, food and gut contents within the peritrophic membrane move posteriorly, after having entered the midgut from the foregut, while fluid containing partially or completely digested food materials outside the peritrophic membrane moves anteriorly between the midgut cell surfaces and the peritrophic membrane (Figure 2.14). The anterior movement of fluids outside the peritrophic membrane is promoted by fluid secreted into the ectoperitrophic space

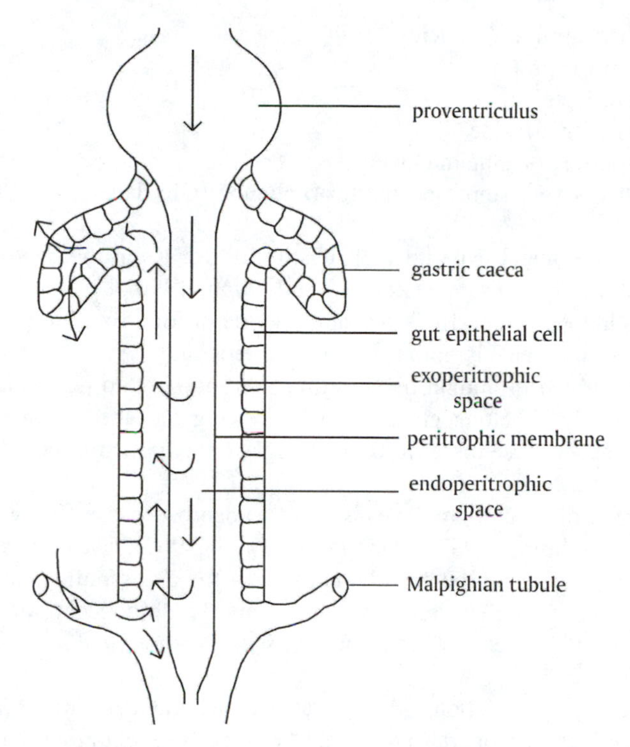

FIGURE 2.14 Diagrammatic illustration of the countercurrent flow that occurs in the midgut of some insects. Fluid may be passed forward from the Malpighian tubules to help create the forward flow that carries small end products of digestion to the gastric caeca, which are very efficient in absorption of fluid and nutrients.

by the posterior midgut or, in some cases, by the Malpighian tubules (Dow, 1981). Fluids and dissolved digestion products may be absorbed all along the midgut by the columnar epithelial cells, but in some insects the gastric caeca rapidly absorb fluids and dissolved nutrients brought to them by the countercurrent flow. Only a few species have been studied critically with respect to this mechanism, and it is not clear how common it is in insects. Dow (1981) suggested four criteria for determining whether an endo-ectoperitrophic flow occurred in insects:

1. A posterior region of the gut (could include Malpighian tubules) should be specialized for secretion of fluid into the posterior midgut.
2. An anterior region of the midgut should be specialized for absorption.
3. There should be a concentration gradient of small molecules between the point of fluid secretion and that of absorption.
4. Metabolites from a meal should remain in the gut longer than the bulk of the solid food originally ingested.

Absorption of amino acids from the midgut has been studied in only a few insects; but in several lepidopterans (*Manduca sexta, Philosamia cynthia, Bombyx mori*), amino acids are actively absorbed by transport proteins specific to particular amino acids (Dow, 1986). Several specific transport proteins have been isolated and identified (Giordana et al., 1989). The driving energy for amino acid absorption from the gut in these lepidopterans is the proton-ATPase pump described in goblet cells (Section 2.4.3). The pump creates the high K^+ concentration in the gut lumen and the high transepithelial potential across the gut wall, both shown to be important for amino acid absorption by columnar cells in lepidopterans. There are at least six transport systems, including:

1. Transporter of neutral amino acids
2. Specific system for proline
3. Specific system for glycine
4. Specific system for L-lysine
5. Specific transporter for glutamic acid
6. Transporter that is very stereospecific for D-alanine (Giordana et al., 1989)

Sodium ions are an efficient experimental substitution for K^+ in the neutral transport system, and in some other transporter systems (Reuveni and Dunn, 1990, 1993; Hennigan et al., 1993a, 1993b), but in the lepidopteran gut, K^+ is probably the major ion involved because it is present in high concentration, whereas Na^+ is not. Transport proteins and systems for leucine and tyrosine also have been demonstrated in midgut tissue of the Colorado potato beetle, *Leptinotarsa decemlineata* (Reuveni et al., 1993; Hong et al., 1995). Blocking the absorption of amino acids and disruption of their transport systems have been sugggested as potential targets for insect control (Hong et al., 1995).

Glucose from digestion of carbohydrates is rapidly absorbed passively by a process known as **facilitated diffusion** (Treherne, 1957, 1958, 1959; Wyatt, 1967). Fat body cells, which can be found attached in small groups on the hemolymph side of the gut, rapidly synthesize the absorbed glucose into the disaccharide trehalose, keeping the hemolymph concentration of glucose low in most insects. Consequently, even low concentrations of glucose in the gut can continue to be absorbed passively.

Fatty acids undergo esterification as they traverse the gut cells and are released into the hemolymph as diacylglycerols for transport to fat body cells (Weintraub and Tietz, 1973, 1978; Turunen, 1975; Turunen and Chippendale, 1977; Chino and Downer, 1979; Thomas, 1984). The diacylglycerols are picked up at the hemolymph side of the gut by lipoprotein complexes, **lipophorins**, that enable transport of the absorbed lipids through the aqueous hemolymph. Lipids are mainly stored in fat body cells as triacylglycerols. After the lipophorin delivers the absorbed lipid to the fat body, it can recirculate to load and transport additional lipids absorbed. (Chino and Kitazawa, 1981; Chino, 1985; Surholt et al., 1991; Van Heusden et al., 1991; Gondim et al., 1992; Blacklock and Ryan, 1994). Additional details on lipophorin and lipid transport are given in Chapter 6, Intermediary Metabolism.

2.11 THE TRANSEPITHELIAL AND OXIDATION-REDUCTION POTENTIAL OF GUT

There is a **transepithelial potential** (**TEP**) across the gut wall, and an oxidation-reduction or **redox potential** within the lumen of various parts of the gut. In most insects it is not known how these potentials are created or controlled. TEP values ranging from lumen negative to lumen positive occur in different insects. In *P. americana* midgut, the TEP varied from –8 to –26 mV (lumen negative to hemolymph), while the TEP in the hindgut ranged from Eo = –84 to –240 mV (Bignell, 1981). High negative values, such as those in the hindgut of insects (termites) with large populations of microorganisms that digest cellulose, indicate anaerobic conditions, while positive or slightly negative values indicate aerobic conditions. Insects such as clothes moths, dermestid beetles, and bird lice that digest keratin, the major protein of wool, fur, and feathers, have strongly negative redox (reducing) potentials in the midgut. The reducing conditions facilitate the breaking of disulfide bonds in the keratin molecule, making keratin more digestible.

The redox potential in various regions of the gut may play a large role in digestion and assimilation of food materials, in detoxication reactions, and in production of toxic metabolites from ingested food materials. The ability to adjust gut redox potentials in plant-feeding insects may be one of the ways that insects adapt in the co-evolutionary race with plants in combating the wide range of allelochemicals common in many plants. Appel and Martin (1990) found reducing conditions in the midguts of two lepidopterans, *M. sexta* and *Polia latex*, which they suggest would make ingested phenolic compounds less likely to be oxidized to highly toxic and reactive quinones than the oxidizing potentials prevailing in the midguts of several other lepidopterans studied. Oxygen levels in the foregut and midgut lumens of ten species of caterpillars and three species of grasshoppers were generally very low, indicating nearly anoxic conditions (O_2 equal to or less than 7.3 mmHg), and the gut was able to deplete oxygen caused by swallowing oxygen with the food or feeding an artificial diet that increased oxygen tension in the gut in some cases (Johnson and Barbehenn, 2000). The authors suggest that low oxygen tension in the gut may be very common in herbivorous insects, and that it is an adaptation to reduce the rate of oxidation of ingested plant allelochemicals that may be more toxic when oxidized.

2.12 GUT pH

Few generalizations about insect **gut pH** can be made, except that it is highly variable in different insects (Table 2.2). The pH of a gut segment greatly influences the action of any enzymes secreted into or carried with the food into that segment. In addition, gut pH may influence the solubility of ingested components, the toxicity of some potential toxins, and the population of gut microorganisms. Cathepsin and trypsin-like enzymes attacking proteins work optimally at acidic and alkaline pHs, respectively. Carbohydrate digesting enzymes usually work best at near-neutral pH or under slightly acidic conditions. Lipases that digest triglycerides and other esters work best at alkaline pHs near 8.

The crop tends to be slightly acidic in most insects, with little or no presence of buffering agents that could alter pH due to organic acids produced when digestion occurs in the crop, as it does in Orthoptera, Dictyoptera, and some other insects. When proteins are the primary food material for the cockroach, *P. americana*, the crop is slightly acidic at pH 6.3, but when sugars (maltose, lactose, sucrose, or glucose) are eaten, the crop has a pH of 4.5 to 5.8 because of the glycolytic cycle acids produced.

Larvae of Lepidoptera and Trichoptera tend to have a very high midgut pH, varying from about 8 to 10, promoted by goblet cells that secrete potassium bicarbonate into the lumen of the midgut. Lepidoptera caterpillars are predominately phytophagous, and Berenbaum (1980) concluded from a survey of published pH values for 60 species in 20 families that midgut pH was related to host plant chemistry. Those larvae that fed on leaves of trees, which typically contain larger quantities of tannins, had an average midgut pH of 8.67, while those that fed mostly upon herbs and forbs had an average midgut pH of 8.29. The higher midgut pH in those feeding on tannin-rich food may have evolved as a protective mechanism to reduce the toxicity of tannins, which tend to complex with proteins but do so less readily at higher pH.

Very acidic conditions prevail in the special hindgut regions of some termites, crickets, and possibly other insects that have hindgut fauna to digest cellulose. The acidic conditions are caused by the anaerobic fermentation of glucose from cellulose digestion, resulting in production of short-chain fatty acids, including acetic, propionic, and butyric acids.

2.13 DIGESTIVE SYSTEM MORPHOLOGY AND PHYSIOLOGY IN MAJOR INSECT ORDERS

Detailed studies of digestion have been made in only a relatively few insects within the major insect orders. Because of the diversity of insects, it is always risky to generalize from studies on a few insects, and the reader should keep in mind that there may be exceptions in minor, or even major, details from these limited studies. More extensive details on food habits and related gut structure and function can be found in reviews by Dow (1986), Terra (1990), and Billingsley (1990).

TABLE 2.2
The pH in Various Parts of the Gut of Selected Insects

Insect	Foregut	Midgut	Hindgut	Ref.[a]
		Orthoptera		
Melanoplus sanguinipes, grasshopper (Acrididae)	5.52	6.75	6.80	1
Photaliotes nebrascensis, grasshopper (Acrididae)	6.03	7.12	6.11	1
Schistocerca gregaria, desert locust (Acrididae)	5.5	6.7–7.0		4
Schistocerca gregaria		5.3		5
Gryllus rubens, cricket (Gryllidae)	5.8–6.0	7.4–7.6	7.6–7.8 (anterior hindgut)	7
Gryllus bimaculatus, cricket (Gryllidae)	5.84 (crop)	8.07	8.50–7.59 (illeum–rectum)	19
Scapteriscus borelli, mole cricket (Gryllotalpidae)	5–7	6–8 (gastric caeca)	7–8 (anterior hindgut)	7
Periplaneta americana, cockroach (Blattidae)		6.3		8
Leucophaea madeirae, cockroach (Blattidae)		9.5 (posterior midgut)		9
		Coleoptera		
Popillia japonica larvae, Japanese beetle (Scarabaeidae)		8.5		2
Exomala orientalis larvae, Oriental beetle (Scarabaeidae)		8.5–9.0		2
Rhizotrogus majalis larvae, European chafer (Scarabaeidae)		9.0–9.5		2
Maladera castanea larvae, Asiatic garden beetle (Scarabaeidae)		8.5		2
Lichnanthe vulpina larvae, cranberry root grub (Scarabaeidae)		8.5		2
Phyllophaga anixia (Scarabaeidae)		8.5–9.0		2
Oryctes nasicornis (Scarabaeidae)		12.2		14
Epilachna varivestis larvae, Mexican bean beetle (Chrysomelidae)		5.8		3
Anthonomus grandis larvae, boll weevil (Curculionidae)		4.6–5.6		3
Tribolium castaneum, red flour beetle (Tenebrionidae)		6.0		18

		Lepidoptera		
Agrotis ipsilon, black cutworm (Noctuidae)		8.5–9.0		2
Manduca sexta, tobacco hornworm (Sphingidae)		9.5–9.7		5
Manduca sexta		6.4 apical folds, anterior midgut		
		8.2 basal folds, anterior midgut		
		7.2 apical folds, posterior midgut		6
		Diptera		
Simulium vitatum, blackfly (Simuliidae)	11.4			15
Tipula abdominalis, cranefly (Tipulidae)	11.6			16
Lucilia cuprina larvae, blowfly (Calliphoridae)		7.4–8 anterior midgut		
		3.3 middle midgut		
		7.4–8 posterior midgut		17
		Tricoptera		
Caddisfly larvae	7			10
		Plecoptera		
Stonefly larvae (Pteronarcyidae)	7			11
		Isoptera		
Termites		>10 anterior midgut		12
5 species of soil-feeding termites (Termitidae: Termitinae)	Slightly acid	Slightly acid to slightly alkaline in different species	11–12.5 most anterior hindgut; >10 in second dilation of hindgut; slightly >7 approaching rectum; 4.8–6 in rectum	13

[a] References for data in Table 2.1; additional data and references can be found in Berenbaum (1980). 1. Barbehann et al. (1996); 2. Broadway and Villani (1995); 3. Murdock et al. (1987); 4. Evans and Payne (1964); 5. Martin et al. (1987); 6. Dow and O'Donnell (1990); 7. Thomas and Nation (1984); 8. O'Riordan (1969); 9. Engelmann and Geraert (1980); 10. Martin et al. (1981a); 11. Martin et al. (1981b); 12. Bignell and Anderson (1980); 13. Brune and Kühl (1996); 14. Bayon (1980); 15. Undeen (1979); 16. Martin et al. (1980); 17. Waterhouse and Stay (1955); 18. Krishna and Saxena (1962); 19. Teo (1997).

2.13.1 Orthoptera

The crop is a major site of digestion in locusts, grasshoppers, and crickets, and possibly in other Orthoptera. Starch digestion is accomplished in the crop of crickets with enzymes secreted forward from the midgut. Salivary enzymes play only a minor role in digestion. Midgut caeca located at the anterior end of the midgut in locusts and crickets rapidly absorb fluids and dissolved nutrients as these enter from the crop. A Type I peritrophic membrane is secreted in the midgut. Cellulase activity is found in the midgut of some grasshoppers, but its origin and the extent of its function are uncertain. Orthoptera may not generally have the endo-ectoperitrophic countercurrent flow in the midgut, although in *Schistocerca gregaria*, countercurrent flow occurs in starved locusts, but not in constantly feeding locusts. Terra (1990) has suggested that the countercurrent flow in a starved individual may be an adaptation to keep food and digestive enzymes in the midgut longer for more complete digestion. This might represent a trade-off of averting starvation vs. keeping potential allelochemicals in the gut longer. The gut of the praying mantis, *Tenodora sinensis*, exhibits extreme modifications to serve the predatory habits and intermittent feeding of this carnivorous insect. The foregut, especially the crop, is long and wide, and occupies nearly the entire length of the body, apparently as an adaptation for storage of opportunistically available prey (Dow, 1986). The midgut, eight gastric caeca, and the hindgut are shortened and compressed into the last three abdominal segments.

2.13.2 Dictyoptera

Cockroaches are scavengers and opportunistic feeders. The crop is large in cockroaches, and is a major organ of digestion. Amylase from the salivary glands, and protease, lipase, and carbohydrate digesting enzymes secreted forward from the midgut contribute to digestion in the crop. Crop emptying is gradual and is regulated by the osmotic pressure created by the small molecules resulting from digestion of crop contents (Englemann, 1968). The higher the osmotic pressure in the crop, the slower the crop empties into the midgut, functionally preventing oversaturation of absorption by gastric caeca and possible loss of poorly absorbed nutrients. Gastric caeca located at the anterior of the midgut are the main sites for absorption. A Type I peritrophic membrane is present, and there may be some countercurrent endo-ectoperitrophic flow. Some final digestion probably occurs in the ectoperitrophic space on the surface of the midgut cells. *P. americana* incorporates ^{14}C into hemolymph trehalose from labeled cellulose (Bignell, 1977), but the digestion of the cellulose occurs in the hindgut (colon) with the aid of cellulases from bacteria located on the gut lumenal wall. The redox potential of the hindgut favors the action of these anaerobic bacteria by varying from –84 to –240 mV, values that are indicative of an anaerobic gut segment (Bignell, 1981). Short-chain fatty acids are produced by bacterial fermentation of glucose liberated from cellulose in the colon. The fatty acids are absorbed through the hindgut wall.

2.13.3 Isoptera

The gut of termites is highly specialized for housing gut microbiota that aid in digestion of cellulose that they obtain from wood, fungi, or other sources, depending on their lifestyle. Gut variation exists among the castes in a colony. For example, soldiers in the family Rhinotermitidae are fed liquid food by the worker caste, and therefore do not have to digest cellulose so they have reduced gut structure. The workers are the social caste responsible for colony construction and nutrition and have highly evolved gut chambers to hold various types of microbiota. Termites hatch without their gut microbiota, but soon receive them by feeding on fluid and excreta from the proctodeum of older nymphs. They lose most of their gut symbionts at each molt and become reinfected by proctodeal feeding.

The so-called lower termites have flagellate protozoans as well as bacteria in the hindgut, and they get their cellulase(s) from their symbionts. Those termites belonging to the “higher termites” in the family Termitidae lack symbiotic protozoa, but have symbiotic bacteria in the hindgut, which is divided into five segments. Many of the higher termites feed on fungi (Anklin-Mühlemann et al., 1995) and some (except fungus-growing Macrotermitinae, which get their cellulases from the conidiophores of a fungus growing in their nests) may be able to secrete their own cellulases (but see Martin, 1983, for a different opinion).

Spirochetes are present in the hindgut of many termites, but their role in digestion is uncertain. They may help to recycle nitrogen, synthesize amino acids, assist in maintaining a low redox potential, and protect from various pathogens (Breznak, 1982; Boucias et al., 1996). The diet of termites tends to be low in protein content, and they acquire protein from the bacterial cells in feces by trophyllaxis and by feeding on the fecal wastes of each other. Symbionts in the hindgut also can fix atmospheric nitrogen and synthesize proteins from the fixed nitrogen. A peritrophic membrane is usually present in termites (Noirot and Noirot-Timothee, 1969). Protein is digested in the midgut, and some termites probably have an endo-ectoperitrophic flow of fluids and enzymes.

Wood-feeding termites tend to have acetogenic bacteria that ferment glucose to acetate, while some fungus-growing and soil-feeding termites evolve methane from anaerobic fermentation (Brauman et al., 1992). The principal metabolite from cellulose digestion is glucose, which is fermented to acetate that is actively absorbed by the hindgut cells through an energy requiring mechanism. For example, cellulolytic organisms in the hindgut of *Reticulitermes flavipes* ferment glucose to acetate, CO_2, and H_2, as shown in Equation (2.1):

$$C_6H_{12}O_6 + 2\ H_2O \rightarrow 2\ CH_3COOH + 2\ CO_2 + 4\ H_2 \tag{2.1}$$

Then, CO_2-reducing acetogenic bacteria convert the free H_2 and CO_2 to an additional acetate, according to Equation (2.2):

$$4\ H_2 + 2\ CO_2 \rightarrow CH_3COOH + H_2O \tag{2.2}$$

Termites absorb the acetate from the hindgut and utilize it for an energy source by metabolism in the Krebs cycle. Some fungus-growing termites convert the H_2 and CO_2 from the initial fermentation of glucose (Equation 2.1, above) into methane (CH_4) rather than into another acetate molecule. Some investigators have suggested that termites are a significant environmental source of methane, a greenhouse gas, but Brauman et al. (1992) caution that much more data are needed about the distribution of the two processes among termites to make accurate estimates.

2.13.4 Hemiptera

R. prolixus (family Reduviidae), the "kissing bug," has been a favorite model hemipteran for the study of digestion. Each instar takes just one blood meal (if allowed to feed to repletion), digests it slowly, utilizes the nutrients to support growth and molting to the next instar, and after molting takes another blood meal. The midgut is divided into two major divisions: the anterior midgut and the posterior midgut. Columnar, cuboidal, regenerative cells, and endocrine cells occur in the midgut. The columnar and cuboidal cells have microvilli on the apical surface and basal infoldings. There is no peritrophic membrane, but a peculiar perimicrovillar membrane composed of two trilaminar membranes forms continuously over the microvilli. The two membranes are held very close, but a constant distance apart, by structural columns or pegs (composition unknown). Little is known about the origin, ultimate fate, and function of the membranes. The anterior midgut functions in carbohydrate and lipid processing, among other functions, but does not secrete enzymes for protein digestion. The posterior midgut has separate functionalities in its anterior and posterior parts. The first part of the posterior midgut secretes cathepsins B and D (active at acidic pH), aminopeptidase, and carboxypeptidase, among other enzymes, and protein digestion is initiated. Digestion is completed in the posterior midgut, where additional protein digesting enzymes are secreted and absorption of products occurs. Endocrine cells are concentrated in this region of the midgut, but practically nothing is known of their function. The excellent review by Billingsley (1990) should be consulted for more details on gut structure and function in *Rhodnius*.

Little is known about digestion in other predatory reduviid bugs, but they probably secrete a complete set of digestive enzymes with their saliva as it is injected into the body of prey. Some additional digestion likely occurs in the midgut as partially digested food is swallowed. Many hemipterans are phytophagous and take plant sap or liquefied plant tissues. Seed-feeding hemipterans secrete enzymes into the seed, where a large amount of digestion takes place, but final digestion occurs in the midgut.

Unequivocal evidence of a peritrophic membrane has not been demonstrated in Hemiptera, most of whom take liquid or semiliquid food not likely to contain rough particles that could abrade the midgut microvilli. If the peritrophic membrane does protect from invading microorganisms, it raises the question of whether hemipterans have other protective means.

2.13.5 Homoptera

Homoptera take xylem or phloem sap, both of which are poor in amino acids and protein, but usually rich in sucrose (150 to >700 m*M*). Homoptera typically excrete a copious, dilute fluid, and in some, such as aphids, the fluid contains so much sugar that it is called honeydew. They have to ingest large volumes of fluid to get the amino acids, and then they have to get rid of the excess water and sucrose. A characteristic evolutionary feature of the gut in Homoptera is the filter chamber in which a loop of the hindgut is in intimate contact with part of the foregut and some fluid passes directly into the hindgut from the foregut without passing through the midgut. The filter chamber is able to concentrate gut fluid up to tenfold in some xylem feeders (Cicadoidea and Cercopoidea), but only about 2.5-fold in members of the Cicadelloidea, which are phloem feeders. Xylem feeders probably need to concentrate xylem fluids more because of the lower amino acid content (xylem, 3 to 10 m*M* amino acids) than do phloem feeders (phloem, 15 to 65 m*M* amino acids).

Heteroptera and Fulguroidea (Homoptera) secrete a lipid "membrane" that does not form a distinct sac like the peritrophic membrane, but nevertheless it does create a perimicrovillar space between the food mass and the microvillar surface of the cells.

2.13.6 Coleoptera

The crop is often absent or only slightly developed in beetle larvae and in adults of the Polyphaga, but is usually present in adult Adephaga. The crop is a site of considerable digestion in the lower (Adephaga) and some higher (Polyphaga) coleopterans by action of enzymes secreted forward from the midgut (Terra et al., 1985). Pre-oral digestion occurs in many of the predacious beetles, and predacious Carabidae complete the process of digestion in the crop by action of enzymes passed forward from the midgut.

Scarabaeid larvae, some of which feed on food containing cellulose, probably digest the cellulose with the aid of bacterial-derived cellulases, and sometimes they ingest the simpler breakdown products from fungal-digested cellulose. Adult coccinellids may have their own cellulases, but see Martin (1983). Cerambycid larvae live in logs and other wood and digest cellulose by ingesting fungal cellulases from their fungus-infected wood habitat. Their nutrition is marginal, and most have a slow growth pattern and long larval development time. Some Coleoptera (Polyphaga) have no, or a reduced, crop and have cathepsin-like proteinases rather than trypsin-like ones. This might be an evolved adaptation to the presence of trypsin inhibitors in some foods.

Tenebrio molitor has been a favorite coleopteran model insect for digestion studies as well as other physiological experiments because of its size and ease of rearing. It should not be assumed to be typical of beetles, however. The larvae produce amylase, cellobiase, and trehalase from the anterior part of the midgut, and trypsin from the posterior of the midgut. These enzymes are secreted by exocytosis into the gut lumen. Less than 5% of the total of some of the major digestive enzymes are excreted by larvae with the feces and other undigested food, which suggests that the larvae probably have an endo-ectoperitrophic countercurrent flow of food and digestion products. The majority of digestion appears to occur within the peritrophic membrane, although an aminopeptidase is bound to the microvillar surface, so some final digestion of smaller polypeptides likely occurs at the microvillar surface (Terra, et al., 1985; Ferreira et al., 1990).

2.13.7 Hymenoptera

The crop is not a major organ of digestion in Hymenoptera, and is reduced in size in many larvae; many hymenopterans also have lost the anterior midgut caeca characteristic of many other groups of insects. The midgut is closed off from the hindgut by a plug of cellular tissue in larval Apocrita (bees and wasps), and the connection does not open until just before pupation. Any undigested residue (for example, the shell of pollen grains in bees) can then be passed into the hindgut and voided with fecal material so that the gut is cleared before pupation. Larvae of the woodwasp (genus *Sirex*, Symphyta, Siricidae) acquire cellulase and xylanase from fungi ingested with the wood on which they feed. Pollen grains ingested by adult bees are not crushed or cracked by mouth or gut action, but the nutrients inside the pollen grains are dissolved and leached from the grains. The nearly empty pollen grain shells are accumulated in the hindgut and are excreted (only) during flight. There may be an endo-ectoperitrophic circulation within the midgut, but evidence is not conclusive.

2.13.8 Diptera

There is a prominent esophageal invagination into the midgut in larval mosquitoes, and midgut cells in a ring between the walls of the invagination secrete a Type II peritrophic membrane. The esophageal invagination thus acts like a chute to channel food into the stocking-like peritrophic membrane (PM). Continued entry of food from the foregut seems necessary to force the lengthening of the PM. There are some differences in the formation of the PM in anopheline and culicine mosquito adults, but in all of them the peritrophic membrane is secreted only after a blood meal is ingested. A PM (or evidence of its formation) may be present as soon as 30 minutes after a blood meal, or only after several hours.

Caeca at the anterior end of the midgut are believed to be the major sites of absorption. Malpighian tubules transfer fluids to the midgut (Stobbart, 1971) and help create a countercurrent endo-ectoperitrophic flow. Adult mosquitoes have an immature midgut upon emergence and do not usually feed for some period of time. Both males and females take nectar, and females (only) also take blood meals for egg maturation.

Nectar taken by male and female mosquitoes is stored in a large, sac-like crop that is a diverticulum from the foregut, but blood meals taken by the females are passed directly into the midgut for the beginning of digestion. The midgut is differentiated functionally into an anterior and a posterior region. The anterior part secretes carbohydrate digesting enzymes, and nectar components are digested as fluid from the crop is passed into the anterior midgut. The arrangement keeps possible trypsin inhibitors that may be present in nectar away from the site of protein digestion, which occurs in the posterior midgut. Simple sugars resulting from digestion, or those already in the nectar, are absorbed in the anterior midgut.

The posterior midgut cells secrete trypsin-like enzymes, and protein (blood) digestion and absorption occur in the posterior midgut. The posterior midgut cells, more so than anterior midgut cells, have extensive microvilli and basal infoldings characteristic of secretion and absorptive processes. The midgut cells in this region get stretched by the large volume of blood that a mosquito takes if it is allowed to feed to repletion. Consequently, the cells have several types of connecting structures between cells to help hold them together and prevent excessive leaking of materials in or out between cells while they are stretched. For example, *Anopheles* species have septate desmosomes connecting the apical (nearest the gut lumen) third of adjacent cells. Culicine females have zonula continua attachments between adjacent cells near the apical apex, and desmosomes between cells in the basal region. Regenerative cells are common in both anterior and posterior midgut regions.

Cells believed to have endocrine function(s) are common in the posterior midgut. *A. aegypti* adults have about 500 such cells concentrated toward the posterior of the midgut (Billingsley, 1990) that, if they are endocrine cells, would make the midgut the largest endocrine organ of adult mosquitoes. Multiple cell types exist, suggesting the possibility of several functions and/or hormone products, but none has been identified yet.

Most larvae of the cyclorrhaphous flies (higher Diptera; including the housefly, *Drosophila* spp., and tephritid fruit flies) are saprophagous. The adults feed mostly on liquids or substances they can solubilize by regurgitating a droplet of fluid on the substance. Nectar, oozing fruit juices, sap, bird droppings, and honeydew on leaves or other substrate are utilized. Such nutrient-rich sources are likely to also contain bacteria, yeast, and possibly other microorganisms or fungi from environmental contamination, and these may be ingested with the fluid content as an additional source of nutrients.

Starch digestion occurs in the crop of houseflies by the action of salivary amylase. Final carbohydrate digestion may occur on the midgut cell surfaces by action of a membrane-bound maltase. Bacteria in the ingested food are likely to be killed by the low pH of the midgut and the action of lysozyme in the gut. Trypsin acts on proteins in the midgut, but final amino acids are liberated at the midgut cell surface by membrane-bound aminopeptidases.

The midgut of *S. calcitrans* (stable fly, a blood feeder) is divided functionally into three parts. The blood meal is stored temporarily in an anterior region of the midgut where no digestion appears to occur because the blood retains its bright red color. As the blood passes into a middle region of the midgut, known as the opaque zone, it changes color and becomes dark red or brown as it encounters the action of the trypsin-like enzyme. Midgut cells in the opaque zone synthesize trypsin-like enzyme as a zymogen (Moffatt and Lehane, 1990) and store it as granules that are released in part by an apocrine mechanism into the gut lumen. The zymogen is converted to the active enzyme when blood enters the opaque zone, but details of the conversion process have not been elucidated. Possible advantages of storing the enzyme as an inactive zymogen may be that the active enzyme can be made available quickly, and autodigestion of the insect's own midgut cells may be reduced when no blood is present. Digestion of the blood meal is completed and absorption occurs from a posterior region of the midgut.

2.13.9 Lepidoptera

Larvae of Lepidoptera have a very short foregut; a large, long, relatively straight midgut; and a short hindgut. There is no storage or digestion in the short, nearly vestigial foregut. Nearly all lepidopterous larvae are phytophagous feeders, and the gut modifications appear to be an adaptation to pass food quickly into the long midgut so that digestion can begin. Feeding is nearly continuous when plenty of food is available, and larvae may ingest more than their body weight in food daily. Food moves rapidly through the relatively straight gut, and frass droppings are frequent in phytophagous caterpillars. A Type I peritrophic membrane is present in the midgut of larvae. Larvae do not have gastric caeca. Digestion and absorption occur along the length of the midgut, with columnar cells secreting the enzymes and performing absorption. Goblet cells secrete K^+ into the midgut lumen, but do not seem to be involved in other gut functions. The presence of an endo-ecto-peritrophic countercurrent flow has been suggested on the basis that digestive enzymes are not rapidly excreted with the steady flow of frass droppings. Clear-cut evidence of such a flow is not available, and a countercurrent flow is to some extent counter-intuitive to the observed rapid movement of food through the gut.

Because the larval and adult forms of Lepidoptera have very different life histories and food habits, the adult gut is quite different from that of the larva. Many adult Lepidoptera feed only on nectar, which is stored in the crop and slowly released into the midgut for digestion to simple sugars. Some adult Lepidoptera have vestigial mouthparts and do not feed at all; they survive and (females) produce eggs at the expense of body substance, and generally live only a few days. In addition to nectar, adults of *Heliconius* butterflies feed on pollen, bird droppings, and other food sources.

In *Erinnyis ello* caterpillars initial digestion occurs in the endoperitrophic space, with final digestion occurring at the midgut cell surface by membrane bound enzymes. Digestion does not occur in the foregut.

An unusual food utilized by *Tineola bisselliella* larvae (clothes moth) is wool, and larvae have a very strong reducing action in the midgut that reduces disulfide bonds to sulfhydryl bonds, which facilitates further protein digestion by proteinases.

2.14 THE INSECT GUT AS A POTENTIAL TARGET FOR POPULATION MANAGEMENT AND CONTROL OF THE SPREAD OF PLANT AND ANIMAL DISEASE ORGANISMS

The gut, and particularly the midgut, has been recognized by many entomologists and disease vector specialists as a potential attack point for insect population control and/or control of transmission of the disease organism. The midgut is one of the principal points of entry for toxins, viruses, hormones, bacteria, and other potential agents that might be introduced into insects for population control. For example, a toxin that acts upon the midgut is produced by a family of bacteria *Bacillus thuringiensis* (Bt) (reviewed by Federici, 1993, 1999; Knowles, 1994). Different strains of the bacteria have variable toxicity levels for different insects, and some insects, including many beneficial ones, are not attacked by Bt. Particularly virulent strains have been discovered that are useful for biological control of some Lepidoptera, Diptera, and Coleoptera. The protoxin consists of a mixture of crystalline proteins, the δ-endotoxins. The δ-endotoxin crystals dissolve in the midgut of susceptible insects, releasing proteinacious toxins that range in size from 27 to 140 kDa. These are further broken into smaller toxic polypeptides by the insect's own protein digesting enzymes. Thus, by its own digestive action, the insect exposes itself to a wide variety of toxins. The CryIA(c) δ-endotoxin of *B. thuringiensis* binds to the brush border membrane-bound aminopeptidase in the midgut of the gypsy moth (Valaitis et al., 1995). Several, or all of the toxins bind to receptors on midgut microvillar surfaces, where they promote the opening of relatively large pores in the midgut cell membranes, with subsequent disruptions in cell osmotic balance, swelling, and eventual lysis. Even susceptible insects have shown the ability to develop resistance to Bt, and strategies are being explored to minimize resistance (McGaughey and Whalon, 1992).

The midgut also plays a role in the transmission of *Leishmania* parasites to humans. *Leishmania* parasites are passed from an infected host to a biting insect, and taken into the midgut of the insect with a blood meal where conditions may be favorable for the parasite to initiate rapid cell division into an early developmental stage, the promastigotes. If the insect is not a suitable host, the promastigotes soon die and are excreted with the fecal wastes of the insect. In susceptible hosts, such as sandflies, specific developmental changes in sugar residues on the surface of the promastigotes enable them to bind to midgut microvilli. As the attached parasites go through further developmental stages, including changes in the surface sugar residues, they are released again into the midgut, where they may be passed to a new host, possibly a human, by regurgitation during a sandfly bite (Pimenta et al., 1992). Agents that could prevent binding of the promastigotes to microvillar surfaces might break the transmission cycle to humans.

The introduction of proteinase inhibitors into commercial plants that would be specific against a particular pest is being investigated. While this is probably technically feasible, a great deal of research must be done to determine if significant control can be achieved, and whether resistance will develop rapidly.

REFERENCES

Anklin-Mühlemann, R., D.E. Bignell, P.C. Vievers, R.H. Leuthold, and M. Slaytor. 1995. Morphological and biochemical studies of the gut flora in the fungus-growing termite, *Macrotermes subhyalinus*. *J. Insect Physiol.,* 41:929-940.

Appel, H.M., and M.M. Martin. 1990. Gut redox conditions in herbivorous lepidopteran larvae. *J. Chem. Ecol.,* 16:3277-3290.

Baines, D.M. 1978. Observations on the peritrophic membrane of *Locusta migratoria* migratorioides (R. & F.) nymphs. *Acridia,* 7:11-22.

Barbehenn, R.V., M.M. Martin, and A.E. Hagerman. 1996. Reassessment of the roles of the peritrophic envelope and hydrolysis in protecting polyphagous grasshoppers from ingested hydrolyzable tannins. *J. Chem. Ecol.,* 22:1911-1929.

Barillas-Mury, C.V., F.G. Noreiga, and M.A. Wells. 1995. Early trypsin activity is part of the signal transduction system that activates transcription of the late trypsin gene in the midgut of the mosquito *Aedes aegypti*. *Insect Biochem. Mol. Biol.,* 25:242-246.

Barrett, A.J., and N.D. Rawlings. 1991. Proteinases: Types and families of endopeptidases. *Biochem. Soc. Trans.,* 19:707-715.

Bayon, C. 1980. Volatile fatty acids and methane production in relation to anaerobic carbohydrate frementation in *Oryctes nasicornis* larvae (Coleoptera: Scarabaeidae). *J. Insect Physiol.,* 26:819-828.

Berenbaum, M. 1980. Adaptive significance of midgut pH in larval Lepidoptera. *Am. Nat.,* 115:138-146.

Bernays, E.A., and D.J. Chamberlain. 1980. A study of tolerance of ingested tannin in *Schistocerca gregaria*. *J. Insect Physiol.,* 26:415-420.

Berridge, M.J. 1970. A structural analysis of intestinal absorption. *Symp. R. Entomol. Soc. London,* 5:135-150.

Beyenback, K.W. 1995. Mechanisms and regulation of epithelial transport across Malpighian tubules. *J. Insect Physiol.,* 41:197-207.

Beyenback, K.W., T.L. Pannabecker, and W. Nagel. 2000. Central role of the apical membrane H^+-ATPase in electrogenesis and epithelial transport in Malpighian tubules. *J. Exp. Biol.,* 203:1459-1468.

Bignell, D.E. 1977. An experimental study of cellulose and hemicellulose degradation in the alimentary canal of the American cockroach. *Can. J. Zool.,* 55:579-589.

Bignell, D.E. 1981. Nutrition and Digestion, pp. 57-86, in K.G. Adiyodi and W.J. Bell (Eds.), *The American Cockroach*, Chapman & Hall, London.

Bignell, D.E., and J.M. Anderson. 1980. Determination of pH and oxygen status in the guts of lower and higher termites. *J. Insect Physiol.,* 26:183-188.

Billingsley, P.F. 1990. The midgut ultrastructure of hematophagous insects. *Annu. Rev. Entomol.,* 35:219-248.

Blacklock, B.J., and R.O. Ryan, 1994. Hemolymph lipid transport. *Insect Biochem. Mol. Biol.,* 24:855-873.

Blakemore, D., S. Williams, and M.J. Lehane. 1995. Protein stimulation of trypsin secretion from the opaque zone midgut cells of *Stomoxys calcitrans*. *Comp. Biochem. Physiol.,* 110B:301-307.

Borovsky, D., and Y. Schlein. 1988. Quantitative determination of trypsin-like and chymotrypsin-like enzymes in insects. *Arch. Insect Biochem. Physiol.,* 8:249-260.

Borovsky, D., D.A. Carlson, P.R. Griffin, J. Shabanowitz, and D.F. Hunt. 1990. Mosquito oostatic factor: A novel decapeptide modulating trypsin-like enzyme biosynthesis in the midgut. *FASEB J.,* 4:3015-3020.

Borovsky, D., D.A. Carlson, P.R. Griffin, J. Shabanowitz, and D.F. Hunt. 1993. Mass-spectrometry and characterization of *Aedes aegypti* trypsin modulating oostatic factor (TMOF) and its analogs. *Insect Biochem. Mol. Biol.,* 23:703-712.

Borovsky, D., Q. Song, M.C. Ma, and D. Carlson. (1994). Biosynthesis, secretion, and immunocytochemistry of trypsin modulating oostatic factor of *Aedes aegypti*. *Arch. Insect Biochem. Physiol.,* 27:27-38.

Borovsky, D., C.A. Powell, J.K. Nayar, J.E. Blalock, and T.K. Hayes. (1994). Characterization and localization of mosquito gut receptors for trypsin modulating oostatic factor using a complementary peptide and immunocytochemistry. *FASEB J.,* 8:350-355.

Borovsky, D., and F. Mahmood. 1995. Feeding the mosquito *Aedes aegypti* with TMOF and its analogs; effect on trypsin biosynthesis and egg development. *Regulatory Peptides,* 57:273-281.

Borovsky, D., I. Janssen, J. Van den Broeck, R. Huybrechts, P. Verhaert, H.L. De Bobdt, D. Bylemans, and A. De Loof. 1996. Molecular sequencing and modeling of *Neobellieria bullata* trypsin. Evidence for translational control by *Neobellieria* trypsin-modulating oostatic factor. *Eur. J. Biochem.,* 237:279-287.

Boucias, D.G., C. Stokes, G. Storey, and J.C. Pendland. 1996. The effects of imidacloprid on the termite *Reticulitemes flavipes* and its interaction with the mycopathogen *Beauveria bassiana*. *Pflanzenschutz-Nachrichten Bayer,* 49:103-144.

Brauman, A., M.D. Kane, M. Labat, and J.A. Breznak. 1992. Genesis of acetate and methane by gut bacteria of nutritionally diverse termites. *Science,* 257:1384-1387.

Breznak, J.A. 1982. Intestinal microbiota of termites and other xylophagous insects. *Annu. Rev. Microbiol.,* 36:323-343.

Broadway, R.M. 1995. Are insects resistant to plant proteinase inhibitors? *J. Insect Physiol.,* 41:107-116.

Broadway, R.M., and M.G. Villani. 1995. Does host range influence susceptibility of herbivorous insects to non-host plant proteinase inhibitors? *Entomol. Exp. Appl.,* 76:303-312.

Brune, A., and M. Kühl. 1996. pH profiles of the extremely alkaline hindguts of soil-feeding termites (Isoptera: Termitidae) determined with microelctrodes. *J. Insect Physiol.,* 42:1121-1127.

Chao, A.C., D.F. Moffett, and A. Koch. 1991. Cytoplasmic pH and goblet cavity pH in the posterior midgut of the tobacco hornworm *Manduca sexta*. *J. Exp. Biol.,* 155:403-414.

Cheeseman, M.T., and C. Gillot. 1987. Organization of protein digestion in *Calosoma calidum* (Coleoptera: Carabidae). *J. Insect Physiol.,* 33:1-18.

Chino, H. 1985. Lipid transport: Biochemistry of hemolymph lipophorin, pp. 115-135, in G.A. Kerkut and L.I. Gilbert (Eds.), *Comprehensive Insect Physiology, Biochemistry and Pharmacology,* Vol. 10, Pergamon Press, Oxford.

Chino, H., and R.G. Downer. 1979. The role of diacylglycerol absorption of dietary glyceride in the American cockroach. *Insect Biochem.,* 9:379-382.

Chino, H., and K. Kitazawa. 1981. Diacylglycerol-carrying lipoprotein of hemolymph of the locust and some insects. *J. Lipid Res.,* 22:1042-1052.

Cohen, A.C. 1995. Extra-oral digestion in predaceous terrestrial Arthropoda. *Annu. Rev. Entomol.,* 40:85-103.

Cristofoletti, P.T., A.F. Ribeiro, and W.A. Terra. 2000. Apocrine secretion of amylase and exocytosis of trypsin along the midgut of *Tenebrio molitor* larvae. *J. Insect Physiol.,* 47:143-155.

Dow, J.A.T. 1981. Countercurrent flows, water movements and nutrient absorption in the locust midgut. *J. Insect Physiol.,* 27:579-585.

Dow, J.A.T. 1984. Extremely high pH in biological systems: A model for carbonate transport. *Am. J. Physiol.,* 246: R633-R635.

Dow, J.A.T. 1986. Insect midgut function. *Adv. Insect Physiol.,* 19:187-328.

Dow, J.A.T., and M.J. O'Donnell. 1990. Reversible alkalinization by *Manduca sexta* midgut. *J. Exp. Biol.,* 150:247-256.

Edwards, M.J., and M. Jacobs-Lorena. 2000. Permeability and disruption of the peritrophic matrix and caecal membrane from *Aedes aegypti* and *Anopheles gambiae* mosquito larvae. *J. Insect Physiol.,* 46:1313-1320.

Englemann, F. 1968. Feeding and crop emptying in the cockroach *Leucophaea maderae*. *J. Insect Physiol.,* 14:1525-1531.

Engelmann, F., and P.M. Geraerts. 1980. The proteases and the protease inhibitor in the midgut of *Leucophaea maderae*. *J. Insect Physiol.,* 26:703-710.

Evans, W.A.L., and D.W. Payne. 1964. Carbohydrases of the alimentary tract of the desert locust, *Schistocerca gregaria* Forsk. *J. Insect Physiol.,* 10:657-674.

Federici, B.A. 1993. Insecticidal bacterial proteins identify the midgut epithelium as a source of novel target sites for insect control. *Arch. Insect Biochem. Physiol.,* 22:357-371.

Federici, B.A. 1999. *Bacillus thuringiensis* in biological control, pp. 575-593, in T.S. Bellows and T.W. Fisher (Eds.), *Handbook of Biological Control,* Academic Press, New York.

Felix, C.R., B. Betschart, P.F. Billingsley, and T.A. Freyvogel. 1991. Post-feeding induction of trypsin in the midgut of *Aedes aegypti* L. (Diptera: Culicidae) is separable into two cellular phases. *Insect Biochem.,* 21:197-203.

Ferreira, C., G.L. Bellinello, A.F. Ribeiro, and W.R. Terra. 1990. Digestive enzymes associated with the glycocalyx, microvillar membranes and secretory vesicles from midgut cells of *Tenebrio molitor* larvae. *Insect Biochem.,* 20:839-847.

Giordana, B., V.F. Sacchi, P. Parenti, and G.M. Hanozet. 1989. Amino acid transport systems in intestinal brush-border membranes from lepidopteran larvae. *Am. J. Physiol.,* 257: R494-R500.

Gondim, K.C., G.C. Atella, J.H. Kawooya, and H. Masuda. 1992. Role of phospholipids in the lipophorin particles of *Rhodnius prolixus*. *Arch. Insect Biochem. Physiol.,* 20:303-314.

Graf, R., and H. Briegel. 1985. Isolation of trypsin isozymes from the mosquito *Aedes aegypti* (L.). *Insect Biochem.,* 15:611-618.

Graf, R., and H. Briegel. 1989. The synthetic pathway of trypsin in the mosquito *Aedes aegypti* L. (Diptera: Culicidae) and *in vitro* stimulation in isolated midguts. *Insect Biochem.,* 19:129-137.

Graf, R., A.S. Raikhel, M.R. Brown, A.O. Lea, and H. Briegel. 1986. Mosquito trypsin: Immunocytochemical localization in the midgut of blood-fed *Aedes aegypti* (L.). *Cell Tissue Res.,* 245:19-27.

Gupta, B.J., and M.J. Berridge. 1966. Fine structural organization of the rectum in the blowfly, *Calliphora erythrocephala* (Meig.) with special reference to connective tissue, tracheae and neurosecretory innervation in the rectal papillae. *J. Morphol.,* 120:23-82.

Harvey, W.R. 1992. The physiology of V-ATPases. *J. Exp. Biol.,* 172:1-17.

Hennigan, B.B., M.G. Wolfersberger, and W.R. Harvey. 1993a. Neutral amino acid symport in larval *Manduca sexta* midgut brush-border membrane vesicles deduced from cation-dependent uptake of leucine, alanine, and phenylalanine. *Biochim. Biophys. Acta,* 1148:216-222.

Hennigan, B.B., M.G. Wolfersberger, R. Parthasarathy, and W.R. Harvey. 1993b. Cation-dependent leucine, alanine, and phenylalanine uptake at pH 10 in brush-border membrane vesicles from larval *Manduca sexta* midgut. *Biochim. Biophys. Acta,* 1148:209-215.

Hong, Y.S., M. Reuveni, and J.J. Neal. 1995. A sodium- and potassium-stimulated tyrosine transporter from *Lepitinotarsa decemlineata* midguts. *J. Insect Physiol.,* 41:527-533.

Hopkins, C.R. 1967. The fine-structural changes observed in the rectal papillae of the mosquito *Aedes aegypti* L. and their relation to the epithelial transport of water and inorganic ions. *J. R. Microsc. Soc.,* 86:235-252.

House, H.L. 1974. Digestion, pp. 63-117, in M. Rockstein (Ed.), *Physiology of the Insecta,* 2nd ed., Academic Press, New York.

Houseman, J.G., A.E.R. Downe, and P.E. Morrison. 1985. Similarities in digestive proteinase production in *Rhodnius prolixus* (Hemiptera: Reduviidae) and *Stomoxys calcitrans* (Diptera: Muscidae). *Insect Biochem.,* 15:471-474.

Hua, Y.-J., D. Bylemans, A. DeLoof, and J. Koolman. 1994. Inhibition of ecdysone biosynthesis in flies by a hexapeptide isolated from vitellogenic ovaries. *Mol. Cell. Endocrinol.,* 104:R1-R4.

Hua, Y.-J., and J. Koolman. 1995. An ecdysiostatin from flies. *Reg. Pept.,* 57:263-271.

Humbert, W. 1979. The midgut of *Tomocerus minor* Lubbock (Insecta, Collembola): Ultrastructure, cytochemistry, ageing and renewal during a moulting cycle. *Cell. Tissue Res.,* 196:39-57.

Johnson, K.S., and R.V. Barbehenn. 2000. Oxygen levels in the gut lumens of herbivorous insects. *J. Insect Physiol.,* 46:897-903.

Jordao, B.P., W.R. Terra, A.F. Ribeiro, M.J. Lehane, and C. Ferreira. 1996. Trypsin secretion in *Musca domestica* larval midguts: A biochemical and immunocytochemical study. *Insect Biochem. Mol. Biol.,* 26:337-346.

Kane, P.M., and K.J. Parra. 2000. Assembly and regulation of the yeast vacuolar H^+-ATPase. *J. Exp. Biol.,* 203:81-87.

Klein, U., G. Löffelmann, and H. Wieczorek. 1991. The midgut as a model system for insect K^+-transporting epithelia: Immunocytochemical localization of a vacuolar-type H^+ pump. *J. Exp. Biol.,* 161:61-75.

Knowles, B.H. 1994. Mechanism of action of *Bacillus thuringiensis* insecticidal δ-endotoxins. *Adv. Insect Physiol.,* 24:275-308.

Krishna, S.S., and K.N. Saxena. 1962. Digestion and absorption of food in *Tribolium castaneum* (Herbst.). *Physiol. Zool.,* 35:66-78.

Lehane, M.J. 1976. Formation and histochemical structure of the peritrophic membrane in the stablefly, *Stomoxys calcitrans. J. Insect Physiol.,* 22:1551-1557.

Lehane, M.J. 1997. Peritrophic matrix structure and function. *Annu. Rev. Entomol.,* 42:525-550.

Lehane, M.J., D. Blakemore, S. Williams, and M.R. Moffatt. 1995. Regulation of digestive enzyme levels in insects. *Comp. Biochem. Physiol.,* 110B:285-289.

McFarlane, J.E. 1985. Nutrition and digestive organs, pp. 59-89, in M.S. Blum (Ed.), *Fundamentals of Insect Physiology,* John Wiley & Sons, New York.

McGaughey, W.H., and M.E. Whalon. 1992. Managing insect resistance to *Bacillus thuringiensis* toxins. *Science,* 258:1451-1455.

McGhie, T.K., J.T. Christeller, R. Ford, and P.G. Allsopp. 1995. Characterization of midgut proteinase activities of white grubs: *Lepidiota noxia*, *Lepidiota negatoria*, and *Antitrogus consanguineus* (Scarabeidae, Melolonthini). *Arch. Insect Biochem. Physiol.,* 28:351-363.

McManus, M.T., D.W.R. White, and P.G. McGregor. 1994. Accumulation of a chymotrypsin inhibitor in transgenic tobacco can affect the growth of insect pests. *Transgenic Res.,* 3:50-58.

Martin, M.M. 1983. Cellulose digestion in insects. *Comp. Biochem. Physiol.,* 75A:313-324.

Martin, M.M., J.S. Martin, J.J. Kukor, and R.W. Merritt. 1980. The digestion of protein and carbohydrate by the stream detritivore, *Tipula abdominalis* (Diptera, Tipulidae). *Oecologia,* 46:360-364.

Martin, M.M., J.J. Kukor, J.S. Martin, D.L. Lawson, and R.W. Merritt. 1981a. Digestive enzymes of larvae of three species of caddisflies (Trichoptera). *Insect Biochem.,* 11:501-505.

Martin, M.M., J.S. Martin, J.J. Kukor, and R.W. Merritt. 1981b. The digestive enzymes of detritus-feeding stonefly nymphs (Plecoptera; Pteronarcyidae). *Can. J. Zool.,* 59:1947-1951.

Martin, J.S., M.M. Martin, and E.A. Bernays. 1987. Failure of tannic acid to inhibit digestion or reduce digestibility of plant protein in gut fluids of insect herbivores: Implications for theories of plant defense. *J. Chem. Ecol.,* 13:605-621.

Merzendorfer, H., R. Gräf, M. Huss, W.R. Harvey, and H. Wieczorek. 1997. Regulation of proton-translocating V-ATPases. *J. Exp. Biol.,* 200:225-235.

Moffatt, M.R., and M.J. Lehane. 1990. Trypsin is stored as an inactive zymogen in the midgut of *Stomoxys calcitrans. Insect Biochem.,* 20:719-723.

Moffatt, M.R., D. Blakemore, and M.J. Lehane. 1995. Studies on the synthesis and secretion of digestive trypsin in *Stomoxys calcitrans* (Insecta-Diptera). *Comp. Biochem. Physiol.,* 110B:291-300.

Moffett, D.F., and A.R. Koch. 1988a. Electrophysiology of K^+ transport by midgut epithelium of lepidopteran insect larvae. I. The transbasal electrochemical gradient. *J. Exp. Biol.,* 135:25-38.

Moffett, D.F., and A.R. Koch. 1988b. Electrophysiology of K^+ transport by midgut epithelium of lepidopteran insect larvae. II. The transapical electrochemical gradients. *J. Exp. Biol.,* 135:39-49.

Murdock, L.L., G. Brookhart, P.E. Dunn, D.E. Foard, S. Kelley, L. Kitch, R.E. Shade, R.H. Shukle, and J.L. Wolfson. 1987. Cysteine digestive proteinases in Coleoptera. *Comp. Biochem. Physiol.,* 87B:783-787.

Nation, J.L. 1983. Specialization in the alimentary canal of some mole crickets (Orthoptera: Gryllotalpidae). *Int. J. Insect Morphol. Embryol.,* 12:201-210.

Noirot, C., and C. Noirot-Timothee. 1969. The digestive system, pp. 49-88, in K. Krishna and R.M. Wheeler (Eds.), *Biology of Termites,* Vol. I, Academic Press, New York.

Noreiga, F.G., X.-Y. Wang, J.E. Pennington, C.V. Barillas-Mury, and M.A. Wells. 1996. Early trypsin, a female-specific midgut protease in *Aedes aegypti*: Isolation, amino-terminal sequence determination, and cloning and sequencing of the gene. *Insect Biochem. Mol. Biol.,* 26:119-126.

Noriega, F.G., K.A. Edgar, W.G. Goodman, D.K. Shah, and M.A. Wells. 2001. Neuroendocrine factors affecting the steady-state levels of early trypsin mRNA in *Aedes aegypti. J. Insect Physiol.,* 47:515-522.

O'Riordan, A.M. 1969. Electrolyte movement in the isolated midgut of the cockroach (*Periplaneta americana*). *J. Exp. Biol.,* 51:699-714.

Parenti, P., B. Giordana, V.F. Sacchi, G.M. Hanozet, and A. Guerritore. 1985. Metabolic activity related to the potassium pump in the midgut of *Bombyx mori* larvae. *J. Exp. Biol.,* 116:69-78.

Pimenta, P.F.P., S.T. Turco, M.J. McConville, P.G. Lawyer, P.V. Perkins, and D.L. Sacks. 1992. Stage-specific adhesion of *Leishmania* promastigotes to the sandfly midgut. *Science,* 256:1812-1815.

Reuveni, M., and P. Dunn. 1990. The use of membrane vesicles as a tool for investigating known and potential pesticides, pp. 523-531, in D.L. Weigmann (Ed.), *Pesticides in the Next Decade: The Challenge Ahead,* Virginia Water Research Center Press, Richmond, VA.

Reuveni, M., and P. Dunn. 1993. Absorption pathways of amino acids in the midgut of *Manduca sexta* larvae. *Insect Biochem. Mol. Biol.,* 23:959-966.

Reuveni, M., Y.S. Hong, P.E. Dunn, and J.J. Neal. 1993. Leucine transport into brush border membrane vesicles from guts of *Leptinotarsa decemlineata* and *Manduca sexta. Comp. Biochem. Physiol.,* 104A:267-272.

Richards, A.G., and P.A. Richards. 1977. The peritrophic membranes of insects. *Annu. Rev. Entomol.,* 22:219-240.

Riemann, J.G., and H.M. Flint. 1967. Irradiation effects on midguts and testes of the adult boll weevil, *Anthonomus grandis,* determined by histological and shielding studies. *Ann. Entomol. Soc. Am.,* 60:298-308.

Rudall, K.M., and W. Kenchington. 1973. The chitin system. *Biol. Rev. Cambridge Philos. Soc.,* 48:597-636.

Santos, C.D., and W.R. Terra. 1984. Plasma membrane associated amylase and trypsin: intracellular distribution of digestive enzymes in the midgut of the cassava hornworm *Erinnyis ello. Insect Biochem.,* 14:587-595.

Santos, C.D., and W.R. Terra. 1986. Midgut alpha-glucosidase and beta-fructosidase from *erinnyis ello* larvae and imagoes. *Insect Biochem.,* 16:819-824.

Santos, C.D., A.F. Ribeiro, and W.R. Terra. 1986. Differential centrifugation, calcium precipitation and ultrasonic disruption of midgut cells of *Erinnyis ello* caterpillars. Purification of cell microvilli and inferences concerning secretory mechanisms. *Can J. Zool.,* 64:490-500.

Sehnal, F., and D. Zitman. 1990. Endocrines of insect gut, pp. 510-515, in A. Epple, C.G. Scanes, and M.H. Stetson (Eds.), *Progress in Comparative Endocrinology,* Wiley-Liss, New York.

Skaer, R.J. 1981. Cellular sieving by a natural, high-flux membrane. *J. Microsc.,* 124:331-333.

Southwood, T.R.E. 1973. The insect/plant relationship — an evolutionary perspective. *Symp. R. Entomol. Soc. London,* 6:3-30.

Stobbart, R.H. 1971. Factors affecting the control of body volume in the larvae of the mosquitoes *Aedes aegypti* (L.) and *Aedes detritus* Edw. *J. Exp. Biol.,* 54:67-82.

Surholt, B., J. Goldberg, T.K.F. Schulz, A.M.Th. Beenakkers, and D.J. Van der Horst. 1991. Lipoproteins act as a reusable shuttle for lipid transport in the flying death's-head hawkmoth *Acherontia atropos. Biochem. Biophys. Acta,* 1086:15-21.

Teo, L.H. 1997. Tryptic and chymotryptic activities in different parts of the gut of the field cricket *Gryllus bimaculatus* (Orthoptera: Gryllidae). *Ann. Entomol. Soc. Am.,* 90:69-74.

Terra, W.R. 1988. Physiology and biochemistry of insect digestion: An evolutionary perspective. *Brazil J. Med. Biol. Res.,* 21:675-734.

Terra, W.R. 1990. Evolution of digestive systems of insects. *Annu. Rev. Entomol.,* 35:181-200.

Terra, W.R., C. Ferreira, and F. Bastos. 1985. Phylogenetic considerations of insect digestion — disaccharidases and the spatial organization of digestion in the *Tenebrio molitor* larvae. *Insect Biochem.,* 15:443-449.

Thie, N.M.R., and J.G. Houseman. 1990. Identification of cathepsin B, D and H in the larval midgut of Colorado potato beetle, *Leptinotarsa decemlineata* Say (Coleoptera: Chrysomelidae). *Insect Biochem.,* 20:313-318.

Thomas, K.K. 1984. Studies on the absorption of lipid from the gut of desert locust, *Schistocerca gregaria. Comp. Biochem. Physiol.,* 77A:707-712.

Thomas, K.K., and J.L. Nation. 1984. Protease, amylase and lipase activities in the midgut and hindgut of the cricket, *Gryllus rubens* and mole cricket, *Scapteriscus acletus. Comp. Biochem. Physiol.,* 79A:297-304.

Treherne, J.E. 1957. Glucose absorption in the cockroach. *J. Exp. Biol.,* 34:478-485.

Treherne, J.E. 1958. The absorption of glucose from the alimentary canal of the locust, *Schistocerca gregaria* (Forsk.). *J. Exp. Biol.,* 35:297-306.

Treherne, J.E. 1959. Gut absorption. *Annu. Rev. Entomol.,* 12:43-58.

Turunen, S. 1975. Absorption and transport of dietary lipid in *Pieris brassicae. J. Insect Physiol.,* 21:1521-1529.

Turunen, S., and G.M. Chippendale. 1977. Lipid absorption and transport: Sectional analysis of the larval midgut of the corn borer, *Diatraea grandiosella. Insect Biochem.,* 7:203-208.

Undeen, A.H. 1979. Simuliid larval midgut pH and its implications for control. *Mosquito News,* 39:391-393.

Valaitis, A.P. 1995. Gypsy moth midgut proteinases: Purification and characterization of luminal trypsin, elastase and the brush border membrane leucine aminopeptidase. *Insect Biochem. Mol. Biol.,* 25:139-149.

Valaitis, A.P., M.K. Lee, R. Rajamohan, and D.D. Dean. 1995. Brush border membrane aminopeptidase-N in the midgut of gypsy moth serves as the receptor for the CryIA(c) δ-endotoxin of *B. thuringiensis. Insect Biochem. Mol. Biol.,* 25:1143-1151.

Van Heusden, M.C., D.J. Van der Horst, J.K. Kawooya, and J.H. Law. 1991. *In vivo* and *in vitro* loading of lipid by artificially lipid-depleted lipophorins: Evidence for the role of lipophorin as a reusable lipid shuttle. *J. Lipid Res.,* 32:1789-1794.

Waterhouse, D.F. 1953. The occurrence and significance of the peritrophic membrane with special reference to adult Lepidoptera and Diptera. *Aust. J. Zool.,* 1:299-318.

Waterhouse, D.F., and B. Stay. 1955. Functional differentiation in the midgut epithelium of blowfly larvae as revealed by histochemical tests. *Aust. J. Biol. Sci.,* 8:253-277.

Weintraub, H., and A. Tietz. 1973. Triglyceride digestion and absorption in the locust, *Locusta migratoria. Biochem. Biophys. Acta.,* 306:31-41.

Weintraub, H., and A. Tietz. 1978. Lipid absorption by isolated intestinal preparation. *Insect Biochem.,* 8:267-274.

Wigglesworth, V.B. 1961. *The Principles of Insect Physiology*, Methuen & Co. Ltd., London, 546 pp.

Wieczorek, H., G. Grüber, W.R. Harvey, M. Huss, H. Merzendorfer, and W. Zeiske. 2000. Structure and regulation of insect plasma membrane H^+ V-ATPase. *J. Exp. Biol.,* 203:127-135.

Wieczorek, H., M. Putzenlechner, W. Zeiske, and U. Klein. 1991. A vacuolar-type proton pump energizes K+/H+ antiport in an animal plasma membrane. *J. Biol. Chem.,* 266:15340-15347.

Wieczorek, H., S. Weerth, M. Schindlbeck, and U. Klein. 1989. A vacuolar-type proton pump in a vesicle fraction enriched with potassium transporting plasma membranes from tobacco hornworm midgut. *J. Biol. Chem.,* 264:11143-11148.

Wolfson, J.L., and L.L. Murdock. 1990. Diversity of digestive proteinase activity among insects. *J. Chem. Ecol.,* 16:1089-1102.

Wyatt, G.R. 1967. The biochemistry of sugars and polysaccharides in insects. *Adv. Insect Physiol.,* 4:287-360.

Yin, C.-M., H. Duan, and J.G. Stoffolano, Jr. 1993. Hormonal stimulation of the brain for its control of oogenesis in *Phormia regina* (Meigen). *J. Insect Physiol.,* 39:165-171.

Yin, C.-M., B.-X. Zou, M.-F. Li, and J.G. Stoffolano, Jr. 1994. Discovery of a midgut peptide hormone which activates the endocrine cascade leading to oogenesis in *Phormia regina* (Meigen). *J. Insect Physiol.,* 40:1-9.

Zeiske, W., W. Van Driessche, and R. Zeigler. 1986. Current-noise analysis of the basolateral route for K^+ ions across a K^+-secreting insect midgut epithelium (*Manduca sexta*). *Pfluegers Arch.,* 407:657-663.

Zieler, H., C.F. Garon, E.R. Fischer, and M. Shahabuddin. 2000. A tubular network associated with the brush-border surface of the *Aedes aegypti* midgut: Implications for pathogen transmission by mosquitoes. *J. Exp. Biol.,* 203:1599-1611.

3 Nutrition

CONTENTS

PREVIEW

Insects need the same basic nutritional components that larger animals need. Balance of nutrients is very critical to most insects studied. Experimentally, some insects are able to self-select among multiple choices of artificial diet formulations to compensate for single-diet deficiencies. This suggests that some oligophagous insects may do the same thing in nature. Immature insects often have different nutritional requirements from adults. Some adults (some Lepidoptera, for example) do not feed as adults and acquire all the nurtritional components needed for development of ovaries and eggs during larval life. Most adults need a nitrogen source to mature ovaries and eggs and a carbohydrate source for energy. Although slight variability is known among different insects, the majority studied need dietary arginine, histidine, isoleucine, leucine, lysine, methionine, phenylalanine, threonine, tryptophan, and valine, the same ten essential amino acids required by larger animals. Some insects need a carbohydrate source for complete development, while others do not. Many adult insects need carbohydrate as an energy source. Insects cannot synthesize sterols, and thus immature insects need a dietary sterol as a precursor that can be transformed into the molting hormone, which has a sterol structure. Eggs contain sterols and the first instar may be able to molt without a dietary source, but subsequent molts may be impossible if dietary sterol is not present. Some adult insects need dietary sterol to produce the normal number and/or hatching of eggs. Immatures of some groups need polyunsaturated fatty acids for normal development. Insects generally need the B vitamins, vitamin A (or a carotenoid), ascorbic acid, and some may need small amounts of other vitamins. Insects do not need vitamin D, and probably not vitamin K. Vitamins, some essential amino acids, and sterols may be supplied by symbionts in the gut or fat body. Development of artificial diets for culture of insects has been stimulated by the desire to learn and compare nutritional requirements, as well as the need to rear large numbers of insects efficiently and economically for commercial and scientific purposes. Procedures to measure growth, digestibility, and conversion of food into body weight and tissues have been devised to evaluate growth and development of insects on artificial diets. Feeding stimulants and deterrents are important in the feeding behavior of insects in their natural environment. Frequently, the presence of natural or similar feeding stimulants and absence of feeding deterrents are important factors in getting insects to eat artificial diets.

3.1 INTRODUCTION

The basic nutritional requirements for growth and reproduction of insects are well known and were largely determined in the middle decades of the 20th century. Insects generally have about the same basic nutritional needs as large animals (Dadd, 1985), although minor variations in both qualitative and quantitative requirements are known in some insects. In general, insects need the ten essential amino acids required by the rat, a model animal for larger vertebrates. One notable difference between vertebrates and insects is the insect requirement for a dietary source of sterol; although some can synthesize squalene, the hydrocarbon backbone needed for the ring structure of a sterol, they cannot form the rings.

Research on insect nutrition stems from:

1. Comparative scientific interest
2. Efforts to achieve increased productivity from desirable insects such as silkworms, honeybees, pollinators, and experimental insects
3. Mass production of parasites, predators, or insects for sterile-insect release programs
4. Development of control strategies that might exploit nutritional requirements
5. Understanding metabolic pathways related to nutritional requirements
6. Understanding nutritional influence on polymorphism

Development of insect diets for ease of rearing and for mass rearing has been an especially active field (Anderson and Leppla, 1992; Singh, 1974, 1976, 1977).

The literature on insect nutrition, diet development, and rearing on artificial or semi-artificial diets is extensive. There are reviews, including but not limited to: House (1965a), Davis (1968), Hsiao (1972), House et al. (1971), Schoonhoven (1972), Gordon (1972), House (1974), Vanderzant (1974), Dadd (1973, 1985), Scriber and Slansky (1981), Slansky (1982), Reinecke (1985), Slansky and Scriber (1985), Waldbauer and Friedman (1991), Anderson and Leppla (1992), Locke and Nichol (1992), and Simpson and Raubenheimer (1995).

Currently, many insects (largely those of some economic importance) can be reared on a synthetic or semi-synthetic diet, including some endoparasitoids (Bracken, 1966; Yazgan, 1972), but few representatives of some groups have been reared on synthetic diets. For example, it appears that only three or possibly four butterflies have been reared on synthetic diets, although many moths are reared relatively easily.

3.2 IMPORTANCE OF BALANCE IN NUTRITIONAL COMPONENTS

One of the strongest principles to come from insect diet studies is that **balance of nutrients** is important to effective growth. Gordon (1959) stressed that balance of nutrients is the most dominant quantitative factor in a diet. Sang (1956) presented detailed quantitative data to document the adverse effect of nutrient imbalance upon growth of *Drosophila melanogaster* larvae, and numerous authors have found similar effects with other insects. House (1965a, 1965b, 1966a, 1969, 1974) found that nutrient imbalance resulted in reduced food intake and that optimum growth depended, among other things, on a proper ratio of amino acids to minerals. In one experiment, 67% of the larvae of *Agria affinis* selected a complete, balanced diet from among choices including a diet deficient in an essential amino acid, a complete but imbalanced diet (improper proportion of essential to nonessential amino acids), and agar (House, 1967). About 45% of the larvae reached maturity in the normal time of 6 days on the complete, balanced diet, while those selecting the other diets remained in the first instar or died. The stress of excreting excess nutrients may be detrimental and wasteful of energy. Optimal balance, however, frequently changes with species, sex, and age or stage of development.

3.3 ABILITY OF INSECTS TO SELF-SELECT NUTRITIONAL COMPONENTS

Many animals, including some insects, demonstrably **self-select dietary components** both from natural foods and from defined diets. The criteria of self-selection are that (1) there is nonrandomness of choice; (2) a uniform cohort of individuals tend to select nutrients, at least the major ones, in consistent proportions; and (3) individuals having a choice to self-select do as well or better than if self-selection is not possible (see review by Waldbauer and Friedman, 1991). Confused flour beetles *Tribolium confusum,* given a choice of 1:1:1 mix of particles of germ, bran, and endosperm selected 81% germ, 2% bran, and 17% endosperm that provided a protein:carbohydrate ratio of 57:43, close to the optimum of 50:50 for these immature beetle larvae (Waldbauer and Bhattacharya, 1973). Corn earworms *Helicoverpa zea* self-selected portions from defined diets providing a protein:carbohydrate ratio of 79:21, a ratio almost identical to the 80:20 ratio of protein:carbohydrate shown to be optimal for growth of tobacco hornworm larvae *Manduca sexta* (Waldbauer et al., 1984; Cohen et al., 1987b). Nymphal brown banded cockroaches *Supella longipalpa* self-selected a ratio of 16:84 protein:carbohydrate when given a choice of diets (Cohen et al., 1987a). In an attempt to rationalize the very different ratios of protein:carbohydrate selected by the two lepidopterans and the cockroach, Waldbauer and Friedman (1991) observed that these particular (and indeed most) lepidopterans have relatively short life cycles, and the high proportion of protein in the self-selected diet of larvae permits rapid growth to the pupal stage, with little expenditure of energy in searching for food. Longer-lived cockroaches, on the other hand, are genetically programmed to grow more slowly (about 256 days required to reach the adult stage), and a high protein diet does not speed growth appreciably. During its long developmental period, however, it needs carbohydrates to provide the fuel for foraging. Food requirements and habits, of necessity, are correlated with life history.

The mechanisms by which insects self-select dietary components is not known. Changes in chemoreceptor sensitivity that correlate with feeding behavior have been observed in locusts (Simpson et al., 1990) and some lepidopterous caterpillars (Schoonhoven et al., 1991), leading to the postulate that changes in peripheral taste receptor sensitivity regulates self-selection through feedback from metabolic and physiological state of various tissues (Simpson and Simpson, 1990; Schoonhoven et al., 1991). Associative learning, association of a specific stimulus with a reward, such as associating a chemical component with a food that promotes growth, also may play a role (Simpson and White, 1990). In an attempt to obtain some evidence for a peripheral receptor mechanism, Ahmad et al. (1993) maxillectomized third instars of tobacco hormworm *Manduca sexta*, a procedure known to alter their ability to discriminate among host plants, and gave them a choice of defined diets lacking carbohydrate or protein. The larvae still self-selected from both diet formulations to obtain a protein:carbohydrate ratio equal to that of sham-operated insects. Thus, the mechanism involved in self-selection of diet by insects remains unknown, and direct evidence for feedback that changes peripheral receptor sensitivity is lacking.

3.4 REQUIREMENTS FOR SPECIFIC NUTRIENTS

To determine the nutrient requirements of insects, it is most desirable, and often necessary, to (1) rear the insects through multiple generations, and (2) rear them under **aseptic** or **axenic** conditions. The reason for the first practice is that small insects require only small amounts of some nutrients, and slight contamination of dietary components with traces of those nutrients, and/or carry-over in the egg and body of the insects, may be sufficient for several generations. Some examples will be enumerated in the following account of specific nutrients. The second practice, rearing under axenic conditions, is necessary because many, and perhaps all, insects contain a microfauna and flora in the gut, or in special bodies called mycetomes, or as bacteroids scattered among fat body cells, and these symbionts usually supply some nutrients to their host. Blood-feeding insects, phloem and xylem sap feeders, stored-products insects, and cockroaches and termites have symbionts that are known to, or may, supply some vitamins, essential amino acids, and sterols (Dadd, 1985).

3.4.1 A Nitrogen Source: Proteins and Amino Acids

Most insects obtain amino acids from their foods by ingesting proteins. Purified proteins such as casein from milk, gluten from wheat, albumin from eggs, and sometimes soybean and peanut protein preparations have been used in artificial diets. A product called wheast, prepared from milk whey and yeast used in the brewery industry, also is available. In artificial diet formulations, investigators commonly add one or more of these protein sources. Unfortunately, no single purified protein is an entirely satisfactory source of amino acids of balanced proportions for all insects. Casein and egg albumin have a good balance of most amino acids and have been the most widely used in insect diets. Each, however, is relatively low in histidine and tryptophan, two of the essential amino acids. Casein is also relatively low in cysteine and glycine, although these are not essential for insects. Most insects probably have an optimum level of protein required in the diet for best growth, but this varies widely for different species. Restricting the dietary protein of several species of cockroaches retarded growth, but prolonged longevity. From 22 to 24% protein in the diet gave fastest growth and lowest nymphal mortality of German and Oriental cockroaches, but 11% protein promoted maximum longevity. The American cockroach grew fastest on 49 to 78% protein, but survived longest on 22 to 24% protein (Haydak, 1953).

Many adult female insects require a source of protein to mature their ovaries and eggs. In part, protein deprivation may manifest itself in the failure to secrete juvenile hormone (JH), which is needed for ovary and egg development; but even if JH or an analog such as methoprene is administered to protein-starved insects, they do not produce the normal complement of eggs simply because they do not have sufficient protein reserves in the body. Male insects usually do not require protein as adults in order to mature sperm.

These examples of sex and developmental requirements illustrate the generalization that optimal nutritional requirements frequently differ with age, sex, and physiological stress. Any attempt to state optimal requirements for protein or amino acids must include a definition of the evaluation criteria.

3.4.2 Essential Amino Acids

The classical method for determining amino acid requirements has been deletion of one amino acid at a time from a diet fed to a group of insects. This is obviously quite time-consuming, requiring rearing many insects on a large series of diets, and it also presupposes that the insects can be grown on a synthetic diet of known composition. Some knowledge of the amino acid composition of the protein sources commonly eaten by a particular insect may be helpful in formulating an amino acid mixture on which the insect can survive. For example, Vanderzant (1958) determined that an amino acid mixture characteristic of proteins from cotton supported growth and development of the pink bollworm *Pectinophora gossypiella* better than a mixture based on casein. Achieving a suitable balance of essential and nonessential amino acids (and sometimes their relationship to other nutrients) is often critical to successful rearing (House, 1965b, 1966a).

The **essential amino acids** of a number of insects from different orders (Table 3.1) have been demonstrated to be **arginine, histidine, isoleucine, leucine, lysine, methionine, phenylalanine, threonine, tryptophan**, and **valine (all in the L-form).** These are the same essential amino acids required by the experimental white rat. In the absence of one of these essential amino acids, growth and development ceases in the insects listed in Table 3.1. In some cases, nonessential amino acids stimulate growth, and this may be related to the optimization of nutrient balance and the efficiency of the biochemical pathways involved in synthesis of the nonessential amino acids. Although some species can be reared on a diet in which the ten essential amino acids are the only amino acid source, others cannot be reared with only the ten essential ones. Most insects that have been studied actually grow better with a balance of essential and nonessential amino acids. Such a mixture undoubtedly saves energy that otherwise must be expended in synthesizing the nonessential amino acids from the essential ones.

The classical deletion method for determining the essential amino acids cannot be used, of course, if the insect cannot be reared on a defined diet. Kasting and McGinnis (1958) demonstrated the value of an indirect method for defining the essential amino acids for such insects based on injecting or feeding the insects with glucose-U-^{14}C (uniformly labeled glucose) or another suitable general precursor compound (Table 3.2). Essential amino acids are expected to have no ^{14}C label, because they are not supposed to be synthesized; and nonessential amino acids should be labeled because they are synthesized. In practice, low label incorporation is often found in some amino acids known to be essential from deletion studies in cases where both methods have been compared. Evidently, slight synthesis of some essential amino acids may occur but the rate is too low to eliminate the need for a dietary supply. The method was compared with the classical deletion procedure in a test with the blowfly, *Phormia regina* (Kasting and McGinnis, 1960). Third instars were injected with 3 to 6 μl of glucose-U-^{14}C containing 5000 counts/min/μl. Sixty-eight hours later, the larvae were homogenized and their body proteins were hydrolyzed to yield free amino acids, which were then separated by ion exchange column chromatography. The amino acids synthesized from the radioactive glucose were expected to contain significant ^{14}C label and these would be considered nonessential. The essential amino acids, which the blowfly larvae could only get from their food, should contain very little or no label. The results suggested several discrepancies between the deletion and radiolabel techniques. For example, the deletion study indicated that proline was essential, while the label incorporation data indicated that proline was synthesized (i.e., it was nonessential). This was later clarified by the finding that some strains of *P. regina* need proline, while other strains do not. Also, label was not incorporated into tyrosine, suggesting it was essential, while the deletion study showed it to be nonessential. In other insects, tyrosine has also been shown to be nonessential. It is probable that phenylalanine is the precursor of tyrosine in the blowfly, as has been demonstrated in a number of other insects. If so, no label would be expected in either phenylalanine or tyrosine. Finally, the labeling technique indicated that both methionine and cysteine were essential, whereas the deletion study had indicated that neither was essential. Subsequent deletion studies showed that when both methionine and cysteine were simultaneously deleted, at least one of them, either one, was essential. Overall, the labeling technique gave results that agreed well with deletion studies for *P. regina* after certain ambiguities were clarified by additional research.

The isotope labeling technique has been used with the prairie grain wireworm *Agrotis orthogonia* (Kasting and McGinnis, 1962), two spotted spider mites, *Tetranchus urticae* (Rodrigues and Hampton, 1966) and *Helicoverpa zea* (Rock and Hodgson, 1971). Rock and Hodgson (1971) compared the labeling technique with the deletion method for *H. zea*. The deletion method indicated that

TABLE 3.1
Insects Known to Require the Ten Essential L-Amino Acids: Arginine, Histidine, Isoleucine, Leucine, Lysine, Methionine, Phenylalanine, Threonine, Tryptophan, and Valine

Species	Ref.[a]
Aedes aegypti	4
Agria affinis	5
Cochliomyia hominivorax	18
Culex pipiens	18
Drosophila melanogaster	6
Phormia regina	8
Stegobium paniceum	19
Bombyx mori	18
Myzus persicae	17
Ctenicera destructor	16
Agrotis orthogonia	10
Tribolium confusum	2
Trogoderma granarium	3
Hylemya antiqua	7
Chilo suppressalis	11
Pectinophora gossypiella	12
Helicoverpa zea	13
Attagenus sp.	1
Argyrotaenia velutinana	14
Anthonomus grandis	15
Apis mellifera	9

[a] References:
1. Moore (1946)
2. Lemonde and Bernard (1951)
3. Pant et al. (1958)
4. Singh and Brown (1957)
5. House (1954)
6. Hinton et al. (1951)
7. Friend et al. (1957)
8. Kasting and McGinnis (1960)
9. DeGroot (1952)
10. Kasting and McGinnis (1962)
11. Ishii and Hirano (1955)
12. Vanderzant (1958)
13. Rock and Hodgson (1971)
14. Rock and King (1968)
15. Vanderzant (1973)
16. Kasting et al. (1962)
17. Dadd and Krieger (1968); Mittler (1971)
18. Dadd (1985)
19. Pant et al. (1960)

arginine, histidine, isoleucine, leucine, lysine, methionine, phenylalanine, threonine, tryptophan, and valine were essential. These amino acids had a low label content, but tryptophan and phenylalanine were labeled strongly enough to be inconclusive. Other amino acids were highly labeled; and in agreement with the deletion method, they were considered nonessential.

The isotope labeling technique is clearly useful when a defined diet is not available, and it is much faster in producing results than the slower deletion method that requires many trials. The previous results indicate, however, that the results must be viewed with caution, and confirmed when possible with deletion techniques. Additionally, neither method can account for the possible contribution of symbionts in the synthesis of amino acids.

It is a common observation that insects often eat their molted exuviae, and Mira (2000) has suggested that one possible benefit from such behavior may be the acquisition of a protein meal. In experiments, Mira found that larval cockroaches (*Periplaneta americana*) usually ate the exuviae

TABLE 3.2
Determination of Essential and Nonessential Amino Acids by Administration of Glucose ^{14}C to Prairie Grain Wireworms and Two-Spotted Spider Mites

	Prairie Grain Wireworm[a]		Two-Spotted Spider Mite[b]	
	c/min/µM	Requirement	c/min/µM	Requirement
Arginine	0.6	+	19	+
Histidine	0.5	+	25	+
Isoleucine	0.11	+	57	+
Leucine	0.2	+	19	+
Lysine	0.01	+	11	+
Methionine	3.5	+	285	+
Phenylalanine	0.7	+	17	+
Threonine	?	?	2989	–
Valine	0.5	+	94	+
Proline	22	·	605	–
Alanine	170	–	7247	–
Aspartic acid	37	·	5500	–
Serine	79	–	1350	–
Glycine	38	–	1256	–
Glutamic acid	24	–	2671	–
Tyrosine	1.3	–	50	+

[a] Data from Kasting and McGinnis (1962).
[b] Data from Rodrigues and Hampton (1966).

during larval life, and adult females ate the exuviae more often than males, possibly because of a greater need for nitrogen for reproduction. Cockroaches reared on high-protein diets most often did not eat the exuviae, while those rendered aposymbiotic always ate their exuviae, both of which tend to support a nutritional role for the exuviae, although other explanations are also possible.

3.4.3 CARBOHYDRATES

Most insects do not have an absolute growth requirement for a specific carbohydrate in the diet, although carbohydrate is a major energy source for most insects. In general, insects can synthesize carbohydrate from amino acids and from lipids. Species in the genera *Tenebrio* (meal worm), *Ephestia* (flour moth), and *Oryzaephilus* (saw-toothed grain beetle) need a carbohydrate source to reach maturity. Other stored-grain insects such as species of *Tribolium* (flour beetles), *Lasioderma* (cigarette beetle), and *Ptinus* (powder post beetles) can be reared to maturity on diets lacking carbohydrates. The adults of the dipterans *Calliphora erythrocephala*, *Lucilia cuprina*, *Anastrepha suspensa,* some other tephritid fruit flies, and probably many adults dipterans require carbohydrate (typically satisfied by sucrose) for an energy source and continued survival. Worker honeybees have a requirement for carbohydrate at the time of pupation. Worker larvae can be reared in the laboratory on worker jelly, but they fail to pupate on the worker jelly (Shuel and Dixon, 1968). Worker larvae fed royal jelly, the food normally fed to developing queen larvae, pupate normally in the laboratory. Worker larvae appear to have a sugar requirement for pupation, and worker jelly, with only 4% carbohydrate content, has too little. Royal jelly contains about 12% carbohydrate content, so the pupation requirement must lie between 4 and 12% carbohydrate. Shuel and Dixon (1968) showed that addition of 40 mg glucose and 40 mg fructose per gram worker jelly (i.e., 8% additional sugar) allowed worker larvae to pupate in the laboratory when fed the altered worker jelly. Worker jelly is a glandular secretion produced by worker bees and fed to worker larvae for the first 3 days by adult bees. In a honeybee colony, the adult bees feed older worker larvae on a modified worker food containing honey and some pollen, so the carbohydrate content of their natural food is high as they approach pupation.

3.4.4 Lipids

"Lipid" is a broad term that includes biological molecules that are soluble in such organic solvents as ether, alcohol, and similar solvents. Typical lipids in biological organisms include free and bound fatty acids, short- and long-chain alcohols, tri-, di-, and monoacylglycerols, steroids and their esters, phospholipids, and several other groups of compounds. Most insects have the necessary metabolic machinery to convert carbohydrate into lipids, and many insects synthesize lipids and store them in the fat body tissue. A specific lipid required in the diet is sterol, and some insects require polyunsaturated fatty acids.

3.4.5 Sterols

Because of their inability to synthesize sterols, insects must obtain sterol(s) from their food and/or from their symbionts. They use sterol as a precursor for synthesis of the ecdysteroid molting hormone (see Chapter 5), and as a component of all cell membranes. Cholesterol, and sparing sterols such as cholestanol, are incorporated into all tissues of *Eurycotis floridana*, a cockroach. A **sparing sterol** is one that can be incorporated into cell membranes but cannot be used to synthesize the molting hormone. The requirement for a dietary sterol was first noted by Hobson (1935) in larvae of the blowfly *Lucilia sericata*, and has since been verified in many different insects. The firebrat *Ctenolepisma* sp., representative of a very primitive group of insects, can synthesize some sterol but probably not enough for its needs.

Cholesterol usually satisfies the sterol requirement. Frequently, several different sterols can replace or spare cholesterol, probably serving in a relatively nonspecific capacity as a cell membrane component in place of cholesterol. In only a few insects is it known that the ecdysteroid molting hormone can be synthesized from a sterol other than cholesterol. Although cholesterol has been detected in small quantities in some plants, phytophagous insects normally ingest β-sitosterol and stigmasterol as major plant sterols, and other sterols that occur in plants. Biochemical pathways for conversion of plant sterols into cholesterol have been demonstrated in many phytophagous insects and some nonphytophagous ones (Svoboda et al., 1975). Cholesterol or β-sitosterol satisfies the sterol requirement in most of the 18 species studied, and ergosterol or 7-dehydrocholesterol can be utilized by about three fourths of them (see Chapter 5 for more details on specific sterol use for synthesis of the molting hormone, and for conversion pathways).

A few species are known to have a requirement for a very specific dietary sterol. *Drosophila pachea* breeds only in the senita cactus *Lophocereus schotti* in the Sonoran Desert of the southwestern United States, and it requires 7-stigmasten-3β-ol, an uncommon sterol found in senita cactus. Only 7-cholesten-3β-ol and 5,7-cholestadien-3β-ol can substitute for the cactus sterol (Kircher et al., 1967). *Xyleborus ferrugineus*, a scolytid beetle, will use cholesterol and lanosterol for egg production and hatching, but larvae fail to pupate unless ergosterol or 7-dehydrocholesterol is available in the diet, presumably indicating a critical need for either of these two sterols for synthesis of its molting hormone. The larvae normally obtain ergosterol from the symbiotic fungus *Fusarium solani* growing on dead trees that are hosts for the scolytid larvae (Norris and Baker, 1967). *Lobesia botrana*, the grape berry moth, has a mutualistic relationship with the fungus *Botrytis cinerea*, and the moth grows faster, survives longer, and has greater fecundity when grown on an artificial diet containing mycelium or purified sterols from the fungus (Mondy and Corio-Costet, 2000).

Sterol deficiency may be manifest in any of the stages of an insect. Newly hatched larvae that lack dietary sterol usually die in the first or second instar because they exhaust the sterol received

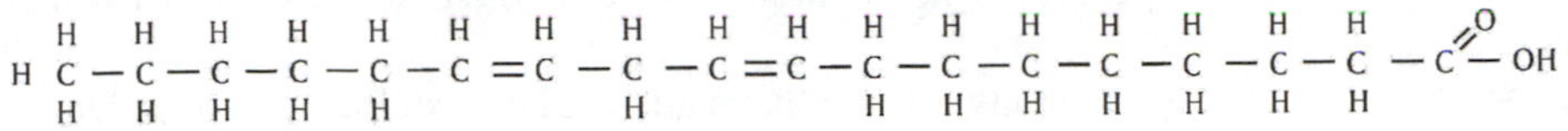

```
   H   H   H   H   H   H   H   H   H   H   H   H   H   H   H   H   H      //O
 H C — C — C — C — C — C = C — C — C = C — C — C — C — C — C — C — C — C — OH
   H   H   H   H   H           H           H   H   H   H   H   H   H
```

(Z,Z) - 9 , 12 - octadecadienoic acid

linoleic acid

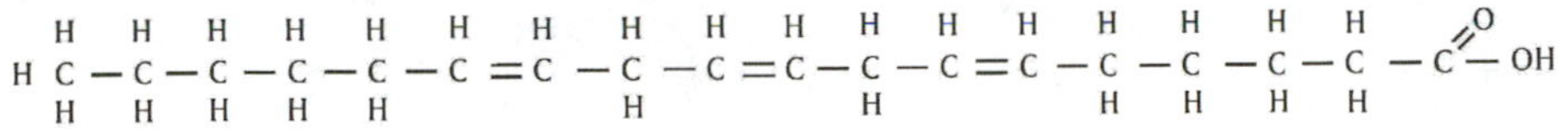

```
   H   H   H   H   H   H   H   H   H   H   H   H   H   H   H   H   H      //O
 H C — C — C — C — C — C = C — C — C = C — C — C = C — C — C — C — C — C — OH
   H   H   H   H   H           H           H           H   H   H   H
```

(Z,Z,Z) - 6 , 9 , 12 - octadecatrienoic acid

linolenic acid

FIGURE 3.1 The structures of the two polyunsaturated fatty acids that are essential for the growth and development of some insects.

in the egg from the mother. Lack of a sterol by adult female houseflies results in an 80% reduction in egg hatch, although the number of eggs laid is not affected. Adult female boll weevils fed a diet in which cholestanol replaces half the cholesterol requirement cannot maintain normal egg production, but eggs that are laid do hatch. Replacement of more than half the cholesterol requirement with cholestanol, however, results in eggs that fail to hatch (Earle et al., 1967).

No general rule can be drawn as to the quantity of sterol needed in an artificial diet. Wide variation in required quantity exists, apparently related to species differences. In general, 0.1% or less of cholesterol is considered to be satisfactory in artificial diets. A few insects, such as *Musca vicinia*, *Dermestes vulpinus*, and *Attagenus piceus,* need as little as 0.01% sterol by weight of diet, while others need about 0.1% by weight.

The ability to synthesize a sterol is mixed among other invertebrates. Some marine annelids can synthesize sterols, but the earthworm *Lumbricus terrestris* cannot. Other invertebrates that cannot synthesize a sterol include the crabs *Astacus astucus* and *Cancer pagurus*, the sea urchin *Paracentratus lividus*, the oyster *Ostrea gryphea*, the mollusk *Mytilus californians*, the tapeworm *Spirometra manosonoides*, and the nematodes *Caenorhabditis briggsae*, *Turbatrix aceti*, and *Panagrellus redivivus*.

3.4.6 Polyunsaturated Fatty Acids

About 50 species of insects from five orders have been shown to require a dietary source of **polyunsaturated fatty acids** (Dadd, 1961; Chippendale et al., 1964; Dadd, 1985). Linoleic acid [(Z,Z)-9,12 octadecadienoic acid] and Linolenic acid [(Z,Z,Z)-9,12,15-octadecatrienoic acid] (Figure 3.1) are effective in relieving the symptoms of deficiency; in some species, one of these is more effective than the other.

The requirement for polysaturated fatty acids was initially discovered in lepidopterans, in which a deficiency is dramatically displayed in failure of pupal or adult ecdysis. Individuals that successfully or partially ecdyse are likely to have misformed wings and lack normal scales on the body. Some hymenopterans show similar difficulty in ecdysis when linoleic and linolenic acids are absent, or very low in the diet. Acridid grasshoppers also tend to produce deformed adults on fatty acid-deficient diets. Coleopterans show slowed growth and decreased adult fecundity in response to deficiency in polyunsaturated fatty acids. Growth of some insects is improved by adding polyunsaturated fatty acids to the diet although an absolute requirement has not been demonstrated. Possibly so little is required in the diet that traces in the test diet, or carry-over from the egg, may support some insects through a generation. It may be necessary to test for a requirement through more than one generation, especially if no apparent defect is noted, and if adult performance is a criterion evaluated (Dadd, 1985).

In general, it appears that dipterans may not have an absolute requirement for polyunsaturated fatty acids although they are unable to synthesize them. Some dipterans, however, show improved growth when polyunsaturated fatty acids are added to the diet. A few insects, including the cricket *Acheta domesticus*, the American cockroach *Periplaneta americana*, and the termite *Zootermopsis angusticollis,* are able to synthesize polyunsaturated fatty acids.

3.4.7 Vitamins

Studies of insect vitamin requirements are particularly subject to ambiguity if axenic conditions are not maintained. Some early work with stored-products insects reared on very dry diets seemed to avoid the interfering effect of microorganisms, which can synthesize many of the vitamins that insects then utilize. Those studies showed that insects need thiamine, riboflavin, pyridoxine, niacinamide, pantothenic acid, biotin, folic acid, and choline. Carnitine, also called vitamin B^T, is a requirement for *Tenebrio molitor* and *Tribolium obscurus*, *T. confusum,* and *T. castaneum*. Different strains of these insects show variable requirements. One of the critical roles for carnitine is as a participant in the passage of fatty acids across mitochondrial membranes in insects and vertebrates.

Houseflies and blowflies are able to use β-methylcholine and γ-butyrobetaine to reduce the need for choline. When these compounds are fed to flies, the phospholipids of most tissues contain β-methylcholine, but acetylcholine in central nervous system tissue is not substituted. *Tenebrio molitor* larvae also incorporate β-methylcholine into body phospholipids, which spares the choline requirement.

There is a demonstrated requirement for water-soluble vitamins in the nutrition of a few insects. Ascorbic acid is required for normal growth and development of some insects, and seems particularly needed by phytophagous insects (Vanderzant and Richardson, 1963; Beck et al., 1968). Boll weevils, *Anthonomus grandis*, grown under aseptic conditions require inositol for normal growth and development, as do *Blattella germanica* and *Periplaneta americana*, German and American cockroaches, respectively. Inositol often seems to improve the growth of many insects but it has not been demonstrated to be essential in most insects.

Carotene and/or vitamin A are required by insects for normal pigmentation and eye function. *Schistocerca gregaria* needs β-carotene for normal body coloration. Vitamin A is required by houseflies, *Musca domestica*, and tobacco hornworms, *Manduca sexta*, for normal structure of the eye. So little carotene or vitamin A is required (and/or small amounts were contaminating the "purified" diet) that houseflies had to be reared for 15 generations on a diet lacking carotenoids and vitamin A to demonstrate conclusively that the vitamin is needed. The 12th and 13th generations had about the same sensitivity in the compound eyes, measured by electroretinograms in response to 340-nm and 500-nm light, as the lst generation. The response from eyes of deficient flies was 2 log units (100×) less sensitive than from eyes of normal flies (Goldsmith et al., 1964). There were changes in rhabdom structure, loss of basement membrane in some places, and degeneration of nervous tissue in the eyes of *M. sexta* reared for several generations on a diet deficient in vitamin A or β-carotene. Moths from the deficient diet showed abnormal orientation to light and the eyes failed to dark-adapt (Carlson et al., 1967). Vitamin A accelerates growth of the fly *Agria affinis* and the silkworm *Bombyx mori*, but it is not clear that it has a metabolic function in growth apart from its visual function.

Vitamin B_{12} stimulates the growth of some insects, but a clear-cut requirement for growth has not been shown. Possibly the small amount that may satisfy an insect requirement can occur as a contaminate of other nutrients or be provided by symbionts. Omission of the vitamin from the diet of *Blattella germanica* results in nonviable eggs, so possibly it plays a biochemical role in at least some insects.

Vitamin E is necessary to the beetle *Cryptolaemus montrousieri* in order for adult females to mature and oviposit eggs. The vitamin also is required for spermatogenesis in male house crickets *Acheta domesticus*. The parasitoid *Agria affinis* needs vitamin E in the larval diet for adult females to produce viable offspring. The vitamin also stimulates growth and development of larvae (House, 1966b). There is no evidence that vitamin D is required by insects. Vitamin K in its several forms has been tested on some insects, usually without any observable effects, but it may have some positive benefit (mechanism unknown) on crickets and may act as a phagostimulant for adult worker honeybees (Dadd, 1985).

3.4.8 Minerals

In general, only major mineral requirements are known for a few insects. Contamination of other food materials with small amounts of minerals, as well as formulation and chemical interactions when minerals are added to a synthetic diet, make determination of trace element requirements very difficult (Dadd, 1968). It stands to reason that insects need small amounts of many minerals because metal ions are required as enzyme co-factors and as constituents of metalloenzymes. For example, molybdenum, is part of the xanthine dehydrogenase involved in purine metabolism of insects. Thus, it seems reasonable to conclude that insects will need traces of Mo in their diet. Insects and vertebrates clearly have some notable differences in quantitative requirements for certain minerals. Vertebrates need large quantities of iron and calcium for hemoglobin and bone formation, respectively. Insects use these elements for neither of those functions (except a few species of insects that do have iron incorporated into hemoglobin) and require only trace amounts of iron and calcium (see review by Locke and Nichol, 1992). Many phytophagous insects need relatively large quantities of potassium and only trace amounts of sodium, while vertebrates need these elements in the reverse order.

Just how much sodium an insect needs is not known, but some Lepidoptera appear to have a need that is met by puddling, or drinking at standing water, usually on the ground. Puddling behavior has been observed primarily in male Lepidoptera, but females also puddle. Smedley and Eisner (1995) experimentally evaluated puddling behavior in male *Gluphisia septentrionis* moths (Notodontidae). The adult moth has its mothparts modified in a way that seems to facilitate rapid sucking up of puddle water while straining out debris that might be present (Figure 3.2A, B). A male moth may imbibe so much water at natural puddles that it ejects an average of 8 μl fluid from the anus about every 3 seconds (Figure 3.3); such behavior was observed to continue for greater than 200 minutes in some individuals. A maximum excretion of fluid equivalent to 600 times the body weight of a moth was observed. The authors showed by quantitative analyses of imbibed fluid and excreted fluid that there is specifically a gain in body sodium by the male moth without a necessary gain of potassium, magnesium, or calcium in test solutions imbibed (Figure 3.4). The time spent in puddling and volume of fluid excreted is inversely related to solutions containing 0.01, 0.1, and 1 m*M* Na. Males transfer the sodium acquired by puddling to females at mating, and the females incorporate it into eggs.

Aphids fed on a liquid diet that can be highly purified have proven useful in mineral studies. Two aphids, *Myzus persicae* and *Aphis fabae*, require trace amounts of Fe, Mn, Zn, and Cu, as well as major quantities of K, Mg, and phosphate (Dadd, 1967, 1968). Potassium and magnesium are major requirements of *D. melanogaster* (Sang, 1956). Zinc is necessary to *Tenebrio molitor* and about 6 μg/g diet satisfies the requirement. Cockroaches grown on artificial diets with very low levels of manganese and zinc tend to lose symbionts from their mycetocytes, but the mechanism for this interaction remains unknown (Brooks, 1960).

The balance of minerals in a salt mixture and the proportion of minerals to other groups of nutrients are important. Wesson's salt mixture is designed for vertebrates (Osborne and Mendel, 1932), and is high in Ca, Fe, and Na. Wesson's salt mixture (often called Salt Mix W) is not

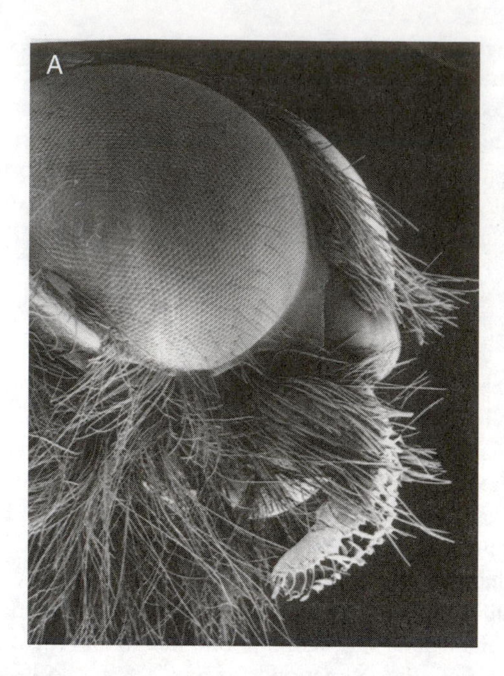

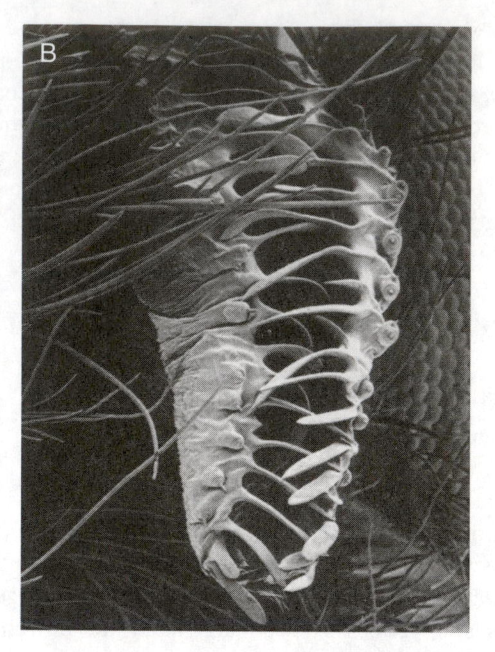

FIGURE 3.2 A: Head of the male moth *Gluphisia septentrionis* showing the stubby, highly modified proboscis that enables the moth to suck up puddle water while straining out debris. B: An enlarged view of the male proboscis illustrating the oral cleft and sieving apparatus. (Reproduced with permission from Smedley and Eisner, 1995. Photographs courtesy of Maria and Tom Eisner.)

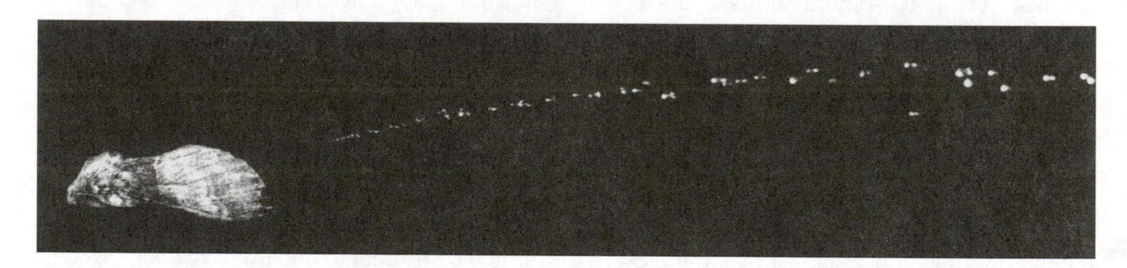

FIGURE 3.3 A male *Gluphisia septentrionis* forcefully excreting excess fluid during puddling behavior. Some males may eject fluid equal to 600 times the body weight, excreting an average of 8 μl every 3 seconds for up to 200 minutes. (Reproduced with permission from Smedley and Eisner, 1995. Photograph courtesy of Tom Eisner and Scott Smedley.)

adequate to support development of the European corn borer on an artificial diet. The Ca level in Wesson's salt mixture is toxic to *B. germanica* (Gordon, 1959). A mixture of salts based on a successful formula for confused flour beetles, *T. confusum* (Medici and Taylor, 1966) supports development of corn borers (Beck et al., 1968).

3.5 TECHNIQUES AND DIETS USED IN INSECT NUTRITION STUDIES

Diets for rearing insects are important not only as a way to study insect nutritional requirements, but in mass rearing programs for sterile releases and augmentation of natural parasites and predators. **Holidic** diets consist of chemicals that have a precisely known chemical structure before the various chemicals are mixed. Holidic diets are sometimes referred to as chemically defined diets, although chemical components may react on mixing to produce new chemical compounds that may not be

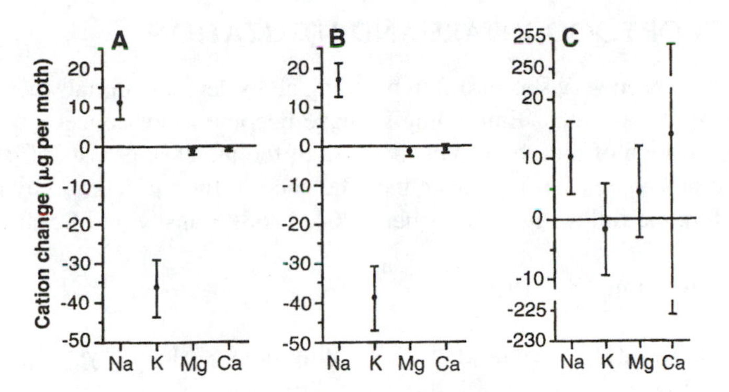

FIGURE 3.4 Net change in cation content of puddling male *Gluphisia*, calculated from concentration differences in imbibed and ejected fluids when 1 m*M* Na is imbibed (A), when a 1-m*M* solution of all four cations is imbibed (B), and when field puddles are imbibed (C). Sodium ions are concentrated from the imbibed fluids and a large percentage (mean of 40%) of potassium is excreted with the first ejected fluid. Magnesium and calcium ions are not accumulated significantly. (Reproduced by permission from Smedley and Eisner, 1995. Copy of figure provided by Tom Eisner.)

known. Holidic diets are important in the study of nutritional requirements. **Meridic** diets contain a holidic base with addition of one (or possibly a few) unknown or poorly defined substance(s). **Oligidic** diets contain complex organic material, such as lettuce for grasshoppers, dog food or chick mash for crickets and cockroaches, or ground pinto beans for some lepidopteran larvae. When insects are reared so that no other species (no bacteria, no fungi, no internal symbionts) are present, the culture is called an **axenic** culture. Axenic rearing is quite difficult to achieve, but precise definition of nutritional requirements for insects demands axenic rearing, and it has been successfully accomplished in a few cases. A **gnotobiotic** culture is one in which all the existing species are known; such a culture may or may not be axenic, depending on how many species exist in the culture. For practical purposes of maintaining laboratory cultures and mass production, insects are usually grown in xenic cultures (unknown number of organisms in the culture) on oligidic diets.

3.6 CRITERIA FOR EVALUATING THE NUTRITIONAL QUALITY OF A DIET

The measurement of growth rate has frequently been used to determine nutritional quality of diets fed to immature stages. Measurements of weight gains, time between molts or to pupation, and time to adult emergence also have been used. The percent successful pupation or emergence of adults can be used. Adult diets can be evaluated by number of eggs laid, percent hatch of eggs, longevity of adults, time to sexual maturity, or other physiological parameters the investigator believes to be influenced by nutrition. It is usually desirable to have more than one criterion for evaluating a diet. Nutritional quality may have little or no effect on one criterion, while causing great changes in another.

Requirements for some nutrients may not be manifest in the first generation; nutrient reserves are frequently stored in body tissues or egg yolk. Hence, in the absence of any effects of an experimental deficiency being tested, several generations should be reared. Axenic rearing conditions should be maintained when possible. Microorganisms present in the gut or in mycetomes frequently contribute to digestion, availability of nutrients, and biosynthesis of some nutrients, particularly vitamins, and sometimes sterols and some essential amino acids.

The purity of dietary components becomes crucial in some experiments. Traces of sterols frequently occur in protein sources such as casein or egg albumin. In general, the requirements for trace elements such as Na, Zn, Fe, Mn, and Cu have not been established for insects because other dietary components contain these elements in sufficient quantity as contaminants. Removal of contaminants, whether sterols, vitamins, or trace minerals, may be tedious and costly.

3.7 MEASURES OF FOOD INTAKE AND UTILIZATION

Animal breeders have been very successful in breeding and selecting animals that maximize weight gains per unit of food consumed. Entomologists have become more concerned with such factors as efficiency of utilization of food by insects because of the increasing costs of large, mass rearing programs. Several procedures for measuring the efficiency of food utilization by insects have been developed, including the following (Waldbauer, 1964, 1968; Slansky and Scriber, 1985).

1. Relative growth rate (R.G.R.):

$$\text{R.G.R.} = (\text{Dry weight gained})/(\text{Feeding days} \times \text{Mean dry weight})$$

2. Approximate digestibility (A.D.):

$$\text{A.D.} = [(\text{Dry weight of food ingested} - \text{Dry weight of feces})/(\text{Dry weight of food ingested})] \times 100$$

3. Efficiency of conversion (E.C.I.) of ingested food to body matter:

$$\text{E.C.I.} = (\text{Weight gained}/\text{Dry weight of food ingested}) \times 100$$

4. Efficiency with which digested food is converted to body matter (E.C.D.):

$$\text{E.C.D.} = [(\text{Weight gained})/(\text{Dry weight of food ingested} - \text{Dry weight of feces})] \times 100$$

Experimental measurements of A.D., E.C.I., or E.C.D. require quantitative data for food ingested and weight of feces excreted. In some cases, weighing the food that remains after the insects have ceased feeding, or at the end of a chosen interval, may be satisfactory. Feces of some insects can be manually separated from uneaten food and weighed.

Addition of chromic oxide to the food ingested and subsequent chemical analysis of the amount of chromic oxide in the feces has been used in vertebrates, and in some insects, to indicate the amount of food consumed (McGinnis and Kasting, 1964). The method works if (1) the chromic oxide is uniformly distributed in the food, (2) it has no toxic effects and does not alter the digestion or physiology of the animal, and (3) it is not absorbed from the gut. In the chromic oxide method, the percentage of ingested food that is utilized is given by the formula:

$$1 - (\text{Weight of chromic oxide}/\text{Unit dry weight of food})/(\text{Weight of chromic oxide}/\text{Unit dry weight of feces}) \times 100$$

Utilizing the chromic oxide method with 5th instars of the pale western cutworm *Agrotis orthogonia,* McGinnis and Kasting (1964) found that 41% of the sprouts of Thatcher wheat were utilized by the larvae, but only 21% of a mixture of equal parts of sprouts and powdered cellulose and 16% of the pith from Rescue wheat were utilized. The method requires that a sample of feces must be separated from the uneaten food, the dry weight obtained, and the concentration of chromic oxide determined. If only a sample of feces, and not the total quantity of feces, is collected for analysis, possible error may occur if the concentration of chromic oxide is not reasonably the same in feces excreted at different ages or times of the day.

The uric acid produced by insects from protein and purine catabolism has also been used as an indicator of food utilization (Bhattacharya and Waldbauer, 1970, 1972). Uric acid, which does not occur in most foods, especially not in stored grains, is easily determined quantitatively. A small

TABLE 3.3
Approximate Digestibility Data from Larval to Pupal Ecdysis of *Heliothis virescens* and *Argyrotaenia velutinana*

	Method for Approximate Digestibility	
Insect	**Manual Separation**	**Uric Acid Procedure**
H. virescens	56.0 ± 1.6	55.4 ± 1.6
A. velutinana	51.0 ± 1.8	50.7 ± 1.8

Note: 20 insects used in each test.

Reproduced with permission from Chow et al., 1973.

TABLE 3.4
Comparisons between Male and Female *Argyrotaenia velutinana* in Food Utilization Determined by the Uric Acid Procedure

	A.D.	E.C.I.	E.C.D.
Male	50.5 ± 3.4	12.2 ± 0.8	26.4 ± 3.2
Female	50.5 ± 2.5	13.4 ± 0.8	27.1 ± 3.0

Reproduced with permission from Chow et al., 1973.

sample of mixed food and feces must be carefully separated manually, the feces weighed, and the quantity of uric acid determined per unit weight of feces. Then the uric acid content of a larger unseparated sample of food-feces is determined. For example, if a carefully separated sample of feces contained 10% uric acid, while a much larger weighed mixture of food and feces contained 1% uric acid, one could estimate that the mixture was 90% uneaten food and 10% feces. This method is subject to error if the quantity of uric acid per unit weight of feces varies with age or stage of development of the insect (and it often does). The potential errors in the chromic oxide and uric acid methods should be investigated prior to indiscriminate use of either method in a particular insect. The use of the uric acid method gave good agreement with the more laborious technique of manually separating all the uneaten food and feces in *Tenebrio molitor*, *Tribolium confusum*, *Argyrotaenia velutinana*, and *Heliothis virescens*. Chow et al. (1973) compared the uric acid method with the manual separation and gravimetric analysis of feces and uneaten food in a study of approximate digestibility in two lepidopterans; there was no difference in results between the two methods (Table 3.3), and no differences in the results from a comparison of female and male *A. velutinana* (Table 3.4).

Although the digestibility of a particular food may be good, it may not be readily converted into body substance (Table 3.5). An imbalance in nutrients resulting from digestion may be one factor that prevents efficient conversion into body substance. About 45% of the food ingested goes into net weight gain of honeybee worker larvae when they are fed royal jelly, but this is not their normal food. An interesting technique that made use of a radioactive nuclide, ^{32}P, was devised to study food consumption in honeybee larvae. The ^{32}P was mixed with royal jelly, and the labeled jelly was hand-fed to larvae growing in incubators. Because honeybee larvae do not have a complete gut until just prior to pupation, larvae only void feces just before they pupate. Thus, the ^{32}P ingested

TABLE 3.5
Effect of Diet on Approximate Digestibility (A.D.) and Efficiency of Conversion of Digested Food to Body Substance (E.C.D.) by *Triboluim confusum* Larvae

Food	A.D.	E.C.D.
Cracked wheat	63.2 ± 0.88	7.2 ± 0.69
Wheat germ	66.5 ± 1.89	15.6 ± 0.94
Ground wheat	55.5 ± 1.79	8.4 ± 0.48
Ground wheat plus 5% brewer's yeast	65.4 ± 0.43	8.1 ± 0.46

Reproduced with permission from Bhattacharya and Waldbauer, 1970.

each day accumulated in the body. By knowing the specific activity of the food, it was possible to calculate the rate and cumulative food intake after removing a larva from the food and obtaining a total body count. For ^{32}P with a half-life of approximately 14 days, some correction should be made if the experiment runs for the 5-day life of a honeybee larva.

3.8 PHAGOSTIMULANTS

Phagostimulants are chemical compounds that induce feeding. Insects are induced to feed by chemosensory stimulation from components in or on their food (Thorsteinson, 1960; Chapman, 1995). Gustatory sensilla (Figure 3.5A, B) are located on the mouthparts, tarsi, and antennae (Schoonhoven et al., 1991; Chyb et al., 1995). Adult western corn rootworms show strong phagostimulatory responses to several L-amino acids (Kim and Mullin, 1998); to γ-amino butyric acid (GABA); and to as little as 0.1 μM cucurbitacin B, a bitter (to humans) substance in cucurbits (Figure 3.6). Caterpillars of the tobacco hornworm *Manduca sexta* have two pairs of myo-inositol-sensitive gustatory sensilla, and the caterpillars respond strongly to inositol by feeding. The compound also counteracts inhibitory effects of some aversive stimuli (Glendinning et al., 2000).

In a few cases, an insect's behavior in the presence of phagostimulants has been divided into a biting response and a swallowing response. Phagostimulants are important to normal insect feeding, and they are very useful in insect diet, nutrition, and mass rearing studies. One commonly encountered frustration suffered by many entomologists has been the finding that a diet formulated on the principles of good nutrition is refused by insects brought in from the field. Part of the problem may be the absence of a normal phagostimulant.

If food recognition mechanisms are acted upon by natural selection, then one might expect a mechanism responsive to specific nutritional requirements to evolve. Major nutritional food materials such as carbohydrates, proteins, lipids, etc. are logical indicators in plant and animal tissues of food and, in fact, virtually all the major nutritional substances normally required by animals serve as phagostimulants for one species of insect or another. Many insects have evolved sensory mechanisms responsive to nonnutritive substances as food recognition signals, apparently as a result of the co-evolution of plants and insects.

Among primary nutrients, sugars, and sucrose especially, are phagostimulants for many insects. Glucosides, which are combinations of glucose with a non-sugar molecule, serve as both phagostimulants and deterrents to feeding in different insects. The aglucone or non-sugar moiety seems to control the role of glucosides.

Combinations of nutrients are important in many cases. Whole proteins, such as wheat gluten for confused flour beetles, are sometimes phagostimulants, but more frequently, amino acids induce feeding. Leucine, methionine, lysine, and isoleucine in phosphate buffer were effective feeding stimulants for female houseflies. In this case, the presence of phosphate ions was important to the

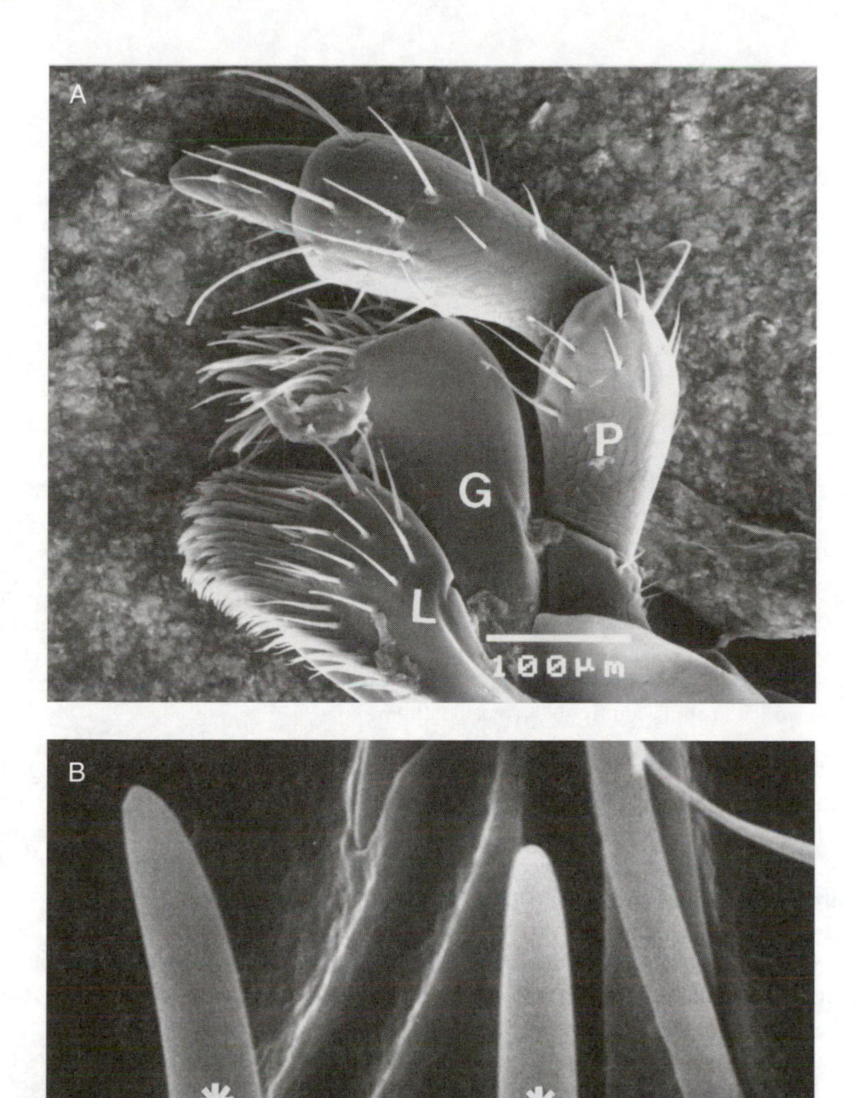

FIGURE 3.5 A: A scanning electron micrograph (SEM) of the maxilla of an adult *Diabrotica virgifera virgifera* illustrating the maxillary palp (P), galea (G), and lacinea (L). The galea contains chemosensilla and (putative) mechanosensilla. B: SEM views of two galeal sensilla, marked with an asterisk. (Reproduced with permission from Chyb et al., 1995. Photographs courtesy of Chris Mullin.)

phagostimulant activity of the amino acids; but in some insects, amino acids may act alone or in combination with other amino acids or derivatives of amino acids (Chen and Henderson, 1996; Hollister and Mullin, 1998). Houseflies are induced to feed by casein and yeast hydrolysates and guanosine monophosphate (GMP) in phosphate buffers.

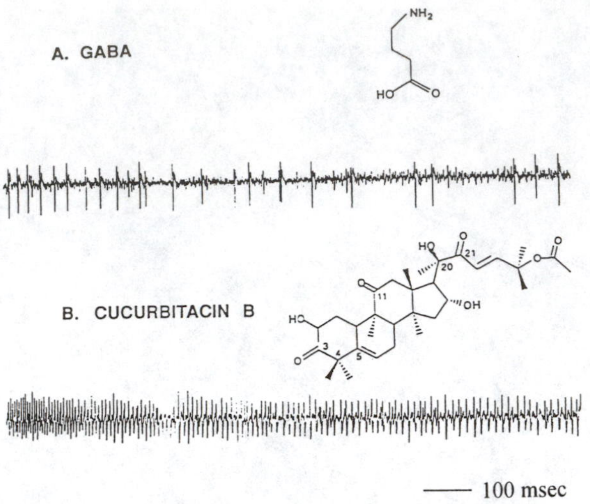

A. GABA

B. CUCURBITACIN B

100 msec

FIGURE 3.6 The structures of γ-aminobutyric acid (GABA) and cucurbitacin B, two feeding stimulants for adult *Diabrotica virgifera virgifera*, and electrophysiological responses recorded from galeal sensilla to 10 μM concentrations of each. The response to GABA is phasic, showing a decrease in firing rate over a second or so time span, while the response to cucurbitacin B is vigorous and only slightly phasic over the same unit of time. (Reproduced with permission from Chyb et al., 1995.)

Reduced glutathione, a tripeptide composed of glutamic acid, cysteine, and glycine, is a phagostimulant for some, and perhaps all, ticks. Adenosine triphosphate (ATP) in the presence of sodium ions is a phagostimulant for the adult mosquito *Aedes aegypti*. Lipids frequently are phagostimulants. Phospholipids, triglycerides, sterols, sterol esters, and fatty acids are important phagostimulants for various insects.

Many secondary plant substances, sometimes referred to as "token factors or stimulants," induce insects to feed. These substances are usually present in small amounts and are frequently restricted to a group of related plants. Insects have no absolute metabolic or nutritional requirement for these token stimulants, so far as is known; they appear to serve as indicators of appropriate food. Combinations of secondary plant substances are frequently more effective stimulants than single compounds. The boll weevil *Anthonomus grandis* is stimulated to feed by a mixture of numerous compounds known to be present in cotton plants, including gossypol. Gossypol by itself, however, is only a weak stimulant to feeding. Secondary plant substances also may signal females to lay eggs. Sinigrin, a mustard oil glucoside, induces the female butterfly *Pieris brassicae* to oviposit, even on foreign or nutritionally sterile substrates.

3.9 FEEDING DETERRENTS

Many substances are known to have an inhibitory effect on insect feeding, and chemoreceptors responsive to specific compounds are involved, as in phagostimulation. For example, (–)-β-Hydrastine and strychnine-HCl are powerful feeding deterrents for adult western corn beetles mediated through the response of chemoreceptors (Figure 3.7) (Chyb et al., 1995), although exactly how the sensory information is translated into behavioral action by the beetles is not known. Cantharidin from the blood of blister beetles is effective at 10^{-5} *M* as a feeding deterrent to carabid beetles, *Calosoma prominens* (Carrel and Eisner, 1974). Ammonium nitrate inhibits feeding of the sweet clover weevil *Sitona cylindricollis*. Feeding by the American cockroach, *Periplaneta americana*, is inhibited by 1,4-naphthoquinone. Food intake by the desert locust, *Schistocerca gregaria*, is decreased by injection of Lom-sulfakinin, a neuropeptide found in the corpus cardiacum of locusts.

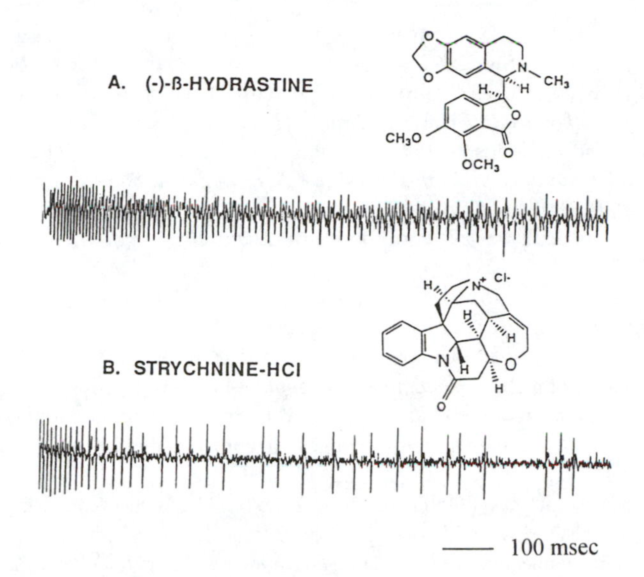

FIGURE 3.7 The electrophysiological responses of chemoreceptors on the galea of adult *Diabrotica virgifera virgifera* to 10 μM concentrations of (–)-β-hydrastine and strychnine-HCl, two alkaloid phagodeterrents. The response to strychnine-HCl is more phasic than that to (–)-β-hydrastine. The data suggest that a peripheral GABA/glycine-type receptor may be important in host plant discrimination in these adult beetles because the two phagodeterrent alkaloids can bind to and inhibit GABA/glycine receptors. (Reproduced with permission from Chyb et al., 1995.)

Although the mechanism has not been elucidated, authors suggest that the sulfakinins may reduce the sensitivity of taste receptors (Wei et al., 2000). For more details and a list of feeding deterrents, see Schoonhoven (1969, 1972) and Dethier (1980).

REFERENCES

Ahmad, I., G.P. Waldbauer, and S. Friedman. 1993. Maxillectomy does not disrupt self-selection by larvae of *Manduca sexta* (Lepidoptera: Sphingidae). *Ann. Entomol. Soc. Am.,* 86:458-463.

Anderson, T.E., and N.C. Leppla. 1992. *Advances in Insect Rearing for Research and Pest Management*, Westview Press, Boulder, CO, 517 pp.

Beck, S.D., G.M. Chippendale, and D.E. Swinton. 1968. Nutrition of the European corn borer, *Ostrinia nubilalis*. VI. A larval rearing medium without crude plant fractions. *Ann. Entomol. Soc. Am.,* 61:459-462.

Bhattacharya, A.K., and G.P. Waldbauer. 1970. Use of faecal uric acid method in measuring the utilization of food by *Tribolium confusum*. *J. Insect Physiol.,* 16:1983-1990.

Bhattacharya, A.K., and G.P. Waldbauer. 1972. The effect of diet on the nitrogenous endproducts excreted by larval *Tribolium confusum* with notes on correction of A.D. and E.C.D. for fecal urine. *Ent. Exp. Appl.,* 15:238-247.

Bracken, G.K. 1966. Role of ten dietary vitamins on fecundity of the parasitoid *Exeristes comstockii* (Cresson) (Hymenoptera: Ichneumonidae). *Can. Entomol.,* 98:918-922.

Brooks, M.A. 1960. Some dietary factors that affect ovarial transmission of symbiotes. *Proc. Helminthol. Soc. Wash.,* 27:212-220.

Carlson, S.D., H.R. Steeves III, J.S. Van de Berg, and W.E. Robbins. 1967. Vitamin A deficiency: Effects on retinal structure of a moth *Manduca sexta*. *Science,* 158:268-270.

Carrel, J.E., and T. Eisner. 1974. Cantharidin: Potent feeding deterrent to insects. *Science,* 183:755-757.

Chapman, R.F. 1995. Chemosensory regulation of feeding, pp. 101-136, in R.F. Chapman and G. de Boer (Eds.), *Regulatory Mechanisms in Insect Feeding*, Chapman & Hall, New York.

Chen, J., and G. Henderson. 1996. Determination of feeding preference of Formosan subterranean termite (*Coptotermes formosanus* Shiraki) for some amino acid additives. *J. Chem. Ecol.,* 22:2359-2369.

Chippendale, G.M, S.D. Beck, and F.M. Strong. 1964. Methyl lineolenate as an essential nutrient for the cabbage looper, *Trichoplusia ni* Hübner). *Nature (London),* 204:710-711.

Chow, Y.M., G.C. Rock, and E. Hodgson. 1973. Consumption and utilization of chemically defined diets by *Argyrotaenia velutinana* and *Heliothis virescens. Ann. Entomol. Soc. Am.,* 66:627-632.

Chyb, S., H. Eichenseer, B. Hollister, C.A. Mullin, and J.L. Frazier. 1995. Identification of sensilla involved in taste mediation in adult western corn rootworm (*Diabrotica virgifera virgifera* LeConte). *J. Chem. Ecol.,* 21:313-329.

Cohen, R.W., S.L. Heydon, G.P. Waldbauer, and S. Friedman. 1987a. Nutrient self-selection by the omnivorous cockroach *Supella longipalpa. J. Insect Physiol.,* 33:77-82.

Cohen, R.W., G.P. Waldbauer, S. Friedman, and N.M. Schiff. 1987b. Nutrient self-selection by *Heliothis zea* larvae: A time-lapse film study. *Entomol. Exp. Appl.,* 44:65-73.

Dadd, R.H. 1961. The nutritional requirements of locusts, V. Observations on essential fatty acids, chlorophyll, nutritional salt mixtures, and the protein or amino acid components of synthetic diets. *J. Insect Physiol.,* 6:126-145.

Dadd, R.H. 1967. Improvement of synthetic diet for the aphid *Myzus persicae* using plant juices, nucleic acids, or trace metals. *J. Insect Physiol.,* 13:763-778.

Dadd, R.H. 1968. Problems connected with inorganic components of aqueous diets. *Bull. Entomol. Soc. Am.,* 14:22-26.

Dadd, R.H. 1973. Insect nutrition: Current developments and metabolic implications. *Annu. Rev. Entomol.,* 18:381-420.

Dadd, R.H. 1985. Nutrition: Organisms, pp. 313-390, in G.A. Kerkut and L.I. Gilbert (Eds.), *Comprehensive Insect Physiology, Biochemistry and Pharmacology,* Vol. 4, Pergamon Press, Oxford.

Dadd, R.H., and D.L. Krieger. 1968. Dietary amino acid requirements of the aphid, *Myzus persicae. J. Insect Physiol.,* 14:741-764.

Davis, G.R.F. 1968. Phagostimulation and consideration of its role in artificial diets. *Bull. Entomol. Soc. Am.,* 14:27-29.

De Groot, A.P. 1952. Amino acid requirements for growth of the honeybee (*Apis mellifica* L.). *Experientia,* 8:192-193.

Dethier, V.G. 1980. Evolution of receptor sensitivity to secondary plant substances with special reference to deterrents. *Am. Nat.,* 115:45-66.

Earle, N.W., A.B. Walker, M.L. Burks, and B.H. Slatten. 1967. Sparing of cholesterol by cholestanol in the diet of the boll weevil *Anthonomus grandis* (Coleoptera: Curculionidae). *Ann. Entomol. Soc. Am.,* 60:599-603.

Friend, W. 1958. Nutritional requirements of phytophagous insects. *Annu. Rev. Entomol.,* 3:57-74.

Friend, W.G., R.H. Backs, and L.M. Case. 1957. Studies on amino acid requirements of larvae of the onion maggot, *Hylemya antiqua* (MG.) under aseptic conditions. *Can. J. Zool.,* 35:535-543.

Glendinning, J.I., N.M. Nelson, and E.A. Bernays. 2000. How do inositol and glucose modulate feeding in *Manduca sexta* caterpillars? *J. Exp. Biol.,* 203:1299-1315.

Goldsmith, T.H., R.J. Barker, and C.F. Cohen. 1964. Sensitivity of visual receptors of carotinoid-depleted flies: A vitamin A deficiency in an invertebrate. *Science,* 146:65-67.

Gordon, H.T. 1959. Minimal nutritional requirements of the German roach, *Blatella germanica* L. *Ann. N.Y. Acad. Sci.,* 77:290-315.

Gordon, H.T. 1972. Interpretation of insect quantitative nutrition, pp. 73-105, in J.G. Rodrigues (Ed.), *Insect and Mite Nutrition: Significance and Implications in Ecology and Pest Management.* Elsevier-North Holland, Amsterdam.

Haydak, M.H. 1953. Influence of the protein level of the diet on the longevity of cockroaches. *Ann. Entomol. Soc. Am.,* 46:547-560.

Hinton, T., D.T. Noyes, and J. Ellis. 1951. Amino acids and growth factors in a chemically defined medium for *Drosophila. Physiol. Zool.,* 24:335-353.

Hobson, R.P. 1935. On a fat-soluble growth factor required by blowfly larvae. II. Identity of the growth factor with cholesterol. *Biochem. J.,* 29:2023-2026.

Hollister, B., and C.A. Mullin. 1998. Behavioral and electrophysiological dose–response relationships in adult western corn rootworm (*Diabrotica virgifera virgifera* LeConte) for host pollen amino acids. *J. Insect Physiol.,* 44:463-470.

House, H.L. 1954. Nutritional studies with *Pseudosarcophaga affinis* (Fall), a dipterous parasite of the spruce budworm, *Choristoneura fumiferana* (Clem). I. A chemically defined medium and aseptic culture technique. *Can. J. Zool.,* 32:331-341.

House, H.L. 1965a. Insect nutrition, pp. 769-813, in M. Rockstein (Ed.), *Physiology of Insecta*, 1st ed., Vol. 2, Academic Press, New York.

House, H.L. 1965b. Effects of low levels of the nutrient content of a food and of nutrient imbalance on the feeding and the nutrition of a phytophagous larva, *Celerio euphorbiae* (Linnaeus) (Lepidoptera: Sphingidae). *Can. Entomol.,* 97:62-68.

House, H.L. 1966a. Effects of varying the ratio between the amino acids and the other nutrients in conjunction with salt mixture on the fly *Agria affinis* [Fall.]. *J. Insect Physiol.,* 12:299-310.

House, H.L. 1966b. Effects of vitamins E and A on growth and development and the necessity of vitamin E for reproduction in the parasitoid *Agria affinis* [Fall.]. *J. Insect Physiol.,* 12:409-418.

House, H.L. 1967. The role of nutritional factors in food selection and preference as related to larval nutrition of an insect, *Pseudosarcophaga affinis* (Diptera, Sarcophagidae) on synthetic diets. *Can. Entomol.,* 99:1310-1321.

House, H.L. 1969. Effects of different proportions of nutrients on insects. *Ent. Exp. Appl.,* 12:651-669.

House, H.L. 1974. Nutrition, pp. 1-62, in M. Rockstein (Ed.), *The Physiology of Insects*, 2nd ed., Academic Press, New York, 648 pp.

House, H.L., P. Singh, and W.W. Batsch. 1971. Lepidoptera, pp. 55-115, in *Artificial Diets for Insects: A Compilation of References with Abstracts*, Information Bulletin No. 7, Research Institute, Canada Department of Agriculture, Belleville, Ontario, 156 pp.

Hsiao, T.H. 1972. Chemical feeding requirements of oligophagous insects, pp. 225-240, in J.G. Rodrigues (Ed.), *Insect and Mite Nutrition: Significance and Implications in Ecology and Pest Management*, Elsevier-North Holland, Amsterdam.

Ishii, S., and C. Hirano. 1955. Qualitative studies on the essential amino acids for the growth of the larva of the rice stem borer, *Chilo simplex* (Butler), under aseptic conditions. *Bull. Natl. Inst. Agric. Sci. (Japan),* Ser. C., No. 5, p. 35-48.

Kasting, R., and A.J. McGinnis. 1958. Use of glucose labeled with carbon-14 to determine the amino acids essential to an insect. *Nature (London),* 182:1380-1381.

Kasting, R., and A.J. McGinnis. 1960. Use of glutamic acid-U-C^{14} to determine nutritionally essential amino acids for larvae of the blowfly, *Phormia regina*. *Can. J. Biochem. Physiol.,* 38:1229-1234.

Kasting, R., and A.J. McGinnis. 1962. Nutrition of the pale western cutworm, *Agrotis orthogonia* Morr. IV. Amino acid requirements determined with glucose-U-^{14}C. *J. Insect Physiol.,* 8:97-103.

Kasting, R., G.R.F. Davis, and A.J. McGinnis. 1962. Nutritionally essential and non-essential amino acids for the prairie grain wireworm, *Ctenicera destructor* Brown, determined with glucose-U-C^{14}. *J. Insect Physiol.,* 8:589-596.

Kim, J.H., and C.A. Mullin. 1998. Structure-phagostimulatory relationships for amino acids in adult western corn rootworm, *Diabrotica virgifera virgifera*. *J. Chem. Ecol.,* 24:1499-1511.

Kircher, H.W., W.B. Heed, J.S. Russell, and J. Groove. 1967. Senita cactus alkaloids: their significance to Sonoran desert *Drosophila* ecology. *J. Insect Physiol.,* 13:1869-1874.

Leclercq, J. 1950. La vitamine T, facteur de croissance pour les larves de "*Tenebrio molitor.*" *Arch. Int. Physiol. LVII,* Fas., 30:350-352.

Lemonde, A., and R. Bernard. 1951. Nutrition des larves de *Tribolium confusum* (Duval). II. Importance des acides amine. *Can. J. Zool.,* 29:80-83.

Locke, M., and H. Nichol. 1992. Iron economy in insects: Transport, metabolism, and storage. *Annu. Rev. Entomol.,* 37:195-215.

McGinnis, A.J., and R. Kasting. 1964. Chromic oxide indicator method for measuring food utilization in a plant-feeding insect. *Science,* 144:1464-1465.

Medici, J.C., and M.W. Taylor. 1966. Mineral requirements of the confused flour beetle, *Tribolium confusum* (Duval). *J. Nutr.,* 88:181-186.

Mira, A. 2000. Exuviae eating: A nitrogen meal? *J. Insect Physiol.,* 46:605-610.

Mittler, T.E. 1971. Dietary amino acid requirements of the aphid *Myzus persicae* affected by antibiotic uptake. *J. Nutr.,* 101:1023-1028.

Moore, W. 1946. Nutrition of *Attagenus* (?) sp. II. (Coleoptera: Dermestidae). *Ann. Entomol. Soc. Am.,* 39:513-521.

Mondy, N., and M.-F. Corio-Costet. 2000. The response of the grape berry moth (*Lobesia botrana*) to a dietary phytopathogenic fungus (*Botrytis cinerea*): The significance of fungus sterols. *J. Insect Physiol.,* 46:1557-1564.

Norris, D.M., and J.K. Baker. 1967. Symbiosis: Effects of a mutualistic fungus upon the growth and reproduction of *Xyleborus ferrugineus*. *Science,* 156:1120-1122.

Osborn, T.B., and L.B. Mendel. 1932. A modification of the Osborne-Mendel salt mixture containing only inorganic constituents. *Science,* 75:339-340.

Pant, N.C., J.K. Nayar, and P. Gupta. 1958. On the significance of amino acids in the larval development of Khapra-beetle, *Trogoderma granarium* Everts. (Coleoptera: Dermestidae). *Experientia,* 14:176-177.

Pant, N.C., B. Gupta, and J.K. Nayar. 1960. Physiology of intracellular symbiotes of *Stegobium paniceum* L. with special reference to amino acid requirements of the host. *Experientia,* 16:311-312.

Reinecke, J.P. 1985. Nutrition: Artificial diets, pp. 391-419, in G.A. Kerkut and L.I. Gilbert (Eds.), *Comprehensive Insect Physiology, Biochemistry and Pharmacology,* Vol. 4, Pergamon Press, Oxford.

Rock, G.C., and K.W. King. 1968. Amino acid synthesis from glucose-U-^{14}C in *Argyrotaenia velutinana* (Lepidoptera: Tortricidae) larvae. *J. Nutr.,* 95:369-373.

Rock, G.C., and E. Hodgson. 1971. Dietary amino requirements for *Heliothis zea* determined by dietary deletion and radiometric techniques. *J. Insect Physiol.,* 17:1087-1097.

Rodrigues, J.G., and R.E. Hampton. 1966. Essential amino acids determined in the two-spotted spider mite, *Tetranychus urticae* (Acarina: Tetranychidae), with glucose-U-^{14}C. *J. Insect Physiol.,* 12:1209-1216.

Sang, J.H. 1956. The quantitative nutritional requirements of *Drosophila melanogaster. J. Exp. Biol.,* 35:45-72.

Schoonhoven, L.M. 1969. Gustation and foodplant selection in some lepidopterous larvae. *Ent. Exp. Appl.,* 12:555-564.

Schoonhoven, L.M. 1972. Secondary plant substances and insects, pp. 197-224, in V.C. Runeckles and T.C. Tso (Eds.), *Structural and Functional Aspects of Phytochemistry,* Academic Press, New York.

Schoonhoven, L.M., M.S.J. Simmonds, and W.M. Blaney. 1991. Changes in the responsiveness of the maxillary styloconic sensilla of *Spodoptera littoralis* to inositol and sinigrin correlate with feeding behavior during the final larval stadium. *J. Insect Physiol.,* 37:261-268.

Scriber, J.M., and F. Slansky, Jr. 1981. The nutritional ecology of immature insects. *Annu. Rev. Entomol.,* 26:183-211.

Shuel, R., and S. Dixon. 1968. The importance of sugar for the pupation of the worker honeybee. *J. Apicult. Res.,* 7:109-112.

Simpson, S.J., and P.R. White. 1990. Associative learning and locust feeding: Evidence for a "learned hunger" for protein. *Anim. Behav.,* 40:506-513.

Simpson, S.J., and D. Raubenheimer. 1995. The geometric analysis of feeding and nutrition: A user's guide. *J. Insect Physiol.,* 41:545-553.

Simpson, S.J., and C.L. Simpson. 1990. The mechanisms of nutritional compensation by phytophagous insects, pp. 111-160, in E.E. Bernays (Ed.), *Insect-Plant Interactions,* Vol. 2, CRC Press, Boca Raton, FL.

Simpson, C.L., S. Chyb, and S.J. Simpson. 1990. Changes in chemoreceptor sensitivity in relation to dietary selection by adult *Locusta migratoria*. *Ent. Exp. Appl.,* 56:259-268.

Singh, K.R.P., and A.W.A. Brown. 1957. Nutritional requirements of *Aedes aegypti* L. *J. Insect Physiol.,* 1:199-220.

Singh, P. 1974. Artificial diets for insects: A compilation of references with abstracts. (1970–72). *N.Z. Dept. Sci. Ind. Res. Bull.* 214, 96 pp.

Singh, P. 1976. Synthetic diets for insects and mites, pp. 131-250, in M. Recheigl (Ed.), *Handbook of Nutrition and Food,* Vol. 2, Sec. G., CRC Press, Boca Raton, FL.

Singh, P. 1977. *Artificial Diets for Insects, Mites, and Spiders.* IFI/Plenum, New York.

Singh, P., and H.L. House. 1970. Antimicrobials: "Safe" levels in a synthetic diet of an insect *Agria affinis*. *J. Insect Physiol.,* 16:1769-1782.

Slansky, F., Jr. 1982. Insect nutrition: An adaptationist's perspective. *Fla. Entomol.,* 65:45-71.

Slansky, F., Jr., and M. Scriber. 1985. Food consumption and utilization, pp. 87-163, in G.A. Kerkut and L.I. Gilbert (Eds.), *Comprehensive Insect Physiology, Biochemistry and Pharmacology,* Pergamon Press, Oxford.

Smedley, S.R., and T. Eisner. 1995. Sodium uptake by puddling in a moth. *Science,* 270:1816-1818.

Svoboda, J.A., J.N. Kaplanis, W.E. Robbins, and M.J. Thompson. 1975. Recent developments in insect steroid metabolism. *Annu. Rev. Entomol.,* 20:205-220.

Taylor, M.W., and J.C. Medici. 1966. Amino acid requirements of grain beetles. *J. Nutr.,* 88:176-180.

Thorsteinson, A.J. 1960. Host selection in phytophagous insects. *Annu. Rev. Entomol.,* 5:193-218.

Vanderzant, E.S. 1958. The amino acid requirements of the pink bollworm. *J. Econ. Entomol.,* 51:309-311.

Vanderzant, E.S. 1973. Axenic rearing of larvae and adults of the boll weevil on defined diets: Additional tests with amino acids and vitamins. *Ann. Entomol. Soc. Am.,* 66:1184-1186.

Vanderzant, E.S. 1974. *Development,* significance, and application of artificial diets for insects. *Annu. Rev. Entomol.,* 19:139-160.

Vanderzant, E.S., and C.D. Richardson. 1963. Ascorbic acid in the nutrition of plant-feeding insects. *Science,* 140:989-991.

Waldbauer, G.P. 1964. The consumption, digestion and utilization of solanaceous and non-solanaceous plants by larvae of the tobacco hornworm, *Protoparce sexta* (Johan.) (Lepidoptera: Sphingidae). *Ent. Exp. Appl.,* 7:253-269.

Waldbauer, G.P. 1968. The consumption and utilization of food by insects. *Adv. Insect Physiol.,* 5:229-288.

Waldbauer, G.P., and A.K. Bhattacharya. 1973. Self-selection of an optimum diet from a mixture of wheat fractions by the larvae of *Tribolium confusum. J. Insect Physiol.,* 19:407-418.

Waldbauer, G.P., and S. Friedman. 1991. Self-selection of optimal diets by insects. *Annu. Rev. Entomol.,* 36:43-63.

Waldbauer, G.P., R.W. Cohen, and S. Friedman. 1984. Self-selection of an optimal nutrient mix from defined diets by larvae of the corn earworm, *Heliothis zea* (Broddie). *Physiol. Zool.,* 57:590-597.

Wei, Z., G. Baggerman, R.J. Nachman, G. Goldsworthy, P. Verhaert, A. De Loof, and L. Schoofs. Sulfakinins reduce food intake in the desert locust, *Schistocerca gregaria. J. Insect Physiol.,* 46:1259-1265.

Yazgan, S. 1972. A chemically defined synthetic diet and larval nutritional requirements of the endoparasitoid *Itoplectis conquisitor* (Hymenoptera). *J. Insect Physiol.,* 18:2123-2141.

4 Integument

CONTENTS

PREVIEW

The integument of insects comprises the cuticle and the epidermal cells beneath that secrete the cuticle. The cuticle is the skeleton of insects, and skeletal muscles are attached to it. The cuticle may be hard and rigid, as is that of adult beetles, or soft and pliable, as is the case in many immature insects and some adults. The head capsule and the thorax, which supports the leg and wing attachments, are usually the most heavily sclerotized (hardened) parts of the body. All stages of insects contain an epicuticular layer, which waterproofs the body. This layer is heavily sclerotized but does not contain chitin. Beneath the epicuticle layer, many insects have an exocuticle layer (hard, sclerotized cuticle) and a layer of soft, relatively unsclerotized endocuticle next to the epidermal cells. Either of the latter two layers may be reduced or absent; there is no endocuticle in very hard cuticle such as the elytra of beetles, and no exocuticle in very soft-bodied insects, such as some larval Diptera. In all immature insects, the exoskeleton gets too small as the insect grows and the cuticle must be molted. Periodically, under hormonal regulation, the old cuticle separates from the epidermal cells, a process called apolysis. Parts of the old cuticle are digested by molting enzymes, and reabsorbed components are used in the synthesis of new cuticle. Secretion of a new cuticle begins beneath the old cuticle even as it is being digested. The first part of the new cuticle to be secreted is the outermost layer, the cuticulin layer of the epicuticle. Additional cuticle, usually called procuticle because it is not sclerotized at this stage regardless of its future destiny, is secreted underneath the cuticulin layer. Cuticle is secreted in thin sheets, with successive sheets pushed up from below. The sheets of cuticle contain a protein matrix with chitin rods embedded in the matrix. Often, each successive sheet is rotated slightly with respect to the long axis of the previous sheet, and successive layers give rise to a helicoid appearance in cross section. For some time interval, lasting from hours to days in different insects, an insect has two cuticular coverings, the old and the new. Muscle attachments to the old cuticle are at last severed, freeing the old cuticle to be shed, and the muscles rapidly attach to the new cuticle. Eclosion, or shedding of the old cuticle, and especially eclosion of adults from the pupal stage, is regulated by a complex of neuropeptide hormones. A period of quiescence is necessary in most cases for the cuticle to harden sufficiently to withstand the strain imposed by muscle contraction. Cuticle sclerotization is regulated by a neurohormone, bursicon, secreted from the nervous system. Chemically, the cuticle contains chitin, a polysaccharide polymer of N-acetylglucosamine, protein, lipids that function in waterproofing, and phenols and quinones that are important in sclerotization, the hardening process in the cuticle. Sclerotization of the cuticle over most of the body is accompanied by darkening, the formation of brown to black melanin pigments, but cuticle can sclerotize without darkening. For example, the compound eyes are usually covered by relatively clear cuticle, and some insects have transparent cuticle over some or even most of the body. The melanin pigments are formed from chemical changes in the polyphenols in the cuticle. Many different proteins have been detected in cuticle prior to sclerotization; but once these proteins are cross-linked, they usually cannot be dissolved from the cuticle. There are some differences, however, in the proteins comprising soft vs. hard regions of cuticle. The hardness of cuticle is a function of sclerotization, not chitin content. Some of the hardest parts of cuticle do not contain chitin. Not only are proteins cross-linked to each other during sclerotization, but the multiple thin sheets of cuticle are cross-linked to each other, giving the cuticle great strength. Lipids on the surface of the epicuticle and within the layers give cuticle excellent waterproofing properties, an important function for nearly all insects because of their large surface area-to-volume ratio. Aquatic insects benefit from the water impermeability of the cuticle by not absorbing large quantities of water via osmosis.

4.1 INTRODUCTION

The integument is composed of the **cuticle** and the underlying **epidermal cells** that secrete the cuticle. The cuticle serves as the exoskeleton of the insect; the site for muscle attachment; and the first line of defense from fungi, bacteria, predators and parasites, and environmental chemicals, including pesticides. The integument functions in some or all insects in locomotion, breathing and respiration, feeding, excretion, protection from desiccation, behavior, osmoregulation, water control, and as a food reserve. The many roles played by the integumentary covering of insects is, in part, reflected in the complexity of its structure and chemistry, and in the special ways it is adapted to function in the ecology of its owner. The surface morphology of the external cuticle is extraordinarily varied, reflecting species specificity and diversity. The beauty, color, shape, and intricate sculpturing on the surface of insects attracts amateurs and professionals alike to collect and study insects. Moreover, taxonomists and systematists have traditionally used the surface sculpturing, setae, and sutures on the cuticle in classification of insect species.

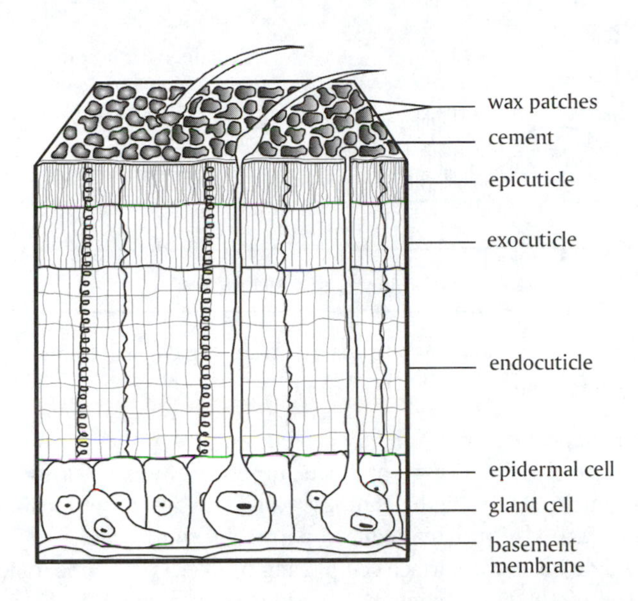

FIGURE 4.1 Diagrammatic representation of a cross-sectional area of the integument illustrating the major layers of the cuticle.

Despite many species-specific features, there are certain common features found in the integument. There is always a single layer of epidermal cells lying immediately beneath the cuticle. These cells secrete the new cuticle at molting and, in some insects at least, continue to secrete cuticle even in the adult. In all insects, there is a thin layer of cuticle with special properties called the epicuticle at the surface of the insect, and beneath this there is additional cuticle that may sometimes be divided into several layers, depending primarily on the degree of sclerotization or cross-linking of the molecules of protein and chitin. This chapter explores the physiology and biochemistry of cuticle and relates these to species similarities and differences in the integument.

4.2 STRUCTURE OF THE INTEGUMENT

The cuticle is composed of several layers, including **epicuticle**, **exocuticle**, and **endocuticle** (Figure 4.1). The single layer of epidermal cells secretes the cuticle at each molt. There is always an epicuticle layer and the epidermal cell layer, but the exocuticle and endocuticle may be greatly reduced or absent in particular parts of the cuticle of the same or different insects. Often, there are additional cuticle layers within one or more of the three principal layers, as evidenced by electron density in cross sections viewed in the transmission electron microscope, but these usually have not been given names. The epidermal cells secrete the overlying cuticle, the lipids (waxes), cement, and often many additional chemical components that occur on or in the cuticle layers. When a new cuticle is secreted at molting, the cuticle below the epicuticle layer is first called procuticle because it is not sclerotized into the distinct layers noted above. Such sclerotization occurs, however, soon after the cuticle is secreted.

4.2.1 EPICUTICLE

Epicuticle is the outermost part of the integument in contact with the environment. Because it is so thin (from 1 to 4 μm in thickness), the detailed chemical structure of the epicuticle has been difficult to discern, but it is known to contain sclerotized proteins impregnated with lipoproteins, lipids, waxes, cement, and minor amounts of various minerals and other chemical components. It does not contain chitin, a major structural carbohydrate in other parts of the cuticle. The proteins

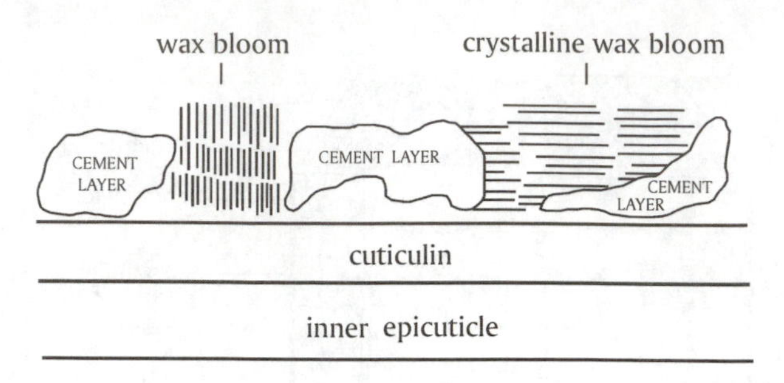

FIGURE 4.2 Diagrammatic illustration of the mosaic of wax (lipids) and cement on the surface of the epicuticle of some insects. (Modified from Locke, 1965.)

and some of the lipids appear to be covalently linked, and the proteins are tanned or sclerotized by phenolic compounds and their oxidized products, quinones. **Sclerotization**, the cross-linking of molecules, gives the epicuticle strength, hardness, and low water permeability. Lipids and the cement layer on the surface also provide reduced permeability to water.

Because the epicuticle bears the surface sculpturing of insects, its appearance is highly variable in different species. It does, however, contain one common layer in all insects, the **cuticulin** layer. Cuticulin is a very thin layer of protein about 0.0075 μm thick, about the thickness of an animal cell membrane. It is impregnated with lipids, some of which may be covalently bound to proteins as lipoproteins. Cuticulin is the first layer of new cuticle to be secreted in preparation for a molt, and new procuticle is secreted beneath it. There may be additional layers in the epicuticle, such as an inner cuticulin layer similar in composition to cuticulin. If present, it is usually thicker than the cuticulin layer and may be up to 1 μm thick. In some insects, numerous layers interior to cuticulin can be differentiated as layers of greater or lesser electron density with transmission electron microscopy.

Cement, often described as a shellac-like substance at the air interface of the cuticle, is secreted by specialized epidermal cells called dermal glands and transported to the surface of the epicuticle. A traditional view is that the cement layer is the outermost layer on the cuticle, with a lipid layer just beneath the cement. Probably in some insects this view is valid, but Locke (1965) (Figure 4.2) suggested, on the basis of electron microscope studies of the cuticle of the lepidopteran caterpillar *Calpodes ethlius*, that the cement layer is not continuous, but is broken by patches of lipids, called **wax blooms**, at the surface. It seems likely in many insects that a mosaic patchwork of cement and lipid exists at the surface of the epicuticle. Some examples of insects that have much or even all the body covered with lipid at the surface will be described in later sections of this chapter. Despite the thin nature of the epicuticle, it nevertheless is extremely important to surface pattern and features, to permeability of the cuticle, and it represents a limitation on expansion of the cuticle in immature insects, necessitating molting during growth.

4.2.2 Exocuticle

The exocuticle, containing both chitin and protein, lies just beneath the epicuticle. It is highly sclerotized and is therefore hard and rigid. Lamellae or layers within the exocuticle may refract light in such a way to produce structural colors in some insects. Many of the iridescent greens and blues of insects are structural colors due to refracted light rather than to pigments. The thickness of the exocuticle is variable and species specific. Adult insects generally have a thicker and more sclerotized exocuticle than larval insects. In particular, the thorax in flying insects has heavily sclerotized exocuticle to support the strong flight muscles. Many larvae have a soft flexible cuticle with little or no exocuticle. As in so many cases with insects, exceptions do exist. There are larval insects with hard sclerotized exocuticle, and soft-bodied adults with little or none. The greater the degree of sclerotization, the harder the cuticle. The content of chitin does not control the hardness of the cuticle, but sclerotization does. Because of the sclerotization, little or none of the exocuticle is digested by molting fluid, and it is shed, along with the epicuticle, at molting.

4.2.3 Mesocuticle

Some insects have a layer of less highly sclerotized cuticle between the exo- and endocuticle called **mesocuticle**. It also contains chitin and protein. Mesocuticle is distinguished on the basis of the degree of sclerotization and by its red staining reaction with acid fuchsin in Mallory's triple stain.

4.2.4 Endocuticle

The endocuticle is soft, flexible cuticle containing both chitin and proteins. It has little sclerotization, which is why it is soft and flexible. Stabilization of the proteins and chitin in the endocuticle occurs through some covalent bonds, hydrogen bonds, and probably occasional quinone cross-links. In general, soft-bodied insects have relatively thick endocuticle, which remains flexible and soft, and thin or no exocuticle. There is, however, always an epicuticle layer at the body surface of soft-bodied insects.

4.2.5 Pore Canals and Wax Channels

Pore canals are passageways from 0.1 to 0.15 μm in diameter, extending from the epidermal cells through the endo- and exocuticle. Larger canals are often flattened, ribbon-like, and may be twisted or straight. Pore canals transport lipids and cement, and sometimes additional chemical components. Formation of the pore canals has not been entirely resolved. Some research has suggested that the passageways are the result of cytoplasmic extensions of the epidermal cells present during cuticle secretion. Usually after a new cuticle has been secreted, the cell extensions are withdrawn, leaving the open canals. Some researchers who have studied cuticle formation do not accept this mode of formation because cell extensions are not always evident, even during new cuticle secretion, but no competing alternative idea has been advanced. The pore canals can be very numerous; for example, as many as $1.2 \times 10^6/mm^2$ in some cockroaches, or as few as $15{,}000/mm^2$ (about 50 to 70 per epidermal cell) in a sarcophagid (a fleshfly) larva.

Pore canals terminate at the junction of the epicuticle; they do not penetrate epicuticle. There are, however, passageways through the epicuticle called **wax channels** (about 0.006 to 0.013 μm in diameter) that are 10 to 20 times smaller than pore canals (Locke, 1965). Wax channels stain with osmium tetroxide, a stain for unsaturated lipids, and this is taken as evidence that the channels continue the transport of lipids, and probably other materials, to the surface. There is no evidence of a 1:1 correspondence of pore canals to wax channels at the junction of the exo- and epicuticle.

Chemicals passing up through the pore canals probably diffuse out of the pore canals laterally to some extent all along their length, and also at the epicuticular interface, and thus impregnate the entire cuticle. Some of the lipids find their way to the surface of the insect, where they provide waterproofing and perhaps other ecological and behavioral functions. Continued secretion of cuticular lipids in many insects is a dynamic process, replacing epicuticular lipids that are volatilized (as semiochemicals, for example) and lipids that no doubt rub or wear off the cuticle in longer-lived insects, and lipids lost with each molt.

4.2.6 Epidermal Cells

The cells underlying the cuticle are arranged in a single layer (Figure 4.3). They are usually called the epidermal cells, but are sometimes referred to as the **cuticular epithelium,** the **epidermis,** and (in older literature) the **hypodermis**. All epidermal cells probably secrete chitin and proteins, and some may secrete lipids. Specially modified glandular epidermal cells may secrete cement. There frequently are specialized glandular cells in the cell layer, either in small groups or as scattered

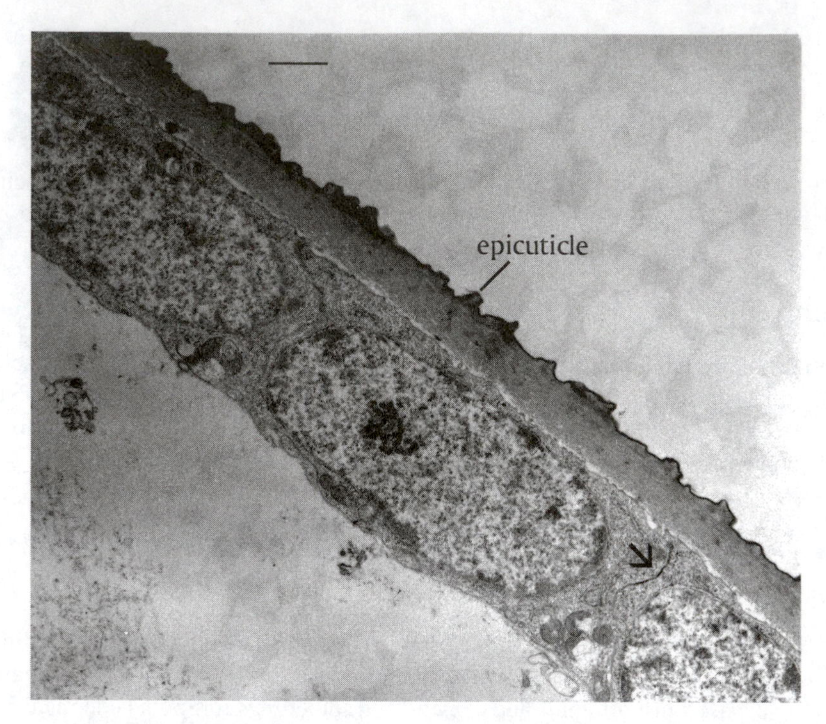

FIGURE 4.3 Epidermal cells and cuticle from a larva of the mosquito *Ochlerotatus* (formerly *Aedes*) *triseriatus*. The arrow points to a row of septate junctions between the membranes of two adjacent cells. The scale bar at the upper left represents 1 μm. (Photograph courtesy of Jimmy Becnel and Alexandra Shapiro.)

isolated glandular cells, that secrete special products (Noirot and Quennedey, 1974). For example, sex pheromones in most female Lepidoptera are secreted by small patches of tall, columnar epidermal cells located beneath the cuticle of the ventral intersegmental membrane in the eighth to ninth segment of the abdomen.

Epidermal cells are separated from the circulating hemolymph by a **basement membrane,** a layer of poorly defined chemical composition, but with pores large enough to permit passage of larger hemolymph proteins and other molecules from hemolymph into the epidermal cells. Small tracer particles of gold or ferritin pass from the hemolymph through the basement membrane, but larger particles are stopped on the hemolymph side. The origin of the basement membrane is not defined for most insects; some evidence supports secretion by hemocytes, while desmosomal attachments to epidermal cells suggest that the epidermal cells themselves secrete it. Tracheae, tracheoles, and nerves pass through the basement membrane to reach the epidermal cells. The basement membrane of epidermal cells may be smooth or may contain many and deep infoldings, depending on function and the stage in which the insect exists. The membrane changes in appearance, and undoubtedly in function, as the insect develops, and there are marked changes in preparation for a molt. **Hemidesmosomes** hold the basement membrane to the epidermal cells.

In the intermolt period, epidermal cells typically have a regular polygonal outline and form a sheet of cells beneath the cuticle. Contiguous epidermal cells, as well as gut and Malpighian tubule epithelium, follicular cells in the ovary, and other cells in insects, are held together by various types of **junctional contacts** (Lane and Skaer, 1980) (Figure 4.4). The junctional contacts not only hold cells together, but depending on the type of contact, may have other functions. **Septate junctions** are close together and numerous, giving a ladder-like appearance between cells (Figure 4.5). Septate junctions tend to occur between the lateral faces of cells, particularly toward the apical (cuticular) surface, and play an important role in preventing inward and outward movement of materials between the cells. **Gap junctions** also occur between the lateral faces of cells. The two cell

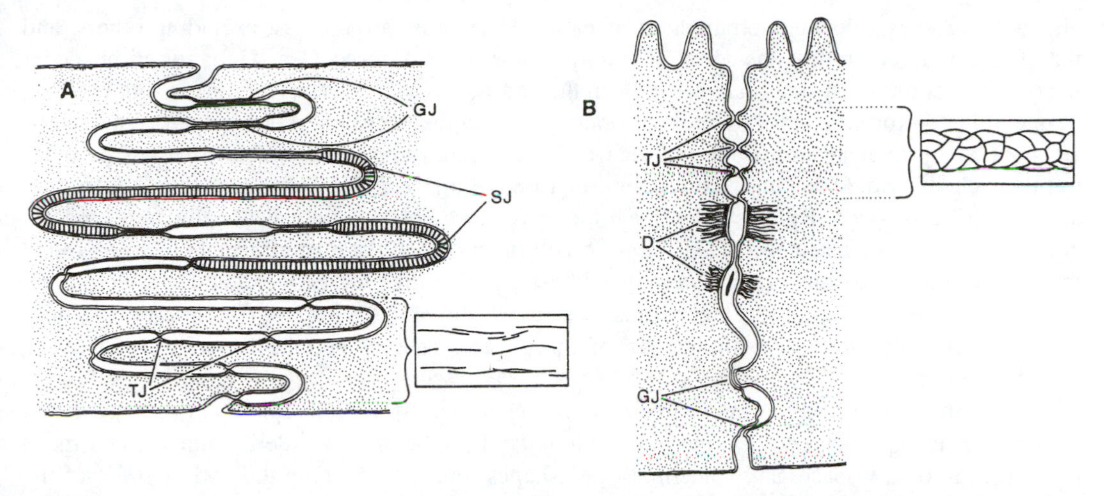

FIGURE 4.4 A variety of cell-to-cell structures holding cells together and, in some cases, preventing passage of chemical substances between cells. The diagram compares the major differences between location along cell boundaries and structure of insect junctions (A) with vertebrate junctions (B). Lateral borders of insect cells exhibit a high degree of interdigitation, with tight junctions (TJ) near the basal cell surface that have a simple linear intramembranous structure indicated in the box on the right of the figure. Vertebrate tight junctions tend to be near the apical surface and adjacent cell membranes exhibit relatively straight, noninterdigitation (box to right of Figure B). In insects, the complex interdigitating membranes between adjacent cells probably aid in making tight junctions resist the passage of substances between cells in many cases. GJ, gap junctions; SJ, septate junctions; TJ, tight junctions; D, desmosomes. (Terminology and figure reproduced with permission from Lane and Skaer, 1980.)

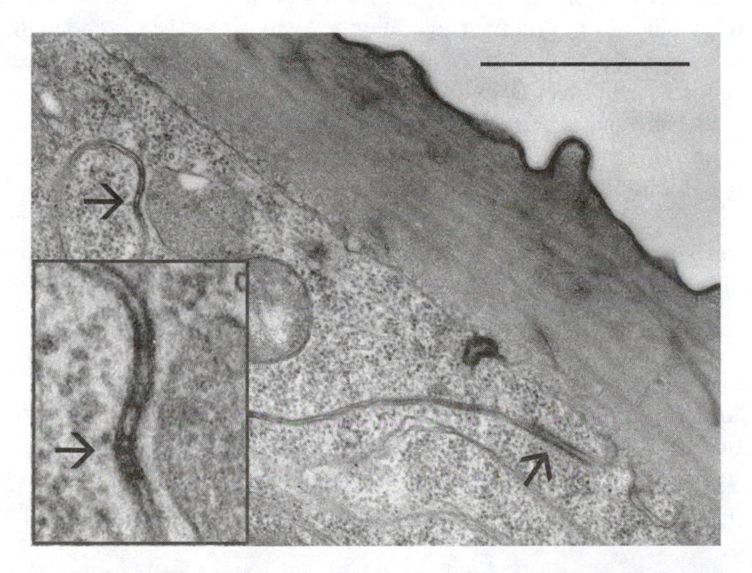

FIGURE 4.5 Septate junctions between adjacent cell membranes in the epidermal cells of a larva of the mosquito *Ochlerotatus* (formerly *Aedes*) *triseriatus*. The arrows in the lower magnification view point to septate junctions, and the inset shows a higher magnification view of the junctions above the inset box. The scale bar in the upper-right corner represents 1 μm. (Photograph courtesy of Jimmy Becnel and Alexandra Shapiro.)

membranes are very close together (a gap of 2 to 4 nm) at gap junctions, and for this reason they also have been called **close junctions**. Gap junctions appear to confer electrical coupling upon cells, can make cells function like a synctium, and play a role in cell-to-cell communication. They have also been called *macula communicans* because of their role in communication. Close junctions also act as sieves, allowing certain sizes of molecule to pass through, but excluding others, and may function in controlling the speed of entry of some types of molecules. **Tight junctions** occur in some insect tissues such as between cells in the compound eyes, testes, rectal pads, and between perineural cells forming part of the blood-brain barrier in the central nervous system. Where cells are bonded by tight junctions, the adjacent cell membranes appear to contain rows of particles tightly packed in ridges that make contact or even fuse the adjacent membranes together, obliterating any intercellular space. Tight junctions seal the passageway between adjacent cells and present a barrier to passage of substances. Depending on cell function and the physiological condition of the insect, there may be sinuses with varying width from time to time between adjacent epidermal cells, with cells being held together primarily at basal and apical surfaces. Cationic ferritin does not penetrate these spaces, an indication of the general lack of permeable pathways between epidermal cells.

Under the influence of hormones secreted in preparation for a molt, epidermal cells generally enlarge and change shape to become more columnar. They begin to divide by mitosis, and move or expand into new space, and sometimes new shapes, that will become the body outline of the next instar as they secrete the new cuticle.

Epidermal cells secrete cuticle at the apical face. The cuticle is attached to the apical face of cells by a very large 2.5-MDa extracellular matrix protein (Wilkin et al., 2000). The protein is encoded in *Drosophila melanogaster* by the *dumpy* (*dp*) gene, a complex locus in excess of 100 kb. The gene is expressed at numerous locations, including epidermal cell-cuticle attachment sites, muscle-tonofibrillae attachment sites, and where a cuticle intima (lining) is attached to cells such as in the tracheae, fore- and hindgut sites. Dumpy protein comprises 308 epidermal growth factor (EGF) modules and 185 of a new class of modules called DPY. Near the carboxy terminal, the protein has a cross-linking *zona pellucida* domain and a transmembrane anchoring sequence. The DPY module forms a β sheet motif that is stabilized by covalent disulfide bonds, and linked end-to-end with EGF modules to form a fiber that may be as much as 0.8 μm in length. Functionally, the protein provides a strong anchor for tissue-cuticle connections, permitting mechanical tension without allowing the tissue to tear away from the cuticle. Such tension sites will occur at several places in the exoskeleton and gut, and most notably at muscle attachment sites.

Epidermal cells have extensive **rough endoplasmic reticulum** (**RER**) where protein synthesis occurs, and areas of **smooth endoplasmic reticulum** (**SER**) for lipid synthesis. Cell nuclei are often polyploid, with multiple nucleoli. During development, nucleoli enlarge and develop multiple lobes when there is evidence of new synthesis of RNA and ribosomes. The **Golgi complex** is prominent in epidermal cells (Locke, 1984), and probably serves several functions, including the following:

1. Processing of secretory substances necessary to synthesize cuticle
2. Production of material for the plasma membrane of the cell
3. Modification of newly synthesized proteins
4. Packaging of cellular components in isolation envelopes for later autophagy
5. Processing and packaging of lysozymes needed for autophagy and heterophagy

Epidermal cells are involved in wound repair and can move from an area of undamaged cells into an area of damaged or destroyed cells. Cells at the leading edge spread over the wound until they cover it and establish contact with another epidermal cell. The population of cells in the peripheral zone around a wound is temporarily reduced as cells migrate toward the wound area, but cell divisions soon repopulate the area.

Oenocytes are large, prominent cells scattered among the epidermal cells, and also clustered at spiracles, near the origin of larger tracheae, and scattered among fat body cells. Oenocytes located among the epidermal cells are considered a type of epidermal cell. Oenocytes are differentiated from epidermal tissues during embryogenesis, as well as later in development. These cells are usually large, polyploid, and always have an extensive tubular smooth endoplasmic reticulum and well-developed plasma membrane reticular system. Their function is not very clear, but their morphology suggests lipid secretion and lipid metabolism.

4.3 MOLTING AND FORMATION OF NEW CUTICLE

The external skeleton gradually becomes too small for the growing body tissues of an immature insect, and it must **molt** its cuticle (Figure 4.6). Molting is a vulnerable time for insects; they are easy prey for predators and subject to environmental hazards, particularly desiccation. The muscles that move the body must be detached from the old cuticle, but they are detached only immediately before the ecdysis and new muscle attachments are made quickly to the new epicuticle (for more detail, see Section 9.2.2). The new cuticle must harden sufficiently to resist the pull of the musculature, or muscle action can cause skeletal deformation and result in permanent restriction of movement, and especially failure of flight ability.

Preparation for molting is under endocrine and nervous control. The events leading to molting in the tobacco hornworm larva represent a described scenario, but this pattern is not typical of all insects. Although covered in more detail in Chapter 5, a brief summary of hormonal control of molting is provided here. A tobacco hornworm fourth instar grows until it reaches a certain size and/or weight. Stretch receptors in the body are probably stimulated by the increasing growth of body tissues; and when it attains the critical size, the brain secretes prothoracicotropic hormone (PTTH). PTTH is released into the circulating hemolymph, circulates around the body, and finds specific receptor proteins on the surface of prothoracic gland (PGL) cells to which it binds. Multiple biochemical reactions are initiated that result in synthesis of ecdysone by the PGL. Ecdysone is released into the circulating hemolymph and is converted into 20-hydroxyecdysone by epidermal cells as well as other tissues. Epidermal cells respond to 20-hydroxyecdysone by separating from the old cuticle (apolysis) (Figure 4.6B) and by mitotic activity that produces new cells to spread over the larger body surface that must be enclosed within the new epidermal cell layer and new cuticle (Figure 4.6C). **Apolysis**, or separation of the epidermal cells from the old cuticle (Jenkin and Hinton, 1966), marks the beginning of a molt and of a new instar, and the animal within the loosened, but not yet shed, cuticle is the **pharate** next instar (or stage if the next form is the pupa or adult).

4.3.1 THE APOLYSIAL SPACE

The **apolysial space** is at first a minute space created by the separation of the epidermal cells from the old cuticle. Soon, molting fluid is secreted into the space and activated; and later, as the molting fluid digests some of the old cuticle, the space widens. Typically, there is the discharge of discontinuous patches of membrane-bound secretion into the apolysial space. The vesicles of secretion, often called apolysial droplets, are secreted by exocytosis from the plasma membrane of epidermal cells. The presence of apolysial droplets seems to precede apolysis in *Calpodes ethlius* and some other insects, and to follow apolysis in others, such as *Galleria mellonella* and *Hyalophora cecropia*. Although it is still open to experimental analysis, it seems reasonable that some secretion, although not directly visible, may be involved in dissolving the attachments of the old cuticle to the epidermal cells.

4.3.2 MOLTING FLUID SECRETION

Molting fluid is first evident as osmiophilic droplets (i.e., droplets likely rich in polar lipids) secreted by the epidermal cells into the apolysial space. An **ecdysial membrane** soon appears and can be observed in histological sections. It may result from coalescence of the droplets in some insects, as reported from both *Rhodnius prolixus* and *Calpodes ethlius*; or it may be formed from inner layers of the old endocuticle, as reported in *Schistocerca gregaria*. The ecdysial membrane persists through the premolt period, and later is shed with the old exuvium.

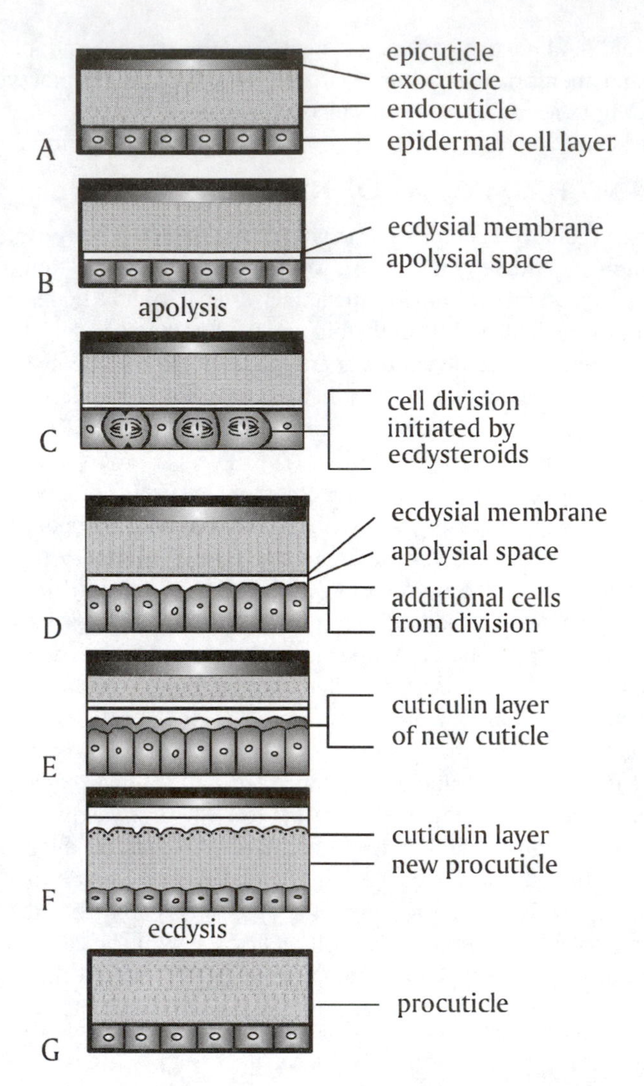

FIGURE 4.6 Diagrammatic illustration of the process of apolysis, secretion of new cuticle, and ecdysis of the old cuticle. A: Old cuticle just before molting begins. B: Formation of the ecdysial membrane and apolysial space. C: Initiation of cell division in epidermal layer in response to molting hormone. D: New epidermal cells, usually developing an irregular apical surface. E: New cuticle secretion begins with secretion of cuticulin layer; digestion of the old endocuticle continues. F: New unsclerotized procuticle is formed. G: The old cuticle shell has been ecdysed and the new cuticle will be covered with a wax and cement layer, and some of the procuticle may be sclerotized into exocuticle, depending on the insect and location on the body.

Molting fluid contains both **proteinases** and **chitinases** that digest the proteins and chitin, respectively, in old endocuticle (and possibly some of the mesocuticle in some insects). The chitin digesting enzymes have received more detailed attention so far. **Chitinase** (EC 3.2.1.14) is an endo enzyme and attacks a chitin chain at random by internal hydrolysis. It produces smaller, soluble oligosaccharides that are attacked at the ends by **N-acetyl-β-D-glucosaminidase** (EC 3.2.1.30), an exo enzyme, yielding free **N-acetylglucosamine**. Chitinase isolated from a *Drosophila* cell line has a pH optimum of about 6. Injection of 20-hydroxyecdysone into 5th instars of the silkworm, *Bombyx mori* and *Manduca sexta* causes secretion of chitinase and N-acetyl-β-D-glucosaminidase, but induction of chitinase requires higher levels of hormone than induction of N-acetyl-β-D-glucosaminidase. Chitinase exists in more than one molecular size (88 and 65 kDa), and may exist as a **zymogen** (215 kDa) that is converted to the active enzyme at the proper time in *B. mori*.

At least ten proteases occur in the molting fluid from tobacco hornworm pharate pupae (Brookhart and Kramer, 1990). Both endo- and exo-cleaving proteolytic enzymes occur and are most active in the neutral to alkaline pH range. Some of the proteases have trypsin-like and chymotrypsin-like activity, but they differ from similar gut enzymes in that they are not affected by some inhibitors of gut trypsin and chymotrypsin. None of the enzymes detected are sulfhydryl or carboxyl proteases. At least some of the enzymes are secreted in a zymogen form.

In general, it appears likely that digestion of the old cuticle starts with proteolytic enzymes acting upon the proteins of the cuticle, and thereby exposing chitin rods or crystallites embedded in the protein matrix. Perhaps there also are places where chitin can be attacked without prior release from surrounding protein.

4.3.3 New Cuticle Formation

New cuticle secretion begins soon after the ecdysial space opens. During new cuticle secretion, an epidermal cell has a series of ridges or knobby projections on its apical face where the proteins and fibers of chitin are secreted. These projections have been called **plasma membrane plaques** by Locke (1984). At the initiation of new cuticle synthesis, cuticulin, the first new cuticular layer secreted, begins to form on these knobby plaques (Figure 4.6D, E). It is secreted initially as small discontinuous patches over the plasma membrane plaques, but the patches enlarge and eventually form a continuous layer of cuticulin. There also may be long microvilli on the apical face of epidermal cells during the molting process. Molting fluid digests the old endocuticle, and the products of digestion are reabsorbed by epidermal cells and used in the synthesis of new cuticle. The manner in which the new cuticulin is protected from digestion is not known; one suggestion is that the ecdysial membrane that is formed represents a barrier between the molting fluid and the new cuticulin layer being secreted just below the membrane. As old endocuticle is digested, new cuticle (unsclerotized) is secreted below the new cuticulin layer, thus pushing the cuticulin upward and outward. The new cuticle secreted at this stage is unsclerotized, and many authors have called it **procuticle** (Figure 4.6F, G). Later, after ecdysis, some or much of the new cuticle, depending on the insect and body part involved, may become heavily sclerotized and very hard (exocuticle), while other layers might be less sclerotized (mesocuticle) or very lightly sclerotized (endocuticle). Locke (2001) suggests describing cuticle layers in terms of the envelope (the cuticulin layer), the epicuticle, and the procuticle, each of which is secreted in a different process. The envelope is laid down at the plasma membrane surface, usually on plaques at the tips of microvilli, as noted above. The newly impermeable envelope then protects the epidermal cell surface from digestive enzymes in the molting fluid, but allows the digestion of the old endocuticle so that the amino acids from protein digestion and glucose from chitin digestion can be reassembled into new procuticle. The epicuticle is secreted on the inner face of the envelope and, with the envelope, forms the outer boundary of the cuticle compartment. Procuticle is then formed at the cell surface until it fills the cuticle compartment. Locke notes that some type of limiting boundary — an envelope (but not a cuticulin envelope) — covers the cells in most invertebrate phyla, including bacteria, protozoa, trematodes, nematodes, mollusks, and arthropods. The envelope, according to Locke, provides a mechanism for extending metabolic control of the extracellular compartment (i.e., the cuticle); limits size; provides protection from environmental chemicals, bacteria, and fungi; regulates permeability of the cuticular compartment; and is involved in surface reflectivity and color.

4.3.4 Reabsorption of Molting Fluid

The molting fluid that accumulates in the apolysial space disappears shortly before ecdysis. Most of it appears to be reabsorbed by the insect; at this critical stage, preventing loss of fluid volume by terrestrial insects may be vital. Some of the molting fluid may be reabsorbed through the epidermal cells, but in *Manduca sexta* pharate pupae (Cornell and Pan, 1983) and in pharate pupae of the skipper *Calpodes ethlius* (Yarema et al., 2000), it primarily flows beneath the old and new cuticles to the mouth and anal openings and is accumulated in the midgut. Similar swallowing of molting fluid through the mouth occurs in the pharate adult *M. sexta* (Miles and Booker, 1998). The fluid probably contains protein (enzymes) and other potentially useful nutrients, and the water may be useful in helping to flush the gut and/or Malpighian tubules of accumulated waste products, particularly in newly eclosed adults, many of which excrete the **meconium** as the accumulated waste from molting. Excretion of any excess water also lightens the body for flight.

4.4 ECDYSIS

To facilitate **ecdysis** of the old cuticle, some insects swallow air — or swallow water if they are aquatic insects — to expand the gut and split the old cuticle. The muscular actions involved in ecdysis are controlled by nervous motor programs. Ecdysis-related motor programs have been identified in a number of insects, and some insects have sequential stages or sequential motor programs that must occur in proper sequence for molting to be successful. Events leading up to and completion of molting (ecdysis) in insects illustrate interactions between the endocrine and nervous systems, for although the nervous system controls muscular movements, in moths, and perhaps in other insects as well, the motor program is initiated by hormonal action. The hormonal controls have been most thoroughly investigated in Lepidoptera (Truman, 1978; Horodyski, 1996).

Ecdysis of the adult from the pupal stage and ecdysis of an immature from one instar to the next appear to be very similar and under similar hormonal controls in Lepidoptera. Ecdysis of the tobacco hormworm, *Manduca sexta*, has been divided into two phases: **pre-ecdysis behavior** and **ecdysis (= eclosion)** (Reynolds, 1980). The motor programs for pre-ecdysis and ecdysis are coordinated by a concert of hormones. The **eclosion hormone (EH)** (Truman and Riddiford, 1970) and the **pre-ecdysis-triggering hormone (PETH)** and **ecdysis-triggering hormone (ETH)** (Žitňan et al., 1996; (Žitňan and Adams, 2000) are involved. Eclosion hormone (EH), a neurosecretory polypeptide of 62 amino acids, is secreted in *M. sexta* by ventrally located neurosecretory cells in the tritocerebrum in response to falling ecdysteroid titer (Truman, 1985; Kingan and Adams, 2000). Its release can be inhibited by injecting 20-hydroxyecdysone, thus preventing acquisition of competence in the Inka cells (Kingan and Adams, 2000). Under falling ecdysteroid levels, it is released from the proctodeal nerve, arising from the terminal abdominal ganglion during larval and pupal ecdysis. The cells secreting EH in the pharate adult are modified during adult development so that the cells send an axon to the corpora cardiaca (CC). EH is transferred through this axon for storage in the CC and released into the circulating hemolymph for pharate adult eclosion. Hemolymph-borne EH causes the release of PETH and ETH from the Inka cells. Within a few minutes after release, low levels of PETH and ETH initiate pre-ecdysis I and II behaviors. Further action of ETH and EH on the tritocerebrum and the subesophageal ganglion sets in motion a neural network in abdominal ganglia by elevating cGMP levels. Ecdysis is delayed by inhibitory factors from the cephalic and thoracic ganglia until an independent clock mechanism in each abdominal ganglion is activated by central release of **crustacean cardioactive peptide** (CCAP) (Gammie and Truman, 1997, 1999; (Žitňan and Adams, 2000). **ETH** is a polypeptide of 26 amino acids released from Inka cells (Žitňan et al., 1996) located segmentally as part of **epitracheal glands** in Lepidoptera larvae. The hormone acts directly on the central nervous system and triggers preecdysis and ecdysis behavior. It is not known whether a second messenger is involved with Mas-ETH. The epitracheal glands consist of several cells, but the most prominent cell is a large opaque cell, named the Inka cell by Žitňan. The opaqueness appears to be due to the protein hormone it is ready to secrete because, after secretion, it is not so large nor opaque. Žitňan et al. (1996) suggest that EH acts to release Mas-ETH, and then Mas-ETH acts directly on the central nervous system to initiate the motor program for pre-ecdysis. Hesterlee and Morton (1996) suggest that a positive feedback cycle occurs in which initially each polypeptide hormone is released in small amounts, with each release stimulating greater release of the other hormone, resulting in the cascade of the two hormones that act on the nervous system to begin the motor program for pre-ecdysis behavior. Extracts of Inka cells (Mas-ETH polypeptide) injected into pharate larvae very near their normal time to molt cause them to begin pre-ecdysis muscular contractions within 2 to 10 minutes, and ecdysis follows after 35 to 65 minutes of pre-ecdysis muscular contractions. When injected into fifth instars 10 to 36 hours prior to the time for normal ecdysis, Mas-ETH induces pre-ecdysis behavior within a few minutes; however, ecdysis does not follow or is incomplete, indicating that the timing of secretion of Mas-ETH is critical to complete ecdysis. Pre-ecdysis behavior includes dorsoventral contractions occurring synchronously in thoracic and abdominal segments. These pre-ecdysis contractions occur every 10 to 12 seconds, with a contraction lasting 5 to 7 seconds. Isolated ventral nerve cords (VNC) of *Hyalophora cecropia*, another moth, are capable of generating the motor program (Truman and Riddiford, 1970). At first, each ganglion along the VNC produces bursts of action potentials alternately from the right and left sides. In an intact animal, these nerve impulses would initiate muscle action in each segment that could cause wiggling and rotatory movements of the abdomen. After about an hour — sometimes less, sometimes more — the pre-ecdysis contractions give way to ecdysis contractions, a series of peristaltic waves of contractions that originate in the most posterior segment and pass anteriorly for about 10 minutes until ecdysis is complete. Monitoring of nervous activity in the lateral nerves indicates that synchronous bursts of action potentials occur from both sides of each ganglion, but alternating in time between successive ganglia. These action potentials cause the peristaltic muscle contractions necessary to push the insect out of the old cuticle. Some minutes after the adult is out of the pupal cuticle, all ganglia simultaneously begin to produce prolonged bursts of action potentials, resulting in a steady tonic contraction of the abdomen that aids the pumping of hemolymph into the wings to inflate them. Eclosion hormone(s) probably exists in most or possibly all insects (Truman and Riddiford, 1970; Truman et al., 1981), but behavioral and biochemical assays for eclosion hormone are most thorough in Lepidoptera.

After ecdysis, the air- or water-filled gut that aided in splitting the old cuticle expands the body size and stretches the new cuticle while it is still pliable and unsclerotized, thus giving the insect some room for growth if it is at an immature stage. Lipids and cement are secreted onto the new epicuticle, muscle reattachments become firmly fixed, and the cuticle begins to tan. In *Calpodes* larvae, wax secretion is under the control of the corpora allata/corpora cardiaca complex, but details have not been elaborated. Formation of layers of endocuticle continues in many insects during intermolt periods, and even during adult life. In some insects, there are two distinctive layers formed daily — one during the day and another at night — so that growth rings can be observed and counted in a cross section of cuticle. These growth layers are present in certain grasshoppers and some cockroaches (*Periplaneta*), and are present but not so clearly demarcated in milkweed bugs for the first 8 days of adult life.

4.5 SCLEROTIZATION OF CUTICLE

Sclerotization is the process of cross-linking protein to protein chains, chitin to chitin chains, and possibly protein to chitin chains [see reviews by Sugumaran (1988) and Hopkins and Kramer 1992)]. Sclerotization is also called **tanning**, sometimes simply described as hardening of the cuticle. Tanning refers to the cross-linking process itself, and not to a color change, although sclerotization is often accompanied by tan, brown, or black colors. The colors are created by a variety of pigments, including melanin. The phenols associated with sclerotization easily undergo auto-oxidation (phenols to quinones) and quinones readily polymerize, processes usually leading to melanin and tan or brown to black colors. Hardening (sclerotization/tanning) and darkening are two different processes, and cuticle can become sclerotized without darkening, for example, over the compound eyes.

Only protein to protein sclerotization occurs in the epicuticle because no chitin occurs there; but in other layers of the cuticle, all the combinations may exist. Sclerotization gives strength and rigidity to the cuticle. Apodemes, the elytra of beetles, and the mandibles of chewing insects are

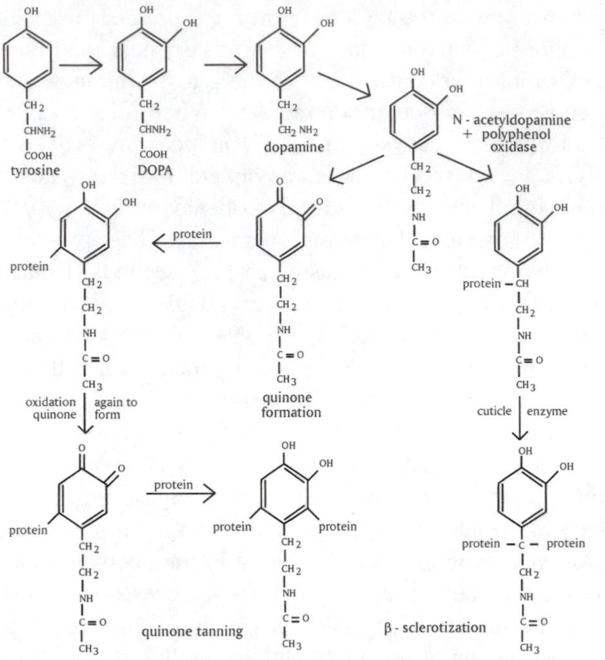

FIGURE 4.7 A generalized biosynthetic pathway for metabolism of tyrosine to N-acetyldopamine, a common sclerotizing agent, and the linking of proteins to the phenolic ring in either quinone tanning or to the beta carbon in β-sclerotization.

examples of heavily sclerotized cuticle. Intersegmental membranes and the cuticle of soft larvae are lightly sclerotized. Hardness of the cuticle is a function of the degree of sclerotization, and is not indicative of the content of chitin in the cuticle, as once thought.

The cross-linking or sclerotizing agents are phenols and their oxidation products, quinones. A number of phenols and quinones exist in various insect cuticles and all probably participate to some extent as sclerotizing agents, but the chemistry has been best elucidated for production of **N-acetyldopamine**, a common and major sclerotizing agent in many cuticles. N-acetyldopamine is formed from tyrosine by a number of enzymatically controlled steps (Figure 4.7). Early instars of Diptera, in which the process has been studied in some detail, metabolize tyrosine to N-acetyltryamine and *p*-hydroxyphenyl propionic acid, which are not involved in sclerotization. Only late in the last instar is there a switch in synthesis to N-acetyldopamine under the influence of the molting hormone.

N-β-alanyldopamine has been implicated as the principal tanning agent in the pupal cuticle of *Manduca sexta*, in which it increases as much as 800-fold during tanning of the pupa. It is a major cuticular constituent of a number of insects in various orders, and may be the typical sclerotizing agent in pupae because the pupal *o*-diphenoloxidase oxidizes it most readily among a variety of potential substrates (Hopkins et al., 1982).

Quinones cross-link protein chains by reacting with free amino groups such those of lysine, tryptophan, arginine, histidine, and the terminal amino group at one end of a protein. Chitin chains are also linked to each other and possibly to protein chains through the amino group of N-acetylglucosamine. The sulfhydryl group (-SH) of the amino acid cysteine may also participate in cross-linking protein chains through formation of disulfide linkages (-S-S-).

When protein chains are linked to the ring of the phenolic cross-linking agent, the process is called **quinone tanning** or **quinone sclerotization**. Proteins may also be linked to the β-carbon (the carbon nearest the ring in the side chain) of N-acetyldopamine. Quinones also are involved in producing this type of sclerotization but the process is called **β-sclerotization** to distinguish it from protein attachment to the ring. How an insect controls the type of sclerotization that occurs is unknown. Some evidence indicates that both quinone tanning and β-sclerotization can occur in the same small region of cuticle. It has been suggested by some workers that β-sclerotization can harden cuticle to produce lighter colored or transparent cuticle, although how this might be controlled is unknown. Transparent cuticle is important in the covering of the compound eyes, for example, and many insects have lightly colored patches of cuticle elsewhere on the body. An important area of integumentary physiology still to be elucidated is the production of transparent cuticle.

The tanning of the ootheca of American cockroaches, *Periplaneta americana,* is a model for sclerotization in the cuticle (Figure 4.8). The ootheca of cockroaches contains proteins but no chitin. When first formed, the ootheca is white and soft; but it soon sclerotizes and darkens to a hard covering for the developing eggs and embryos. The sclerotizing process involves secretions from two accessory or collateral glands that are part of the reproductive tract in female cockroaches. The left gland contains the enzyme **diphenoloxidase**, and the glucosides of 3,4-dihydroxybenzoic acid and of 3,4-dihydroxybenzyl alcohol. The right gland contains **β-glucosidase**. When the secretions from the two glands are poured over the newly formed ootheca, β-glucosidase cleaves the two glycosides to free glucose and the diphenol **3,4-dihydroxy benzoic acid**. Diphenoloxidase oxidizes the phenolic acid to its quinone form. The quinone reacts without enzymatic help with a free amino group from a protein, hooking the protein to the ring of the compound (i.e., quinone sclerotization) and simultaneously becoming reduced again to the phenol form. In the presence of excess free quinone, the protein-phenol complex is again oxidized to a quinone, which may then react with another free amino group of a protein, with this protein also becoming hooked to the phenolic ring. These reactions may be repeated until several protein chains have been linked to the phenolic ring in quinone sclerotization. In the process, the ootheca becomes a hard, tough, waterproof case covering the eggs and developing embryos.

4.5.1 Hormonal Control of Sclerotization: Bursicon

Bursicon is a **neuropeptide** that promotes sclerotization and specifies how much cross-linking of molecules will occur. This hormone is secreted by the nervous system. It has not been isolated in pure enough form for a complete amino acid determination, but it is a small polypeptide of about 40,000 Da. It has been found in various ganglia of the central nervous system of many insects and is now believed to occur in most, or possibly all, insects.

Bursicon was first discovered in a newly formed adult fly (Cottrell, 1962; Frankel and Hsiao, 1962, 1963; Frankel et al., 1966). Soon after an adult fly emerges from the puparium, its peripheral nervous system sends signals to the brain to secrete bursicon (from the Greek *bursikos*, tanning or pertaining to tanning). Bursicon is released from neurosecretory cells (NSC) in the pars intercerebralis of the brain and from NSC in the large combined abdominal and thoracic ganglion of cyclorrhapha dipterans, in which bursicon is present at even higher concentrations than in the brain. Bursicon has also been demonstrated from the nervous system and corpora cardiaca of the cockroach *Periplaneta americana*. In ways that have not been elucidated in detail, bursicon promotes hardening or sclerotization of the cuticle. Bursicon may promote production of some of the sclerotizing enzymes, control access to quinone precursors, or control penetration and permeability of the cuticle to phenols and quinones.

4.6 CHEMICAL COMPOSITION OF CUTICLE

4.6.1 Chitin

One of the important constituents of cuticle is **chitin**, a polymer of **N-acetyl-β-D-glucosamine** (2-acetamido-2-deoxy-β-D-glucose) residues held together by **β-(1-4)-glycosidic linkages** (Figure 4.9). Enzymatic hydrolysis of chitin with chitinase and chitobiase releases N-acetylglucosamine, and also some free glucosamine. From these and other experiments, it has been suggested

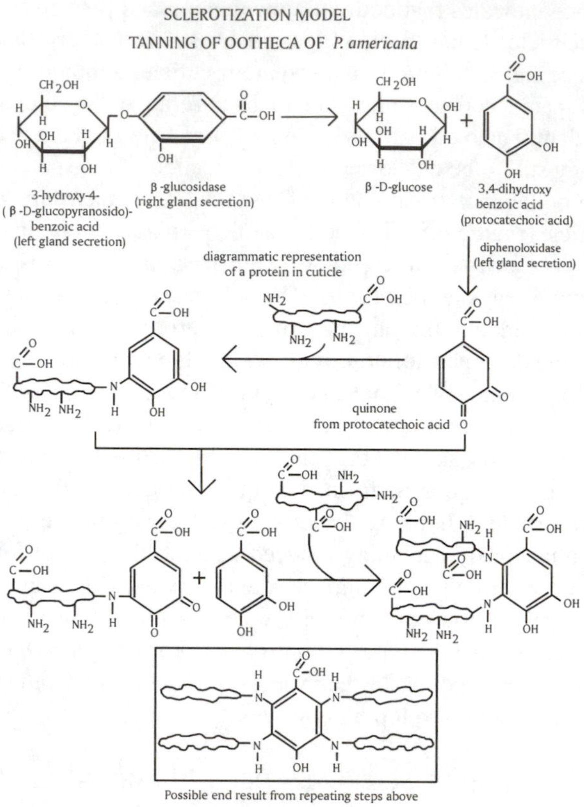

SCLEROTIZATION MODEL
TANNING OF OOTHECA OF *P. americana*

3-hydroxy-4-(β-D-glucopyranosido)-benzoic acid (left gland secretion)
β-glucosidase (right gland secretion)
β-D-glucose
3,4-dihydroxy benzoic acid (protocatechoic acid)
diagrammatic representation of a protein in cuticle
diphenoloxidase (left gland secretion)
quinone from protocatechoic acid
Possible end result from repeating steps above

FIGURE 4.8 A general sclerotization model based on sclerotization of the ootheca in the American cockroach, *Periplaneta americana*.

that every sixth or seventh residue in chitin may be glucosamine, with the remainder being N-acetylglucosamine. Chitin cannot be extracted directly from cuticles with any solvent, but it is left behind when the proteins, mineral deposits (if present), lipids, and other chemicals have been removed. However, the procedures that remove the other constituents almost always cause varying degrees of degradation of the chitin that is left. For example, treatment with hot alkali (KOH) removes protein from insect cuticles (cold alkali does the same thing, but requires more time; certain cuticular structures, such as genitalia, are "cleared" by allowing the tissue to stay in cold alkali for several days to weeks). Both cold and hot alkali remove some of the acetyl groups from chitin, leaving a less acetylated product called chitosan. Transparent and flexible, chitosan reacts with iodine (van Wisslingh's test) to give a dark purple color, and this often has been used to test for the presence of chitin, although it may not be infallible. In light of the vigorous procedures required to purify chitin from insect cuticles, it is not surprising that isolated chitin does not show all the same properties of the original insect cuticle.

Chitin, one of the most widely occurring polysaccharides in nature, is found in the cuticle of crustaceans and insects, many other invertebrates, nematode eggs, and as a structural cell wall component of fungi. As previously noted, chitin does not occur in the epicuticular layer of cuticle, and it may not be the major constituent in other parts of the cuticle. In some cuticles, protein is present in greater percentage by weight than chitin (Table 4.1).

The inertness and insolubility of chitin in the cuticle serve insects well, but make it difficult to characterize the molecular arrangement of chitin within cuticle. X-ray diffraction studies indicate that chitin exhibits a crystalline structure, but unit crystal dimensions and number of chitin chains per unit crystal vary with the source of the cuticle. Three types of chitin, named α-, β-, and γ-chitin,

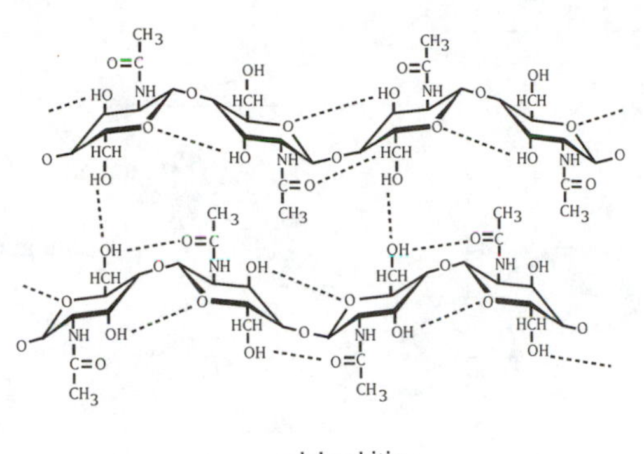

FIGURE 4.9 An illustration of the chemical structure of α-chitin chains and the anti-parallel arrangement of (two) chains that is typical in insect cuticles. Hydrogen bonds, some of which are intrachain and some interchain bonds, are represented by the dotted lines. Not shown are hydrogen bonds between chains in adjacent layers of chitin.

TABLE 4.1
The Proportions of Protein, Chitin, Phenolic Compounds, and Lipids in or on the Exuviae of Some Species of Cockroaches, as Determined by ^{13}C-NMR Analyses

	Relative Percentage in/on Exuviae			
Species	**Protein**	**Chitin**	**Phenolics**	**Lipids**
Periplaneta americana	49	38	11	2
Blatella germanica	59	30	9	2
Gromphadorhina portentosa	53	38	8	1
Blaberus cranifer	52	42	5	1
Leucophaea maderae	61	35	4	1

Reproduced with permission from Kramer et al., 1991.

have been described, and all three occur in some form in insects. The three types of chitin have differences in crystal cell size, number of chitin chains per unit cell, and degree of hydration (Rudall, 1963). Differences in orientation of the three chitin chains in chitinized structures lead to differences in physical packing of chains and in overall physiological properties of structures. In **α-chitin**, adjacent chains run antiparallel to each other, which allows them to pack closer together and maximizes the number of within- and between-chain hydrogen bonds (Figure 4.9). Typically, 18 to 20 α-chitin chains are packed together in a roughly circular (about 3 nm in diameter) rod or "**crystallite**" that is embedded in a protein matrix (Figure 4.10). The extensive **hydrogen bonding** within chains, between chains, and between adjacent sheets of cuticle contributes to the rigidity, strength, and water-proofing of the cuticle, leaving few hydration sites for water. Water molecules form hydrogen bonds with appropriate partners, and most of the potential sites for hydration in α-chitin are already involved in hydrogen bonds. The high content of water of hydration not only weakens the tightness of chain packing, but it allows more water movement across the cuticle, disrupting the high degree of impermeability typical of the cuticle. The tight packing and intra- and interchain hydrogen bonds provide strength and stability, and contribute to the impermeability of the cuticle to water.

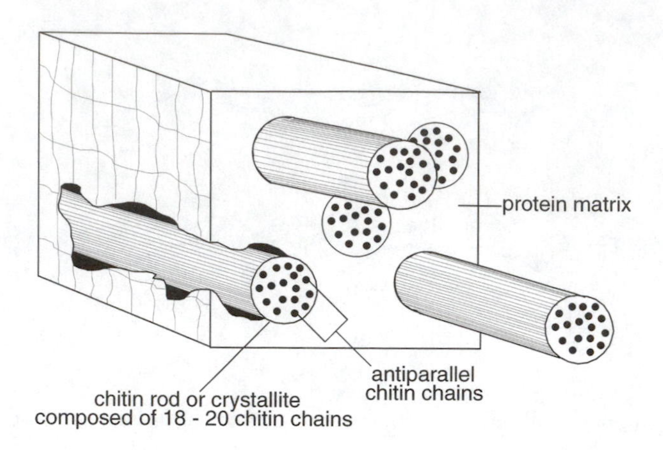

FIGURE 4.10 Chitin chains (about 20 typically in antiparallel arrangement) held together by hydrogen bonds to form chitin rods or crystallites embedded in a protein matrix in cuticle. (Modified from a model presented in Giraud-Guille and Bouligand, 1986.)

Adjacent chains in **β-chitin** run parallel to each other, and the relatively large N-acetyl groups projecting from the chain act like spacers to hold chains further apart, thus reducing tightness of packing and the number of hydrogen bonds that can form between chains. Chains in **γ-chitin** may be oriented in various ways, but one common orientation is a repeating pattern of two parallel chains adjacent to an antiparallel chain, again reducing packing tightness and interchain hydrogen bonds. Chains in β- and γ-chitin have more free groups that can form hydrogen bonds with water of hydration. β-Chitin and γ-chitin occur in cocoons of some beetles, and both are found in some other non-insect invertebrates. γ-Chitin has been identified in the peritrophic membrane of some insects. Greater hydration and less packing of chains allow chitinous structures with large amounts of β-and γ-chitins to be flexible and soft.

Cuticle is secreted as thin lamellae or sheets, like sheets of paper stacked on top of each other. The rods or crystallites of chitin are embedded in the protein matrix of a sheet (Giraud-Guille and Bouligand, 1986) and provide strengthening in much the same way as that provided by steel rods (reinforcement bars commonly called "rebar" in the construction industry) that are embedded in concrete columns and walls. Adjacent sheets of cuticle are stabilized by quinone tanning agents and by hydrogen bonds between chitin rods in adjacent cuticle sheets when the rods are near the surface of each sheet. It remains uncertain if quinone tanning agents directly link chitin rods to protein.

In successive sheets of cuticle, the chitin rods are often shifted slightly in orientation relative to the sheet above (the older sheet), and this gives rise to **Bouligand helicoids** (Figure 4.11) in thin transmission electron micrographs sections. One structural model suggests that chitin rods are embedded parallel to each other in the protein matrix in a plane or sheet of cuticle only a few nanometers thick (Bouligand, 1972), and in each successive sheet of cuticle, the model suggests that rods are reoriented slightly through a small angle relative to rods in the plane lying above. The shift in orientation of the chitin rods produces a helicoid pattern when oblique sections are cut through the cuticle. When the rotation has passed through 180°, the result is a lamella of cuticle.

Cross sections and freeze-fracture of insect cuticles often show a "plywood-like" arrangement (Figure 4.12A, B). The "plywood" structure occurs when chitin rods do not shift during the formation of many overlying layers of cuticle, then shift rapidly through 90° in a few thin lamella, and finally do not shift again while another thick layer of cuticle is laid down. Manufactured plywood sheets are designed in this way for added strength, and one can assume that similar strength is imparted to cuticle with this arrangement.

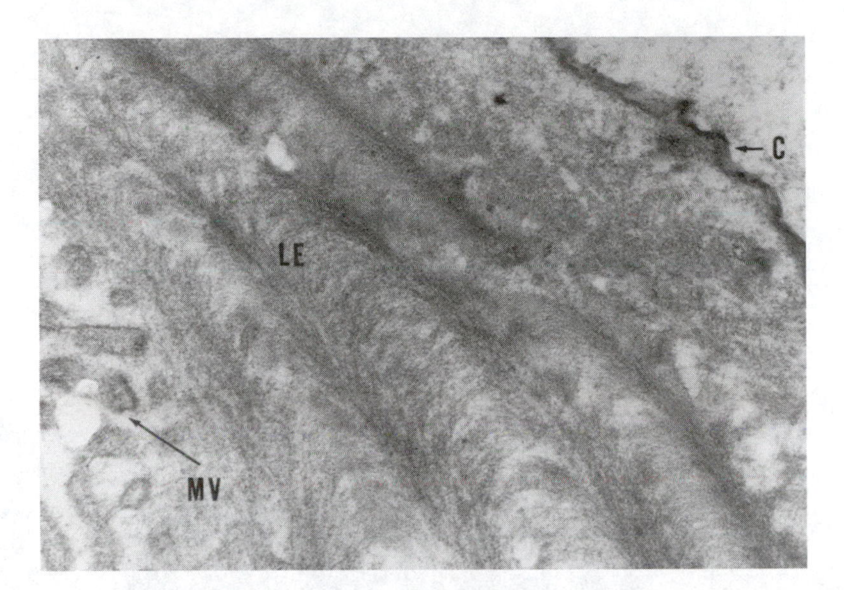

FIGURE 4.11 Bouligand helicoids in the cuticle of wing discs cultured *in vitro* from the Indian meal moth *Plodia interpunctella*. C, cuticulin layer; LE, lamellate endocuticle showing Bouligand helicoid patterns; MV, microvillae of underlying epidermal cell. (Photograph courtesy of Herbert Oberlander.)

4.6.2 Biosynthesis of Chitin

Information as to how chitin is synthesized has been obtained from a variety of invertebrates, including insects, and from fungi. The process remains incompletely understood in all of these organisms. The starting point for synthesis of chitin is β-D-glucose (Figure 4.13), although the β-D-glucose may come from a storage form such as trehalose or glycogen. The latter two compounds are formed from α-D-glucose, but a rapid and dynamic isomerization of glucose in aqueous solution occurs so that at any given moment there is 37% α-D-glucose and 63% β-D-glucose. Any removal of one form, for example, by synthesis into a polysaccharide, rapidly leads to a new equilibrium. A general review of biosynthesis has been presented by Cohen (1987). Initially, glucose is phosphorylated at the expense of ATP (Figure 4.14) and then isomerized to fructose-6-phosphate. An amino group is transferred from glutamine to fructose-6-phosphate to form glucosamine-6-phosphate. The latter molecule is acetylated, probably with acetyl CoA contributing the acetyl group, to form N-acetyl-glucosamine-6-phosphate. A transfer of the phosphate group from carbon-6 to carbon-1 is necessary to form N-acetyl-glucosamine-1-phosphate, which reacts with uridine triphosphate and forms uridine diphospho-N-acetyl-glucosamine (UDPGlcNAc). The detailed steps (likely there are several) between UDPGlcNAc and the linking of N-acetyl-β-D-glucosamine units together with β-1,4 linkages by the enzyme chitin synthase are poorly elucidated. Difficulties have been encountered in purifying and maintaining the stability of the enzymes needed in the final polymerization step.

The steps up to and including the formation of UDPGlcNAc have been verified in a number of insects (Kramer and Koga, 1986), and are similar in crustaceans, insects, and fungi. Little is known about the specifics of the final steps in chitin synthesis in any organism. Probably one intermediate step involves transfer of the N-acetylglucosamine residue to a lipid to form dolicyl-diphosphate-N-acetylglucosamine (Turnbull and Howells, 1982, 1983), as has been described in integumentary tissue from the sheep blowfly *Lucilia cuprina*. A similar process has also been described from crustaceans. Chitin synthase, necessary for the final step(s), has been prepared from insects as a large complex of proteins, some of which are very unstable in isolated form. The

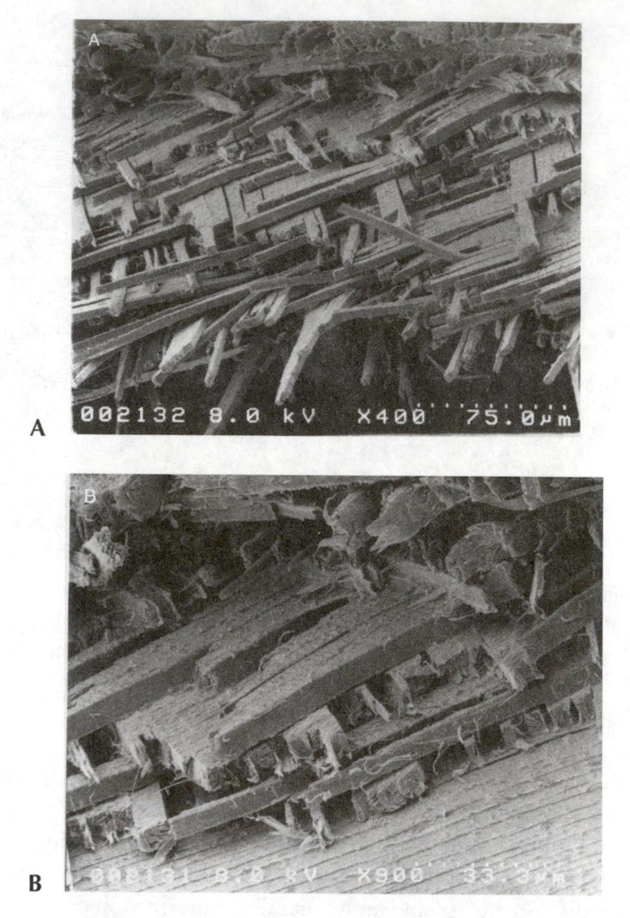

FIGURE 4.12 Freeze-fractured break in the thoracic cuticle of the weevil *Rhynchophorus cruentatus* showing plywood-like arrangement of cuticle layers that gives the cuticle added strength. (Magnification: ×400 (A), ×900 (B).) (Photographs courtesy of Robin Giblin-Davis.)

enzyme is present in microsomal preparations from epidermal cells, integument and gut; multiple forms of chitin synthase may occur in different tissues and insects (Kramer and Koga, 1986). It may exist as a zymogen that requires activation when chitin synthesis is under way (Mayer et al., 1980).

4.6.3 Cuticular Proteins

Cuticular proteins are synthesized in the epidermal cells and are secreted from the apical surface of the cells into the cuticle. In the cuticle, they fill in the matrix around chitin rods or crystallites. In hard cuticles, the proteins are stabilized (sclerotized, cross-linked) by phenolic and quinone compounds that form covalent bonds and cross-link proteins to each other (and possibly to chitin), forming a very hard, rigid structure. Even in very soft cuticles, there is some degree of stabilization, but probably relatively few cross-links. Cuticle proteins in proximity to chitin rods may be bound to chitin by quinones or hydrogen bonds, or both. Protein bound to chitin has never been extracted from cuticle, however, and the degradative procedures necessary to extract proteins or chitin from cuticle have been blamed for the failure to find bound protein-chitin complexes. Sclerotized cuticular proteins are difficult or impossible to extract without extensive degradation resulting from the treatments needed to break the phenolic cross-links, which also break the peptide bonds within proteins. These difficulties of solubilization have hindered analyses of cuticular proteins in general, and with the exception of resilin, few specific cuticular proteins have been characterized.

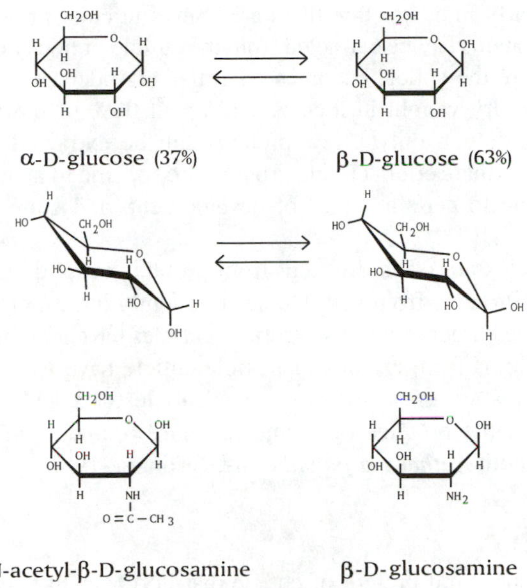

α-D-glucose (37%) β-D-glucose (63%)

N-acetyl-β-D-glucosamine β-D-glucosamine

FIGURE 4.13 A illustration of two ways of representing the structure of α-D-glucose and β-D-glucose, the proportions of each in the dynamic equilibrium that always exists in aqueous solution, and the structures of N-acetyl-β-D-glucosamine and β-D-glucosamine, all precursors of chitin.

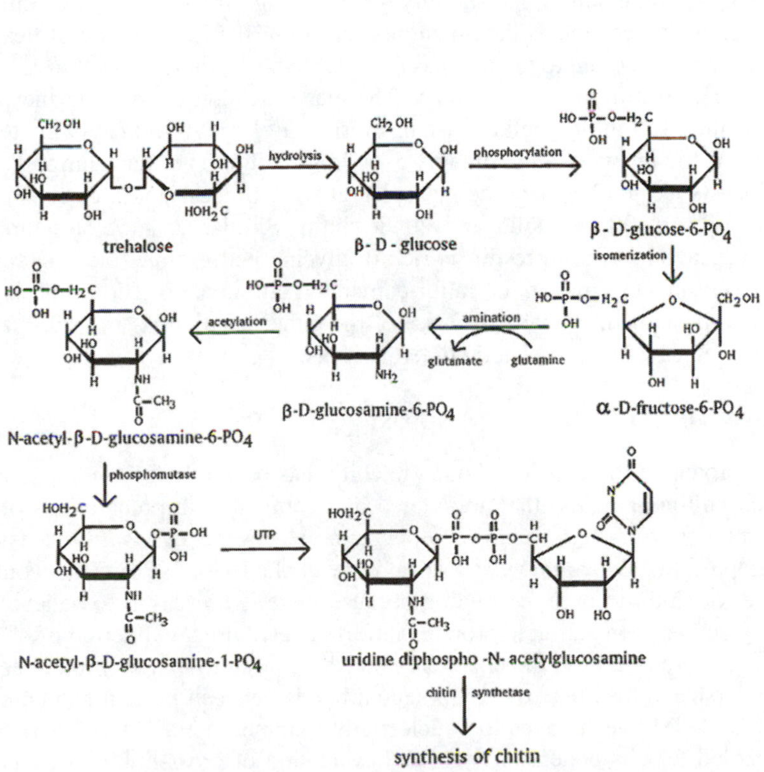

FIGURE 4.14 A general and tentative scheme for the biosynthesis of chitin.

Current research clearly indicates that there are many cuticular proteins. At least 100 electrophoretically separable proteins can be extracted from the cuticle of newly eclosed migratory locusts, *Locusta migratoria*, before the cuticle becomes sclerotized (Andersen et al., 1986). Proteins make up to 70% of the cuticle dry weight in locusts, and about 90% of the proteins can be extracted prior to sclerotization, although only a few proteins can be extracted from hard or sclerotized cuticle. Some studies show that certain cuticle proteins are specific to anatomical structures or body regions, some are specific to certain stages of development, and some are even specific to age within a stage.

Proteins of soft cuticles are often different from proteins in hard cuticles in ways other than degree of sclerotization. Proteins from soft cuticles frequently have a high content of polar amino acids, while proteins in the harder or more sclerotized cuticles have a higher content of hydrophobic amino acids. Locust proteins from regions of flexible cuticle have isoelectric points between 4.4 and 5.0, and these proteins are missing from harder cuticle (Cox and Willis, 1985). In contrast, only minor differences in the proteins of segmental (hard) cuticle and intersegmental (flexible) cuticle occur in the silkmoth *Antheraea polyphemus* (Sridhara, 1983).

4.6.4 Resilin

Resilin is an important structural protein of cuticle that is rubber-like, colorless, transparent, and insoluble in water. It has remarkable properties of elasticity, like rubber, but shows less deformation upon prolonged stretching than rubber. A stretched rubber band gets longer upon prolonged stretching; but even after extended periods of stretching, resilin returns to about 97% of its original length. To be flexible, resilin obviously cannot be very sclerotized or have many cross-links with other protein chains, but dityrosine and trityrosine residues (Figure 4.15) provide a few internal cross-links, giving resilin chains some stability, yet allowing high elasticity. Resilin occurs in the wing hinges of some insects and in the hinge mechanism of the jumping leg of fleas. The prealar arm connecting the mesotergum to the first basalar sclerite of the thoracic wall in *S. gregaria* wings contains about 50 µg resilin and 15 µg chitin. The elastic tendon connecting the pleuro-subalar muscle to the ventral wall in dragonflies contains resilin. In large *Aeshna* spp. the tendon is about 0.7 mm long by 0.15 mm wide and contains 5 to 7 µg resilin. The main hinge ligament of the forewings of locusts located between the mesopleural wing process and the second axillary wing sclerite contains about 100 µg resilin and 20 µg chitin. Similar to such structural proteins as collagen, elastin, and silk fibroin, resilin is rich in glycine and proline, but contains no hydroxyproline, hydroxylysine, tryptophan, or sulfur-containing amino acids. Biosynthesis of resilin has not been extensively studied, probably because of the rather small areas in which it occurs, but it is believed to be secreted by specialized epidermal cells.

4.6.5 Do Protein-to-Chitin Bonds Exist?

Whether or not protein in the cuticle is bound to chitin has been a long-standing question. **Glycoproteins** (called **chitinoproteins**) that are altered by treatment with preparations of chitinase are found in cell cultures from *Aedes aegypti*, *Bombyx mori*, several *Drosophila* species, and in all stages of development of *D. melanogaster* (Kramerov et al., 1986, 1990). This would seem to be strong evidence for chitin-to-protein bonding because there is no reason to believe that chitinase should have any action upon a simple protein. Unfortunately, direct extraction of chitin-to-protein units cannot be achieved because the chemical procedures used to extract protein and chitin from cuticle are too harsh and they destroy any potential bonds between protein and chitin. The recent application of solid NMR techniques to cuticle study (Kramer et al., 1991, 1995) may provide a noninvasive method for component analysis and evaluation of cross-linking of components, and has already provided evidence that quinones derived from N-acetylcatecholamines form cross-links and adducts with functional groups of proteins.

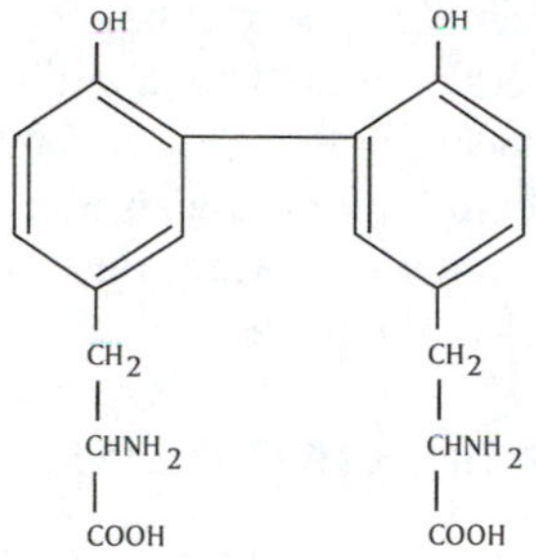

dityrosine

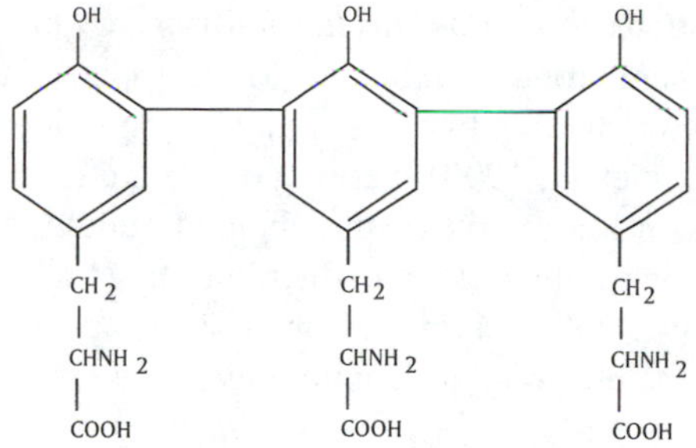

trityrosine

FIGURE 4.15 The chemical structure of dityrosine and trityrosine, two unique amino acids in resilin that are believed to be important in holding the proteins chains together, but allowing for elasticity.

4.6.6 Stage-Specific Differences in Cuticle Proteins

Developmental biologists have frequently raised the question of whether or not each stage in the development of an insect is controlled by specific genes or gene sets. There might be a larval set of genes, a pupal set, and an adult set. Somehow, each set would be activated and then, at the appropriate time, deactivated. The identification of specific cuticular proteins at a particular stage of development might provide an opportunity for isolating and identifying the genes that controlled the proteins, and ultimately of resolving the questions of whether there are stage-specific genes. Regional-specific, stage-specific, and temporal differences in the proteins of *Hyalophora cecropia* (Cox and Willis, 1985, 1987) have been found, but a *cecropia* silk moth larva also conserves proteins from some stages to the next, as well as synthesizes new proteins specific to its current stage of development. A protein band has been found in extracts of head and thoracic cuticle of *Drosophila* that does not occur in abdominal cuticle (Chihara et al., 1982).

Stage-specific proteins occur in *Drosophila melanogaster*, *Manduca sexta*, *Antheraea polyphemus*, *Bombyx mori*, and *Tenebrio molitor*. cDNA, mRNA, and electrophoretic analyses have been used to determine that there are different proteins in pupal and adult wing cuticle of *Antheraea polyphemus* (Sridhara, 1983, 1985). Cuticular proteins from larval, pupal, and adult cuticle of *Tenebrio* are specific for each stage (Lemoine and Delachambre, 1986). Electrophoretic and immunoblot analyses have been used to demonstrate different cuticular proteins in pupal and larval cuticles of the silkworm *Bombyx mori* (Nakato et al., 1990).

In an apparent contradiction to much of the above evidence of the stage specificity of proteins, the cotton boll weevil, *Anthonomus grandis*, has a high degree of antigenic similarity among cuticular proteins of various stages (Stiles and Leopold, 1990). Many of the proteins are glycosylated, however, which may promote antigenic cross-reactivity and lead to a false indication of similarity (Stiles, 1991). Thus, it remains unclear whether or not the proteins in the boll weevil are stage specific.

In summary, some proteins in the cuticle appear to be stage specific, but there also is evidence that some proteins are common to several stages. The data do not cast much light on the original question of whether or not there are specific gene sets governing each developmental stage. The same protein in more than one stage could simply mean that the gene coding for it is present in one or more sets. If there were no proteins that were the same in, for example, a larva and the pupa, then there would be a stronger case for two sets of genes.

4.6.7 Protective Functions of Cuticle Proteins

Cuticular proteins in some or all insects may function in protection against invading bacteria or other organisms (Marmaras et al., 1993). When the surface of the epicuticle of silkworm larvae is lightly abraded and treated with bacteria (*Bacillus lichenformis* or *Enterobacter cloacae*) or bacterial cell wall compounds, mRNAs are produced that code for cecropin, a large protein with antibacterial properties. The response is systemic, and cecropin appears in epidermal cells and fat body cells remote from the abrasion (Brey et al., 1993). Subsequently, and as early as 8 hours post-abrasion, antibacterial activity appears in the matrix of the abraded cuticle, but not in unabraded cuticle. More intense abrasion of cuticle results in a wider distribution of the antibacterial activity, including activity in fat body and hemolymph. It is not clear how surface abrasion of the cuticle and the introduction of bacteria into the abrasion are communicated to epidermal and fat body cells.

4.6.8 Cuticular Lipids

The lipids on the cuticle of insects have received a great deal of attention, not only because they promote water conservation, but because they often have additional behavioral, pheromonal, ecological, and taxonomic significance (Howard and Blomquist, 1982; Liepert and Dettner, 1996; Doi et al., 1997). Among the lipids on the cuticle are true chemical waxes, hydrocarbons, alcohols, fatty acids, glycerides, sterols, ketones, aldehydes, and esters. Cuticular lipids are relatively easy to separate, identify, and quantitatively measure with gas chromatography combined with mass spectrometry (Bagnères and Morgan, 1990).

The lipid layer on the cuticle has often been described as a wax layer. In most insects, only a small percentage of the lipids are true waxes, which are chemically defined as esters between long-chain alcohols and long-chain fatty acids. In most cases, "wax layer" is clearly a misnomer. The quantity of waxes on the cuticle varies among insects and stage of development (Lockey, 1988). Only about 3% of the cuticular lipids of a weevil, *Ceutorrhynchus assimilis*, are true wax esters; but up to 74% of the lipids from the cuticle of the burrowing cockroach *Arenivaga investigata* are wax esters. Wax esters made up 24.4, 29.2, and 3.9% of the cuticular lipids of the larva, pupa, and adult, respectively, of the bean beetle *Epilachna varivestis*. Honeybees have wax esters (34% of total lipids) on their cuticle and secrete several types of wax esters to make beeswax, which typically contains about 50% wax esters.

The lipid layer may lie beneath the cement layer, as many introductory entomology books suggest, but this is not always true. There is a mosaic pattern consisting of patches of lipids and cement on the cuticle of *Calpodes ethlius* (Locke, 1965). A tenebrionid beetle, *Cryptoglossa verrucosa*, that lives in the Sonoran Desert in the American Southwest has a "wax bloom" on its surface. The wax bloom serves as camouflage, reduces water evaporation, and provides some thermal protection by restricting air movement at the cuticular surface (Hadley, 1979). The beetles change color, depending on how much wax is secreted to the surface. The wax is secreted mainly at low humidity from the tips of miniature tubercles at the cuticle surface, and the beetles look blue in low humidity and black at high humidity (Figure 4.16). The filaments of wax spread over the cuticle to form a fibrous meshwork about 20 µm thick, and light reflected from the surface makes the beetles appear light bluish-white in color. Under high humidity conditions, the lipid filaments are not secreted and the beetles are black. The layering of the filaments and their thickness,

FIGURE 4.16 Color phases of the beetle *Cryptoglossa verrucosa* (Le Conte) (Coleoptera: Tenebrionidae) in response to humidity. The beetle is dark, nearly black (right) at higher humidity and blue (left) at low humidity. Wax filaments secreted during low humidity by tubercles on the elytra disperse the light and give the beetles a beautiful blue color. The photographs were taken in Bahia de Kino, Sonora, Mexico. (Photographs courtesy of J. Nathaniel Holland, Department of Ecology and Evolutionary Biology, University of Arizona, Tucson.)

and the boundary air layer trapped between the meshwork and the surface of the cuticle, probably retard transcuticular water loss and possibly reduce the rate of body heating by acting as a reflective shield. Bluish beetles lose 0.109 ± 0.032 mg/cm^2/h at 40°C and 0% RH, while black beetles lose 0.140 ± 0.026 mg/cm^2/h (Hadley, 1979). Another insect with lipid at the surface of the cuticle is the eri silkworm, which covers itself with a white powder composed of two long-chain alcohols (Bowers and Thompson, 1965).

The type and quantity of lipids on the cuticle seem likely to be one of the many ways insects are adapted for the environment in which they live. For example, young larvae of *Sarcophaga bullata* (Diptera, Sarcophagidae), which live in very wet environments, have only small quantities of cuticular hydrocarbons on the cuticle. Pupae, representing a closed system to water, and adults subject to the drying influence of the air have greater quantities of surface lipids (Armold and Regnier, 1975). Diapausing pupae of *Manduca sexta* have a thicker lipid layer on the cuticle than non-diapausing ones (Bell et al., 1975), apparently providing greater protection from desiccation during the long diapause. Nymphs of the desert cicada *Diceroprocta apache* live underground and have smaller quantities of cuticular hydrocarbons than adults, which fly and live at a temperature close to 50°C (Hadley, 1980). The underground burrows or chambers in which nymphs live have lower and more stable temperatures, and probably high humidity, thus affording more protection from water transpiration than the air in which adults live. Season, and apparently temperature, are correlated with the type and quantity of hydrocarbons on the cuticle of the desert tenebrionid *Eleodes armata* (Hadley, 1977). Summer beetles have a greater quantity of hydrocarbons and more long-chain ones than winter beetles, but longer-chain molecules can be induced by holding winter beetles at 35°C for 5 to 10 weeks.

Hydrocarbons are often the major components on the cuticle, and many different compounds are usually present, including straight-chain saturated compounds (alkanes), olefins (unsaturated alkenes), and alkanes with methyl branches (2-methyl, 3-methyl, and various internally branched alkanes are common) (Lockey, 1988). Molecules with complex branching and multiple double bonds occur on some cuticles. In many insects, cuticular hydrocarbons are species specific, leading to the use of cuticular hydrocarbon analyses in taxonomy and systematics (Lockey, 1988, 1991; Page et al., 1990, 1997; Haverty et al., 1996, 1997). Some caution must be exercised in interpreting hydrocarbon composition as species indicators, however, because there may be sexual, seasonal, stage-specific, or dietary-related variations in the cuticular hydrocarbons (Espelie et al., 1994). Cuticular lipids are "worn-off" the cuticle and periodically replaced, especially in long-lived insects, and are lost with the epicuticle at each molt. At ecdysis, new lipids are secreted again just before and/or just after ecdysis of the old cuticle. Hydrocarbons are synthesized by abdominal oenocytes and or epidermal cells in female *Blatella germanica* cockroaches, and transported by lipophorins in the hemolymph to distribution sites in the body, including the cuticle (Gu et al., 1995).

Insects have a large surface-to-volume ratio, and desiccation is a potential hazard for many insects. One of the major functions of the cuticle is to protect insects from losing excess water from the body by transevaporation across the cuticle, and from imbibing water and flooding the body in the case of aquatic insects (Hadley, 1984). All parts of the integument, including the epidermal cells, are important in maintaining the impermeable nature of the cuticle, but the lipids and wax bloom at the surface of the cuticle are especially important in providing waterproofing (Noble-Nesbitt, 1991). Although the cuticle is relatively impermeable to water, 85% or more of the water lost from the body is lost through the cuticle (see Table 12.2 in Chapter 12).

Insects experience sudden, rapid water loss through the cuticle at a critical temperature (Ramsay, 1935; Wigglesworth, 1945), a "transition temperature" that varies with the species. The mechanisms involved in this sudden shift in water permeability are not yet understood, but it may be due in part to a reorientation of the lipid molecules on the cuticle surface. Although this phenomenon has been suspected to be related to changes in the lipids on the cuticle, experimental details have, until recently, been lacking. It has now been documented that increasing temperature does indeed modify the physical state (induces melting) of epicuticular lipids and disrupts packing of molecules on the cuticle as a water proof barrier (Gibbs, 1995, 1998; Gibbs and Pomonis, 1995; Rourke and Gibbs, 1999). The critical transition temperature T_m was defined by Gibbs (1995) as the temperature causing 50% melting of cuticular hydrocarbons and subsequent disruption of molecular orientation. Melting and orientation depend on carbon chain length, degree of branching, location of branching, and saturation (alkanes) or unsaturation (alkenes) in the mixture of hydrocarbons. Longer-chain alkanes melt at higher temperatures than alkenes of the same carbon number, and *n*-alkanes (straight chains) melt at higher temperatures than branched chains with the same number of carbons. Internally branched hydrocarbons melt at lower temperatures than terminally branched alkanes (2- or 3-methylalkanes) (Gibbs, 1995; Gibbs and Pomonis, 1995). Rourke and Gibbs (1999) determined that the transition temperature for rapid water loss from the cuticle of the grasshopper *Melanoplus sanguinipes* occurs when the cuticular lipids are about 30% melted, and Rourke (2000) found that increased quantity and higher melting points of cuticular lipids correlate with lower rates of body water loss. Increased water loss through the cuticle of the German cockroach *Blattella germanica* L. occurs even when cuticular hydrocarbons are only about 5% melted (Young et al., 2000). Although the T_m does not correlate well with typical environmental temperature exposure of some insects, the type and packing of hydrocarbons on the cuticle do influence water loss, and in some cases possibly at only a few degrees higher than the typical environmental exposure (Young et al., 2000).

4.7 MINERALIZATION OF INSECT CUTICLES

Insect cuticles generally do not incorporate significant quantities of minerals into their cuticles. One insect that does is the face fly, *Musca autummnalis*. About 63% of its puparium dry weight is ash, and the major mineral ions are calcium, phosphorus, and magnesium. The related housefly, *M. domestica*, and the stable fly, *Stomoxys calcitrans*, have only 3.65 and 2.31% ash, respectively, as dry weight of their puparia, but the major ions are still calcium, phosphorus, and magnesium, although in much lower quantity than in the face fly (Roseland et al., 1985).

4.8 CAPTURE OF ATMOSPHERIC WATER ON CUTICULAR SURFACES

A few insects have specialized regions of cuticle upon which water condensation from the atmosphere can occur, with the result that the insect can absorb the water into the body. The desert cockroach, *Arenivaga investigata* can capture water vapor on the cuticular lining in the mouth, and the mealworm *Tenebrio molitor* and the thysanuran firebrat *Thermobia domestica* can absorb water across the rectal cuticle from moist air. *Tenebrio molitor* can absorb water across its rectal cuticle at a rate of 0.4 $mg/cm^2/h$ in high humidities, but larvae lose water no faster than 0.08 $mg/cm^2/h$ even at low humidity. A Namib desert tenebrionid, *Onymacris unguicularis*, captures water over its entire body surface, as it orients in a head-down position at or near the top of a sand dune where water condenses on its body from wind-driven fogs at night. The vertical orientation of the body causes the water to run in droplets toward its mouth, and the beetle ingests the water.

REFERENCES

Andersen, S.O., P. Hojrup, and P. Roepstorff. 1986. Characterization of cuticular proteins from the migratory locust, *Locusta migratoria*. *Insect Biochem.,* 16:441-447.

Armold, M.T., and F.E. Regnier. 1975. A developmental study of the cuticular hydrocarbons of *Sarcophaga bullata*. *J. Insect Physiol.,* 21:1827-1833.

Bagnères, A.G., and E.D. Morgan. 1990. A simple method for analysis of insect cuticular hydrocarbons. *J. Chem. Ecol.,* 16:3263-3275.

Bell, R.A., P.R. Nelson, T.K. Borg, and D.L. Caldwell. 1975. Wax secretion in non-diapausing and diapausing pupae of the tobacco hornworm, *Manduca sexta*. *J. Insect Physiol.,* 21:1725-1729.

Bouligand, Y. 1972. Twisted fibrous arrangements in biological materials and cholesteric mesophases. *Tissue Cell,* 4:189-217.

Bowers, W.S., and M.J. Thompson. 1965. Identification of the major constituents of the crystalline powder covering the larval cuticle of *Samia cynthia ricini* (Jones). *J. Insect Physiol.,* 11:1003-1011.

Brey, P.T., W.-J. Lee, M. Yamakawa, Y. Koizumi, S. Perrot, M. Francois, and M. Ashida. 1993. Role of the integument in insect immunity: epicuticular abrasion and induction of cecropin synthesis in cuticular epithelial cells. *Proc. Natl. Acad. Sci. U.S.A.,* 90:6275-6279.

Brookhart, G.L., and K.J. Kramer. 1990. Proteinases in molting fluid of the tobacco hornworm, *Manduca sexta*. *Insect Biochem.,* 20:467-477.

Chihara, C.J., D.J. Silvert, and J.W. Fristrom. 1982. The cuticle proteins of *Drosophila melanogaster*: Stage specificity. *Dev. Biol.,* 89:379-388.

Cohen, E. 1987. Chitin biochemistry: Synthesis and inhibition. *Annu. Rev. Entomol.,* 32:71-93.

Cornell, J.C., and M.L. Pan. 1983. The disappearance of moulting fluid in the tobacco hornworm, *Manduca sexta*. *J. Exp. Biol.,* 107:501-504.

Cottrell, C.B. 1962. The imaginal ecdysis of blowflies. *J. Exp. Biol.,* 39:395-449.

Cox, D.L., and J.H. Willis. 1985. The cuticular proteins of *Hyalophora cecropia* from different anatomical regions and metamorphic stages. *Insect Biochem.,* 15(3):349-362.

Cox, D.L., and J.H. Willis. 1987. Analysis of the cuticular proteins of *Hyalophora cecropia* with two dimensional electrophoresis. *Insect Biochem.,* 17:457-486.

Doi, M., T. Nemoto, H. Nakanishi, Y. Kuwahara, and Y. Oguma. 1997. Behavioral response of males to major sex pheromone component, (Z,Z)-5,25-hentriacontadiene, of *Drosophila ananassae* females. *J. Chem. Ecol.,* 23:2067-2078.

Espelie, K., R.F. Chapman, and G.A. Sword. 1994. Variation in the surface lipids of the grasshopper, *Schistocerca americana* (Drury). *Biochem. Syst. Ecol.,* 22:563-575.

Frankel, G., and C. Hsiao. 1962. Hormonal and nervous control of tanning in the fly. *Science,* 138:27-29.

Frankel, G., and C. Hsiao. 1963. Tanning in the adult fly: A new function of neurosecretion in the brain. *Science,* 141:1057-1058.

Frankel, G., C. Hsiao, and M. Seligman. 1966. Properties of bursicon: An insect protein hormone that control cuticular tanning. *Science,* 151:91-93.

Gammie, S.C., and J.W. Truman. 1997. Neuropeptide hierarchies and the activation of sequential motor behaviours in the hawkmoth, *Manduca sexta*. *J. Neurosci.,* 17:4389-4397.

Gammie, S.C., and J.W. Truman. 1999. Eclosion hormone provides a link between ecdysis-triggering hormone and crustacean cardioactive peptide in the neuroendocrine cascade that controls ecdysis behaviour. *J. Exp. Biol.,* 202:343-352.

Gibbs, A. 1995. Physical properties of insect cuticular hydrocarbons: model mixtures and lipid interactions. *Comp. Biochem. Physiol.,* 112B:667-672.

Gibbs, A.G. 1998. Waterproofing properties of cuticular lipids. *Am. Zool.,* 38:471-482.

Gibbs, A.G., and J.G. Pomonis. 1995. Physical properties of insect cuticular hydrocarbons: model mixtures and interactions. *Comp. Biochem. Physiol.,* 112B:243-249.

Giraud-Guille, M.M., and Y. Bouligand. 1986. Chitin-protein molecular organization in Arthropods, pp. 29-35, in R. Muzzarelli, Ch. Jeuniaux, and G.W. Gooday (Eds.), *Chitin in Nature and Technology,* Plenum Press, New York.

Gu, X., D. Quilici, P. Juarez, G.J. Blomquist, and C. Schal. 1995. Biosynthesis of hydrocarbons and contact sex pheromone and their transport by lipophorin in females of the German cockroach (*Blatella germanica*). *J. Insect Physiol.,* 41:257-267.

Hadley, N.F. 1977. Epicuticular lipids of the desert tenebrionid beetle, *Eleodes armata*: Seasonal and acclimatory effects on composition. *Insect Biochem.,* 7:277-283.

Hadley, N.F. 1979. Wax secretion and color phase of the desert tenebrionid beetle *Cryptoglossa verrucosa* (LeConte). *Science,* 203:367-369.

Hadley, N.F. 1980. Cuticular lipids of adult and nymphal exuviae of the desert cicada, *Diceroprocta apache* (Homoptera, Cicadidae). *Comp. Biochem. Physiol. [B],* 65:549-553.

Hadley, N.F. 1984. Cuticle: Ecological significance, pp. 685-693, in J., Bereiter-Hahn, A.G. Matoltsy, and K. Sylvia Richards (Eds.), *Biology of the Integument, 1. Invertebrates,* Springer-Verlag, New York.

Haverty, M.I., B.L. Thorne, and L.J. Nelson. 1996. Hydrocarbons of *Naustitermes acajutlae* and comparison of methodologies for sampling cuticular hydrocarbons of Caribbean termites for taxonomic and ecological studies. *J. Chem. Ecol.,* 22:2081-2109.

Haverty, M.I., M.S. Collins, L.J. Nelson, and B.L. Thorne. 1997. Cuticular hydrocarbons of termites of the British Virgin Islands. *J. Chem. Ecol.,* 23:927-964.

Hesterlee, S., and D.B. Morton. 1996. Insect physiology: The emerging story of ecdysis. *Curr. Biol.,* 6:648-650.

Horodyski, F.M. 1996. Neuroendocrine control of insect ecdysis by eclosion hormone. *J. Insect Physiol.,* 42:917-924.

Howard, R.W., and G.J. Blomquist. 1982. Chemical ecology and biochemistry of insect hydrocarbons. *Annu. Rev. Entomol.,* 27:149-172.

Hopkins, T.L., T.D. Morgan, Y. Aso, and K.J. Kramer. 1982. N-β-alanyldopamine: Major role in insect cuticle tanning. *Science,* 217:364-366.

Hopkins, T.L., and K.J. Kramer. 1992. Insect cuticle sclerotization. *Annu. Rev. Entomol.,* 37:273-302.

Jenkin, P.M., and H.E. Hinton. 1966. Apolysis in arthropod moulting cycles. *Nature (London),* 211:871.

Kingan, T.G., and M.E. Adams. 2000. Ecdysteroids regulate secretory competence in Inka cells. *J. Exp. Biol.,* 203:3011-3018.

Kramer, K.J., and D. Koga. 1986. Insect chitin, physical state, synthesis, degradation and metabolic regulation. *Insect Biochem.,* 16:851-877.

Kramer, K.J., A.M. Christensen, T.D. Morgan, J. Schaefer, T.H. Czapla, and T.L. Hopkins. 1991. Analysis of cockroach oothecae and exuviae by solid-state ^{13}C-NMR spectroscopy. *Insect Biochem.,* 21:149-156.

Kramer, K.J., T.H. Hopkins, and J. Schaefer. 1995. Applications of solids NMR to the analysis of insect sclerotized structures. *Insect Biochem. Mol. Biol.,* 25:1067-1080.

Kramerov, A.A., D.V. Mukha, E.V. Metakovsky, and V.A. Gvozdev. 1986. Glycoproteins containing sulfated chitin-like carbohydrate moiety are synthesized in an established *Drosophila melanogaster* cell line. *Insect Biochem.,* 16(2):417-432.

Kramerov, A.A., Y.M. Rozovsky, N.A. Baikova, and V.A. Gvozdev. 1990. Cognate chitinoproteins are detected during *Drosophila melanogaster* development and in cell cultures from different insect species. *Insect Biochem.,* 20(8):769-775.

Lane, N.J., and H. leB. Skaer. 1980. Intercellular junctions in insect tissues. *Adv. Insect Physiol.,* 15:35-213.

Lemoine, A., and J. Delachambre. 1986. A water-soluble protein specific to the adult cuticle in *Tenebrio*. Its use as a marker of a new programme expressed by epidermal cells. *Insect Biochem.,* 16(3):483-489.

Liepert, C., and K. Dettner. 1996. Role of cuticular hydrocarbons of aphid parasitoids in their relationship to aphid-attending ants. *J. Chem. Ecol.,* 22:695-706.

Locke, M. 1965. Permeability of insect cuticle to water and lipids. *Science,* 147:295-298.

Locke, M. 1984. Epidermal cells, pp. 502-522, in J. Bereiter-Hahn, A.G. Matoltsy, and K. Sylvia Richards (Eds.), *Biology of the Integument, 1. Invertebrates,* Springer-Verlag, New York.

Locke, M. 2001. The Wigglesworth Lecture: Insects for studying fundamental problems in biology. *J. Insect Physiol.,* 47:495-507.

Lockey, K.H. 1988. Lipids of insect cuticle: Origin, composition and function. *Comp. Biochem. Physiol.,* 89B:595-645.

Lockey, K.H. 1991. Insect hydrocarbon classes: Implications for chemotaxonomy. *Insect Biochem.,* 21:91-97.

Marmaras, V.J., S.N. Bournazos, P.G. Katsoris, and M. Lambropoulou. 1993. Defense mechanisms in insects: Certain integumental proteins and tyrosinase are responsible for nonself-recognition and immobilization of *Escherichia coli* in the cuticle of developing *Ceratitis capitata. Arch. Insect Biochem. Physiol.,* 23:169-180.

Mayer, R.T., A.C. Chen, and J.R. DeLoach. 1980. Characterization of a chitin synthase from the stable fly, *Stomoxys calcitrans* (L.). *Insect Biochem.,* 10:549-556.

Miles, C.I., and R. Booker. 1998. The role of the frontal ganglion in the feeding and eclosion behavior of the moth *Manduca sexta. J. Exp. Biol.,* 20:1785-1798.

Nakato, H., M. Toriyama, S. Izumi, and S. Tomino. 1990. Structure and expression of mRNA for pupal cuticle protein of the silkworm, *Bombyx mori. Insect Biochem.,* 20:667-678.

Noble-Nesbitt, J. 1991. Cuticular permeability and its control, pp. 252–283, in K. Binnington and A. Retnakaran (Eds.). *Physiology of the Insect Epidermis,* CSIRO Publications, Melbourne, Australia.

Neville, C. 1986. *The Biology of the Arthropod Cuticle,* Carolina Biology Reader 103, 16 pp.

Noirot, C., and A. Quennedey. 1974. Fine structure of insect epidermal glands. *Annu. Rev. Entomol.,* 19:61-80.

Page, M., L.J. Nelson, M.I. Haverty, and G.J. Blomquist. 1990. Cuticular hydrocarbons as chemotaxonomic characters for bark beetles: *Dendroctonus ponderosae, D. jeffreyi, D. brevicomin,* and *D. frontalis* (Coleoptera: Scolytidae). *Ann. Entomol. Soc. Am.,* 83:892-901.

Page, M., L.J. Nelson, G.J. Blomquist, and S.J. Seybold. 1997. Cuticular hydrocarbons as chemotaxonomic characters of pine engraver beetles (*Ips* spp.) in the *grandicollis* subgeneric group. *J. Chem. Ecol.,* 23:1053-1099.

Ramsay, J.A. 1935. The evaporation of water from the cockroach. *J. Exp. Biol.,* 12:373–383.

Reynolds, S.E. 1980. Integration of behavior and physiology in ecdysis. *Adv. Insect Physiol.,* Vol. 15:475-595.

Roseland, C.R., M.J. Grodowitz, K.J. Kramer, T.L. Hopkins, and A.B. Broce. 1985. Stabilization of mineralized and sclerotized puparial cuticle of muscid flies. *Insect Biochem.,* 15:521-528.

Rourke, B.C. 2000. Geographic and altitudinal variation in water balance and metabolic rate in a California grasshopper, *Melanoplus sanguinipes. J. Exp. Biol.,* 203:2699-2712.

Rourke, B.C., and A.G. Gibbs. 1999. Effects of lipid phase transitions on cuticular permeability: Model-membrane and *in situ* studies. *J. Exp. Biol.,* 202:3255-3262.

Rudall, K.M. 1963. Chitin protein complexes of insect cuticle. *Adv. Insect Physiol.,* 1:257-311.

Sridhara, S. 1983. Cuticular proteins of the silkmoth, *Antheraea polyphemus. Insect Biochem.,* 13:665-676.

Sridhara, S. 1985. Evidence that pupal and adult cuticular proteins are coded by different genes in the silkmoth, *Antheraea polyphemus. Insect Biochem.,* 15:333-339.

Stiles, B. 1991. Cuticle proteins of the boll weevil, *Anthonomus grandis,* abdomen: Structural similarities and glycosylation. *Insect Biochem.,* 21:249-258.

Stiles, B., and R.A. Leopold. 1990. Cuticle proteins from the *Anthonomus grandis* abdomen: Stage specificity and immunological relatedness. *Insect Biochem.,* 20:113-125.

Sugumaran, M. 1988. Molecular mechanisms for cuticle sclerotization. *Adv. Insect Physiol.,* 21:179-231.

Truman, J.W. 1978. Hormonal release of stereotyped motor programmes from the isolated nervous system of the Cecropia silkmoth. *J. Exp. Biol.,* 74:151-174.

Truman, J.W. 1985. Hormonal control of ecdysis, pp. 109-151, in G.A. Kerkut and L.I. Gilbert (Eds.), *Comprehensive Insect Physiology, Biochemistry and Pharmacology,* Pergamon Press, Oxford.

Truman, J.W., and L.M. Riddiford. 1970. Neuroendocrine control of ecdysis in silkmoths. *Science,* 167:1624-1626.

Truman, J.W., P.H. Taghert, P.F. Copenhaver, N.J. Tublitz, and L.M. Schwartz. 1981. Eclosion hormone may control all ecdyses in insects. *Nature,* 291:70-71.

Turnbull, I.F., and A.J. Howells. 1982. Effects of several larvicidal compounds on chitin biosynthesis by isolated larval integuments of the sheep blowfly *Lucilia cuprina. Aust. J. Biol. Sci.,* 35:491-503.

Turnbull, I.F., and A.J. Howells. 1983. Integumental chitin synthase activity in cell-free extracts of larvae of the Australian sheep blow-fly, *Lucilia cuprina* and two other species of Diptera. *Aust. J. Biol. Sci.,* 36:251-262.

Wigglesworth, V.B. 1945. Transpiration through the cuticle of insects. *J. Exp. Biol.,* 21:97-114.

Wilkin, M.B., M.N. Becker, D. Mulvey, I. Phan, A. Chao, K. Cooper, H.-J. Chung, I.D. Campbell, M. Baron, and R. MacIntyre. 2000. *Drosophila* dumpy is a gigantic extracellular protein required to maintain tension at epidermal-cuticle attachment sites. *Curr. Biol.,* 10:559-567.

Yarema, C., H. McLean, and S. Caveney. 2000. L-Glutamate retrieved with the moulting fluid is processed by a gluamine syntetase in the pupal midgut of *Calpodes ethlius. J. Insect Physiol.,* 46:1497-1507.

Young, H.P., J.K. Larabee, A.G. Gibbs, and C. Schal. 2000. Relationship between tissue-specific hydrocarbon profiles and lipid melting temperatures in the cockroach *Blattella germanica. J. Chem. Ecol.,* 26:1245-1263.

Žitňan, D., and M.E. Adams. 2000. Excitatory and inhibitory roles of central ganglia in initiation of the insect ecdysis behavioural sequence. *J. Exp. Biol.,* 203:1329-1340.

Žitňan, D., T.G. Kingan, J.L. Hermesman, and M.E. Adams. 1996. Identification of ecdysis-triggering hormone from an epitracheal endocrine system. *Science,* 271:88-91.

5 Hormones and Development

CONTENTS

PREVIEW

Insects have an external skeleton; and as they grow, it becomes too small. Consequently, all insects periodically secrete a new, more flexible exoskeleton that they can "grow into" inside the old one, and then shed (molt, ecdyse) the old skeleton. The majority of insects also metamorphose into an adult form at the last molt. Molting and metamorphosis are under the control of hormones, with the brain as a master control gland. A few brain neurosecretory cells (NSC) secrete prothoracicotropic hormone (PTTH) at an appropriate time in each instar to set in motion further hormonal and physiological events necessary for molting. The cues that stimulate NSC to secrete PTTH have been identified in only a few insects, but at least three types of stimuli are known in different insects, including an environmental stimulus (cold exposure in a diapausing insect), attaining a certain weight or size, and stretching of the abdomen in response to a large blood meal. PTTH passes down the axons of the secreting NSC to the corpora cardiaca, small paired (and sometimes fused) masses of tissue of ectodermal origin just behind the brain. In the tobacco hornworm and possibly other Lepidoptera, NSC axons terminate in the corpora allata, small paired structures just posterior to the corpora cardiaca. PTTH is released from the corpora cardiaca (or corpora allata in some or possibly all Lepidoptera) and picked up by the circulating hemolymph. The corpora cardiaca and corpora allata are neurohemal organs where neurosecretions are passed into the hemolymph. PTTH binds to receptors on the outer cell membrane of the prothoracic glands, and adenyl cyclase on the inner side of the cell membrane is activated. Adenyl cyclase converts ATP into cAMP, the second messenger that sets in motion the cascade of reactions resulting in synthesis of ecdysone from cholesterol or one of the C28 or C29 plant sterols. Ecdysone is not stored in the PGL but is secreted into the hemolymph as it is produced. It generally is considered a prohormone, and a 20-monooxygenase enzyme (present in many tissues but not in the prothoracic gland cells) requiring cytochrome P450 rapidly converts ecdysone into the active hormone 20-hydroxyecdysone by adding the hydroxyl group at the C-20 position in the β-configuration (which is the rationale for the older name of β-ecdysone for 20-hydroxyecdysone). 20-Hydroxyecdysone is the molting hormone, although it cannot be absolutely concluded that ecdysone itself does not have hormonal activity. Several molecular structures similar to ecdysone and to 20-hydroxyecdysone with hormonal activity are known from different insects, and frequently all the steroid hormones are described as ecdysteroids. Receptors on the epidermal cells are the targets for ecdysteroids in immature insects. A number of actions are initiated by ecdysteroids, including mitosis and cell division of epidermal cells, apolysis or separation of the old cuticle from the cells, secretion of molting fluid, and secretion of a new cuticle. Later, when holometabolous insects are about to pupate, many tissues express ecdysteroid receptors and become targets for reorganization into pupal, and finally adult, structures. Juvenile hormone (JH) is secreted in each instar prior to the peak of ecdysteroid secretion, and modifies cuticle secreted so that an additional juvenile type cuticle is secreted. When the insect is large enough to pupate, JH is present only in very small quantities, and the ecdysteroid molting hormone then causes a pupal cuticle to be secreted, with appropriate changes in various internal tissues as well. Subsequently, ecdysteroid secretion with little or no JH allows the epidermal cells to secrete an adult cuticle, and internal organs and tissues are also reorganized to reflect the adult stage. Each time ecdysteroids are secreted, PTTH is secreted first; and in the immature stages; JH is also secreted ahead of the ecdysteroid peak. The exact stimuli and controls upon secretion of JH still are not well defined, but nervous control is believed to be a major influence. Ecdysteroids act at the gene level by regulating or modifying expression of genes. The hormone binds to a receptor in the nucleus, and zinc fingers on the receptor bind the receptor-hormone complex to DNA. Several different receptor isoforms are known, and expression and number of receptors on the cell surface may be one of the ways that some cells respond to ecdysteroid while others do not respond, or respond only at certain times, such as at pupation. JH may also act at the gene level, but the evidence is not so clear as that for ecdysteroids.

5.1 INTRODUCTION

Two critical and important physiological events in the life of insects are **molting** and **metamorphosis.** All insects molt periodically in order to grow, and all but a very few go through either gradual (no pupal stage) or complete (with pupal stage) metamorphosis to become an adult. How are these events in the life of all insects controlled? Molting and metamorphosis are not rapid changes in the same sense that response to many other daily encountered stimuli cause rapid movement away from, or attraction to, the source of the stimulus. The nervous system controls the latter type of rapid responses, but the hormonal system is better suited to control the slower physiological and biochemical changes requiring sustained stimulation needed in molting and metamorphosis. Nevertheless, as in vertebrates, the nervous system exerts control by secretion of neurohormones (Gilbert et al., 1988), and by nervous feedback over many and possibly all endocrine functions. Thus, nervous control of endocrine function antedates the split between vertebrate and invertebrate lines of evolution.

This chapter describes endocrine controls of growth, molting, and metamorphosis. Many other functions are under endocrine regulation, and will be described in the appropriate subject chapters. Additional details on developmental hormones and hormones regulating other functions in insects can be found in books by Raabe (1982), Downer and Laufer (1983), Nijhout (1994), Gilbert et al. (1996b), and in recent reviews by Riddiford and Truman (1993), Riddiford (1994), Jones (1995), and Gilbert et al. (2000).

5.2 HISTORICAL BEGINNINGS FOR THE CONCEPT OF HORMONAL CONTROL OF MOLTING AND METAMORPHOSIS

Stefan Kopeč (1917) published that the brain of gypsy moth caterpillars is necessary for successful pupation. His experiments involved surgically removing the brain from some larvae while performing sham surgery (incision made but brain not removed) on control larvae. A high percentage of the sham-operated larvae pupated, while brainless larvae usually failed to form pupae, although they continued to live. Kopeč found that he could also isolate the posterior body region from brain influence by tying a silk ligature tightly around the body at various points posterior to the head. Regions posterior to a ligature failed to show the cuticular changes associated with pupation, while anterior to the ligature the cuticle changed to look more like pupal cuticle. If he removed the brain late in last instars, the brainless larvae pupated anyway, leading Kopeč to suggest that the brain was necessary for successful pupation for only a short period of time. The latter experiments led to the concept of a **critical period**, a time period when the brain is necessary for its hormonal influence to be exerted.

Thus, although the idea that the brain controlled metamorphosis was current during the 1920s and 1930s, not much attention was given to this concept. Experiments by Fukuda (1940, 1944) on *Bombyx mori* led him to the conclusion that a secretion from the prothoracic region was necessary for pupation. Kopeč and Fukuda were each partially correct; both the brain and a gland in the prothorax are now known to be necessary for molting from one instar to the next and for successful pupation and transformation to the adult.

Kopeč and Fukuda were each looking at different halves of a two-step endocrine mechanism regulating molting and metamorphosis. However, it turns out that a third critical step is also involved — the secretion of a third hormone, the **juvenile hormone**, from glands in the head modifies the type of molt. Identification of the **corpora allata** as the source of this third hormone stems from classical extirpation and reimplantation experiments conducted by Wigglesworth on the reduviid blood-feeding bug *Rhodnius prolixus* in the 1930s. Wigglesworth (1936) first called the hormone from the corpora allata an inhibitory hormone. When multiple corpora allata were implanted into last instars of *Rhodnius*, the bugs molted into **supernumerary larvae** rather than changing into adults as expected. In this sense, it did inhibit metamorphosis. Later, Wigglesworth (1940) called the hormone the **juvenile hormone** as it became clearer that it functioned in other insects and generally had a juvenilizing effect rather than a strictly inhibitory effect.

Finally, the dichotomy surrounding the roles of the brain and prothoracic glands was resolved by Carroll Williams at Harvard University in a series of experiments. Williams designed experiments to test the idea that the **brain hormone** might activate the **prothoracic glands** to produce a **molting hormone**. Williams used pupae of a native silkmoth *Hyalophora cecropia* for his experiments. These large pupae have an obligatory pupal diapause in which they survive the winter in the soil and leaf litter. Following a period of cold exposure at 5 to 10°C for at least 6 weeks, pupae will molt into adults after a few weeks at warm temperatures. Williams found that diapausing pupae could be induced to complete development even without chilling when an "active" brain (the brain from a pupa that had been chilled) was implanted. Williams was able to slice these large pupae in half, seal the abdominal half with a glass coverslip and wax, and implant either an active brain, bits of prothoracic gland, or both into the abdomen. Only when both brain and prothoracic gland tissue were implanted did these isolated abdomens metamorphose into adult abdomens with moth scales and adult reproductive structures. Thus, Williams (1947) established that both brain and prothoracic glands were necessary for adult development, and that the secretion from the brain activated the prothoracic glands.

5.3 THE INTERPLAY OF PTTH, ECDYSTEROIDS, AND JUVENILE HORMONE CONTROLS DEVELOPMENT

It is now established that developmental changes such as molting and metamorphosis are under the control of three major hormones: **PTTH** from **brain neurosecretory cells**, **ecdysteroids** from the **prothoracic gland**, and **juvenile hormone** (**JH**) from the **corpora allata**. As described in subsequent chapter sections, there are molecular variants of each of these three hormones; but in the following general discussion, each is used as a generic name.

Overall, the brain is in control. By secreting PTTH, the brain stimulates the prothoracic glands to synthesize and secrete ecdysteroids. Ecdysteroids combine with a receptor protein in the nucleus of cells, and the ecdysteroid-receptor complex binds to DNA and induces transcription of a few master genes. Transcripts from these few genes turn on a cascade of gene activity, ultimately resulting in cell division in epidermal cells, secretion of molting fluid, secretion of a new cuticle, and (depending on the stage) may result in numerous structural changes in morphology and physiology of internal organs such as the nervous system, gut, and reproductive organs. The timing of secretion and quantity of JH at target cells modulates the action of ecdysteroid by influencing the nature of the molt, whether larval-larval, larval-pupal, or pupal-adult (Gilbert et al., 1996a). JH also determines whether major changes will occur in internal organs; usually little or no changes in internal morphology occur between larval molts, but major changes occur during transformation into pupa or adult. Although ecdysteroids act at the gene level, the mode of action by which JH modifies ecdysteroid-induced gene switching leading to molting and metamorphosis remains unknown.

JH is secreted in advance of the rise of ecdysteroid secretion in early instars of hemimetabolous insects, such as *Nauphoeta cinerea* (Figure 5.1A), and falls toward the end of the instar, allowing secretion of another nymphal cuticle. JH titers are very low or not measurable in the last instar, and this appears to be important in allowing the molt into an adult. In locusts, small pre-molt pulses of ecdysteroid secretion near the end of the last instar are important in inducing mitosis in wing pads, initiating growth of future flight muscles, and starting changes in male accessory glands before the molt actually starts. Subsequently, a major pulse of ecdysteroid in the continued absence of JH causes secretion of an adult cuticle, with concomitant changes in internal structure and physiology characteristic of adult locusts.

JH also tends to be high, relative to ecdysteroid, during the early part of an instar in holometabolous insects, such as *Manduca sexta*, and falls only moderately just prior to each molt

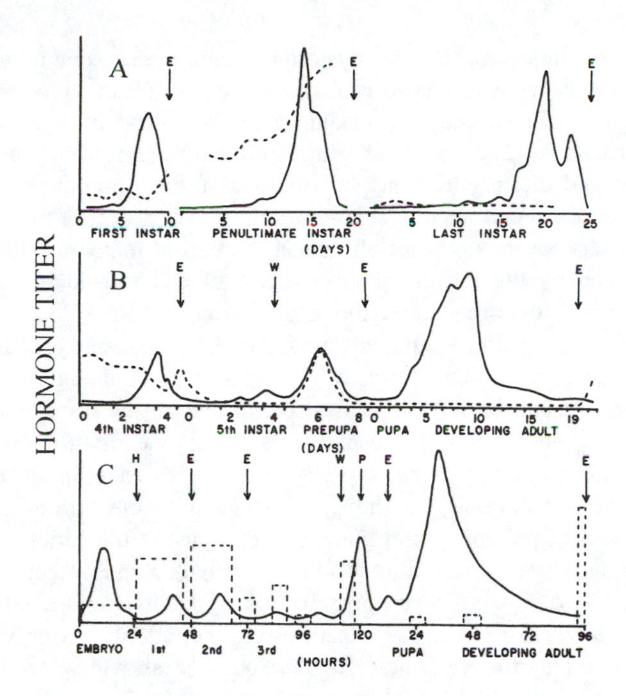

FIGURE 5.1 Ecdysteroid and juvenile hormone titers during development of model insects. A: The cockroach, *Nauphoeta cinerea* with gradual metamorphosis. B: *Manduca sexta*, the tobacco hornworm, with complete metamorphosis. C: *Drosophila melanogaster*, also with complete metamorphosis. In each case, the solid line indicates ecdysteroid titer and the broken line indicates juvenile hormone titer. E, ecdysis; W, wandering larva; P, pupation. (Reproduced with permission from Riddiford, 1994.)

(Figure 5.1B). As a consequence of this hormonal interplay, the epidermal cells secrete another larval cuticle in the early instars. In the last instar, however, JH falls to a low level before the molt, partly in response to a decline in synthesis due to a decreased level of methyl transferase (Bhaskaran et al., 1986), the enzyme that adds the methyl group to JH acid, and to an increase in JH esterase (Roe and Venkatesh, 1990), the enzyme that hydrolyzes JH. As a consequence, the JH level drops below detectable levels early in the last instar.

In *M. sexta,* there is only one peak of PTTH secretion in the penultimate instar, but two peaks in the last instar. The first peak of PTTH induces a small peak of ecdysteroid that reprograms the larval tissues and causes the epidermal cells to become committed to pupal development. This reprogramming peak of ecdysteroid can occur only when the JH titer falls below a critical level because JH appears to act directly on the brain to prevent PTTH secretion (Nijhout and Williams, 1974; Roundtree and Bollenbacher, 1986). Physiological changes in the nervous system cause cessation of feeding, induce wandering behavior, and metabolic changes occur in the fat body. When a larva finds a suitable pupation site, wandering behavior ceases, and a large release of ecdysteroid (in response to another release of PTTH) and a rise in JH cause the epidermal cells to secrete a pupal cuticle. Later, a rise in the level of JH esterase removes JH, allowing the large pupal pulse of ecdysteroid to promote adult development. Each level of ecdysteroid — the rising phase, peak, and falling phase — may be important to the responsiveness of certain cells and tissues in some or even all insects. For example, the DOPA decarboxylase gene in *Manduca* is regulated by decreasing ecdysteroid titer (Hiruma and Riddiford, 1993), and low ecdysteroid levels may play a role in decreasing the titer of JH at metamorphosis (Gu and Chow, 1993). Experimental elevation of ecdysteroid levels in some insects at times when the titer should be low or falling has detrimental effects (see Riddiford, 1994, for an excellent review).

There are six clearly defined pulses of ecdysteroid secretion during development of *Drosophila melanogaster* (Figure 5.1C) (Handler, 1982), each preceded by a pulse of PTTH secretion. JH is secreted in conjunction with each peak of ecdysteroid except the last one in which the pupa molts to the adult. The first pulse of ecdysteroid in *D. melanogaster* occurs about 10 h into embryonic development. A second pulse of ecdysteroid during the 1st instar initiates secretion of a larval cuticle and the molt into the 2nd instar. The third pulse of ecdysteroid causes the secretion of another larval cuticle and molting into the 3rd (the last) instar. In contrast to the situation in a hemimetabolous insect, JH is produced in *Drosophila* in the early part of the last instar, but it falls late in the instar to a very low or nondetectable level. As part of many physiological and morphological changes induced by the fourth secretory pulse of ecdysteroid in the last instar, body-shortening and pupariation to a prepupa are the most obvious. About 12 h after pupariation, a fifth pulse of ecdysone induces secretion of a pupal cuticle. Finally, a sixth and broad pulse of ecdysteroid is secreted in the presence of very low levels of JH in the pupa, and adult development begins.

Exogenous JH administered prior to a critical period late in the penultimate or final instar of Hemiptera, Coleoptera, and Lepidoptera usually results in one or more supernumerary molts (Riddiford, 1996). For example, in *Tenebrio molitor,* two genes encoding proteins specific to the adult cuticle are expressed normally in the pupal stage with the falling ecdysteroid titer. JH application to the newly ecdysed pupa prevents the appearance of the cuticle-specific proteins and a second pupa is formed instead of an adult (Riddiford, 1996). Some insects, such as Diptera and Siphonaptera, are unable to undergo supernumerary molts, and exogenous JH applied at a critical time late in the last instar usually results in death of the individual. Not only does JH interact in some unknown way to modify the type of cuticle secreted, there is evidence that JH can act directly on the brain to inhibit secretion of PTTH (Nijhout and Williams, 1974; Roundtree and Bollenbacher, 1986), and depending on the stage of *M.sexta*, JH can both stimulate and inhibit synthesis of ecdysteroids by the prothoracic glands (Gilbert et al., 1996a). For example, in the lepidopteran *M. sexta*, there are two pulses of PTTH in the last instar: the first is a small pulse and the second is a larger, more sustained pulse. The first small peak of PTTH can occur only when the JH titer falls below a critical level. A number of ideas have been advanced as to how JH might specify the nature of a molt, and more than one may be correct. For example, one possibility is that the JH-JHreceptor interacts with some transcription factors important in the regulation of continuing larval gene transcription during the intermolt. Another possibility is that JH might be involved in protein-protein interactions that stabilize chromatin configuration. When JH is low or absent, ecdysteroid may be able to destabilize the chromatin and open regions for new gene expression. Not all JH action may be nuclear; JH might act on post-translational processing or on translation (Nijhout, 1994; Riddiford, 1994; Riddiford, 1996).

Although the interplay of ecdysteroid and JH holds the fate of cells and tissues, not all cells respond in the same way or at the same time. Imaginal disks that give rise to adult structures do not grow in synchrony (Riddiford, 1994). In *Drosophila* and in tephritid fruit flies, the eye disks that make the compound eyes of adults are large and contain many cells even in the 1st instar, whereas the leg disks are too small to be located easily in early instars, but the latter grow rapidly in the last instar. Some cells and tissues, such as those in larval organs, are destined to die as the imaginal disks replace them with adult structures. Application of exogenous JH within the critical period, when some cells and tissues are still sensitive to its influence while others are not, can cause mosaic insects (Wigglesworth, 1940) that have a mixture of morphological and physiological characteristics representing two stages.

5.4 BRAIN NEUROSECRETORY CELLS AND PROTHORACICOTROPIC HORMONE: PTTH

5.4.1 SOURCE AND CHEMISTRY

Although the hormone from the brain that induces prothoracic glands to secrete molting hormone was initially called **brain hormone**, the more appropriate name is the **prothoracicotropic hormone (PTTH)**. This avoids confusion because, in addition to PTTH, the brain produces a number of

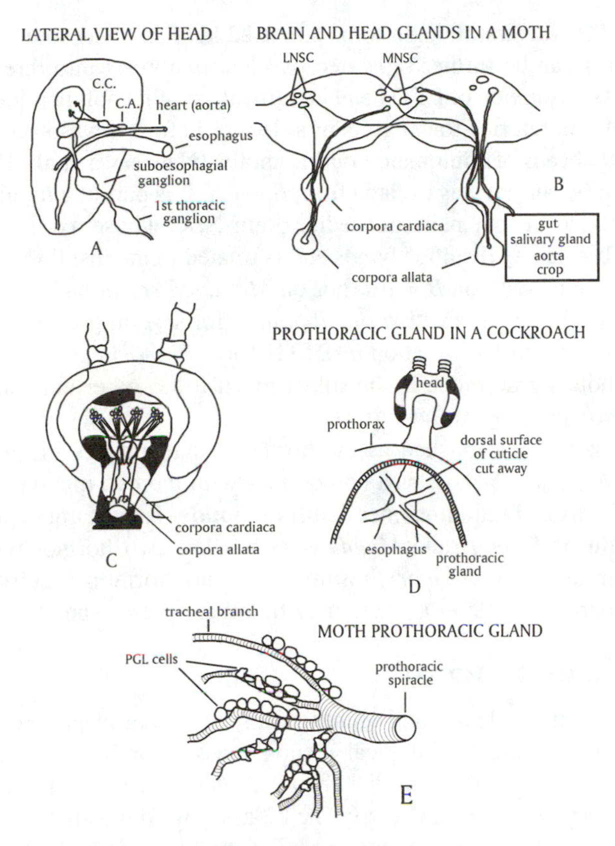

FIGURE 5.2 Representations of neuroendocrine structures in different insect groups. A: Generalized concept of a lateral view of the three regions of the brain (protocerebrum, deutocerebrum, and tritocerebrum) with neurosecretory cells (NSC) in the protocerebrum terminating in the corpora cardiaca and corpora allata. B: Cross-sectional view of the protocerebrum showing general positioning of medial and lateral NSC and their axons to corpora cardiaca and corpora allata typical of Lepidoptera larvae. C: Cutaway view of the head of an American cockroach to show the fused corpora cardiaca and lateral corpora allata lying on top of the aorta, with diagrammatic representation of neurosecretory cells in the protocerebrum of the brain with axons leading to CC and CA. D: The "X" shaped, nearly transparent prothoracic gland typical in an American cockroach nymph. E: Prothoracic glands cells clustered around the larger tracheal branches near the prothoracic spiracle in a Lepidoptera larva.

hormones that regulate other functions in insects. PTTH is produced by **neurosecretory cells (NSC)** in the brain (Figure 5.2A, B, C). In some insects, only a few large NSC in the brain are involved in producing PTTH; only two NSC in each hemisphere of the protocerebrum of *B. mori* were pinpointed as producers with immunofluorescent antibodies to PTTH. PTTH is not released directly from the brain, but passes down monopolar axons (Figure 5.2B) as electron-dense granules to a **neurohemal organ**, the general name for a structure from which neurosecretory hormones are released into the circulating hemolymph. The neurohemal organs that release PTTH in most insects are the bilaterally paired **corpora cardiaca (CC)**, although in some insects these organs are fused. In *M. sexta*, and perhaps in other or even all Lepidoptera, PTTH is released from the paired **corpora allata (CA)** (Agui et al., 1980). The corpora allata are paired bilaterally in Lepidoptera and some other insects, but are fused in some groups of insects. Serving as a release site for hormones made elsewhere does not preclude a neurohemal organ from also synthesizing hormones. The corpora allata, for example, release PTTH made in the brain, and parts of the gland also synthesize and release JH.

PTTH is known to exist in two different molecular sizes: a relatively small polypeptide of about 4.4 kDa and a large form of about 30 kDa. These often are designated as **4 K PTTH** and **30 K**

PTTH. The 30 K PTTH was first thought to be about 22 kDa and was so described in some early reports. The 4 K PTTH can be further separated by electrophoresis into three species designated as **PTTH-I**, **-II**, and **-III**. Isolation of PTTH and identification of its polypeptide molecular structure occupied a number of laboratories for more than a decade before it was successfully isolated and identified from 648,000 heads of adult male *Bombyx* moths (Nagasawa et al., 1984). The 4 K PTTH is known as **bombyxin** because it was isolated from *B. mori*. It is part of a family of related peptides coded by genes expressed in four pairs of medial brain NSC whose axons terminate in the CA. Bombyxin is a dimer bonded by disulfide bonds and is related to the insulin family of polypeptides. Bombyxin does not have activity on *B. mori,* nor on *Manduca sexta*, but does have molt-inducing activity in the bioassay with *Samia cynthia ricini*. *Samia*, for several practical reasons, was usually used as the bioassay animal in the isolation of PTTH form *B. mori* extracts. It remains unclear if bombyxin is a molt-inducing hormone in the silkworm, if it has other unidentified activity, and if it is an artifact of the preparation of the extracts.

The 30 K PTTH is now established as the hormone that induces the prothoracic glands to produce ecdysteroid hormone. It exists as a homodimer with the two monomer units **glycosylated** and held together by inter- and intramonomer **disulfide bonds** (Kawakami et al., 1990). It has been isolated from the brain of *B. mori* and *Manduca sexta*. The two hormones are not particularly similar to each other, and the *B. mori* hormone has no hormonal activity in *M. sexta* or *D. melanogaster* (Gilbert et al., 1996a). There may be other species-specific PTTHs.

5.4.2 Bioassay for PTTH Activity

Each step in any successful isolation and identification of a natural product must be monitored with an assay. Initially, at least, a biological response assay, or **bioassay**, is used because the chemistry is usually unknown. Even when the chemical identity is known, the biological response is usually more sensitive than a chemical assay. The bioassay used in isolating PTTH was induction of adult development in a brainless (**Dauer**) pupa of another silkmoth, *S. cynthia ricini,* after injection of active material. Dauer pupae were prepared by surgical removal of the brain from newly pupated *S. cynthia ricini*. This rendered them incapable of producing PTTH, of course, but left them alive for long periods of time in a suspended state of development (Dauer is a German word that roughly translates as "continuing without change"). These pupae needed a stimulus to the prothoracic glands to cause them to produce ecdysone in order to continue toward adult development. Even with exogenously added PTTH to stimulate the prothoracic glands, these pupae could not produce a new brain in the developing adult, so of course, they would not be normal adults. Often they could successfully metamorphose into an adult, sometimes needing a little help from the investigator to completely get loose from the pupal cuticle. For assay of PTTH, however, complete development is not necessary and the wait is undesirable; it is enough to see the beginning secretion of an adult cuticle, and particularly wing cuticle, as an indication that a preparation contains PTTH activity. In another assay, the sample to be assayed is added to a culture of *Samia* prothoracic glands. Approximately 1×10^{-11} *M* PTTH, equal to about 40 ng of PTTH, activates the glands to produce ecdysone, which can be measured in several ways.

5.4.3 Stimuli for Secretion of PTTH

What stimuli induce the brain to secrete PTTH and initiate the molting process? The answer is only partially known for a few insects, but those examples illustrate the diversity, adaptiveness, and complexity that characterize insect biology. In addition, they serve to stimulate ideas for investigating other possible control mechanisms in insects. In the known examples the brain gets its cue from (1) the filled gut activating **stretch receptors** that send a message to the brain; (2) from some measurement of having attained a **critical size**; and (3) from cold exposure, and probably other **environmental stimuli**, that can stimulate the brain. An example of each of these three mechanisms is described in the following chapter subsections.

5.4.4 PTTH Secretion after Brain Activation by Stretch Receptors

Activation of stretch receptors in the gut and abdomen as a stimulus for PTTH secretion was first defined from *R. prolixus*, a blood feeding reduviid bug. *Rhodnius* typically takes one large blood meal (if allowed to feed successfully to repletion) in each instar. It spends the rest of the instar digesting the meal and getting ready to molt into the next instar. By decapitating insects at various times after a blood meal and sealing the wound with wax, Wigglesworth and subsequent workers showed that *Rhodnius* needed to retain its head (i.e., the brain) for about 3 days after feeding in the 4th instar, and for about 5 to 7 days in the last (5th) instar for a successful molt to be initiated. Thus, PTTH secretion was not an instantaneous event, but one that had to continue for a critical period of time. Apparently, the prothoracic glands need a sustained stimulus in order to secrete enough ecdysone to cause epidermal cells to begin to produce a new cuticle. The period during which the brain must secrete PTTH is called the **critical period** and was defined by Nijhout (1994) as "the point in time at which half the animals are able to complete a normal molt in the absence" of the brain. The duration of the critical period probably varies in different insects, and as evidence from *Rhodnius* suggests, in different stages of the same species. Clearly, the critical period cannot be as long in housefly pupae, which emerge as adults in about 3 days, as it is in *Rhodnius*. That the stimulus is stretching, and not nutritional, was shown by giving a non-fed *Rhodnius* an enema of saline. The saline enema stretched the gut and initiated the process of molting, while an intermittent series of small blood meals that never really stretched the abdomen did not allow molting to occur. The stretch receptors are **tonic receptors** located in one of the abdominal nerves. These stretch receptors do not adapt, but continue to fire bursts of nerve impulses while pressed against the abdominal wall by the filled gut. Abdominal stretching is also the activator of the brain in another rediviid bug, *Dipetalogaster maximus*, and the receptors are located in abdominal nerves that became stretched up to 1½ times the normal length during feeding (Nijhout, 1984). Stretching is important in the milkweed bug, *Oncopeltus fasciatus,* but other undefined factors also influence molting. An injected saline enema can induce molting in milkweed bugs only after a sharply defined **critical weight** is attained (Nijhout, 1979).

5.4.5 Gated PTTH Secretion in Tobacco Hornworm

Attainment of a **critical body mass** or weight by *Manduca sexta* larvae is necessary for PTTH secretion. Secretion of PTTH occurs only during a well-defined temporal window, or gate, near the beginning of the **scotophase** (dark period) after a larva reaches a critical weight. Approximately one third of 4th instars reared at 25°C with a 12:12 L:D cycle reach the critical size necessary to secrete PTTH about 36 h into the 4th instar. Secretion of PTTH in these larvae begins in the first hours after the beginning of thc scotophase; they are designated as **Gate I larvae**. The remaining two thirds of the larvae that have not grown to the critical size to make Gate I must wait 24 h, even if they reach the critical size before the next scotophase. They cannot begin to secrete PTTH until the scotophase, and these larvae are described as **Gate II larvae**.

The stimulus for PTTH secretion is probably nervous, and may be due to stretching of some part of the body, although this has not been proven. The secretion of PTTH is clearly related to internal physiology and the growth rhythm of the larva. Release of PTTH activates the prothoracic glands to produce ecdysone, and larvae molt into the 5th instar at the beginning of a photophase about 50 h after Gate I or II. During the 50 h between PTTH secretion and ecdysis, the epidermal cells are active in cell division, apolysis occurs, and new cuticle secretion begins as some of the old cuticle is digested.

5.4.6 Secretion of PTTH after Brain Activation by Cold Exposure

The brain is "made ready" to secrete PTTH by **exposure to low temperature** for a required minimum of days in a diapausing larva of the *cecropia* silkmoth, *H. cecropia*, but secretion of

PTTH begins only some weeks after return of pupae to room temperature. How the cold activates or primes the brain has not been elucidated. Carroll Williams, who discovered this process in *cecropia*, and his students took advantage of the need for chilling by keeping cecropia diapausing pupae that were collected in the late summer in the refrigerator until needed for experiments. The requirement for a prolonged period of chilling to activate the brain to produce PTTH, followed by a period of warm exposure, is clearly adaptive for *H. cecropia*, which must get through the winter as an inactive pupa in the soil or leaf litter, and emerge as an adult with the coming warm weather of late spring. Activation of the brain after only a few nights of cold exposure, followed by a few warm days, could have the moths emerging in the late fall, long before there would be any new leaves on the host trees for its caterpillars. Probably, many other insects that must pass a period of dormancy experience similar environmentally induced stimuli.

5.4.7 Regulation of Tissue and Hemolymph Levels of PTTH

Tissue and hemolymph levels of PTTH are currently difficult to measure. The availability of very sensitive radioimmunoassay (RIA) techniques should allow more data to be accumulated on regulation; but at present, little is known about how the hemolymph and tissue levels of PTTH are regulated. Probably there are enzymes, such as proteinases, that degrade PTTH that is not bound to a receptor at the target tissue (prothoracic glands), and tissue sequestration and excretion may occur.

5.4.8 Mode of Action of PTTH

PTTH binds with a **receptor** at the outer surface of **prothoracic glands** (Figure 5.3) and sets in motion a cascade of reactions resulting in synthesis of ecdysteroid hormone. One of the first actions is elicitation of an increase in **cyclic adenosine monophosphate** (**cAMP**) in the cells (Smith et al., 1984). Cyclic AMP, a **second messenger**, is produced by **adenylate cyclase** acting upon **ATP** inside the cell. At least part of the function of cAMP is to regulate Ca^{2+} ions in cells and promote the conversion of inactive cellular protein kinase *b* into active protein kinase *a* (Smith, 1993; Smith et al., 1985, 1986, 1993). **Kinases** are enzymes that are involved in phosphorylation reactions, a process that often activates proteins and enzymes. Activation of protein kinase activity in PTTH-stimulated PGL reaches a maximum in about 5 min (Smith et al., 1986), correlating well with rapid generation of cAMP. There is some evidence that PTTH may modulate or influence the growth of the prothoracic glands and may play a role in regulating ecdysone synthesizing enzymes (Gilbert et al., 1996a).

5.5 THE PROTHORACIC GLANDS AND ECDYSTEROIDS

The **prothoracic glands** secrete ecdysone, or a closely related compound. The glands were described by Lyonet in 1762, who apparently had no idea they were involved in molting or metamorphosis. Tomaya redescribed the glands in 1902 in silkworm larvae, and Ke suggested the name prothoracic glands about 1930. Cells of the prothoracic glands are derived from ectodermal tissue in the embryo. Prothoracic glands have a variety of shapes and names in different insects, and they have been called thoracic glands, or peritracheal glands in some insects, and ventral glands in Ephemeroptera and Odonata. The glands in Lepidoptera consist of loose clusters of cells that are widely scattered in the prothorax, but may even reach into the head region. In *H. cecropia* larvae, the glands consist of loose clusters of large cells (about 47 × 22 μm in 4th instars) scattered along the major tracheal branches near the prothoracic spiracle (Figure 5.2E). The cells receive neurons from the prothoracic ganglion and also from the sub-esophageal ganglion. There are 220 cells in the prothoracic glands of *M. sexta*. In cockroaches, the glands form the figure of an "X" in the prothoracic segment (Figure 5.2D). When activated by PTTH, the prothoracic glands secrete

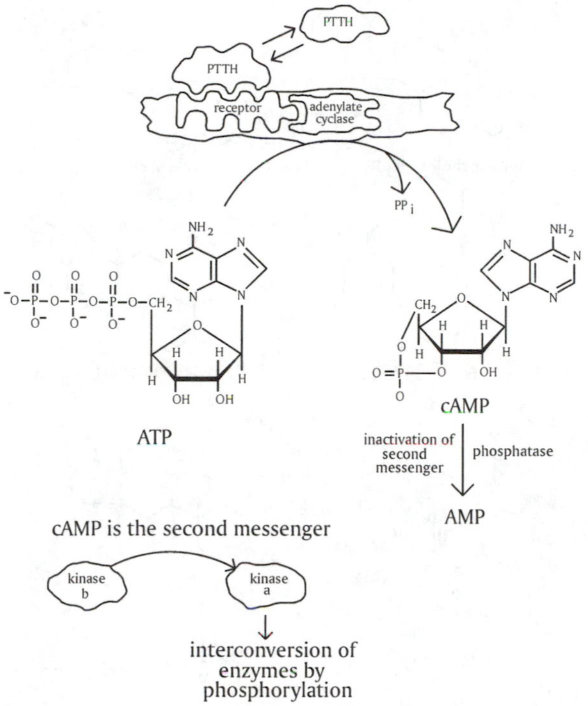

FIGURE 5.3 Conceptual model for PTTH-receptor interaction at the prothoracic gland cell membrane, with activation of adenylate cyclase to form cAMP, a second messenger within the cell cytoplasm. The second messenger sets in motion a cascade of intracellular reactions.

ecdysone or a closely related ecdysteroid. The prothoracic glands in the hemipteran *Rhodnius prolixus* have an endogenous photosensitive circadian oscillator that regulates ecdysteroid synthesis after a blood meal causes the release of PTTH, which acts as an entraining agent (Vafopoulou and Steel, 1999). The prothoracic gland cells of the cockroach *Periplaneta americana* do not have an endogenous circadian oscillator, but rhythmicity of ecdysteroid secretion during the photophase is controlled by secretion of PTTH during the scotophase (Richter, 2001). The brain in larvae of the blowfly *Calliphora vicina* contains extractable prothoracicotropic and prothoracicostatic compounds, and ecdysteroid synthesis seems likely to be under the control of a complex of several factors with interacting and opposing activity (Hua et al., 1997). The prothoracic glands of most insects degenerate during or soon after metamorphosis, and possibly they, like some other tissues, are JH dependent in larval life (Gilbert et al., 1996a). Ecdysteroids are produced by adult female insects, but the ecdysteroids are produced by the follicular epithelial cells of the ovary.

5.5.1 Biosynthesis of Ecdysone

The prothoracic glands sequester **cholesterol** from the circulating hemolymph and convert it into **ecdysone** or a closely related ecdysteroid (Figure 5.4). Cholesterol must be obtained from the diet; it cannot be synthesized by insects. The first step in the synthesis of ecdysone is the conversion of cholesterol to 5,7-dehydrocholesterol. In *Locusta migratoria* and *M. sexta* the enzyme responsible is a microsomal cytochrome P450 monooxygenase requiring NADPH (Kappler et al., 1988; Grieneisen et al., 1993). The 7-dehydrocholesterol is shuttled from the cytoplasm into the mitochondria where oxidative and hydroxylation steps occur that produce a diketol or a ketodiol, depending on the sequence of reactions at carbon-3. A mitochondrial membrane shuttle then moves the resulting trideoxyecdysteroid back to the endoplasmic reticulum for hydroxylation at carbon-25 by a microsomal cytochrome P450 enzyme. Finally, the shuttle returns the steroid to the mitochondrial compartment

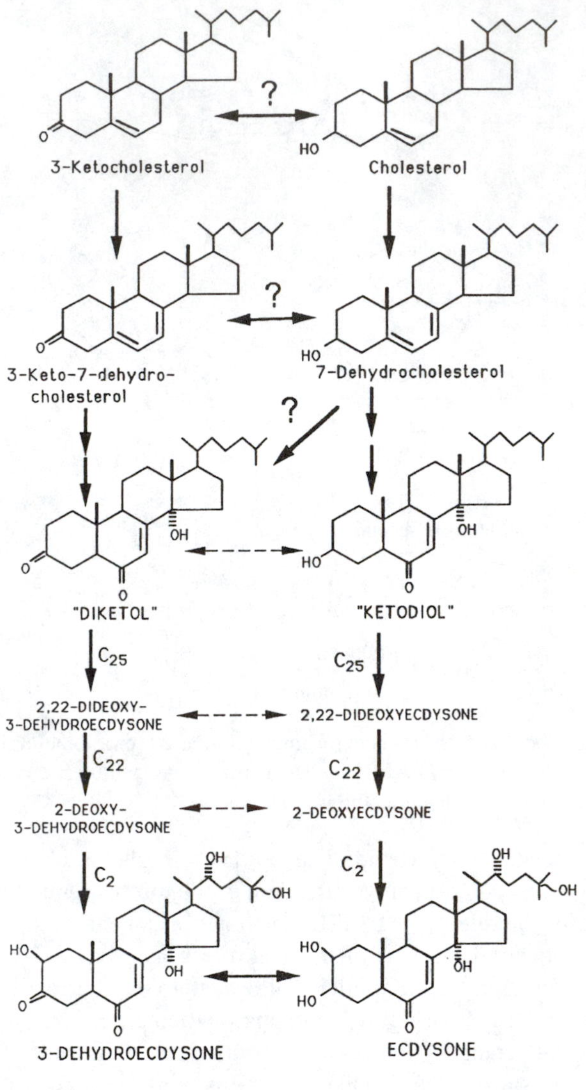

FIGURE 5.4 Possible biosynthetic routes to ecdysone and analogs in the prothoracic glands. (Reproduced with permission from Grieneisen et al., 1993.)

where the C-22 and C-2 hydroxyl groups are added by mitochondrial cytochrome P450 enzymes (Kappler et al., 1988; Grieneisen et al., 1993). The frequent shuttling between the endoplasmic reticulum and the mitochondria requires expenditure of ATP energy, and is facilitated by a sterol carrier protein (Grieneisen et al., 1993). The end products, depending on the chemistry at C-3, are either 3-dehydroecdysone or ecdysone. Both compounds have been found in the prothoracic glands of several lepidopterans and in the Y-organ of some crustaceans. 3-Dehydroecdysone can be converted to ecdysone by reduction of the 3-keto group. In *M. sexta*, 3-dehydroecdysone is released from the prothoracic glands (Warren et al., 1988a, 1988b) and it is reduced in the hemolymph to ecdysone by a ketoreductase (Sakurai et al., 1989).

5.5.2 Conversion of Ecdysone into 20-Hydroxyecdysone

The prothoracic glands do not store ecdysone, but secrete it into the hemolymph as it is made. Ecdysone appears to have low hormonal activity itself, although this is difficult to gauge because

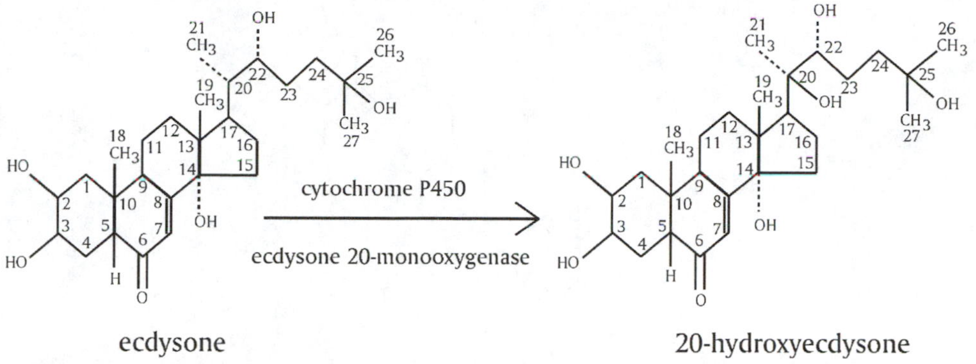

FIGURE 5.5 Structures of ecdysone, the principal prohormone produced by the prothoracic glands, and 20-hydroxyecdysone, the active hormone. Ecdysone is converted to 20-hydroxyecdysone by the enzyme ecdysone 20-monooxygenase, with participation of cytochrome P450. The reaction is characteristic of many tissues, but does not occur in prothoracic gland cells.

it is rapidly converted to **20-hydroxyecdysone** (Figure 5.5; note that ecdysone and 20-hydroxyecdysone each have 27 carbons in their structure) by the **enzyme 20-hydroxymonooxygenase** present in most insect tissues. The Malpighian tubules, gut, and fat body are especially rich in 20-hydroxymonooxygenase. The enzyme, however, is not present in the prothoracic glands. In older literature, 20-hydroxyecdysone is also known as **β-ecdysone**, **ecdysterone**, and **crustecdysone**. 20-Hydroxyecdysone was first called β-ecdysone because the hydroxyl group at carbon-20 is in the β-configuration. The name "crustecdysone" came from its isolation and description from crustaceans at a time when it was not known that its structure was identical to 20-hydroxyecdysone.

5.5.3 Molecular Diversity in the Structure of the Molting Hormone

Although cholesterol is the sterol obtained from the food by carnivorous insects, about half of all insects are plant feeders and thus have access to plant sterols that typically have 28 or 29 carbons in the molecule while cholesterol has 27 (Figure 5.6). Phytophagous insects that have been studied usually dealkylate plant sterols to cholesterol. A few phytophagous insects (only a few have been studied) do not dealkylate the plant sterols, but synthesize a C28 or C29 ecdysteroid molting hormone. **Makisterone A,** or **20-hydroxy-24-α-methylecdysone** (Figure 5.7), is the principal molting hormone of honeybees, and radiolabeled tracers have shown that they synthesize it from the plant sterol **campesterol**, a C28 sterol with the C24 methyl in the α-configuration. Makisterone A also has been detected in embryos of the milkweed bug, *Oncopeltus fasciatus*, and in other hemipterans. A sensitive enzyme immunoassay (detection limit about 3 pg) to detect Makisterone A has been developed (Royer et al., 1993).

The leaf cutting ant, *Acromyrmex octospinosus*, makes **24-epi-Makisterone A** (**20-hydroxy-24-β-methylecdysone**) (Figure 5.8) from a sterol in a fungus they farm. The ants cut leaves, take them into their underground nest, and eat the fungus that grows on the leaves. The fungus contains a C28 sterol with the C24 methyl group in the β-configuration, from which the ants synthesize 24-epi-Makisterone A.

The cotton stainer bug, *Dysdercus fasciatus*, feeds on plant sap and ingests the plant sterol **sitosterol**, a C29 sterol that it uses to make **Makisterone C**, a C29 ecdysteroid, as its molting hormone (Figure 5.9).

Drosophila melanogaster, although not a phytophagous insect, can convert campesterol and sitosterol into cholesterol when reared aseptically on defined diets (3.3 and 8.1% cholesterol in tissues of insects raised on campesterol and sitosterol, respectively). Thus, the biochemical machinery to make the conversion of plant sterols may be widespread in insects (Feldlaufer et al., 1995).

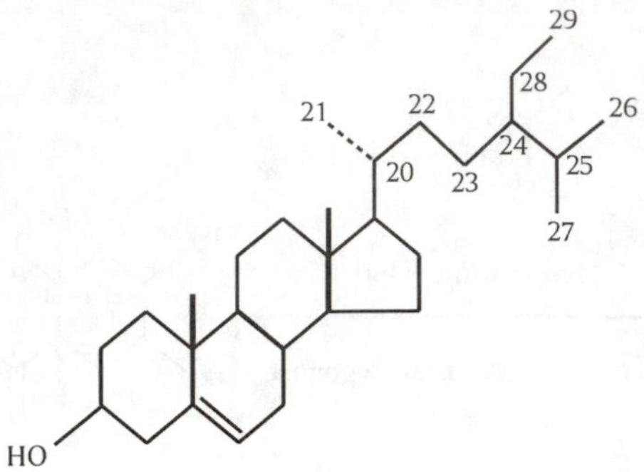

β-sitosterol

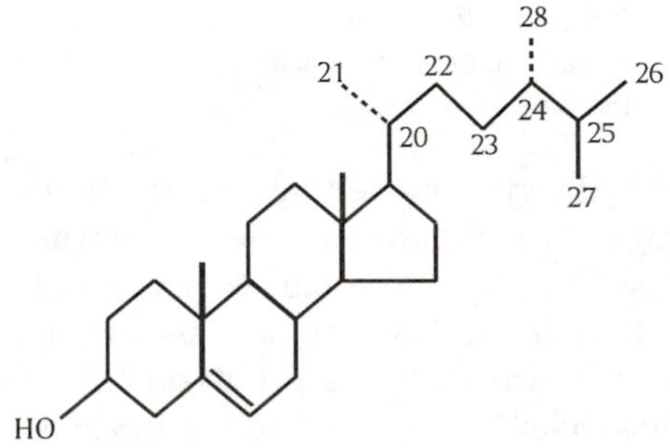

campesterol

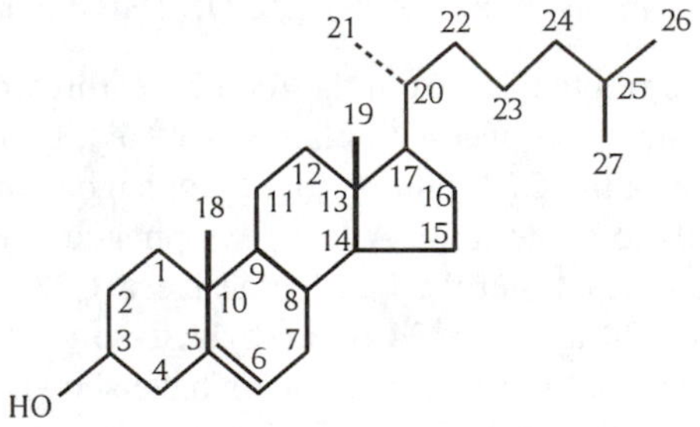

cholesterol

FIGURE 5.6 Two typical plant sterols — β-sitosterol, a 29-carbon sterol, and campesterol, a 28-carbon sterol — and the typical animal sterol, cholesterol, with 27 carbons.

D. melanogaster contains small amounts of 3-dehydro-20-hydroxyecdysone (3-dehydroecdysone) but the principal ecdysteroids in flies are ecdysone and 20-hydroxyecdysone. A small amount of Makisterone A, probably formed from campesterol, is present in pupae. *M. sexta* embryos have small amounts of 20,26-hydroxyecdysone and 26-hydroxyecdysone. There is evidence of the synthesis of ecdysteroids in isolated abdomens of *M. sexta* pupae.

5.5.4 THE *CALLIPHORA* ASSAY FOR ECDYSTEROIDS

The oldest bioassay for ecdysteroids is the ***Calliphora* assay** developed by Fraenkel (1935). In this assay with late 3rd instars of *Calliphora erythrocephala*, a blowfly, a ligature is placed around the body about one third of the body length from the anterior end, so that the brain is isolated from chemical communication with the posterior part of the body. The ligature must be placed before

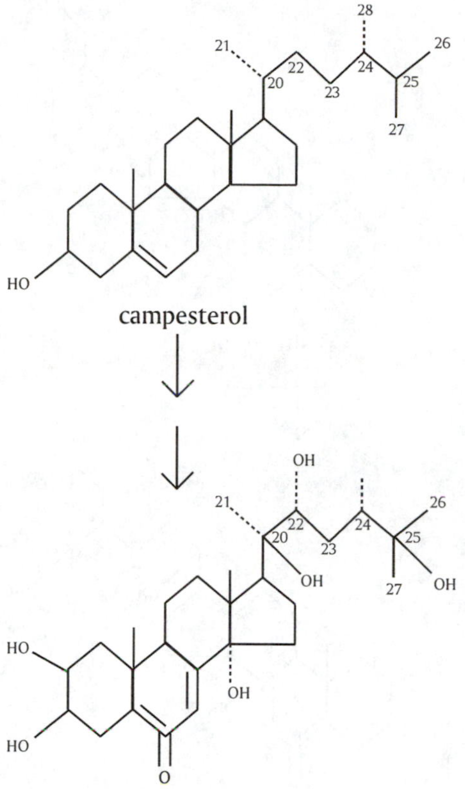

campesterol

makisterone A
20-hydroxy-24-α-methylecdysone

FIGURE 5.7 Conversion of campesterol to form the molting hormone 20-hydroxy-24-α-methylecdysone, or Makisterone A in honeybees, *Apis mellifera* L.

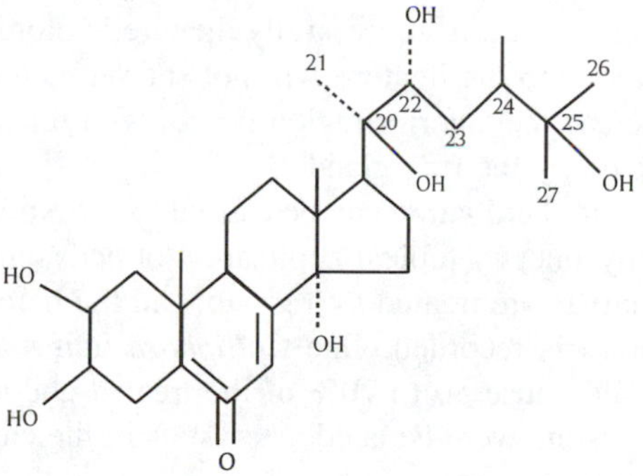

24-epi-makisterone A
20-hydroxy-24-β-methylecdysone

FIGURE 5.8 24-epi-Makisterone (20-hydroxy-24-β-methylecdysone), the molting hormone of leaf cutting ants *Acromyrmex octospinosus*.

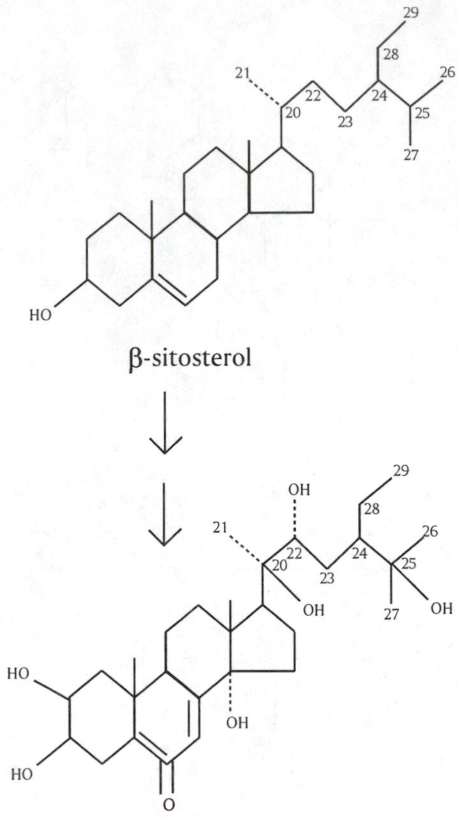

β-sitosterol

makisterone C

20-hydroxy-24-ethylecdysone

FIGURE 5.9 Conversion of β-sitosterol to the molting hormone 20-hydroxy-24-β-ethylecdysone (Makisterone C) in the cotton stainer bug *Dysdercus fasciatus*.

the brain secretes PTTH, or at least before the ring gland secretes the molting hormone. The **ring gland** is a tissue in dipterous larvae that contains both ecdysone-secreting and juvenile hormone secreting cells (Figure 5.10). If larvae are successfully ligatured before secretion of ecdysone, then the portion of the larva posterior to the ligature will not shorten and form a puparium because it receives no ecdysteroid, whereas the anterior region proceeds to form a modified puparial cuticle because of ecdysteroid secreted by the ring gland.

The body region posterior to the ligature can be caused to form puparial cuticle by dipping the abdomen in a solution of ecdysone, by topical application of ecdysone, or by injecting ecdysone. In a typical assay, 20 to 30 larvae are treated as a group, and the percentage forming the puparial cuticle posterior to the ligature is recorded. One *Calliphora* unit was originally defined as that amount of hormone that would cause 50 to 70% of the treated abdomens to pupariate; but after ecdysone and 20-hydroxyecdysone were isolated as crystals in the early 1950s, a *Calliphora* unit was redefined as 0.01 μg pure ecdysone.

5.5.5 Radioimmunoassay for Ecdysone and Related Ecdysteroids

The most widely used method for determining ecdysteroid levels in tissue and hemolymph is a **radioimmunoassay (RIA)** (Borst and O'Connor, 1972). Improved RIA methods for ecdysone are sensitive to about 1 to 10 pg ecdysone, or about 1000-fold the sensitivity of the *Calliphora* assay.

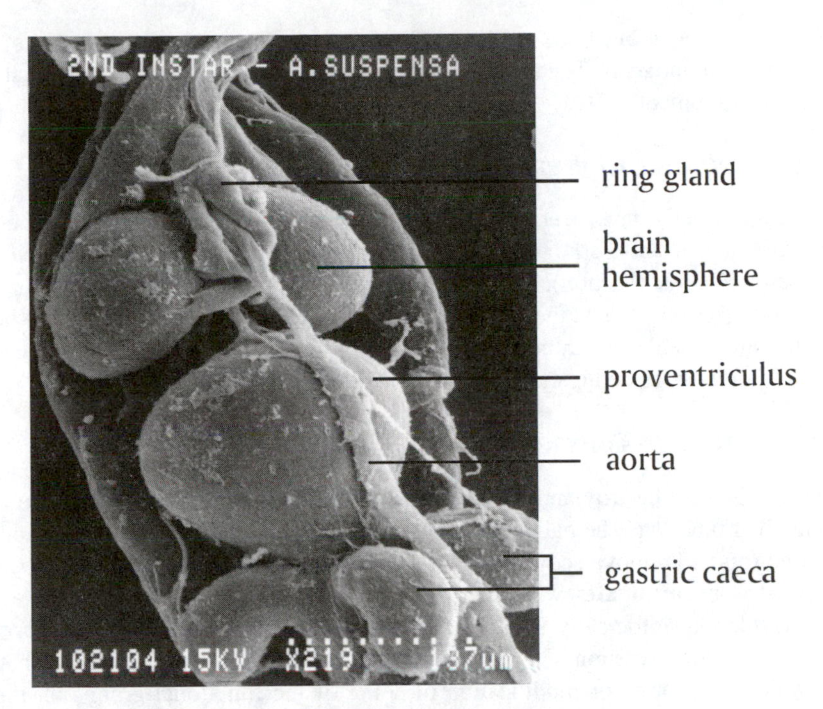

FIGURE 5.10 A scanning electron micrograph of the larval ring gland and brain in a 2nd instar of the tephritid fruit fly *Anastrepha suspensa*. Note the aorta passing through the ring gland.

Although the antibody molecules used in the original RIA method were not completely specific for either ecdysone or 20-hydroxyecdysone, more specific antibody preparations and refined procedures have increased specificity for particular ecdysteroids (Porcheron et al., 1989; Royer et al., 1993). Nevertheless, because there are several steroidal molecules with hormonal activity in insects, analyses are often reported as total ecdysteroids detected rather than attempting to attribute activity to specific ecdysteroids.

5.5.6 Assay by Physicochemical Techniques

Thin layer chromatography (**TLC**), **high performance liquid chromatography** (**HPLC**), and **gas chromatography** coupled with **mass spectroscopy** (**GC-MS**) are methods that have the advantage of allowing separation of the various ecdysteroids present in a biological sample before quantitation. TLC is currently used mainly as a preliminary clean-up method because its ability to separate closely related steroids is limited. HPLC is probably the most widely used of the physicochemical methods (Lafont and Wilson, 1990; Wainwright et al., 1997), providing good separation of the various ecdysteroids and quantitation. The most definitive method for identification of ecdysteroids is the use of GC-MS or HPLC-MS (Wainwright et al., 1997). The ecdysteroids are too polar to chromatograph directly by GC, however, and therefore suitable derivatives must be made. Several problems can be encountered in derivatization, including difficulty of reaction, uncontrolled chemical changes in the ecdysteroids, and very large molecular weight compounds after derivatization that require high temperatures for successful elution. Successful separation of derivatized ecdysteroids by GC coupled to a mass spectrometer, however, then allows characterization and quantitation from the mass spectrum. A further refinement of mass spectroscopy, known as **coupled mass spectroscopy-mass spectroscopy** (**MS-MS**), involving two tandemly operated mass spectrometers gives even greater resolution. This technique can be coupled with GC or HPLC (HPLC-MS-MS, for example). Mauchamp et al. (1993) used GC-MS-MS and HPLC-MS-MS to demonstrate the application of the technique to ecdysteroid analysis and found Makisterone A and

Makisterone C in the eggs of *D. fasciatus*, confirming a similar analysis by Royer et al. (1993), who used enzyme immunoassay. The most obvious disadvantage of the physicochemical techniques is the cost of the equipment.

5.5.7 Tissues and Cell Cultures Used in Assays

Cultured tissues and cells have been used for many years to assay the activity of ecdysteroids (reviewed by Oberlander and Ferkovich, 1994), and were very important in the elucidation of an ecdysone receptor (see later section). Cell lines from several lepidopterans (Sohi et al., 1995) have been used as model systems to test the ecdysone agonists RH-5849 and RH-5992, two nonsteroidal compounds that are capable of binding to the ecdysone receptor and can induce precocious and incomplete molting when added exogenously to some insects.

5.5.8 Degradation of Ecdysone

There are mechanisms to destroy and remove ecdysteroids rapidly after the hormone has accomplished its intended function. The half-life of most ecdysteroids is very short, a few hours at most. Excretion is one route to remove ecdysteroids from the body, and ecdysteroids have been detected in the fecal wastes as **conjugates** with glucose, glucuronic acid, and sulfates. These conjugates render the molecule physiologically inactive, and although the ecdysteroids are remarkably water soluble (a necessity for excretion via the Malpighian tubules), the conjugates probably further facilitate excretion. Oxidation or modification of some of the functional groups on the molecule also reduces or destroys biological activity. An active enzyme in the larval midgut of *M. sexta* converts ecdysteroids to inactive compounds by 3-epimerization; that is, the conversion of the 3β-hydroxy group to a 3-oxo-group or to a 3α-hydroxy-group (Weirich et al., 1993). Conversion of 20-hydroxyecdysone to 20,26-hydroxyecdysone is also a degradation pathway. These conversion products have only very low molting hormone activity. Remarkably enough, the high-level pulses of molting hormone secretion characteristic of immature insects in preparation for each molt may induce the enzymes necessary for hormone removal, a kind of self-destruct mechanism. There is recent evidence for this in the cotton leafworm, *Spodoptera littoralis*, in which treatment of last instars with ecdysone, 20-hydroxyecdysone, or the ecdysone agonist RH-5849 induced 26-hydroxylase activity (Chen et al., 1994). The induction required new mRNA (gene transcription) and new protein synthesis (gene translation). The degradation pathway (Figure 5.11) was 20-hydroxyecdysone → 20,26 hydroxyecdysone → 20-hydroxyecdysone-26-aldehyde → 20-hydroxyecdysone-26-oic acid (Chen et al., 1994).

5.5.9 Virus Degradation of Host Ecdysteroids

A remarkable adaptation has evolved in a baculovirus, the *Autographa californica* nuclear polyhedrosis virus (AcMNPV), that enables the virus to produce a **uridine 5′-diphosphate (UDP)-glucosyl transferase** that catalyzes the transfer of glucose from UDP-glucose to ecdysteroids, thus inactivating the hormone (O'Reilly and Miller, 1989). Although enzymes of this type typically transfer glucuronic acid to various drugs and carcinogens in vertebrates, glucose is normally transferred in insects. Virus expression of the *egt* gene that controls this enzyme allows the virus to interfere with normal larval development and prevents molting of the host, the fall armyworm, *Spodoptera frugiperda*. The evolutionary advantage gained by the virus is not entirely clear, but as a model it introduces the fascinating possibility that other pathogens and parasites may manipulate the hormonal milieu of their host. Indeed, Palli et al. (2000) demonstrate another case of virus modulation of its host's hormonal titers: larvae of the spruce budworm, *Choristoneura fumiferana* infected with an entomopoxvirus, (*Choristoneura fumiferana* entomopoxvirus, CfEPV) feed and grow in size, but eventually die without metamorphosing, apparently because the infected larvae have increased JH titer and decreased ecdysteroid titer.

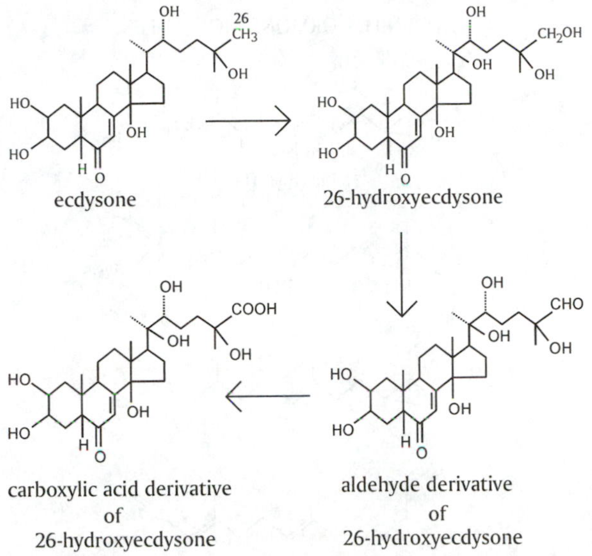

ecdysone → 26-hydroxyecdysone

↓

carboxylic acid derivative of 26-hydroxyecdysone ← aldehyde derivative of 26-hydroxyecdysone

FIGURE 5.11 One of the ways that the molting hormone is inactivated by the cotton leafworm *Spodoptera littoralis*. In this scheme, ecdysone is converted to the molting hormone 20-hydroxyecdysone. Hormone not bound to receptors is converted rapidly to 20,26-hydroxyecdysone with low hormonal activity. 20,26-hydroxyecdysone is further metabolized to the 26-aldehyde derivative, with the aldehyde group subsequently oxidized to a carboxyl group. The latter two compounds do not have hormonal activity and are rapidly excreted. (Modified from Chen et al., 1994.)

5.5.10 Dependence of Some Parasitoids on Host Ecdysteroids

Some parasitoids depend on the host's hormones for their own development and pupation. The parasitoid *Biosteres longicaudatus*, an endoparasite of larvae of the Caribbean fruit fly (a tephritid), typically oviposits into early 3rd instar larvae where the parasitoid 1st instar feeds on host tissues and grows to a critical size ready to molt (Lawrence, 1986). The parasitoid does not produce its own molting hormone, but depends on its host's hormone. The parasitoid does not molt until the host secretes ecdysone to initiate its own larval-to-pupal transformation. The parasitoid then molts to the next instar and continues to feed on the pupa of the fly. Eventually, the parasitoid kills the host by feeding on critical tissues, and then pupates and completes its transformation into an adult wasp inside the puparial shell of the host. *Diapetimorpha introita* is an ichneumonid ectoparasitoid of the fall armyworm *S. frugiperda*, and is capable of producing its own ecdysteroids when grown on artificial diet. Hemolymph ecdysteroid titers were higher, however, in host-reared than in diet-reared individuals, suggesting a significant nutritional role for most effective ecdysteroid biosynthesis in these ectoparasitoids (Gelman et al., 2000).

5.6 THE CORPORA ALLATA AND JUVENILE HORMONES

5.6.1 Glandular Source and Chemistry

Juvenile hormones (**JHs**) are **sesquiterpenoid** compounds produced in the cells of the **corpora allata** (**CA**), which are bilaterally paired structures in Lepidoptera, but are often fused into one mass of tissue in other groups of insects. The CA, derived embryologically from ectodermal tissue, lie beneath the aorta and posterior to the brain and corpora cardiaca. A nerve tract connects the CA with the corpora cardiaca and brain and brain neurosecretory cells. In Orthoptera, Thysanura, and Ephereroptera, there is a nerve tract to the subesophageal ganglion. Five JH variants (six, if

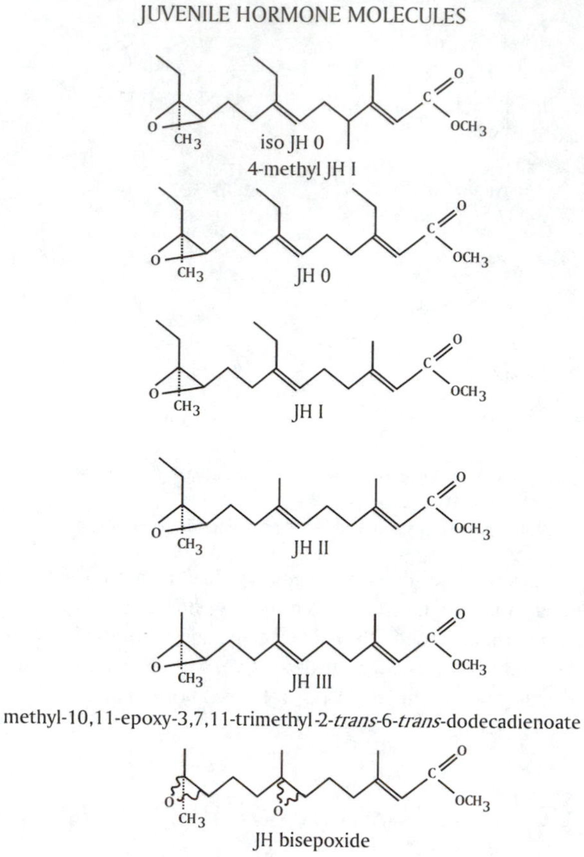

JUVENILE HORMONE MOLECULES

iso JH 0
4-methyl JH I

JH 0

JH I

JH II

JH III

methyl-10,11-epoxy-3,7,11-trimethyl-2-*trans*-6-*trans*-dodecadienoate

JH bisepoxide

FIGURE 5.12 Naturally occurring molecules with juvenile hormone (JH) activity. JH III may be the most common JH of insects. JH I, II, III, JH 0, and Iso JH 0 have been found in some Lepidoptera. JH bisepoxide is synthesized by the ring gland of some Diptera.

one includes methylfarneosate, which is the immediate precursor of the JH molecules and which is secreted by some insects and has JH activity on its own) are known. The structures of **JH I**, **JH II**, **JH III**, **JH 0**, and **iso JH 0** (also called **4-methyl JH I**) are shown in Figure 5.12. Although JH III has most often been found as the principal or only JH molecule in many insects, more detailed analyses with GC-MS (Bergot et al., 1981) have shown multiple JHs in some insects, particularly Lepidoptera. JH III is only detectable as a trace or sometimes not at all in some Lepidoptera (Edwards et al., 1995; Ramaswamy et al., 1997; Park et al., 1998), and JH I and JH II are often the principal JHs. JH I, II, and III are released from isolated CA of 10-day-old *Actebia fennica* moth females, but JH II is the principal JH (Everaerts et al., 2000). JH II is the principal JH in the 4th and 5th instars of the tomato moth larvae *Lacanobia oleracea,* and in the pupa, along with some JH I (Edwards et al., 1995), but Audsley et al. (2000) found that 90% of the JH released by isolated CA is JH II. In addition, **JH II acid** and **JH I acid** are also released from the CA, but Audsley et al. (2000) attribute the acid analogs to the presence of JH esterase activity in the CA and breakdown of already-formed JH, rather than the failure of methylation of farnesoic acid in biosynthesis. It has not been determined if the multiple forms of JH have unique functional roles in all the insects in which they have been found, but in some cases they do have specific functional roles (Gilbert et al., 2000).

Although JH I is the principal JH of early instars of *M. sexta* (Schooley et al., 1984), production of the JH acids seems to be the normal process in the last instar of *M. sexta*, which loses the enzymatic ability to methylate the final step in synthesis of the various JHs (Bhaskaran et al., 1986; Baker et al., 1987). JH acid production continues into the pupal and adult stages. The JH acids have little biological activity, but they can be methylated slowly in imaginal disc tissues (Sparagana et al., 1985), and the slow methylation may be important to overall function of JH and metamorphosis in *M. sexta*, and possibly other Lepidoptera (Gilbert et al., 1996a). Enough evidence has accumulated, however, to show that the JH acids themselves have hormonal activity in some insects (Gilbert et al., 2000).

JH bisepoxide (JHB3) is the principal JH of *D. melanogaster* (Richard et al., 1989a, 1989b; Yin et al., 1995), and occurs also in *Phormia regina* (Richard et al., 1989a), *Calliphora vomitoria* (Cusson et al., 1991), the sheep blowfly, *Lucilia cuprina* (Lefevere et al., 1993), and in four species of mosquitoes (Borovsky et al., 1994). In addition to JHB3, *L. cuprina* also produces some JH III (East et al., 1997).

The JH molecules have chiral centers, and some evidence indicates that the natural enantiomers may be more active and less rapidly degraded than unnatural enantiomers (Peter et al., 1979; King and Tobe, 1993). JH III has a chiral center at carbon 10 and the other JH molecules have chiral centers at C-10 and C-11. JHB3 has chiral centers at C-6, C-7, and C-10.

5.6.2 Assays for JH Activity

The oldest assay is the ***Galleria* wax test**, in which a small hole about 1 mm^2 is cut through the cuticle of a newly molted pupa, a test material is applied, and the wound is sealed with molten wax. It is more effective to administer samples potentially containing JH in an agent, such as peanut oil, that protects it from JH esterases (Gilbert and Schneiderman, 1960). If JH is present in the sample applied to the pupal wound, the adult that emerges about 10 days later has a small patch of pupal cuticle devoid of scales, or with few scales, at the wound site. The size and quality of the pupal patch is scored to give a rough measure of JH present in the test solution. The *Galleria* assay has the disadvantages that it is slow, requiring about 2 weeks to determine if a sample has JH activity; considerable experience is required to obtain reliable data; and biological variability requires that a large sample of pupae must be tested. Currently, most of the determinations of JH activity are performed by selective ion monitoring with GC-MS. It is highly specific for the different JH molecules (Figure 5.13 and Table 5.1) and is a good quantitative method (Bergot et al., 1976, 1981; Teal et al., 2000). An RIA technique for JH is also available (Hunnicutt et al., 1989; Huang et al., 1994; Borst et al., 2000).

A very sensitive method for measuring biosynthesis of JH is based on transfer of the radiolabeled methyl group from methionine to JH during the biosynthesis of JH. This method, however, is not suitable for the measurement of total tissue or hemolymph titer of JH.

5.6.3 Regulation of the Tissue and Hemolymph Levels of JH

The titer of JH in the hemolymph and tissues is a function of biosynthesis by the CA balanced against degradation in the tissues. JH is secreted continuously by CA during larval life, but there are stage-specific fluctuations (Tobe and Stay, 1985). JH is not stored in the CA or in other tissues. In some insects, there is evidence that nervous connections from the brain influence biosynthesis of JH in the CA, either through stimulation or inhibition. These examples generally come from studies of endocrine control of reproduction in adults. Nervous suppression of JH synthesis is found in virgin female cockroaches, *Diploptera punctata*, which normally do not mature oocytes or ovary until mating releases the inhibition. Mating often stimulates the CA to begin JH production, possibly by relieving the inhibition that the nervous system may have on CA activity (Tobe and Feyereisen, 1983). Experimentally severing nerve connections between brain and CA allows oocytes to mature and promotes a cycle of JH secretion without mating. In *Schistocerca* sp., however, severing the nerve connections to the CA causes a decline in JH production.

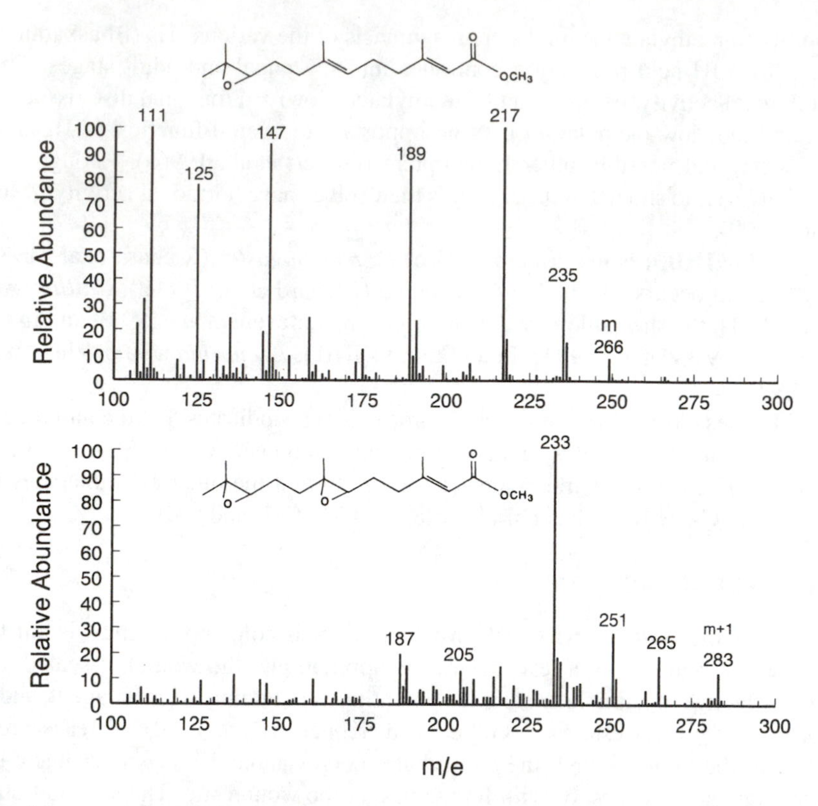

FIGURE 5.13 Mass spectra of JH III (top) and JH III bisepoxide (bottom) extracted from the hemolymph of sexually mature, 12-day-old tephritid fruit fly *Anastrepha suspensa* males. Samples were analyzed in a Finnigan Magnum Ion Trap mass spectrometer interfaced with a Varian Star 3400 gas chromatograph (GC) containing a 30-m × 0.25-mm × 0.1-μm film thickness J&W DB5-MS analytical column. The Ion Trap was operated in chemical ionization mode with isobutane as the reagent gas. A 10-m × 0.25-mm uncoated, deactivated fused silica retention gap column in the GC and a 10-cm × 0.5-mm length of uncoated, deactivated fused silica column in the injector allowed large volumes of sample to be injected without loss of resolution. The conditions of chromatography were as follows: initial injector temperature of 40°C for 30 s; injector temperature increased at 170°C/min to 270°C; initial column temperature 40°C for 5 min; column temperature increased at 5°C/min to 210°C. He carrier gas with linear flow velocity equal to 24 cm/s. GC-MS transfer line temperature was 230°C. Ion Trap operating conditions were as follows: multiplier voltage, 1900 volts; manifold temperature, 130°C; emission current, 16 μa; mass acquisition range, 60 to 350 amu; 1 scan per second. (MS record courtesy of Peter Teal.)

The brain clearly has some control over the secretion of JH by the corpora allata (reviewed by Stay et al., 1994). There are numerous reports of changes in the corpora allata and JH titers in reaction to severing nervous connections between the CA and brain, and after total removal of the brain. Some of the brain's control is likely to be through the intact nerve connections and nerve impulses, but some studies have demonstrated that extracts of the brain also influence CA activity and JH levels. A number of **allatotropins (ATs)** and **allatostatins (ASTs)**, peptides from the nervous system of several different species and from several orders of insects, have been isolated and bioassayed for their effect on JH biosynthesis. ATs stimulate JH synthesis while ASTs inhibit synthesis. In general, these peptides have mainly been studied in adult insects, in which JH is necessary in most insects for female reproduction (reviewed by Stay et al., 1994). Audsley et al. (2000) found, however, that Mas-AT (isolated from *M. sexta* and the only AT that has been completely sequenced) seems to be the principal regulator of corpora allata activity in the larval tomato moth *L. oleracea*. These authors suggest that the larval corpora allata are activated by Mas-AT as

TABLE 5.1
Description of Cleavage Assignments Resulting in Diagnostic Ions Used for Quantitation of JH Compounds

Ion #		Mass to Charge (m/e)	
		JH III	JH III Bisepoxide
1	M + 1		283
	Ion 1 — HOH (from ring cleavage of epoxide)		265
2	Ion 1 — CH_3OH (from methyl ester)	235	251
3	Ion 2 — HOH (from ring cleavage of epoxide)	217[a]	233[a]
4	Ion 3 — CO (from methyl ester)	189	205
5	Ion 4 — HOH (from ring cleavage of second epoxide)		187
6	Ion 1 — $C_2H_4O_2$ (from methyl ester) — C_3H_8O (from epoxy terminus)	147	
7	M — $C_8H_{13}O_2$ (cleavage between C-6 and C-7)	125	
8	$C_7H_{11}O$ (scission between C-6 and C-7 after loss of CH_3OH)	111	

[a] Parent ion.

Data courtesy of Peter Teal.

needed, and that removal of the peptide when JH is not needed stops JH synthesis, with no or little need for Mas-AST to turn off synthesis. Additional information on ATs and ASTs is given in Chapter 7.

Additional agents have been shown to stimulate or inhibit corpora allata in some insects, including neurotransmitters such as dopamine and octopamine; second messenger systems, including cAMP, inositol trisphosphate, diacylglycerol, and calcium signaling; location, kind, and number of potential receptors; and ecdysteroid action (Granger et al., 1996, 2000; Woodring and Hoffmann, 1994; Gilbert et al., 2000).

The main degradative pathways for JH involve specific and nonspecific **JH esterases** (**JHEs**) described from numerous insects, and **JH epoxide hydrolases** (**JHEHs**) reported from some insects. The esterases attack the ester linkage, while epoxide hydrolases open the epoxide ring and create a diol (Figure 5.14). Only one, or both actions, occur in some insects. The metabolic changes not only eliminate all or most of the hormonal activity of the molecules, but the molecules become more water soluble and can be excreted by the Malpighian tubules.

JH is transported through the hemolymph bound to a protective lipoprotein. The proteins are known as **JH binding proteins** (**JHBPs**), and proteins with high and low specific binding have been isolated. Lipophorins are primary JH carriers in hemimetabolous insects (Kanost et al., 1990). In two cockroach species, lipophorin and vitellogenin are the two principal JHBPs (Engelmann and Mala, 2000). In *Manduca sexta,* a 32-kDa JH binding protein has been purified. JH binds in a hydrophobic pocket in the protein, so it is well protected from esterase attack (Touhara and Prestwich, 1992). A major factor contributing to degradation of JH in *M. sexta* is a JH-specific esterase that is synthesized in large amount during the latter part of each larval molt, and in the final instar. Although the JHBPs protect JH from esterase attack, the high titer of esterase just prior to a molt destroys any JH that dissociates. Hydrolysis of undissociated JH continues to pull the equilibrium of bound JH toward dissociation and degradation, allowing the molt to the next instar (Riddiford, 1996). In *L. migratoria*, JHBP protects the natural enantiomer, (10*R*)-JH III, from hemolymph esterase activity better than it protects the unnatural enantiomer, (10*S*)-JH III (Peter et al., 1979, 1983). In *Nauphoeta cinerea* cockroaches, however, natural (10*R*)-JH III is degraded by hemolymph esterase more rapidly than a racemic mixture of the enantiomers (Lanzrein et al., 1993). Thus, fluctuations in the level of JHBP in the hemolymph at different developmental times

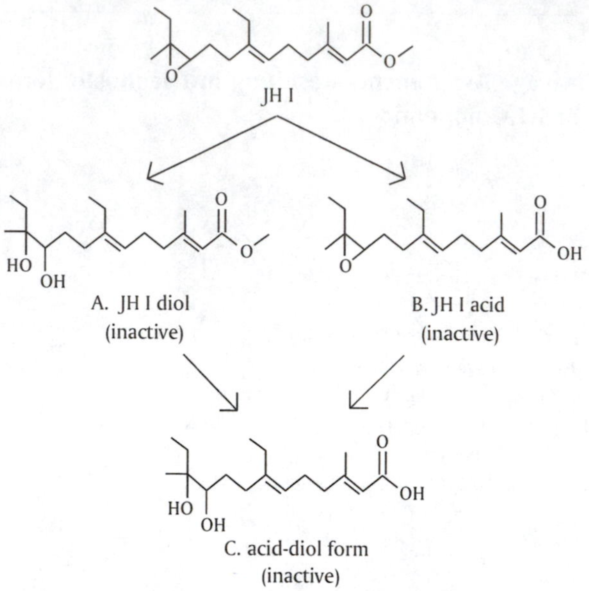

JH I

A. JH I diol (inactive)

B. JH I acid (inactive)

C. acid-diol form (inactive)

FIGURE 5.14 Metabolic pathways for the degradation of JH. The epoxide ring can be opened with hydrolysis and production of two hydroxyl groups, as in A, or the ester group can be hydrolyzed to the free acid as in B. Both A and B are inactive, and either or both can occur in most insects.

are a factor in JH titer. And, although JH titer is influenced by (1) rate of synthesis, (2) rate of breakdown, (3) sequestering in some target tissues, (4) presence of JHBP, and (5) excretion, the two main processes appear to be the rate of synthesis and the rate of breakdown.

JH biosynthesis occurs through an isoprenoid pathway in which acetyl CoA units are used to build farnesyl pyrophosphate, and finally methyl farnesoate. In the final enzymatically controlled step occurring in corpora allata, methyl farnesoate epoxidase, a cytochrome P450 containing monooxygenase, epoxidizes the 10,11-double bond. Photoaffinity labels have been used to label a protein (presumably methyl farnesoate epoxidase) of about 55 kDa in the corpora allata of the cockroach *Diploptera punctata* (Andersen et al., 1995). The label binds to the heme iron of cytochrome P450 and to the hydrophobic substrate binding pocket of the enzyme. These or similar probes can be useful in defining the exact chemistry of the epoxidase enzyme and the final epoxidation step, as well as demonstrating exactly where within the CA cells the reaction occurs.

5.6.4 Insect Growth Regulators and Compounds That Are Cytotoxic to the Corpora Allata

For reasons still not understood, JH is an extremely easy molecule to mimic in all or part of its physiological function(s). Thousands of compounds (about 5000) and extracts of many tissues, including those of vertebrates, have JH activity to varying degrees. Compounds with JH activity are generally called insect growth regulators (IGRs). A few IGRs have enough JH activity (examples are kinoprene, hydroprene, and methoprene) to be used commercially as insecticides. Methoprene is very effective against the late larval stage of mosquitoes and fleas. Kinoprene is most active in Lepidoptera, and hydroprene in Orthoptera.

Some plants contain substances that are active against the corpora allata. For example, chromene compounds, given the trivial names **Precocene I** and **Precocene II** (Figure 5.15), occur in various vegetative parts of *Ageratum houstonianium*, a common bedding ornamental (Bowers et al., 1976). These compounds have a powerful effect on milkweed bugs, *Oncopeltus fasciatus*, and numerous other insects, but they are usually most effective on Hemiptera. The compounds cause 2nd and 3rd

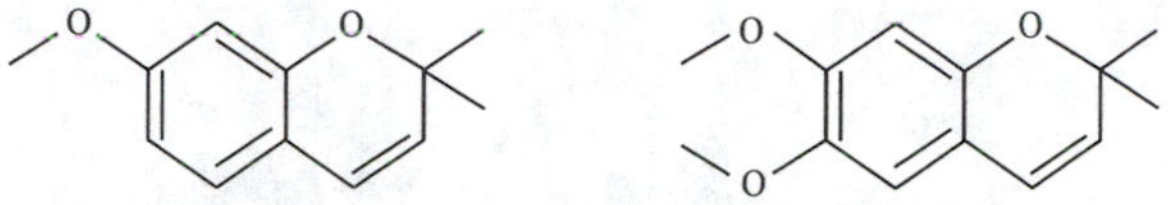

precocene I precocene II

FIGURE 5.15 Naturally occurring compounds precocene I and precocene II discovered in *Ageratum houstonianium* ornamental bedding plants. The compounds cause precocious molting in some insects, most notably Hemiptera.

instars of milkweed bugs to precociously metamorphose into small, imperfectly formed adults. The small adults are incapable of reproducing because the ovaries do not develop. Considerable excitement ensued from this discovery, and scientists in many laboratories around the world synthesized derivatives of the precocenes and/or started searches for other naturally occurring compounds with similar hormonal effects. Neither the precocenes nor derivatives, and no other naturally occurring compounds with similar action, have been commercialized.

The precocenes do not directly antagonize the action of JH at target sites. Treated insects can sometimes be rescued with large doses of JH or an effective JH analog. The compounds have a cytotoxic effect on the cells of the corpora allata, causing them to atrophy and fail to produce JH. Lack of JH allows ecdysteroids to catapult young instars into precocious adult development.

5.6.5 Cellular Receptors for JH

A number of studies have shown that various JH molecules are bound to hemolymph components, and to cytoplasmic and nuclear proteins (reviewed by Goodman and Chang, 1985). A 29-kDa nuclear protein has been isolated from larval epidermal and fat body cells of *M. sexta* with high specificity for binding JH I and JH II (Palli et al., 1990, 1994; Riddiford, 1990; Riddiford and Truman, 1993). This nuclear binding protein is not present in the nuclei of epidermal cells when no high-affinity JH binding sites are present, such as in wandering larvae or in those larvae that are allatectomized. The protein is considered a **putative JH receptor**. It does not appear to be very similar to any of the known DNA-binding protein families, which may reflect the fact that non-arthropods lack a hormone that is comparable to JH.

A JH receptor has been purified from fat body cells of both adult sexes of the cockroach *Leucophaea maderae* (Engelmann, 1995). The receptor is a binding protein of about 64 kDa composed of two 32-kDa subunits. This JH receptor appears to be more related to egg production in the adult than to the development of immature stages. At present, the receptor has only been detected in the last instar and adult, both stages capable of responding to exogenous JH or JH analog by synthesizing vitellogenin (in females) for incorporation into eggs.

5.7 MODE OF ACTION OF ECDYSTEROIDS AT THE GENE LEVEL

5.7.1 Chromosomal Puffs

The **polytene chromosomes** in both *Chironomus tentans,* a midge (Chironomidae: Diptera), and *D. melanogaster* have been important for understanding the mode of action of ecdysteroids. Polytene chromosomes are the result of chromosomal replication without mitosis, and polytene chromosomes in a single cell may consist of up to 2^{13} chromatids (Lezzi, 1996). Alternating bands of condensed and decondensed chromatin occur along the length of the polytene chromosomes, producing the characteristic banding pattern observed under the microscope. Work with *Drosophila* chromosomes was aided greatly by the extensive genetic and developmental background for *D. melanogaster.* The polytene chromosomes (Figure 5.16) in salivary glands of *D. melanogaster* and *C. tentans* show characteristic **puff patterns** (Figure 5.17) that enlarge and regress during development (Becker, 1959; Clever and

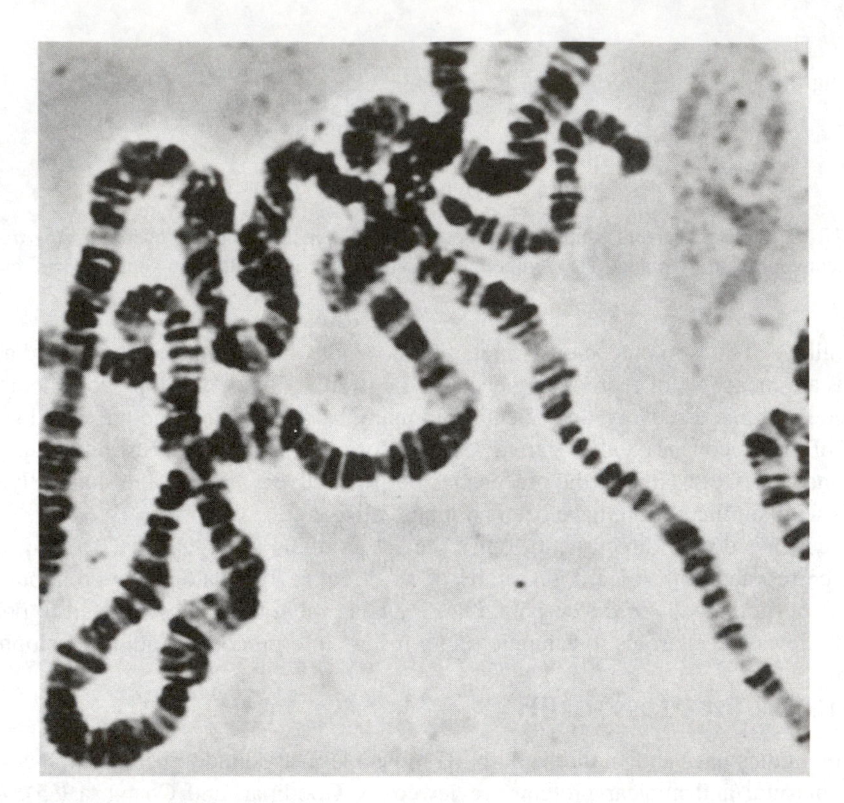

FIGURE 5.16 A polytene chromosome from *Drosophila melanogaster.* (Photograph courtesy of Marie Nation Becker.)

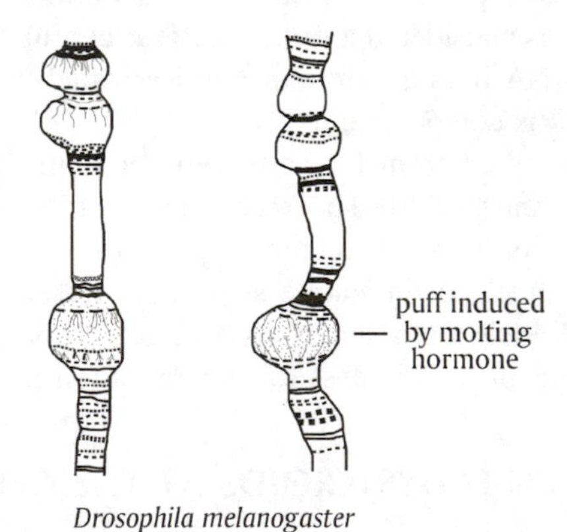

FIGURE 5.17 A drawing of a chromosomal puff from a polytene chromosome of *Drosophila melanogaster* induced by molting hormone. (Reproduced with permission from Becker, 1959.)

Karlson, 1960). The puffs are sections of the DNA (i.e., genes) in the act of transcribing the genetic code to new mRNA. Clever and Karlson (1960) took advantage of the availability of the isolated and identified molting hormone 20-hydroxyecdysone (Butenandt and Karlson, 1954) to show that the hormone injected into last instars of *C. tentans* induces chromosomal puffing within 2 h, and the puff regions are the same as those observed during normal initiation of pupariation. They ventured the seminal suggestion that "the primary effect of ecdysone is to alter the activity of specific genes." Numerous studies of the effect of 20-hydroxyecdysterone on chromosomes of *C.*

tentans and *D. melanogaster* were soon underway, with significant improvement in developing and standardizing the use of organ cultures of *Drosophila* salivary glands and a testable model scheme for how the genes worked (Ashburner, 1972). Within a few minutes after the addition of 20-hydroxyecdysone to a culture medium containing isolated 3rd instar *Drosophila* salivary glands, **six "early" puffs** develop at sites designated as 22B4-5, 23E, 63F, 74EF, 75B, and 74C (Ashburner, 1972). The early puffs grow larger over a period of about 1 hour (h), persist for about 4 h, then begin to regress, and disappear in about 6 h. The six puff regions are the same ones that occur at the initiation of pupariation of late third instars, thus providing satisfying evidence that isolated salivary glands respond in the same way as intact glands in whole insects. A few hours after the early puffs regress, up to 100 puffs can be observed in other chromosomal regions; these are called **"late" puffs**, and have been divided into "early-late" and "late-late" puffs in reference to their time of puffing.

The puffed regions of polytene chromosomes are indicative of a loosening of the DNA strands of polytene chromosomes, and **transcription** of genes into mRNA. Experiments showed that the six early puffs appeared when cycloheximide, a general inhibitor of protein synthesis, was added with 20-hydroxyecdysone to the culture medium. These results indicated that no new protein synthesis was required to induce the early genes to transcribe new mRNA. Ashburner and Richards (1976) suggested that 20-hydroxyecdysone acted directly on the early genes, probably, they speculated, after combining with a receptor protein.

The early puffs begin to regress after about 4 h, and many new puffs appear. These actions occur even in the continued presence of 20-hydroxyecdysone, but regression of early puffs and the development of late puffs can be prevented in the presence of cycloheximide. This indicated that new protein synthesis was necessary for gene regression and appearance of late puffs. Utilizing the idea of negative feedback, Ashburner postulated that when enough of some new gene product(s) had been made, it (they) acted to inhibit the early genes and induce the late genes.

Inhibition of early genes and induction of late genes require exposure of the salivary gland chromosomes to hormone for a critical period of time, as shown by washing added 20-hydroxyecdysone from gland cultures after various exposure periods. After only a short exposure to 20-hydroxyecdysone, early puff regression and late puff development are aborted. Ashburner and Richards (1976) suggested that the early gene transcription products have to be made in sufficient quantity to compete with the 20-hydroxyecdysone-receptor complex for binding sites on the DNA.

5.7.2 Isolation of an Ecdysteroid Receptor

The search for an **ecdysone receptor** led to attempts to demonstrate whether ^{14}C- or ^{3}H-labeled ecdysone accumulated at the site of chromosomal puffs. While some workers found tantalizing suggestions that ecdysone accumulated in the nucleus, and at puffs, the miniscule amount of ecdysone at a puff site, and the lack of very high specific radiolabeling of ecdysone with ^{14}C or ^{3}H were drawbacks to conclusive proof. A major development occurred with the discovery that an analog of ecdysone, ponasterone, has good hormonal activity after iodination to form 26-[^{125}I]-iodoponasterone (Figure 5.18). The radioactive iodine gives iodoponasterone very high specific labeling, thus facilitating detection in the nucleus. Labeled iodoponasterone shows powerful hormonal effects on cultured *Drosophila* Kc cells, an embryonic cell line derived from *D. melanogaster*, and *in vivo* activity, and it is bound in the nucleus to DNA (Cherbas et al., 1988, 1991).

A gene *EcR* from *D. melanogaster*, that codes a protein with ecdysteroid-binding properties has been isolated and sequenced (Koelle et al., 1991). The receptor protein that it encodes, designated EcR, shows high specific binding to DNA and to labeled ecdysteroids. EcR is localized in the nucleus and can be detected with anti-EcR monoclonal and polyclonal antibodies in a variety of *Drosophila* tissues, including imaginal discs, fat body, tracheae, salivary glands, central nervous system tissue, gut, ring gland, and epidermal cells, and in 13- to 16-hour-old embryos, older embryos, and in late 3rd instars.

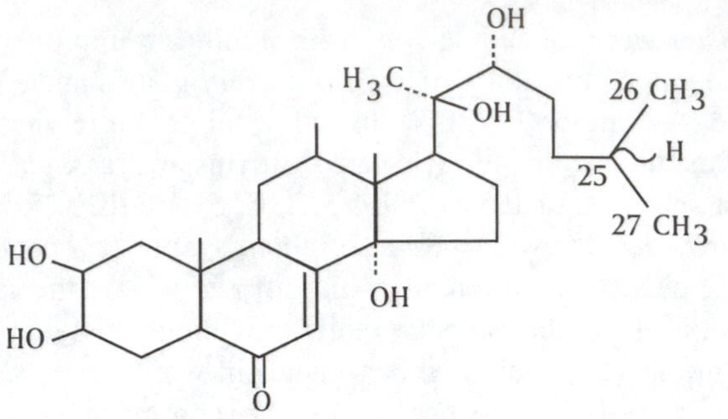

ponasterone

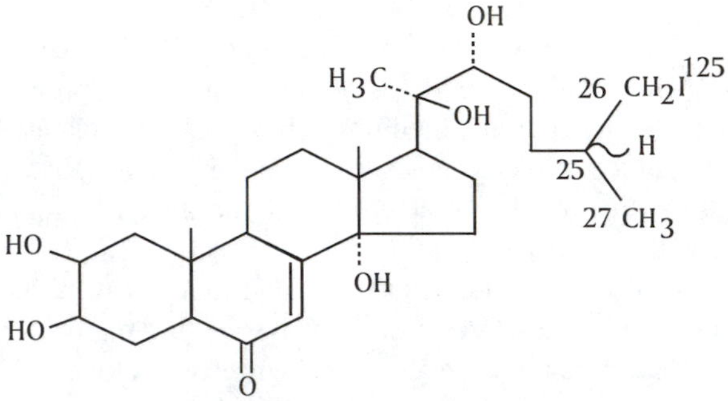

26-iodoponasterone

FIGURE 5.18 Structure of ponasterone, a synthetic ecdysteroid with low hormonal activity *in vivo* and in *Drosophila* cell lines, and after iodination with ^{125}I to form 26-iodoponasterone, which has relatively high hormonal activity and high specific activity due to the ^{125}I.

The EcR protein contains two highly conserved domains characteristic of vertebrate steroid receptors. The N-terminal portion of the molecule contains the **DNA-binding domain**, while the **ecdysteroid-binding domain** is near the C-terminus. Near the C-terminus there is a dimerization sequence, again a similarity to previously discovered vertebrate steroid receptors. EcR forms dimers that are stable only when binding between protein and ecdysteroid occurs; in contrast, some of the vertebrate receptors form stable dimers even when not bound to hormone (Ozyhar et al., 1991).

When the sequence of amino acids in the hormone-binding domain of the EcR protein is compared with sequences of amino acids in vertebrate steroid receptors, EcR is most similar to a subfamily of steroid receptors that includes the human thyroid receptor, human vitamin D receptor, and human retinoic acid receptor, as well as some other hormone receptors. All of the most highly conserved amino acids in the DNA-binding domain of the vertebrate steroid receptors are present in EcR, including nine cysteine residues. Eight of the cysteine residues coordinate two zinc atoms, as in the vertebrate receptors, resulting in the folding of the receptor chain into two zinc fingers (Figure 5.19). The zinc fingers are involved in binding the hormone-receptor complex to DNA at base sequences known as **ecdysone response elements** (**EREs**). In the case of vertebrates, in which there are a number of different steroid hormones, the binding response elements are simply called hormone response elements (HREs).

EcR binds to the heat-shock protein gene *hsp27*, a gene that at present is not known to have a defined role in molting. It codes for heat-shock protein 27 (MSP27), and is activated both by heat shock and by 20-hydroxyecdysone. EcR specifically binds to the *hsp27* gene promoter region at the EREs with the **imperfect palindromic base sequence GGTTCA-TGCACT**, a sequence that resembles the known vertebrate steroidal HREs (Ozyhar et al., 1991). The *hsp27*-DNA-bound EcR has a molecular mass of about 270 kDa (Ozyhar et al., 1991).

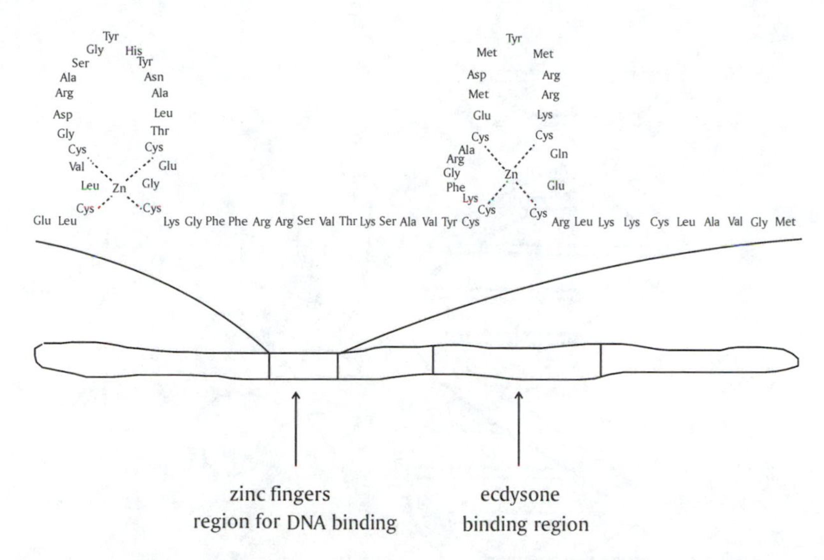

FIGURE 5.19 Representation of the amino acid sequence and Zn atoms in the two zinc fingers in the DNA binding region of the ecdysone receptor isolated from *Drosophila* cultured K_c cells. Additional portions of the receptor molecule (bottom diagram) are involved with actual binding of the ecdysteroid hormone and possibly with transactivation after binding to the DNA.

5.7.3 Differential Tissue and Cell Response to Ecdysteroids

A major challenge is to explain the basis for differential cell and tissue response to ecdysone. Ecdysteroid action on genes may involve activation of some genes, repression of some genes, and indirect action through transcription products. Clearly, the interaction with JH must also be considered, but its interaction with ecdysteroid hormone to control development is much less clear than the present state of ecdysteroid action. Not all tissues and cells respond to ecdysone and JH in the same way, yet presumably all are exposed to the same hormonal stimulation at each molt. For example, epidermal cells respond at each molt and secrete a cuticle that may be larval, pupal, or adult in structure, depending on the interaction of ecdysteroids and JH. On the other hand, some tissues, such as nervous system and imaginal discs, change little or not at all during some of the ecdysteroid pulses, but may respond later in development, or even during adult life (e.g., reproductive organs). How can these tissues, all exposed to the same hormonal stimuli, respond so differently? The answer, at least in part, must lie in (1) the presence of different molecular forms of the ecdysteroid receptor, (2) different number and/or combinations of receptors in different tissues, (3) dimerization and/or heterodimerization of the receptor, and (4) the presence and interaction of specific tissue factors and hormone-induced transcription factors acting in conjunction with the hormone-receptor complex.

Multiple ecdysone receptors are known to exist, but tissue distribution and the number of the different ecdysone receptors remain uncertain. A single gene has been identified and cloned from *D. melanogaster* that codes for three different ecdysteroid receptors (Talbot et al., 1993). Only the B1 receptor is predominant in larvae. The imaginal discs primarily contain the A form (Talbot et al., 1993). The three receptor forms — EcR-A, EcR-B1, and EcR-B2 — have common domains for binding DNA and ecdysteroids, but each has its own unique N-terminal domain, the region believed to direct or modify the type of response a cell makes to a steroid hormone. Thus, one way that differential cell and tissue response to ecdysteroid secretion can be mediated is through the number, type, and distribution of different receptors.

During larval development of *Drosophila* and *Manduca*, the central nervous system (CNS) contains little ecdysone receptor; but at pupariation, all cells have high levels of EcR and, in

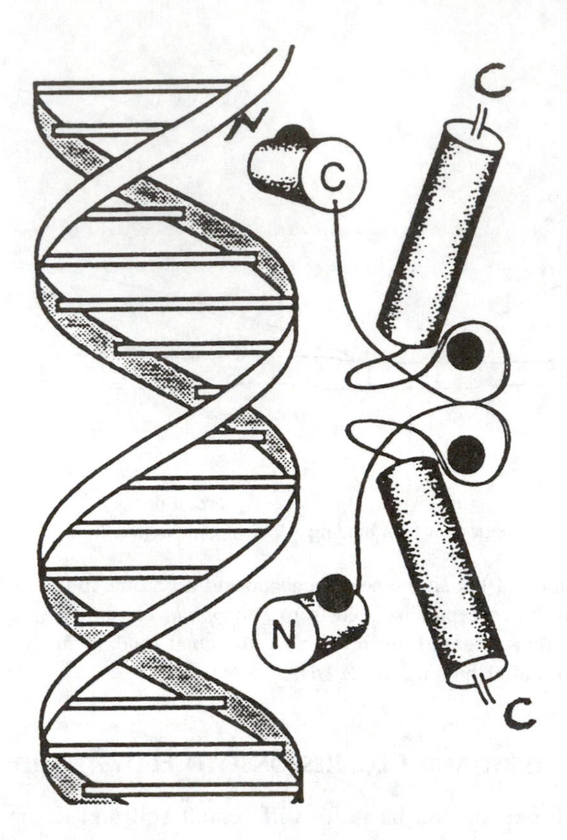

FIGURE 5.20 A model representing the potential binding of a dimer composed of two vertebrate steroid receptor molecules. The zinc fingers, the DNA-binding region of each receptor, may bind to hormone response elements (HREs) situated at adjacent major grooves located 34 Å apart in the DNA helix. (Reproduced with permission from Schwabe and Rhodes, 1991.)

Drosophila, the receptor is the B1 form. The B1 receptor disappears after pupariation; and in adult development, the type and amount of EcR varies with whether a particular cell is a new cell or a cell from the larva that has metamorphosed (Truman et al., 1994). Application of JH at the appropriate time can prevent the appearance of EcR in cells of the CNS of *Manduca* (Riddiford, 1996), but JH does not prevent the appearance of 20-hydroxyecdysone-induced EcR mRNA in epidermal cells during larval or pupal molt in *Manduca.* In the latter tissue, EcR is present in larval epidermis throughout larval life, with higher levels occurring at molts.

Another possible mechanism for differential response is in the **dimerization** of the ecdysteroid receptor. Steroid receptors in vertebrates, and apparently in insects, typically form dimers before they bind to the hormone response elements of DNA (Figure 5.20). Both homodimers and heterodimers are known in vertebrates. In **homodimers**, two receptor molecules bind together, with each also binding a steroid hormone molecule. **Heterodimers** are composed of one molecule of a receptor combined with a different receptor, with each receptor binding a steroid molecule. Heterodimers and homodimers may confer upon the receptor complex different or variable gene regulatory properties. The nature of the dimerization may be one way that tissues respond differently to ecdysone exposure, and the way that the same tissue responds differently to ecdysone at different times. An important role for heterodimer formation in vertebrate hormone signaling has been demonstrated, but its importance remains to be shown in insects. *Drosophila* EcR, however, is known to form a heterodimer with the *Drosophila* protein Ultraspiracle, a product of the *ultraspiracle* (*usp*) gene locus (Yao et al., 1992; Thomas et al., 1993). Heterodimerization conferred DNA binding and functional activity on the complex in the presence of 20-hydroxyecdysone in co-transfected CV1 monkey kidney cells. It is not known, however, if Ultraspiracle is a normal part of the *in vivo* dimerization of EcR.

Tissue and cell response may also be related to the presence of **tissue-specific factors** and **transcription products** that can modify hormone-receptor action. For example, the transcription factor designated E74 occurs in *D. melanogaster* in two forms appearing sequentially during pupariation. The two forms may occur in response to different transcription times and possibly in response to different levels of 20-hydroxyecdysone in the tissues in the time leading up to and during pupariation (Thummel, 1990; Thummel et al., 1990). When details of these and perhaps other control mechanisms are elucidated, it may become clear why some late 3rd instar tissues exposed to the same strong pulse of ecdysone undergo histolysis (e.g., gut, salivary glands), while others (imaginal disks) grow and form adult structures.

Finally, all DNA binding sites for ecdysteroid-receptor complex may not function as response elements (i.e., all may not allow the hormone-receptor complex to promote, inhibit, or modify gene action). Additional transcription factors, possibly specific to particular cells, may be required to turn a binding site into a response element. Because it is believed that the ecdysone receptor must form a dimer in order to bind to DNA, as all vertebrate steroid receptors do, then formation of and interactions between homo- and heterodimers may influence how the hormonal message is applied in each cell and tissue.

5.8 A POSSIBLE TIMER GENE IN THE MOLTING PROCESS

The early gene *E74A* unit in *D. melanogaster* is large, consisting of 60 kb, and it takes about 1 h to transcribe the gene, with elongation of the RNA-transcript taking place at a rate of about 1.1 kb/min (Thummel et al., 1990). In agreement with this time frame, *E74A* mRNA is not detectable in the cell cytoplasm sooner than about 60 min after gene puffing action. These data provide an explanation of why *E74* puffs continue to expand during the first hour after exposure to ecdysone; they expand because new transcripts of mRNA, beginning at the 5′ end, are being formed at the puff site; as more transcript is made, the puff gets bigger. After about an hour, the first complete transcripts are released from the end of the 60-kb unit, and the rate of new transcript formation reaches an equilibrium with the release of transcripts. This agrees with observations that the puff slows in expansion after about an hour and reaches a stable size until it begins to regress at about 4 h post-treatment. Regression in size after about 4 h presumably reflects the time it takes for enough of the new mRNA transcript to be translated into protein products, or for some other product(s) of transcription to be made, which act to repress the *E74A* promoter. The half-life of *E74A* mRNA appears to be about 1 h (Thummel et al., 1990).

At least three of the early puff genes — *E74A*, *E75*, and *2B5* — are large and also might serve as timer genes. If the genes were small, on the order of only a few kilobases in size, then transcripts would be pushed off the end in a few minutes, and the genes would likely be repressed by the products of their transcripts in a much shorter time than the 4 to 6 h that actually occurs. Whether the large size of several of the early genes is indeed related to a significant timing of events is not known, but the **timer gene concept** is important as a model. One can now imagine that some physiological, morphogenetic, or biochemical event may be timed by the size of a gene unit and the time required for transcripts to be made.

5.9 ECDYSONE-GENE INTERACTION IDEAS STIMULATED VERTEBRATE RESEARCH

It took about 30 years of research to prove the astute guess made by Clever and Karlson in 1960 that ecdysone alters "…the activity of specific genes." The idea that steroid hormones might work at the gene level was actively pursued by vertebrate biologists and geneticists, and they were successful much sooner than insect biologists in demonstrating that vertebrate steroid hormones bind to DNA and work at the gene level. There are many more details available at the molecular level on the receptors and binding of vertebrate steroid hormones than is known about ecdysteroid function in insects and other invertebrates. One important aspect of the comparative vertebrate and invertebrate work is the revelation that the basic mechanism of steroid hormone action has been very conserved, and some functional aspects of steroid hormone action clearly predates the separation of vertebrate and invertebrate lines of evolution.

Vertebrate steroid hormones first bind either to a cytoplasmic receptor (the glucocorticoid steroids) or a nuclear receptor (estrogen) receptor. The receptors are proteins. After binding occurs, the hormone-receptor complex translocates to the nucleus (if the receptor is not already located in the nucleus), where binding to nuclear DNA occurs. Genes may be turned on or off by the action of the bound hormone-receptor, and transcription of mRNA is regulated.

Vertebrate steroid receptors bind the steroid hormone near the carboxy-terminal end of the protein molecule. Near the amino-terminal end, the vertebrate steroid receptors contain a highly conserved region with a sequence of amino acids that recognizes and binds to a specific sequence of bases in DNA, the hormone response element (HRE). The sequence of amino acids in several vertebrate steroid receptors has been determined, and they share a common feature in which the DNA-binding region of the protein chain is folded into two zinc fingers. Each zinc finger structure is maintained by a single zinc atom that forms coordinate bonds to four cysteine residues.

The hormone-receptor complex typically binds to DNA as a dimer, and it is the carboxy-terminal region of the receptor that contains the dimerization sequence. Both ends of the receptor protein contain regions that have transactivation and possibly transcription regulatory action. Further discussion of vertebrate mechanisms is not appropriate here, but much more extensive reviews can be found in Carson-Jurica et al. (1990) and Schwabe and Rhodes (1991).

REFERENCES

Agui, N., W. Bollenbacher, N. Granger, and L.I. Gilbert. 1980. Corpus allatum is release site for the insect prothoracicotropic hormone. *Nature (London),* 285:669-670.

Andersen, J.F., M. Ceruso, G.C. Unnithan, E. Kuwano, G.D. Prestwich, and R. Feyereisen. 1995. Photoaffinity labeling of methyl farnesoate epoxidase in cockroach corpora allata. *Insect Biochem. Mol. Biol.,* 25:713-719.

Ashburner, M. 1972. Patterns of puffing activity in the salivary gland chromosomes of *Drosophila. Chromosoma,* 38:255-281.

Ashburner, M., and G. Richards. 1976. Sequential gene activation of ecdysone in polytene chromosomes of *Drosophila melanogaster. Dev. Biol.,* 54:241-255.

Audsley, N., R.J. Weaver, and J.P. Edwards. 2000. Juvenile hormone biosynthesis by corpora allata of larval tomato moth, *Lacanobia oleracea*, and regulation by *Manduca sexta* allatostatin and allatotropin. *Insect Biochem. Mol. Biol.,* 30:681-689.

Baker, F.C., L.W. Tsai, C.C. Reuter, and D.A. Schooley, 1987. *In vivo* fluctuation of JH, JH acid, and ecdysteroid titer, and JH esterase activity, during development of fifth stadium *Manduca sexta. Insect Biochem.,* 17:989-996.

Becker, H.J. 1959. Die Puffs der Speicheldrusenchromosomen von *Drosophila melanogaster. Chromosoma,* 10:654-678.

Bergot, B.J., D.A. Schooley, G.M. Chippendale, and C.-M. Yin. 1976. Juvenile hormone titre determinations in the southwestern corn borer, *Diatraea grandiosella* by electron capture-gas chromatography. *Life Sci.,* 18:811-820.

Bergot, B.J., M. Ratcliff, and D.A. Schooley. 1981. Method for quantitative determination of the four known juvenile hormones in insect tissue using gas chromatography-mass spectroscopy. *J. Chromatogr.,* 204:231-244.

Bhaskaran, G., S.P. Sparagana, P. Barrera, and K.H. Dahm. 1986. Change in corpus allatum function during metamorphosis of the tobacco hornworm *Manduca sexta*. Regulation at the terminal step in juvenile hormone biosynthesis. *Arch. Insect Biochem. Physiol.,* 3:321-338.

Borovsky, D., D.A. Carlson, R.G. Hancock, H. Rembold, and E. Van Handel. 1994. *De novo* biosynthesis of juvenile hormone III and I by the accessory glands of the male mosquito. *Insect Biochem. Mol. Biol.,* 24:437-444.

Borst, D.W., and J.D. O'Connor. 1972. Arthropoid molting hormone: radioimmune assay. *Science,* 178:418-419.

Borst, D.W., M.R. Eskew, S.J. Wagner, K. Shores, J. Hunter, L. Luker, J.D. Hatle, and L.B. Hecht. 2000. Quantification of juvenile hormone III, vitellogenin, and vitellogenin-mRNA during the oviposition cycle of the lubber grasshopper. *Insect Biochem. Mol. Biol.,* 30:813-819.

Bowers, W.J., T. Ohta, J.S. Cleere, and P.A. Marsella. 1976. Discovery of insect anti-juvenile hormones in plants. *Science,* 193:542-547.

Butenandt, A., and P. Karlson. 1954. Über die Isolierung eines Metamorphose-Hormone der Insekten in kristallierten Form. *Z. Naturforsch.,* 9b:389-391.

Carson-Jurica, M.A., W.T. Schrader, and B.W. O'Malley. 1990. Steroid receptor family: structure and functions. *Endocrine Rev.,* 11:210-220.

Chen, J., M. Kabbouh, M.J. Fisher, and H.H. Rees. 1994. Induction of an inactivation pathway for ecdysteroid in larvae of the cotton leafworm, *Spodoptera littoralis. Biochem. J.,* 301:89-95.

Cherbas, L., K. Lee, and P. Cherbas. 1991. Identification of ecdysone response elements by analysis of the *Drosophila Eip28/29* gene. *Genes Dev.,* 5:120-131.

Cherbas, P., L. Cherbas, S. Lee, and K. Nakanishi. 1988. 26-[^{125}I]Iodoponasterone A is a potent ecdysone and a sensitive radioligand for ecdysone receptors. *Proc. Natl. Acad. Sci. U.S.A.,* 85:2096-2100.

Clever, U., and P. Karlson. 1960. Induktion von Puff-Veranderugen in den Speicheldrusenchromosomen von *Chironomus tentans* durch Ecdyson. *Exp. Cell Res.,* 20:623-626.

Cusson, M., K.J. Yagi, Q. Ding, H. Duve, A. Thorpe, J.N. McNeil, and S.S. Tobe. 1991. Biosynthesis and release of juvenile hormone and its precursors in insects and crustaceans: the search for a unifying arthropod endocrinology. *Insect Biochem.,* 21:1-6.

Daneholt, B., and H. Hosick. 1972. The transcription unit in Balbiani ring 2 of *Chironomus tentans.* pp. 629-635, in *Cold Spring Harbor Symposia on Quantitative Biology,* Vol. XXVIII, Cold Spring Harbor Laboratory, Cold Spring Harbor, NY.

Downer, R.G.H., and H. Laufer (Eds.). 1983. *Endocrinology of Insects*, Alan R. Liss, New York, 707 pp.

East, P.D., T.D. Sutherland, S.C. Trowell, A.J. Herlt, and R.W. Richards. 1997. Juvenile hormone synthesis by ring glands of the blowfly *Lucilia cuprina. Arch. Insect Biochem. Physiol.,* 34:239-253.

Edwards, J.P., T.S. Corbitt, H.F. McArdle, J.E. Short, and R.J. Weaver. 1995. Endogenous levels of insect juvenile hormones in larval, pupal, and adult stages of the tomato moth, *Lacanobia oleracea. J. Insect Physiol.,* 41:641-651.

Engelmann, F. 1995. The juvenile hormone receptor of the cockroach *Leucophaea maderae. Insect Biochem. Mol. Biol.,* 25:721-726.

Engelmann, F., and J. Mala. 2000. The interactions between juvenile hormone (JH), lipophorin, vitellogenin, and JH esterases in two cockroach species. *Insect Biochem. Mol. Biol.,* 30:793-803.

Everaerts, C., M. Cusson, and J.N. McNeil. 2000. The influence of smoke volatiles on sexual maturation and juvenile hormone biosynthesis in the black army cutworm, *Actevia fennica* (Lepidoptera: Noctuidae). *Insect Biochem. Mol. Biol.,* 30:855-862.

Feldlaufer, M.F., G.F. Weirich, R.B. Imberski, and J.A. Svoboda. 1995. Ecdysteroid production in *Drosophila melanogaster* reared on defined diets. *Insect Biochem. Mol. Biol.,* 25:709-712.

Fraenkel, G. 1935. A hormone causing pupation in the blowfly *Calliphora erythrocephala. Proc. R. Soc. London, Ser. B,* 118:1-12.

Fukuda, S. 1940. Hormonal control of moulting and puation in the silkworm. *Proc. Imp. Acad. Japan,* 16:417-420.

Fukuda, S. 1944. The hormonal mechanism of larval molting and metamorphosis in the silkworm. *J. Fac. Sci. Tokyo Univ.,* 4:477-532.

Gelman, D.B., J.E. Carpenter, and P.D. Greany. 2000. Ecdysteroid levels/profiles of the parasitoid wasp, *Diapetimorpha introita,* reared on its host, *Spodoptera frugiperda* and on an artificial diet. *J. Insect Physiol.,* 46:457-465.

Gilbert, L.I., and H. Schneiderman. 1960. The development of a bioassay for the juvenile hormone of insects. *Trans. Am. Micros. Soc.,* 79:38-67.

Gilbert, L.I., W.L. Combest, W.A. Smith, and V.H. Meller, V.H. 1988. Neuropeptides, second messengers and insect molting. *BioEssays,* 8:153-157.

Gilbert, L.I., N.A. Granger, and R.M. Roe. 2000. The juvenile hormones: historical facts and speculations on future research directions. *Insect Biochem. Mol. Biol.,* 30:617-644.

Gilbert, L.I., R. Rybczynski, and S.S. Tobe. 1996a. Endocrine cascade in insect metamorphosis, pp. 59-107, in L.I. Gilbert, J.R. Tata, and B.G. Atkinson (Eds.), *Metamorphosis: Postembryonic Reprogramming of Gene Expression in Amphibian and Insect Cells*, Academic Press, New York, 687 pp.

Gilbert, L.I., J.R. Tata, and B.G. Atkinson (Eds.). 1996b. *Metamorphosis: Postembryonic Reprogramming of Gene Expression in Amphibian and Insect Cells*, Academic Press, New York, 687 pp.

Goodman, W.G., and E.S. Chang. 1985. Juvenile hormone cellular and hemolymph binding proteins, pp. 491-510, in G.A. Kerkut and L.I. Gilbert (Eds.), *Comprehensive Insect Physiology, Biochemistry and Pharmacology*, Vol. 7, Pergamon Press, Oxford.

Granger, N.A., S.L. Sturgis, R. Ebersohl, C. Geng, and T.C. Sparks. 1996. Dopaminergic control of corpora allata activity in the larval tobacco hornworm, *Manduca sexta. Arch. Insect Biochem. Physiol.,* 32:449-466.

Granger, N.A., R. Ebersohl, and T.C. Sparks. 2000. Pharmacological characterization of dopamine receptors in the corpus allatum of *Manduca sexta* larvae. *Insect Biochem. Mol. Biol.,* 30:755-766.

Grieneisen, M.L., J.T. Warren, and L.I. Gilbert. 1993. Early steps in ecdysteroid biosynthesis: Evidence for the involvement of cytochrome P450 enzymes. *Insect Biochem. Mol. Biol.,* 23:13-23.

Gu, S.-H., and Y.-S. Chou. 1993. Role of low ecdysteroid levels in the early last larval instar of *Bombyx mori. Experientia,* 49:806-809.

Handler, A.M. 1982. Ecdysteroid titers during pupal and adult development in *Drosophila melanogaster. Dev. Biol.,* 93:73-82.

Hiruma, K., and L.M. Riddiford. 1993. Molecular mechanisms of cuticular melanization in the tobacco hornworm, *Manduca sexta* (L.) (Lepidoptera: Sphingidae). *Int. J. Insect Morphol. Embryol.,* 22:103-117.

Hua, Y.-J., R.-J. Jiang, and J. Koolman. 1997. Multiple control of ecdysone biosynthesis in blowfly larvae: Interaction of ecdysiotropins and ecdysiostatins. *Arch. Insect Biochem. Physiol.,* 35:125-134.

Huang, Z.-H., G.E. Robinson, and D.W. Borst. 1994. Physiological correlates of division of labor among similarly aged honeybees. *J. Comp. Physiol. A,* 174:731-739.

Hunnicutt, D., Y.C. Toong, and D.W. Borst. 1989. A chiral specific antiserum for juvenile hormone. *Am. Zool.,* 29:48a.

Jones, G. 1995. Molecular mechanisms of action of juvenile hormone. *Annu. Rev. Entomol.,* 40:147-169.

Kanost, M.R., J.K. Kawooya, J.H. Law, R.O. Ryan, M.C. Van Heusden, and R. Ziegler. 1990. Insect haemolymph proteins. *Adv. Insect Physiol.,* 22:299-396.

Kappler, C., M. Kabbouh, C. Hetru, F. Durst, and J.A. Hoffmann. 1988. Characterization of three hydroxylases involved in the final steps of biosynthesis of the steroid hormone ecdysone in *Locusta migratoria* (Insecta, Orthoptera). *J. Steroid Biochem.,* 31:891-898.

Kawakami, A., H. Kataoka, T. Oka, A. Mizoguchi, M. Kimura-Kawakami, T. Adachi, M. Iwami, H. Nagasawa, A. Suzuki, and H. Ishizaki. 1990. Molecular cloning of the *Bombyx mori* prothoracicotrophic hormone. *Science,* 247:1333-1335.

King, L.E., and S.S. Tobe. 1993. Changes in the titre of a juvenile hormone III binding lipophorin in the haemolymph of *Diploptera punctata* during development and reproduction: Functional significance. *J. Insect Physiol.,* 39:241-251.

Koelle, M.R., W.S. Talbot, W.A. Segraves, M.T. Bender, P. Cherbas, and D.S. Hogness. 1991. The *Drosophila* EcR gene encodes an ecdysone receptor, a new member of the steroid receptor superfamily. *Cell,* 67:59-77.

Kopeč, S. 1917. Experiments on metamorphosis of insects. *Bull. Int. Acad. Cracovie,* (B):57-60.

Lafont, R., and I.D. Wilson. 1990. Advances in ecdysteroid high performance liquid chromatography, pp. 79-94, in A.R. McCaffery and I.D. Wilson (Eds.), *Chromatography and Isolation of Insect Hormones and Pheromones*, Plenum Press, New York.

Lanzrein, B., R. Wilhelm, and R. Reichsteiner. 1993. Differential degradation of racemic and 10*R*-juvenile hormone-III by cockroach (*Nauphoeta cinerea*) haemolymph and the use of lipophorin for long-term culturing of corpora allata. *J. Insect Physiol.,* 39:53-63.

Lawrence, P.O. 1986. The role of 20-hydroxyecdysone in the moulting of *Biosteres longicaudatus*, a parasite of the Caribbean fruit fly, *Anastrepha suspensa. J. Insect Physiol.,* 32:329-337

Lefevere, K.S., M.J. Lacey, P.H. Smith, and B. Roberts. 1993. Identification and quantification of juvenile hormone biosynthesized by larval and adult Australian sheep blowfly *Lucilia curpina* (Diptera: Calliphoridae). *Insect Biochem. Mol. Biol.,* 23:713-720.

Lezzi, M. 1996. Chromosome puffing: Supramolecular aspects of ecdysone action, pp. 145-173, in L.I. Gilbert, J.R. Tata, and B.G. Atkinson (Eds.), *Metamorphosis: Postembryonic Reprogramming of Gene Expression in Amphibian and Insect Cells*, Academic Press, New York, 687 pp.

Mauchamp, B., C. Royer, L. Kerhoas, and J. Einhorn, 1993. MS/MS analyses of ecdysteroids in developing eggs of *Dysdercus fasciatus*. *Insect Biochem. Mol. Biol.,* 23:199-205.

Nagasawa, H., H. Kataoka, A. Isogai, S. Tamura, and A. Suzuki. 1984. Amino-terminal amino acid sequence of the silkworm prothoracicotrophic hormone: homology with insulin. *Science,* 226:1344-1345.

Nijhout, H.F. 1979. Stretch-induced moulting in *Oncopeltus fasciatus*. *J. Insect Physiol.,* 25:277-281.

Nijhout, H.F. 1984. Abdominal stretch reception in *Dipetalogaster maximus* (Hemiptera: Reduviidae). *J. Insect Physiol.,* 30:629-633.

Nijhout, H.F. 1994. *Insect Hormones*, Princeton University Press, Princeton, NJ, 267 pp.

Nijhout, H.F., and C.M. Williams. 1974. Control of moulting and metamorphosis in the tobacco hornworm *Manduca sexta* (L.): Cessation of juvenile hormone secretion as a trigger for pupation. *J. Exp. Biol.,* 61:493-501.

O'Reilly, D.R., and L.K. Miller. 1989. A baculovirus blocks insect molting by producing ecdysteroid UDP-glucosyl transferase. *Science,* 245:110-1112.

Oberlander, H., and S.M. Ferkovich. 1994. Physiological and developmental capacities of insect cell lines, pp. 127-140, in K. Maramorosch and A.H. McIntosh (Eds.), *Insect Cell Technology*, CRC Press, Boca Raton, FL.

Ozyhar, A., M. Strangmann-Diekmann, H. Kiltz, and O. Pongs. 1991. Characterization of a specific ecdysteroid receptor-DNA complex reveals common properties for invertebrate and vertebrate hormone-receptor/DNA interactions. *Eur. J. Biochem.,* 200:329-335.

Palli, S.R., E.O. Osir, W.-S. Eng, M.F. Boehm, M. Edwards, P. Kulcsar, I. Ujvary, K. Hiruma, G.D. Prestwich, and L.M. Riddiford. 1990. Juvenile hormone receptors in insect larval epidermis: Identification by photoaffinity labeling. *Proc. Natl. Acad. Sci. U.S.A.,* 87:796-800.

Palli, S.R., K. Touhara, J. Charles, B.C. Bonning, J.K. Atkinson, S.C. Trowell, K. Hiruma, W.G. Goodman, T. Kyraikides, G.D. Prestwich, B.D. Hammock, and L.M. Riddiford. 1994. A nuclear juvenile hormone-binding protein from larvae of *Manduca sexta*: A putative receptor for the metamorphic action of juvenile hormone. *Proc. Natl. Acad. Sci. U.S.A.,* 91:6191-6195.

Palli, S.R., T.R. Ladd, W.L. Tomkins, S. Shu, S.B. Ramaswamy, Y. Tanaka, B. Arif, and A. Retnakaran. 2000. *Choristonerua fumiferana* entomopoxvirus prevents metamorphosis and modulates juvenile hormone and ecdysteroid titers. *Insect Biochem. Mol. Biol.,* 30:869-876.

Park, Y.I., S. Shu, S.B. Ramaswamy, and A. Srinivasan. 1998. Mating in *Heliothis virescens*: Transfer of juvenile hormone during copulation by male to female and stimulation of biosynthesis of endogenous juvenile hormone. *Arch. Insect Biochem. Physiol.,* 38:100-107.

Peter, M.G., S. Gunawan, G. Gellisen, and H. Emmerich. 1979. Differences in hydrolysis and binding of homologous juvenile hormones in *Locusta migratoria* haemolymph. *Z. Naturforsch.,* 34:558-598.

Peter, M.G., H.P. Stupp, and K.U. Lentes. 1983. Reversal of the enantioselectivity in the enzymatic hydrolysis of juvenile hormone as a consequence of the protein fractionation. *Angew. Chem.,* 95:773-774.

Porcheron, P., M. Moriniere, J. Grassi, and P. Pradelles. 1989. Development of an enzymo-immunoassay for ecdysteroids using acetylcholinesterase as label. *Insect Biochem.,* 19:117-122.

Raabe, M. 1982. *Insect Neurohormones*, Plenum Press, New York, 352 pp.

Ramaswamy, S.B., S. Shu, Y.I. Park, and F. Zeng. 1997. Dynamics of juvenile hormone-mediated gonadotropism in the Lepidoptera. *Arch. Insect Biochem. Physiol.,* 35:539-558.

Richard, D.S., S.W. Applebaum, T.J. Sliter, F.C. Baker, D.A. Schooley, C.C. Reuter, V.C. Henrich, and L.I. Gilbert. 1989a. Juvenile hormone bisepoxide biosynthesis *in vitro* by the ring gland of *Drosophila melanogaster*: A putative juvenile hormone in the higher Diptera. *Proc. Natl. Acad. Sci. U.S.A.,* 86:1421-1425.

Richard, D.S., S.W. Applebaum, and L.I. Gilbert. 1989b. Developmental regulation of juvenile hormone biosynthesis by the ring gland of *Drosophila melanogaster*. *J. Comp. Physiol. B,* 159:383-387.

Richter, K. 2001. Daily changes in neuroendocrine control of moulting hormone secretion in the prothoracic gland of he cockroach *Periplaneta americana* (L.). *J. Insect Physiol.,* 47:333-338.

Riddiford, L. 1990. Juvenile hormone receptors in insect larval epidermis: Identification by photoaffinity labeling. *Proc. Natl. Acad. Sci. U.S.A.,* 87:796-800.

Riddiford, L.M. 1994. Cellular and molecular actions of juvenile hormone I. General considerations and premetamorphic actions. *Adv. Insect Physiol.,* 24:213-274.

Riddiford, L.M. 1996. Molecular aspects of juvenile hormone action in insect metamorphosis, pp. 223-251, in L.I. Gilbert, J.R. Tata, and B.G. Atkinson (Eds.), *Metamorphosis: Postembryonic Reprogramming of Gene Expression in Amphibian and Insect Cells*, Academic Press, New York, 687 pp.

Riddiford, L.M., and Truman, J.W. 1993. Hormone receptors and the regulation of insect metamorphosis. *Am. Zool.,* 33:340-347.

Roe, R.M., and K. Venkatesh. 1990. Metabolism of juvenile hormones: Degradation and titer regulation, pp. 125-179, in A.P. Gupta (Ed.), *Morphogenetic Hormones of Arthropods*, Vol. 1, Rutgers University, New Brunswick, NJ.

Roundtree, D.B., and W.E. Bollenbacher. 1986. The release of prothoracicotropic hormone in the tobacco hornworm, *Manduca sexta*, is controlled intrinsically by juvenile hormone. *J. Exp. Biol.,* 120:41-58.

Royer, C., P. Porcheron, P. Pradelles, and B. Mauchamp. 1993. Development and use of an enzymatic tracer for new enzyme immunoassay for Makisterone A. *Insect Biochem. Mol. Biol.,* 23:193-197.

Sakurai, S., J.T. Warren, and L.I. Gilbert. 1989. Mediation of ecdysone synthesis in *Manduca sexta* by a hemolymph enzyme. *Arch. Insect Biochem. Physiol.,* 10:179-197.

Schooley, D.A., F.C. Baker, L.W. Tsai, C.A. Miller, and C.G. Jamieson. 1984. Juvenile hormones 0, I, and II exist only in Lepidoptera, pp. 373-383, in J.A. Hoffmann and M. Porchet (Eds.), *Biosynthesis, Metabolism and Mode of Action of Invertebrate Hormones*, Springer-Verlag, Heidelberg.

Schwabe, J.W.R., and D. Rhodes. 1991. Beyond zinc fingers: steroid hormone receptors have a novel structural motif for DNA recognition. *Trends Biochem. Sci.,* 16:291-296.

Smith, W.A. 1993. Second messengers and the action of prothoracicotropic hormone in *Manduca sexta*. *Am. Zool.,* 33:330-339.

Smith, W.A., L.I. Gilbert, and W.E. Bollenbacher. 1984. The role of cyclic AMP in the regulation of ecdysone synthesis. *Mol. Cell. Endocrinol.,* 37:285-294.

Smith, W.A., L.I. Gilbert, and W.E. Bollenbacher. 1985. Calcium-cyclic AMP interactions in prothoracicotropic hormone stimulation of ecdysone synthesis. *Mol. Cell. Endocrinol.,* 39:71-78.

Smith, W.A., W.L. Combest, and L.I. Gilbert. 1986. Involvement of cyclic AMP-dependent protein kinase in prothoracicotropic hormone-stimulated ecdysone synthesis. *Mol. Cell. Endocrinol.,* 47:25-33.

Smith, W.A., A.H. Varghese, and K.J. Lou. 1993. Developmental changes in cyclic AMP-dependent protein kinase associated with increased secretory capacity of *Manduca sexta* prothoracic glands. *Mol. Cell. Endocrinol.,* 90:187-195.

Sohi, S.S., S.R. Palli, B.J. Cook, and A. Retnakaran. 1995. Forest insect cell lines responsive to 20-hydroxyecdysone and two nonsteroidal ecdysone agonists, RH-5849 and RH-5992. *J. Insect Physiol.,* 41:457-464.

Sparagana, S.P., G. Bhaskaran, and P. Barrera. 1985. Juvenile hormone acid methyltransferase activity in imaginal discs of *Manduca sexta* prepupae. *Arch. Insect Biochem. Physiol.,* 2:191-202.

Stay, B., S.S. Tobe, and W.G. Bendena. 1994. Allatostatins: Identification, primary structures, functions and distribution. *Adv. Insect Physiol.,* 25:267-337.

Talbolt, W.S., E.A. Swyryd, and D.S. Hogness. 1993. *Drosophila* tissues with different metamorphic response to ecdysone express different ecdysone receptor isoforms. *Cell,* 73:1323-1337.

Teal, P.E.A., Y. Gomez-Simuta, and A.T. Proveaux. 2000. Mating experience and juvenile hormone enhance sexual signaling and mating in male Caribbean fruit flies. *Proc. Natl. Acad. Sci. U.S.A.,* 97:3708-3712.

Thomas, H.E., H.G. Stunnenberg, and F.A. Stewart. 1993. Heterodimerization of the *Drosophila* ecdysone receptor with retinoid X receptor and ultraspiracle. *Nature,* 362:471-475.

Thummel, C.S. 1990. Puffs and gene regulation-molecular insights into the *Drosophila* ecdysone regulatory hierarchy. *BioEssays,* 12:561-568.

Thummel, C.S., K.C. Burtis, and D.S. Hogness. 1990. Spatial and temporal patterns of E74 transcription during *Drosophila* development. *Cell,* 61:101-111.

Tobe, S.S., and R. Feyereisen. 1983. Juvenile hormone biosynthesis: Regulation and assay, pp. 161-178, in R.G.H. Downer and H. Laufer (Eds.), *Endocrinology of Insects,* Alan R. Liss, New York.

Tobe, S.S., and B. Stay. 1985. Structure and regulation of the corpus allatum. *Adv. Insect Physiol.,* 18:305-432.

Touhara, K., and G.D. Prestwich. 1992. Binding site mapping of a photoaffinity-labeled juvenile hormone binding protein. *Biochem. Biophys. Res. Commun.,* 182:466-473.

Truman, J.W., W.S. Talbot, S.E. Fahrbach, and D.S. Hogness. 1994. Ecdysone receptor expression in the CNS correlates with stage-specific responses to ecdysteroids during *Drosophila* and *Manduca* development. *Development,* 120:219-234.

Vafopoulou, X., and C.G.H. Steel. 1999. Daily rhythm of responsiveness to prothoracicotropic hormone in prothoracic glands of *Rhodnius prolixus*. *Arch. Insect Biochem. Physiol.,* 41:117-123.

Wainwright, G., M.C. Prescott, L.L. Lomas, S.G. Webster, and H.H. Rees. 1997. Development of a new high-performance liquid chromatography-mass spectrometric method for the analysis of ecdysteroids in biological extracts. *Arch. Insect Biochem. Physiol.,* 35:21-31.

Warren, J.T., S. Sakurai, D.B. Roundtree, and L.I. Gilbert. 1988a. Synthesis and secretion *in vitro* of ecdysteroids by the prothoracic gland of *Manduca sexta*. *J. Insect Physiol.,* 34:571-576.

Warren, J.T., S. Sakurai, D.B. Roundtree, L.I. Gilbert, S.-S. Lee, and K. Nakanishi. 1988b. Regulation of the ecdysteroid titer of *Manduca sexta:* Reappraisal of the role of the prothoracic glands. *Proc. Natl. Acad. Sci. U.S.A.,* 85:958-962.

Weirich, G.F., M.F. Feldlaufer, and J.A. Svoboda. 1993. Ecdysone oxidase and 3-oxoecdysteroid reductases in *Manduca sexta*: Developmental changes and tissue distribution. *Arch. Insect Biochem. Physiol.,* 23:199-211.

Wigglesworth, V.B. 1936. The function of the corpus allatum in the growth and reproduction of *Rhodnius prolixus* (Hemiptera). *Quart. J. Micr. Sci.,* 79:91-121.

Wigglesworth, V.B. 1940. The determination of characters at metamorphosis in *Rhodnius prolixus*. *J. Exp. Biol.,* 17:201-222.

Williams, C.M. 1947. Physiology of insect diapause. II. Interactions between the pupal brain and prothoracic glands in the metamorphosis of the giant silkworm, *Platysamia cecropia*. *Biol. Bull. Woods Hole,* 93:86-98.

Woodring, J., and K.H. Hoffmann. 1994. The effects of octopamine, dopamine and serotonin on juvenile hormone synthesis, *in vitro*, in the cricket *Gryllus bimaculatus*. *J. Insect Physiol.,* 40:797-802.

Yao, T., W.A. Segraves, A.E. Oro, M. McKeown, and R.M. Evans. 1992. *Drosophila* ultraspiracle modulates ecdysone receptor function via heterodimer formation. *Cell,* 71:63-72.

Yin, C.-M., B.-X. Zou, M. Jiang, M.-F. Li, W. Qin, T.L. Potter, and J.G. Stoffolano, Jr. 1995. Identification of juvenile hormone III bisepoxide (JHB_3), juvenile hormone III and methyl farnesoate secreted by the corpus allatum of *Phormia regina* (Meigen), *in vitro* and function of JHB_3 either applied alone or as a part of a juvenoid blend. *J. Insect Physiol.,* 41:473-479.

6 Intermediary Metabolism

CONTENTS

PREVIEW

Metabolism is the sum of all the chemical reactions occurring in an organism. This chapter deals primarily with catabolism, reactions that break down molecules to release energy. The most intense and rapid energy demands made by insects come with flight. Within seconds, the available high-energy phosphates in the body are used in flight, and energy must be made available rapidly and for long periods for flight to continue. The tracheal system of insects is able to supply oxygen to mitochondria even during flight, and insects do not incur an oxygen debt in flight. As a consequence of the efficiency of the tracheal system, and a fast glycerol-3-phosphate shuttle that regenerates NAD^+ for use in glycolysis, virtually all metabolic glucose can go directly to pyruvate, and pyruvate can go directly into mitochondria for metabolism to release large amounts of energy. Flight muscle mitochondria are highly specialized to support a high rate of metabolism. They are extraordinarily large (up to 4 µm long) and irregular in shape. The cristae of flight muscle mitochondria are numerous, like the pages of a book, and there is relatively little open (matrix) space within flight muscle mitochondria. Up to 40% of the wet weight of flight muscle from *Phormia regina*, a blowfly, is mitochondrial mass, and half the muscle protein is mitochondrial protein. Just 1 mg of flight muscle from a blowfly can contain 1.1×10^8 mitochondria. The biochemist Albert Lehninger estimated that flight muscle mitochondria have as much as 400 m^2 surface per gram mitochondrial protein. By way of comparison, rat liver mitochondria have about 40 m^2/g protein. The outer membrane of insect mitochondria is permeable to most soluble components, but the inner membrane is very selectively permeable. Cytochrome C reductase and hexokinase, among other enzymes, are located on the outer membrane. The space between the outer and inner membrane contains adenylate kinase and nucleoside diphosphokinase activities. The outer surface of the inner membrane contains glycerol-3-phosphate dehydrogenase, proline dehydrogenase, and trehalase. The inner membrane contains the respiratory chain enzymes, ATP-synthesizing enzymes, and α-ketoglutarate dehydrogenase. The inner side of the inner membrane contains succinic dehydrogenase and NADH dehydrogenase. The matrix contains citrate synthetase; aconitase; isocitrate dehydrogenase; fumarase; malate dehydrogenase; alanine and aspartate amino transferase; and carnitine, acetyl, and palmityl transferases. Most of the Krebs cycle intermediates do not readily cross the inner membrane, and are not usually metabolized when added exogenously to isolated mitochondria. Knob-like structures about 8 to 9.5 nm in diameter are connected to the cristae by stalks that are 3 to 4 nm in diameter and 4 to 5 nm in length, and it is within these knob-like structures that ATP is actually synthesized by a chemiosmotic gradient.

Some groups (Lepidoptera, Orthoptera, and some others) burn lipids (fatty acids) as flight fuels. Fatty acids, which must be metabolized within the mitochondria and hence require availability of oxygen, release large amounts of energy per unit weight of substrate metabolized. The ability to rapidly mobilize and transport lipids from fat body, and the availability of oxygen from the tracheal system, are major adaptations in those insects that burn fatty acids for flight. Some insects that metabolize lipids are able to fly continuously for hours and undertake long-distance migration. A few insects use proline as flight fuel. Its complete metabolism yields much less energy per unit weight metabolized, and only a few insects have evolved to depend on it as a major flight fuel.

6.1 INTRODUCTION

The day-to-day activities of an insect require a constant supply of energy. Most adults need an intake of food to support activities such as dispersal, reproduction, and flight. Flight, in particular, is a very energy-intensive activity, requiring rapid mobilization of energy sources, transport, and transformation of food energy into energy of ATP. Metabolism involves all the biochemical reactions occurring in an organism, coverage that clearly is not possible in one chapter. Thus, those metabolic reactions most directly involved in mobilizing stored energy reserves, and in releasing that energy for flight, are the subject of this chapter. The same processes support general maintenance activities as well, but at a less intense level.

In the animal kingdom, only birds, insects, and bats fly with their own muscle power. Flight enables insects to disperse rapidly and widely, and seek new areas to colonize. It also enables them to seek new and/or sparsely distributed food resources, locate potential mates, and search for oviposition sites.

In some insects, such as blowflies (Diptera) and some Hymenoptera, flight is the most energy-intensive biological process known per unit weight of tissue. Perhaps because of the unique position it holds, flight metabolism has been studied extensively, and many reviews are available, including Sacktor (1974), Bailey (1975), Candy (1985), Friedman (1985), Downer (1985), and Beenakkers et al. (1986). Blacklock and Ryan (1994) present an excellent review of lipid transport and metabolism.

This chapter presents a basic introduction to metabolic pathways that are important to the release of energy for general cell and body maintenance and growth, and for intense muscular activity of flight.

6.2 THE ENERGY DEMANDS FOR INSECT FLIGHT

A honeybee in continuous flight burns up to 2400 cal/g muscle/h (Weis-Fogh, 1952). Contrast this with the recorded metabolic rate of 215 cal/g muscle/h for hummingbirds during hovering flight, (Hainsworth, 1981), one of the highest rates of metabolism known among vertebrates. The mass-specific metabolic rates for flying honeybees are about three times greater than those measured for hovering hummingbirds, and 30 times those of human athletes in maximum exercise activity (Suarez et al., 2000, and references therein). Not only do some flying insects have high oxygen and calorie consumption values, but they can reach these high metabolic rates within a few seconds after taking flight. Upon cessation of flight, the metabolic rate returns almost instantly to a low "resting" rate. An oxygen debt does not have to be paid after intense activity because flight metabolism in insects is aerobic, in contrast to largely anaerobic work accomplished in vertebrate muscles during intense muscular activity.

How insects control the rapid "turn-on" and "turn-off" of flight metabolism has been of great interest. Biochemists describe the adjustment in metabolic rate from rest to activity as the control value, calculated as the ratio of oxygen consumption rate during intense muscular activity divided by the resting rate. Control implies, of course, that an animal regulates its oxygen consumption and metabolic processes to support the intense activity, and then scales down the processes when the activity ceases. Upon initiation of flight, oxygen consumption rate in many insects jumps within seconds to values as much as 100 times the resting rate. A blowfly, *Lucilia sericata*, consumes 33 to 50 µl oxygen/min/g tissue while resting, but almost instantly increases that to as much as 1625 to 3000 µl oxygen/min/g tissue upon taking flight (Davis and Fraenkel, 1940). A simple calculation shows the control value to be at least 50, and possibly as much as 100 times the resting value. A variety of moths, which are not particularly fast flyers, have oxygen consumption values of 7 to 12 µl/min/g muscle at rest and 700 to 1660 µl/min/g muscle in flight, again yielding control values of approximately 100 (Zebe, 1954).

There are many dynamic changes occurring in an insect upon initiation of flight, including changes in various metabolites and ions, increased nerve firing, flight muscle contractions, mobilization of components from the fat body, transport through the hemolymph, and the release of hormones. All these events contribute to the physiological control ability of flying insects to achieve the very rapid 50- to 100-fold increase in metabolic activity and oxygen consumption during flight. Hummingbirds in flight have only about a fivefold control from resting metabolism to flight (Pearson, 1950), and trained, conditioned human sprinters also have control values of about 5 during a sprint.

Insects are able to use several different substrates as fuel during flight. As energy is released, it is trapped in the universal metabolic currency, ATP. As in all other organisms, ATP is present in relatively small amounts in cells, and more is synthesized as needed. The ATP concentration in cells is one of the regulators of metabolism, with "large" amounts inhibiting some key enzymes involved in ATP synthesis, while decreased amounts stimulate new synthesis.

Probably in a typical insect the level of **ATP** in the flight musculature is sufficient for only about 1 second of flight; a **phosphagen** reserve of **arginine phosphate**, sufficient for an additional 2 to 4 seconds of flight, can be used rapidly to synthesize ATP (Candy, 1989), as shown in the following equation.

$$\text{Arginine phosphate} + \text{ADP} \xrightarrow{\text{Arginine phosphokinase}} \text{ATP} + \text{Arginine}$$

Clearly, metabolism of some additional substrates and synthesis of new ATP must start in the first 1 or 2 seconds if flight is to continue. All insects appear to initially metabolize carbohydrates and a little bit of proline to "prime the Krebs cycle" upon taking flight. Some, such as Diptera and Hymenoptera, can sustain flight only as long as carbohydrates are available to metabolize; while

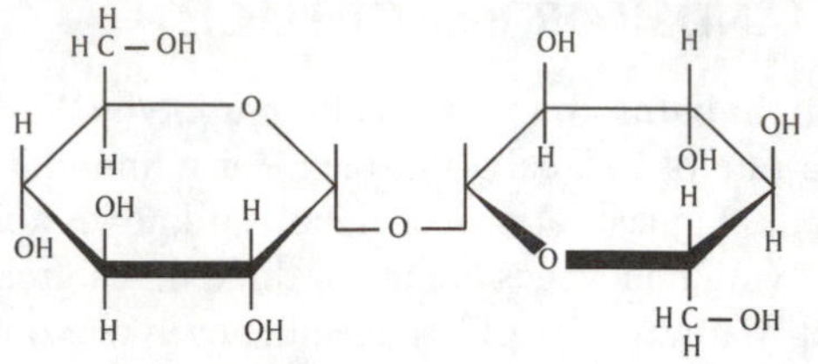

TREHALOSE

α - D - 1,1 - glucopyranoside

FIGURE 6.1 Trehalose (α-D-glucopyranosyl-α-D-glucopyranoside), a principal disaccharide storage and transport sugar in most insects.

Lepidoptera, Orthoptera, and a number of other insect groups rapidly switch to metabolism of lipids before their carbohydrates are gone. A few insects metabolize the amino acid proline as a major fuel supply to support flight.

6.3 METABOLIC STORES

6.3.1 CARBOHYDRATE RESOURCES

The two most common carbohydrate stored reserves of insects are the disaccharide **trehalose** and the polysaccharide **glycogen**. The hemolymph, fat body, and gut tissue are major sources of stored carbohydrates, but small amounts of trehalose and glycogen occur in muscles. Trehalose is usually present in large quantity in the hemolymph and is rapidly hydrolyzed to two glucose molecules for muscles or other tissues to use. Glycogen stored in fat body cells and gut cells must be hydrolyzed to release glucose units, which are then converted to trehalose and transported by the hemolymph to active tissues. Carbohydrate reserves are usually sufficient in a well-fed dipteran or hymenopteran to support continuous flight for 30 min to perhaps 2 h, depending on the species, size of the insect, size of fat body (which varies considerably in insects), and trehalose content of the hemolymph (also variable).

6.3.1.1 Trehalose Resources

When energy is needed, trehalose is usually the first metabolite used, and its hydrolysis yields two molecules of glucose for each trehalose molecule hydrolyzed. Trehalose is the principal storage sugar of insects, and from 200 mg to as much as 1.5 g per 100 ml hemolymph occur in the hemolymph of various insect species. Trehalose is a disaccharide (**α-D-glucopyranosyl-α-D-glucopyranoside**), with the two glucose units linked α-1,1 (Figure 6.1). As a consequence of the 1,1 linkage of the two glucose units, trehalose is a nonreducing sugar, perhaps an important feature because a reducing sugar that occurred in such large concentration as trehalose in the hemolymph might interact with and reduce other components in the hemolymph or tissues. Additional stores of trehalose occur in muscle cells and fat body cells.

Trehalose is rapidly synthesized from glucose as it is absorbed from the midgut. The absorbed glucose may have several fates, as shown in Figure 6.2, but with few exceptions, glucose is not stored as such in insects and is not usually present in any appreciable quantity in the hemolymph. Rapid synthesis of the absorbed glucose into trehalose keeps the hemolymph level of glucose very low, and glucose absorption occurs without an energy-requiring membrane transport mechanism. The process is called **facilitated diffusion**, and even when the gut concentration is low, glucose is still effectively absorbed.

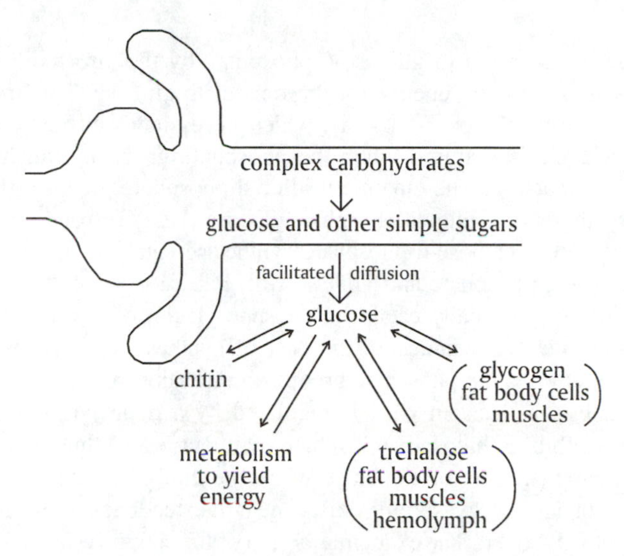

FIGURE 6.2 Possible fates for glucose absorbed from the gut.

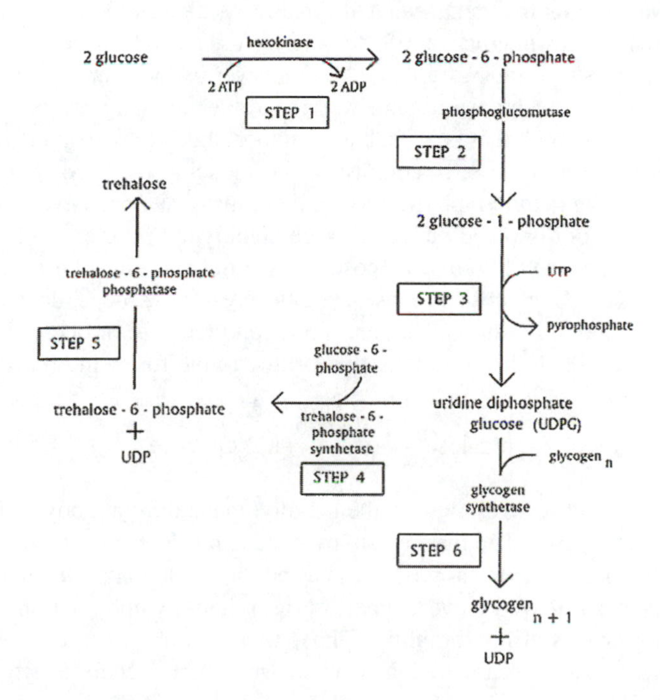

FIGURE 6.3 Biosynthetic pathway for the formation of trehalose and glycogen from glucose.

The conversion of [^{14}C]glucose into trehalose has been demonstrated in tissue preparations from a number of insects, including the orthopterans *Schistocerca gregaria*, *Locusta migratoria*, the dipteran *Phormia regina*, the dictyopteran (cockroach) *Leucophaea maderae*, and the lepidopterans *Bombyx mori* and *Hyalophora cecropia*. Most of the synthesis of trehalose occurs in fat body cells, with small amounts synthesized in other tissues, such as gut and muscle cells.

Trehalose is a costly sugar for insects to synthesize, requiring several enzymes, steps, and input of high-energy phosphate from ATP. The pathway for trehalose synthesis in fat body (Figure 6.3) is based on the work of investigators in several different laboratories. Immediately after absorption from the gut, glucose is converted to glucose-6-phosphate by the enzyme hexokinase, with ATP supplying the phosphate group and energy for its transfer to glucose. Two molecules of glucose-6-phosphate are needed for trehalose synthesis, which necessitates an investment of 2 ATPs. No significant energy exchange is required when the phosphate group is transferred to carbon-1 of glucose in a subsequent reaction. The enzyme uridine diphosphate glucose-glucose pyrophosphorylase catalyzes the synthesis of uridine diphosphate glucose (UDPG) from glucose-1-phosphate and uridine triphosphate (UTP). Trehalose-6-phosphate synthetase catalyzes the formation of trehalose-6-phosphate from glucose-6-phosphate and UDPG, with release of UDP. Regeneration of UTP for future reactions requires enzymatically catalyzed phosphorylation of uridine diphosphate (UDP) by ATP, so one more ATP must be counted in the cost of synthesis of trehalose. Finally, trehalose-6-phosphate synthetase removes the phosphate group from trehalose-6-phosphate to form trehalose.

Synthesis of trehalose, at least in part, is regulated by a **negative feedback mechanism** in which free trehalose inhibits trehalose-6-phosphate synthetase and thus acts like a brake on the system. By slowing new synthesis of trehalose when the concentration of trehalose is high and little metabolic need for an energy supply exists, negative feedback presumably helps shift the synthesis of glucose into glycogen. The exact interaction of these two systems for storage of glucose, however, has not been worked out in detail in insects.

Even if no attempt is made to account for the energy required to synthesize the enzymes of the trehalose pathway, or for maintenance of the pathway, at least 3 moles ATP is required to synthesize 1 mole trehalose from glucose. Why do insects use a costly storage form for sugar when so many other organisms store sugar as glucose? The answer is not clear, but various guesses have been offered. Possibly one selection pressure was the value of a large reserve of an immediate energy source in the hemolymph that is available to support the energy demand of sustained flight. The nonreducing nature of trehalose, in contrast to the reducing nature of glucose, may prevent undesirable reactions in the hemolymph. Moreover, there may have been evolutionary selection to reduce the osmotic effect of dissolved nutrients in the hemolymph. For example, 10 m*M* trehalose has the equivalent energy value of 20 m*M* glucose, but the trehalose in solution will have only half the osmotic value of 20 m*M* glucose because osmotic pressure is dependent on the number of chemical particles in solution or suspension, and not upon their size or chemical nature.

Glucose is readily available from trehalose according to the following reaction:

$$\text{Trehalose} \xrightarrow{\text{Trehalase}} 2\ \text{Glucose}$$

Most insects have high levels of **trehalase** in the hemolymph and in fat body cells, but the enzyme exists as an inactive proenzyme. The mechanism by which an insect converts the proenzyme to the active form under normal physiological processes has not been elucidated. Trehalase can be rapidly activated by the simple act of wounding and collection of hemolymph for trehalose assay, and by disruption of other tissues during sampling. Thus, to measure the true level of trehalose in hemolymph or tissues requires care to inactivate or minimize trehalase activity during tissue collection and processing.

6.3.1.2 Metabolic Stores of Glycogen

Glycogen is a second form of storage energy. Insect flight muscles contain glycogen, but most muscles are too small to store very much. Glycogen is present at 10 to 15 mg/g thorax wet weight, which is mostly muscle tissue, in the blowfly *Phormia regina* (Childress et al., 1970), and this is sufficient for a few minutes of flight. To sustain flight, additional fuels must be brought to the flight muscles. Glucose can be released from glycogen stored in fat body by **glycogen phosphorylase**. This enzyme is present in muscle tissue and fat body as inactive phosphorylase *b*, and it must be activated to phosphorylase *a*. Activation is under control of the **hypertrehalosemic hormone**

(**HTH**), a peptide hormone formerly called the **hyperglycemic hormone** (**HGH**). Initiation of flight activates the corpora cardiaca, probably through nervous control, to secrete the hormone (Steele, 1961, 1980, 1985). HTH requires the participation of a second messenger, **cAMP**, at the fat body cell membrane surface (Hanaoka and Takahashi, 1977) to activate phosphorylase *b* kinase, which converts inactive phosphorylase *b* to active phosphorylase *a*.

$$\text{Phosphorylase } b \xrightarrow[\text{HTH, cAMP}]{\text{Phosphorylase } b \text{ kinase}} \text{Phosphorylase } a$$

Glycogen phosphorylase *a* then releases glucose from glycogen as follows:

$$\text{Glycogen}_n + \text{PO}_4 \xrightarrow[+\ \text{Ca}^{2+},\ \text{PO}_4]{\text{Phosphorylase } a} \text{Glycogen}_{n\text{-}1} + \text{Glucose-1-PO}_4$$

Because glucose-1-PO_4, rather than glucose, is released from glycogen, the investment of 1 ATP is saved in the initial stages of the glycolytic pathway. Free Ca^{2+} at concentrations as low as 10^{-8} *M* and inorganic phosphate stimulate phosphorylase *b* kinase, and stimulation is near maximum at 10^{-6} *M* Ca^{2+} (Chaplain, 1967; Hansford and Sacktor, 1970). Both free Ca^{2+} and inorganic PO_4 are increased as a result of the initiation of flight. The reverse reaction that inactivates glycogen phosphorylase by conversion of phosphorylase *a* to phosphorylase *b* is catalyzed by phosphorylase *a* phosphatase, but little is known about how this enzyme functions in insects.

6.3.1.3 Glycogen Synthesis

Storage glycogen occurs mainly in fat body cells, but some glycogen is stored in gut epithelial cells, and to a slight extent in muscle cells. Synthesis of glycogen is catalyzed by the **enzyme UDP-glucose-glycogen transglycosylase**, also known as **glycogen synthetase**, according to the following equation:

$$\text{UDP-glucose} + \text{Glycogen}_n \xrightarrow{\text{Glycogen synthetase}} \text{UDP} + \text{Glycogen}_{n+1}$$

The precise regulatory controls determining the synthesis of trehalose vs. glycogen are not clear in insects, but one factor known to stimulate glycogen synthesis in insect tissues is the accumulation of glucose-6-phosphate. Glucose-6-phosphate can accumulate slowly as the rate of trehalose synthesis declines due to the feedback inhibition of free trehalose upon trehalose-6-phosphate synthetase. Declining synthesis of trehalose is likely to shift the UDP-glucose pool toward synthesis of glycogen.

6.4 HORMONES CONTROLLING CARBOHYDRATE METABOLISM

The principal hormone controlling carbohydrate metabolism is the peptide hormone **HTH**, hypertrehalosemic hormone. A related peptide hormone, **adipokinetic hormone** (**AKH**), may supplant the action of HTH in some insects. For example, *Manduca sexta*, the tobacco hornworm, utilizes AKH for controlling carbohydrate metabolism during larval growth and development, but adults use AKH to mobilize lipids for flight fuel (Zeigler et al., 1990; Nijhout, 1994).

HTH and AKH have been purified and sequenced. The two compounds are closely related chemically and are considered to be members of the same family of peptide hormones (Gäde, 1990; Nijhout, 1994). HTH is a polypeptide of ten amino acids, and the sequence of amino acids varies slightly from species to species. AKH, also isolated from several different insects, may have from eight to ten amino acids in its structure.

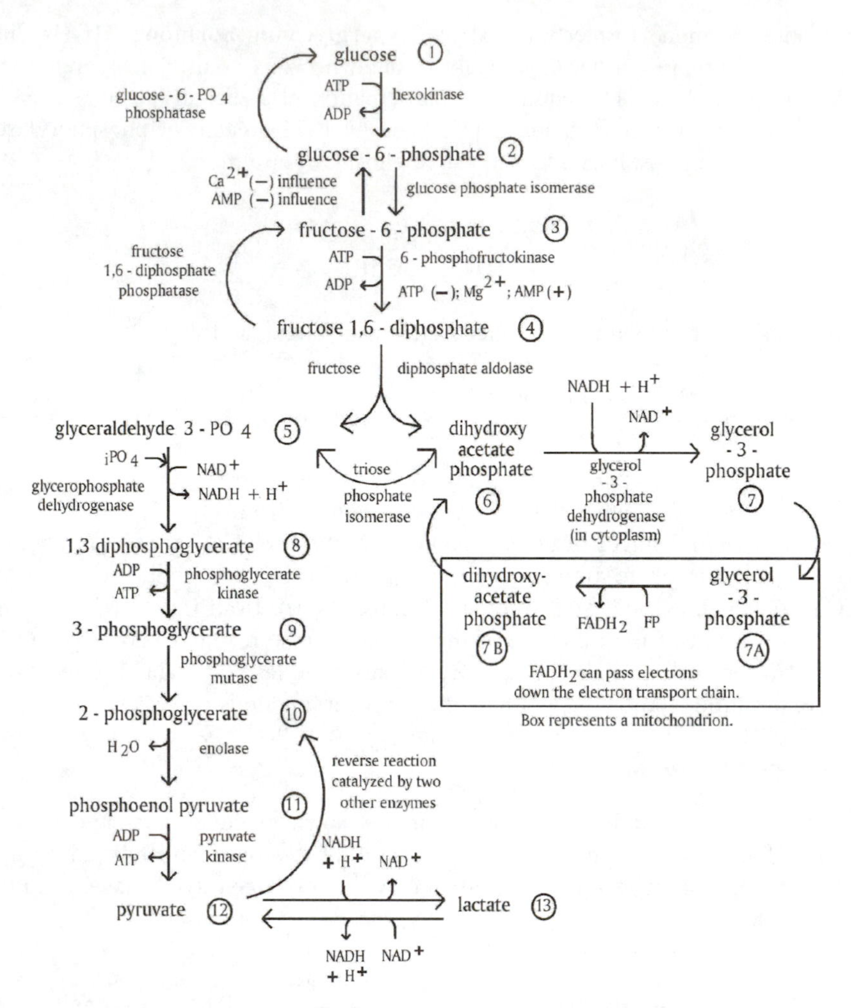

FIGURE 6.4 The glycolytic pathway for metabolism of glucose in insects.

6.5 PATHWAYS OF METABOLISM SUPPORTING INTENSE MUSCULAR ACTIVITY SUCH AS FLIGHT

6.5.1 GLYCOLYSIS

All insects tested metabolize carbohydrates first upon taking flight. For some insects, such as dipterans and hymenopterans, carbohydrate is the only fuel they can mobilize fast enough to support flight. Other insects metabolize carbohydrate at the initiation of flight, but if flight continues, they switch to another fuel such as proline or fatty acids. **Glycolysis** (Figure 6.4), the process by which insects start to metabolize glucose, is similar to the process in vertebrates and other organisms, with the exception that glycolysis in insect flight muscle is always aerobic, never anaerobic as in actively working vertebrate muscle. The tracheal supply to insect flight muscle is extensive (see Chapter 12), and capable of supplying sufficient oxygen for totally aerobic oxidation during flight. The enzymes of glycolysis function equally well under aerobic conditions or anaerobic conditions.

Another specialization in insect flight muscle glycolysis is the way in which NADH in the cytoplasm is oxidized to NAD^+. The quantity of cytoplasmic NAD^+ is limited in the flight muscles of insects, just as it is in vertebrate muscles, and in order for glycolysis to continue, NAD^+ must be constantly regenerated. Cytoplasmic NAD^+ in insect flight muscle is regenerated through the

glycerol-3-phosphate shuttle, not through conversion of pyruvate to lactate as in vertebrates. Some slower working skeletal muscles in insects, such as leg muscles, may oxidize NADH to NAD^+, however, by the pyruvate to lactate step.

Because glucose in most insects is not present in significant quantities in the cell cytoplasm or hemolymph, glucose entering the glycolytic process will be derived first from the hydrolysis of trehalose and slightly later from glycogen. Glucose derived from trehalose must be phosphorylated with participation of ATP and hexokinase. This investment of ATP to get the process started must be subtracted later from the total number of ATPs produced as the result of complete glucose metabolism. Paradoxically, for an insect initiating flight and needing energy from glucose, insect muscle hexokinase activity is easily inhibited by the product of its action, glucose-6-phosphate, as it is in other animal systems. The inhibition is countered by other products, however, such as inorganic phosphate that accumulates from use of ATP to power the sudden intense muscle activity of flight. Initially, the situation is somewhat analogous to driving a car with one foot on the brake and the other on the accelerator. During sustained flight, however, a steady state is soon reached so that glycolysis proceeds smoothly. Glucose is released from glycogen in a phosphorylated state, without the expenditure of a high-energy phosphate such as ATP. The phosphate group comes from inorganic phosphate.

In glycolysis, glucose-1-phosphate is changed to glucose-6-phosphate by phophoglucomutase, and glucose-6-phosphate is changed to fructose-6-phosphate, with neither reaction requiring the expenditure of ATP. Conversion of fructose-6-phosphate to fructose-1,6-diphosphate requires another phosphorylation, with ATP supplying the energy and phosphate group. Thus, depending on the source of glucose at the start, either 1 ATP (if the glucose is derived from glycogen) or 2 ATPs (if it is from trehalose) must be invested to get glycolysis under way.

Conversion of fructose-6-phosphate to fructose-1,6-diphosphate is one of the major control points for carbohydrate metabolism in insects, as it is in other organisms. Excess ATP inhibits phosphofructokinase isolated from blowfly insect flight muscle, and acts as a brake on glycolysis when demand for ATP drops. Phosphofructokinase is stimulated by AMP, inorganic phosphate, and cyclic AMP (Walker and Bailey, 1969), products expected to accumulate from initiation of flight and the use of available ATP in muscle contractions. Although ATP decreased in the blowfly *Phormia* upon initiation of flight, it fell only slightly from 6.9 m*M* to 6.2 m*M* (Sacktor and Hurlbut, 1966), a drop that is unlikely to relieve inhibition of phosphofructokinase because the lower concentration of 6.2 m*M* ATP still inhibited isolated phosphofructokinase *in vitro*. There is also a concomitant rise in AMP level from 0.12 m*M* at rest to 0.30 m*M* in flight, but again a magnitude of change that seems insufficient to account for the large increase in flight metabolism. Additional factors, perhaps relating to compartmentalization, and other as yet unidentified agents acting upon this control point are probably involved.

6.5.1.1 The Glycerol-3-Phosphate Shuttle and Regeneration of NAD^+

Fructose-1,6-diphosphate is split into two three-carbon products, glyceraldehyde-3-phosphate and dihydroxyacetone phosphate. These two compounds are interconvertible and the enzyme for conversion is **triosephosphate isomerase**. The oxidation of glyceraldehyde-3-phosphate to 1,3-diphosphoglycerate is a very important step because it is dependent on the availability of inorganic phosphate and the oxidized form of **nicotinamide adenine dinucleotide** (NAD^+). Inorganic phosphate (possible forms might be $NaHPO_4$ or $KHPO_4$) is unlikely ever to be a limiting factor in the reaction, but only small amounts of NAD^+ are present in the cytoplasm, and the oxidized form must be regenerated as rapidly as it is used in order for this cytoplasmic reaction to continue. In the reaction, two electrons and two protons are removed from glyceraldehyde-3-phosphate, thereby oxidizing it to 1,3-diphosphoglycerate. The two electrons and one proton are accepted by NAD^+, reducing it to **NADH** (Figure 6.5), and one proton is buffered by the cytoplasmic medium. If NADH resulting from this reaction could get into the mitochondria, which as noted above always have available oxygen in flight muscles, it could be reoxidized to NAD^+, but flight muscle mitochondria are relatively impermeable to NADH, NADPH, NAD^+, and $NADP^+$ (Sacktor, 1961; Sacktor and

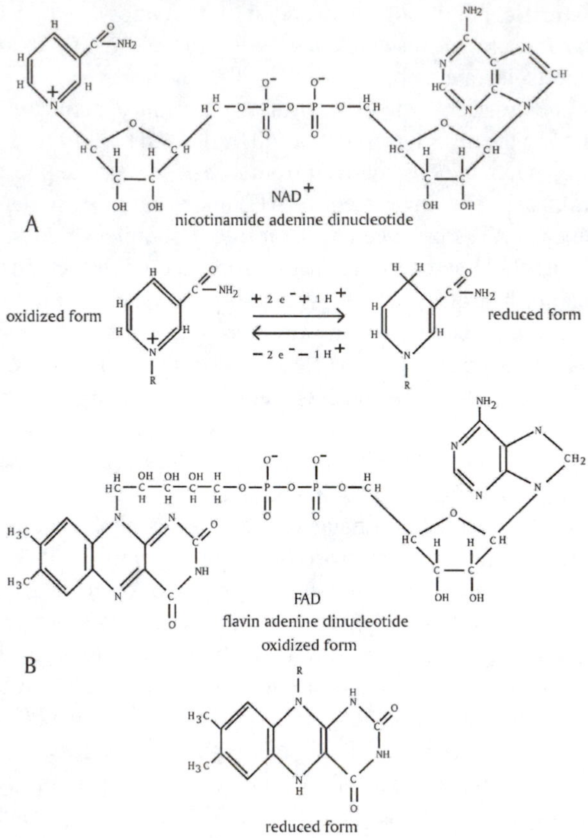

A

NAD^+
nicotinamide adenine dinucleotide

oxidized form
$+ 2e^- + 1H^+$
$- 2e^- - 1H^+$
reduced form

B

FAD
flavin adenine dinucleotide
oxidized form

reduced form

FIGURE 6.5 Reduced and oxidized forms of (A) nicotinamide adenine dinucleotide (NAD^+) and (B) flavin adenine dinucleotide (FAD).

Dick, 1962). Thus, a cytoplasmic mechanism is necessary to regenerate NAD^+. The common cytoplasmic mechanism for regenerating NAD^+ in working vertebrate muscle is the transfer of the two electrons and one proton from NADH (and a cytoplasmic proton, H^+) to pyruvate, thereby reducing it to lactate. Although lactic dehydrogenase, the catalyst for this reaction, occurs in insect walking leg muscles and other muscles that perform slower movements, its activity in flight muscle is very low and is unable to regenerate NAD^+ fast enough to allow carbohydrate metabolism to continue at a high rate. Flight muscles have high levels of another enzyme, **cytoplasmic glycerol-3-phosphate dehydrogenase** (Table 6.1), that catalyzes the regeneration of NAD^+ in the cytoplasm much faster by the cytoplasmic half of the glycerol-3-phosphate shuttle reactions, as shown here.

$$\text{Dihydroxyacetone phosphate} + \text{NADH} + \text{H}^+ \xrightarrow{\substack{\text{Cytoplasmic} \\ \text{glycerol-3-phosphate} \\ \text{dehydrogenase}}} \text{Glycerol-3-phosphate} + \text{NAD}^+$$

This cytoplasmic reaction is the first of a two-step shuttle for transferring electrons from the cytoplasm to mitochondria. Regeneration of cytoplasmic NAD^+ allows continued oxidation of glyceraldehyde-3-phosphate, with substrate-level production of 1 mole of ATP/3-carbon fragment oxidized to pyruvic acid. Pyruvate rapidly enters mitochondria and leads to further oxidation and production of ATP through the Krebs cycle.

Glycerol-3-phosphate (G-3-P) from the reaction above does not accumulate in the cytoplasm of flight muscle tissues, as lactate does in a working vertebrate muscle. About the same concentration

TABLE 6.1
Glycerol 3-Phosphate Dehydrogenase Activity

Tissue Source	μmoles/g wet wt./min
Blowfly flight muscle	1230
Honeybee flight muscle	700
Locust flight muscle	167
Cockroach flight muscle	48
Cockroach leg muscle	32
Locust leg muscle	33
Rat skeletal muscle	50
Beef smooth muscle	0.1

Adapted from Bailey (1975).

(2 m*M*) is present in both resting and working flight muscle (Sacktor and Wormser-Shavit, 1966). As fast as it is produced, G-3-P crosses the outer membrane of flight muscle mitochondria; and at the outer surface of the inner mitochondrial membrane, it is rapidly oxidized to dihydroxyacetone phosphate by an FAD-linked **mitochondrial glycerol-3-phosphate dehydrogenase** bound to the inner membrane, according to the following reaction.

$$\text{Glycerol-3-phosphate} + \text{FAD} \xrightarrow{\substack{\text{Mitochondrial} \\ \text{glycerol-3-phosphate} \\ \text{dehydrogenase}}} \text{Dihydroxyacetone phosphate} + \text{FADH}_2$$

The flavoprotein accepts the two electrons and two protons, which ultimately get rapidly transferred through the electron transport system in mitochondria to molecular oxygen as the final acceptor. Several evolutionary adaptations have made this shuttle possible, including (1) high activity of the cytoplasmic glycerol-3-phosphate dehydrogenase, (2) localization of an active glycerol-3-phosphate dehydrogenase on the outer surface of the inner membrane of flight muscle mitochondria, and (3) availability of oxygen and a fully functional electron transport system in mitochondria of working muscle. Dihydroxyacetone phosphate at the inner membrane surface rapidly diffuses out of mitochondria into the cytoplasm. The overall process can be succinctly summarized as follows: NADH is oxidized to NAD^+ in the cytoplasm and its electrons and proton (plus another cytoplasmic proton) are shuttled across the outer mitochondrial membrane via a carrier, glycerol-3-phosphate. At the outer surface of the inner membrane, the carrier is oxidized to dihydroxyacetone phosphate, which returns to the cytoplasm to repeat the process, and the electrons and protons pass down the electron transport chain with the generation of 2 ATP/cytoplasmic NADH oxidized (see Figure 6.7 and Section 6.5.3).

Evidence for the importance of the G-3-P shuttle in flight comes from mutants of *Drosophila melanogaster* that are deficient in cytoplasmic glycerol-3-phosphate dehydrogenase (Bewley et al., 1974; Collier et al., 1976) and are incapable of flight, presumably because they have no effective way to rapidly regenerate ATP in the cytoplasm.

6.5.1.2 Significance and Control of the Glycerol-3-Phosphate Shuttle

The significance of the glycerol-3-phosphate shuttle hinges on the assumption that the shuttle is self-generating when catalytic amounts of dihydroxyacetone phosphate are introduced (Sacktor and Dick, 1962). A small number of dihydroxyacetone phosphate (DHAP) molecules converted to

G-3-P in the cytoplasm, with subsequent conversion of G-3-P to DHAP in the membranes of mitochondria, may cycle over and over during flight and keep the cytoplasmic level of NAD^+ high. This would allow nearly all the DHAP produced from the splitting of fructose-1,6-diphosphate to be converted to glyceraldehyde-3-phosphate, and ultimately converted to pyruvate. In this scenario, all the initial glucose can be converted to two pyruvate molecules that enter the Krebs cycle; thus, ATP production in glycolysis and in the Krebs cycle is maximized. The shuttle itself should produce 4 moles ATP per mole glucose metabolized (2 ATPs for each cytoplasmic NAD^+ regenerated; or said another way, 2 ATPs for each FADH2 produced within the mitochondria as a result of the shuttle action); and if the shuttle makes it possible for nearly all of the glucose to be converted to pyruvate, then 4 moles ATP moles/mole glucose would be produced by substrate oxidations in the reactions of glycolysis. Thus, in this scenario, as many as 8 moles ATP could be produced per mole of glucose metabolized during glycolysis. If glucose is derived from trehalose, then 2 ATPs will be required in early phosphorylations (to form glucose-6-phosphate and fructose-1,6-diphosphate), so the net production would be 6 ATP. If glycogen provides the glucose, then only 1 ATP is needed in an early phosphorylation, and the net production of ATP/glucose metabolized is seven. The important point is that flight can continue for long periods supported by aerobic metabolism, which provides much more ATP than anaerobic metabolism.

The importance of the shuttle in insect flight muscle suggests that there must be control points in the shuttle mechanism, and indeed there are. Free Ca^{2+}, and possibly Mg^{2+}, are important in stimulating the metabolism of glycerol-3-phosphate. Ethylene diamine tetraacetic acid (EDTA), a sequestering agent for divalent cations, inhibits oxidation of glycerol-3-phosphate, but the inhibition can be reversed by adding additional Ca^{2+} or Mg^{2+}, thus implicating one or both of these ions as potential control factor in the shuttle reactions (Estabrook and Sacktor, 1958). Although Ca^{2+} is bound to the sarcoplasmic reticulum (SR), a network of membranes in muscle tissue, arrival of nerve impulses causes its release. Under these conditions concentrations of free Ca^{2+} ions at 10^{-6} *M* to 10^{-7} *M* occur in the sarcoplasm. Ca^{2+} released in muscle tissue stimulates G-3-P dehydrogenase and antagonizes resting inhibition of the dehydrogenase (Sacktor and Wormser-Shavit, 1966), with 10^{-7} *M* Ca^{2+} stimulating G-3-P dehydrogenase to about half-maximal activity (Hansford and Chappel, 1967; Donnellan and Beechey, 1969; Carafoli and Sacktor, 1972).

6.5.2 The Krebs Cycle

The reactions of the **Krebs cycle** are shown in Figure 6.6. The two products from glucose metabolism in the glycolytic pathway, pyruvate and glycerol-3-phosphate, rapidly enter mitochondria (Sacktor and Wormser-Shavit, 1966). Cytoplasmic pyruvate accumulates very briefly for the first few seconds after flight begins, probably due to the need to prime the Krebs cycle with intermediates, particularly oxaloacetate as the acceptor for acetate resulting from the oxidation of pyruvate. Priming of the cycle may result from proline metabolism. Proline decreases initially at the start of flight, and it may be converted by proline dehydrogenase to glutamate, which in turn undergoes transamination with pyruvate to form α-ketoglutarate and alanine. α-Ketoglutarate, a normal component of the Krebs cycle, is oxidized through several steps to oxaloacetate. In any event, the delay in oxidation of pyruvate is rapidly relieved.

Most substrates from the Krebs cycle are not readily oxidized when added exogenously to isolated mitochondria, apparently because they cannot penetrate the mitochondrial membranes (Van den Bergh and Slater, 1962). In preparations of isolated mitochondria from some insects, added succinate is metabolized, apparently after transport by a carrier in the mitochondrial membrane. The carrier is rapidly saturated by inorganic phosphate buffer, and succinate oxidation is best demonstrated in non-phosphate buffer systems. A few examples of rapid oxidation of other Krebs cycle intermediates have emerged, and it is probable that some variability exists among insect species as to which, if any, of the Krebs cycle intermediates can be metabolized by isolated mitochondria. In intact insects, of course, Krebs cycle intermediates are produced from within and

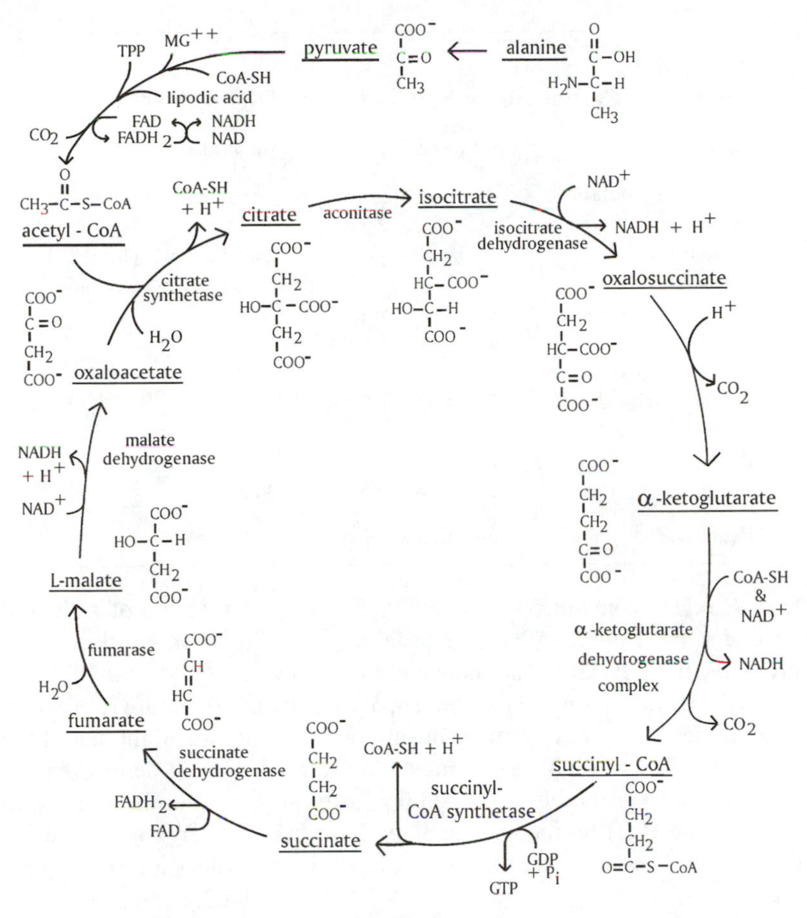

FIGURE 6.6 The reactions of the Krebs cycle in insect mitochondria. The principal source of pyruvate entering the Krebs cycle is glycolysis, but insects can convert alanine to pyruvate as one way to metabolize amino acids for energy. Amino acids can also enter the Krebs cycle at other points.

do not have to penetrate mitochondria. Lack of membrane permeability to the Krebs cycle intermediates serves to keep internal concentrations high, because they do not leak out.

The enzyme **pyruvate dehydrogenase** in insect flight muscle tissue exists in an inactive form (phosphorylated form) and an active form (dephosphorylated), with activation controlled by a phosphatase enzyme activated by free Ca^{2+}. Entry of pyruvate into mitochondria and its conversion to acetyl-Coenzyme A is likely to involve a multi-step sequence in insects, as it does in other organisms. A complex five-step sequence for passage of pyruvate across the mitochondrial membranes and into the mitochondrial matrix has been described for some non-insect mitochondrial preparations (Lehninger, 1975).

6.5.2.1 Control of Krebs Cycle Metabolism and Regulation of Carbohydrate Metabolism in Flight Muscles

Control points in the Krebs cycle enable insects to rapidly increase metabolic activity upon taking flight. **Isocitrate dehydrogenase**, an NAD-linked enzyme, is a major **control point** (Goebell and Klingenberg, 1964; Zahavi and Tahori, 1972; Hansford, 1972). It is inhibited by high levels of ATP, and stimulated by isocitrate, ADP, and inorganic PO_4. The latter two compounds accumulate from the use of ATP to initiate and support flight. The concentrations and relative ratios of ATP, ADP, and AMP play an important role in regulating metabolism in insects, as they do in other organisms.

TABLE 6.2
Control of Carbohydrate Metabolism in Flight Muscles

Enzyme	Activators	Inhibitors
Trehalose synthetase		Trehalose-6-phosphate
Phosphorylase *b* to *a*	CA^{2+}, P_i, AMP	
Hexokinase	P_i, Alanine	Glucose-6-phosphate
Aldolase		Citrate, palmitoylcarnitine
Phosphofructokinase	AMP, cyclic AMP, P_i, fructose-2,6-DiPO4	ATP
Fructose *bis*phosphatase		AMP
Glycerol-3-phosphatase (mitochondrial)	Ca^{2+} at 10^{-7} *M*	
Pyruvate carboxylase	Acetyl-CoA	
Isocitrate dehydrogenase	ADP, P_i, isocitrate	ATP

Reproduced with permission from Candy (1989).

Flight muscle of *P. regina* contains 6.9, 1.5, and 0.13 μmol/g wet weight of ATP, ADP, and AMP, respectively (Sacktor and Hurlbut, 1966). Upon initiation of flight, the level of ATP falls rapidly, while concentrations of ADP, AMP, and inorganic PO_4 increase.

Many steps in the mobilization and metabolism of carbohydrates require regulatory mechanisms to enable the rapid increase in rate of metabolism upon initiation of flight, and the equally rapid decrease in metabolism when flight stops. Table 6.2 summarizes well-defined enzyme regulators acting through stimulation or inhibition. In addition, carbohydrate mobilization for metabolism is under the control of the HTH hormone, as previously noted. The large mass and high metabolic rate of flight muscles and associated mitochondrial data in honeybees are shown in Table 6.3.

6.5.3 The Electron Transport System

The **electron transport system** in insect mitochondria is similar to that in other animals (Sacktor, 1974). The components of the respiratory chain are arranged in a sequence on the inner mitochondrial membrane so that electrons flow down the chain, as illustrated in Figure 6.7. Electrons pass sequentially to a component with a lower oxidation/reduction potential (a more positive value) until molecular oxygen accepts two electrons and two protons to form water. Flight muscle mitochondria are not permeable to NAD^+ or NADH, so cytoplasmic nucleotides do not readily enter mitochondria, nor do those inside the mitochondria pass outside. The dehydrogenase enzyme-NAD^+ complex is often tightly bound to the inner membrane. An exception is the mitochondrial glycerol-3-phosphate dehydrogenase involved in the shuttle described in glycolysis; this dehydrogenase is located on the outer part of the inner membrane and is linked to a flavoprotein other than NAD^+. Insect mitochondria have non-heme iron containing coenzyme Q, also called ubiquinone, between the flavoprotein and cytochrome *b* in the chain sequence. Like the cytochromes, one molecule of coenzyme Q accepts only one electron and no protons. Coenzyme Q has an isoprene side chain of varying length in different insects.

The **cytochromes** are proteins (enzymes) that contain iron (Fe) held in a heme-porphyrin structure (Figure 6.8). One molecule of any of the cytochromes can accept only one electron (or give up one upon oxidation). The atom of iron in the heme structure is the part of the molecule that accepts one electron (thereby becoming reduced Fe^{2+}). Fe^{2+} becomes oxidized to Fe^{3+} when the electron is passed to the next cytochrome in the sequence. No protons can be accepted by the cytochromes; the protons removed from FPH_2 are pumped into the mitochondrial matrix, while the two electrons are passed to two molecules of cytochrome *b*. Later these protons will return through

TABLE 6.3
Metabolic Rate Data and Morphometric Measurements of Mitochondrial Characteristics of Honeybees at 22°C

Body mass (mg)	82.61 ± 5.99
Thorax mass (mg)	28.55 ± 0.44
Metabolic rate	
Individual (ml O_2 h^{-1})	9.13 ± 0.25
Mass-specific (ml O_2 g^{-1} h^{-1})	114.8 ± 10.6
Thoracic (ml O_2 g^{-1} h^{-1})	320.2 ± 9.1
Muscle (ml O_2 cm^{-3} min^{-1})	8.38 ± 0.24
Myofibrils (% of muscle fiber volume)	53.8 ± 0.5
Mitochondria (% of muscle fiber volume)	43.0 ± 0.5
Surface density of mitochondria (m^2 cm^{-3})	
Outer membrane	3.0 ± 0.1
Inner membrane	48.0 ± 2.4
Mitochondrial respiration rate	
Total volume (ml O_2 cm^{-3} min^{-1})	19.49 ± 0.49
Cristae surface area (μl O_2 m^{-2} min^{-1})	414.1 ± 27.0

Note: Values are the means ± SE from eight individuals.

Data from Tables 1 and 2 in Suarez et al. (2000). Reproduced with permission.

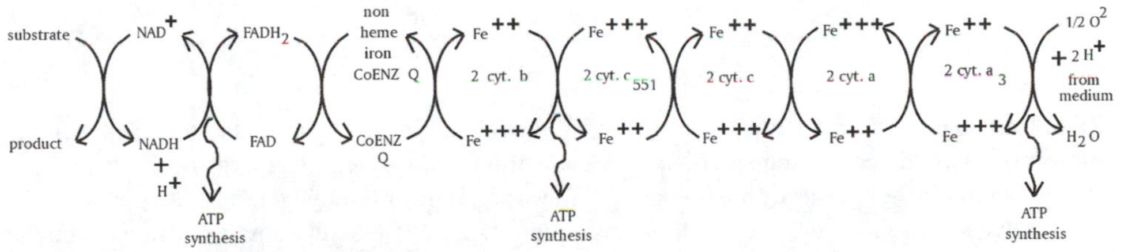

FIGURE 6.7 The electron transport system in insect mitochondria.

ATP synthase complexes in the inner membrane, enabling the formation of new ATP, and the protons join with two electrons in the structure of H_2O. All the cytochromes have absorption characteristics in the reduced state determined by the heme-porphyrin rings, the Fe, and the protein chain. Thus, cytochrome c_{551} is a characteristic cytochrome of insects and its absorption maximum in the reduced state is at 551 nm. In vertebrates, the cytochrome in this position in the chain is called cytochrome c_1, and its absorption spectrum has a maximum at 554 nm, indicating that it is slightly different in some way, perhaps only in the folding of the protein backbone of the cytochrome. Cytochrome *c*, the next cytochrome in the sequence, has been isolated and purified from many organisms, including different families of insects. The last two cytochromes in the chain make up a unit known as cytochrome oxidase. An electron is transferred from cytochrome *a* to cytochrome a_3 and subsequently to an atom of oxygen. The two necessary protons for water formation are taken from the mitochondrial matrix. Water formed through transfer of substrate electrons and protons to oxygen is called **metabolic water**. It is a very important source of water for all insects, and especially those that live in very dry environments.

At three points in the electron transfer chain enough energy is released in a single step to enable the synthesis of an ATP molecule from ADP and inorganic phosphate (a minimum of about 7.5

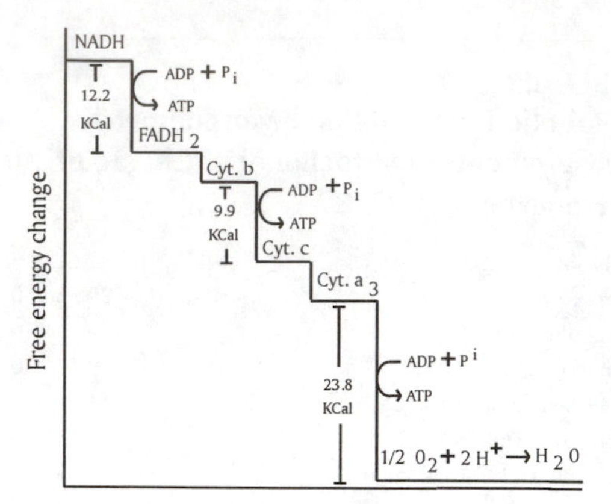

FIGURE 6.8 Changes in free energy and points at which enough energy is available to enable synthesis of ATP as electrons pass down the transport chain. A cytochrome, illustrated diagrammatically here, is a large enzymatic protein attached to an iron heme group. The iron atom in the cytochromes accepts one electron (becoming reduced) or gives up one electron (becoming oxidized) as electrons flow from component to component in the chain. Electron flow is only one way, always toward the component with the higher redox potential, and finally to molecular oxygen.

kcal is required) (Figure 6.8). Thus, when two electrons are passed through the complete electron transfer pathway, three ATPs can be formed. The first synthesis of ATP occurs when two electrons are passed from NADH to FAD. If this step is bypassed, as when FAD directly accepts the electrons from substrate oxidation (for example, in the mitochondrial reaction of glycerol-3-phosphate to dihydroxyacetone as part of the glycerol-3-phosphate shuttle), then the first ATP is not formed, and only two remaining ATPs per two electrons transferred are formed. The second ATP is formed when two cytochrome *b* molecules each transfer one electron to two cytochrome *c* molecules. The final ATP is formed when two cytochrome a_3 molecules transfer two electrons to molecular oxygen to form water (with the two protons coming from the general buffers of the mitochondrial matrix). In every step in which ATP is formed, more energy is released than can be captured in one ATP synthesized; but in this step in particular, a large release of energy occurs, equal to about 23.8 kcal/mole of substrate. This is sufficient energy to synthesize nearly three ATPs, but only one is actually formed. The remainder of the energy is dissipated as heat. Insects, like other biological organisms, are capable of capturing only about 40% of the energy in a glucose molecule as ATP.

Although fine details of ATP production in insects have not been elucidated, the **chemiosmotic hypothesis** proposed by Peter Mitchell (1979, and additional references therein) seems the likely

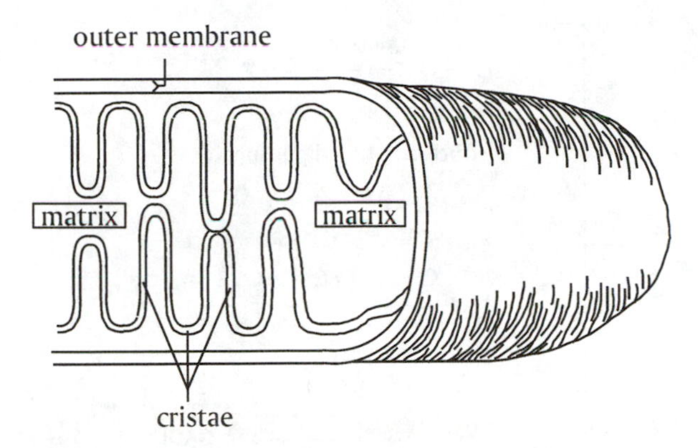

FIGURE 6.9 A diagrammatic cut-away view of a mitochondrion to illustrate the cristae typical of flight muscle mitochondria where the electron transport reactions occur, and the matrix where most of the citric acid cycle reactions occur. Cristae are much more numerous than this diagram shows.

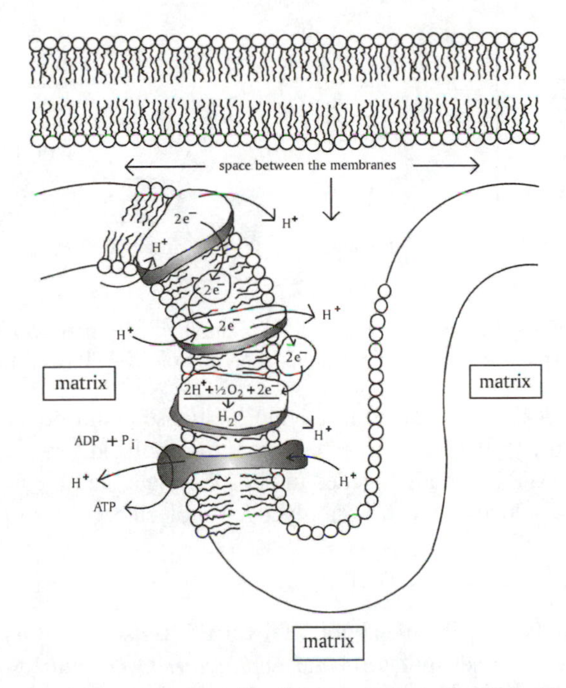

FIGURE 6.10 An illustration of the steps in transfer of electrons down the electron transfer chain, the pumping of protons into the space between the membranes, and the passage of protons back to the matrix through the ATP synthetase complex causing the formation of ATP.

mechanism. In the chemiosmotic process, energy from the transfer of electrons is used to pump protons into the space between the inner and outer mitochondrial membranes (Figures 6.9 and 6.10). This leaves the inner compartment of the mitochondrion negatively charged relative to the intermembrane space, which has a positive charge because of the accumulating H^+. Thus, a small battery results. Mitochondria utilize the potential difference between the membranes to supply the energy for ATP synthesis from ADP as the protons pass down the chemiosmotic gradient by returning to the inner compartment through F_1 membrane complexes, also called the **F_1 ATP synthase complex**. The complex is a **membrane ion channel** composed of proteins that allow the passage of protons, and ATP synthesizing enzymes are part of the complex.

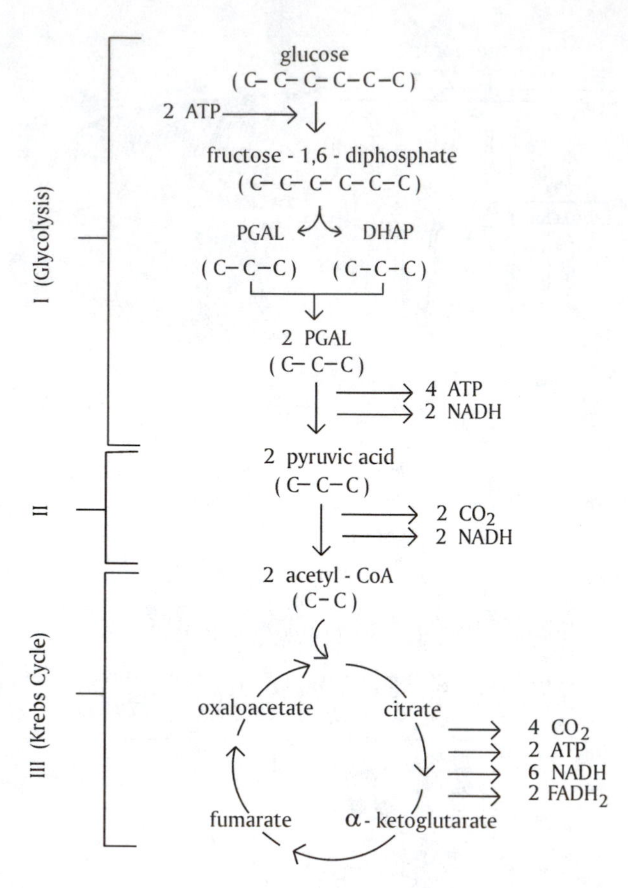

FIGURE 6.11 A summary of the major stages in the metabolism of glucose in flight muscle, with CO_2, reduced co-factors, and a small amount of substrate level formation of ATP resulting.

The energy yield from the metabolism of 1 mole glucose could be as much as 36 net moles ATP (Figures 6.11 and 6.12). If glycogen is the source metabolized, the number of ATPs could be higher because one less ATP is initially needed to phosphorylate the glucose-1-phosphate resulting from the cleavage of one glucose unit from a glycogen molecule.

6.5.4 Proline as a Fuel for Flight

The amino acid **proline** is a major metabolic fuel for the tsetse fly (Bursell, 1963, 1965, 1966, 1981); for adults of the Colorado potato beetle *Leptinotarsa decemlineata* (DeKort et al., 1973; Khan and De Kort, 1978; Mordue and De Kort, 1978; Weeda et al., 1980), some beetles in the family Scarabaeidae (Pearson et al., 1979), including *Melolontha melolontha*, *Heliocopris dilloni*, and *Popillia japonica*, and some beetles in the family Cerambycidae (Gäde and Auerswald, 2000). In the South African longhorned beetle *Phryneta spinator*, about 50% of the carbohydrates and 40% of the proline in the hemolymph were metabolized to support 5 min of flight, and alanine increased. An increase of alanine is expected when proline is metabolized for energy because of the transamination reaction in which the amino group from glutamic acid is transferred to pyruvic acid, creating alanine and α-ketoglutarate, as follows:

$$\text{Glutamic acid} + \text{Pyruvic acid} \rightarrow \alpha\text{-Ketoglutaric acid} + \text{Alanine}$$

Complete proline metabolism releases up to 14 moles ATP per mole proline and is a mitochondrial process, so flight and proline metabolism are linked to and dependent on a rich supply of

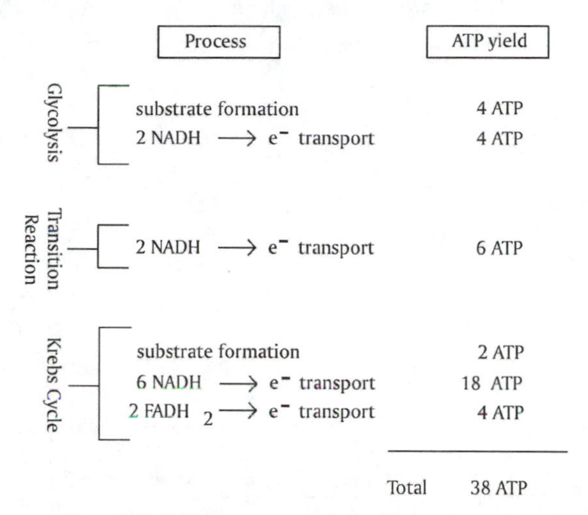

FIGURE 6.12 A summary of the energy yield in the several stages of glucose metabolism in flight muscle. Electron transport (e^- transport) occurs through the electron transport chain.

oxygen to flight muscles. The pathway for metabolism of proline, proposed in part by Bursell (1967), is shown in Figure 6.13. Proline readily enters mitochondria and is first oxidized to glutamate (Bursell, 1967) by a very active **proline dehydrogenase** located in tsetse fly flight muscle mitochondria. A flavoprotein accepts the two electrons and two protons that are removed. Glutamate then undergoes a transamination reaction with pyruvate to produce α-ketoglutarate and alanine. The α-ketoglutarate formed is a normal component of the Krebs cycle, and it is readily metabolized by the Krebs cycle pathway.

One of the products from proline metabolism, alanine, is rapidly removed from the muscle and transported to the fat body where, by addition of a two-carbon unit derived from fatty acids, it is converted into proline again. It can then be transported to the muscles to repeat the proline cycle reactions. Thus, the proline pathway represents a shuttle to transfer two-carbon units from the fat body to the muscles for metabolism (Candy, 1989). The sum of the reactions is as follows:

$$\text{Proline} \rightarrow \text{Alanine} + 2\ CO_2 + 3\ \text{NADH} + 2\ FPH_2 + 1\ \text{GTP}$$

No satisfactory explanation has been offered as to why some insects use proline, while others use glucose or lipid metabolism, both of which produce larger amounts of ATP per mole substrate metabolized than proline. Possibly, many insects metabolize small amounts of proline at the beginning of flight to prime the Krebs cycle. Evidence for proline use in this way has been obtained for the housefly *Musca domestica*, the blowflies *Phormia regina* and *Sarcophaga nodosa*, and *Schistocerca gregaria* locusts (Sacktor and Wormser-Shavit, 1966). In these insects, there is an initial disappearance of proline during the first minute of flight, with all of them switching rapidly to carbohydrate, and the locust ultimately switching to lipids if flight continues for an hour or so. Thus, perhaps a sort of preadaptation of the necessary enzymes may have been present in the early evolution of insects. A few insects have evolved much more dependence on the proline pathway for energy release. Because the pathway produces much less ATP per mole initial substrate metabolized, however, it probably was selected against by very strong and longer-distance fliers. Some attempts to relate proline metabolism to the blood-feeding behavior of tsetse flies no longer seem tenable in light of the several beetles now known to utilize proline.

Proline metabolism is subject to control by the cellular level of ADP. An increase in ADP stimulates oxidation of proline, while glutamate resulting from oxidation inhibits through a feedback

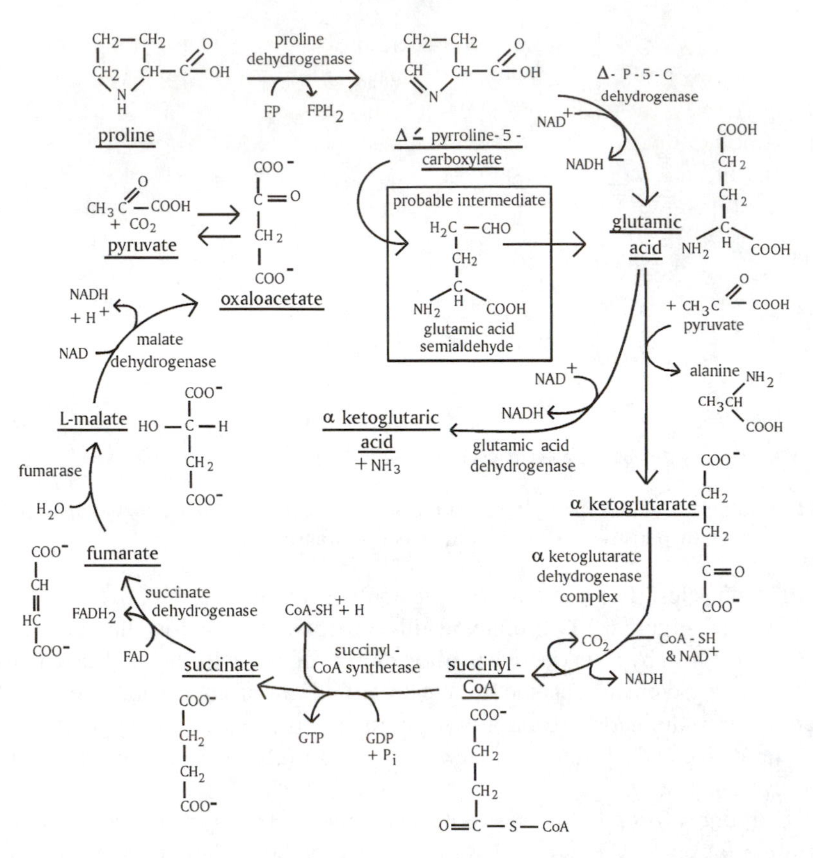

FIGURE 6.13 The pathway for proline metabolism in mitochondria in support of flight in the tsetse fly and some other insects. Proline enters mitochondria and is metabolized to glutamate and thence to α-keotglutarate by a transamination reaction. The remainder of the pathway is identical to the Krebs cycle. There is little of the enzyme glutamic acid dehydrogenase in tsetse fly mitochondria, so very little of the glutamic acid is converted to α-keotglutarate in that way.

mechanism. Isolated mitochondria from a blowfly, *P. regina*, are stimulated by addition of ADP (Hansford and Sacktor, 1970), which allosterically lowers the apparent K_m of proline dehydrogenase for proline from 33 to 6 m*M*. This is probably significant for the blowfly because the concentration of proline in flight muscle tissue was found to be between 6 and 7 m*M* (Sacktor and Wormser-Shavit, 1966). Similar effects of ADP on mitochondrial metabolism of proline were shown for tsetse flies, (another) blowfly, and houseflies, but not for locusts. The stimulatory action of ADP is probably a control point for the oxidation of proline in intact insects because ADP would be expected to accumulate upon initiation of flight. Glutamate formed from proline oxidation inhibits proline dehydrogenase by negative feedback. ADP can counter the inhibition of glutamate which aids the tsetse fly in using it as its main flight fuel.

6.5.5 Mobilization and Use of Lipids for Flight Energy

Adult Lepidoptera, some Orthoptera, and some other insects can use fatty acids as a flight fuel, and well-fed ones generally have sufficient lipids in the body to support flight for much longer periods than can be supported by the available supply of carbohydrates. The metabolism of fatty acids to support flight, however, presents several problems. Nearly all the fatty acids are stored in fat body cells as triacylglycerols, and thus before they can be metabolized, they must be released from fat body cells and transported to the muscles. Release of lipid from fat body cells is under the control of a peptide hormone, **adipokinetic hormone (AKH)**, secreted from the corpora cardiaca. AKH action, mediated through participation of cAMP at the fat body cell membrane and subsequent activation of a lipase, causes the release of diacylglycerol from fat body cells. The released diacylglycerol is transported through an aqueous medium, the hemolymph, to the thoracic flight muscles. Another muscle outer membrane lipase releases the two fatty acids, which now must have help in crossing the mitochondrial membrane. Metabolism of fatty acids is a mitochondrial process, and the efficient delivery of oxygen to flight muscles was a preadaptation to evolution of lipid metabolism by flight muscles.

6.5.5.1 Hormonal Control of Lipid Mobilization

Adipokinetic hormone (AKH), a decapeptide hormone synthesized in neurosecretory cells in one part of the **corpora cardiaca (CC)** and released from the CC (Beenakkers et al., 1986), promotes rapid release of diacylglycerol from fat body cells. The stimulus for secretion of AKH from the corpora cardiaca is not well defined, but probably nervous control associated with activation of the thoracic musculature is involved. AKH peptides have been isolated from a number of insects and the amino acid sequence indicates they are all members of the same family of neurohormones, and they generally have cross-reactivity (Nijhout, 1994). The peptides have eight to ten amino acids, depending on the species from which the different AKHs have been isolated. A typical structure, that of Locust AKH-1, is H_2N-Thr-Gly-Trp-Asn-Pro-Thr-Phe-Asn-Leu-PCA.

PCA, at the carboxy terminal end, is pyrrolidone carboxylic acid formed from the amino acid glutamine, also sometimes shown as pGLU, standing for cyclized pyroglutamate. Both names and symbols stand for the same chemical structure, and all the AKHs isolated have this modified amino acid at the carboxy terminal end. The amino acid at the amino terminal varies with the species. A specific radioimmunoassay for AKH has been developed (Fox and Reynolds, 1990).

AKH circulating through the hemolymph binds to a receptor on the surface of fat body cells, and activates adenylcyclase to produce cAMP as a second messenger. In turn, cAMP activates a lipase that removes one fatty acid residue from triacylglycerol, the storage form of lipid in fat body. The resulting diacylglycerol (DAG) is released from fat body cells, to be picked up by high-density (unloaded apoLp-A) lipoprotein particles at the hemolymph/fat body cell surface interface. The lipophorin particle also loads a small circulating hemolymph lipoprotein (apoLp-III) to form low-density lipophorin particles (LDLp). These LDLp particles are transported to the flight muscles. The process is summarized in Figure 6.14.

6.5.5.2 Transport of Lipids by Lipophorin

Insects transport lipids through the aqueous hemolymph as lipoprotein complexes called **lipophorins**. A variety of lipids have been found in lipophorin particles, including hydrocarbons, phospholipids, and tri- and diacylglycerols, but the one destined for metabolism in flight muscles is diacylglycerol. There are three identified proteins associated with lipophorin: apoLp-I, apoLp-II, and ApoLp-III. ApoLp-III associates reversibly with diacylglycerol-loaded lipophorin and dissociates from lipophorin in the unloaded state. It appears to be necessary to stabilize the loaded lipophorin-diacylglycerol particle. ApoLp-I and II remain in the structure of lipophorin in both the loaded and unloaded state.

Although the structure of lipophorins may vary, it appears that in general they have a hydrophobic core of hydrocarbons, a middle layer of diacylglycerol, and a surface monolayer composed of phospholipids and apolipoproteins. In contrast to the situation in vertebrates, insect lipophorin is not degraded after it delivers a lipid load to muscles (or to other tissues), but circulates in the hemolymph. The unloaded lipophorin may pick up more dietary lipids at the midgut or return to

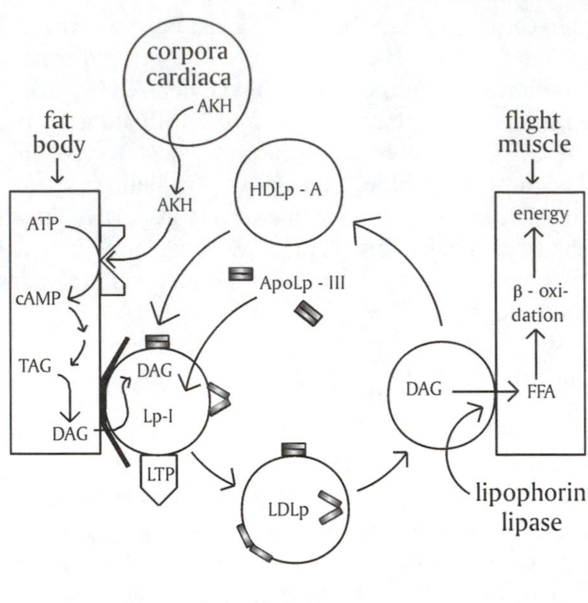

FIGURE 6.14 A diagram of the role of AKH in promoting DAG release from the fat body, loading of DAG into a lipoprotein transport complex, and delivery of it to muscle cells in an insect that can utilize lipid for flight metabolism. (Modified from Blacklock and Ryan, 1994.)

the fat body for a new load of lipids (Chino and Kitazawa, 1981; Chino, 1985; Surholt et al., 1991; Van Heusden et al., 1991; Gondim et al., 1992). Loading of an already existing lipophorin particle, as opposed to having to synthesize a new particle, is one of the adaptations evolved by insects to facilitate rapid mobilization of lipids for flight (Blacklock and Ryan, 1994). A detailed review of insect lipophorin and associated physiology has been presented by Blacklock and Ryan.

6.5.5.3 Activation of Fatty Acids, Entry into Mitochondria, and β-Oxidation

At the hemolymph/flight muscle cell interface, DAG is unloaded under the influence of AKH (Wheeler, 1989), the apoLp-III protein dissociates from the lipophorin, and the remaining high-density lipoprotein particle (HDLp-A) is shuttled back through the hemolymph to transport DAG again, either from the gut to the fat body, or from fat body to the muscles. A membrane-bound lipase in flight muscle tissue has high affinity for diacylglycerol, releasing fatty acids and glycerol from diacylglycerol bound to lipoprotein A^+, the transport lipoprotein complex (Wheeler and Goldsworthy, 1985; Van Heusden et al., 1986). Glycerol resulting from the hydrolysis can be phosphorylated and metabolized for energy (as glycerol-3-phosphate), or returned to the fat body where it can be used again to form triacylglycerols. The free fatty acids are bound to an intracellular protein in the locust *Schistocerca gregaria* (Haunerland and Chisholm, 1990), which helps to move them through the aqueous medium of the cell cytoplasm to the site of metabolism (the mitochondria). The fatty acids must be activated to cross mitochondrial membranes.

1. *Activation of fatty acids in the cytoplasm of muscle fibers and entry into mitochondria.* The sequence of reactions leading to β-oxidation within mitochondria is shown in Figure 6.15. Fatty acids in the cytoplasm must be complexed with the vitamin **carnitine** in order to cross mitochondrial membranes. In the cytoplasm, the fatty acid is activated by reaction with Coenzyme A (CoA) and ATP to form a fatty acyl CoA derivative. The reaction costs the equivalent of two ATPs per molecule of fatty acid activated, because

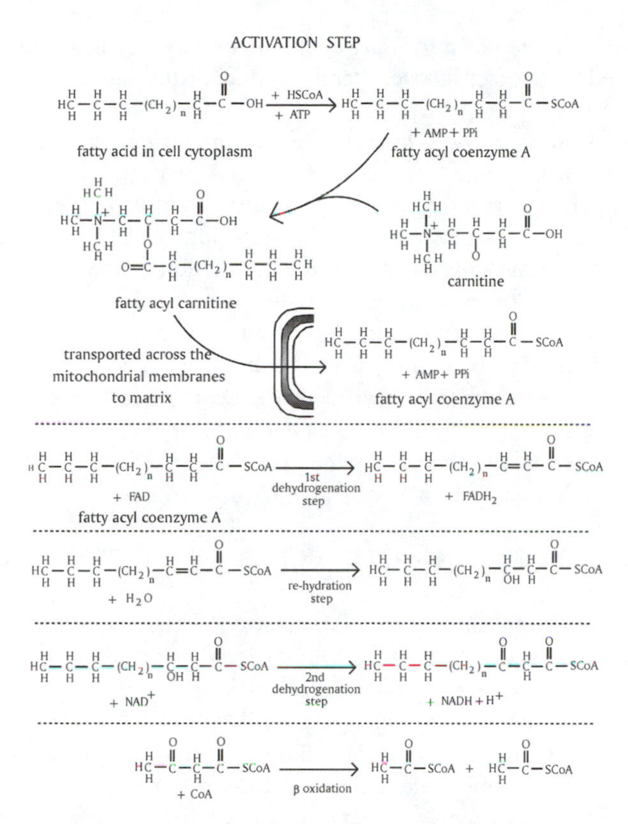

FIGURE 6.15 A summary of the metabolic reactions leading to β-oxidation of a fatty acid and release of acetyl CoA units within flight muscle mitochondria in insects that utilize lipids for flight energy. Carnitine aids in transferring fatty acids into the mitochondria. Subsequent reactions occur within muscle mitochondria, and the acetyl CoA generated is further metabolized by the Krebs cycle. An even-numbered fatty acid such as C18:3DB (oleic acid with three double bonds, which is a common fatty acid in insect lipids) will yield 9 acetyl CoA units, each of which yields 12 ATPs in the Krebs cycle. Palmitic acid (C16) would yield 8 acetyl CoA units. Thus, fatty acid metabolism yields a large number of ATPs.

the reaction results in AMP and PPi as by-products rather than ADP. Two ATPs will be needed to phosphorylate AMP in the process of replenishing the supply of ATP for additional reactions.

2. Carnitine is complexed in the presence of carnitine acyl transferase to the activated fatty acyl CoA, with release of CoA. CoA can now participate in activation of another fatty acid in the cytoplasm, while the fatty acyl carnitine derivative passes across the outer and inner mitochondrial membranes. Inside mitochondria at the inner surface of the inner membrane, carnitine is removed and acetyl CoA reacts with fatty acid to activate it again. Carnitine returns to the cytoplasm and is available to assist in the entry of another molecule of activated fatty acid. The fatty acyl CoA molecule, now in the matrix of the mitochondrion, undergoes further reactions leading to **β-oxidation** and removal of acetyl CoA units that readily enter the oxidative reactions of the Krebs cycle. The following reactions are necessary.
3. *Dehydrogenation with FAD as the cofactor: the first matrix reaction.* The initial matrix reaction is a **dehydrogenation** (an oxidation) in which a double bond is introduced at the β-carbon. FAD is the electron and proton acceptor. The reduced $FADH_2$ transfers its electrons through the electron transport system, which will result in the production of two ATPs per two electrons transferred.

4. *Addition of water to the activated fatty acid molecule.* Water is added across the double bond introduced in the step above, thereby reducing the fatty acid, but no significant energy input or release occurs.
5. *Dehydrogenation to yield the β-keto form of the molecule.* The activated fatty acid is oxidized again, but this time the two electrons and two protons are removed from the same carbon, carbon number 3, leaving a β-keto group at this position (Figure 6.15). NAD^+ participates in this reaction, and the resulting NADH passes its two electrons through the electron transport chain with the production of three ATPs.
6. *β-Oxidation.* Acetyl CoA is cleaved from the fatty acid chain, with a second CoA participating in the reaction so that the fatty acid (now minus two carbons) is left in an activated form. This new, shorter fatty acyl CoA molecule repeats the reactions of Steps 3, 4, 5, and 6. The shortened fatty acyl CoA molecule continues to repeat these steps, becoming shorter by two carbons each time another acetyl CoA is removed from it, until in its final passage through the steps, when only four carbons of the original molecule remain, it will yield two molecules of acetyl CoA as follows:

$$CH_3COCH_2COSCoA + CoA \rightarrow CH_3COSCoA + CH_3COSCoA$$

The energy derived from fatty acid metabolism. The metabolism of fatty acids yields many more ATPs per mole of fatty acid metabolized than can be derived from metabolism of glucose. Each $FADH_2$ from the oxidation in Step 3 yields 2 ATPs as the electrons pass down the electron transport chain; and similarly, each NADH from Step 5 yields 3 ATPs. Each acetyl CoA released from a fatty acid molecule results in 12 ATPs from oxidation through the Krebs cycle. Because a fatty acid must pass successively through Steps 3, 4, 5, and 6 each time a two-carbon unit is released by β-oxidation, 5 ATPs will be produced with each pass. A fatty acid will have to make $(n/2 - 1)$ passes (n = number of carbons in the original fatty acid). The scheme is based on metabolism of fatty acids with an even number of carbons, but generally these are the ones found in both plant and animal fats. Thus, the total ATPs produced for palmitic acid with 16 carbons, as an example, when it has been completely metabolized to carbon dioxide and water can be calculated from the following equations:

$$\text{Palmitoyl CoA} + 7\ CoA + 7\ FAD + 7\ NAD^+ \rightarrow 8\ CH_3COSCoA + 7\ FADH_2 + 7\ NADH$$

When 7 $FADH_2$ and 7 NADH transfer their electrons through the electron transport system, 35 ATPs will be produced. Each acetyl CoA metabolized through the Krebs cycle will yield 12 ATPs, and a total of 96 ATPs can be produced from the metabolism of 8 acetyl CoA molecules by the TCA cycle, as follows:

$$8\ CH_3COSCoA + 16\ O_2 + 96\ PO_4 + 96\ ADP \rightarrow 8\ HSCoA + 96\ ATP + 104\ H_2O + 16\ CO_2$$

The total ATP derived from complete metabolism of 1 molecule of palmitic acid will then be 35 ATPs + 96 ATPs = 131 ATPs. The net production of ATP will be 129 ATPs because 2 ATPs were needed in Step 1 to activate the fatty acid in the cytoplasm.

The ATP produced in these reactions will be used by a flying moth or locust to work the flight muscles, to supply energy to the nervous system, and for all the other physiological processes of the body that must go on even in flight. The metabolic water produced as a result of metabolism of just one fatty acid (108 molecules H_2O per molecule palmitic acid) is indicative of the large amounts of metabolic water that can be produced when an insect is able to metabolize lipid for flight. Metabolic water is a valuable resource to insects. Pupae of many insects contain large amounts of lipids accumulated during larval life, and these large stores of lipids can be metabolized slowly during pupal transformation to the adult to provide energy for new syntheses and cellular changes, and also supply water to the closed system.

REFERENCES

Bailey, E. 1975. Utilization of fuels by muscle, pp. 3-87, in D.J. Candy and B.A. Kilby (Eds.), *Insect Biochemistry and Function*, Chapman & Hall, London.

Beenakkers, A.M.Th., D.J. Van der Horst, and W.J.A. van Marrewijk, 1986. Insect lipids and lipoproteins, and their role in physiological processes. *Prog. Lipid Res.,* 24:19-67.

Bewley, G.C., J.M. Rawls, Jr., and J.C. Lucchesi. 1974. α-Glycerophosphate dehydrogenase in *Drosophila malanogaster*: Kinetic differences and developmental differentiation of the larval and adult isozymes. *J. Insect Physiol.,* 20:153-165.

Blacklock, B.J., and R.O. Ryan. 1994. Hemolymph lipid transport. *Insect Biochem. Mol. Biol.,* 24:855-873.

Bursell, E. 1963. Aspects of the metabolism of amino acids in the tsetse fly, *Glossina* (Diptera). *J. Insect Physiol.,* 9:439-452.

Bursell, E. 1965. Oxaloacetic carboxylase in flight musculature of the tsetse fly. *Comp. Biochem. Physiol.,* 16:259-266.

Bursell, E. 1966. Aspects of the flight metabolism of tsetse flies (*Glossina*). *Comp. Biochem. Physiol.,* 19:809-818.

Bursell, E. 1967. The conversion of glutamate to alanine in the tsetse fly (*Glossina morsitans*). *Comp. Biochem. Physiol.,* 23:825-829.

Bursell, E. 1981. The role of proline in energy metabolism, pp. 135-154, in R.G.H. Downer (Ed.), *Energy Metabolism in Insects*, Plenum Press, New York.

Candy, D.J. 1985. Intermediary metabolism, pp. 1-41, in G.A. Kerkut and L.I. Gilbert (Eds.), *Comprehensive Insect Physiology, Biochemistry and Pharmacology*, Vol. 10, Pergamon Press, Oxford.

Candy, D.J. 1989. Utilization of fuels by the flight muscles, pp. 305-319, in G.J. Goldsworthy and C.H. Wheeler (Eds.). *Insect Flight.* CRC Press, Boca Raton, FL.

Carafoli, E., and B. Sacktor. 1972. The effects of ruthenium red on reactions of blowfly flight muscle mitochondrial with calcium. *Biochem. Biophys. Res. Commun.,* 49:1498-1503.

Chaplain, R.A. 1967. The effect of Ca^{2+} and fibre elongation on the activation of the contractile mechanism of insect fibrillar flight muscle. *Biochim. Biophys. Acta,* 131:385-392.

Childress, C.C., B. Sacktor, I.W. Grossman, and E. Bueding. 1970. Isolation, ultrastructure, and biochemical characterization of glycogen in insect flight muscle. *J. Cell. Biol.,* 45:83-90.

Chino, H. 1985. Lipid transport: Biochemistry of hemolymph lipophorin, pp. 115-135, in G.A. Kerkut and L.I. Gilbert (Eds.), *Comprehensive Insect Physiology, Biochemistry and Pharmacology*, Vol. 10, Pergamon Press, Oxford.

Chino, H., and K. Kitazawa. 1981. Diacylglycerol-carrying lipoprotein of hemolymph of the locust and some insects. *J. Lipid Res.,* 22:1042-1052.

Collier, G.E., D.T. Sullivan, and R.J. MacIntyre. 1976. Purification of α-glycerophosphate dehydrogenase from *Drosophila melanogaster. Biochim. Biophys. Acta,* 429:316-323.

Davis, R.A., and G. Fraenkel. 1940. The oxygen consumption of flies during flight. *J. Exp. Biol.,* 17:402-407.

DeKort, C.A.D., A.K.M. Bartelink, and R.R. Schuurmans. 1973. The significance of L-proline for oxidative metabolism in the flight muscles of the Colorado potato beetle, *Leptinotarsa decemlineata. Insect Biochem.,* 3:11-17.

Donnellan, J.F., and R.B. Beechey. 1969. Factors affecting the oxidation of glycerol-1-phosphate by insect flight-muscle mitochondria. *J. Insect Physiol.,* 15:367-372.

Downer, R.G.H. 1985. Lipid Metabolism, pp. 77-113, in G.A. Kerkut and L.I. Gilbert (Eds.), *Comprehensive Insect Physiology, Biochemistry and Pharmacology*, Vol. 10, Pergamon Press, Oxford.

Estabrook, R.W., and B. Sacktor. 1958. α-Glycerophosphate oxidase of flight muscle mitochondria. *J. Biol. Chem.,* 233:1014-1019.

Fox, A.M., and S.E. Reynolds. 1990. Quantification of *Manduca* adipokinetic hormone in nervous and endocrine tissue by a specific radioimmunoassay. *J. Insect Physiol.,* 36:683-689.

Friedman, S. 1985. Carbohydrate metabolism, pp. 43-76, in G.A. Kerkut and L.I. Gilbert (Eds.), *Comprehensive Insect Physiology, Biochemistry and Pharmacology*, Vol. 10, Pergamon Press, Oxford.

Gäde, G. 1990. The adipokinetic hormone/red pigment-concentrating hormone peptide family: Structures, interrelationships and functions. *J. Insect Physiol.*, 36:1-12.

Gäde, G., and L. Auerswald. 2000. Flight substrates and their regulation by a member of the AKH/RPCH family of neuropeptides in Cerambycidae. *J. Insect Physiol.*, 46:1575-1584.

Goebell, H., and M. Klingenberg. 1964. DPN-spezifische Isocitrat-dehydrogenase der Mitochondrien. I. Kinetische Eigenschaften, Vorkommen und Function der DPN-spezifischen Isocitrat-dehydrogenase. *Biochem. Z.*, 340:441-464.

Gondim, K.C., G.C. Atella, J.H. Kawooya, and H. Masuda. 1992. Role of phospholipids in the lipophorin particles of *Rhodnius prolixus*. *Arch. Insect Biochem. Physiol.*, 20:303-314.

Hainsworth, F.R. 1981. Energy regulation in hummingbirds. *Am. Sci.*, 69:420-429.

Hanaoka, K., and S.Y. Takahashi. 1977. Adenylate cyclase system and the hyperglycaemic factor in the cockroach, *Periplaneta americana*. *Insect Biochem.*, 7:95-99.

Hansford, R.G. 1972. Some properties of pyruvate and 2-oxoglutarate oxidation by blowfly flight-muscle mitochondria. *Biochem. J.*, 127:271-283.

Hansford, R.G., and J.B. Chappel. 1967. The effect of Ca^{2+} on the oxidation of glycerol phosphate by blowfly flight-muscle mitochondria. *Biochem. Biophys. Res. Commun.*, 27:686-692.

Hansford, R.G., and B. Sacktor. 1970. The control of the oxidation of proline by isolated flight muscle mitochondria. *J. Biol. Chem.*, 245:991-994.

Haunerland, N., and J.M. Chisholm. 1990. Fatty acid binding protein in flight muscle of the locust *Schistocerca gregaria*. *Biochim. Biophys. Acta*, 1047:233-238.

Khan, M.A., and C.A.D. DeKort. 1978. Further evidence for the significance of L-proline for flight in the Colorado potato beetle, *Leptinotarsa decemlineata*. *Comp. Biochem. Physiol. B*, 60:407-411.

Lehninger, A.L. 1975. *Biochemistry*, 2nd ed., Worth Publishers, New York, 1104 pp.

Mitchell, P. 1979. Keilin's respiratory chain concept and its chemiosmotic consequences. *Science*, 206:1148-1159.

Mordue, W., and C.A.D. DeKort. 1978. Energy substrates for flight in the Colorado potato beetle, *Leptinotarsa decemlineata*. *J. Insect Physiol.*, 24:221-224.

Nijhout, H.F. 1994. *Insect Hormones*, Princeton University Press, Princeton, NJ, 267 pp.

Pearson, O.P. 1950. The metabolism of hummingbirds. *Condor*, 52:145-152.

Pearson, D.J., M.O. Imbuga, and J.B. Hoek. 1979. Enzyme activities in flight and leg muscles of the dung beetle in relation to proline metabolism. *Insect Biochem.*, 9:461-466.

Sacktor, B. 1961. The role of mitochondria in respiratory metabolism of flight muscle. *Annu. Rev. Entomol.*, 6:103-130.

Sacktor, B. 1974. Biological oxidations and energetics in insect mitochondria, pp. 271-353, in M. Rockstein (Ed.), *Physiology of Insecta*, 2nd ed., Vol. 4, Academic Press, New York.

Sacktor, B., and A. Dick. 1962. Pathways of hydrogen transport in the oxidation of extramitochondrial reduced diphosphopyridine nucleotide in flight muscle. *J. Biol. Chem.*, 237:3259-3263.

Sacktor, B., and E.C. Hurlbut. 1966. Regulation of metabolism in working muscle *in vivo*. II. Concentrations of adenine nucleotides, arginine phosphate, and inorganic phosphate in insect flight muscle during flight. *J. Biol. Chem.*, 241:632-634.

Sacktor, B., and E. Wormser-Shavit. 1966. Regulation of metabolism in working muscle *in vivo*. I. Concentrations of some glycolytic, tricarboxylic acid cycle, and amino acid intermediates in insect flight muscle during flight. *J. Biol. Chem.*, 241:624-631.

Steele, J.E. 1961. Occurrence of a hyperglycaemic factor in the corpus cardiacum of an insect. *Nature*, 192:680-681.

Steele, J.E. 1980. Hormonal modulation of carbohydrate and lipid metabolism in the fat body, pp. 253-271, in M. Locke and D.S. Smith (Eds.), *Insect Biology in the Future*, Academic Press, New York.

Steele, J.E. 1985. Control of metabolic processes, pp. 99-145, in G.A. Kerkut and L.I. Gilbert (Eds.), *Comprehensive Insect Physiology, Biochemistry and Pharmacology*, Pergamon Press, New York.

Suarez, R.K., J.F. Staples, J.R.B. Lighton, and O. Mathieu-Costello. 2000. Mitochondrial function in flying honeybees (*Apis mellifera*): Respiratory chain enzymes and electron flow from complex III to oxygen. *J. Exp. Biol.*, 203:905-911.

Surholt, B., J. Goldberg, T.K.F. Schulz, A.M.Th. Beenakkers, and D.J. Van der Horst. 1991. Lipoproteins act as a reusable shuttle for lipid transport in the flying death's-head hawkmoth *Acherontia atropos*. *Biochem. Biophys. Acta,* 1086:15-21.

Van den Bergh, S.G., and E.C. Slater. 1962. The respiratory activity and permeability of housefly sarcosomes. *Biochem. J.,* 82:362-371.

Van Heusden, M.C., D.J. Van der Horst, J.M. Van Doorn, J. Wes, and A.M.Th. Beenakkers. 1986. Lipoprotein lipase activity in the flight muscle of *Locusta migratoria* and its specificity for haemolymph lipoproteins. *Insect Biochem.,* 16:517-523.

Van Heusden, M.C., D.J. Van der Horst, J.K. Kawooya, and J.H. Law. 1991. *In vivo* and *in vitro* loading of lipid by artificially lipid-depleted lipophorins: Evidence for the role of lipophorin as a reusable lipid shuttle. *J. Lipid Res.,* 32:1789-1794.

Walker, P.R., and E. Bailey. 1969. A comparison of the properties of the phosphofructokinase of the fat body and flight muscle of the adult male desert locust. *Biochem. J.,* 111:365-369.

Weeda, E., C.A.D. DeKort, and A.M.Th. Beenakkers. 1980. Oxidation of proline and pyruvate by flight muscle mitochondria of the Colorado potato beetle, *Leptinotarsa decemlineata*. *Insect Biochem.,* 10:305-311.

Weis-Fogh, T. 1952. Fat combustion and metabolic rate of flying locusts (*Schistocerca gregaria* Forskål). *Phil. Trans. R. Soc. B,* 237:1-36.

Wheeler, C.H. 1989. Mobilization and transport of fuels to the flight muscles, pp. 273-303, in G.J. Goldsworthy and C.H. Wheeler (Eds.), *Insect Flight,* CRC Press, Boca Raton, FL.

Wheeler, C.H., and G.J. Goldsworthy. 1985. Specificity and localisation of lipoprotein lipase in the flight muscles of *Locusta migratoria*. *Biol. Chem. Hoppe-Seyler,* 366:1071-1077.

Zahavi, M., and A.S. Tahori. 1972. Activity of mitochondrial NAD-linked isocitric dehydrogenase in alatiform and apteriform larvae of *Myzus persicae*. *J. Insect Physiol.,* 18:609-614.

Zebe, E. 1954. Über den Stoffwechsel der Lepidopteren. *Z. Vergl. Physiol.,* 36:290-317.

Zeigler, R., K. Eckart, and J.H. Law. 1990. Adipokinetic hormone controls lipid metabolism in adults and carbohydrate metabolism in larvae of *Manduca sexta*. *Peptides,* 11:1037-1040.

7 Neuroanatomy

CONTENTS

PREVIEW

The central nervous system (CNS) comprises the brain, ventral nerve cord, and ventral ganglia. The brain consists of fused ganglia that make up the protocerebrum, deutocerebrum, and tritocerebrum. The protocerebrum is a major integrative center and receives sensory input from the compound eyes. The deutocerebrum receives sensory input from the antennae and sends motor output to the antennae. The tritocerebrum sends motoneurons to muscles in the labrum and pharynx, and innervates the stomato-gastric (foregut) nervous system controlling foregut muscles. In some insects, sensory axons from sensory receptors on the head terminate in the tritocerebrum; and in some insects the tritocerebrum receives sensory input from receptors on the mouthparts. The subesophageal ganglion has sensory and motor connections to sensory structures and muscles of the mouthparts, salivary glands, neck receptors in some insects, and neck muscles. Axons from neurons in the subesophageal ganglion project forward to the brain and posterior to the thoracic ganglia. The subesophageal ganglion has influence over motor patterns involved with walking, flying, and breathing, although those motor patterns originate in other ganglia. Typically, there are three thoracic ganglia, the pro-, meso-, and metathoracic ganglion, located, respectively, in the pro-, meso-, and metathoracic segments. Each thoracic ganglion sends motor axons to the leg muscles of its segment and receives sensory axons from sensory receptors in the tarsi and leg joints. The meso- and metathoracic ganglia send motor nerves to the wing muscles. Although the primitive evolutionary condition seems to have been that each abdominal ganglion innervated, and received sensory information from, structures in its segment, in all existing insects fusion of some abdominal ganglia has occurred. Some Apterygota have eight abdominal ganglia, some Odonata larvae have seven, and Orthoptera have five or six. In some highly evolved dipterans and hemipterans, all abdominal ganglia have fused with thoracic ganglia. Nerves radiate from fused ganglia to organs and muscles representing the evolutionary segmental origin of the ganglia that have fused. The central region of all ganglia is an area of synaptic connections called the neuropil, and cell bodies of motor neurons and interneurons tend to be located peripherally. The cell bodies of sensory neurons are located in peripheral parts of the body near the sensory site (e.g., in the antennae or tarsi) and many other internal and cuticular sites. The brain, ventral connectives and ganglia, and large lateral nerves are protected from direct contact with the hemolymph by a selectively permeable barrier, the hemolymph-brain barrier, consisting of an outer acellular layer called the neurolemma and an inner cellular layer called the perineurium. Nerve cells have a high demand for oxygen and nutrients. The CNS receives a rich supply of tracheae delivering oxygen, and hemolymph bathes the nerves and ganglia in the open circulatory system. Neurons are classified as sensory, or afferent, if they deliver signals to the CNS, and motor, or efferent, if they deliver output to muscles, glands, and organs. Interneurons mediate between sensory and motor neurons, and generally are located within the CNS. Neurosecretory cells are present in all ganglia, and play a major role in producing neuropeptides that regulate a variety of physiological and behavioral functions. Motor programs originate in various ganglia and control repetitive muscular actions, such as tracheal ventilation, walking, and ecdysis.

7.1 INTRODUCTION

So little was known about the anatomy and function of nervous systems just a little more than a century ago that the prevailing idea then was that the nervous system was composed of anastomosing cells, and individual cells were thought not to exist. The very small size of neurons, their long extended processes, and histological stains and procedures that did not clearly differentiate individual neurons led some early anatomists including Golgi, Weigert, and Nissl, all of whom nevertheless made major contributions to neuroanatomy, to make or support this erroneous conclusion. Despite the difficulty of studying the detailed internal anatomy of the nervous system, studies of gross anatomy continued to reveal fine details. With the development of better stains, a great deal of progress in defining the anatomy of nervous systems occurred in the last half of the 19th century. Golgi developed the staining procedure in 1873 that bears his name and is still used today. It allowed him to see evidence of individual neurons within a ganglion. The Spanish anatomist Ramon y Cajal used Golgi staining with improvements he devised, and published in 1888 his "neuron doctrine," in which he declared that nervous systems of animals were composed of individual cells, just as all other organs were. Cajal and Golgi later shared the Nobel prize for their pioneering work in neuroanatomy. Each advance stimulated another, and many anatomical details of the nervous system in animals, including insects, were published in the late 19th and early 20th centuries. Most of the major details concerning insect nervous system anatomy were known by the early part of the 20th century. Indeed, as early as 1762 Lyonet had published a highly detailed and accurate drawing of the entire nervous system of a caterpillar, *Cossus* (cited in Strausfeld, 1976).

7.2 THE CENTRAL NERVOUS SYSTEM (CNS)

The **central nervous system** (**CNS**) of insects consists of the **brain**, the **ventral ganglia**, and the **ventral nerve cord** (Figure 7.1). In the earliest insects to evolve, a ganglion probably occurred in each segment, and nerves from it controlled the muscles and glands in that segment. Aristotle believed that the brain in insects resided somewhere between the head and the tail (cited in Strausfeld, 1976, p. 3); he may have been uncertain because a great deal of autonomy for control resides in each ganglion. The brain is in the head in most insects (but it is located several body segments posterior to the anterior end of the body in dipterous larvae). Some insects experimentally rendered headless, or merely brainless, can live for long periods of time and may even walk, mate, and lay eggs. Removal of the brain, entire head, or other parts of the nervous system, while keeping the insect alive, has been a useful approach in the study of hormones produced by the nervous system.

Fusion of ganglia has occurred in all insects, sometimes to the extent that there are no ganglia in the abdominal segments. The ventral nerve connectives between ganglia still show pairing indicative of bilateral symmetry of the system, but ganglia from the two sides of a segment have fused along the midline in all extant insects.

The ventral ganglia and connecting nerve cord usually lie close to the cuticle on the ventral side of the body. Typically, but not invariably, nerves from a ganglion innervate muscles and organs within the segment where the ganglion resides. Nerves from fused ganglia project to the various body segments and structures in the posterior of the body.

A ganglion typically contains a mass of neuronal cell bodies of inter- and motoneurons at the periphery and a central region, the **neuropil**, where synapses occur. The cell bodies of sensory neurons of insects are usually near the site of sensory stimulus reception, and consequently, many sensory neuron cell bodies are located peripherally in the cuticle and in or on internal organs.

7.3 THE BRAIN

The brain consists of three fused ganglionic masses: the **protocerebrum**, the **deutocerebrum**, and the **tritocerebrum** (Figure 7.2). These three ganglionic masses typically rest on top of the esophagus, which passes posteriorly between the connectives to the subesophageal ganglion. Together, the three parts are sometimes called the supraesophageal ganglion. Two excellent books have been written detailing the anatomy and function of the brain and other parts of the central nervous system in insects; Strausfeld (1976) deals primarily with Diptera, and Burrows (1996) concentrates on locusts.

7.3.1 PROTOCEREBRUM

The protocerebrum is the site of major integrative centers that process incoming information from many sensory sources. The optic lobes, which process information from the compound eyes, are

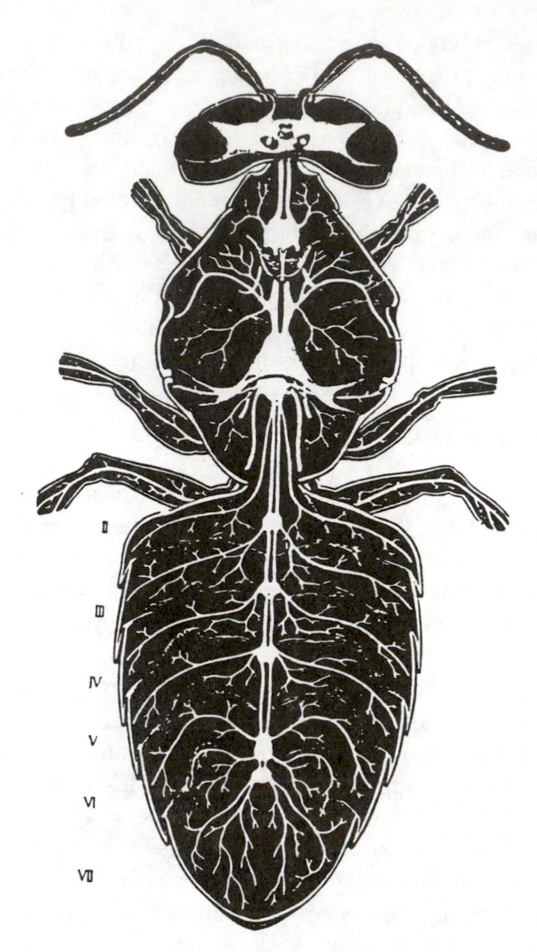

FIGURE 7.1 A drawing of the brain, ventral ganglia, ventral connectives, and some of the major nerves in a worker honeybee. The ganglia and nerve cord lie beneath the ventral cuticle, so this drawing depicts the system as a dissection from the dorsal surface, with all organs removed except the nervous system. In the earliest insects to evolve, there probably was a ganglion in each segment, but coalescence of ganglia has occurred in all living insects. The bilateral symmetry of the nervous system is still evident in the paired ventral connectives between ganglia, although in some insects the connectives are also fused together. There has been a tendency during the evolution of some insect groups for abdominal ganglia to become fused with thoracic ganglia, and only nerves pass into the abdomen to innervate the various muscles and organs. (From *The Hive and the Honey Bee,* 1975, Dadant & Sons, editors and publisher. Permission granted with acknowledgment.)

part of the protocerebrum. The optic lobes contain several neuropil regions related to visual processing (see Chapter 10, Sensory Physiology). The protocerebrum also receives input from ocelli via the ocellar nerves. Small paired nerves, the nervus corporis cardiaci I and II, link the protocerebral neurosecretory cells with the corpora cardiaca and corpora allata.

The **corpora pedunculata**, the mushroom bodies, are large, bilateral integrative centers in the protocerebrum (Figure 7.3). The size of the protocerebrum dedicated to the mushroom bodies varies in insects, with estimates of about 50,000 cells in locusts and as many as 1.2×10^6 in honeybees. These integrative centers are believed to be involved in olfactory learning through connections and input from the olfactory lobe integrative centers in the deutocerebrum, the region that receives olfactory input from the antennae. The mushroom bodies are divided into the peduncle (or stalk) and the calyx, the cap part of the mushroom shape. The peduncle contains fibers of neurons going to and from the calyx, a synaptic region. The mushroom shape of the corpora pedunculata is not so evident in all insects.

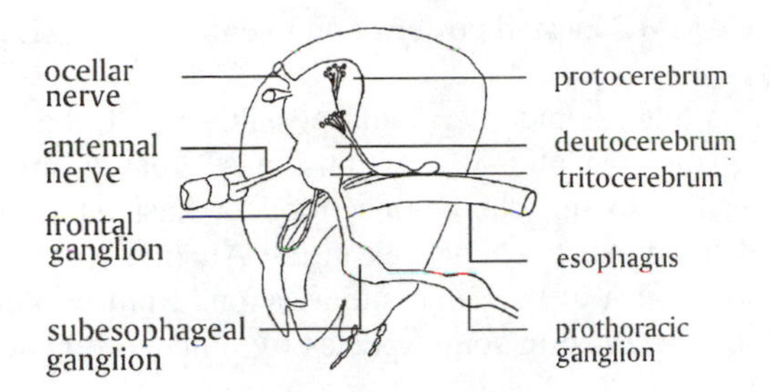

FIGURE 7.2 A lateral view of the three major parts of the brain (protocerebrum, deutocerebrum, and tritocerebrum) with associated head connections and ventral connectives to the subesophageal ganglion. (Modified from Snodgrass, 1935 and Jenkin, 1962.)

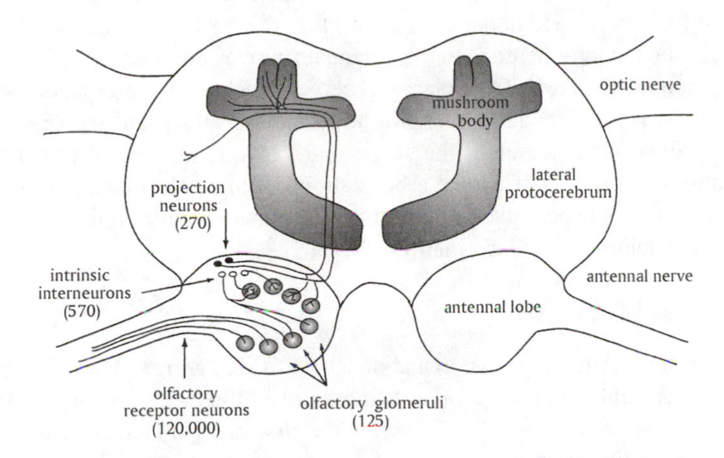

FIGURE 7.3 A frontal section through the brain of a female cockroach illustrating the antennal lobes and mushroom bodies. Sensory neurons from the antennae pass into the deutocerebrum, where they synapse in glomeruli with interneurons projecting to the protocerebrum. The numbers indicate the approximate numbers of neurons in various parts of the olfactory system in a cockroach. (Reproduced with permission from Lemon and Getz, 1999.)

Another major neuropil region in the central part of the protocerebrum is the **central body complex,** which is located between the bases of the stalks of the mushroom bodies. Although not a great deal is known about central body complex functions, it appears to be involved in "arousal" behavior, and it mediates between the two sides of the brain through fibers connecting both sides of the protocerebrum. It also receives input from the optic lobes. There are also paired lateral neuropil regions in the protocerebrum but their functions are poorly known. Internal commissures or fiber tracts connect parts of the protocerebrum and deutocerebrum with each other.

7.3.2 Deutocerebrum

The deutocerebrum receives sensory input from mechano- and chemosensory receptor neurons on the antennae (Hildebrand, 1996, 1995; Homberg et al., 1989; Stocker, 1994; Rodrigues and Pinto, 1989; Hösl, 1990), and sends motor signals to muscles of the antennae. There are separate neuropil regions in the deutocerebrum that process the information from the chemosensory and mechanosensory neurons; that is, the axons from the two types of receptors project (send axons) to separate areas within the deutocerebrum. Chemosensory input goes to the **antennal lobe** (**AL**) neuropil, while the **antennal mechanosensory and motor center** (**AMMC**) receives the mechanoreceptor input and sends motor information out. Each of these centers is represented on the left and right sides of the brain, with the AMMC located posterior and ventral to the AL (Marion-Poll and Tobin, 1992; Hildebrand, 1995).

At least in some insects (e.g., Lepidoptera) and possibly in all, the chemosensory inputs are further partitioned into separate synaptic sites within the AL, based on whether the information comes from receptors sensitive to sex pheromone, food or host odors, or carbon dioxide. The antenna is not reproduced in an exact spatial way in the AL (i.e, receptor axons from the distal portion of the antenna may project to the same site as axons from proximal antennal receptors), but directional information is retained in some species by unilateral input from an antenna to the ipsilateral side of the brain.

7.3.2.1 The Antennal Mechanosensory and Motor Center (AMMC) Neuropil

Relatively little is known at present about the AMMC, but as the name indicates, it contains arborizations of both motor and sensory neurons. Motor centers controlling muscles and glands of the head are in the deutocerebrum (with some additional ones controlling these structures also in the tritocerebrum). The deutocerebrum sends motor neurons to the antennal muscles and muscles of the labrum.

As a sensory center, the AMMC primarily receives mechanoreceptor neuron terminals from Böhm's organ, Johnston's organ, Janet's organ, and other mechanosensory structures located on the two basal segments of the antenna (the scape and pedicel) of various insects. In addition to terminal arborizations in the AMMC, some arborizations from mechanosensory cells project to the protocerebrum, the subesophageal ganglion, and into thoracic ganglia, indicating widespread distribution of some mechanosensory information.

7.3.2.2 The Antennal Lobe (AL)

In contrast to the meager information known about the AMMC, a great deal of information is now known about the **antennal lobe neuropil** and its interconnections with sensory structures and other parts of the nervous system (Figure 7.4). The AL is the first-order olfactory center in insects. Each AL receives sensory information from chemoreceptors on the flagellum of the antenna on the ipsilateral (same) side of the body. Within the chemosensory neuropil of the AL, there are two neuropil divisions in some insects: (1) one concerned with food, host, and perhaps other general environmental odors; and (2) a **macroglomerular complex (MGC)** that is sex specific in males of Dictyoptera (cockroaches) and Lepidoptera (moths) as the sensory neuropil for the sex pheromone receptor neurons. Axons from the antennal receptor cells pass through the antennal nerve and terminate in AL neuropil regions in structures called **glomeruli**.

7.3.2.2.1 Organization of Glomeruli in the Antennal Lobe (AL)

Glomeruli are somewhat cup-shaped masses of axonal terminals. Typically, a glomerulus is approximately 50 to 100 μm in diameter, and each is separated from other glomeruli by layers of glial cell membranes. The number of glomeruli is species specific, with approximately 10 in *Aedes aegypti* mosquitoes, 200 in *Formica pratensis* (an ant), and 60 in *Manduca sexta*. There is a great deal of convergence of the sensory axons: from 10^5 to 10^8 receptor neurons converge upon ten to a few hundred glomeruli in different insects. The glomeruli house axonal terminals of antennal olfactory receptor cells, neurites of local and output (projection) neurons, and terminals of centrifugal neurons projecting to the antennal lobe from other sites in the brain. A given glomerulus appears to be associated with a group of axons related to particular odor identification rather than representing a strict morphological array on the antenna; that is, in a given glomerulus, the converging sensory axons may come from various sites on the antenna where there are receptors sensitive to a specific odor.

Inside the glomeruli are the soma of one to several interneurons, whose terminals make synaptic connections with the incoming sensory axons. Three classes of interneurons are associated with the glomeruli:

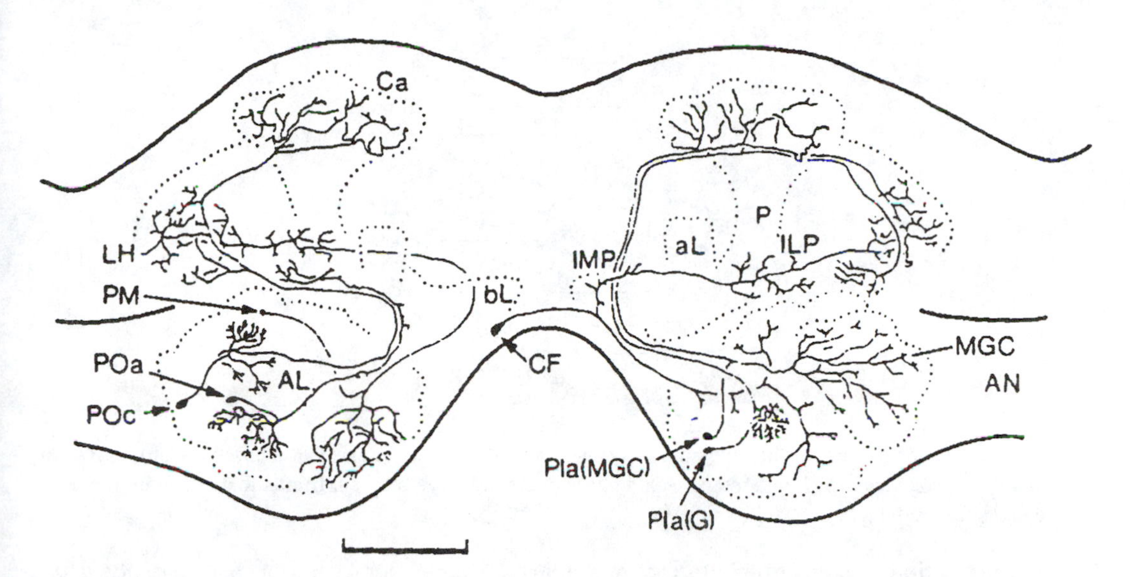

FIGURE 7.4 Horizontal cross-section through the brain in an adult *Manduca sexta*, the tobacco hornworm moth, showing a variety of neurons from the antennal lobe projecting to parts of the protocerebrum. The antennal lobes receive sensory input from the antennae. AL, antennal lobe; CF, centrifugal neuron; MGC, macroglomerular complex; bL, beta lobe; aL, alpha lobe; ILP, inferior lateral protocerebrum; P, pedunculus of mushroom body; IMP, inferior medial protocerebrum; Ca, calyces of mushroom bodies; LH, lateral horn of protocerebrum; Pla(G), Pla(MGC), POa, POc, various projection neurons. (Reproduced with permission from Homberg et al., 1989.)

1. Local interneurons (LNs) that interconnect areas of the AL but remain within the AL
2. Projection neurons (PNs) that have dendrites located in the AL with axons projecting into the mushroom bodies and lateral parts of the protocerebrum to which they relay information
3. Centrifugal neurons (CNs) whose axons project into the AL but cell bodies and dendrites are located outside the AL (some, but not exclusively, in the protocerebrum, for example).

Somata of the LNs and of most PNs are located in groups in the peripheral part of the AL. Somata of some PNs and of most CNs are located in the protocerebrum instead of the deutocerebrum.

Fiber tracts connect each AL with the protocerebrum and the subesophageal ganglion. In some Diptera, the ALs on opposite sides of the brain are interconnected by fiber tracts. ALs on each side of the brain can also communicate with each other through commissures in several groups of insects, especially Diptera (commissures are larger pathways between parts of the brain). Through synaptic connections with LN, PN, and CN interneurons, whose terminals project to many parts of the brain, sensory information is distributed to higher centers in the brain.

In male *M. sexta,* there are three sexually dimorphic glomeruli comprising the MGC (Rospers and Hildebrand, 1992). These glomeruli receive axonal arborizations from pheromone receptors on male antennae, and in addition to interneuronal connections noted above, glomeruli also receive terminals from an identified 5-hydroxytryptamine (5-HT)-immunoreactive neuron (Sun et al., 1993). The magnitude and duration of neuronal potentials from pheromone receptors on the antenna are increased by 5-HT added to the bathing saline, suggesting that 5-HT is a neuromodulator of pheromone signals relayed from the MGC to higher-order integrative centers in the protocerebrum (Kloppenburg and Heinbockel, 2000).

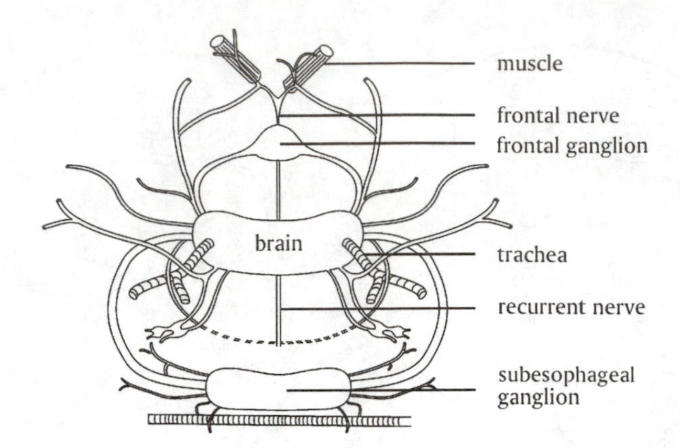

FIGURE 7.5 A dorsal view of the brain and associated head ganglia and nerves of a larval tomato hornworm *Manduca quinquemaculata* (Lepidoptera: Sphingidae). Note the large tracheal trunks that penetrate the brain tissue and provide for gas exchange in the brain.

The projection of axons from olfactory receptors to discrete glomeruli is a characteristic feature of both invertebrate and vertebrate olfactory systems (Hildebrand, 1995), and the organizational pattern may be as much as 500 million years old (Dethier, 1990, cited in Hildebrand, 1995). The role of glomeruli in odor perception has been reviewed by Galizia and Menzel (2001).

7.3.3 Tritocerebrum

The tritocerebrum sends motoneurons to muscles in the labrum and pharynx, and innervates the **stomato-gastric nervous system**, a system of several small ganglia, including the frontal ganglion, hypocerebral ganglion, and ingluvial ganglia having control of foregut muscles. In some insects, sensory axons from sensory receptors on the head terminate in the tritocerebrum; and in *Manduca sexta* it is known to receive projection neurons from sensory receptors on the mouthparts (Kent and Hildebrand, 1987). A commissural connective from each side of the tritocerebrum passes around the esophagus and provides cross-communication between the two halves, and lateral connectives connect each half of the tritocerebrum with the subesophageal ganglion.

The unpaired frontal ganglion (Figures 7.2 and 7.5), lying on top of the esophagus and anterior to the brain, is connected to the tritocerebrum by lateral connectives. The recurrent nerve and other small nerves arise from the frontal ganglion and carry motoneurons to muscles of the gut wall. Nerves from the frontal ganglion innervate the pharynx. Posteriorly, the median recurrent nerve runs along the surface of the esophagus, passes beneath the brain, and connects behind the brain with the hypocerebral ganglion, also lying on the surface of the esophagus. The small, unpaired **hypocerebral ganglion** innervates the corpora cardiaca. The small, paired **ingluvial ganglia** send nerves to the posterior foregut. The tritocerebrum has lateral nerve cord connections to the **sub-esophageal ganglion.** The brain typically rests on top of the gut, which passes between the lateral nerve cord connectives to the subesophageal ganglion. The subesophageal ganglion is formed from the fusion of three pairs of ganglia. It has sensory and motor connections to sensory structures and muscles of the mouthparts, salivary glands, neck receptors in some insects, and neck muscles. Axons from neurons in the subesophageal ganglion project forward to the brain and posterior to the thoracic ganglia. The subesophageal ganglion has influence over motor patterns involved with walking, flying, and breathing, although those motor patterns originate in other ganglia.

7.4 Ventral Ganglia

Many insects have three thoracic ganglia — the **pro**-, **meso**-, and **metathoracic ganglia** — located, respectively, in each of the thoracic segments (Figure 7.6), and such an arrangement was probably

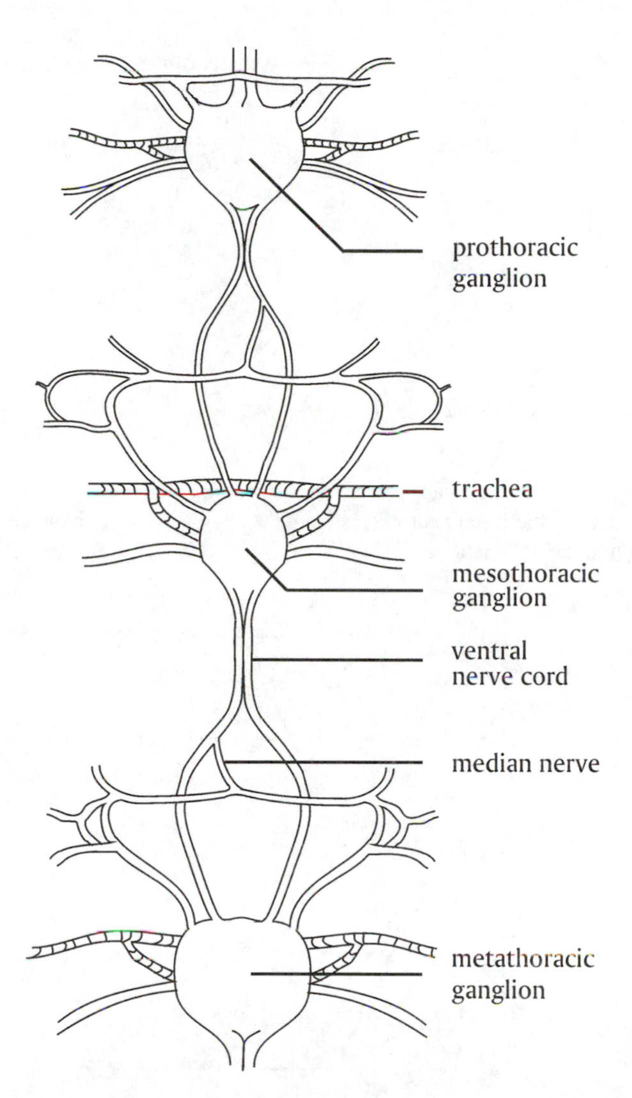

FIGURE 7.6 A dorsal view of the thoracic ganglia of a larval tomato hornworm. Note the median nerve that arises from the ganglion in front, travels with the connective for a distance, and then separates from the ventral connective and splits into two branches passing to each side of the posterior segment. A branch of the median nerve innervates the spiracular muscles that control opening and closing. Large tracheal trunks supply numerous smaller branches to the ganglia.

an early evolutionary one. In some insect groups (particularly Hemiptera and Diptera), the meso- and metathoracic ganglia have fused, often with fusion also of some of the abdominal ganglia into a large thoracic ganglion (Figure 7.7). Each thoracic ganglion sends motor axons to the leg muscles of its segment, and receives sensory axons from sensory receptors in the tarsi and leg joints. The meso- and metathoracic ganglia supply motor nerves to the wing muscles, respectively, of the meso- and metathoracic segments.

7.4.1 Abdominal Ganglia

The number of abdominal ganglia is highly variable in different orders. One ganglion in each segment was the evolutionarily primitive condition; but in all living insects, some fusion of abdominal ganglia has occurred. Some Apterygota still have eight abdominal ganglia, some Odonata larvae have seven, and Orthoptera have five or six. There may be fewer than five in some insects;

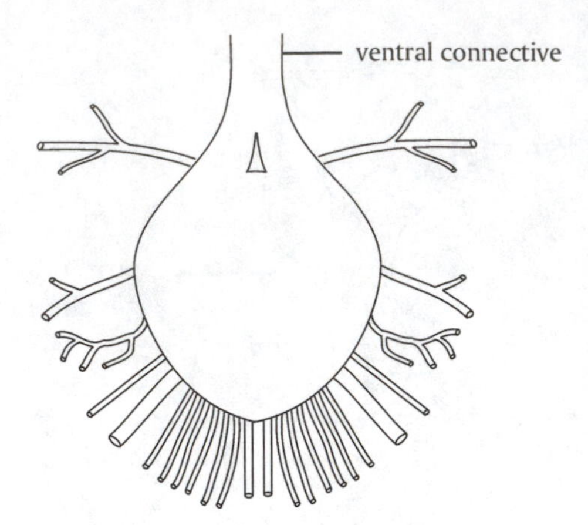

FIGURE 7.7 The large second thoracic ganglion of the hemipteran, *Oncopeltus fasciatus*, the large milkweed bug. The ganglion contains all the fused neuromeres from the abdomen. Note the many nerves that arise from the ganglion and pass into the abdomen.

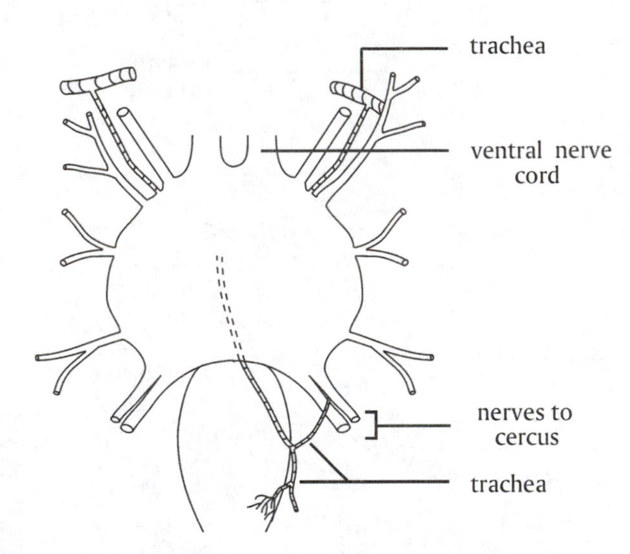

FIGURE 7.8 The sixth and terminal abdominal ganglion (TAG) of the American cockroach *Periplaneta americana*. The TAG contains fused neuromeres from several posterior abdominal segments. Nerves from the TAG pass into the cerci, and sensory neurons from the cerci synapse in the TAG with large (giant) axons that project forward to the thoracic ganglia and brain.

and in some highly evolved dipterans and hemipterans, all abdominal ganglia have fused with thoracic ganglia into the one large metathoracic ganglion. In locusts, *Schistocerca gregaria*, for example, the first three abdominal ganglia are fused in the adult with the large metathoracic ganglion. The large 6th abdominal ganglion in the American cockroach is the **terminal abdominal ganglion** (**TAG**) and is a fusion of neuromeres (neuromeres are ganglionic masses associated with a particular segment) from the posterior segments (Figure 7.8). Numerous nerves radiate out from the TAG to the posterior structures in the body, and it is well supplied with tracheae. Cercal nerves from the cerci carry large numbers of sensory axons into the TAG and synapse with giant axons, functioning as a fast escape mechanism when the cockroach is threatened. The giant axons pass forward through abdominal ganglia without synapsing to eventually synapse with interneurons connecting to leg motoneurons in the thoracic ganglia. Thus, information can be sent from the cerci to the legs very rapidly.

Fused ganglionic masses (i.e., fused neuromeres) send nerves to the various muscles and glands of the body segments according to their evolutionary origin, and also carry sensory axons back to the fused neuromeres. According to Snodgrass (1935), the best morphological indication of the composition of fused ganglia (or of a ganglion that has migrated out of the primitive segmentation pattern) is the distribution of nerves to various segments and segmental muscles or glands.

Thoracic and abdominal ganglia tend to be divided into three anatomical and functional divisions: a **dorsal motor neuropil**, a **middle integrative neuropil**, and a **ventral sensory neuropil**. Sensory information tends to come into the ventral portion of a ganglion from lateral nerves and from nerve tracts in the ventral nerve cord. Motor output more often occurs from the dorsal portion of a ganglion after associative interneurons allow the two regions to communicate. Internally, the middle layer (or associative zone) appears to contain the most complex array of variously shaped interneurons. Intersegmental ventral nerve cord connectives also tend to maintain the pattern of dorsal motor nerves and ventral sensory nerves. Burrows (1996) does not attribute any special significance to the anatomical and functional divisions of ganglia.

7.4.2 Lateral Nerves

Ganglia give rise to variable numbers of nerve tracts in different groups of insects. Most of the nerve tracts that arise from the brain or ventral ganglia contain mixed nerves; that is, the tract carries both sensory and motor neurons. An exception is the ocellar nerve connecting the ocelli with the protocerebrum. It is purely sensory, carrying interneurons with sensory information inward to the protocerebrum. The antennal nerves are mixed, carrying sensory fibers from olfactory and mechano-receptors inward, and motoneurons leading to tentorial ptilinial muscles and tentorial frontal muscles. The labro-frontal nerves and maxillary-labellar nerves are mixed, bringing sensory information from receptors on the mouthparts and carrying motoneurons to the muscles of the mouthparts.

Lateral nerves from the thoracic ganglia innervate the wing musculature and return sensory information from receptors associated with wing orientation. Motor neurons from the thoracic ganglia innervate the musculature of the legs and tarsi, and thoracic ganglia receive mechanosensory information from the legs and probably mechano- and chemosensory information from the tarsi in most insects. Motor neurons and sensory neurons do not directly synapse with each other, but each makes synaptic connections with interneurons (probably with many interneurons). Interneurons make synaptic connections with many other interneurons, and thus every part of the nervous system is potentially in communication. Incoming sensory information is received in various neuropils, interpreted, and motor commands are formulated to send to muscles or glands.

7.5 OXYGEN AND GLUCOSE SUPPLY TO THE BRAIN AND GANGLIA

Nerve cells typically have a high rate of metabolism, requiring a steady supply of nutrients and efficient gas exchange. Glucose is delivered (as trehalose) by the hemolymph, and gas exchange occurs through the tracheal system. Large tracheal trunks penetrate ganglia, branch into smaller tracheae and tracheoles, and bring air to within a few micrometers of each neuron or glial cell. Carbon dioxide is delivered back to the outside via the spiracles. The trachea and tracheoles have been estimated to take up 4 to 8% of the neuropil volume in the brain of the housefly (Strausfeld, 1976).

7.6 THE NEUROPIL

The main tissue mass of any ganglion is a central region of synaptic connections between arborizations of sensory, motor, and interneurons called the **neuropil**. The largest and most complex neuropil occurs in the brain in such associative centers as the mushroom bodies, central body, and optic lobes, but all ganglia contain neuropil regions. The neuropil is concentrated in the center of a

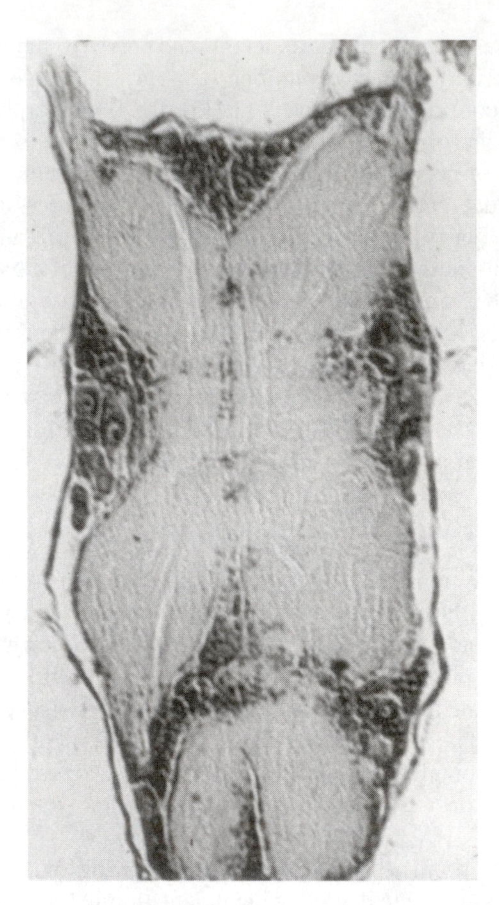

FIGURE 7.9 A light microscope view of a cross-section through the large thoracic ganglion of a dipteran to show the central neuropil and surrounding shell of somata. The ganglion contains the fused abdominal ganglia.

ganglion and is surrounded with a shell of cell bodies, the **somata** (**perikarya** of some authors) of motoneurons, interneurons, and glial cells (Figure 7.9). The neuropil continues to grow in size with new synaptic connections in the milkweed bug *Oncopeltus fasciatus*, in *Drosophila melanogaster*, and possibly in most or all insects, even after the nervous system as a whole stops growth. This suggests the importance of new information processing and integration that an insect uses to acquire food, mate, lay eggs, and survive in its environment.

7.7 HEMOLYMPH-BRAIN (CNS) BARRIER

At the hemolymph interface, the brain, ventral connectives, ventral ganglia, and large nerves are protected from direct contact with the hemolymph by a selectively permeable barrier, the **hemolymph-CNS barrier**, or **blood-brain barrier** (Scharrer, 1939; Strausfeld, 1976; Abbott and Treherne, 1977; Treherne, 1985). The barrier consists of an outer acellular layer called the **neurolemma** (also neural lemma, neural lamella) and an inner cellular layer, the **perineurium**, consisting of the perineural cells (Figure 7.10).

The open circulatory system with high hemolymph concentrations of trehalose and amino acids, some of which may function as neurotransmitters (Iversen et al., 1975), and the high potassium to sodium ratio in some phytophagous insects may have greatly influenced the evolution of the blood-brain barrier. The blood-brain barrier protects not only the brain but all ganglia and major nerves from direct contact with the hemolymph.

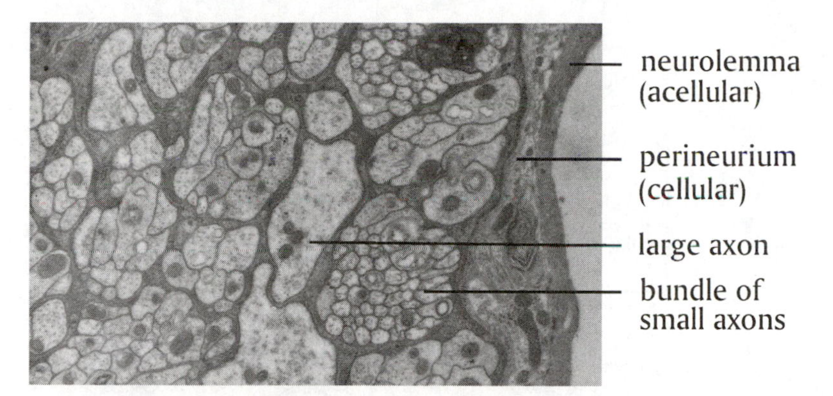

FIGURE 7.10 A transmission electron micrograph of a portion of the ventral nerve cord of a mosquito, *Culex* species. The brain, ventral nerve cord, ganglia, and major nerves are protected from contact with the hemolymph by the hemolymph-brain barrier. The barrier consists of an acellular neurolemma and a cellular perineurium made up of various types of glial cells. The neurolemma is rather permeable and leaky, and the perineurium is the main barrier.

The entire acellular and cellular barrier is thin, varying from 7 μm to about 15 μm thick around the brain of the housefly *Musca domestica*, for example. Some insects have a continuous fatty sheath of variable thickness composed of fat body cells exterior to the neurolemma that covers the entire brain and ventral nerve cord (Treherne, 1985). When present, the fatty envelope also acts as a barrier to ions and osmolites in the hemolymph.

The acellular neurolemma of some insects is composed of several layers identifiable in the electron microscope, and contains fibrils of collagen-like material that are embedded in a matrix of glycoaminoglycans (Ashhurst and Costin, 1971a, 1971b). The fibrous matrix is tough and elastic, and is the first barrier that a neurophysiologist must push a microelectrode through in order to record from within the CNS. The neurolemma is relatively permeable to large molecules, such as inulin and methylene blue, and in itself does not appear to be much of a barrier to hemolymph substances. This acellular neurolemma is probably secreted by the outermost layer of cells, sheath cells, located just below the surface of the neurolemma (Burrows, 1996).

The cellular layers beneath the neurolemma, the perineural cells, constitute the main hemolymph-CNS barrier. The perineural layer is several cells in thickness. Wigglesworth (1959) classified the perineural cells as glial cells, but that is not generally followed now (Strausfeld, 1976). The perineural cells have extensive couplings between the cells via tight and gap junctions (Lane and Skaer, 1980), which probably account for much of the impermeability of the cell layer; large molecules and ions would have great difficulty in passing between the cells because of the between-cell barriers (Strausfeld, 1976). The physiology of these cells is poorly known, but it has been suggested that the perineural cells are nutritive as well as protective. They probably participate in nutrient transfer from the hemolymph to underlying glial and nerve cells.

7.8 NEURONS: THE BUILDING BLOCKS OF A NERVOUS SYSTEM

Nerve cells (neurons) have many shapes, and no single shape can be said to be characteristic. Neurons have one or more **dendrites**, a cell body called the **soma** (perikaryon of some authors) that contains the nucleus, and an **axon** (Figure 7.11). Sometimes, the axon has collateral branches. Extensive arborization of axonal and dendritic processes is typical (Figure 7.12). Within the CNS, the arborizations enable synaptic contact with many other neurons, thus making possible communication throughout the nervous system. A dendrite is defined as any process conducting electrical excitation from the site of stimulation toward the soma, and the axon is the process conducting

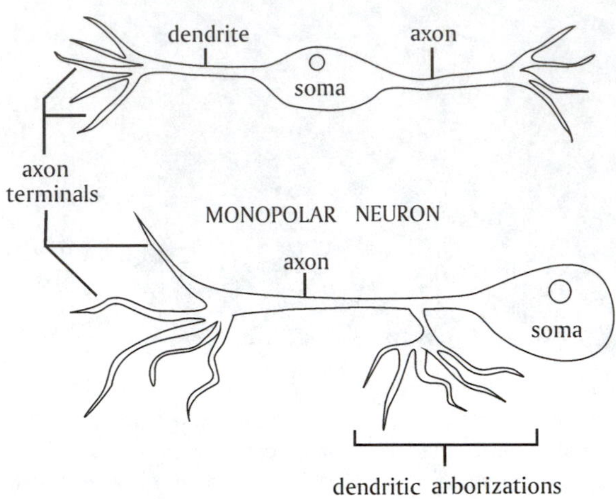

FIGURE 7.11 A schematic drawing of a sensory and motor neuron to illustrate axon, dendrite, soma, and arborizations.

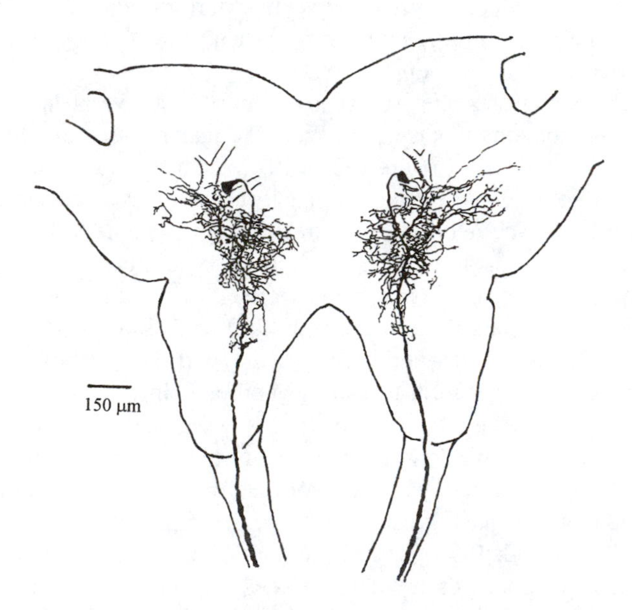

FIGURE 7.12 A drawing of cobalt stained neurons that illustrates the complex structure of neurons. (Reproduced with permission from Rind, 1990.)

excitation away from the soma. An axon may conduct excitation toward the CNS (sensory neuron) or away from the CNS (motoneuron) toward an effector, such as a muscle or gland.

The use of dendrite as a term for the extensive arborizations of neurons in the neuropil of insects has been criticized as borrowed and defined from vertebrate neurobiology (Burrows, 1996). Some of the arborizations from certain neurons in insects are known to receive input in the neuropil (i.e., act like dendrites), while others deliver output (act like axons), and morphology cannot be used to discriminate between these. A few motoneurons have been identified with input and output synapses on the same nerve branch. As a consequence, the word "dendrite" for arborizations of central neurons is often avoided, and different authors have used **neurites**, branches, and arborizations to describe branching without having to specify whether it functions in input or output (Burrows, 1996). When a clearly defined neurite emerges from the neuron in question and transmits spikes along its length, it is called an axon. There are a number of morphological types of cells in the nervous system, but they can be grouped functionally into **sensory neurons**, **motoneurons**, **interneurons**, **neurosecretory cells**, and **glial cells**.

7.8.1 Afferent or Sensory Neurons

Sensory neurons carry nerve impulses toward the CNS. There is a major morphological difference between insects and vertebrates in the location of the somata (singular = soma) of sensory neurons. In vertebrates, the somata of afferent neurons rest in the paravertebral chain of ganglia near the dorsal spinal cord; the dendrites are long and the axons relatively short. The somata of insect sensory neurons, with few exceptions, are located very near the site of stimulus detection, that is, at the cuticular surface for external receptors and as part of various internal organs. The dendritic processes are very short, and the axonal processes tend to be relatively long. Insect sensory neurons tend to be **bipolar**. The short dendritic processes of insect receptor cells probably do not generate or conduct spikes, as they must in the long dendrites of vertebrates, but develop graded electrical activity that spreads from the site of the stimulus. When the graded electrical activity is strong, it is likely to spread to the region where spikes are generated. The number of sensory neurons may change with molts and metamorphosis. For example, mechanoreceptors (innervated hairs) on the abdominal cerci of *Acheta domesticus* increased from 50 to 750 during growth and molting.

7.8.2 Efferent or Motor Neurons

Motoneurons have their somata located in the peripheral region of a ganglion and are usually monopolar in insects. The somata of motoneurons are usually very large, up to 100 μm in diameter. Motoneurons are usually paired on each side of a ganglion, and neurites and axons from a motoneuron usually remain on the side, or exit in case of an axon, from the side of the ganglion that houses the soma. **Dorsal unpaired medial (DUM) motoneurons** are not paired, but have neurites in each half of a ganglion and an axon on each side of the ganglion that emerges through a lateral nerve of the ganglion to innervate an effector gland or muscle.

A few motoneurons have branches on both sides of a ganglion, and the axon may even emerge from the contralateral side. Usually, the single process that arises from the peripheral soma of a motoneuron passes into the neuropil of its ganglion, branches off a network of arborizations (neurites) that make many synaptic connections with neurites of other interneurons, and the axon exits the ganglion on the ipselateral side through a large nerve and continues to its target site of gland or muscle. Some axons exit into the ventral connectives between ganglia and terminate in another part of the central nervous system, or eventually pass out through the lateral nerve of a different ganglion than that of its origin.

Motoneurons, when individually identified, are named (or numbered) after the muscle they innervate, a system started by Snodgrass (1929). The number of motoneurons is a very small percentage of the total nerve cells in an insect. Typically, only about 100 or so metathoracic motoneurons are involved with control of wing and leg muscles, and other thoracic musculature on each side of the body in a locust (Burrows, 1996). Often only two or three, and only in a few cases more, motoneurons go to large muscles. For example, the extensor tibiae of the third pair of legs in a locust receives four neurons, including a fast axon, a slow axon, an inhibitory neuron, and a DUMETi (dorsal unpaired median, extensor tibiae) neuron to the extensor tibiae muscle.

Axon-to-soma synaptic connections, a common type of synapse in vertebrate motoneurons, have not been found in insects. All synapses occur within the neuropil between neurites of neurons. The axon conducts a spike, but the soma and fine neurites of a motoneuron do not ordinarily conduct the spike, although somata of DUM neurons can conduct spikes. Under certain experimental conditions, even the soma of a motoneuron that usually does not conduct a spike can be shown to be excitable, and to possess voltage-sensitive Na^+ and Ca^{2+} channels that may be sensitive to neuromodulators that could create new dimensions in integration of signals (Burrows, 1996).

A motoneuron is generally bombarded with synaptic input from the many fine neurites in the neuropil; the electrical activity initiated in the neurites, while spike-like, is not transmitted over the many fine neurites and to the soma without decrement, as it typically is once it enters an axon. There are changes in rise time, amplitude, and duration of the excitability wave as it is conducted toward the soma and to the axon. If the excitation is strong enough, true spikes propagated without decrement will be generated at a zone in the axon sometimes called the axon hillock. Speed of transmission in axons is variable, depending on the size of the axon (faster transmission occurs in larger axons, such as the so-called giant axons) and other axonal characteristics.

7.8.3 Interneurons

Interneurons, also called **association neurons** and **internuncials**, may be located entirely within a ganglion or they may send intersegmental (axonal) processes through the ventral nerve cord to make synaptic connections in other ganglia. Interneurons make extensive synaptic connections with other interneurons, with incoming afferent (sensory) neurons, and with motoneurons. With their vast array of neurites and interconnections, interneurons are extremely important in coordinating communication between sensory and motor systems and within the central nervous system. The somata of interneurons lie in the peripheral region of a ganglion.

Local interneurons are those that make connections within a ganglion and do not exit from it. Interneurons that exit a ganglion to make connections in other ganglia are called **intersegmental interneurons**. Local interneurons are further classified into spiking and nonspiking interneurons, depending on whether the neuron transmits a spike or only graded electrical activity. Spiking local interneurons are most common in the optic neuropils, mushroom bodies of the protocerebrum, and in the antennal lobes of the deutocerebrum, but they can be found in thoracic ganglia and a few in other parts of the nervous system. In general, local interneurons are paired in the two halves of a ganglion, but the dorsal unpaired median (DUM) local interneurons, as the name indicates, are not paired. The somata of these DUM neurons are located in the periphery of a ganglion near the midline, and extensive neurites from each DUM spread into both halves of the ganglion. DUM neurons may be local, intersegmental, or efferent DUM neurons (Burrows, 1996). Intersegmental DUM neurons have neurites in both halves of a ganglion, and an axon arises from among the neurites on each side of the ganglion and passes through the intersegmental connectives into the next ganglion. Efferent DUM neurons were described in Section 7.8.2.

There may be several different neurotransmitters involved with spiking interneurons. Some show inhibitory action and evidence (immunoreactivity) of release of **γ-aminobutyric acid (GABA)**, a common neurotransmitter at inhibitory synaptic endings. Other spiking interneurons have excitatory action at the postsynaptic connections, and they must release a stimulatory neurotransmitter, but the chemical nature has not been identified. Efferent DUM interneurons appear to be octopaminergic (releasing octopamine) and probably exert neurohormonal control over muscle or gland action.

Nonspiking interneurons have only a vast array of neurites projecting into the neuropil of a ganglion and no identifiable axons. They are contained within a single ganglion. They release their neurotransmitter at the synaptic junctions without transmitting spikes. The monopolar neurons in the neural cartridges of the lamina ganglionaris in the optic lobe are nonspiking interneurons. They also occur in the neural network controlling walking behavior in cockroaches (Pearson and Fourtner, 1975) and other insects, and in the terminal abdominal ganglion (TAG) where they process signals from certain mechanoreceptors on the cerci (Kondoh et al., 1991, 1993). The dynamics of the electrical excitation and its movement over these nonspiking interneurons is not well-defined. There is evidence for K^+ and possibly Ca^{2+} ion channels in the membrane (Laurent, 1990, 1991), but no evidence for fast Na^+ currents (Burrows, 1996). The nature of neurotransmitters involved with nonspiking interneurons is not known.

7.8.4 Glial Cells

Glial cells are always present in nervous systems and are in intimate contact with neurons. Glial cells ensheath the axonal and neurite processes, as well as the somata, of neurons. Glial cells serve neurons and the nervous system in a number of ways; they provide structural support, nutritive and metabolic functions, and protection from outside chemical and ionic influence. They also may provide regenerative guidance for regrowth of severed or damaged neuronal processes, repair and maintenance of neurons, and may possibly function in signaling and information processing (Burrows, 1996). Multiple layers of glial cells help provide the blood-brain barrier that protects the CNS from direct contact with chemical components in the hemolymph.

Typically, in insects (and sometimes in vertebrates as well), many axons may be ensheathed by the same glial cell. For example, in *Acheta domesticus* (the house cricket), from 10 to 35 contiguous, small-diameter axons may share a glial cell in the dorsolateral bundles of the metathoracic nerve, as well as in the cercal nerves of adult crickets. Although a glial cell may wrap around an axon several times, it is not considered among neurophysiologists to be equivalent to the myelin sheath characteristic of vertebrate axons because the wrappings are not as numerous nor as tight in the case of insects. Insects do not have nodes of Ranvier, nor do they show saltatory conduction. In general, an axon must be large to have a glial wrapping not shared with other axons. In the house cricket, an axon must have a diameter greater than 1 μm to have its own glial sheath. Glial cells serve protective functions by isolating neurons or their processes from each other, and probably have a role in nutritive support of neurons, although direct evidence for this is not so clear in insects. They may remove and degrade excess neurotransmitter, or transmitter precursors, and possibly other toxins, and store nutritive macromolecules (Smith and Treherne, 1963; Treherne and Pichon, 1972; Strausfeld, 1976).

Large numbers of glial cells are to be found in the peripheral parts of every ganglion where they surround the cell bodies of neurons and the neurites or axons arising from the cell body. Glial cell types have been described primarily on a morphological basis. Wigglesworth (1959) described four types of glia from *Rhodnius*, and included the perineural cells that probably secrete the acellular neurolemma of the blood-brain barrier as one type. Other authors have excluded the perineural cells from their classification of glial cells. Strausfeld (1976) describes four types, based on electron microscopy, as follows:

1. **Type 1** neuroglia occur just inside the perineural cell layer and form tight junctions with perineural cells. The cellular processes from these neuroglia do not appear to enter the neuropil, but wrap around neuronal somata located in the perimeter of the ganglion. Multiple wrappings of motoneuron somata are common, and motoneurons are more heavily wrapped than dorsal unpaired median (DUM) neurons (Burrows, 1996). Extensive interdigitation occurs between glial cell processes and neuronal processes, especially at points where a neurite branches off from a soma to enter the neuropil.
2. **Type 2** glial cells have somata located interior to the Type 1 cells but still lie mainly at the peripheral edge of the ganglion. They isolate somata of neurons and surround neuronal processes passing into the neuropil.
3. **Type 3** cells are more interior in the ganglion and their somata lie just at the interface of the central neuropil. Type 3 cells send extensive lamellate, mossy, or spinniform processes into the neuropil, and mainly insulate neuronal processes within the neuropil. Some Type 3 somata have extensions reaching far back into the periphery of a ganglion where they make tight junctional contacts with perineural cells.

TABLE 7.1
Characteristics of Giant Axons of Selected Insects

	Species		
Characteristic	***Periplaneta americana***	***Locusta migratoria***	***Anax* spp.**
No. giant axons	6–8	4	6–7
Axon diameters (μm)	20–60	8–15	12–16
Location of somata	TAG[a]	TAG	Not determined
Sensory connections	Cerci	Cerci	Paraprocts
Connections to CNS	Thoracic MN[b]	Thoracic MN	Abdominal and thoracic MN

[a] TAG = Terminal abdominal ganglion.
[b] MN = Motoneurons.

4. **Type 4** neuroglia are the sheath cells that enclose single axons (large or giant axons) or groups of smaller axons in the brain, ganglia, ganglionic connectives, and in lateral nerves. When groups of small axons are enclosed by a single sheath cell, the enclosed axons may lie naked, and adjacent to each other; or in other cases the sheath cell invaginates a little way between the axons, partially isolating them (Strausfeld, 1976).

Synaptic junctions are free from glia, and some electron micrographs have shown indications that not even all neuronal processes in the neuropil may be isolated by glia. The number of glial cells in the nervous system of insects is not known in any insect, and may vary widely among insects. A gene, *repo*, is expressed in most glial cells but not in neurons (Halter et al., 1995) of *Drosophila melanogaster* and may prove to be a good marker of glial cells and facilitate counting.

7.9 GIANT AXONS IN THE INSECT CENTRAL NERVOUS SYSTEM

In contrast to the very small diameter of most neurons (the typical soma diameter is only a few micrometers, and the axon diameter is even less), some insects have **"giant" axons** of varying size running through the abdominal ganglia and into the thoracic ganglia. Table 7.1 shows data on some known giant axons. The CNS giant axons of insects develop by anastomosis of adjacent segmental neurons. The ganglionic junctional synapses are electrical rather than chemical, and much faster conducting than chemical synapses. The large diameter of giant axons also promotes more rapid rate of impulse conduction. A few such giant axons may be common in many, if not all, insects, but only a few have been studied. In the cockroach, the somata of the giant axons are located in the last abdominal ganglion, and the giants make synaptic contacts within the neuropil of the ganglion with sensory neurons from the paired cerci, the short appendages from the last segment of the insect. The giant axons run without synapsing through the abdominal ganglia and synapse in the thoracic ganglia with motoneurons going to the legs. These axons provide a rapid pathway for sensory information from the cerci to activate the legs in an escape behavior. A puff of air on the cerci causes a cockroach to make rapid escape movements. Some of the smaller giants also pass through the thorax and synapse in the brain (Hess, 1958). The giant axons in the CNS of dragonfly larvae (*Anax* spp.) run from the last abdominal ganglion to the thoracic ganglia, but these giants have synapses in each abdominal ganglion with motoneurons to abdominal muscles as well as with motoneurons in the thoracic ganglia to the legs (Fielden, 1960). The escape reaction in dragonfly larvae involves raising the legs, releasing the hold on the substrate, and simultaneous contracting abdominal muscles to force water out of the large rectum in a jet propulsion mechanism that propels the insect forward. Insect giant axons are hardly "giants" in comparison with the giant fibers of mollusks, some of which can be up to 1 mm in diameter. The very large axons in mollusks were very useful in determining the physiological properties of nerve impulse generation and transmission (Hodgkin and Huxley, 1952).

Giant axons have been found in the functioning of very fast trap jaws of ponerine ants in the genus *Odontomachus* (Gronenberg et al., 1993). These ants capture small prey in their jaws, which they lock open in a cocked position. Each of a number of mechanoreceptor hairs on the inner edges of the mandibles contains a large sensory neuron (15 to 20 μm in diameter) that passes through the mandibular nerve to the subesophageal ganglion. When the mechanoreceptor trigger hairs are touched by prey, the jaws snap closed in only 0.33 to 1 ms.

7.10 NERVOUS SYSTEM CONTROL OF BEHAVIOR: MOTOR PROGRAMS

Motor programs are neural mechanisms for coordinating and regulating repetitive behaviors. Although motor programs control and support many, perhaps all, of the repetitive behavioral actions of insects, few have been defined in any detail, and very few identified neurons in the pathways are known. Several examples of motor programs are described here to illustrate their nature and complexity. Motor programs originating in the CNS allow an animal to conduct rhythmic behavior without requiring timing signals from rhythmically stimulated sensory organs (Delcomyn, 1980). Some motor neurons may function in more than one motor program, for example, in programs controlling walking, running, and posture, all involving the legs. A motor program controlling **ecdysis** was described in Chapter 4.

7.10.1 A Motor Program That Controls Walking

One of the better defined motor programs (Figure 7.13) controls muscles involved in walking by the American cockroach, *Periplaneta americana* (Pearson, 1972, 1976; Pearson and Fourtner, 1975; Fourtner, 1976; Fourtner and Pearson, 1976; Graham, 1985; Delcomyn, 1985). There are six control centers in the thoracic ganglia of a cockroach, one for each leg, with a segmental ganglion controlling the pair of legs attached to its segment. A small number of **central command (COM) neurons** provides coordination among the centers and ensures that only one leg at a time is moved. The COM neurons send output to a group of interneurons, the **flexor burst generator (FBG) neurons**. The FBG neurons produce rhythmic **excitatory** output to motoneurons whose axons synapse with flexor leg muscles, and **inhibitory** output to neurons whose axons connect with extensor muscles of a leg. Thus, the leg can be flexed for the next step. The flexor muscles bend the leg and swing it forward. As it nears its full forward swing, hair sensilla near the coxal base are activated and send negative feedback to the FBG neurons, which inhibits their output to the flexor motoneurons and reduces inhibition of extensor motoneurons. The hair sensilla also have positive feedback to the extensor motoneurons, thus helping to make them ready for the leg to contact the substrate.

As the leg contacts the surface on which the cockroach is walking, the extensor muscles, now receiving activation from the COM neurons and the hair sensilla, extend the leg, push it backward, and move the body forward. As the leg takes some of the weight of the body, dome sensilla near the femoral-tibial joint are stimulated and their input provides additional positive feedback to the extensor motor neurons while simultaneously inhibiting the FBG neurons. With the leg in this extended position, the hair sensilla at the coxal joint are not active, so their previous inhibition of the FBG neurons is relieved and the FBG neurons now send impulses that flex the leg and swing it forward for the next step, and the cycle repeats.

Both hair sensilla and dome sensilla have positive and negative feedback loops so that when activated, they simultaneously stimulate one set of neurons and inhibit another set. Such positive

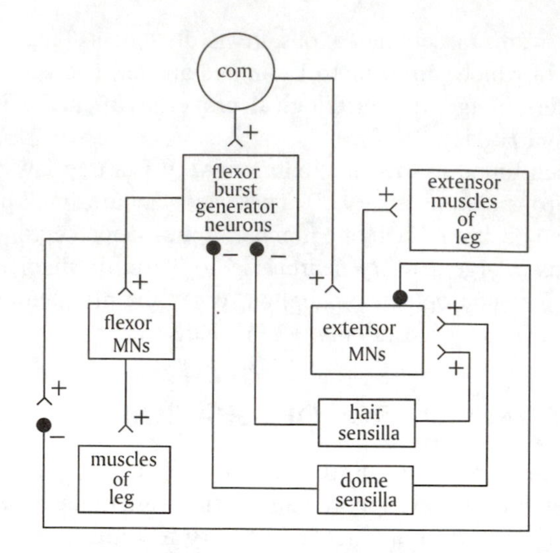

FIGURE 7.13 A schematic diagram of the major components in a motor program to control walking in the American cockroach, *Periplaneta americana*. (Based on Pearson, 1972; Pearson and Fourtner, 1975; Fourtner, 1976; Fourtner and Pearson, 1976.)

and negative feedback loops are very common in control of antagonistic sets of muscles in all animals.

There are four types of sensory organs on the legs, including a femoral chordotonal organ, campaniform sensilla, hair plates, and hair sensilla (see Chapter 10, Sensory Physiology, for more specific details about these sensory structures). Sensory input from one or more of the leg structures superimposes on the CNS output to provide proper timing of the start of leg movements and stance. Experimentally preventing input from only one of the four structures causes only minor alteration in leg movements, and the absence of all sensory input from the legs makes leg movement abnormal but does not destroy the ability of the insect to walk (Pearson and Iles, 1973).

7.10.2 A Motor Pattern for Rhythmic Breathing

Large insects (generally larger than a *Drosophila* fruit fly) must have pumping or ventilatory movements of the abdomen in order to force air through the tracheal system and create exchange of tissue gases. A motor pattern has been observed in several large insects, and the pattern in the locusts *Schistocerca gregaria* and *Locusta migratoria* has been well described.

The driving force for the motor program originates in the metathoracic ganglion. The evidence for this location is that the (experimentally) isolated metathoracic ganglion maintains its rhythmical output of spike activity, whereas no other isolated thoracic or head ganglia display the pattern. Isolated abdominal ganglia have a rhythm, but it is slow and somewhat abnormal. The regularity and frequency of the output from the metathoracic ganglion indicate that it drives the rhythm in the abdominal ganglia. Only abdominal segments 3 to 8 participate in the ventilatory movements. Segments 1 and 2 have no dorsoventral muscles to function in the inspiratory phase, and segments 9 to 11 are highly modified to bear the genitalia. In an active, deeply ventilating locust, the participating abdominal segments contract together; but during very shallow ventilation, there is a delay of about 80 to 400 ms between activation of segments so that a sort of ripple movement runs from anterior segments toward the posterior of the insect. The thorax, of course, is too rigid to participate in routine ventilatory movements; but in flight, the flight muscle contractions as well as some movement of the thoracic wall help to create a faster rate of air flow through the large tracheae.

Air flow through the large longitudinal tracheal trunks is a directed flow from anterior to posterior. There are two spiracles in the thorax and eight in the abdomen of locusts. Spiracles 1 to 4 are open during inspiration and the remainder are closed. During expiration, the pattern is reversed; and in shallow ventilation, fewer spiracles are open in inspiration and only spiracle 10 may be open for expiration. Each spiracle opening is guarded by two valves. Some spiracles have both an opener and a closer muscle, while others have only a closer muscle.

Ventilatory movements in each segment involve 13 muscles receiving innervation from two pairs of lateral nerves and a single median nerve from each ganglion. Lateral nerve 1 contains axons of about 30 motoneurons, some of which terminate on dorsal longitudinal muscles. Contraction of these muscles pulls the segments closer together; in stressed ventilation, there is marked telescoping of the abdomen. These movements result in expansion of the abdomen, and air is sucked in through the open anterior spiracles. Lateral nerve 2 contains axons of about 13 motoneurons, some of which innervate expiratory dorsoventral muscles. These axons display bursts of spike activity during expiration. Contraction of the dorsoventral muscles lifts the sternites upward and compresses the body cavity, forcing air out through open posterior spiracles. The single median nerve contains axons of four motoneurons that divide and innervate the spiracular valve muscles and muscles on each side of the body involved in inspiration in the next posterior segment. The axons in the median nerve spike during inspiration, indicating that the median nerve output is related to the inspiration phase of the cycle.

Lewis et al. (1973) proposed that an interneuron (IN 1) in the metathoracic ganglion produces a burst of spikes and is the central command neuron. It receives feedback stimulatory input and inhibitory input from receptors responding to carbon dioxide and oxygen, and possibly other factors (e.g., neuromodulators) at sites in the CNS. In the model, IN 1 output acts as a brake on two coordinating interneurons (IN 2), one in each ventral nerve cord connective. Axons of IN 2 run the length of the ventral nerve cord and synapse in each ganglion with a small interneuron (IN 3) in each ganglion that controls expiratory motoneurons (MN 4s) for that segment. IN 3 also sends inhibitory input to inspiratory motoneurons (MN 5s) in each ganglion. IN 2s directly sends weak inhibitory signals to MN 5s. The inspiratory MN 5s have a spontaneous firing rhythm, and when released from inhibition by IN 3s, they promote inspiration while simultaneously inhibiting the expiratory MN 4s through a sixth interneuron (IN 6). Feedback to IN 1 from CO_2 and O_2 receptors in the tissues determines how much it inhibits IN 2s, and thus can determine the rate of ventilatory movements.

Coordination of these ventilatory movements by the rhythm driven from the metathoracic ganglion results in alternate contraction and expansion of the abdomen, sucking air in through open anterior spiracles and forcing it out through open posterior ones. Opening and closing of the spiracles is under control of the median nerve, but the spiracle muscles receive rhythmic nerve input linked to the ventilatory rhythm.

7.11 NEUROSECRETORY CELLS (NSC) AND NEUROSECRETION PRODUCTS FROM THE CNS

7.11.1 Neurosecretory Cells (NSC)

Neurosecretory cells usually have a very large soma, are usually monopolar, and occur in all ventral ganglia and the brain of insects. The somata are located peripherally in a ganglion. They are usually characterized by their large size and staining properties. Axonal processes from neurosecretory neurons often project to the periphery of the body, and staining suggests that they carry the neurosecretory products to functional sites. Except for a few neurotransmitter and modulating chemicals, most neurosecretory hormones (in both insects and vertebrates) are peptides or small proteins.

Neurosecretion, the secretion and release of products that may function as hormones and as neuromodulators, is one of the major functions of the nervous system. Neurosecretion is ideally suited for control of physiological and biochemical processes in which sustained stimulation is needed, such as secretion of PTTH over several days in some insects in order to stimulate the prothoracic glands to begin to produce the molting hormone.

All known, physiologically active molecules secreted by the nervous system (of all animals) have been peptides or small proteins (except neurotransmitters, such as acetylcholine, γ-aminobutyric acid, and some biogenic amines), and they are usually called neuropeptides. Immunocytochemistry, in which an internal secretory component reacts with antibodies prepared to identify specific neurosecretory products, has enabled cytologists to rapidly determine the location of cells that secrete specific products. Thus, more than 100 neuropeptide sequences from insects have been described, but few have proven functions because the products have not been isolated, nor a functional bioassay developed. Usually, the neuropeptides found in insects have been small, composed of 10 to 15 or fewer amino acids. The sequence of a number has been determined and the molecules can be synthesized for bioassay tests. Nevertheless, a clearly defined function has been demonstrated for only a few of the isolated peptides and, more often, a described function is tenuous and imprecise, often said to be "AKH-like" or "proctolin-like," or having properties similar to some other well-defined neuropeptide. Immunoreaction to rabbit antiserum is one of the favorite methods of peptide detection. The antiserum reagent has often been used in an ELISA reaction or complexed with a fluorescent dye, as is usually the case in immunohistochemistry. Current work to isolate and sequence neuropeptides is a very active research field, and new natural products and synthetically modified peptides based on a natural structure with physiological functions appear in the literature regularly. The greatest knowledge gap in neurosecretion at present is understanding the function of the many neuropeptides and the characterization of receptors for the described neuropeptides.

Many regions of the CNS of insects have been mapped for the presence of various neuropeptide-reactive neurons, with detection of neuropeptides in interneurons, motoneurons, and neurosecretory cells. A few of the major neuropil regions in the insect brain receive neuronal connections from many different neuropeptide-containing neurons. For example, the fan-shaped body in the protocerebrum is innervated by neurons that react to antisera to FaRPs, proctolin, AKH, leucokinins, locustatachykinin, and several other known neuropeptides. The pars intercerebralis in the protocerebrum and the medulla in the optic lobe show similar diversity in contacts with neuropeptide-containing neurons. This diversity of innervation is further argument for the great diversity in function and behavior modulating activity of the insect CNS.

The identified peptides have been grouped into families based on structure similarity. A family does not, however, necessarily indicate similarity of function. There are about 20 such families. Some of the neuropeptides have described functions as hormones (such as PTTH, AKH, eclosion hormone, and diuretic hormone), but it is highly likely that some neuropeptides function as neurotransmitters and as neuromodulators that could modify the input or output from neural connections. There is isolated evidence in insects of the co-localization of neuropeptides and neurotransmitters in the same nerve terminals, and neuropeptides are known in some cases to be released simultaneously with neurotransmitters. If some neuropeptides do indeed work in this way, a single neuron in a network may be able to regulate many variations on a basic behavior by modulation with neuropeptides. As an example, the cardioacceleratory peptides (CAPs) in *Manduca sexta* modulate four different behaviors at different periods in the life of the tobacco hornworm, with two modifying feeding, ingestion, and nutrition, and two others regulating heart activity in relation to wing inflation and flight (Tublitz et al., 1991). Neuromodulators might alter response characteristics of neurons, including such activities as feed-back, feed-forward, motor output, and muscle or gland response to nervous activity. Neuropeptides may have roles in embryonic development (Hokfelt, 1991; Li et al., 1991, 1992) and as cytokines in non-self recognition and response (Scharrer, 1991; Johnson et al., 1992).

Included here are a few of the neurosecretory peptides that have been found in insects to illustrate the diversity and function of neurosecretion. Additional details regarding the function of some of these compounds can be found in other chapters and in a review by Nassel (1993).

7.11.2 Adipokinetic Hormone (AKH)

AKH, first isolated from the CC of *Locusta migratoria* (Stone et al., 1976) is a decapeptide with the amino acid (coded) sequence pQLNFTPNWGTamide. It is now known from a large number of insects and crustaceans, and has been described by a number of names. The identity of many of these with AKH was only recognized later. For example, substances with AKH functions were described from crustaceans as "red pigment concentrating hormone," and from cockroaches as periplanetins CCI-II, MI, MII, and as neurohormone D. About 20 members in this family have been sequenced from seven orders of insects. Some are octapeptides instead of decapeptides, but usually they begin with pyroglutamate and have an amide function at the carboxy-terminal end. Locust AKH is synthesized as two inactive prohormones, pro AKH-I and AKH-II requiring two different messenger RNAs, followed by complex processing that eventually results in AKH-I and AKH-II, and three dimeric peptides with as yet undescribed functions. In various bioassay preparations, AKH has functional activity on mobilization of lipids or carbohydrates, acceleration of heart beat, myoactivity, and inhibition of fatty acid and protein synthesis (Orchard, 1987; Gäde, 1990). AKH-like immunoreactive (AKH-LI) cells are present in the CC, brain, and subesophageal ganglion of many insects. The brain of the blowfly (*Calliphora*) contains about 50 AKH-LI neurons scattered in proto-, deuto-, and tritocerebrum, and several hundred in the medulla of each optic lobe. One very prominent AKH-LI neuron on each side of the protocerebrum of *Calliphora* has arborizations that cover most of the superior protocerebrum on the ipsilateral side.

7.11.3 Proctolin

Proctolin was the first insect neuropeptide to be sequenced (Starrat and Brown, 1975). It contains the amino acid sequence arginine-tyrosine-leucine-proline-threonine (amino acid code RYLPT). Proctolin has action on skeletal, heart, and visceral muscle. In crustaceans, it controls central pattern-generating nerve networks that regulate feeding and ventilatory behaviors. There are about 40 cell bodies that are proctolin-immunoreactive in the brains of cockroaches, a similar number in Colorado potato beetle, from 80 to 90 in blowfly brain, and about 100 in each lobulus (a specific neuropil region) of the optic lobe in blowflies. Proctolin in the insect brain may act as a neurohormone and as a modulator of responses within central synaptic neuropil (Nassel and O'Shea, 1987). In *M. sexta,* NSC that are proctolin immunoreactive send branches and arborizations to the CA, and in Colorado beetle such terminals are found in the CC.

7.11.4 FMRFamide-Related Peptides (FaRPs)

Neuropeptides with the general structure of **FMRFamides** are widely represented throughout the Metazoa, and are characterized by the amino acid sequence phenylalanine-methionine-arginine-phenylalanine amide (FMRFamide) at the carboxy-terminal end. Because of their wide distribution and important actions in many groups, the FaRPs are the best studied of all the neuropeptides. FaRPs have been isolated and sequenced from *Leucophaea cinerea*, *M. sexta*, *S. gregaria*, *Aedes* sp., and *Calliphora erythrocephala*. There is striking diversity in the distribution and number of FMRFamides within insects, as there is in other groups. In *Calliphora,* 13 different sequences are found, of which CaliFRMFamide 5 is the most abundant and is located in the ventral nerve cord. *D. melanogaster* has 13 known FMRFamides, while a closely related species, *D. virilis,* has only 10; five are shared by the two species. One of those in *D. melanogaster*, a heptapeptide, is the same as CalliFMRFamide 11 in *Calliphora*. The distribution of neurons with sequences that show a positive FMRFamide immunoreaction is very widespread in the brain of a number of insects, including the Colorado potato beetle, *Drosophila*, a blowfly, *M. sexta*, and the honeybee. About 240 cell bodies that react positive are located in the proto-, deuto-, and tritocerebrum and subesophageal ganglion of *Drosophila*. Few functions of FMRFamide peptides have been determined in insects, but CalliFMRFamide 1, 2, and 3 induce salivation from blowfly salivary glands at nanomolar concentrations. FMRFamide peptides may have multiple physiological effects in different tissues.

7.11.5 Tachykinins, Including Locustatachykinins and Leucokinins

Tachykinins are a large family of peptides in lower vertebrates and mammals, and now found in some insects. The group in vertebrates is best represented by **Substance P**. There are four *Locusta* tachykinins (**LomTK I, II, III, and IV**) isolated from the brain and CC of *L. migratoria*. There is about 50% homology of the sequence of LomTK I with a vertebrate tachykinin, physalaemin. Another group of insect kinins are the **leucokinins**, first discovered in the cockroach *Leucophea maderae*, but now known also from a cricket *Acheta domesticus* and a locust. The leucokinins show myotropic action on visceral muscle and influence ion transport in Malpighian tubules. Eight known leucokinins are octapeptides (**LK I** to **VIII**); there are five known **achetakinins** and one **locustakinin**. Neurons that show immunoreactivity for these peptides are present in the brains of the insects indicated by the names of the peptides. It is thought that both tachykinins and kinins act as neuromodulators, and the kinins may also have important roles as neurohormones. In *Drosophila melanogaster*, two different tachykinin receptors have been isolated by recombinant DNA procedures. There is evidence for leucokinin receptors in the gut of some insects. One tachykinin receptor protein has been identified and cloned from *Drosophila*, and it was expressed when put into mouse NIH-3t3 cells, causing the cells to increase synthesis of inositoltrisphosphate (IP3) in response to locustatachykinin II (Monnier et al., 1992).

7.11.6 Pigment-Dispersing Factors

A number of octadecapeptides that have the ability to disperse pigment in chromatophores of some crustaceans have been isolated from insects and crustaceans. Although the bioassay utilizes the pigment-dispersing action of the compounds, pigment dispersing is not the normal physiological role of these peptides in insects, because pigment dispersion is not a typical mechanism in insects as it is in some crustaceans. Highly characteristic groups of neurons that are immunoreactive for these peptides are associated with the visual system and seem to be similar in a number of insect species. These peptides may be involved in regulatory activity within the visual system, possibly regulating a circadian pacemaker system. A neurohormonal role cannot be excluded. Neurons with these neuropeptides are not so widely distributed within the nervous system.

7.11.7 Vasopressin-Like Peptide (Locust F2 Peptide)

Vasopressin is a peptide in vertebrates with activity on smooth muscle of blood vessels; it can elevate blood pressure by causing constriction of the vessels. The role of this neurohormone in insects is not vasoconstriction, however. Two neurons in *L. migratoria* react with antisera against vasopressin, and these neurons have extensive axonal connections throughout the brain and optic lobes. Two locust peptides, F1 and F2, were isolated. F1 is a monomer and is inactive, but F2 is an antiparallel dimer of two F1s and has diuretic activity in the locust assay. Recent reports indicate that F2 neurons have some input from the visual system, and it may indicate that F2 is secreted in response to a light-driven circadian rhythm.

7.11.8 Allatotropins and Allatostatins

Allatotropins (**ATs**) and **allatostatins** (**ASTs**) are neuropeptides isolated from nervous and some non-neural tissues (reviewed by Gilbert et al., 2000) that either stimulate or inhibit, respectively,

TABLE 7.2
Functions of Allatotropin (AT) and Allatostatin (AST) Apparently Unrelated to Their Effects on Corpora Allata

Peptide	Function	Tissue	Species
Manduca sexta AT	Myostimulatory	Heart	*M. sexta*
	Ion transport inhibition	Midgut	*M. sexta*
YXFGL-amide[a] AST	Myoinhibitory	Hindgut muscle	*Diploptera punctata*
			Calliphora vomitoria
			Euborellia annulipes
		Foregut	*Leucophea maderae*
			Cydia pomonella
		Oviduct	*Schistocerca gregaria*
		Heart	*Blatella germanica*
	Inhibition of vitellogenin synthesis	Fat body	*B. germanica*
	Neuromodulatory	Stomatogastric ganglion	*Cancer borealis*[b]

[a] Tyr-Xxx-Phe-Gly-Leu-NH_2.
[b] Crustacea (crab).

From Stay (2000), with permission.

the corpora allata. Although several allatotropins have been discovered based on bioassays showing stimulation of corpora allata (Gilbert et al., 2000; Stay, 2000), the only one of known structure is Mas-AT from *M. sexta*. Mas-AT has the structure Gly-Phe-Lys-Asn-Val-Glu-Met-Met-Thr-Ala-Arg-Gly-Phe-NH_2 (Kataoka et al., 1989). It also has been isolated from the lepidopterans *Spodoptera frugiperda* (fall armyworm) (Oeh et al., 2000) and *Lacanobia olerace* (tomato moth) (Audsley et al., 2000). Physiological stimulation of juvenile hormone synthesis by ATs has generally been demonstrated in adult females, but Mas-AT stimulates, and Mas-AST inhibits, the larval corpora allata of the tomato moth (Audsley et al., 2000).

There are three identified allatostatins: Mas-AST from *M. sexta* (pGlu-Val-Arg-Phe-Arg-Gln-Cys-Tyr-Phe-Asn-Pro-Ile-Ser-Cys-Phe-OH), Dip-AST from the cockroach *Diploptera punctata* (a pentapeptide with amidated C-terminal sequence and variable number of amino acids at the N-terminus in different cockroach species), and an AST from the cricket *Gryllus bimaculatus* that is similar to the cockroach family of ASTs and also a different AST. Mas-AST and Dip-AST have physiological action on larvae and adults, and in the case of Dip-AST, in the embryo. Action of the cricket AST has been demonstrated only in adults (Stay, 2000, and references therein). The *Diploptera* ASTs are a family of 13 allatostatins, each of which has physiological action in inhibiting the corpora allata, but with strikingly different effectiveness. Tobe et al. (2000) suggest from experimental studies with mixtures of the peptides that they likely act in concert to regulate juvenile hormone biosynthesis by interacting with receptors in the corpora allata.

Allatotropins and allatostatins are widely distributed in various tissues of insects and in other invertebrates, and they have physiological action on tissues other than the corpora allata (Table 7.2). It appears highly likely that they have fundamental actions unrelated to juvenile hormone synthesis (Gilbert et al., 2000; Stay, 2000; Truesdell et al., 2000).

7.11.9 Crustacean Cardioactive Peptide (CCAP)

A **cardioactive peptide** with the sequence PFCNAFTGCamide has been isolated from *L. migratoria* and a crab. Similar or possibly identical peptides appear to occur in *Tenebrio molitor* and in *M. sexta*. As the name implies, one action may be to stimulate the heart, but its true function(s) in insects is not known. **Corazonin** is another cardioactive peptide that has been isolated from the cockroach

Periplaneta. Corazonin antiserum D reacts with lateral neurosecretory cells in the protocerebrum and in two descending neurons in the blowfly, suggesting the same or a very similar molecule.

7.11.10 Pheromone Biosynthesis Activating Neuropeptide (PBAN)

PBAN, a neuropeptide, controls the biosynthesis of the pheromone in glands of some female moths, the best documented of which is *Helicoverpa* (formerly *Heloiothis*) *zea* (Raina et al., 1989). PBAN-immunoreactive neurons also have been demonstrated in the CNS of several other species. More detailed discussion of its function can be found in Chapter 14, Pheromones.

REFERENCES

Abbott, N.J., and J.E. Treherne. 1977. Homeostasis of the brain microenvironment: A comparative account, pp. 481-510, in B.L. Gupta, R.B. Moreton, J.L. Oschman, and B.J. Wall (Eds.), *Transport of Ions and Water in Animals*, Academic Press, New York.

Ashhurst, D.E., and N.M. Costin. 1971a. Insect mucosubstances. II. The mucosubstances of the central nervous system. *Histochem. J.,* 3:297-310.

Ashhurst, D.E., and N.M. Costin. 1971b. Insect mucosubstances. III. Some mucosubstances of the nervous systems of the wax-moth (*Galleria mellonella*) and the stick insect (*Carausius morosus*). *Histochem. J.,* 3:379-387.

Audsley, N., R.J. Weaver, and J.P. Edwards. 2000. Juvenile hormone biosynthesis by corpora allata of tomato moth, *Lacanobia oleracea,* and regulation by *Manduca sexta* allatostatin and allatotropin. *Insect Biochem. Mol. Biol.,* 30:681-689.

Burrows, M. 1996. *The Neurobiology of an Insect Brain*, Oxford University Press, Oxford, 682 pp.

Dethier, V.G. 1990. Five hundred million years of olfaction, pp. 1-37, in K. Colbow (Ed.), *Frank Allison Linville's R.H. Wright Lectures on Olfactory Research*, Simon Fraser University, Burnaby, B.C., Canada.

Delcomyn, F. 1980. Neural basis of rhythmic behavior in animals. *Science,* 210:492-498.

Delcomyn, F. 1985. Factors regulating insect walking. *Annu. Rev. Entomol.,* 30:239-256.

Fielden A. 1960. Transmission through the last abdominal ganglion of the dragonfly, *Anax imperator. J. Exp. Biol.,* 37:832-844.

Fourtner, C.R. 1976. Central nervous control of cockroach walking, pp. 519-537, in R.M. Herman, S. Grillner, P.S.G. Stein, and D.G. Stuart (Eds.), *Neural Control of Locomotion*, Plenum Press, New York.

Fourtner, C.R., and K.G. Pearson. 1976. Morphological and physiological properties of motor neurons innervating insect leg muscles, pp. 87-99, in G. Hoyle (Ed.), *Identified Neurons and Behavior of Arthropods*, Plenum Press, New York.

Gäde, G. 1990. The adipokinetic hormone/red pigment-concentrating hormone peptide family: Structures, interrelationships and functions. *J. Insect Physiol.,* 36:1-12.

Galizia, C.G., and R. Menzel. 2001. The role of glomeruli in the neural representation of odours: Results from optical recording studies. *J. Insect Physiol.,* 47:115-130.

Gilbert, L.I., N.A. Granger, and R.M. Roe. 2000. The juvenile hormones: historical facts and speculations on future research directions. *Insect Biochem. Mol. Biol.,* 30:617-644.

Graham, D. 1985. Pattern and the control of walking in insects. *Adv. Insect Physiol.,* 18:31-140.

Gronenberg, W., J. Tautz, and B. Holldobler. 1993. Fast trap jaws and giant neurons in the ant *Odontomachus. Science,* 262:561-563.

Halter, D.A., J. Urban, C. Rickert, S.S. Ner, K. Ito, A.A. Travers, and G.M. Technau. 1995. The homeobox gene *repo* is required for the differentiation and maintenance of glia function in the embryonic nervous system of *Drosophila melanogaster. Development,* 121:317-332.

Hess, A. 1958. The fine structure of nerve cells and fibres, neuroglia, and sheaths of the ganglion chain in the cockroach (*Periplaneta americana*). *J. Biophys. Biochem. Cytol.,* 4:731-742.

Hildebrand, J.G. 1995. Analysis of chemical signals by nervous systems. *Proc. Natl. Acad. Sci. U.S.A.,* 92:67-74.

Hildebrand, J.G. 1996. Olfactory control of behavior in moths: Central processing of odor information and the functional significance of olfactory glomeruli. *J. Comp. Physiol. A,* 178:5-19.

Hodgkin, A.L., and A.F. Huxley. 1952. Currents carried by sodium and potassium ions through the membrane of the giant axon of *Loligo*. *J. Physiol. (London),* 116:449-472.

Hokfelt, T. 1991. Neuropeptides in perspective: the last ten years. *Neuron,* 7:867-879.

Homberg, U., T.A. Christensen, and J.G. Hildebrand. 1989. Structure and function of the deutocerebrum in insects. *Annu. Rev. Entomol.,* 34:477-501.

Hösl, M. 1990. Pheromone-sensitive neurons in the deutocerebrum of *Periplaneta americana*: Receptive fields on the antenna. *J. Comp. Physiol. A,* 167:321-327.

Iversen, L.L., S.D. Iversen, and S.H. Snyder (Eds.). 1975. *Handbook of Psychopharmacology,* Vol. 4, Amino Acid Transmitters, Plenum Press, New York.

Jenkin, P.M. 1962. *Animal Hormones; A Comparative Survey,* Pergamon Press, Oxford, New York.

Johnson, H.M., M.O. Downs, and C.H. Pontzer. 1992. Neuroendocrine peptide hormone regulation of immunity. *Chem. Immunol.,* 52:49-83.

Kataoka, H., A. Toschi, J.P. Li, R.L. Carney, D.A. Schooley, and S.J. Kramer. 1989. Identification of an allatotropin from adult *Manduca sexta*. *Science,* 243:1481-1483.

Kent, K.S., and J.G. Hildebrand. 1987. Cephalic sensory pathways in the central nervous system of larval *Manduca sexta* (Lepidoptera: Sphingidae). *Phil. Trans. R. Soc. London B,* 315:1-36.

Kloppenburg, P., and T. Heinbockel. 2000. 5-Hydroxytryptamine modulates pheromone-evoked local field potentials in the macroglomerular complex of the sphinx moth *Manduca sexta*. *J. Exp. Biol.,* 203:1701-1709.

Kondoh, Y., H. Morishita, T. Arima, J. Okuma, and Y. Hasegawa. 1991. White noise analysis of graded response in a wind-sensitive, nonspiking interneuron of the cockroach. *J. Comp. Physiol.,* 168A:429-443.

Kondoh, Y., T. Arima, J. Okuma, and Y. Hasegawa. 1993. response dynamics and directional properties of nonspiking local interneurons in the cockroach cercal system. *J. Neurosci.,* 13:2287-2305.

Lane, N.J., and H. le B. Skaer. 1980. Intercellular junctions in insect tissues. *Adv. Insect Physiol.,* 15:35-213.

Laurent, G. 1990. Voltage-dependent nonlinearities in the membrane of locust nonspiking local interneurons, and their significance for synaptic integration. *J. Neurosci.,* 10:2268-2280.

Laurent, G. 1991. Evidence for voltage-activated outward currents in the neuropilar membrane of locust nonspiking local interneurons. *J. Neurosci.,* 11:1713-1726.

Lewis, G.W., P.L. Miller, and P.S. Mills. 1973. Neuro-muscular mechanisms of abdominal pumping in the locust. *J. Exp. Biol.,* 59:149-168.

Lemon, W.C., and W.M. Getz. 1999. Neural coding of general odors in insects. *Ann. Entomol. Soc. Am.,* 92:861-872.

Li, X.J., W. Wolfgang, Y.N. Wu, R.A. North, and M. Forte, M. 1991. Cloning, heterologous expression and developmental regulation of a *Drosophila* receptor for tachykinin-like peptide. *EMBO J.,* 10:3221-3229.

Li, X.J., Y.N. Wu, R.A. North, and M. Forte. 1992. Cloning, functional expression and developmental regulation of a neuropeptide Y receptor from *Drosophila melanogaster*. *J. Biol. Chem.,* 267:9-12.

Marion-Poll, F., and T.R. Tobin. 1992. Temporal coding of pheromone pulses and trains in *Manduca sexta*. *J. Comp. Physiol.,* 171:505-512.

Monnier, D.J., F. Cloas, P. Rosay, R. Hen, E. Borelli, and L. Maroteaux. 1992. NKD, a developmentally regulated tachykinin receptor in *Drosophila*. *J. Biol. Chem.,* 267:1298-1302.

Nassel, D.R. 1993. Neuropeptides in the insect brain: a review. *Cell Tissue Res.,* 273:1-29.

Nassel, D.R., and M. O'Shea. 1987. Proctolin-like immunoreactive neurons in the blowfly central nervous system. *J. Comp. Neurol.,* 265:437-454.

Oeh, U., M.W. Lorenz, H. Dyker, P. Lösel, and K.H. Hoffmann. 2000. Interaction between *Manduca sexta* allatotropin and *Manduca sexta* allatostatin in the fall armyworm *Spodoptera frugiperda*. *Insect Biochem. Mol. Biol.,* 30:719-727.

Orchard, I. 1987. Adipokinetic hormone: An update. *J. Insect Physiol.,* 33:451-463.

Pearson, K.G. 1972. Central programming and reflex control of walking in the cockroach. *J. Exp. Biol.,* 56:173-193.

Pearson, K. 1976. The control of walking. *Sci. Am.,* 235:72-86.

Pearson, K.G., and C.R. Fourtner. 1975. Nonspiking interneurons in the walking system of the cockroach. *J. Neurophysiol.,* 38:33-52.

Pearson, K.G., and J.F. Iles. 1973. Nervous mechanisms underlying intersegmental co-ordination of leg movements during walking in the cockroach. *J. Exp. Biol.,* 58:725-744.

Raina, A.K., H. Jaffe, T.G. Kempe, P. Keim, R.W. Blacher, H.M. Fales, C.T. Riley, J.A. Klun, R.L. Ridgway, and D.K. Hayes, 1989. Identification of a neuropeptide hormone that regulates sex pheromone production in female moths. *Science,* 244:796-798.

Rind, F.C. 1990. A directionally selective motion-detecting neurone in the brain of the locust: Physiological and morphological characterization. *J. Exp. Biol.,* 149:1-19.

Rodrigues, V., and L. Pinto. 1989. The antennal glomerulus as a functional unit of odor coding in *Drosophila melanogaster*, pp. 387-396, in R.N. Singh and N.J. Strausfeld (Eds.), *Neurobiology of Sensory Systems*, Plenum Press, New York.

Rospars, J.P., and J.G. Hildebrand. 1992. Anatomical identification of glomeruli in the antennal lobes of male sphinx moth *Manduca sexta. Cell Tissue Res.,* 270:205-227.

Scharrer, B. 1991. Neuroimmunology: the importance and role of a comparative approach, pp. 1-6, in G.B. Stefano and E.M. Smith (Eds.), *Advances in Neuroimmunology*, Vol. 1. Manchester University Press, Manchester, U.K.

Scharrer, B.C.J. 1939. The differentiation between neuroglia and connective tissue sheath in the cockroach (*Periplaneta americana*). *J. Comp. Neurol.,* 70:77-88.

Smith, D.S., and J.E. Treherne. 1963. Functional aspects of the organization of the insect nervous system. *Adv. Insect Physiol.,* 1:401-484.

Snodgrass, R.E. 1929. The thoracic mechanism of a grasshopper, and its antecedents. *Smithson. Misc. Collect.,* 82:1-111.

Snodgrass, R.E. 1935. *Principles of Insect Morphology,* McGraw-Hill, New York.

Starrat, A.N., and B.E. Brown. 1975. Structure of the pentapeptide proctolin, a proposed neurotransmitter in insects. *Life Sci.,* 17:1253-1256.

Stay, B. 2000. A review of the role of neurosecretion in the control of juvenile hormone synthesis: a tribute to Berta Scharrer. *Insect Biochem. Mol. Biol.,* 30:653-662.

Stocker, R.F. 1994. The organization of the chemosensory system in *Drosophila melanogaster*: A review. *Cell Tissue Res.,* 275:3-26.

Stone, J.V., W. Mordue, K.E. Betley, and H.R. Morris. 1976. Structure of locust adipokinetic hormone, a neurohormone that regulates lipid utilization during flight. *Nature,* 265:207-211.

Strausfeld, Nicholas J. 1976. *Atlas of an Insect Brain*, Springer-Verlag, New York, 214 pp.

Sun, X.J., L.P. Tolbert, and J.G. Hildebrand. 1993. Ramification pattern and ultrastructural characteristics of the serotonin immunoreactive neuron in the antennal lobe of the moth *Manduca sexta*: A laser scanning confocal and electron microscopic study. *J. Comp. Neurol.,* 338:5-16.

Tobe, S.S., J.R. Zhang, P.R.F. Bowser, B.C. Donly, and W.G. Bendena. 2000. Biological activities of the allatostatin family of peptides in the cockroach, *Diploptera punctata*, and potential interactions with receptors. *J. Insect Physiol.,* 46:231-242.

Treherne, J.E. 1985. Blood-brain barrier, pp. 115-137, in G.A.Kerkut and L.I. Gilbert (Eds.), *Comprehensive Insect Physiology Biochemistry and Pharmacology, Vol. 5, Nervous System: Structure and Motor Function,* Pergamon Press, Oxford.

Treherne, J.E., and Y. Pichon. 1972. The insect blood-brain barrier. *Adv. Insect Physiol.,* 9:257-313.

Tublitz, N., D. Brink, K.S. Broadie, P. Loi, and A.W. Sylwester. 1991. From behavior to molecules: An integrated approach to the study of neuropeptides. *Trends Neurosci.,* 14:254-259.

Truesdell, P.F., P.M. Koladich, H. Kataoka, K. Kojima, A. Suzuki, J.N. McNeil, A. Mizoguchi, S.S. Tobe, and W.G. Bendena. 2000. Molecular characterization of a cDNA from the true armyworm *Pseudaletia unipuncta* encoding *Manduca sexta* allatotropin peptide. *Insect Biochem. Mol. Biol.,* 30:691-702.

Wigglesworth, V.B. 1959. The histology of the nervous system of an insect, *Rhodnius prolixus* (Hemiptera). II. The central ganglia. *Quart. J. Microsc. Sci.,* 100:299-313.

8 Neurophysiology

CONTENTS

PREVIEW

Neurons function like batteries; they develop and store a potential difference across the cell membrane. With appropriate stimulation, a neuron discharges a flow of electricity along its axonal or dendritic processes. Afferent axons synapse with interneuronal processes, which enable the stimulation to be passed on to many other neurons, including motor neurons. Axons from motor neurons synapse directly with glands or muscles. In nearly all cases, the transmission from neuron-neuron, or from neuron-tissue, is by chemical transmission. A few electrical synapses occur in the central nervous system of insects in which neuronal processes have physically fused so that chemical transmission is not necessary. Graded neuronal responses occur at synapses and receptor neuron endings. Graded responses develop relatively slowly, are not self-propagated, and are proportional in strength to the stimulus intensity. Sufficiently strong graded potentials typically lead to generation of spikes or all-or-none potentials at the axon hillock, a region of the axon where spikes can be produced. All-or-none potentials are called action potentials. Action potentials are not proportional to the stimulus strength, provided the stimulus is above the threshold for spike generation; they rise extremely rapidly, last a few milliseconds, and are propagated along the axon without decrement. The resting potential is the potential difference across the cell membrane when the cell is not being stimulated. Typically in insects, the resting potential across the axon membrane is about 70 mV, inside negative to the outside. The resting potential depends on ion distribution. Ion distribution is a result of a Na^+/K^+ exchange pump that pumps Na^+ out of the cell and brings K^+ into the cell. The resting membrane is extremely impermeable to Na^+ reentry. Other ions involved include negatively charged proteins inside the neuron, and Cl^- on both sides of the neuronal membrane. The outside of the neuron refers to the very small space (called the mesaxon) between the neuronal membrane and the surrounding glial cell; all neurons are surrounded by glial cell membranes. Thus, it is the distribution of ions between the inside of the neuron and its mesaxon space that determines the resting potential. Stimulation above the characteristic threshold for the neuron causes an action potential, in which Na^+ channels rapidly open, allowing an influx (picomolar amounts) of Na^+. This influx of positive ions reverses the potential so that the inside of the neuron is briefly positive to the outside. Na^+ channels close in a few milliseconds, and an outward flow of K^+ ions (again picomolar amounts) repolarizes the neuron and restores the resting potential. The Na^+/K^+ pump only works in long-term maintenance, not in restoration of the resting potential after a stimulus. Transmission across synapses is by chemical diffusion, a slower process than the flow of electrical current represented by the transmission of an action potential. If the synaptic transmitter chemical is acetylcholine or L-glutamic acid, the synapse is a stimulatory synapse, and the postsynaptic potential is called an excitatory postsynaptic potential (EPSP). At inhibitory synapses, the neurotransmitter is γ-aminobutyric acid (GABA) and the potential is called an inhibitory postsynaptic potential (IPSP). Acetylcholine is the stimulatory neurotransmitter at neuron-neuron synapses in the CNS, and L-glutamic acid, and possibly L-aspartic acid are stimulatory transmitters at neuromuscular junctions. The only inhibitory transmitter known from insects is GABA.

8.1 INTRODUCTION

Neurons are composed of a cell body, the **soma**, and **axonal** and **dendritic processes**. **Integration**, the alteration or modification of electrical signals, can occur at a number of levels and sites within a single neuron and in the neuropil of ganglia. The corpora pedunculata (mushroom bodies) and the central body in the protocerebrum are examples of major integrative centers.

This chapter illustrates the functioning of neurons with examples from insect biology whenever possible. The reader should be aware, however, that most of the experiments that first elucidated nerve function were conducted on organisms other than insects, mostly on mollusks and some crustaceans because they have very large, giant axons (some approaching 1 mm in diameter) and in the case of marine mollusks, sea water was a sufficient saline in which to study the properties of neurons. Enough of the major neurophysiological experiments have been repeated on insects to confidently verify that the basic pattern of nerve cell function in insects agrees with principles derived from other groups (Pichon and Ashcroft, 1985). Indeed, experiments on all major groups of animals show that the physiological and biochemical principles of nerve function evolved early in the evolution of animals and these principles have been highly conserved in the course of evolution.

Detailed studies of single nerve cells began in the early 1930s, and three outstanding leaders were A.L. Hodgkin, A.F. Huxley, and J.C. Eccles. They and their colleagues conducted innovative experimentation in nerve function, mostly with mollusks and crustaceans, that culminated in the Nobel prize being awarded to Hodgkin, Huxley, and Eccles in 1963. The physiological model (Hodgkin, 1964; Huxley, 1964; Eccles, 1964) that grew out of these studies came to be called the **Hodgkin and Huxley model**, the basis for understanding how neurons function in insects.

8.2 NERVE CELL RESPONSES TO STIMULI

A number of different types of electrical responses (potentials) from a neuron are possible in response to a stimulus. When an experimental electrical stimulus (electrical stimulation is the usual means of stimulating neurons in the laboratory because the stimulus characteristics can be controlled

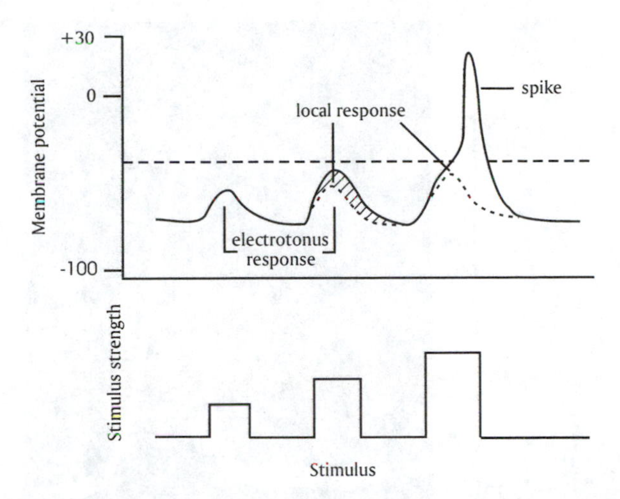

FIGURE 8.1 A conceptual diagram to illustrate the difference between passive electrotonic response, graded potential response, and spike response of a nerve cell membrane subjected to stimuli by increasingly large square-wave current pulses.

and repeated) is delivered to a nerve cell membrane, even if it is too weak to excite the cell into action, it causes a passive alteration in the membrane potential called **electrotonus** (Figure 8.1) because cells are conductors of electricity. The electrotonic effect passively spreads along the length of the cell as far as the natural tissue resistance and capacitance will allow. If the tissue is electrically excitable, however, and the electrical stimulation is strong enough, the neuron responds with a **graded membrane response**, often called a **local potential**. An even stronger stimulus can cause the neuron to respond by changing the graded response into an **all-or-none spike**, the **action potential** (Figure 8.1). A graded response always precedes a spike. In some neurons, any part of the neuron may conduct a spike; but typically in insect neurons, spikes are conducted by axonal processes.

Electrically excitable cells have a characteristic **membrane threshold** that must be exceeded in order to generate a response. In general, a stimulus must be strong enough to cause a change of about 10 to 15 mV in the axon membrane potential to exceed the threshold and thereby generate a spike. An oscilloscope or computer is usually used to visualize nerve activity, and permanent recordings are usually made on tape for playback and evaluation. Software is currently available that allows a computer to store and process data from stimulation of nerve cells. Action potentials similar to that shown in Figure 8.2 have been recorded from axons of many insects (Pichon and Boistel, 1966; Treherne and Maddrell, 1967; Gwilliam and Burrows, 1980; Tanouye et al., 1981).

8.2.1 Graded Responses

Graded responses are very important responses that nerve cells make, and some neurons (e.g., some interneurons in the ventral nerve cord) only make graded responses. Graded responses are often named after the site of their occurrence, such as synaptic potentials, receptor potentials, pacemaker potentials, or local potentials. Graded potentials are always localized and usually do not spread very far from their origin; they are propagated decrementally, becoming weaker the farther away they travel from their origin, and they soon are extinguished by resistance of the tissue to current flow. Their intensity is proportional to the strength of the stimulus. They rise and fall slowly in comparison with the 1- to 3-ms time duration of a spike. Because they rise and fall slowly, graded potentials are sometimes called **slow potentials**. Some defining characteristics of graded potential compared with characteristics of spikes are given in Table 8.1.

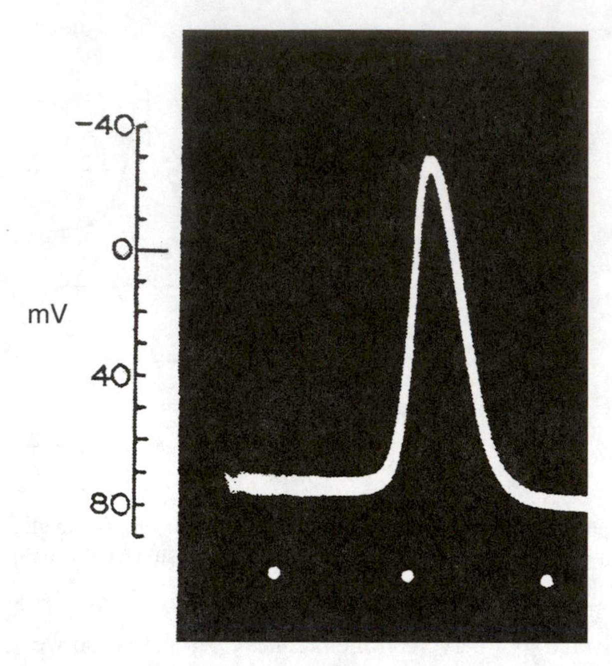

FIGURE 8.2 An action potential recorded from a cockroach giant axon. The overshoot is indicated by the height of the spike above the "0" potential. The dots along the *x*-axis indicate time in ms. (Reproduced with permission from Yamasaki and Narahashi, 1959.)

TABLE 8.1
Comparison of Graded Potential and Spike Potential Characteristics

	Type of Potential	
Response Characteristic	**Graded**	**Spike**
1. Rate of Rise	Slow, related to stimulus strength	Rapid, not related to stimulus strength[a]
2. Threshold	No threshold	Definite threshold
3. Magnitude	Related to stimulus strength	Characteristic of the neuron; not related to stimulus
4. Propagation	Decremental, usually over a few mm only	Self-generating, non-decremental
5. Refractory period	None	Definite relative and absolute refractory period
6. Ability to summate	Yes	No

[a] The stimulus is assumed to be above the threshold for the neuron.

8.2.2 Spike Potentials

All-or-none spike or action potentials rise and fall very rapidly, self-generate along the axon, and are propagated without decrement. The size of the spike is not proportional to the stimulus strength, provided the stimulus exceeds the spike threshold. Within the same neuron over a short period of time, spikes may be about the same size, but different neurons develop spikes of different size. Investigators are often interested in a train or burst of spikes arising from stimulation of a receptor, and can sometimes determine how many neurons may be responding in the receptor by the differential size of the recorded spikes.

8.3 THE PHYSIOLOGICAL BASIS FOR NEURONAL RESPONSES TO STIMULI

8.3.1 MEMBRANE ION CHANNELS: BIOELECTRIC POTENTIALS

Stimuli cause changes in membrane permeability of excitable cells. Potential differences across the cell membrane of about 70 mV are common in insect neurons, but lower and higher values have been recorded. The potential difference is usually written with a negative sign to connote that the inside is negative to the outside of the cell. The potential difference results from the unequal distribution of ions (both inorganic and organic) on the two sides of the membrane. Moreover, there is differential permeability to diffusible ions between the inside and outside.

The major ions involved in the transmembrane potential of nerve cells are K^+, Na^+, Cl^-, Ca^{2+}, and large negatively charged organic ions (primarily $proteins^-$). The membrane has different permeabilities to each of these, and the permeabilities are very different in a resting neuron than in one undergoing stimulation. There is virtually no permeability to large, negatively charged protein ions at any time in a healthy neuron, and permeability to the other ions is variable and depends on the physiological state and voltage across the membrane. The permeabilities to K^+ and Na^+ in a resting neuron can change extremely rapidly upon stimulation. Chloride ions tend to follow the dictates of the distributions of the positive ions. In nerve cells calcium is primarily involved with the release of neurotransmitter at the presynaptic terminals, and will be considered in that respect later.

The distribution of sodium, potassium, chloride, and negatively charged proteins that influence the membrane potential can be represented as follows:

$$K^+_{inside} + Na^+_{inside} = Protein^-_{inside} + Cl^-_{inside} \quad (8.1)$$

$$K^+_{outside} + Na^+_{outside} = Protein^-_{outside} + Cl^-_{outside} \quad (8.2)$$

It is important to note that although there is a potential difference *across* the cell membrane, there is electrical neutrality on a side. The total negatively charged ions on a given side equals the total positively charged ions on the same side. The "outside" in Equation 8.2 is the small space, the **mesaxon** between the nerve cell membrane and the membrane of the protective glial cell. The mesaxon space, and representative ion concentrations in the axon, mesaxon space, and in the surrounding hemolymph of a cockroach neuron are shown in Figure 8.3. The concentration of protein is normally very low in the mesaxon space, so most of the negatively charged ions in the mesaxon space are chloride ions. Thus, Cl^- concentration is normally greater outside the neuron than inside because the inside negative charge is shared by Cl^- and large, nondiffusible negative ions, such as $proteins^-$.

During depolarization and repolarization of a nerve cell, ions move through microscopic **pores** in the membrane. **Transmembrane proteins** (Figure 8.4) control the movement of the ions by forming narrow, hydrophilic channels through the cell membrane. These pores, which are capable of opening and closing rapidly, are frequently described as **gates, channels, gated channels,** and **ion channels**. These terms are used interchangeably. Some ion channels are very selective for a particular ion, while others are relatively nonselective. Potassium channels and sodium channels in nerve cell membranes are generally very selective for the named ion. Numerous ion channels occur in biological tissues (Stevens, 1984), but the most important ones for nerve cell function (and muscle cells) are channels for Na^+, K^+, Cl^-, and Ca^{2+}.

Channels are **ligand gated** (Figure 8.5) if a neurotransmitter or other molecule controls the opening of the gates, or **voltage gated** if the membrane voltage controls the opening of the channels. Acetylcholine, for example, and other neurotransmitters at nerve and muscle synapses are ligands that bind to one or more of the channel proteins composing the gate, and cause the gate to open.

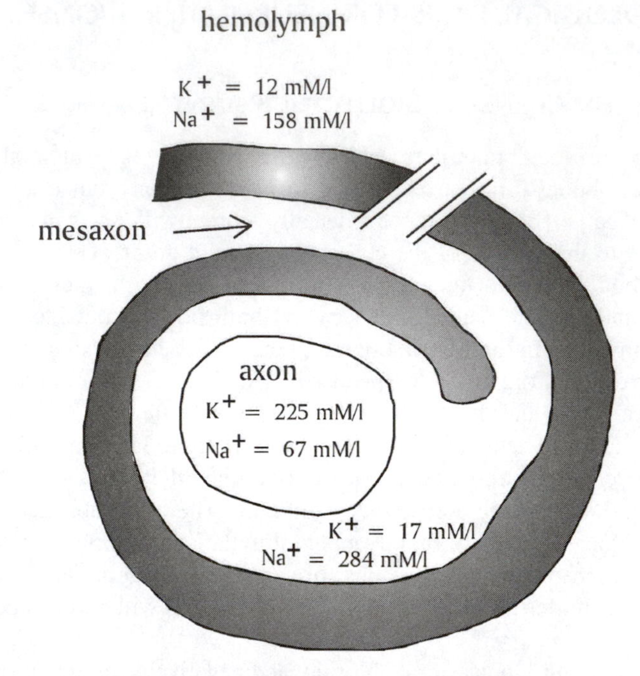

FIGURE 8.3 The mesaxon and the importance of the glial cell that protects a nerve cell from hemolymph ion concentrations and the sodium-potassium pump that maintains the distribution of ions within the mesaxon channel necessary for normal nerve cell function. The ion concentrations shown come from data in Narahashi and Yamasaki (1960).

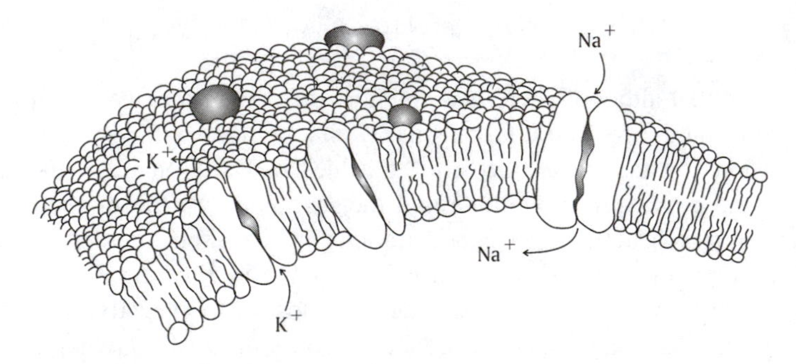

FIGURE 8.4 A conceptual illustration of transmembrane proteins that function as ion channels. Some channels are ion selective, while others may allow passage of several different ions. The opening of ion channels is controlled by binding of a ligand, by neuromodulators secreted from nerve cells, by neurotransmitters, and by voltage across the cell membrane.

Some potassium gates (those involved in membrane repolarization) are voltage gated, but others in *Drosophila melanogaster* muscle are known to be calcium activated (i.e., ligand gated, with calcium as the ligand) (Ganetzky et al., 1993).

Sodium channels along an axon are voltage gated and are opened by the flow of weak **local currents** flowing out from the main site of depolarization. Once an all-or-none depolarization develops, it gives rise to the local currents that depolarize the region ahead of the action potential, thus leading to the "self-generating" movement of the depolarization along the neuron. Sodium channels, as well as other channels, can be demonstrated by use of antibodies to specific peptide sequences in channel proteins. In the thoracic ganglion of a cockroach, an appropriate antibody intensely stains axons in tracts, commissures, and nerves, indicating many voltage-gated sodium

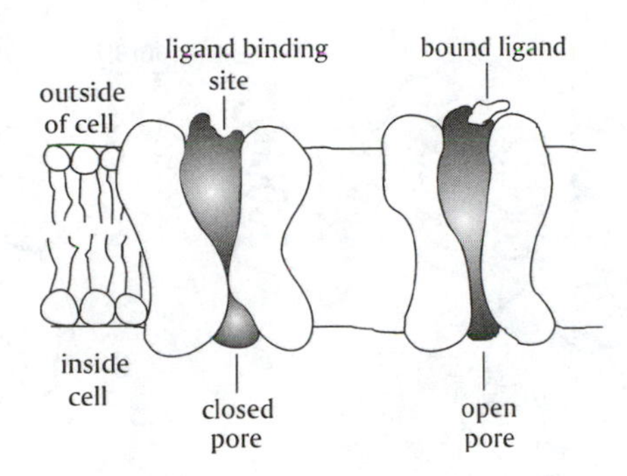

FIGURE 8.5 Illustration of an ion channel controlled by the binding of a ligand. The hypothetical channel in this diagram is composed of four transmembrane proteins, but only three are shown; the fourth protein forming the front of the channel has been omitted to allow a view of the channel. The ligand is shown being bound to the stippled transmembrane protein and opening the channel.

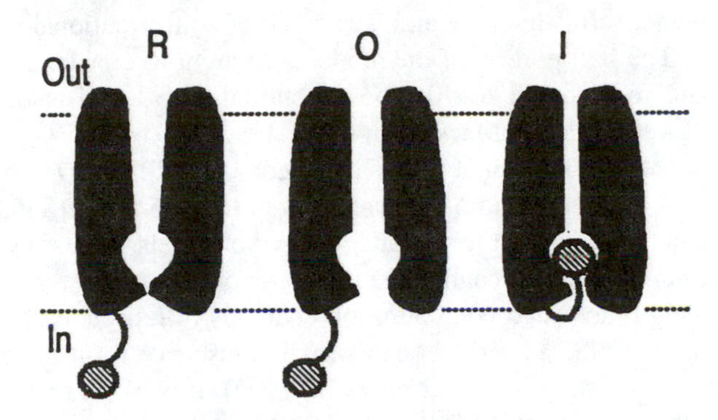

FIGURE 8.6 The ball and chain model illustrated as a mechanism for opening or closing the *shaker* potassium channel in *Drosophila melanogaster.* R, resting state of the channel; O, open state; I, inactivated state. (Reproduced with permission from Miller, 1991.)

channels. There is very little staining of the cell body membranes, indicating few sodium channels in the soma (French et al., 1993). In this particular study of the cockroach, the density of channels in the membrane was about 90 channels/μm^2, and this is similar to the density in some non-insect axons. An analysis of sodium channels in larvae and adult moths of *Heliothis virescens* shows that the channels generally exhibit similar characteristics to sodium channels in vertebrates, but the moth channels are more sensitive to scorpion toxin than vertebrate channels (Lee and Adams, 2000).

The precise mechanisms by which transmembrane proteins control movement of ions through cell membranes are still not well known, but evidence is accumulating that conformational changes in some of the transmembrane proteins making up a channel open a pore between adjacent proteins. A conceptual model (Armstrong and Bezanilla, 1977) for ion channel function is known as the **ball and chain model** (Figure 8.6). The model consists of several transmembrane protein domains, and an intracellular ball and chain sequence of amino acids. A molecular conformation allows the ball portion to swing in to block the channel, or swings out to open it. Channel proteins may be composed of a variable number of transmembrane segments and protein subunits.

Drosophila has been important in the study of potassium ion channels. The *Shaker (Sh)* gene codes for the proteins that make the potassium channel in *D. melanogaster*, and it was the first potassium channel gene to be cloned from any biological tissue and studied in detail (Timpe et al., 1988). The *shaker* potassium channel polypeptide includes seven hydrophobic transmembrane segments (Miller, 1991) that loop back and forth across the membrane (Figure 8.7). Six segments of the molecule exist as alpha-helices that span the membrane, while one segment, a beta-hairpin

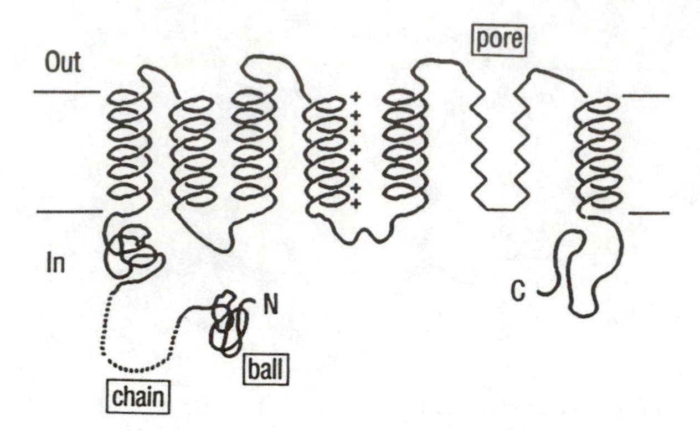

FIGURE 8.7 A conceptual model of one of the shaker K^+ channel polypeptides forming the K^+ channel in nerve tissue. Amino acid residues 1 to 20 at the N terminus represent the ball, residues 23 to 40 the chain, and residues 431 to 449 the conduction pore. (Reproduced with permission from Miller, 1991.)

loop within the membrane, forms the pore and is capable of conformational changes that allow or occlude passage of K^+. The ball portion of the model is thought to consist of amino acid residues 1 to 20, and the chain, residues 23 to 40. The functional gate consists of a tetramer of four polypeptide subunits (Li et al., 1992; Mackinnon et al., 1993).

A number of other potassium channel genes have been cloned from *Drosophila melanogaster*, including three genes, *Shaw*, *Shab*, and *Shal,* that are similar to *Shaker* (Salkoff et al., 1992), all of which code for channel proteins that form voltage-gated channels. Ganetzky et al. (1993) cloned two additional *Drosophila* potassium channel genes, *slowpoke (slo)* and *ether à go-go (eag)*, two genes that code for calcium-activated potassium ion channels in muscle cells. The gene for *slo* is expressed in neurons of the CNS and peripheral system, in muscle cells, in some cells in the midgut, and in tracheal cells of *D. melanogaster* (Becker et al., 1995). It is likely that the gene controls ion movements in cells with very different functionalities. *Drosophila* gene probes have been used to isolate potassium channel genes from other insects and from vertebrates. For example, a homolog to *slo* was isolated from a mosquito, *Aedes* sp., with >90% similarity to the *Drosophila* gene in coding for amino acid sequence. Both *slo* (Ganetzky et al., 1993) and *eag* (Warmke and Ganetzky, 1993) homologues have been cloned from mouse and human tissues. The proteins coded by the mouse and human genes have 71 and 48% identity, respectively, with *Drosophila eag* protein. The Slo proteins from mice and humans have about 70% identity to the *Drosophila* proteins coded by *slo*. Defining how the different potassium channels function is a major research effort for the future. Sodium and potassium channels close on a time-dependent basis, and neither ligands nor voltage directly control closing. Certain drugs, insecticides, and naturally occurring poisons, however, can block various ion gates open or closed.

8.3.2 The Resting Potential

The **resting potential** is the potential difference between the inside and the outside of the cell at rest; the inside of the cell is negative relative to the outside. At rest, the cell membrane is nearly impermeable to sodium ions and to the large charged protein molecules, but relatively permeable to potassium and chloride ions. The space outside a nerve cell is the mesaxon space between the neuron membrane and the protective glial cell membrane, and potassium ions are typically in low concentration here, but sodium ion concentration is usually high. Thus, both an electrical gradient and a concentration gradient act on any movement of sodium and potassium ions.

The negative charge on the inside of a resting neuron will attract positively charged ions, such as K^+ and Na^+. The concentration gradient acting on K^+ will tend to force it outward until the attraction of the negative charge (the electrical gradient) on the inside just balances the concentration force. Any change in the membrane potential immediately results in a redistribution of K^+ ions until a new equilibrium is achieved. Although both a concentration gradient and an electrical gradient act on Na^+ to promote inward movement when a channel is open, the membrane at rest is nearly impermeable to Na^+ entry. In a membrane at rest, Na^+ only very slowly leaks in with time.

An energy-requiring membrane pump, the sodium-potassium exchange pump works continuously to pump K^+ from the mesaxon space into the nerve cell and to pump Na^+ out of the cell and into the mesaxon space. It is important to understand, however, that this pump serves a long-term maintenance function, and it does not move ions fast enough to cause the repolarization process, which usually requires only a few milliseconds following a stimulus response. Ion pumps in some cases can induce s slow potential change, but the important point here is that repolarization is not a function of the Na^+-K^+ exchange pump. Repolarization is based on different physiological properties of a neuron, which are explained below.

In general the presence and distribution of chloride has little effect on resting or action potentials, and chloride ions tend to distribute according to the dictates of Equations (8.1) and (8.2). The presence of the large nondiffusible organic ions, mostly proteins, that carry a net negative charge at physiological pH are extremely important to overall nerve cell charge maintenance, enabling the inside to hold its negative charge relative to the outside. The mesaxonal space contains very few of these large molecules.

The **Nernst equation**, a physical chemistry model derived from studies with solutions of ions separated by a semipermeable membrane in laboratory situations, gives a fairly accurate prediction of membrane potentials of a neuron. The resting potential, influenced most strongly by potassium distribution inside and outside the cell, is calculated as a potassium equilibrium potential, as follows,

$$E_m = (RT/\eta F) \ln ([K^+]_o/[K^+]_i) \quad (8.3)$$

where E_m is the membrane potential in volts, R is the universal gas constant (0.082 liters × atmospheres per mole per degree Kelvin temperature scale, or 8.314 joules per mole per degree Kelvin), T is temperature on the Kelvin or absolute scale (°C + 273), η is the valance of the ion involved (for potassium, the valence is 1), F is the Faraday unit (96,496 coulombs of electricity per gram-equivalent of ion moving), ln is the logarithm to the base e (natural log as opposed to log base 10), and $[K^+]_i$ and $[K^+]_o$ refer to ionic activity of potassium on the inside and outside, respectively, of the membrane. The equation asks for ionic activities of potassium (indicated by the brackets in the equation) rather than concentration, but the more conveniently measured concentrations are often used in the equation to obtain a good approximation of the voltage across a nerve cell membrane.

When a neuron membrane is resting and its permeability to Na^+ is very low (as is normal), then the magnitude of the resting membrane potential is directly influenced by the K^+_o/K^+_i ratio, shown in Figure 8.8. The graph shows the magnitude of the resting potential when the external K^+ concentration in the saline bathing the nerve is varied. High levels of potassium outside the cell reduce, and may destroy, the resting membrane potential, as predicted by the Nernst equation. The equation predicts that the membrane resting potential will decrease as the K^+_o:K^+_i ratio approaches a value of 1; and when the value is 1, there is no transmembrane potential. This experimental effect of manipulating the external potassium concentration reinforces the importance of glial cells, which shield all parts of the nervous system from direct exposure to the hemolymph. Phytophagous insects often have a very high level of potassium ions in the hemolymph and a low level of sodium ions, conditions that would be deleterious to nerve cell function.

The Nernst equation was expanded into the **Goldman constant field equation** to include the three major ions moving during an action potential in an attempt to make even better predictions. The expanded equation is:

$$Em = (RT/\eta F) (\ln (P_k[K^+]_i + P_{Na}[Na^+]_i + P_{Cl}[Cl^-]_o/P_k[K^+]_o + P_{Na}[Na^+]_o + P_{Cl}[Cl^-]_i) \quad (8.4)$$

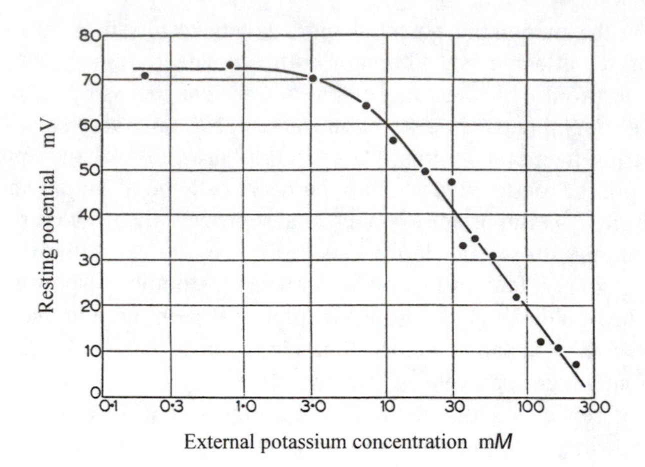

FIGURE 8.8 The magnitude of the resting membrane potential as a function of the potassium in experimentally controlled external bathing solution. Experimentally raising the external K^+ concentration in the bathing solution progressively reduces the magnitude of the resting potential. The normal bathing solution of a neuron is the fluid in the mesaxon channel. (Reproduced with permission from Yamasaki and Narahashi, 1959.)

where P is the membrane permeability value for each of the major ions that move during an action potential. The Goldman field equation gives slightly better agreement with actual observations, but absolute values for P are not available for very many neuronal preparations. *Relative* permeability values for K^+, Na^+, and Cl^-, equal to 1:0.04:0.45, respectively, were used for the squid giant axon to test the equation (Hodgkin and Katz, 1949). The Nernst equation was valuable when it was first applied to neuron physiology because it provided numerous testable predictions that generated many experiments, and it is a classical case of the benefits of mathematical modeling of a biological phenomenon.

8.3.3 The Action Potential: Sodium Activation

Upon stimulation that causes depolarization, a nerve cell undergoes remarkable permeability changes to both Na^+ and K^+. The most dramatic change is that the excited membrane becomes explosively permeable to Na^+, and a small number of sodium ions, acted upon by both electrical and concentration gradients, rush into the membrane causing **depolarization** of the membrane. The sudden increase in Na^+ influx is called **sodium activation**. The rush of Na^+ for 2 to 3 ms into the neuron when the sodium channels are completely open dominates membrane physiology and its electrical properties. The sodium ions carry an inward current, and this is usually recorded on an oscilloscope or similar device as a **spike** or **action potential**, and described as the **all-or-none response** of a neuron. A partial action potential does not occur (except in special experimental situations controlled by the investigator).

The rate at which the membrane potential changes (i.e., the rise of the spike potential) is very fast. For example, the rise of the spike has been measured at 1370 mV/ms in a giant axon of the cockroach *Periplaneta americana*. At this rate, it takes much less than 10 μs to depolarize the resting membrane potential from its typical resting value of about –70 mV in a cockroach giant axon.

Relatively few sodium ions move across the membrane to cause the depolarization. Although the number of ions moving per square centimeter of membrane surface has not been determined for insect axons, experiments with squid giant axons bathed in sea water containing radioactive Na^+ demonstrated that an average of 3.7×10^{-12} moles of Na^+ ions moved across 1 cm^2 membrane surface with each stimulus (Keyes, 1951). Thus, movement of only a few picomoles of sodium ions causes the spike, and the concentration of sodium ions in the mesaxon space is hardly changed.

The rapid inward movement of sodium ions during an action potential allows the membrane potential to overshoot zero, and the inside becomes positive to the outside over that portion of the membrane surface affected by the action potential. The reversal is called the **overshoot potential**. In a cockroach giant axon, overshoot potentials of about +35 mV have been recorded (the plus sign indicates that now the inside is positive to the outside). The total magnitude of the action potential is stated as the sum of the absolute values of the resting potential and the overshoot potential. Thus, if a neuron has a resting potential of –70 mV and an overshoot potential of +35 mV, then the action potential is 105 mV.

During the spike, the membrane is absolutely refractory (the **absolute refractory period**) to further stimulation. The sodium channels are open and Na^+ is entering at a maximal rate, and a stimulus of even great magnitude cannot cause more to enter or the gates to open any wider. The membrane is not capable of making any greater response. Thus, a second stimulus delivered within 1 to 2 ms of the first one will not elicit a response from an axon. Furthermore, in addition to the absolute refractory period, a neuron is partially or **relatively refractory** for a further few to many milliseconds, and only a very strong stimulus will elicit a new response during the relatively refractory period. Thus, although the absolute refractory period and the relative refractory period are short, they set an upper limit (typically about 100 impulses/s) on how many separate nerve impulses a neuron can transmit in 1 s.

The Nernst equation also predicts the magnitude of the overshoot potential, which is a sodium equilibrium potential in which sodium movements dominate the membrane. The form for the equation describing a sodium equilibrium potential is as follows: $E_m = (RT/\eta F)(\ln [Na]_o/[Na]_i)$. Sodium concentration in the mesaxonal space is very important to the action potential, and low sodium concentrations cause small action potentials. Although the concentration of sodium ions is relatively high in the mesaxon space, only a small volume of solution containing the ions occurs in this restricted space. The clefts in the mesaxon channel around a cockroach giant axon contain only enough free sodium ions for about 20 to 30 action potentials if no corrective pump action occurs (Treherne and Schofield, 1981). Thus, without some way to restore the ionic composition of this microchannel around an axon, it would soon fail to fire. In reality, axons are capable of firing continuously for many minutes (Narahashi and Yamasaki, 1960; Parnas et al., 1969) because at the inner membrane surface of the glial cell (i.e., the surface nearest the axon), coupled Na^+-K^+ pumps pump Na^+ from the glial cell into the mesaxon channel, and pump K^+ from the mesaxon channel into the glial cell (Treherne and Schofield, 1981). Accumulation of K^+ in the mesaxon would also be detrimental to continued function. The axon membrane also has a coupled Na^+-K^+ pump that works to pump Na^+ out of the axon and bring K^+ back into the axon. It is important to understand that the pumps provide for long-term maintenance by keeping Na^+ in the mesaxon channel high and the concentration of K^+ low (Treherne, 1985); the pumps do not repolarize the membrane after a depolarization.

Figure 8.9 graphically shows the magnitude of the action and overshoot potential as a function of progressive loss of external sodium concentration in the case of a giant axon of the American cockroach (Yamasaki and Narahashi, 1959). As sodium concentration in the bathing saline is reduced, the action potential is reduced in size, and at very low sodium concentration, the action potential is abolished. Similar experiments with similar results have been conducted on nerves from *Blaberus craniifer* (Pichon and Boistel, 1966), *Carausius morosus* (Treherne and Maddrell, 1967, and *Manduca sexta* (Pichon et al., 1972). The importance of sodium ions to the action potential was further demonstrated when desheathed (removal of the protective fat body and perineural layers) crural nerves of *P. americana* and *Locusta migratoria* bathed in sodium-free saline failed to develop action potentials (Pichon and Treherne, 1973) because there were no sodium ions to carry an inward depolarizing current.

8.3.4 Sodium Inactivation and Repolarization

The excited membrane state is normally a very transitory event, and the sodium channels have a time-dependent closing mechanism, called **sodium inactivation**. The time required for complete

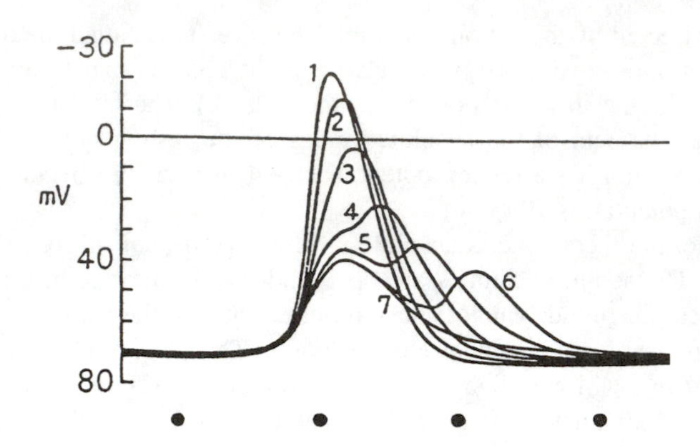

FIGURE 8.9 The effect of experimentally altering external sodium concentration on the magnitude of the action potential. Experimentally eliminating external Na^+, which carries the inward current, progressively reduces the size of the neuron response to stimulation; and finally at very low external sodium concentration, no spike can be developed. The dots along the *x*-axis indicate time in ms. (Reproduced with permission from Yamasaki and Narahashi, 1959.)

sodium inactivation to occur is variable, but can be a few milliseconds or several hundred milliseconds. During sodium inactivation, the sodium channels close. The spike falls rapidly (a falling rate of 640 mV/ms was recorded from a cockroach giant axon), and the rate of fall largely reflects the closing of the sodium channels. In general, as the spike falls, there is a slow positive after potential (slight **hyperpolarization**) and shortly thereafter, an even smaller and slower negative after potential. The after-potentials are graded or slow potentials, and are caused by the transitory displacement of ions in the mesaxon and across the membrane.

Membrane permeability to potassium changes immediately after the spike develops, and potassium starts to move outward across the axonal membrane, but it moves slowly at first, and the membrane is dominated by the inwardly directed sodium ion movements for a few milliseconds. Maximum potassium flux was measured at 440 mV/ms in a cockroach giant axon at the peak of the overshoot potential. The outwardly directed potassium current is counter to the inwardly directed current flow carried by sodium. Only when the sodium channels have partially closed, thus restricting the inward flow of Na^+, does the outward flow of potassium begin to bring the membrane potential back toward the resting value. Repolarization is a much slower process, relatively speaking, than depolarization, and total recovery of a neuron may take from about 10 ms to many tens of milliseconds, depending on the neuron.

As K^+ continues to move out and the sodium channels close, the membrane potential begins to return to its resting condition in which the inside is negative to the outside. The negative pole of this small biological battery now attracts the positively charged potassium ions and slows their outward movement. Net outward flux ceases when the membrane potential has become sufficiently negative enough to attract potassium and counterbalance the concentration gradient that drives it outward. The neuron has recovered its resting value; it is repolarized and ready to respond to a new stimulus.

The few picomoles of sodium ions that enter a neuron during an action potential do not have to be removed from the cell for repolarization to occur. Repolarization occurs when approximately the equivalent number of positively charged potassium ions exit from the neuron. Experiments with a squid giant axon that had been injected with radioactive potassium ions demonstrated that an average of 4.3×10^{-12} mole radioactive K^+/cm^2 exited into the bathing saline with each stimulus (Keyes, 1951), approximately equal to the Na^+ per cm^2 that entered with each impulse. The Na^+-K^+ exchange pump works to restore the normal distribution of ions, but it **does not** account for repolarization of the neuron membrane. Repolarization and continued nerve cell function without the pump was conclusively shown by selectively poisoning the pump in a squid giant axon, which

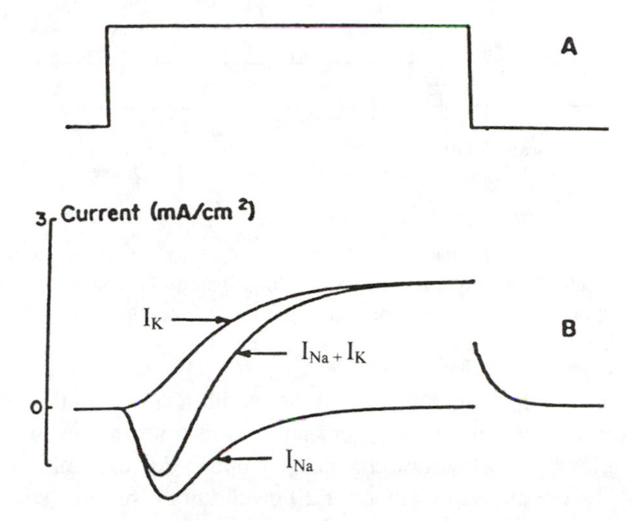

FIGURE 8.10 The partitioning of the membrane current flows (conductance) during an action potential into an inward current carried by Na^+ (I_{Na}) and an outward current carried by K^+ (I_K). In the diagram, inwardly directed current is represented below the baseline at point 0, and outwardly directed current is represented above the baseline. $I_{Na} + I_K$ represents the normal situation in a stimulated neuron in which both ions are moving simultaneously, and can be visualized in this diagram as a sort of inverted spike. (Reproduced with permission from Pichon, 1976.)

nevertheless continued to develop spikes and repolarize repeatedly for hours before the redistribution of Na^+ and K^+ became physiologically limiting. The pump serves a long-term maintenance function to keep Na^+ outside and the majority of K^+ inside. Nerve cell membranes are leaky and allow Na^+ to leak across the membrane, necessitating further action for the membrane pumps. The pumps work slowly and continuously, and require a constant supply of energy. Thus, nervous tissue has high metabolic demands in an insect, as in all other organisms.

8.3.5 Measurement of Ion Fluxes: The Voltage Clamp Technique

How can something that happens in 2 to 3 ms be observed? Even with the oscilloscope, it is impossible to follow all the details of the ion fluxes because of the extremely rapid and transitory nature of the nerve response to a stimulus. What is needed is some way to prevent the sudden explosive changes of the action potential — a way to stop an action potential at a given point and measure what ions are moving and in what direction. An ingenious technique, the **voltage clamp**, was devised independently by Cole (1949) and Marmont (1949), and it continues to be a useful tool (Trudeau et al., 1995).

The voltage clamp technique uses a feedback amplifier in the recording circuit to feed into the membrane just enough current in an opposing direction to counter the action of the ion currents induced by a stimulus. With this technique, the membrane could be stabilized (clamped) at any membrane potential desired by the investigator. For example, the membrane potential can be held at –20 mV, inside negative. This would be equivalent to stopping a depolarization about half-way, something that never occurs naturally. By measuring the magnitude of the current needed to hold the membrane at a given potential, the investigator can get a measure of the strength of the ion current at that potential. Furthermore, instead of the experiment being over in 1 to 2 ms, a sustained membrane response is possible. Figure 8.10 illustrates reconstructed data from a voltage clamp experiment in which inwardly directed current carried by Na^+ is separated from outwardly directed current carried by K^+. The net sum of the two currents represents the spike.

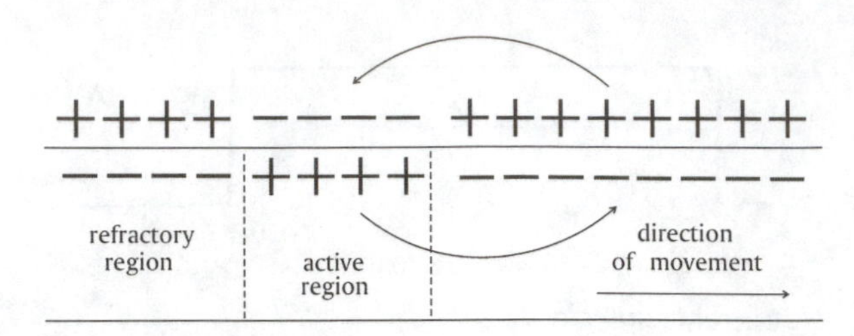

FIGURE 8.11 Local currents and an illustration of how they function to propagate a spike along the axon. Small currents flowing ahead of the active region open voltage-gated Na^+ channels in the axon. The currents also flow back into the region over which the impulse has just passed, but that region is still partially refractory, and the local currents are not strong enough to open Na^+ channels.

By using the voltage clamp technique to hold the membrane potential near, but not exceeding, the threshold value for a spike, an investigator can demonstrate that even slight changes in the membrane potential allow a few sodium channels to open. For example, a squid axon voltage clamped at only 8 mV below its resting value for 20 ms did not result in a spike (because the spike generation threshold was not reached), but it did cause as much as 40% reduction in the spike or sodium current upon subsequent depolarization. Hodgkin and Huxley explained this experiment as showing that even the slight drop in membrane potential had activated the time-dependent inactivation mechanism controlling the sodium channels. When the membrane was finally depolarized, the timing mechanism did not allow the sodium channels to stay open long enough for the spike to be normal in size. This led directly to the concept of a leaky membrane and helped explain why the action potential is usually not as large as predicted by the Nernst equation. Thus, in a typical neuron, the sodium gates allow some leaking. Upon stimulation of a leaky neuron, the sodium conductance values are lower than expected, and sodium conductance does not continue as long as expected. Conversely, hyperpolarization, even by a few millivolts (raising the inside potential to a greater negative value, e.g., from –70 to –90 mV), increases the sodium current and the size of the spike upon depolarization. Hyperpolarization also raises the threshold necessary to cause a neuron to fire, and inhibition in the nervous system often works through hyperpolarization of one or more neurons in a circuit, making it more difficult for other stimulating synaptic connection to fire the circuit.

8.4 CONDUCTION OF THE ACTION POTENTIAL: THE LOCAL CIRCUIT THEORY

After a spike arises, it is self-propagating and travels rapidly along the length of the axon. Local electrical disturbances (Figure 8.11) in the nerve set up a pattern of local currents or local circuits around the excited region of membrane. The propagated impulse does not go backward over the same length of axon it has just passed because that part of the membrane is still too refractory to be depolarized again by the local currents.

According to the **local circuit theory**, the currents flowing into the axon membrane just ahead of the active region open voltage-gated Na^+ channels leading to depolarization. The depolarization then allows local currents to flow into the next part of the membrane and cause a spike to develop in the that new location, and these actions are repeated along the neuron as the depolarized region moves along the axon. The membrane immediately behind the active region also receives the local circuits, but this part of the membrane is in a state of incomplete recovery (in the relative refractory period) and the local circuits are not strong enough to cause a new spike; thus, nerve transmission is normally unidirectional. The local circuit theory of spike propagation is based on experimental evidence from studies on frog sciatic nerve (Hodgkin, 1937a, 1937b). In those experiments, a small section (about 1 mm in length) of the nerve was frozen in order to block transmission across the frozen section because the frozen sodium gates could not open. Local electrical currents, however, would not be stopped by the frozen section because the tissue would still be a conductor of electricity. The intent of the experiment was to freeze a sufficiently long section so that the currents, although attenuated, would flow through the frozen section and do some of the work of lowering the threshold beyond the block. Stimulating electrodes placed just past the frozen section were used to deliver a stimulus just as a wave of depolarization reached the blocked section. The strength of stimulus needed to initiate a new spike just beyond the frozen section was about 10% of normal, and Hodgkin interpreted the experiment as showing that local currents from the wave of depolarization did 90% of the work of exceeding the threshold beyond the block. Without the new stimulus just as the local currents arrived at the post-block site, the nerve impulse would not have continued because the threshold would not have been reached. He reasoned that the local currents should be strong enough in a normal axon with no blocked portion to exceed the threshold and open the sodium channels immediately ahead of the spike. In such a case, the wave of excitation would be self-propagating, as observations indicated. Hodgkin also concluded from this experiment that the action potential and local currents extend over about 1 mm of axon surface. The exact surface of an axon that is excited at any given moment in an insect has not been measured in a similar manner, but the excited state probably spreads over an axon in a larger insect, such as a locust, grasshopper, or cockroach, in much the same way.

8.5 PHYSIOLOGY AND BIOCHEMISTRY AT THE SYNAPSE: EXCITATORY AND INHIBITORY POSTSYNAPTIC POTENTIALS

A **synapse** is any site where one nerve cell influences another neuron. In insects, synapses typically occur between the axon of one cell and a neurite of another neuron within the neuropil of a ganglion. A single neuron can have synaptic contacts with many other neurons. The axon-to-muscle contact is also sometimes described as a synapse. Some synaptic contacts, in addition to the axon to dendrite (neurite), are known in other animals, but they are either not known in insects or are very uncommon; for example, axon-to-soma synapses are very common in vertebrates, but none have been observed in insects. The transmitter chemical is contained within membrane-bound synaptic vesicles, usually from 200 to about 400 Å in diameter, near the presynaptic membrane (Figure 8.12). The arrival of spikes at the presynaptic terminals opens calcium channels and Ca^{2+} enters the presynaptic membrane. The entry of Ca^{2+} promotes the fusion of transmitter vesicle membranes with the presynaptic membrane at the synaptic cleft, thus releasing the transmitter chemical into the synaptic cleft. The released chemical diffuses across the synaptic cleft, a distance of some 100 to 200 Å, and binds to highly specific receptor proteins on the postsynaptic membrane surface. The binding of the transmitter to receptor opens ion channels in the postsynaptic membrane, and a postsynaptic membrane potential (a graded potential) develops. A particular synapse is stimulatory or inhibitory, depending on the neurotransmitter released. Stimulatory synapses give rise to **excitatory postsynaptic potentials** (**EPSP**s), while inhibitory synapses generate **inhibitory postsynaptic potentials** (**IPSP**s). EPSPs and IPSPs are graded potentials, and their magnitude, rate of rise, duration, and spread will depend on the amount of transmitter chemical released, which in turn might depend on the number of spikes arriving per second at the presynaptic terminals. Increasing release of neurotransmitter causes larger postsynaptic potentials. If the postsynaptic potential is a stimulatory potential and is strong enough to spread over the postsynaptic membrane surface and reach the spike generation region of the postsynaptic neuron, then spikes may be generated. If the transmitter chemical is an inhibitory chemical, spike generation will be suppressed or prevented in the postsynaptic neuron. This could mean reduction or cessation of spike generation in a postsynaptic neuron that spontaneously is active.

There is evidence that **acetylcholine** (**ACh**) is a synaptic transmitter at stimulatory synapses in the central nervous system (CNS), and **γ-aminobutyric acid** (**GABA**) is an inhibitory transmitter

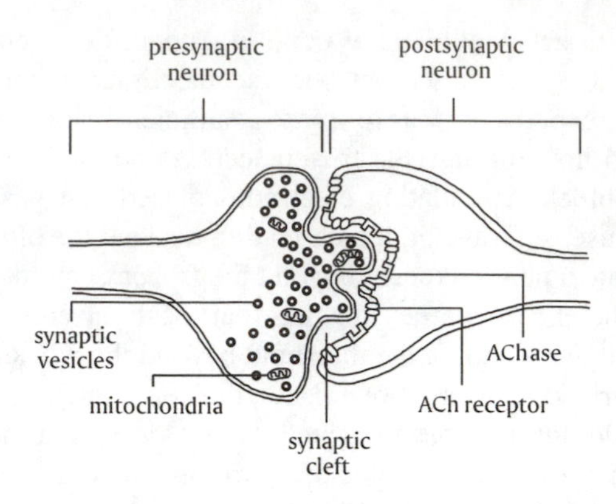

FIGURE 8.12 A schematic diagram for a synapse in an insect. Synaptic vesicles store the synaptic transmitter near the presynaptic membrane. Arrival of a nerve impulse opens voltage-gated calcium channels; calcium enters the presynaptic membrane and promotes the fusion of synaptic vesicles with the presynaptic membrane at the synaptic cleft region. This part of the membrane has been shown to lengthen in a heavily stimulated neuron as the vesicles fuse with the membrane, thereby lengthening it. In a stimulatory neuron in which the transmitter chemical is acetylcholine (ACh), the released ACh diffuses across the synaptic cleft and has about equal probability of encountering an acetylcholine receptor (ACh receptor) or the enzyme acetylcholine esterase (AChase).

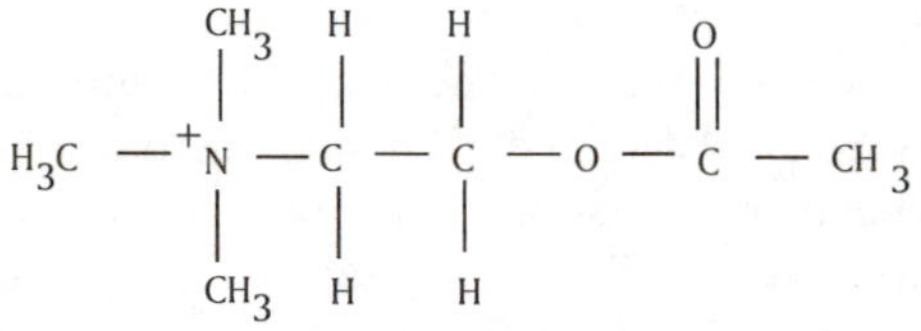

$$H_3C-\overset{+}{N}(CH_3)_2-CH_2-CH_2-O-\overset{\overset{O}{\|}}{C}-CH_3$$

acetylcholine

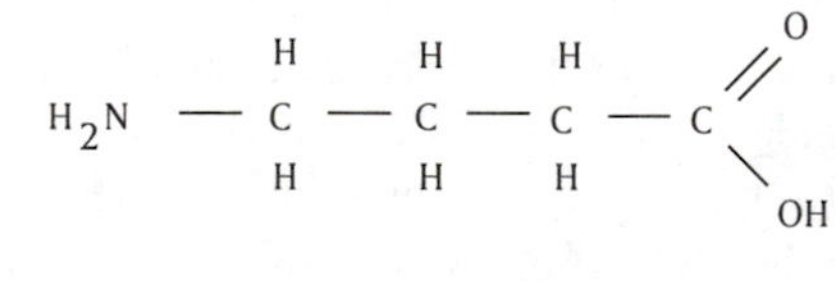

$$H_2N-CH_2-CH_2-CH_2-C\begin{matrix}\nearrow O \\ \searrow OH\end{matrix}$$

γ-aminobutyric acid (GABA)

FIGURE 8.13 The structures for acetylcholine, the stimulatory transmitter at synapses in the central nervous system (CNS) of insects, and γ-aminobutyric acid (GABA), the inhibitory transmitter in the CNS.

in the CNS of insects (Figure 8.13). These two transmitter chemicals are not likely to be the only neurotransmitters in the insect CNS, but others remain to be positively identified. The transmitter at nerve-muscle junctions (neuromuscular synapses) in insects is **L-glutamic acid** and possibly **L-aspartic acid**. There is fragmentary evidence from insects for a number of additional putative transmitter chemicals, including 5-hydroxytryptamine, catecholamines, octopamine, and some peptides. Callec (1985) outlined a number of criteria needed to demonstrate a transmitter function for a putative transmitter chemical.

Inhibition is a vital process in nervous systems and acts like a brake on the system. Synapses in the 6th abdominal ganglion of *P. americana*, for example, exhibit a high degree of sensitivity to applied GABA, and 1.05×10^{-13} *M* is sufficient to hyperpolarize postsynaptic membranes (Kerkut et al., 1969a, 1969b). When GABA is released in response to a wave of excitation, it selectively increases permeability to chloride ions, allowing more of these ions with a negative charge to enter the postsynaptic membrane. This causes the inside negative potential to become slightly more negative (e.g., from –70 mV to as much a –80 mV, or even more). A neuron that has been hyperpolarized requires a greater than normal stimulus to change the membrane potential enough to exceed the characteristic membrane threshold for spike generation.

GABA is synthesized in insects from glutamate by action of the enzyme L-glutamic acid decarboxylase, an enzyme widely distributed in high titers in insect nervous tissue. Following secretion at the nerve endings, GABA is probably rendered inactive by uptake at the synaptic terminal and/or by glial cells. It also may be oxidized to succinic acid.

8.6 ACETYLCHOLINE-MEDIATED SYNAPSES

Numerous studies have shown the presence of the necessary components of a cholinergic system in the CNS of insects; that is, acetylcholine (ACh); choline acetyltransferase that synthesizes ACh, **acetylcholinesterase (AChase)** that breaks down ACh after its secretion into the synapse; and receptors in the CNS for ACh. Cholinergic receptors are located only within the CNS in insects, and are not found at neuromuscular junctions as in vertebrates. Applications of ACh in isolated insect preparations have not always produced spike activity at physiologically relevant ACh concentrations, possibly because the CNS is protected from ACh applied in a bathing saline by the hemolymph-CNS barrier and fatty sheath surrounding the brain, ventral ganglia, and connectives. A very active acetylcholinesterase also acts quickly to destroy applied ACh. However, the high sensitivity of insect neurons to ACh is revealed by application of the ACh with a microsyringe, or iontophoretically, to neurons inside a ganglion (Callec, 1985). For example, application of 5×10^{-6} *M* ACh by microsyringe to dorsal unpaired median cell bodies in the neuropil of the 6th abdominal ganglion of *P. americana* depolarized the cells and produced a volley of spikes (Callec and Boistel, 1967). Acetylcholine administered iontophoretically into the 6th abdominal ganglion of *P. americana* was stimulatory at a dilution of 1.31×10^{-13} *M* ACh (Kerkut et al., 1969a, 1969b). Pretreatment of nervous tissue with inhibitors of AChase further increases sensitivity of desheathed ganglia to ACh (Narahashi, 1971; Shankland et al., 1971).

8.6.1 Action of Acetylcholine at the Synapse

A volley of spikes arriving at the presynaptic terminal increases permeability of the presynaptic membrane to calcium, which diffuses into the terminal, and through a second messenger sets up a cascade of actions that facilitate attachment of synaptic vesicles to the synaptic membrane. The synaptic vesicles fuse with the membrane and release quanta or packets of ACh into the synaptic cleft. Electron micrographs have shown that the presynaptic membrane expands slightly with the incorporation of vesicular membranes. When released into the synaptic cleft, the ACh molecules diffuse randomly, with some contacting and attaching to acetylcholine receptors and others encountering acetylcholinesterase. ACh is rapidly released from the receptor and may randomly collide with another receptor to repeat the process. When ACh is bound to its receptor, Na^+ channels are opened. When large numbers of Na^+ channels are opened in the postsynaptic membrane, the inward movement of Na^+ depolarizes the postsynaptic membrane with production of an EPSP that is conducted decrementally away from the site of origin. If the stimulation is strong enough, the excitation may spread to the region of the axon that generates spikes.

ACh molecules seem to have about the same probability of encountering the enzyme acetylcholinesterase, which is also bound to the postsynaptic membrane, as they do to encounter a receptor molecule. An ACh molecule encountering acetylcholinesterase is hydrolyzed to acetic acid and choline, neither of which has any physiological action at the synapse. Both breakdown products diffuse out of the synapse and/or are taken up by the presynaptic neuron, and may be used to synthesize new ACh through action of the enzyme choline acetyltransferase. Acetylcholinesterase is protective in function, and poisoning it with molecules such as organophosphate insecticides results in prolonged stimulation at synaptic sites throughout the CNS of insects. Poisoned insects

TABLE 8.2
Examples of Types and Binding Capacity of Cholinergic Receptors in Nervous Tissue

Organism	Tissue	Nicotinic Receptors	Muscarinic Receptors
		Binding (fmol/mg protein)	
Locusta	Ganglia	1775	116
Drosophila	Head	800	65
Periplaneta	Nerve cord	910	138
Mouse	Brain	180	570
Rat	Hippocampus	60	1000

Reproduced with permission from Breer et al., 1987.

typically show uncontrolled leg tremors, buzzing of the wings without control for flight, and eventual death. Probably many other physiological and biochemical processes are disrupted, such as release of neurohormones and depletion of energy reserves in uncontrolled muscular actions, and all of these contribute to the death of a poisoned insect.

8.6.2 Nicotinic and Muscarinic Cholinergic Receptors in Insects

In insects, as in vertebrates, more than one cholinergic receptor type exists, and the cholinergic receptors in insects have been characterized as **nicotinic**, **muscarinic**, and **mixed** receptors (Callec, 1985). ACh is the neurotransmitter at all these cholinergic receptors, and the function of the different receptor types is not clear either in vertebrate or invertebrate systems. At very low concentrations, nicotine and muscarine mimic the action of ACh; but at higher concentrations, they block the receptor. The ACh-mediated skeletal muscle receptors in vertebrates are of the nicotinic type, while vertebrate heart and gut muscle have muscarinic-type ACh receptors. The brain tissue of vertebrates has both nicotinic and muscarinic receptors, with muscarinic receptors generally outnumbering nicotinic ones in the brain of vertebrates. In contrast, insect central nervous tissue has more nicotinic receptors that muscarinic ones (Table 8.2), and ACh is not the synaptic mediator at insect neuromuscular junctions, so insect muscle tissue has neither type of ACh receptor.

Nicotinic cholinergic receptors in insects, as in vertebrates, are sensitive to the inhibitory action of a very potent toxin, α-bungarotoxin, derived from the venom of snakes of the family Elapidae (snakes in Southeast Asia). The toxin binds irreversibly to nicotinic-type receptors, producing a block of the synapse. It is often used pharmacologically to characterize and study the properties of nicotinic receptors. α-Bungarotoxin binding data indicate that nicotinic ACh receptors are present in the CNS of *D. melanogaster*, *P. americana*, *Musca domestica*, and *Manduca sexta* (reviewed by Callec, 1985). Membrane-bound nicotinic receptors that have very high binding capacity (Bmax) of 8926 fmol/mg protein are highly localized in the neuropil regions of the brain of the American cockroach (Orr et al., 1990).

ACh receptors that specifically bind muscarine, a very potent poison obtained from an *Amanita* sp. of mushroom, are classified as muscarinic-type receptors. Muscarinic receptors have been demonstrated in the head of *D. melanogaster* (Haim et al., 1979) and in the last abdominal ganglion of the cricket, *Acheta domesticus* (Meyer and Edwards, 1980). Insects appear to have a much higher concentration of nicotinic receptor types than muscarinic receptor types (Lummis and Sattelle, 1985; Breer et al., 1987). Pyrrolizidine alkaloids (PAs), present in many plant families, bind to muscarinic cholinergic receptors (Schmeller et al., 1997) and may exert some of their toxicity through this mode of action.

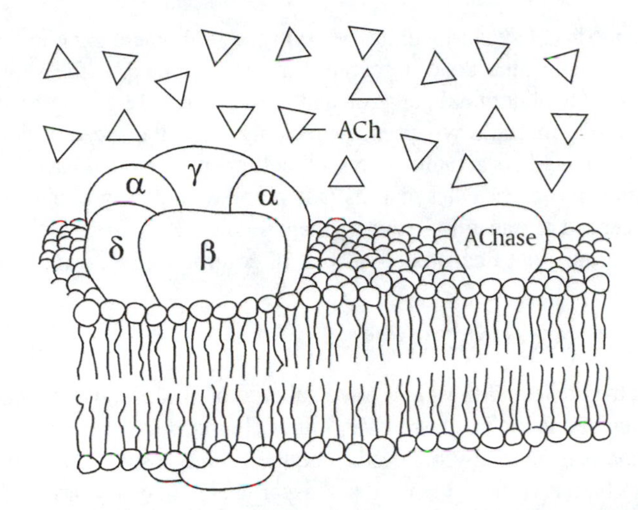

FIGURE 8.14 A conceptual model for the postsynaptic membrane in which acetylcholine (ACh triangles) may encounter its receptor (the α subunits of the sodium channel, schematically represented at the left) or acetylcholine esterase molecules (AChase, at the right) at the postsynaptic membrane. The ACh receptor modeled here is that shown to occur in a vertebrate and consists of five protein subunits, two α plus one β, δ, and γ. The structure of the receptor in an insect has not been completely elucidated. When ACh binds to each of the α subunits, the channel formed by the proteins opens and sodium ions enter the postsynaptic neuron, resulting in a postsynaptic (graded) potential. The ACh receptor quickly releases a bound ACh molecule and it may bind to another receptor and repeat the action. ACh encountering AChase is hydrolyzed into acetic acid and choline, both of which are inactive as far as nerve response is concerned.

8.6.3 Acetylcholine Receptor Structure

The complete structure of insect ACh receptors has not been elucidated, but there is evidence that the ACh receptor comprises part of the sodium channel, as it does in vertebrates. In contrast to the vertebrate receptor, however, the nicotinic receptor isolated from locust nervous tissue appears to be composed of identical polypeptide subunits (Breer et al., 1987). The electric fish ACh receptor at the neuromuscular junction is composed of five polypeptide subunits: two alpha, and one each of beta, gamma, and delta polypeptide units. The five subunits form a barrel-shaped transmembrane protein with the sodium channel through the middle (Figure 8.14) (Changeux et al., 1984, 1992; Changeux, 1993). One molecule of ACh binds to each of the two alpha subunits to open the channel to entry of Na^+.

8.7 ELECTRIC TRANSMISSION ACROSS SYNAPSES

Transmission of impulses across some synapses is electrical. In insects, some or possibly all the synapses within giant fiber systems of the ventral nerve cord are electrical. The spike crosses the electrical synapse without involvement of a chemical transmitter. Electrical synapses allow faster transmission of spikes (a message) than a network containing several synapses. The cercal nerve-giant axon complex provides the circuitry for a startle and escape reaction in cockroaches. An escape reaction starts with the reception of stimuli at the mechanoreceptors on the cerci. Strong stimuli result in large receptor potentials and a series of spikes that travel over the cercal nerve to the 6th abdominal ganglion. In the neuropil of the 6th ganglion, ACh is released at synapses with one or more of the giant axons. There is a delay in giant axon response of 0.68 ms to the released ACh, followed by a slow rise time of EPSPs over about 2 to 3 msec. The EPSPs are only about 2 to 5 mV in amplitude, but they give rise to spikes in the giant axon that do not have to cross additional chemically mediated synapses until they synapse in thoracic ganglia with mononeurons to the leg muscles. Much of the time delay in escape can be attributed to the slowness of chemically mediated synapses in the 6th abdominal ganglion and in the thorax. In general, synaptic transmission, and especially a circuit with multiple synapses, appreciably slows the speed of communication within the nervous system. The giant fibers actually represent multiple neurons that have anastomosed together, and the points of fusion are electrical synapses that a spike crosses without a chemical mediator. Probably, selection for speed of transmission was a major evolutionary force acting on the development of electrical synapses in the giant fiber system as part of control for an escape mechanism.

8.8 NEUROMUSCULAR JUNCTIONS

The junction between the nerve and muscle is a special type of synapse, usually called the **neuromuscular junction.** L-glutamate is the excitatory transmitter chemical at the neuromuscular junction in a few insects studied, and it is generally believed to be the typical insect neuromuscular transmitter at stimulatory junctions. There is some evidence that L-aspartic acid may also act as a transmitter at some neuromuscular synapses. GABA is the transmitter chemical at inhibitory nerve-muscle junctions.

REFERENCES

Armstrong, C.M., and F. Bezanilla. 1977. Inactivation of the sodium channel II. Gating current experiments. *J. Gen. Physiol.,* 70:567-590.

Becker, M.N., R. Brenner, and N.S. Atkinson, 1995. Tissue-specific expression of a *Drosophila* calcium-activated potassium channel. *J. Neurosci.,* 15(9):6250-6259.

Breer, H., D. Benke, W. Hanke, R. Kleene, M. Knipper, and L. Wieczorek, 1987. Identification, reconstitution and expression of neuronal acetylcholine receptor polypeptides from insects, pp. 95-105, in J.H. Law (Ed.), *Molecular Entomology,* Alan R. Liss, New York.

Callec, J.J. 1985. Synaptic transmission in the central nervous system, pp. 139-179, in G.A. Kerkut and L.I. Gilbert (Eds.), *Comprehensive Insect Physiology, Biochemistry and Pharmacology,* Vol. 5, Pergamon Press, Oxford.

Callec, J.J., and J. Boistel, 1967. Les effets de l'acetylcholine aux nivaux synaptique et somatique dans le cas du dernier ganglion abdominal de al blatte, *Periplaneta americana* L. *C.R. Soc. Biol.,* 161:442-446.

Changeux, J.-P. 1993. Chemical signaling in the brain. *Sci. Am.,* 1993:58-62.

Changeux, J.-P., A. Devillers-Thiéry, and P. Chemouilli, 1984. Acetylcholine receptor: An allosteric protein. *Science,* 225:1335-1345.

Changeux, J.-P., J.L. Galzi, A. Devillers-Thiéry, and D. Bertrand, 1992. The functional architecture of the acetylcholine nicotinic receptor explored by affinity labeling and site-directed mutagenesis. *Quart. Rev. Biophysics,* 25:395-432.

Cole, K.S. 1949. Dynamic electrical characteristics of the squid axon membrane. *Arch. Sci. Physiol.,* 3:253-258.

Eccles, J.C. 1964. Ionic mechanism of postsynaptic inhibition. *Science,* 145:1140-1147.

French, A.S., E.J. Sanders, E. Duszyk, S. Prasad, P.H. Torkkeli, J. Haskins, and R.A. Murphy. 1993. Immunocytochemical localization of sodium channels in an insect central nervous system using a site-directed antibody. *J. Neurobiol.,* 24:939-948.

Ganetzky, B., J.W. Warmke, G. Robertson, N. Atkinson, and R. Drysdale. 1993. Genetic and molecular analysis of potassium channels in *Drosophila,* pp. 9-22, in A.B. Borkovec and M.J. Loeb (Eds.), *Insect Neurochemistry and Neurophysiology,* CRC Press, Boca Raton, FL.

Gwilliam, G.F., and M. Burrows, 1980. Electrical characteristics of the membrane of an identified insect motor neurone. *J. Exp. Biol.,* 86:49-61.

Haim, N., S. Nahum, and Y. Dudai. 1979. Properties of a putative muscarine cholinergic receptor from *Drosophila melanogaster. J. Neurochem.,* 32:543-552.

Hodgkin, A.L. 1937a. Evidence for electrical transmission in nerve. Part I. *J. Physiol.,* 90:183-210.

Hodgkin, A.L. 1937b. Evidence for electrical transmission in nerve. Part II. *J. Physiol.,* 90:211-232.

Hodgkin, A.L. 1964. The ionic basis of nerve conduction. *Science,* 145:1148-1154.

Hodgkin, A.L., and B. Katz. 1949. The effect of sodium ions on the electrical activity of the giant axon of the squid. *J. Physiol.* (*London*), 108:37-77.

Huxley, A.F. 1964. Excitation and conduction in nerve: Quantitative analysis. *Science,* 145:1154-1159.

Kerkut, G.A., R.M. Pitman, and R.J. Walker. 1969a. Sensitivity of neurones of the insect central nervous system to iontophoretically applied acetylcholine or GABA. *Nature,* 222:1075-1076.

Kerkut, G.A., R.M. Pitman, and R.J. Walker. 1969b. Iontophoretic application of acetylcholine and GABA onto insect central neurones. *Comp. Biochem. Physiol.,* 31:611-633.

Keyes, R.D. 1951. The ionic movements during nervous activity. *J. Physiol.,* 114:119-150.

Lee, D., and M.E. Adams. 2000. Sodium channels in central neurons of the tobacco budworm, *Heliothis virescens*: Basic properties and modification by scorpion toxins. *J. Insect Physiol.,* 46:499-508.

Li, M., Y.N. Jan, and L.Y. Jan. 1992. Specification of subunit assembly by the hydrophilic amino-terminal domain of the *Shaker* potassium channel. *Science,* 257:1225-1229.

Lummis, S.C.R., and D.B. Sattelle. 1985. Binding of N-[propionyl-^{3}H] propionylated α-bungarotoxin and L-[benzilic-4,4′-^{3}H] quinuclidinyl benzelate to CNS extracts of the cockroach *Periplaneta americana*. *Comp. Biochem. Physiol.,* 80C:75-83.

Mackinnon, R., R.W. Aldrich, and A.W. Lee. 1993. Functional stoichiometry of *Shaker* potassium channel inactivation. *Science,* 262:757-759.

Marmont, G. 1949. Studies on the axon membrane. I. A new method. *J. Cell. Comp. Physiol.,* 34:351-382.

Meyer, M.R., and J.S. Edwards. 1980. Muscarinic cholinergic binding sites in an orthopteran central nervous system. *J. Neurobiol.,* 11:215-219.

Miller, C. 1991. 1990: Annus mirabilis of potassium channels. *Science,* 252:1092-1096.

Narahashi, T. 1971. Effects of insecticides on excitable tissues. *Adv. Insect Physiol.,* 8:1-93.

Narahashi, T., and T. Yamasaki. 1960. The mechanism of after-potential production in the giant axons of the cockroach. *J. Physiol.* (*London*), 151:75-88

Orr, G.L., N. Orr, and R.M. Hollingworth. 1990. Localization and pharmacological characterization of nicotinic-cholinergic binding sites in cockroach brain using α- and neuronal bungarotoxin. *Insect Biochem.,* 20:557-566.

Parnas, L., M.E. Spira, R. Werman, and F. Bergmann. 1969. Non-homogeneous conduction in giant axons of the nerve cord of *Periplaneta americana*. *J. Exp. Biol.,* 50:635-649.

Pichon, Y., and F.M. Ashcroft. 1985. Nerve and muscle: Electrical activity, pp. 85-113, in G.A. Kerkut and L.I. Gilbert (Eds.), *Comprehensive Insect Physiology, Biochemistry and Pharmacology,* Vol. 5, Pergamon Press, Oxford.

Pichon, Y., and J. Boistel. 1966. Application aux fibres géantes de Blattes (*Periplaneta americana* L. et *Blaberus craniifer* Bürm.) d'une technique permettant l'introduction d'une microélectrode dans le tissu nerveux sans résection préalable de la gaine. *J. Physiol.* (Paris), 58:592.

Pichon, Y., D.B. Sattelle, and N.J. Lane. 1972. Conduction processes in the nerve cord of the moth, *Manduca sexta,* in relation to its ultra-structure and haemolymph ionic composition. *J. Exp. Biol.,* 56:717-734.

Pichon, Y., and J.E. Treherne. 1973. An electrophysiological study of the sodium and potassium permeabilities of insect peripheral nerves. *J. Exp. Biol.,* 59:447-461.

Salkoff, L., K. Baker, A. Butler, M. Covarrubias, M.D. Pak, and A. Wei. 1992. An essential "set" of K^+ channels conserved in flies, mice and humans. *Trends Neurosci.,* 15:161-166.

Schmeller, T., A. El-Shazly, and M. Wink. 1997. Binding of pyrrolizidine alkaloids to acetylcholine, serotonin, and dopamine receptors. *J. Chem. Ecol.,* 23:399-416.

Shankland, D.L., J.A. Rose, and C. Donniger. 1971. The cholinergic nature of the cercal nerve-giant fiber synapse in the sixth abdominal ganglion of the American cockroach, *Periplaneta americana* L. *J. Neurobiol.,* 2:247-262.

Stevens, C.F. 1984. Biophysical studies of ion channels. *Science,* 225:1346-1350.

Tanouye, M.A., A. Ferrus, and S.C. Fujita. 1981. Abnormal action potentials associated with the *Shaker* complex locus of *Drosophila*. *Proc. Natl. Acad. Sci. U.S.A.,* 78:6548-6552.

Timpe, L.C., Y.N. Jan, and L.Y. Jan. 1988. Four cDNA clones from the *Shaker* locus of *Drosophila* induce kinetically distinct A-type potassium currents in *Xenopus* oocytes. *Neuron,* 1:659-667.

Treherne, J.E. 1985. Blood-brain barrier, pp. 115-137, in G.A. Kerkut and L.I. Gilbert (Eds.), *Comprehensive Insect Physiology, Biochemistry and Pharmacology,* Vol. 5, Pergamon Press, Oxford.

Treherne, J.E., and S.H.P. Maddrell. 1967. Membrane potentials in the central nervous system of a phytophagous insect (*Carausius morosus*). *J. Exp. Biol.,* 46:413-421.

Treherne, J.E., and P.K. Schofield. 1981. Mechanisms of ionic homeostasis in the central nervous system of an insect. *J. Exp. Biol.,* 95:61-73.

Trudeau, M.C., J.W. Warmke, B. Ganetzky, and G.A. Robertson. 1995. *HERG*, a human inward rectifier in the voltage-gated potassium channel family. *Science,* 269:92-95.

Warmke, J.W., and B. Ganetzky. 1993. A novel potassium channel gene family: *EAG* homologs in *Drosophila*, mouse and human. *Biophys. J.,* 64: A340 (abstract).

Yamasaki, T., and T. Narahashi. 1959. The effects of potassium and sodium ions on the resting and action potentials of the cockroach giant axon. *J. Insect Physiol.,* 3:146-158.

9 Muscles

CONTENTS

PREVIEW

Insect muscles are composed of cells that have anastomosed into multinucleate fibers of myofibrils. Myofibrils are divided into sarcomeres, which are the contractile units of muscle. Although wing muscles and jumping leg muscles in some insects are relatively large, muscles in small insects and in small appendages are necessarily small, and often are composed of only a few fibers. Skeletal muscles are attached to the cuticle, typically by tonofibrillae that pass through the endo- and exocuticle and attach to the inner layer of the epicuticle. Mitochondria, sometimes called sarcosomes because of their large and irregular size, are the powerhouses for muscle function. Only a few motor neurons are allocated to innervate most insect muscles, and typically there is a fast axon producing a rapid, twitch-like response in the muscle, and a slow axon that produces a slower, but more sustained contraction. The fast axon innervates each fiber in a muscle, while the slow axon innervates only about 30 to 40% of the fibers. Some muscles also receive an inhibitory neuron. A few large muscles receive multiple motor neurons. Graded contractions are achieved in some muscles by activating the slow or the fast axon, depending on the degree of muscle action needed, and perhaps by combining these with the action of the inhibitory axon. Each motor nerve breaks into many terminals that make contact with the muscle fibers at intervals of 40 to 80 μm apart. In general, an action potential is not conducted by the muscle fiber itself, but contractions occur around the nerve terminals and thus sum over the entire muscle. The transmitter chemical at excitatory motor endings is L-glutamic acid or L-aspartic acid, and the transmitter at inhibitory neurons is γ-aminobutyric acid (GABA). Most skeletal muscles of insects are synchronous muscles that require nerve input for each contraction, but wing muscles in some insects are asynchronous, and multiple contractions for each motor nerve input can be obtained. The ability of fibrillar muscles to yield multiple contractions is based on anatomical arrangement in the thorax, internal anatomy, and physiology. During the active state induced in muscles by the arrival of nerve impulses, calcium ions bound to the sarcoplasmic reticulum are released and then bind to a subunit of troponin. This induces a conformation change that pulls tropomyosin away from an active site on actin. Myosin binds to the active site and pulls actin into a new position. Myosin releases from actin when ATP binds to it and is split, releasing energy for return to the original state of myosin. Binding, sliding, release, and repeat are very rapid events, occurring in about 0.1 ms. Contraction is terminated by rapid binding of calcium to the sarcoplasmic reticulum, necessitating more nerve input to free Ca^{2+} for any additional contraction. Sarcoplasmic reticulum sequestering of calcium is slower in asynchronous muscle than in synchronous muscle, which is part of the explanation for multiple contractions per nerve input in asynchronous muscle. All muscles in insects are striated, including visceral muscle. Although insects first evolved as flightless animals, flight ability evolved as one of the most intense muscular activities of insects. How the wings evolved is uncertain, and numerous theories have been proposed. The ability to fly has been a major factor in the success of insects, enabling them to fill many ecological niches and to disperse in searching for food and mates. Some insects have special adaptations in their skeletal anatomy and muscle function for jumping and singing.

9.1 INTRODUCTION

The microstructural units of muscles that are most clearly identifiable with low magnification are muscle fibers. Each muscle fiber in skeletal and wing muscles is composed of many cells that have anastomosed so that cell membranes are no longer distinct. The nuclei from these cells, however, are still evident, and muscle fibers are multinucleate in histological sections. In insect gut muscles, the individual cells are more distinct and uninucleate. All insect muscle is striated, including gut muscle. Typically, the muscle fibers are as long as the muscle itself. In some muscles, fibers are grouped into bundles; while in others, especially fibrillar muscles, they are only loosely held together. Muscles are usually divided into broad categories such as skeletal, flight, heart, alary, and gut muscles based on their location within the body, structure, and function. Skeletal muscles are a mixture of fast and slow contracting muscles. Fibrillar muscles are exclusively fast contracting muscles. Heart, alary, and gut muscle are slower contracting muscles.

9.2 BASIC MUSCLE STRUCTURE AND FUNCTION

9.2.1 MACRO- AND MICROSTRUCTURE OF MUSCLE

A muscle can be subdivided into fibers, and a fiber into **myofibrils**. Myofibrils are composed of **sarcomeres**, each of which contains **actin**, **myosin**, and other proteins involved in the contraction

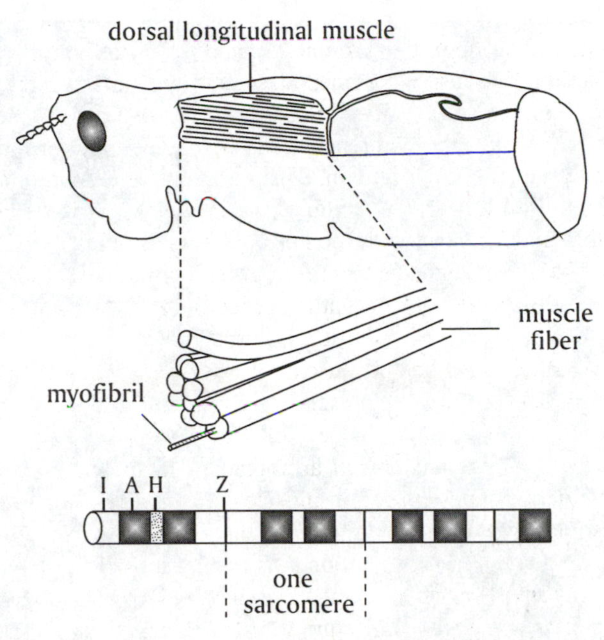

FIGURE 9.1 An illustration of progressively smaller units that compose muscles. Muscle is composed of muscle fibers that in turn are composed of myofibrils. At still higher magnification, myofibrils can be seen to be composed of sarcomere units, which are the contractile units of the muscle. The distribution of muscle proteins within the sarcomeres, which make the light and dark areas in a sarcomere, are designated as I, A, and H bands.

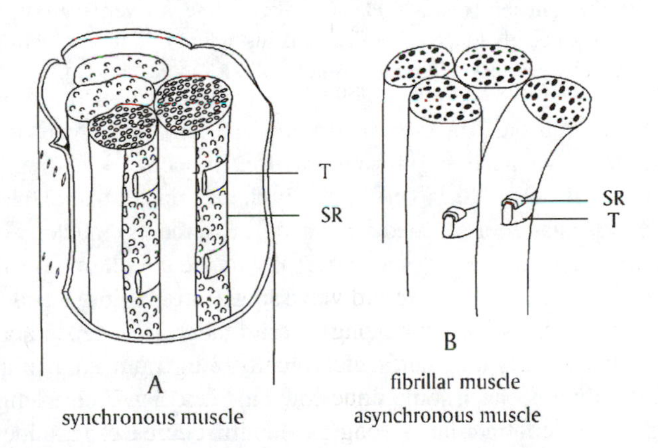

FIGURE 9.2 Diagrammatic illustration of the sarcoplasmic reticulum (SR) and transfer (T) tubules in synchronous muscle (A) and in asynchronous muscle (B).

mechanism (Figure 9.1). In addition, muscle contains **sarcoplasmic reticulum (SR)**, which is much reduced in fibrillar muscles but well developed in fast synchronous muscles (Figure 9.2). The SR is an extensive network of internal membranes broken into vesicles that run longitudinally on the surface of the muscle fibers. The SR plays a major role in the contraction process as a storehouse of calcium ions. Transverse (T) tubules penetrate the muscle from the outside, originating usually, but not always, at the Z bands. The network of T tubules and SR membranes do not open to each other, but they do intersect at closed junctions believed to be the major sites of calcium storage. These junctions are called **dyad** or **triad** junctions, depending on whether a T tubule intersects with one or two SR vesicles, respectively. The T tubules carry the electrical wave of excitation

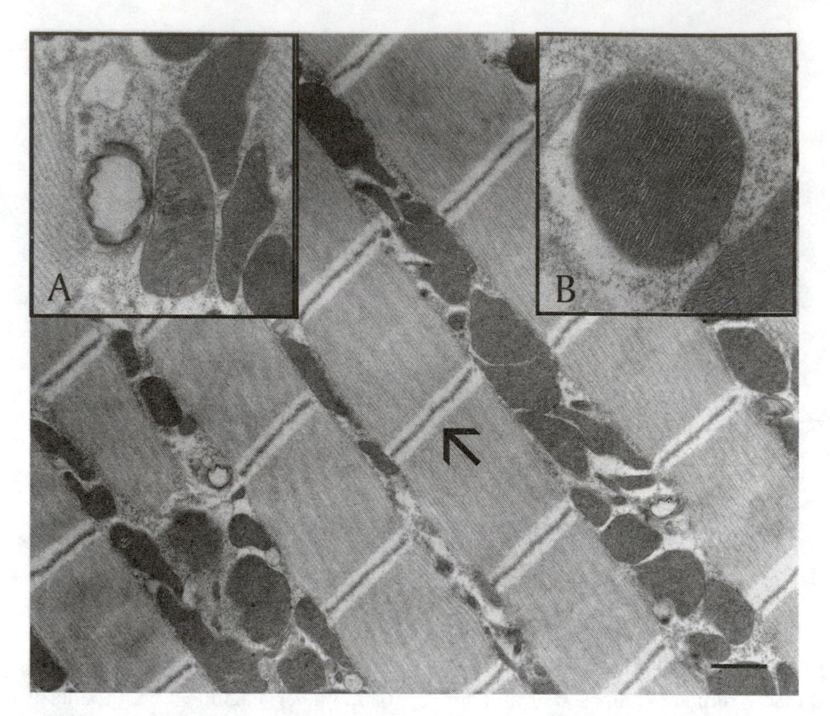

FIGURE 9.3 A transmission electron micrograph of muscle from *Tachinaephagus zealandicus* (Hymenoptera) showing well-defined Z lines (arrow), numerous, large, irregular mitochondria between the myofibrils, and intracellular tracheoles. Inset A shows an enlarged view of one of the intracellular tracheoles and several mitochondria. Inset B shows an enlargement of a mitochondrion. Membranes of the cristae in mitochondria can be seen in the enlargements. The I bands (the light areas on each side of the dark Z line) are very narrow in this muscle, indicating that the myosin filaments extend nearly to the Z line. There is just a hint of a lighter, narrow H zone at the middle of the sarcomeres. (Photographs courtesy of Jimmy Becnel and Alexandra Shapiro.)

arriving at the surface of a muscle (via a nerve) inward where it also spreads to the SR and releases bound calcium as the free ions necessary for contraction to occur.

Muscles contain abundant and often large, irregularly shaped mitochondria (also called **sarcosomes**, especially in thoracic musculature associated with wing movements), nuclei, and **intracellular tracheoles** (Figure 9.3). Intracellular tracheoles are not really inside the plasma membrane of the muscle, but have merely pushed into the muscle interior, like a finger pushed into a soft balloon.

The myofibrils (called fibrils by some authors) are made up of repeating sarcomere units. A sarcomere, the region between two Z bands, is typically 2 to 3 μm long in a muscle at rest, but sarcomeres up to 10 μm long occur in some very slow muscles. Sarcomere length is shorter in fast-contracting muscles. Upon contraction, sarcomere length decreases, as does the entire muscle length. The Z line is a plate-like sheet of protein to which actin (thin filaments, about 5 nm in diameter) and some other muscle proteins are attached. The thin filaments extend on either side of the Z line (about two thirds) to nearly the mid-point of a sarcomere. Thick filaments of myosin, about 20 nm in diameter, lie between the thin filaments (Ashhurst, 1967). The thick filaments extend across the middle of a sarcomere but usually do not extend to the Z line. The various overlapping regions of thick and thin filaments give muscle in thin histological sections a banded appearance (Figure 9.3) as light passes through regions of different density. The A band appears dark because light must pass through the overlapping regions of the actin and myosin filaments, while the H zone in the middle of the sarcomere and the I band near the Z line transmit more light because those regions contain only myosin or actin filaments, respectively. The M line across the middle of the H zone is created by cross-links between myosin filaments that help hold the myosin filaments in place. The various zones are of variable length in different muscles, depending on the degree of

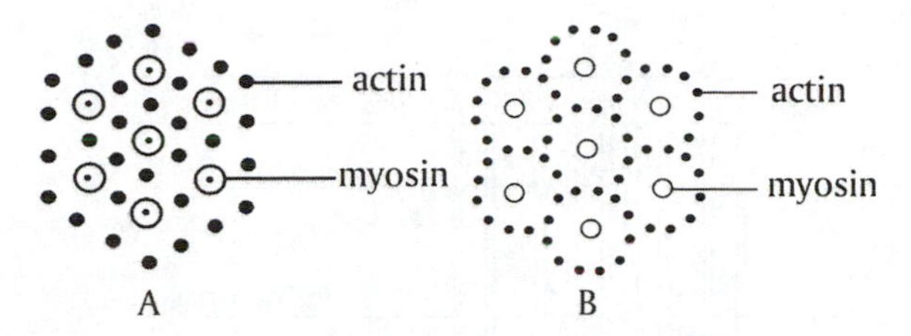

FIGURE 9.4 The arrangement of actin and myosin filaments in different muscles. A: Typically, in the flight muscle of most insects there are six actin filaments spaced such that each is midway between two myosin filaments, producing a 3:1 ratio of thin to thick filaments. B: An arrangement in which 12 actin filaments are arranged between myosin filaments to give a 6:1 ratio. (Reproduced with permission from Toselli and Pepe, 1968.)

overlapping of filaments. The width of the H zone bears some relationship to how much the sarcomere (and thus the muscle) will shorten upon stimulation and how fast it can accomplish its shortening. Muscles with very narrow H and I zones, such as fibrillar muscles that cause the wing movements in Diptera and Hymenoptera, shorten only a small amount upon contraction (2 to 3%), while some muscles may shorten much more. In general, in skeletal and flight muscles the striations or bands of adjacent myofibrils are aligned side by side. Thus, the light and dark bands appear evenly lined up in a large section of muscle, but there are exceptions to this in some insect muscles; in particular, there is less alignment in gut muscles.

Cross sections of fibrillar flight muscle myofibrils viewed with the electron microscope typically show each thick filament surrounded by six thin filaments, with the thin filaments positioned about equally between two thick filaments (Figure 9.4) to give a ratio of three thin filaments to one thick filament. Other ratios are found in some muscles, including up to 12 thin filaments arranged in such a way as to present a 6 thin: 1 thick filament arrangement in intersegmental muscles of the cockroach *Periplaneta americana* (Smith, 1966) and *Rhodnius prolixus* (Hemiptera: Reduviidae) (Toselli and Pepe, 1968), and in wing muscles of large saturniid silkmoths (Lepidoptera: Saturniidae) that have slow wing beats of five to six beats per second (Carnevali and Reger, 1982).

The physiology and biochemistry of muscle contraction appear to be the same in insects as in other organisms. The sliding filament theory in which actin and myosin filaments slide over each other, drawing Z bands closer together, and thus shortening, adequately explains insect muscle contraction. There are, however, many specific adaptations in insect muscle to serve the small size, flight ability, and unique behavior of many insects that are described in this chapter.

9.2.2 Muscle Attachments to the Exoskeleton

One end of a skeletal and wing muscle is anchored to a relatively non-moveable part of the exoskeleton. This is called the **origin** of the muscle, and the opposite end that is attached to a movable part of the body, such as a wing hinge, an appendage, or a part of the exoskeleton is called the **insertion**. Skeletal muscles are usually anchored to the epicuticle layer of the exoskeleton (Figure 9.5). A large 2.5-MDa extracellular matrix protein (**Dumpy protein**) encoded by the *dumpy* (*dp*) gene in *Drosophila melanogaster* connects the outer membrane of muscle fibers to the basal membrane of epidermal cells through a crosslinking zona pellucida domain and a transmembrane anchoring sequence (Wilkin et al., 2000). Dumpy protein provides a strong attachment for muscles to cells and cuticle, allowing high mechanical tension and preventing the tearing of muscle away from the cuticle (see also Chapter 4). Bundles of intracellular microtubules originate at the junction with the epidermal cells, pass through the epidermal cells, and, at the cuticle, are cemented to fibers of chitin (**tonofibrillae**) that are formed extracellularly and embedded in the cuticle (Hinton, 1973). Tonofibrillae are very resistant to the action of molting fluid, and allow muscles to remain attached to the exoskeleton after apolysis and secretion of new cuticle begins. In some insects, apolysis occurs hours and even days before ecdysis, and movement during the interval between apolysis and ecdysis may be critical to feeding and predator escape. The

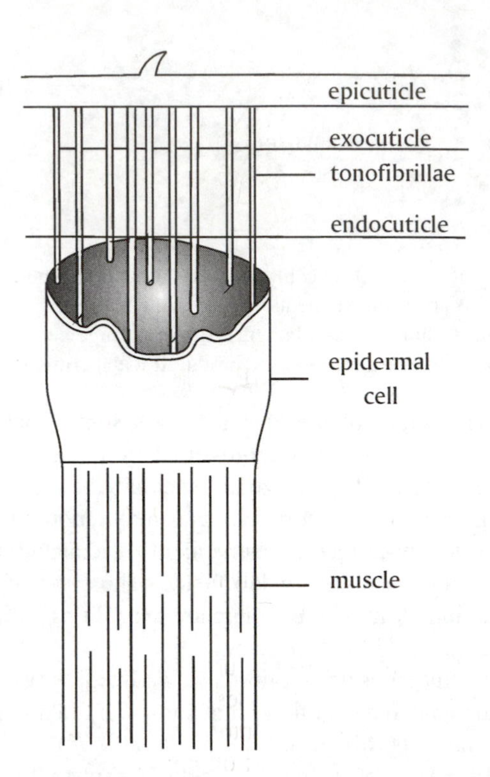

FIGURE 9.5 A schematic illustration of tonofibrillae connecting muscle fibers with the epicuticle. (Modified from Elder, 1975.)

final factors that enter into the breaking of the tonofibrillae to allow the old cuticle to be ecdysed have not been elucidated. After the old attachments of tonofibrillae are dissolved, new attachments to the epicuticle layer occur quickly. It should be remembered that the new cuticle present at ecdysis is largely unsclerotized, and the new epicuticle is probably the most stabilized part of the new cuticle, and the best place to anchor the muscles. The new cuticle must sclerotize adequately before the muscles are used or the soft cuticle will be distorted in shape by the pull of powerful muscles, particularly those used in flight and jumping. Most insects rest quietly for some minutes or hours immediately after molting until the cuticle has hardened.

9.2.3 Skeletal Muscle

Skeletal muscles are usually organized in antagonistic pairs. One muscle of the pair (the **flexor**) bends an appendage, while the second muscle of the pair (the **extensor**) straightens the appendage. A few muscles (e.g., the tymbal muscles of cicadas) are not antagonistically paired, but depend on the natural elasticity of the cuticle to stretch the muscle to the precontraction condition. Skeletal muscles are typically **synchronous muscles** in which the rate of contraction is in 1:1 proportion to the incoming nerve impulses. Peak tension and speed of contraction are influenced by sarcomere length, the degree of overlap of myosin and actin filaments, and the degree of development of the sarcoplasmic reticulum (Elder, 1975).

9.2.4 Polyneuronal Innervation and Multiterminal Nerve Contacts

Because of the small size of insects, only a few neurons are allocated to control each muscle. Pringle (1939) showed that the leg muscles (and many other muscles, it is now known) typically receive two stimulatory neurons, commonly described as the **slow neuron** and the **fast neuron**

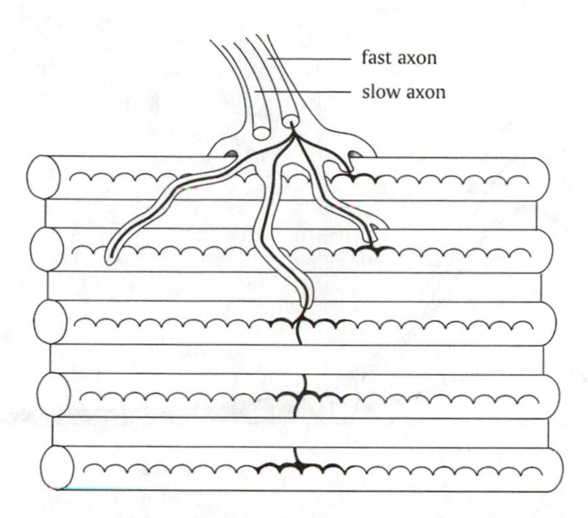

FIGURE 9.6 Polyneuronal innervation of muscle by fast and slow axons and the multiterminal junctional contacts with the muscle fibers. For clarity, only the contacts from the fast axon, which makes multiple contacts with each muscle fiber, are shown. The slow axon usually makes contact with 30 to 40% of the muscle fibers. Both axons may share the same glial sheath and contact points on the muscle fibers.

(Figure 9.6). At the muscle surface, the two nerves usually share a glial sheath and lie in a shallow groove on the muscle fiber surface, or in some cases they are invaginated within the muscle fiber outer membrane. Both fast and slow axons break into a number of arborizations, with arborizations making multiple contacts a few micrometers apart on muscle fibers. This is called **multiterminal innervation**, and it is common in invertebrates. In vertebrates, a neuronal branch makes only one contact with a muscle fiber although the motoneuron typically sends axonal branches to many fibers to form a motor unit.

The fast axon sends multiple terminals to all or most of the muscle fibers in locust jumping leg muscle, but the slow axon makes junctional contacts with only 30 to 40% of the muscle fibers (Hoyle, 1955). In muscles of non-jumping legs, the percentage of fibers innervated by the slow axon is usually greater than 40; and in some small muscles, such as some spiracular muscles, every fiber is dually innervated. Junctional locations on the muscle fibers are generally shared by the fast and slow axons. The distance between the multiterminal contacts on a muscle fiber varies with different species, but the contacts are always close together. In flight muscle of *Geotrupes* sp. (a beetle), junctional contacts are about 80 μm apart, in *Musca domestica* 50 μm apart, 40 μm apart in cockroach leg muscle, and about 60 μm apart in locust and grasshopper leg muscle.

Insect skeletal muscle usually does not develop a propagated action potential (an all or none response similar to nerve) as does vertebrate skeletal muscle, although resting potentials of 40 to 60 mV, inside negative, have been recorded from insect muscles (Aidley, 1985, 1989). Instead, graded potentials are produced around each of the junctional endings, and local contraction of the muscle fiber occurs around each ending. Because junctional terminals are very close together, the net result is a nearly simultaneous contraction of the whole fiber without a propagated muscle potential. After very careful dissection, some investigators have reported that jumping-leg muscle fibers will give a spike instead of the summed graded responses around end plates; but in a living insect, the summed graded responses are probably typical.

The fast neuron, produces a fast, twitch-type response in the muscle, while the slow axon typically produces a much slower, graded response. The slow neuron junctional contacts at the muscle show facilitation, and a characteristic frequency of nerve impulses must arrive at the junction before the muscle contracts. Typically, a muscle innervated by a fast axon contracts rapidly in response to each arriving nerve impulse. The designations "fast" and "slow" indicate the speed of the muscle contraction, not the rate of nerve impulse conduction. Both fast and slow neurons are

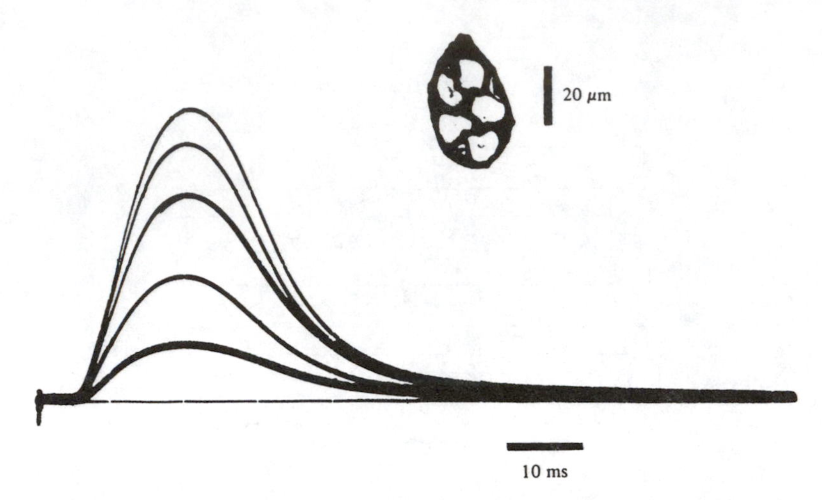

FIGURE 9.7 Mechanical responses of the metathoracic dorsal longitudinal muscles (DLM) from the katydid *Neoconocephalus robustus* to single shocks of gradually increasing intensity at 25°C. The five increases in muscle response suggest that the electrical stimuli elicited a response from five different neurons. Each neuron has its own characteristic threshold, and additional neurons are stimulated (and hence additional muscle fibers are activated) as the stimulus strength is increased. The histological cross-section of the nerve to the DLM shows five large axons. (Reproduced with permission from Josephson and Stokes, 1982.)

about the same size and conduct the nerve impulse at the same rate. They usually share the same glial sheath. This dual (or triple if an inhibitory neuron is present) innervation of a muscle in insects is referred to as **polyneuronal innervation**. A third neuron, an inhibitory one that causes hyperpolarization, was described by Hoyle (1955) from the jumping-leg muscle of the locust, and also from the grasshopper *Romalea microptera* (Usherwood and Grundfest, 1964; Usherwood, 1968).

Only a few large muscles get more than two stimulatory neurons. A few cases have been reported in which there are multiple axons to a large muscle; for example, seven to nine axons to the basalar flight muscle of large beetles (Darwin and Pringle, 1959; Ikeda and Boettiger, 1965). The locust dorsal longitudinal muscles (DLM), which are large, powerful muscles involved in flight, receive five fast motoneurons (Neville, 1963; Burrows, 1977). The mesothoracic DLM in a katydid, *Neoconocephalus robustus*, receives four fast axons, while DLM in the metathorax receives five fast axons (Figure 9.7) (Josephson and Stokes, 1982). There is no evidence for inhibitory neurons to the DLM in *N. robustus*, and the possibility of slow fibers is uncertain. There are at least five motor neurons to the basalar flight muscle of the scarab beetle *Cotinus mutabilis* (Josephson et al., 2000a). Thus, although the flight muscles of some insects receive multiple axons, and skeletal muscles may receive both slow and fast axons, indirect flight muscles of Hymenoptera and Diptera receive only fast axons. Slow flapping of the wings may be unnecessary and even impossible in some insects. The tymbal muscles of cicadas also receive only fast axons.

A few muscles may receive no nervous connections. For example, the long, thin muscle that spirals round the Malpighian tubules of some insects has its own myogenic rhythm and does not receive a nerve supply. At least some of the accessory heart muscles located at the base of legs and antennae of some insects appear to have no nerve supply, while others receive neurosecretory neurons and may be a neurohemal organ site.

The ability to achieve graded muscle contractions is important to the behavior of all animals, and insects appear to have an adequate repertoire of mechanisms available to them to achieve graded contractions. In addition to the simple arithmetic of how many muscle fibers may be activated depending on the use of the fast or slow axon, **facilitation** and **summation** commonly occur at the junctional endings when the slow axon is activated. In the leg, typically about 15 neuronal impulses per second must arrive at the junctional endings of the slow axon to produce contraction.

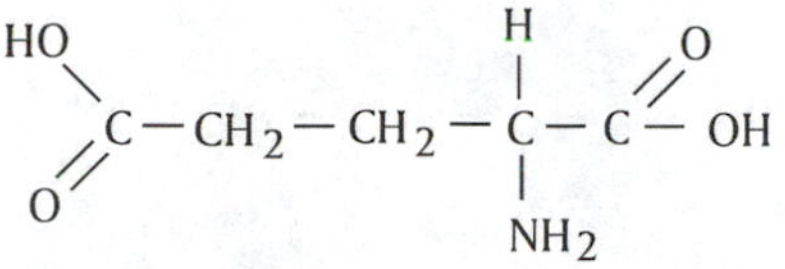

$$HO(O{=})C-CH_2-CH_2-C(H)(NH_2)-C({=}O)-OH$$

L-glutamic acid

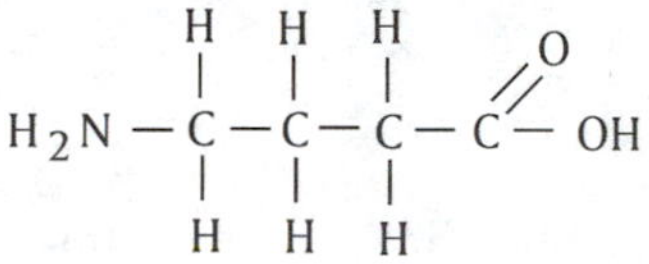

$$H_2N-CH_2-CH_2-CH_2-C({=}O)-OH$$

γ-aminobutyric acid

FIGURE 9.8 The chemical structure of L-glutamic acid, a stimulatory neurotransmitter at nerve-muscle junctions in insects, and γ-aminobutyric acid, an inhibitory transmitter at nerve-muscle junctions.

Temporal facilitation and summation can occur when several small volleys of nerve impulses arrive within one second. Thus, an insect that has some combination of fast, slow, and inhibitory axons to the same muscle may realize a number of degrees of graded muscle responses from the neuronal system or systems utilized, and response may be further modified by neuropeptides secreted at some nerve endings.

9.2.5 The Transmitter Chemical at Nerve-Muscle Junctions

The synaptic cleft between nerve ending and muscle membrane is about 30 nm wide (Aidley, 1989), and a transmitter chemical must diffuse across this gap from axonal ending to muscle sarcolemma. The stimulatory transmitter chemical at fast and slow nerve-muscle junctions is **L-glutamic acid** (Irving and Miller, 1980) and possibly **L-aspartate**, also. At inhibitory junctions, it is **γ-aminobutyric acid (GABA)** (Figure 9.8). The transmitter chemicals are contained in large numbers of vesicles that are from 20 to 60 nm in diameter. Although vertebrates utilize GABA as an inhibitory transmitter in the central nervous system, they use acetylcholine as the stimulatory transmitter at neuromuscular junctions in skeletal muscles.

Several types of glutamate receptors (GluR) have been characterized in insect muscles (and also in the insect CNS), based on binding to indicator compounds including quisqualate (quisqualate glutamate receptor, qGluR), ibotenate (iGluR), and aspartate (aspGluR); but how these different receptors function in muscle activity is unknown. Multiple receptors may be involved in different degrees of graded responses from the muscle. In *Schistocerca gregaria* and *Tenebrio molitor*, where the most detailed studies have been done, populations of these receptors are present at nerve-muscle junctions, as well as at extrajunctional sites (Usherwood, 1994). The fate of L-glutamic acid and GABA at the muscle receptors is unclear, but the neuron may reabsorb the transmitter to deactivate it. The transmitter at visceral muscle junctions has not been definitely determined. There is evidence, however, that some neuropeptides may play a role in gut muscle function, either as neurotransmitters or as neuromodulators that modify the response to a transmitter.

9.3 SYNCHRONOUS AND ASYNCHRONOUS MUSCLES

Muscles of insects can be divided into two groups based on the contraction rate per nerve input. Contraction frequency in **synchronous muscles** is directly controlled in a 1:1 manner by the output

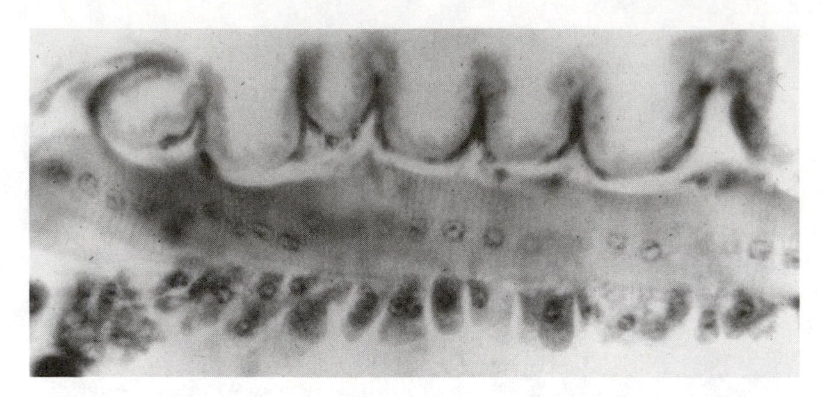

FIGURE 9.9 A photograph of tubular muscle from a tephritid fruit fly adult (*Anastrepha suspensa*) showing the nuclei lying down the middle of the muscle. (Photograph by the author.)

of nerve impulses from the central nervous system. In **asynchronous muscles**, the contraction frequency is a property of the muscle itself and of its anatomical arrangement in the musculoskeletal system. Asynchronous muscles oscillate and produce several to many contractions in response to a single neuronal stimulus.

General skeletal muscles of all insects, some tymbal and stridulatory muscles, and muscles that move the wings of some insects (Table 9.1) are synchronous. Synchronous muscles typically have a well-developed SR and T systems with regularly occurring triad junctions (Figure 9.2). Slowly contracting synchronous muscles are an exception, and they have a poorly developed SR. For example, the SR composes only about 1% of the fiber volume in the slow extensor tibia muscle of *Schistocerca gregaria* (desert locust: Orthoptera) (Cochrane et al., 1972), whereas the SR fills about 30% of the muscle volume in the fast synchronous tymbal muscle of a cicada, *Platypleura capitata* (Homoptera: Cicadidae) (Josephson and Young, 1985).

Pringle (1957, 1968, 1976) and others have made detailed studies of the function of synchronous muscles and suggested that a contraction rate greater than about 100 times per second is unlikely because of the time needed for neuronal repolarization. A firing rate of 100 times per second allows 10 ms for the neuron to repolarize. The contraction frequency of most synchronous muscle is well below 100 Hz (contractions per second), but some stridulatory and tymbal muscles with characteristic internal anatomy of synchronous muscles have rates of contraction two or more times the expected 100-Hz limit. Exactly what is going on in these muscles is not known for sure; they may be synchronous most of the time but capable of going into the oscillatory behavior of asynchronous muscles (Josephson and Young, 1985) (see Section 9.7.1 for more details). Two types of synchronous muscles, **tubular** and **close-packed**, have been well characterized and are found in both general skeletal and flight musculature of some insects.

The fibers of tubular muscle are multinucleate, with nuclei located along the center of the fiber (Figure 9.9) and slab-like myofibrils typically radiating from the center like spokes of a wheel. These slab-like myofibrils run the length of the muscle and shorten in the plane of the long axis of the muscle. Numerous large, somewhat elongated mitochondria are arranged radially between myofibrils. Both longitudinal and transverse tubules of the sarcoplasmic system are well developed in tubular muscles (Smith, 1961). Tubular muscles are the typical skeletal muscles of several orders, including Diptera and Hymenoptera, and are involved in movement of legs and other appendages, spiracle muscle control, and in such activities as compressing the abdomen (the tergosternal muscles) dorsoventrally as part of the breathing rhythm. Tubular muscles (also called radial fibers flight muscle by some authors) are found in the flight musculature of many insect groups, including both direct and indirect flight muscles in some insects (Table 9.1). These flight muscles, which are the typical flight musculature in the lower Orthoptera (such as the Blattidae) and in Odonata, may be most similar to the evolutionarily primitive flight muscles.

Other synchronous muscles occurring in the flight musculature of many groups of insects (Table 9.1) are called **close-packed fibers** (= micro-fibrillar and mosaic fibers of various authors). The muscle fibers are typically 10 to 100 μm in diameter. Nuclei are numerous, peripherally located, and somewhat flattened. Small myofibrils of 0.5 to 1 μm in diameter are interspersed with columns of large mitochondria. The muscle fibers may be circular in shape or have a polygon shape with many angles. The flight muscles of some Orthoptera, Trichoptera, and Lepidoptera are close-packed muscle. In some Orthoptera, strap-like, radially arranged myofibrils of close-packed muscle look somewhat similar to the arrangement in tubular muscle of lower Orthoptera. Both longitudinal and transverse tubules of the sarcoplasmic reticulum are extensively developed.

Asynchronous muscles, also called **fibrillar muscles**, have arisen about ten times in different groups of insects (Cullen, 1974), and the majority of insects that fly do so with asynchronous muscles. Asynchronous muscles are more efficient than synchronous muscles because they reduce the repetitive cycling of Ca^{2+} and thus minimize the energy cost of such cycling (Josephson et al., 2000a). Moreover, asynchronous muscles likely have been selected during evolution of flying insects because they allow a greater power output at high contraction frequencies typical of the flight of many insects (Josephson et al., 2000a, 2000b). Asynchronous flight muscles occur in both direct and indirect muscles associated with wing movements in Diptera, Hymenoptera, Coleoptera, some Hemiptera, and in a number of other insect groups (Table 9.1). Asynchronous muscles allow an insect to beat the wings multiple times for each volley of nerve impulses, and wing beat frequencies greater than 100 Hz are common in insects with asynchronous flight muscles. One very small midge (a dipteran) may beat the wings up to 1000 times per second (Sotavalta, 1953). Although skeletal muscles are not asynchronous muscles, the tymbal muscles of some cicadas are asynchronous. Asynchronous muscles require rhythmical input from the central nervous system for continued contraction, but the contraction frequency is a property of the muscles and their anatomical orientation and is not proportionally related to the nerve input. There may be three, four, or many repeated contractions for each volley of nerve impulses. Wing loading and thoracic resonance properties in insects with asynchronous wing muscles influence the number of wing beats per second. Wing beat frequency can be increased experimentally by reducing wing loading (by cutting off small portions of the wings). In contrast, wing loading does not have much influence on wing beat frequency in insects with synchronous muscles because the frequency is determined by the rate of nerve impulses coming to the muscles.

Fibrillar muscles are so-named because the muscle fibers easily separate from each other upon even slight shearing or tearing action (Figure 9.10). The muscle fibers are very large, ranging from about 100 μm up to 1 mm in diameter. In larger insects, the fibrillar muscles consist of bundles of fibers separated by tracheae, which often push into the plasma membrane of individual fibers to become intracellular tracheoles. The muscle fibers are cylindrical in shape and multinucleate, with nuclei usually lying peripherally in neat rows. Sarcomeres vary from 1.7 to 2.5 μm in length. The A band covers about 90% of the sarcomere length, and I bands are very narrow. These anatomical arrangements mean that maximum contraction can occur with very little shortening of the sarcomere and, consequently, with little shortening of the total muscle length, another property that enables rapid contraction rates. The myofibrils of fibrillar muscles are large, varying from 1 to 5 μm in diameter (as large as muscle fibers in some other types of muscles) and are as long as the muscle itself. The sarcoplasmic reticulum is poorly developed and not continuous along the long axis of the muscle fiber so that junctions between the SR and T system of tubules are reduced to dyad junctions (Figure 9.2B). The transverse tubules are well developed, penetrating usually at the Z lines and ramifying among myofibrils. Six thin filaments (actin) surround the thick filament (myosin) in a very regular order so that the ratio of thick to thin filaments is 1:3. Large, irregular mitochondria lie between myofibrils and may occupy up to 30% of the volume of the muscle. Glycogen deposits and lipid droplets have been observed around the mitochondria in some insects (Ashurst and Cullen, 1977).

TABLE 9.1
Principal Muscles Controlling Flight in Various Orders of Insects

Muscle	Location	Synchronous Tubular	Synchronous Close-Packed	Asynchronous Fibrillar	Function
Indirect Muscles That Attach to the Thoracic Cuticle, But Do Not Insert on the Wing Hinges					
Dorsal longitudinal	Pre- to postphragma	Dictyoptera Odonata	Orthoptera Ephemeroptera Dermaptera Psocoptera (*Troglus*) Neuroptera Mecoptera Trichoptera Lepidoptera Homoptera[a] Hymenoptera (Symphyta except *Xyela*)	Thysanoptera Heteroptera Psocoptera (*Psocus*) Diptera Coleoptera Hymenoptera (Apocrita, *Xyela*) Homoptera (Jassidae, Aphididae, Psyllidae)	Power-producing downstroke of the wings
Dorsoventral	Scutum to sternum, coxa, or trochanter	Diptera	Same orders as above for dorsal longitudinal muscle		Upstroke of wings; power producing
Oblique dorsal	Lateral scutum to postphragma	Orthoptera Dictyoptera	Phasmida and orders above for dorsal longitudinal muscle		Downstroke in Phasmida; upstroke in other orders

Direct Muscles Attached to the Wing Hinges

Basalar	Basalar sclerite or episternum to sternum or coxa	Dictyoptera	Orthoptera	Homoptera (Jassidae, Aphididae, Psyllidae)	Pronates wing; fibrillar and close-packed are power producing
		Diptera	Odonata	Heteroptera (Belostomatidae, Naucoridae, Notonectidae)	
		Hymenoptera (Aculeata)	Lepidoptera	Hymenoptera (Ichneumonidae)	
			Homoptera (same families as for dorsal longitudinal muscles)	Coleoptera	
Subalar	Subalar sclerite or episternum to sternum or coxa	Dictyoptera	Orthoptera	Diptera (Nematocera)	Supination of wing; fibrillar and close-packed are power producing
		Heteroptera (Geocoridae)	Odonata	Coleoptera	
		Diptera (Brachycera, Cyclorrhapha)			
		Hymenoptera (Apocrita)			

[a] Aleurodidae, Cercopidae, Cixidae, Coccidae, Flatidae, Membracidae.

Reproduced with permission from Pringle (1976).

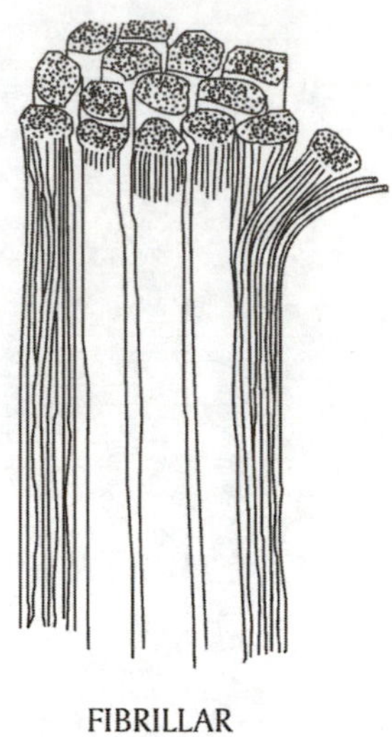

FIBRILLAR
MUSCLE

FIGURE 9.10 A diagrammatic illustration of fibrillar muscle with large myofibrils that tend to separate from each other with slight shearing action.

9.4 MUSCLE PROTEINS AND PHYSIOLOGY OF CONTRACTION

The muscle proteins and contraction physiology are essentially the same in insects as in muscles of other animals (see review by Maruyama, 1985). The major proteins in muscle fibers are **actin**, **tropomyosin**, **troponin** (all part of the thin filaments), and **myosin** (the thick filaments) (Figure 9.11). Shortening of a muscle occurs by the sliding of actin and myosin filaments over each other, pulling the Z bands closer together, and ultimately shortening the muscle toward the point of its fixed attachment, its origin. A single myosin molecule has the shape of a double-headed golf club (Figure 9.11A). Myosin molecules are arranged with their long tails forming the core of the thick filament with heads projecting from the filament. The globular heads have calcium-dependent ATPase activity and contain the binding sites for attachment to actin during contraction. The myosin heads form the structures called **crossbridges** (Squire, 1977).

The smallest unit of actin is a globular polypeptide named **G actin**. G actin subunits are linked together by polypeptide bonds to form a long chain of filament actin (**F actin**) (Figure 9.11C) and a thin filament consists of two chains of F actin twisted around each other in an α-helix. Each G actin subunit in the chain has an active site where a myosin head can attach. Another protein, tropomyosin, is associated with the thin filaments, and it consists of a long filamentous chain running along each of the two grooves created by the F actin helix. Each tropomyosin chain is composed of two α-helical units in a coiled-coil pattern, and the coiled chain runs along the groove created by the F actin helix. Tropomyosin filaments cover the active sites where a myosin head can attach. Troponin is a globular protein composed of three subunits: TnT (**tropomyosin binding**), TnC (**calcium binding**), and TnI (**actin binding**) (Figure 9.11C, D). It functions during nerve stimulation to the muscle by changing shape and pulling tropomysin away from the myosin binding sites. A troponin unit is associated with each G actin unit.

9.4.1 THE ACTIVE STATE: BINDING OF MYOSIN HEADS TO ACTIN AND THE SLIDING OF FILAMENTS

The signal for initiating a contraction is the arrival of a nerve impulse at the junctional ending on the muscle surface, resulting in the release of neurotransmitter at the polyneuronal neuromuscular

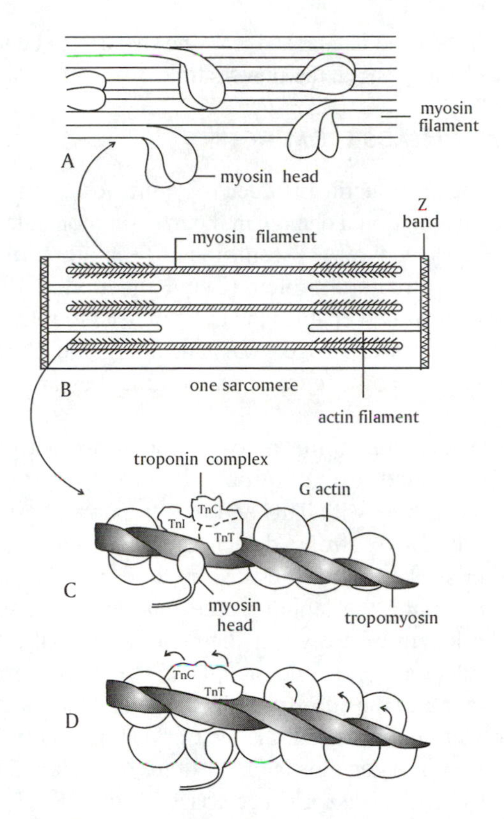

FIGURE 9.11 The arrangement of the major proteins involved with contraction in each sarcomere of a myofibril. A: A diagram of myosin molecules to show the double heads of each myosin filament. B: A sarcomere unit illustrating actin and myosin filaments. C: Helically entwined strands of F actin with a helical coil of tropomyosin lying in the groove of the F actin molecules. Troponin, consisting of three polypeptide units, is bonded through one polypeptide to a G actin molecule and by another subunit to tropomyosin, leaving one subunit free to bind to calcium ions. D: When the TnC part of troponin binds calcium ions, a change in shape of troponin occurs and tropomyosin is pulled away from the active site on G actin, allowing the myosin heads to bind to actin and initiate the power stroke of contraction.

junctions. The nerve impulse sets in motion a cascade of reactions mediated by G proteins and results in the release of Ca^{2+} from binding sites in the SR at the triad (synchronous muscle) or dyad (asynchronous muscle) junctions in the muscle fiber. The threshold of free Ca^{2+} ions for activation of the contraction process is about 10^{-7} M, with maximum contraction occurring at about 10^{-5} *M* (Aidley, 1989). The free Ca^{2+} ions in the sarcoplasm and availability of ATP create an "**active state**" in each myofibril. Once initiated, the active state can persist as long as free Ca^{2+} ions and ATP are available. When the concentration of free Ca^{2+} ions falls below about 10^{-7} *M* in the sarcoplasm because of the sequestering of Ca^{2+} by the sarcoplasmic reticulum, the active state is abolished, and a further contraction is not possible until a new volley of nerve impulses releases more Ca^{2+}. The active state persists much longer in asynchronous muscle than in synchronous muscle because the poorly developed SR in asynchronous muscle sequesters Ca^{2+} ions only slowly.

During the active state, free Ca^{2+} ions bind to the TnC subunit of troponin. This promotes a change in shape of the troponin complex, causing the TnT subunit to pull tropomyosin away from the active site where a myosin head can attach to actin. Attachment sites covered by troponin occur on each of the globular or G actin subunits, spaced about 38.5 nm apart. This allows myosin heads to bind to actin at many sites during the active state. When myosin binds to actin, a conformational change occurs in the myosin molecule in a "hinge-like" region of the myosin head and at another hinge site in the arm to which the head is attached, causing the arm to bend similar to the bending of a person's arm. These shape changes create the power stroke that causes sliding of actin over myosin.

9.4.2 Release of Myosin Heads from Actin

The myosin head must release from actin in order to return to its original state and prepare for another power stroke. The conformational change in the myosin molecule associated with the power stroke exposes a site on the head where ATP can bind. ATP binds to this head site, and myosin, which is an ATPase as well as a structural protein (Lymn and Taylor, 1970), splits the bound ATP into ADP and PO_4. The energy liberated enables the myosin head to release from actin and to return to its original shape, ending the power stroke. ADP and inorganic phosphate remain attached to the head until a new attachment to actin occurs, at which time they are also released from the head and a new power stoke begins.

The binding of ATP, detachment of the myosin head from actin, and reattachment occur extremely rapidly, and are estimated to require only about 0.1 ms (Lymn and Taylor, 1970). If calcium ions are available to combine with TnC and pull tropomyosin away from the active sites, myosin can bind, accomplish its power stroke, detach, and bind again many times per second. The movement caused by a power stroke has been estimated to be 5 to 10 nm (Harrington, 1981), and thousands of such movements along the length of the muscle fiber create total fiber shortening. The myosin heads along the length of a myosin filament are not all attached or detached at the same time, and their independent action results in a steady tension exerted to pull the Z bands of a sarcomere closer together, and thus the entire muscle shortens. After death of an insect, the myosin heads bind to actin and contraction occurs, but the lack of ATP prevents them from detaching from actin, resulting in the condition of rigor mortis, as in other animals. ATP must be available at all times for normal muscle functions because it is necessary for the SR to actively sequester calcium, and for binding to the myosin heads so that the heads detach from actin.

The very large fibrillar muscles of the giant waterbug *Lethocerus* spp. (Hemiptera: Belostomidae) provided important data for developing the concept of the conformation change in the myosin head region when it is attached to actin and for flexing or movement of the head on its arm. Electron micrographs of *Lethocerus* muscle in a state of extended contraction (rigor) showed cross-bridges (the myosin heads) fixed at an angle, which suggested they had to flex during contraction (Reedy et al., 1965). X-ray diffraction studies of active muscle suggested that the heads actually move during contraction (Tregear and Miller, 1969).

9.5 THORACIC STRUCTURE, WING HINGES, AND MUSCLE GROUPS INVOLVED IN FLIGHT

The thorax, although heavily sclerotized to withstand the pull of the flight musculature, is composed of plates joined by sutures that allow some flexibility and movement in several planes relative to the body axis. The wings are hinged to the thoracic plates by a number of small, hard sclerites at the junction between the tergum and the pleuron. The wings pivot up and down over the **pleural wing process**, a heavily sclerotized finger-like fulcrum of cuticle that is part of the pleuron (Figure 9.12). In addition, the hinge points let the thorax move inward and outward during a stroke cycle, which aids in snapping the wing rapidly over the pleural wing process.

The forewings attached to the mesothorax are not used in flight by some insects, such as beetles. In other insects in which both fore- and hindwings are used in flight, the wings may beat together, or the fore- and hindwings may be slightly out of phase. When forewings and hindwings are used in flight, the frontal edges of the wings are reinforced with larger, tubular veins for strength. The frontal edge of the wing leads in both the upstroke and the downstroke. Airflow hits the lower side of the wing during the downstroke, which generates the principal lift forces, and hits the upper

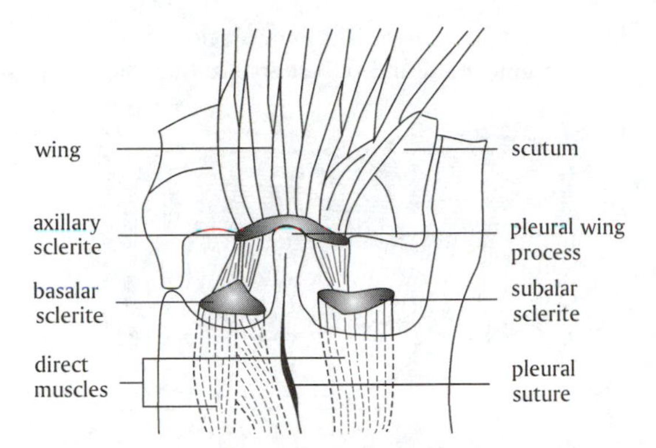

FIGURE 9.12 Thoracic structure showing the subalar and basalar wing hinge sclerites, and heavily sclerotized pleural wing process over which the wing pivots. Subalar and basalar muscles, attached to their respectively named wing hinges, are important in controlling wing orientation movements in all insects, and they also produce the downstroke in dragonflies and damselflies, but not in most other insects.

side during the upstroke, which can also produce lift forces (Nachtigall, 1989). Like a mechanical toggle switch, the wings are only in a stable position when positioned either up or down. As every insect collector knows, the horizontal position of the wings of butterflies and moths preserved in an insect collection can be achieved only by pinning the wings in this position until the thorax dries.

At least ten pairs of muscles are involved in flight, wing orientation, and steering (Pringle, 1976) (Table 9.1). The **indirect flight muscles** power wing movements by changing the shape of the elastic thorax. These are power muscles and include the **dorsal longitudinal muscles** (the wing depressors) that arch the tergum and the **dorsoventral** and **oblique dorsal** muscles (the wing elevators). The **direct wing muscles** insert directly on the wing hinge sclerites, or on axillary sclerites or movable sclerites of the pleuron of the thorax. These muscles are the **basalar** (wing depressors), **subalar** (wing depressors and twisting or supinators of the wing), and **third axillary** (wing-folding) muscles. A third group of muscles in the thorax are accessory indirect muscles consisting of the pleurosternal, anterior tergopleural, posterior tergopleural, and intersegmental muscles. They modify the way in which the power-producing muscles move the wings and the angle of attack. For example, contraction of a small muscle, the pleuroalar (also called the pleuroaxillary) muscle, in the locust *Locusta migratoria* reduces pronation of the forewings during the downstroke and supination during the upstroke, thus helping to control the angle of attack and the development of additional lift forces (Wolf, 1990). The muscle inserts on the third pleuroaxillary sclerite and has a broad, fan-shaped anchor on the pleural wing process. The angle of attack is important to generation of lift forces and to speed of flight and hovering (Nachtigall, 1989).

9.5.1 The Wing Strokes

The wing **downstroke** is produced by contraction of dorsal longitudinal muscles (in all insects except dragonflies and damselflies, in which the mechanism is different; see Section 9.5.3). The dorsal longitudinal muscles are indirect muscles that do not attach directly to the wing hinge sclerites. These large, powerful muscles are attached to the phragma, which are invaginated and hardened cuticular processes at the anterior and posterior of each of the meso- and metathoracic segments (Figures 9.13 and 9.14). When they contract, they shorten the thoracic segments by arching the tergum, slightly lifting the attachment base of the wings at the tergo-pleural junction, and forcing the wings downward. As the wings approach the unstable horizontal position, they suddenly pivot downward over the pleural wing process. This releases the load on the dorsal longitudinal muscles and they cease to shorten, but the new position of the wings introduces the load to the antagonistic set of muscles, the dorsoventral muscles.

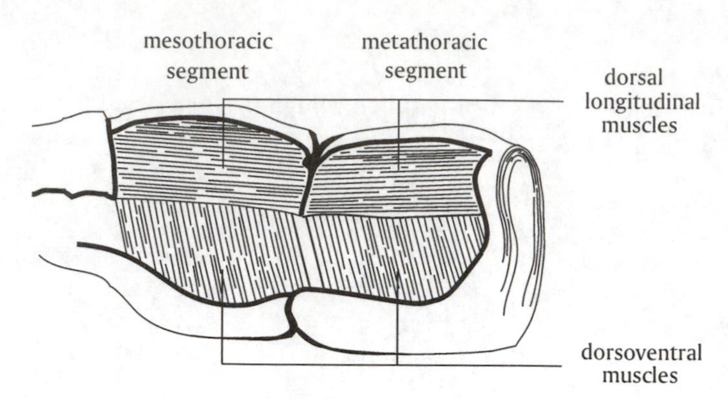

FIGURE 9.13 A diagrammatic illustration of the dorsal longitudinal muscles (DLM) and dorsoventral muscles (DM). The DLM are attached to phragma near the dorsum of the meso- and metathoracic segments in most insects. Upon contraction, these muscles arch the thorax and cause the downstroke of the wings. The DM have their origin on heavily sclerotized cuticle on the ventrum and insert on the thoracic cuticle. When they contract, they depress the thoracic cuticle and pivot the wings upward.

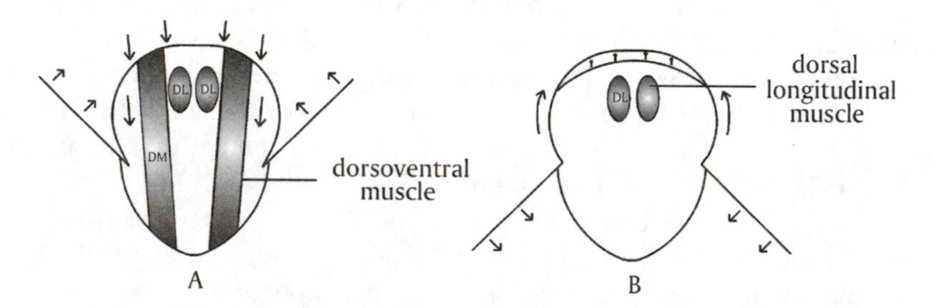

FIGURE 9.14 A crosssection through the thorax to illustrate the antagonistically arranged dorsal longitudinal muscles (DL) and dorsoventral muscles (DM) that produce the downstroke and upstroke, respectively, of the wings in most insects, and allow some insects to beat the wings several times for each nerve input. A: When dorsoventral muscles contract, they pull upon the tergum of the thorax and depress the arching of the thorax. This causes the wings to pivot upward over the pleural wing process. B: Contraction of the dorsal longitudinal muscles, attached at the front and back of each of the meso- and metathoracic segments (see Figures 9.1 and 9.13), shorten the length of the segments slightly and arch the thorax, and lift the dorsal thoracic region so that the wings pivot downward. Contraction of one pair of muscles stretches the antagonistic pair, and serves as a stimulus for a repeated contraction. Thus, several contractions may occur for each nerve impulse received.

The dorsoventral muscles cause the **upstroke** of the wings in all insects, including the Odonata. These powerful muscles are anchored on the heavily sclerotized, relatively rigid ventral thoracic cuticle. They insert on the dorsum of the thorax. When they contract, they pull on the tergum and reduce the arching of the thorax (Figure 9.14A). This causes the wings to pivot upward; and when they again reach the unstable position on the pleural wing process, they snap into the up position. This reduces the load on the dorsoventral muscles and they cease to shorten.

In insects with synchronous muscles, nerve impulses must arrive at the dorsal longitudinal and dorsoventral muscles to evoke each successive contraction. The rhythm for the repeated nerve impulses and continued contraction of the flight muscles in locusts is based in groups of interneurons in the thoracic ganglia that generate a motor program (Wilson, 1968). Similar motor programs originating in the thorax probably exist for other insects. The wingbeat frequency of selected insects is shown in Table 9.2.

TABLE 9.2
Selected Insects to Show Representative Wingbeat Frequencies

Insect	Wingbeat Frequency (Hz)	Reference
Drosophila hydei (Diptera)	170–180	Dickinson and Lighton (1955)
Apis mellifera (Hymenoptera)	180	Pringle (1983)
Anastrepha suspensa (Diptera)	145	Webb et al. (1976)
Forcipomyia sp. (Diptera)	800–950	Sotavalta (1953)
Chironomus sp. (Diptera)	600–650	Sotavalta (1953)
Blowfly (Diptera)	120	Pringle (1983)
Hawkmoth (Lepidoptera)	40	Pringle (1983)
Swallowtail butterfly (Lepidoptera)	5	Pringle (1983)
Schistocerca gregaria (Orthoptera)	17	Pringle (1983)
Neoconocephalus robustus (Orthoptera)		Josephson (1984)
Flight	20	
Stridulation	200	
Musca domestica (Diptera)	~150	Roeder (1951)

9.5.2 Multiple Contractions from Each Volley of Nerve Impulses to Asynchronous Muscles

Insects with asynchronous muscles can achieve several to many contractions of the two antagonistic sets of muscles with one delivery of nerve impulses to each set. The reduced extent of the SR in asynchronous muscles does not rapidly sequester calcium ions released by the arrival of nerve impulses, and this allows the muscle fibers to remain in the active state for a (relatively) long time. Each time the wings snap into the up or down position, the load on one set of antagonistic muscles is released and reintroduced with a stretching stimulus to the other set of muscles. Stretching the muscles acts as a stimulus, and, with Ca^{2+} and ATP available, the stretched muscles start to shorten again without another volley of nerve impulses. The time elapsed during a single wing stroke is very short because the muscles shorten only 2 to 3% of resting length before the wings snap into a new position. Minimal shortening, a prolonged active state with free Ca^{2+} ions available, and reintroduction of a load with stretching allow asynchronous muscles to oscillate and produce several contractions for each burst of nerve impulses received. A control center in the thorax sends out periodic nerve impulses to keep a rhythm going.

9.5.3 Flight in Dragonflies and Damselflies

In dragonflies and damselflies (Odonata), which have synchronous wing muscles, the downstroke is produced by the basalar and subalar muscles (Figure 9.12) attached directly to the wing hinge sclerites. The basalar muscles insert on the basalar sclerite at the anterior base of the wing in front of the pleural wing process. In the Odonata, they pull the wing down and twist downward, or pronate, the anterior leading edge of the wing. The subalars insert on the subalar sclerite at the base of the wing posterior to the pleural wing process, and also pull the posterior edge of the wing downward, with some twisting of the trailing edge. The basalar and subalar muscles are large, powerful muscles in dragonflies and damselflies (Nachtigall, 1989); they work on the short end of the lever to pivot the relatively long wings over the pleural wing process against the resistance of the air. They are anchored to parts of the heavily sclerotized ventral and pleural cuticle. This direct muscle arrangement for the downstroke of the wings only occurs in Odonata according to Nachtigall

(1989), but Pringle (1976) states that the basalar and subalar muscles are also power-producing muscles for the wings in some other groups (see Table 9.1). Pringle (1983) indicates that one basalar muscle group in locusts aids in pronating and pulling the wings down.

9.5.4 The Aerodynamics of Lift Forces Produced by Wings

To remain airborne, an insect must generate **lift** forces at least equal to its weight; and to move forward, the horizontal thrust vector must exceed the drag of air resisting forward motion (Pringle, 1983). The dynamics of flight are complex, and the lift force needed is related to factors such as body weight, wing size, wing shape, speed of air movement over the wings (i.e., wingbeat frequency), and angle of attack of the wings. Smaller insects have to beat the wings faster than larger insects to gain the lift forces required to keep them in the air. There is a popular myth, recounted by McMasters (1989), that someone is supposed to have calculated that the wings of bumblebees are too small to produce enough lift for the insect to fly. The calculations, if they ever really existed, would have been based on steady-state aerodynamic calculations, which predict sufficient lift forces for some insects (locusts, for example; Jensen, 1956), but not for most insects. **Steady-state aerodynamics** are based largely upon calculations derived to explain lift in fixed-wing aircraft. Insects do not have fixed wings, and the flapping of the wings presents special problems, such as changing wing shape during a wing-stroke, acceleration, and deceleration of wings as they change direction of movement. Moreover, insect wings do not present a smoothly contoured airfoil typical of an airplane wing (McMasters, 1989). Unsteady-state conditions, in which there are momentary very high lift forces followed by lower lift forces, or even negative ones, describe insect flight more adequately than steady-state calculations (Weis-Fogh, 1973).

Weis-Fogh (1973) initially described an example of an **unsteady lift** condition that occurs in very small chalcid wasps as a "**clap**" and "**fling**" wing motion. In this small insect, the wings "clap together" at the top of the upstroke, and then twisting motions controlled by some of the small muscles fling them apart at the start of the downstroke. The rapid flinging motion sets up air movements above the wings that increase the lift force of the downstroke, and may aid very small insects in generating enough lift for flight despite having little airfoil surface on their tiny wings. The clap and fling mechanism seems to work mostly in small insects, although *Drosophila melanogaster* also uses it at the start of the downstroke. Somps and Luttges (1985) measured large, transient lift forces 15 to 20 times the body weight of tethered dragonflies, with time-averaged lift values equal to 2 to 3 times body weight created by the turbulent flow of air generated by the independent movement of the front and rear wings. Similar analyses of unsteady lift forces are described by Brodsky (1994).

Tethered tobacco hornworm moths (*Manduca sexta*) generate unsteady lift forces equal to at least 1.5 times the body weight during the downstroke of the wings (Ellington et al., 1996). Coincident with the downstroke movement, intense leading-edge vortices of low-pressure air above the wings are created by the pronation of the wing (tilting downward of the leading edge) during the downstroke. The vortices form first over the leading edge of the wing, move out toward the wing tips, and finally extend behind the insect in a ring of turbulent air (Figure 9.15) (Alexander, 1996; Ellington et al., 1996). These low-pressure vortices increase the lift of the downstroke. The high angle of attack of the wing, a condition that would rapidly create a stall in a fixed-wing aircraft, creates the vortices. This condition of dynamic or delayed stall can be tolerated in an insect for the brief interval of one downstroke, at the end of which the stall conditions are eliminated as the wings change direction (Ellington et al., 1996).

Using a dynamically scaled model of *Drosophila melanogaster* with built-in sensors, Dickinson et al. (1999) describe three interacting mechanisms that provide the lift forces for flight in the fruit fly, and likely in most other insects, perhaps with some degree of variation in importance of some of the components. The three mechanisms they describe are: (1) the upstroke and downstroke of the wings with a high angle of attack (delayed stall), (2) rotational circulation of air eddies above

FIGURE 9.15 A diagrammatic illustration of vortices above the wings and behind the body created by the downstroke of the wings. The model used to illustrate the vortices is a butterfly, the painted lady *Vanessa cardui*, but the concept is based on illustrations and data obtained from experiments with the moth *Manduca sexta* flying in a smoke chamber (Alexander, 1996; Ellington et al., 1996). The vortices contribute to the lift generated by the downstroke, the principal lift-generating wing movement. The size, shape, and trailing nature of the vortices may vary with wing shape and flight speed, and a butterfly has not been subjected to experimental analysis.

the wings, and (3) wake capture, with the latter two mechanisms promoted by the pronation and supination of the wings as they rapidly rotate and change direction at the end of each half stroke. Larger insects, such as sphingid moths, generate lift forces from the development of a leading-edge vortex that produces transient aerodynamic lift forces to keep the insect in the air (Dudley, 1999). The high angle of attack of the wings as they move through the up- and downstrokes generate high unsteady lift forces during the fast movement and short stroke amplitude of an insect wing, but come close to stall conditions. Just prior to stall, relief is provided by reversal of wing direction in movement. The rotation of the wings at the end of each half-stroke (pronation at beginning of downstroke and supination at beginning of upstroke) generates an upwardly directed lift force initially, followed by a downward force. Although the turbulence or wake following a moving object in a fluid medium typically produces drag, by rotating the wings at the end of each half-stroke, the wing encounters its own wake in such a way to generate momentary positive lift (Dudley, 1999). The magnitude of the lift generated by wake capture varies from positive to negative, depending on exactly when in time the rotation occurs relative to the beginning of the next half-stroke; wing rotation that is delayed until the start of a new stroke direction produces negative lift. Dickinson et al. (1999) postulate that the small but significant lift forces associated with the rotational nature of the wings is a powerful means that insects use in steering maneuvers during flight.

Dragonfly wings act as ultra-light aerofoils during gliding flight (Kesel, 2000). In a cross-sectional view, the wings have well-defined corrugations in which rotating vortices develop. The corrugations might be expected to lead to high drag values; but, in fact, the drag from air flowing over the wings is low. The vortices filling the profile valleys smooth out the profile geometry, resulting in low drag similar to that of air flowing over flat plates (Kesel, 2000), while lift forces are much higher than expected from flat plates.

9.5.5 Control of Pitch and Twisting of Wings

Pitch and twisting of the leading and trailing edges of the wings during a wingstroke are important in generating lift forces. The basalar and subalar muscles inserted directly on the wing hinges are mainly involved in controlling the wing angle during a wingstroke (Table 9.1). The basalar muscles attached to hinge points in front of the pleural process pull down (pronate) the leading edge of the wings in a downstroke, while the subalar muscles attached to hinge points behind the pleural process supinate or pull down the posterior edge of the wings and cause the leading edge to tilt upward during an upstroke. These movements cause air to flow faster over the upper surface of the wing than over the lower surface and a lift force is produced. Thrust forces push the insect forward through the air, and the wing tip tends to describe a figure eight during flight.

In flies, bees, and wasps, a different mechanism for pitch and twist of the wings occurs because of the way in which the thorax is constructed. The tergum of the thorax is divided by the scutal cleft into two plates, the scutum and the scutellum. Sclerites at the posterior of the wing base join to the posterior plate (the scutellum). When the scutellum is pulled slightly forward by the contraction of the dorsal longitudinal muscles, the wing tilts slightly forward, or pronates, as it pivots down over the pleural process. Relaxation of the dorsal longitudinal muscles at the end of the downstroke allows the tergal plates to move further apart, and the wing tilts slightly backward (supinates) as it is forced upward by the contraction of the dorsoventral muscles.

9.5.6 Power Output of Flight Muscles

The inertia of stopping the movement and changing the up-and-down direction of the wings is energy demanding, but insects do not have super-efficient muscles. Flight, the most energy-intensive activity in the life of a flying insect, no doubt requires near-maximum mechanical power output by working flight muscles. The power output of the flight muscle of a katydid, *Neoconocephalus triops* (Orthoptera: Tettigoniidae), is equal to about 37 W/kg^{-1} muscle when the (synchronous) muscle is stimulated to give a single twitch, and maximum output of 76 W/kg^{-1} muscle during contractions at 25 Hz and at 30°C (Josephson, 1985a). Weis-Fogh (1973, 1977) estimated the mechanical power of flight muscle at 60 to 360 W/kg^{-1} based on an assumed 20% metabolic to mechanical efficiency of flight muscle. Ellington (1984a, 1984b) estimated the mechanical power output of wing muscle of an insect in hovering flight to be 70 to 190 W kg^{-1}, but questioned whether the metabolic to mechanical efficiency was as high as 20%. Metabolic conversion to mechanical muscle efficiency values of only about 10% were calculated for flight muscle of the fruit fly *Drosophila hydei* (Dickinson and Lighton, 1995), and only 3% efficiency of metabolic energy to mechanical power conversion was calculated for *N. triops* during stridulatory singing (Josephson, 1985b). Assuming higher efficiency makes power output calculations higher than they may really be; and according to Josephson (1985a), the comparative value of power output measurements from muscles of various animals is very limited anyway because the measurements have been made under highly variable conditions, techniques, and assumptions. With that caveat in mind, the values for *N. triops* fall within the range of measurements from active muscle in other animals.

Weis-Fogh (1964) suggested that insects store kinetic energy as elastic energy during one wing movement (either up or down) and then release the stored energy as kinetic energy when the wing movement reverses. They may store the energy in several different ways, including in the muscle system as in dragonflies (*Aeshna* spp.), in the elasticity of the cuticle of the thorax (Sphinx moths), and in the resilin-forming elastic wing hinges (Weis-Fogh, 1964; Dickinson and Lighton, 1995). The natural elasticity of the thoracic cuticle that is distorted by a wing stroke and the compressed resilin in the wing hinge absorb energy and tend to spring back after a wing stroke. Additional reviews and discussion of flight and of wing morphology are presented by Wootton (1992).

9.5.7 Metabolic Activity of Wing Muscles

Working insect flight muscles have the highest rate of metabolism per gram of muscle tissue of all biological tissues, and also the highest control values, ranging from 50 to 100 times resting values. The control value is the ratio of the oxygen consumption in active flight (or other exercise) divided

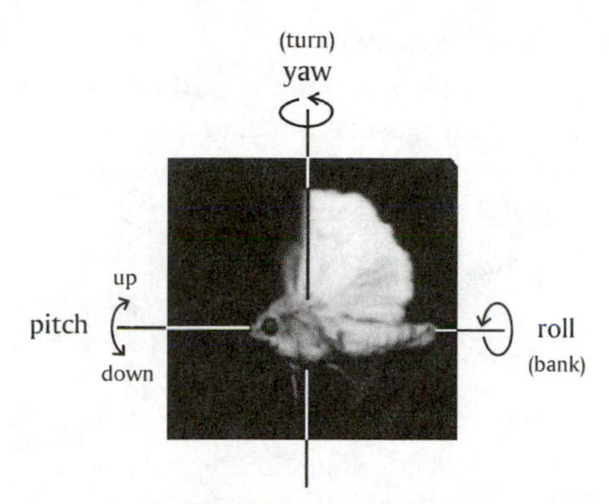

FIGURE 9.16 Insects in flight use wing movements to maneuver the body in a roll around the long axis of the body, pitch up or down around the perpendicular axis of the body, and yaw or turn the body in flight. The model used in this illustration is a flying moth, *Heliothis virescens.* (Photograph courtesy of Peter Teal.)

by the oxygen consumption at rest. Intense energy demands of flight are supported by an extensive tracheal and tracheolar network that can supply oxygen to the flight muscles as they need it, so they do not develop an oxygen debt and do not have to resort to anaerobic glycolysis. Within the wing muscles there are very large and numerous mitochondria (the sarcosomes) with many cristae, like leaves in a book. Flight muscles have very active metabolic pathways for metabolizing carbohydrates for energy release. Muscles of Orthoptera, Lepidoptera, and some other groups can metabolize lipids rapidly enough to support flight.

9.5.8 The Evolution of Wings and Flight

Insects were the first animals to evolve wings. In a recent book on insect flight, Dudley (2000) states that "the ability to fly has been central to the evolutionary diversification of insects." Flight enabled wide dispersal, search for food in diverse locations, and colonization of nearly every available niche. The ability to take flight and maneuver during flight (Figure 9.16) enables insects to escape enemies.

Insects began as wingless creatures, documented by fossils dating back some 400 million years ago; but sometime during the next 100 million years, forms with primitive extensions of cuticle from the thorax appeared (Marden and Kramer, 1994). A number of theories have been advanced in an attempt to explain the selection forces acting on the evolution of wings (reviewed by Kingsolver and Koehl, 1994).

Flight probably evolved only once within the Insecta (Brodsky, 1994), and probably first in an aquatic ancestor. The paranotal lobe theory (reviewed by Wootton, 1976) proposed that wings evolved from rigid extensions of the thoracic terga. Another idea that has received considerable support is that wings evolved from movable gill flaps whose original function was respiration. Marden and Kramer (1994) have suggested that wing-like appendages were first used in the air by surface-skimming aquatic insects in much the same way that some stoneflies (Plecoptera: Taeniopterygidae) and some subadult mayflies (Ephemeroptera) still skim across the water. It is not possible to say with certainty what factors were at work to selectively promote the evolution of wings. It may be that multiple factors were important (Kingsolver and Koehl, 1994).

If wings did evolve from gill flaps, did those gill flaps originally occur only on thoracic segments or also on abdominal segments? Wigglesworth (1976) supported the latter idea, and there may be neurophysiological support (Robertson et al., 1982). Large interneurons in the meso- and metathoracic ganglia of *Locusta migratoria* have rhythmic nervous output in phase with the motoneuron output to the dorsal longitudinal muscles (the muscles that control the downstroke of the wings), and to motoneurons controlling the dorsoventral muscles that raise the wings. Those particular interneurons, however, have their cell bodies located in the third abdominal ganglion (for dorsal

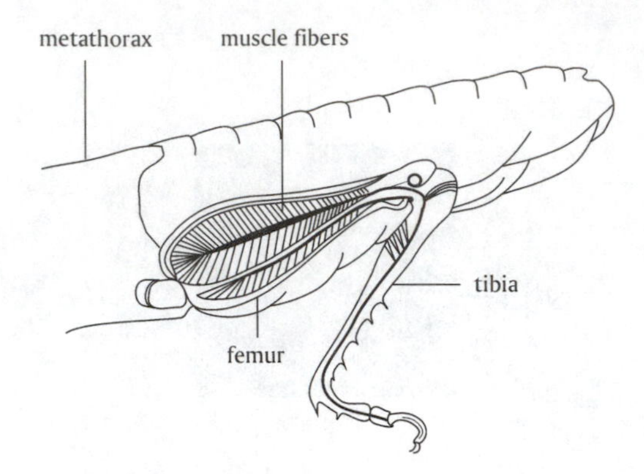

FIGURE 9.17 The arrangement of muscle fibers in the femur of the jumping leg (the metathoracic leg) of a grasshopper. Many very short muscle fibers are anchored on the epicuticle and insert on long cuticular tendons attached to the tibia in such a way as to flex or extend the leg, depending on the muscles activated.

longitudinal muscles) and first abdominal ganglion (for dorsoventral muscles), indicative that their original function might have been to control appendages on the abdomen. In the earliest insects, neurons from a segmental ganglion probably controlled the structures within its own segment. Although these three abdominal ganglia are fused with the metathoracic ganglion in modern adult locusts, embryonically the tissue housing the cell bodies is abdominal ganglionic tissue (Robertson et al., 1982). The question remains open as to what appendages or structures those particular neurons might have controlled early in the evolution of insects, but they might have controlled abdominal gill flaps.

9.6 MUSCLES INVOLVED IN JUMPING

The ability of some insects to jump many times their body length and height might suggest that insect muscles are stronger than vertebrate muscles. Actual measurements of the force developed per unit cross-sectional area for insect muscles reveal about the same force per unit cross-sectional area as for other animals. It is both the anatomical arrangement of muscles and levers as well as physiology that enable some insects to jump so far.

The main muscle in the femur or jumping leg of a grasshopper or locust is composed of many short muscle fibers with origins on the epicuticle of the femur. The muscle fibers insert on the tendon or apodeme from the tibia that runs the length of the femur (Figure 9.17). Thus, each muscle fiber shortens very little, but the many fibers along the length of the long femur apply tension that straightens the leg and propels the locust into the air in a jump. Furthermore, the long femur and tibia of the metathoracic leg (which is much longer than either of the pro- or mesothoracic legs) give the locust something of a "pole-vaulting" advantage, similar to a human using a vaulting pole as an extension of the arms.

Very massive muscles whose fibers must shorten only a small percentage of the muscle length are a common adaptation in fleas that make extraordinary jumps. Jumping fleas (not all fleas are specialized for jumping) have an extraordinarily large coxa, in contrast to most insects, containing a large muscle whose fibers are anchored to the thorax. The coxal muscle inserts on the femur, which is also very large and long. Overall, the jumping leg is about 80% of the length of the body of a cat flea and contains large muscles inserting on the tibia. In preparation for a jump, a flea utilizes a cocking mechanism in which the femur is pulled up to overlap the large coxa. In this cocked position, both coxa and femur are perpendicular to the body axis and to the substrate (Rothschild et al., 1988). The cocking action compresses an arch of resilin in the pleural arch near where the coxa is attached to the thorax. In addition, the cocking action clamps the three thoracic segments together by engagement of "catch" mechanisms between the three thoracic segments. The sternum of the mesothorax, in particular, is rigidly held against the metathorax, and this allows

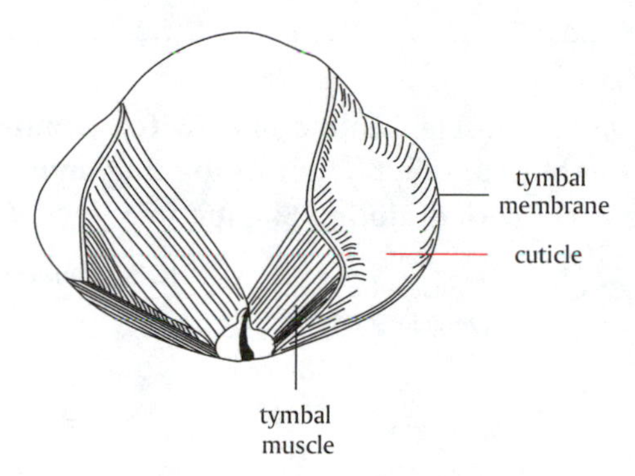

FIGURE 9.18 A cross-sectional view of the tymbal muscles that contract and flex the tymbal cuticular surface inward to produce the loud singing sounds of a cicada. The muscle on each side is unpaired, and the elastic nature of the cuticle stretches the muscle and returns the tymbal to its resting shape after a contraction, usually producing another sound with the outward snap of the cuticle.

the large muscles involved in raising the femur and locking the catches to relax, resisting tiring and reducing energy requirements. In this crouched position, a flea is poised for its leap with trochanter and tibio-tarsal joints and most of the tarsi resting on the substrate. The flea propels itself forcefully into the air by relaxing the levator muscle holding the femur in the perpendicular position and relaxing the ventral longitudinal muscles holding the catches in the cocked position. The compressed resilin, similar to a compressed spring, forcefully drives the trochanter against the substrate to provide the initial leverage off the substrate. Powerful muscles extend the joint between the trochanter and the femur, and extend the leg, providing further thrust as the terminal tarsal segments and tarsal claws press against the substrate. Additional details on the jump of the flea can be found in an illustrated and informative paper by Rothschild et al. (1988).

9.7 SOUND PRODUCTION: TYMBAL AND STRIDULATORY MUSCLES

9.7.1 Tymbal Morphology and Physiology

Tymbals are thin, often ribbed patches of cuticle in male cicadas (Homoptera), some other homopterans, some Hemiptera, and some Lepidoptera (moths in the Families Arctiidae, Ctenuchidae, and Pyralidae, and some nymphalid butterflies) (Ewing, 1989, p. 34) that are used for sound production and communication. The tymbals of cicadas have received the most detailed analysis. Two paired tymbals, each occurring on the lateral surface of the first abdominal segment of male cicadas are used by the male to produce both calling and protest songs. The tymbal membrane is a convex layer of cuticle, often bearing a series of ribs, covering an air-filled cavity that acts as a sound enhancer. The large tymbal muscle, the sound-producing muscle, is anchored on the sternal cuticle and inserted dorsolaterally on the tymbal membrane (Figure 9.18), sometimes by means of a thickened rod of chitin. Contraction of the muscle is initially isometric against the resistance of the convex tymbal, but suddenly the tymbal membrane buckles inward into an unstable, concave shape with a loud clicking sound. The inward buckling of the tymbal releases the load on the tymbal muscle and it stops developing tension and shortening. The sudden reduction of muscle tension allows the natural elasticity of the tymbal cuticle to click back into the resting convex shape. Sounds may be produced by either, or both, inward and outward movement of the tymbal membrane. Some species have ribs in the convex tymbal surface that buckle progressively inward from posterior to anterior, producing a series of sound pulses as the successive ribs buckle, although only a single muscle twitch is involved (Young, 1972; Young and Josephson, 1983b).

The tymbal muscles of some cicadas are synchronous, while others have asynchronous muscles (Pringle, 1981). For example, *Cyclochila australasiae* have synchronous tymbal muscles and

TABLE 9.3
Comparison of Tymbal Muscle Contraction, Twitch Duration, and Time from Onset of One Muscle Contraction to the Beginning of the Next Contraction (the cycle period) during Singing in Selected Cicadas

Cicada Species	Tymbal Muscle Contraction Frequency (Hz)	Twitch Duration (Milliseconds, ms)	Cycle period (Milliseconds, ms)
Okanagana vanduzeei	550	~6	—
Psaltoda claripennis	224	6.6	4.46
P. harrisii	~150	7.7	6.7
P. agentata	192	7.9	5.51
Tamasa tristigma	82	8.7	12.2
Magicicada cassini	~90	9.8	11.1
M. septendecim	~68	11.8	14.7
Abricta curvicosta	72	12.6	13.9
Arunta perulata	31	14.9	32.3
Chlorocysta viridis	56	15.2	17.9
Cystosoma saundersii	~40	22.0	25.0
Cyclochila australasiae	~116	12.3	8.55

Data from Young and Josephson (1983a) and Josephson and Young (1985).

Platypleura capitata have asynchronous ones, but the flight muscles in both are synchronous (Josephson and Young, 1981). Each cicada tymbal muscle is innervated by a single motor axon with multiterminal branches to all the muscle fibers in the muscle (Pringle, 1954a, 1954b; Hagiwara, 1955; Simmons, 1977; Josephson and Young, 1981). A number of cicadas that have synchronous tymbal muscles have contraction frequencies less than 100 Hz (Table 9.3) (Young and Josephson, 1983a), as might be expected of synchronous muscle, but several have high contraction frequencies well above 100 Hz. *Okanagana vanduzeei* has a contraction frequency of 550 Hz. In part, the rapid contractions are enabled by very short twitch duration times (Table 9.3) in the fastest synchronous muscles (Josephson, 1984). However, a contraction rate of 550 Hz in a synchronous muscle means that the nerve must repolarize and be ready to fire again in less than 2 ms. As Table 9.3 shows, only 4 to 6 ms are available for repolarization in several of the cicadas. In the cicadas with very fast synchronous muscles, the muscle may be delicately balanced on the edge of oscillatory instability, and may possibly go into oscillatory behavior (Josephson and Young, 1985). The inward click of the tymbal cuticle releases the load against which the muscle works, and the natural cuticular elasticity that causes it to click back into the convex shape could then stretch and reintroduce the load to the activated muscle, inducing another contraction if the concentration of free calcium ions is still high enough in the cytoplasm. Thus, these tymbal muscles may work like the asynchronous wing muscles.

9.7.2 Stridulatory Muscle Physiology

Another sound-producing mechanism, **stridulation**, involves the rubbing together of two parts of the body. Stridulation is widespread among insect orders, with examples known from Odonata, Orthoptera, Coleoptera, Mecoptera, Lepidoptera, Siphonaptera, and Hymenoptera (Ewing, 1989). As in the tymbal muscles of some cicadas, synchronous muscles involved in stridulation by some katydids (Orthoptera: Tettigoniidae) show unusually high contraction frequencies. Male katydids stridulate to attract females by rubbing their forewings (mesothoracic wings) together. In the mesothoracic segment, the same muscles used to power the downstroke of the wings during flight, the mesothoracic dorsal longitudinal muscles (DLM), are used in stridulation. The wing-stroke frequency of the DLM in the mesothoracic segment during stridulation of a katydid, *N. robustus* (Orthoptera: Tettigoniidae), has been measured at 200 Hz (Josephson and Stokes, 1982). The DLM in both segments are used in flight, but then the wing-stroke frequency is 20 Hz. In a related katydid, *N. triops*, wing-stroke frequency during stridulation is 100 Hz and wing-stroke frequency during flight is 22.9 Hz (Josephson, 1984). The tropical bush cricket *Hexacentrus unicolor* (Orthoptera: Conocephalidae) produces a song with a stridulatory frequency of 320 to 415 Hz using synchronous muscles (Heller, 1986). Thus, it is clear that morphologically synchronous muscle can produce contraction frequencies higher than previously expected. The evolution of asynchronous muscle, which has occurred as many as ten times in different groups of insects (Cullen, 1974), may be related more to operating efficiency, economy in calcium recycling between the cytoplasm and SR, and overall structural economy than to the potential for high operating frequency (Josephson and Young, 1985).

9.8 MORPHOLOGY AND PHYSIOLOGY OF NON-SKELETAL MUSCLE

9.8.1 Visceral Muscles

Antagonistic muscle sets in the gut are arranged as bands of longitudinal and circular muscle. They create peristaltic action in the gut that mixes food with enzymes and aids digestion, and physically macerate food particles in the powerful proventriculus of some insects. Visceral muscles in the foregut and midgut are innervated by nerves from the stomatogastric nervous system, consisting of several small ganglia lying on top of the esophagus and crop. Some visceral muscles in insects are **myogenic** and may or may not have a nerve supply.

Visceral muscle fibers are usually short and small (1 to 5 μm in diameter), and, in contrast to skeletal and flight muscle, uninucleate. Long sarcomere lengths (7 to 10 μm long) are characteristic of, and consistent with, the slow contractions of visceral muscles. Z lines are irregular and often not well lined up in adjacent fibers. In visceral muscles, myosin filaments are surrounded by up to ten or twelve actin filaments, whereas in skeletal muscle the ratio is usually one myosin to six actin filaments (Elder, 1975). Muscles in the gastric caeca of *Aedes* mosquitoes have eight or nine actin filaments for each myosin filament (Jones and Zeve, 1968). The SR is usually poorly developed and the transverse tubules are irregularly located. Mitochondria are small and few in number. The neurotransmitter at gut muscle has not been conclusively identified; neuropeptides are probably important in regulating the rate and force of contraction in gut muscles. Proctolin, a neurohormone from various parts of the nervous system in different insects, has profound action on the muscles in the hindgut.

9.8.2 Heart Muscle

The heartbeat in insects is **myogenic**, but heart rate is influenced by nerves that innervate the heart in most insects (McCann, 1970) and by neurohormones. Sarcomeres tend to be short in heart muscle, with an A-band of 1.8 μm, a short I band, and no H band. There are numerous, but small, mitochondria and extensive tracheal connections, characteristics suggesting moderately high energy demands. Myosin filaments are surrounded by ten to twelve actin filaments. The SR is usually not well developed, and transverse tubules, while well developed, are irregularly spaced. Dyad junctions are common, and some triad junctions occur. Intercalated discs, regions of extensive interdigitation of plasma membranes between adjacent fibers, described from heart muscle of *Blatella germanica* and *Hyalophora cecropia* may be important in spreading the wave of contraction from fiber to fiber (Edwards and Challice, 1960; Sanger and McCann, 1968a).

9.8.3 Alary Muscles

Alary muscles are thin sheets of wing-shaped muscles that help support the heart. The muscle fibers often branch and are variable in length (from 1 to 20 μm in length) (Sanger and McCann, 1968b). Sarcomere lengths are long, with the A band measuring about 5.5 μm. Few mitochondria are present, suggesting a very low metabolic activity, and the SR is poorly developed. Myosin filaments are surrounded by ten to twelve actin filaments. Some alary fibers have intercalated disc junctions with heart muscle fibers, suggesting intercommunication of alary and heart muscle, but this has not been shown experimentally. Little is known about specific neurotransmitters or neurohormonal effects on alary muscles.

Long sarcomeres, a high actin:myosin filament ratio, and a poorly developed SR and transverse tubule system are general characteristics of tonic or slow contracting muscles. Visceral, heart, and alary muscles are slow contracting muscles, as opposed to the twitch or rapid contractions more typical of skeletal muscles.

REFERENCES

Aidley, D.J. 1985. Muscle contraction, pp. 407-437, in G.A. Kerkut and L.I. Gilbert (Eds.), *Comprehensive Insect Physiology, Biochemistry and Pharmacology*, Vol. 10, Pergamon Press, Oxford.

Aidley, D.J. 1989. Structure and function in flight muscle, pp. 31-49, in G.J. Goldsworthy and C.H. Wheeler (Eds.), *Insect Flight*, CRC Press, Boca Raton, FL.

Alexander, R. McNeil. 1996. Smokescreen lifted on insect flight. *Nature,* 384:609-610.

Ashhurst, D.E. 1967. The fibrillar flight muscles of giant water-bugs: an electron-microscope study. *J. Cell Sci.,* 2:435-444.

Ashhurst, D.E., and M.J. Cullen. 1977. The structure of fibrillar flight muscle, pp. 9-14, in R.T. Tregear (Ed.), *Insect Flight Muscle*, North-Holland Publishing Co., New York.

Brodsky, A.K. 1994. *The Evolution of Insect Flight,* Oxford University Press, New York, 229 pp.

Burrows, M. 1977. Flight mechanisms of the locust, pp. 339-356, in G. Hoyle (Ed.), *Identified Neurons and Behaviour of Arthropods*, Plenum Press, New York.

Carnevali, M.D.C., and J.F. Reger. 1982. Slow-acting flight muscles of saturniid moths. *J. Ultrastruct. Res.,* 79:241-249.

Cochrane, D.G., H.Y. Elder, and P.N.R. Usherwood. 1972. Physiology and ultrastructure of phasic and tonic skeletal muscle fibres in the locust, *Schistocerca gregaria. J. Cell. Sci.,* 10:419-441.

Cullen, M.J. 1974. The distribution of asynchronous muscle in insects with particular reference to the Hemiptera: An electron microscope study. *J. Entomol.,* 49A:17-41.

Darwin, F.W., and J.W.S. Pringle. 1959. The physiology of insect fibrillar muscle. I. Anatomy and innervation of the basalar muscle of lamellicorn beetles. *Proc. R. Soc. London B,* 151:194-203.

Dickinson, M.H., and J.R.B. Lighton. 1995. Muscle efficiency and elastic storage in the flight motor of *Drosophila. Science,* 268:87-90.

Dickinson, M.H., F.-O. Lehmann, and S.P. Sane. 1999. Wing rotation and the aerodynamic basis of insect flight. *Science,* 284:1954-1960.

Dudley, R. 1999. Unsteady aerodynamics. *Science,* 284:1937-1939.

Dudley, R. 2000. *The Biomechanics of Insect Flight: Form, Function, Evolution*, Princeton University Press, Princeton, NJ, 476 pp.

Edwards, G.A., and C.E. Challice. 1960. The ultrastructure of the heart of the cockroach, *Blattella germanica. Ann. Entomol. Soc. Am.,* 53:369-383.

Elder, H.Y. 1975. Muscle structure, pp. 1-74, in P.R.N. Usherwood (Ed.), *Insect Muscle*, Academic Press, New York.

Ellington, C.P. 1984a. The aerodynamics of hovering flight. III. Kinematics. *Phil. Trans. R. Soc. Ser. B,* 305:41-78.

Ellington, C.P. 1984b. The aerodynamics of hovering flight. VI. Lift and power requirements. *Phil. Trans. R. Soc. Ser. B,* 305:145-181.

Ellington, C.P., C. van den Berg, A.P. Willmott, and A.L.R. Thomas. 1996. Leading-edge vortices in insect flight. *Nature,* 384:626-630.

Ewing, A.W. 1989. *Arthropod Bioacoustics — Neurobiology and Behaviour,* Comstock Publishing Associates, Cornell University Press, Ithaca, NY, 260 pp.

Hagiwara, S. 1955. Neuromuscular mechanism of sound production in the cicada. *Physiol. Comp. Oecol.,* 4:142-153.

Harrington, W.F. 1981. *Muscle Contraction,* Carolina Biological Supply Co., Burlington, NC, 31 pp.

Heller, K-G. 1986. Warm-up and stridulation in the bushcricket, *Hexacentrus unicolor* serville (Orthoptera, Conocephalidae, Listroscelidinae). *J. Exp. Biol.,* 126:97-109.

Hinton, H.E. 1973. Neglected phases in metamorphosis: A reply to V.B. Wigglesworth. *J. Entomol. (A),* 48:57-68.

Hoyle, G. 1955. Neuromuscular mechanisms of a locust skeletal muscle. *Proc. R. Soc. B,* 143:343-367.

Ikeda, K., and E.G. Boettiger. 1965. Studies on the flight mechanisms of insects. III. The innervation and the electrical activity of the basalar fibrillar muscles of the beetle, *Oryctus rhinoceros. J. Insect Physiol.,* 11:791-802.

Irving, S.N., and T.A. Miller. 1980. Aspartate and glutamate as possible transmitters of the 'slow' and 'fast' neuromuscular junctions of body wall muscles of *Musca* larvae. *J. Comp. Physiol.,* 135:299-314.

Jensen, M. 1956. Biology and physics of locust flight. III. The aerodynamics of locust flight. *Phil. Trans. R. Soc. London B,* 239:511-552.

Jones, J.C., and V.H. Zeve. 1968. The fine structure of the gastric caeca of *Aedes aegypti* larvae. *J. Insect Physiol.,* 14:1567-1575

Josephson, R.K. 1984. Contraction dynamics of flight and stridulatory muscles of tettigoniid insects. *J. Exp. Biol.,* 108:77-96.

Josephson, R.K. 1985a. Mechanical power output from striated muscle during cyclic contraction. *J. Exp. Biol.,* 114:493-512.

Josephson, R.K. 1985b. The mechanical power output of a tettigoniid wing muscle during singing and flight. *J. Exp. Biol.,* 117:357-368.

Josephson, R.K., J.G. Malamud, and D.R. Stokes. 2000a. Power output by an asynchronous flight muscle from a beetle. *J. Exp. Biol.,* 203:2667-2689.

Josephson, R.K., J.G. Malamud, and D.R. Stokes. 2000b. Asynchronous muscle: A primer. *J. Exp. Biol.,* 203:2713-2722.

Josephson, R.K., and D.R. Stokes. 1982. Electrical properties of fibres from stridulatory and flight muscles of a tettigoniid. *J. Exp. Biol.,* 99:109-125.

Josephson, R.K., and D. Young. 1981. Synchronous and asynchronous muscles in cicadas. *J. Exp. Biol.,* 91:219-237.

Josephson, R.K., and D. Young. 1985. A synchronous insect muscle with an operating frequency greater than 500 Hz. *J. Exp. Biol.,* 118:185-208.

Kesel, A.B. 2000. Aerodynamic characteristics of dragonfly wing sections compared with technical aerofoils. *J. Exp. Biol.,* 203:3125-3135.

Kingsolver, J.G., and M.A.R. Koehl, 1994. Selective factors in the evolution of insect wings. *Annu. Rev. Entomol.,* 39:425-451.

Lymn, R.W., and E.W. Taylor. 1970. Transient state phosphate production in the hydrolysis of nucleoside triphosphates by myosin. *Biochemistry,* 9:2975-3983.

McCann, F.V. 1970. Physiology of insect hearts. *Annu. Rev. Entomol.,* 15:173-200

McMasters, J.H. 1989. The flight of the bumblebee and related myths of entomological engineering. *Am. Scientist,* 77:164-169.

Marden, J.H., and M.G. Kramer. 1994. Surface-skimming stoneflies: A possible intermediate stage in insect flight evolution. *Science,* 266:427-430.

Maruyama, K. 1985. Biochemistry of muscle contraction, pp. 487-498, in G.A. Kerkut and L.I. Gilbert (Eds.), *Comprehensive Insect Physiology, Biochemistry and Pharmacology,* Vol. 10, Pergamon Press, Oxford.

May, M. 1991. Aerial defense tactics of flying insects. *Am. Sci.,* 79:316-328.

Nachtigall, W. 1989. Mechanics and aerodynamics of flight, pp. 1-29, in G.J. Goldsworthy and C.H. Wheeler (Eds.), *Insect Flight,* CRC Press, Boca Raton, FL.

Neville, A.C. 1963. Motor unit distribution of the locust dorsal longitudinal flight muscles. *J. Exp. Biol.,* 40:123-136.

Pringle, J.W.S. 1939. The motor mechanism of the insect leg. *J. Exp. Biol.,* 16:220-231.
Pringle, J.W.S. 1954a. A physiological analysis of cicada song. *J. Exp. Biol.,* 31:525-560.
Pringle, J.W.S. 1954b. The mechanism of the myogenic rhythm of certain insect striated muscles. *J. Physiol.,* 124:269-291.
Pringle, J.W.S. 1957. *Insect Flight.* Cambridge University Press, Cambridge, England.
Pringle, J.W.S. 1968. Comparative physiology of the flight motor. *Adv. Insect Physiol.,* 5:163-227.
Pringle, J.W.S. 1976. The muscles and sense organs involved in insect flight, pp. 3-15, in R.C. Rainey (Ed.), *Insect Flight*, Royal Entomological Society and Blackwell Scientific Publications, Oxford.
Pringle, J.W.S. 1981. The evolution of fibrillar muscle in insects. *J. Exp. Biol.,* 94:1-14.
Pringle, J.W.S. 1983. *Insect Flight,* Carolina Biological Supply Co., Burlington, NC, 16 pp.
Reedy, M.K., K.C. Holmes, and R.T. Tregear. 1965. Induced changes in the orientation of the cross-bridges of glycerinated insect flight muscle. *Nature (London),* 207:1276-1280.
Robertson, R.M., K.G. Pearson, and H. Reichert. 1982. Flight interneurons in the locust and the origin of wings. *Science,* 217:177-179.
Roeder, K.D. 1951. Movements of the thorax and potential changes in the thoracic muscles of insects during flight. *Biol. Bull. Mar. Biol. Lab. Woods Hole,* 100:95-106.
Rothschild, M., Y. Schlein, K. Parker, C. Neville, and S. Sternberg. 1988. The flying leap of the flea. *Sci. Am.,* 229:92-100.
Sanger, J.W., and F.V. McCann. 1968a. Ultrastructure of the myocardium of the moth, *Hyalophora cecropia. J. Insect Physiol.,* 14:1105-1111.
Sanger, J.W., and F.V. McCann. 1968b. Ultrastructure of the moth alary muscles and their attachment to the heart wall. *J. Insect Physiol.,* 14:1539-1544.
Simmons, P.J. 1977. Neuronal generation of singing in a cicada. *Nature,* 270:243-245.
Smith, D.S. 1961. The organization of the flight muscle in a dragonfly, *Aeshna* sp. (Odonata). *J. Biochem. Biophys. Cytol.,* 11:119-145.
Smith, D.S. 1966. The structure of intersegmental muscle fibers in an insect, *Periplaneta americana* L. *J. Cell Biol.,* 29:449-459.
Somps, C., and M. Luttges. 1985. Dragonfly flight: novel uses of unsteady separated flows. *Science,* 228:1326-1329.
Sotavalta, O. 1953. Recordings of high wing-stroke and thoracic vibration frequency in some midges. *Biol. Bull.,* 104:439-444.
Squire, J.M. 1977. The structure of insect thick filaments, pp. 91-111, in R.T. Tregear (Ed.), *Insect Flight Muscle*, North-Holland Publishing Co., New York.
Toselli, P.A., and F.A. Pepe. 1968. The fine structure of the ventral segmental abdominal muscles of the insect *Rhodnius prolixus* during the molt cycle. I. Muscle structure at molting. *J. Cell Biol.,* 37:445-461.
Tregear, R.T., and A. Miller. 1969. Evidence of cross-bridge movement during contraction of insect flight muscle. *Nature (London),* 222:1184-1185.
Usherwood, P.N.R. 1968. A critical study of the evidence for peripheral inhibitory axons in insects. *J. Exp. Biol.,* 49:201-222.
Usherwood, P.N.R., and H. Grundfest. 1964. Inhibitory postsynaptic potentials in grasshopper muscle. *Science,* 143:817-818.
Usherwood, P.N.R. 1994. Insect glutamate receptors. *Adv. Insect Physiol.,* 24:309-341.
Webb, J.C., J.L. Sharp, D.L. Chambers, J.J. McDow, and J.C. Benner. 1976. The analysis and identification of sounds produced by the male Caribbean fruit fly, *Anastrepha suspensa* (Loew). *Ann. Entomol. Soc. Am.,* 69:415-420.
Weis-Fogh, T. 1965. Elasticity and wing movements in insects, pp. 186-188, in P. Freeman (Ed.), *Proceedings XII Int. Congress of Entomology,* 8-16 July 1965, Royal Entomological Society, London.
Weis-Fogh, T. 1973. Quick estimates of flight fitness in hovering animals, including novel mechanisms for lift production. *J. Exp. Biol.,* 59:169-230.
Weis-Fogh, T. 1977. Dimensional analysis of hovering flight, pp. 405-420, in T.J. Pedley (Ed.), *Scale Effects in Animal Locomotion*, Academic Press, London.
Wigglesworth, V.B. 1976. The evolution of flight, pp. 255-269, in R.C. Rainey (Ed.), *Insect Flight*, Royal Entomological Society and Blackwell Scientific Publications, London.

Wilkin, M.B., M.N. Becker, D. Mulvey, I. Phan, A. Chao, K. Cooper, H.-J. Chung, I.D. Campbell, M. Baron, and R. MacIntyre. 2000. *Drosophila* dumpy is a gigantic extracellular protein required to maintain tension at epidermal-cuticle attachment sites. *Curr. Biol.,* 10:559-567.

Wilson, D.M. 1968. The flight-control system of the locust. *Sci. Am.,* 218:83-90.

Wolf, H. 1990. On the function of a locust flight steering muscle and its inhibitory innervation. *J. Exp. Biol.,* 150:55-80.

Wootton, R.J. 1976. The fossil record and insect flight, pp. 235-254, in R.C. Rainey (Ed.), *Insect Flight*, Royal Entomological Society and Blackwell Scientific Publications, London.

Wootton, R.J. 1992. Functional morphology of insects wings. *Annu. Rev. Entomol.,* 37:113-140.

Young, D. 1972. Neuromuscular mechanism of sound production in Australian cicadas. *J. Comp. Physiol.,* 79:343-362.

Young, D., and R.K. Josephson. 1983a. Mechanisms of sound-production and muscle contraction kinetics in cicadas. *J. Comp. Physiol.,* 152:183-195.

Young, D., and R.K. Josephson. 1983b. Pure-tone songs in cicadas with special reference to the genus *Magicicada. J. Comp. Physiol.,* 152:197-207.

10 Sensory Systems

CONTENTS

PREVIEW

Sensory structures transduce many different kinds of internal and external stimuli into electrical signals, and then feed these signals into the central nervous system. Sensory receptors are classified in several different ways based on morphology, but morphology is not always a sure indication of physiological function. Sensory receptors on insects are often small, and many receptors have been described from transmission electron microscopy studies without proven physiological functions. A single sensory neuron with its sheath cells is called a sensillum; frequently, a sensory structure consists of many sensilla, that is, many neurons each enclosed in one or more sheath cells. Mechanoreceptors located at many sites on the body monitor body or appendage orientation in space, and serve as wind-speed indicators, tympanal organs, simple contact receptors, and environmental vibration receptors. Thermo-, hydro-, and infrared receptors are also mechanoreceptors. Mechanoreceptors do not have pores opening on the cuticular surface. Proprioceptors located internally are usually mechanoreceptors that monitor stretching, filling of the gut, and other internal movements. Chemoreceptors can be divided into olfactory and gustatory receptors. Olfactory receptors tend to have multiple pores at the cuticular surface, while gustatory receptors tend to have a single pore, usually at the tip of a hair. Olfactory receptors are often concentrated on the antennae, and gustatory receptors are located on the palps, on other mouthparts, and sometimes on the tarsi. Chemoreceptors (probably functioning as contact or gustatory receptors) are often located on the ovipositor of females and enable them to sample an oviposition site. Some olfactory receptors are relatively specialized, as for example, receptors for the sex pheromone of the species, while others may be responsive to a number of chemicals. Gustatory receptors tend to have varying sensitivity to a number of chemicals, and the firing pattern (number and frequency of action potentials, and rate of firing) that several gustatory receptors send into the central nervous system after exposure to a particular chemical compound has been called across-fiber patterning. To paraphrase the late Vincent Dethier, a noted sensory biologist, across-fiber patterning is the way in which a paucity of receptors can detect a surfeit of stimuli. Considerable data have accumulated in support of the stereochemical theory for the interaction of chemicals at the receptor site. In this theory, the chemical combines with a receptor at the dendritic membrane, and this leads to a receptor potential in the dendritic membrane. Insects have several types of light receptors, including compound eyes, ocelli, stemmata, and simple dermal light receptors. Compound eyes form images. While compound eyes of many insects are known to be sensitive to blue, green, and UV wavelengths, color vision has been demonstrated behaviorally in only a few insects. A rigorous test for color vision requires behavioral demonstration that an insect has discriminated between two colors (i.e., two wavelengths of light); and on this basis, color vision has been demonstrated in honeybees, some dipterans, and a few other insects. The visual process and visual cascade in insect compound eyes appears to be essentially the same as that in eyes of vertebrates. One exception is that rhodopsin does not split away from 11-*cis*-retinal in insect eyes after receiving a photon of light and becoming excited to the metarhodopsin state. By absorbing another photon of light, the metarhodopsin can be transformed back into rhodopsin, ready to repeat the visual process all over again. It has been demonstrated that a source of vitamin A or a carotenoid is needed by several species for visual acuity and normal structure of compound eyes. A number of different species detect and use plane-polarized light in behavioral orientation.

10.1 INTRODUCTION

Insects have a surprisingly diverse array of sensory receptors that feed them information about their internal and external environments. The sensory neurons of insects have their cell bodies, with only rare exceptions, located very near the stimulus site, rather than in or near the central nervous system (CNS) as in vertebrates. Many receptors detect changes occurring at the cuticular surface, and the cell bodies are located peripherally just beneath the cuticle. Most sensory neurons are bipolar, with a few multipolar ones, and the dendritic terminals are usually very short compared to the relatively long axon leading to the CNS. The axons from many sensory neurons pass into the brain prior to synapsing and are classified as **primary** or **type I** sensory neurons. **Secondary** or **type II** sensory neurons synapse prior to entering the brain.

A common characteristic of all types of sensory neurons is that they transduce the stimulus energy (such as light, heat, chemical, or mechanical energy) into a slow (or graded) electrical potential. The receptor process can be divided into three steps:

1. Absorption of the stimulus energy
2. Transduction into the receptor potential
3. Repetitive impulse discharge from the axon portion of the receptor neuron.

Repetitive discharge occurs only if the receptor potential is of sufficient magnitude to exceed the threshold for spike generation in the axon.

The input energy can have an excitatory effect on the receptor neuron (depolarization), or it can have an inhibitory effect (hyperpolarization). Sensory neurons are sensitive to change in a stimulus; thus, a receptor cell will make an initial response (depolarization or hyperpolarization) when the stimulus starts (the "on" response), and then it makes the reverse response when the stimulus ceases (the "off" response). The upward deflection in Figure 10.1 indicates a depolarizing stimulus, and the receptor cell membrane has become less negative on the inside because a receptor potential has been produced.

Adaptation, illustrated in Figure 10.1, to a steady stimulus is a characteristic feature of many receptor neurons. During adaptation, the receptor potential falls from its initial response level to some lower level, or perhaps even to a silent state. Receptors that rapidly adapt to continuing steady stimuli are **phasic receptors**, while those that slowly adapt are **tonic receptors**. The foregut stretch receptor in *Phormia regina* is a tonic receptor that maintains a relatively sustained and uniform rate of firing when a constant stretch is applied (Gelperin, 1967). Two bipolar neurons connect the recurrent nerve with the foregut, and the neurons function as stretch receptors, indicating peristalsis and fullness of the gut. Severing the branch of the recurrent nerve carrying the neurons results in the failure of a fly to stop feeding, hyperphagia, and a grossly expanded abdomen (Dethier and Gelperin, 1967).

With the appropriate equipment and technique, the receptor potential can be measured, but sometimes it is easier and more convenient to measure the number of spikes produced from the receptor as an indicator of receptor action. A large receptor potential will produce spikes in rapid succession. As the receptor adapts, the frequency of spike generation falls. The chemoreceptor cell in Figure 10.1 shows another feature of many receptors. It is **spontaneously active**, firing about 5 spikes/s in the absence of any applied stimulus. Such spontaneously active receptors are probably never silent. They may play important roles in information coding by reducing spike frequency or increasing it, respectively, in response to inhibitory or stimulatory stimuli. Most sensory neurons in insects are organized into a complex morphological unit containing associated sheath cells, and the entire structure is called a **sensillum** (plural = **sensilla**). Sensory organs, such as the compound eye, tympanum, and Johnston's organ, are composed of many units or sensilla.

10.2 EXTERNAL AND INTERNAL RECEPTORS MONITOR THE ENVIRONMENT

Receptors can be broadly classified as providing information about the external environment or the internal environment. Receptors that monitor the external environment are usually given descriptive names such as compound eyes, ocelli, tympanum, or Johnston's organ, but simple tactile hairs are also common. Receptors providing information about internal body conditions are called **proprioceptors**. Proprioceptors are present in the connective tissue of the body, among muscles, and

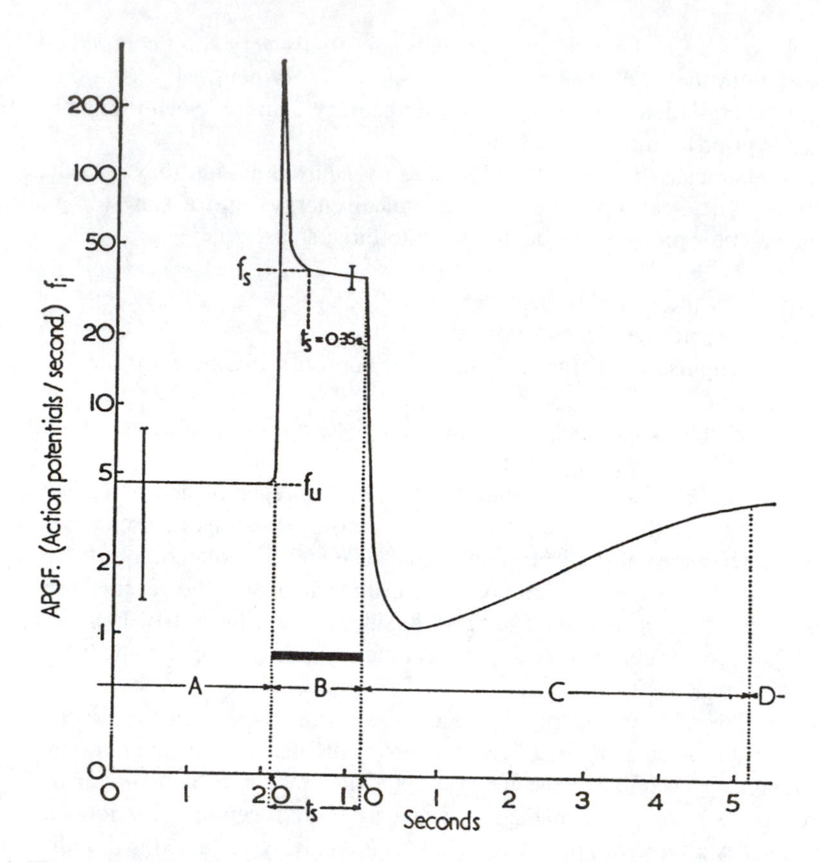

FIGURE 10.1 The response of a chemoreceptor on the labelum of the blowfly *Phormia regina* to a solution of 0.4 *M* KCl. The stimulus duration is indicated by the dark bar at B. Periods A and D show the unstimulated spontaneous activity (about 5 spikes/s) of the receptor. The receptor neuron initially makes a strong response to the stimulus by increasing spike output to a rate greater than 200 spikes/s, but adaptation occurs rapidly; and after about 0.35 s, the output rate falls to a tonic output of about 40 impulses/s. The post-stimulation response or "off" response is shown in C. (Reproduced with permission from Rees, 1968.)

along the surface of the alimentary canal of all insects. The sensory neurons are frequently multipolar. Some types of proprioceptors function as mechanoreceptors or stretch receptors to indicate gut filling, muscle tension, and general body movements, while others act as chemoreceptors, relaying information about the chemical composition of the body. Proprioceptors generally adapt slowly to a constant stimulus, which clearly is adaptive if they must indicate body orientation, equilibrium, limb positioning, or fullness of the gut.

Many proprioceptors monitor stress and strain in the cuticle and provide information about body and limb movements. For example, head movement relative to the long axis of the body is indicated by movement against hair plates on the back of the head and on the prothorax of some insects. The campaniform sensilla in Table 10.1 are proprioceptors indicating cuticle stresses due to movement, but the only external cuticular evidence of these sensilla is a slightly raised dome of cuticle where they are attached.

10.3 GENERAL FUNCTIONAL CLASSIFICATION OF SENSORY RECEPTORS

Receptors can be classified functionally with respect to the type of energy they transduce. For example, they might be classified as light and/or visual detectors, mechanoreceptors (including tactile, vibration, and sound detectors), chemoreceptors (including contact receptors [gustatory or

TABLE 10.1
A Classification Scheme Often Used by Insect Biologists in Describing Sensory Structures

1. *Sensilla trichoidea* are sensory hairs or setae ("Sinneshaare") and their elaborations, including hair plates. These sensilla are widely distributed among insects and other arthropods. Many of the hairs on insects are innervated by sensory neurons. The hair may not have any openings (tactile hairs, hair plates), may be perforated by only one pore (gustatory contact chemoreceptors), or by many pores (olfactory receptors).
2. *Sensilla chaetica* are sensory spines or bristles ("Sinnesborsten") that are stouter than *S. trichoidea* and are often located singly rather than in groups. *S. chaetica* consist of an innervated hair in a flexible socket. The hair may have a single pore (gustatory and some mechanoreceptors), or pores may be absent (most mechanoreceptors).
3. *Sensilla squamiformia* are flattened hairs or sensory scales ("Sinnesscchuppen"). They are common on the wings of Lepidoptera, although not every wing scale is innervated as a sensory organ.
4. *Sensilla basiconica* ("Sinneszapfen") are sensory pegs, cones, or stumpy hairs. They may be either thick walled or thin walled, and lack pores (thermo- and hygroreceptors), or have a single pore (contact chemoreceptors and gustatory receptors), or have many pores (olfactory receptors). *S. basiconica* may contain only a few or many neurons.
5. *Sensilla coeloconica* ("Grubenkegel") are cones or pegs set in small depressions or pits in the cuticle. Both thick-walled and thin-walled sensilla occur. Some have multiple pores and serve as olfactory receptors, and some have no pores and appear to serve as thermo- and hygroreceptors.
6. *Sensilla styloconica* consist of elevated cones that may be located in pits or at the cuticle surface. *S. styloconica* typically serve a gustatory function and have a single pore at the tip.
7. *Sensilla ampullacea* are sensory tubes ("Sinnesflaschen"). They may be cone-shaped and rest on long tubes in sunken pits. They occur on the antennae of bees and other Hymenoptera.
8. *Sensilla placodea* are multiporous plate structures ("Sinnesplatten") with an olfactory function. *S. placodea* have many neurons with dendritic terminals ending on thin plates covering a fluid-filled canal. These sensilla occur on the antennae of several insect orders.
9. *Sensilla campaniformia* include structures variously called campaniform sensilla, Hick's organs, cupola organs, and sensory pores ("Sinneskuppeln"). *S. campaniformia* are innervated by a single neuron that terminates in a small dome 20 to 30 μm in diameter. Sometimes there is a scolopale at the center of the dome where nerve contact occurs. These sensilla may be recessed in cuticular depressions or elevated. In general, they convey information about mechanical strain in the cuticle, for example, information about joint movements, movements of appendages, or other strain or pressure at the cuticular surface. Campaniform sensilla are often arranged in small groups along the long axis of a limb. It appears likely that they are important in maintaining the stance of an insect, in geotaxis, and in coordinating walking or running.

taste] and olfactory receptors), humidity receptors, temperature receptors, magnetic receptors, or geodetectors.

Anatomical classification of insect receptors has been a common practice (Table 10.1) (Horridge, 1965). Given the diversity of insects, probably all the sensory structures that have been, or will be, described from SEM studies will not fit the described categories. Unfortunately, for physiology, it is much easier to obtain excellent SEM photos of structure than to obtain physiological data, and functional information about many sensilla on the surface of insects is sparse or nonexistent. It is important to remember that it is not possible to assign function with absolute certainty to all insect sensory structures based on morphology.

Altner and Prillinger (1980), Zacharuk (1980), and Frazer (1985) recommend a simplified classification scheme based primarily on the presence or absence of pores, and the number of pores in external sensilla, with general functional significance when known. They recommend three major groupings: (1) receptors with multiple pores, (2) receptors with a single pore, and (3) receptors without pores.

10.3.1 Receptors with Multiple Pores

Receptors with multiple pores in the external cuticular structure tend to be **olfactory** receptors that detect airborne chemicals. The external cuticular structure may take the form of a hair (*Sensilla* *trichoidea*), plate (*S. placodea*), peg (*S. basiconica*), or peg-in-a-pit (*S. coeloconica*). Olfactory receptors are probably present on the antennae of all insects, and occur elsewhere on the body of some insects. Functionally, the openings or pores through the cuticular covering allow airborne molecules to enter the sensillum. The external structure tends to be thin-walled and the cuticular socket inflexible.

10.3.2 Receptors with a Single Pore

Receptors that have a single pore near or at the tip of the cuticular structure usually have a **gustatory** function (i.e., taste receptor) and detect chemical substances in solution. Gustatory receptors are also called contact chemoreceptors. The external appearance may be that of a hair (*S. chaetica* or *S. trichoidea*), peg (*S. basiconica*), or dome (*S. styloconica*). The external cuticular structure is thick-walled, and sockets may be flexible or inflexible. Gustatory receptors are numerous on the mouthparts, as well as on the tarsi and ovipositor of some insects.

10.3.3 Receptors without Pores

The lack of any pore in the external cuticular part of a sensillum is typical of mechanoreceptors that detect vibrations in the air (insect "ears" and vibration receptors), water, or substrate on which an insect rests. Mechanoreceptors may consist of a dome (*S. campaniformia*) or hair (*S. chaetica*) set in a flexible socket. Humidity and temperature receptors, which may have the form of a peg (*S. basiconica)* or peg-in-a-pit (*S. coeloconica*) set in an inflexible socket, also usually lack a pore.

Frazer (1985) cautions that although the structural unit is the sensillum, the functional units are the neurons within the sensillum. Multiple neurons in the same sensillum can, and sometimes are known to, serve multiple physiological roles. For example, one neuron may function as a chemoreceptor, while another neuron in the sensillum functions as a mechanoreceptor. Dethier (1955) and Hodgson (1956, 1958) described such combinations in the labellar hairs of blowflies, and subsequent studies have shown that multimodal sensilla occur on many appendages and other parts of the body of insects.

10.4 MECHANORECEPTORS

The anatomy and basic physiology of mechanoreceptors were reviewed by Dethier (1963) and Horridge (1965). Mechanoreceptors are involved in detection of airborne sounds, substrate vibrations, appendage movement and orientation (proprioceptors), flight speed, gravity, and possibly heat detection. Although all mechanoreceptors contain one or more bipolar sensory neurons and associated sheath cells, there are many anatomical variations. Some are located entirely internally, but many have external components.

10.4.1 Structure of a Simple Tactile Hair: A Mechanoreceptor Sensillum

The simplest mechanoreceptor, a sensory hair, contains a minimum of three cells, all derived from a common epidermal mother cell that divides to give rise to the **trichogen**, the **tormogen**, and the bipolar sensory neuron. The trichogen and tormogen are sheath cells; the trichogen is the inner sheath cell enclosing the soma and parts of the dendrites and axon of the sensory neuron, with the tormogen cell then enclosing both trichogen and neuron — a double sheath arrangement. Sometimes, other sheath or specialized glial cells are also present, and the inner one in contact with the neuron in some receptors has been called a **thecogen** cell by some authors. As in other parts of the nervous system, the sheath cells insulate the neuron, may provide it with nutrients, and may help control concentrations of ions necessary for nerve function. The elements of a bipolar neuron and sheath cells in a tactile hair are the common elements of all insect sensilla, although some sensilla contain more complicated structures.

Single tactile hairs, and hairs grouped together in a hair plate, are common tactile mechanoreceptors on the body surface of insects. Tactile hairs are numerous on the antennae (especially on those insects that spend part or all their lives in darkness, such as bees, ants, cockroaches, and cave dwellers) and on the cerci of Orthoptera and Dictyoptera. Cercal receptors detect a range of vibrations in the air and substrate, and can act as sound receptors and vibration receptors. In cockroaches, and possibly in other insects, cercal receptors function in an escape mechanism in which the tactile hairs respond to sudden vibrations or loud sounds by sending spikes through the cercal nerve to connect with giant axons at synapses in the 6th abdominal ganglion. The giant axons pass uninterrupted to the thoracic ganglia, where synaptic connections are made with motoneurons to the leg muscles. The system results in rapid transmission of stimuli that give rise to escape maneuvers. Many caterpillars also have single hairs that detect air vibrations and/or sounds; and caterpillars make behavioral responses to loud sounds and other airborne vibrations. Hairs are relatively insensitive to sound, however, in contrast to more complex tympanal organs that often are very sensitive.

Depending on the way in which the tactile hair is set in its socket, it may bend in only one direction, and thus can indicate the direction of the bending energy, while others are omnidirectional. Tactile hairs usually occur in multiples on appendages, however, and directionality is often possible from the combination of stimuli and receptors responding.

10.4.2 Hair Plates

Hair plates are common at leg joints and at points of limb articulations with the body. They respond to touch, bending, and to joint flexion with output of nerve spikes. They adapt slowly, a characteristic of static receptors that indicate body orientation. These tactile structures enable an insect to know the position of its limbs with respect to the body, and probably function in locomotion. Greater numbers of tactile hairs typically occur on the coxa and trochanter, probably enabling more precise movement of these large, heavily muscled parts of the leg. Hair plates are also common on sclerites at the back of the head and/or neck and on the anterior parts of the prothorax in mantids, locusts, and bees, and act as proprioceptors enabling the insect to know its head orientation with respect to the body. They may be important, in some cases at least, in flight ability because destroying the hairs on the cervical plates of locusts influences their equilibrium in flight. Sensory neurons, like other parts of the nervous system, have a high demand for oxygen; the mechanoreceptors of honeybee cervical hair plates were shown to be very sensitive to oxygen deficiency, with spikes ceasing after 2 minutes and the receptor potential after 10 minutes.

10.4.3 Chordotonal Sensilla

Chordotonal sensilla are anatomically more complex than tactile hairs. They occur at most exoskeletal joints, limb joints, and body segment joints. Field and Matheson (1998) presented a comprehensive review of chordotonal sensilla. There are many morphological variations but, basically, the sensory neuron is enclosed (ensheathed) within parts of two or three other cells, including a characteristic **scolopale** cell and cap cell (Figure 10.2). There may be additional sheath cells. Instead of the dendritic terminals being directly attached to the cuticle, they terminate within the cap cell, and the cap cell is attached to some internal structure or to the cuticle. Any stress or pull on the cap cell is then transmitted to the neuronal endings as a stimulus. Often, the cap cell extends well down over the scolopale sheath cell. Authors have not applied the same terminology, unfortunately, and the scolopale cell may be referred to as the sense rod or scolopale body. A single sensory unit with the scolopale and cap structure is called a **scolopidium** or, alternatively, a chordotonal sensillum. Complex chordontonal organs such as a tympanum or Johnston's organ

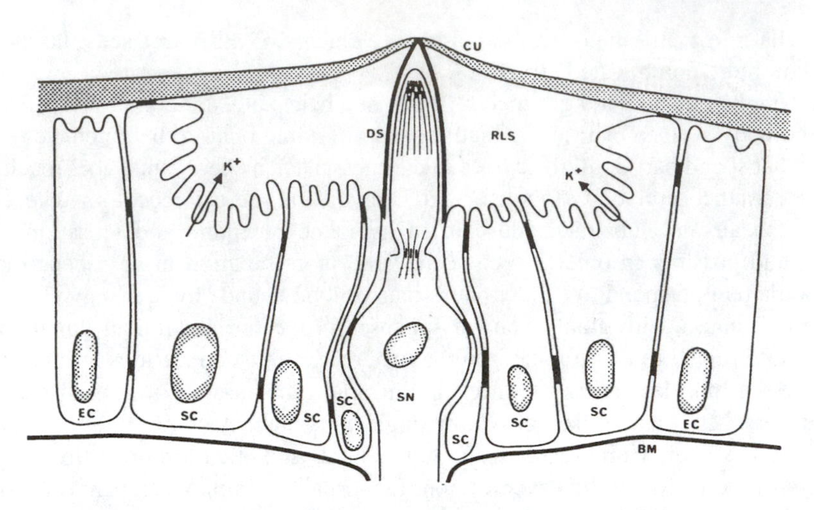

FIGURE 10.2 Diagram of a hypothetical chordotonal sensillum or mechanoreceptor, showing the scolopale and cap cell. Any stress or strain at the cuticular surface where the scolopale cap cell is attached is transmitted to the sensory neuron beneath. BM, basement membrane; CU, cuticle; DS, dendrite sheath or scolopale; EC, epidermal cells; RLS, receptor lymph space; SC, sheath cells; SN, sensory neuron. In the diagram, potassium ions (K^+) are shown being pumped into the receptor lymph space. The short dark bars between adjacent epidermal cells are cell junctions that prevent ion movement between cells and provide high electrical resistance. (Reproduced with permission from French, 1988.)

contain many scolopidia, and both external receptors and proprioceptors contain scolopidia as the morphological unit.

Examples of sensory organs containing scolopidia include (1) subgenual organs found just below the epidermis at the femero-tibial joints of most adult insects; these allow an insect to know where its limbs are, and whether they are flexed or extended, in relation to the body; (2) tympanal organs involved in detection of substrate vibrations and sound detection; (3) Johnston's organ, also a vibrational/sound detector located in the pedicel of the antennae of most adult insects, and in some larvae; and (4) simple structures, each with only a few scolopidia, that occur in various parts of the body of larvae and adults (Horridge, 1965). Sometimes the simple structures of type (4) are simply called chordotonal organs, while the more complex structures, such as subgenual organs or Johnston's organ, have other names; but all are chordontonal organs composed of multiple scolopidia. Even a single sensory (tactile) hair can contain a scolopale, but many of these do not (see above).

10.4.4 Subgenual Organs

The **subgenual organ** is a complex chordotonal organ composed of multiple scolopidia. The term "subgenual" means below the knee, from Latin for knee (*genu*); and this complex chordotonal organ usually is located near the joint between the femur and tibia (Figure 10.3). The organ contains as few as three scolopidia in some earwigs (*Forficula* spp.), but contains more in most insects. It acts as a proprioceptor, detects vibrations of the substrate, and it has become specialized as a tympanal organ in some insects. The subgenual organ is especially well developed in crickets (Gryllidae) and katydids (Tettigoniidae) and is associated with the tympanal organ, with both organs located on the tibia. The two organs have separate innervation, however, and probably have separate functions (Haskell, 1961). In some insects, the scolopidia vary in length, suggesting that different scolopidia might respond to vibrations of different amplitude according to length. The subgenual organ of the American cockroach *Periplaneta americana* is sensitive to vibrations that would displace the foot of the insect by as little as 10^{-9} to 10^{-7} cm (Autrum and Schneider, 1948).

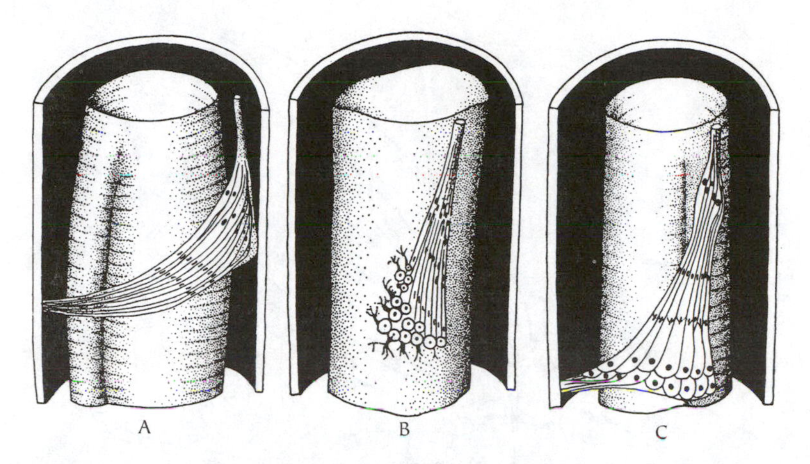

FIGURE 10.3 Structure of subgenual organs from an orthopteran (A), a lepidopteran (B), and a hymenopteran (C). (Reproduced with permission from Autrum and Schneider, 1948.)

The subgenual organ is less well-developed in Lepidoptera, Hymenoptera, and Hemiptera than in the Orthoptera, and some Hemiptera, Coleoptera, and Diptera do not have subgenual organs. Those without a subgenual organ display only low sensitivity to high-frequency substrate vibrations.

Probably all insects have additional chordotonal sensilla on the legs, particularly at or near the leg joints (Haskell, 1961). Some insects lacking a subgenual organ have a similar organ at the distal end of the tibia that may serve much the same function as the subgenual organ (Horridge, 1965).

10.4.5 Tympanal Organs: Specialized Organs for Air-borne Sounds

Tympanal organs are chordotonal organs or "insect ears" that are specialized for high-frequency sound detection as opposed to low-frequency vibration detection. Tympanal organs are located at various places on the body of insects (Figure 10.4) (Yack and Fullard, 1993; Hoy and Robert, 1996). Examples of locations are near the sternum of the first abdominal segment of Acrididae (grasshoppers) and Cicadidae (cicadas), on the tibia of Tettigoniidae (long-horn grasshoppers) and Gryllidae (crickets), on the thorax of Notonectidae (aquatic hemipterans), and on the thorax or abdomen of some Lepidoptera.

Many insects produce sounds, are sensitive to sounds, and utilize sounds in courtship, mating, prey location, and predator avoidance. Haskell (1961) has reviewed and tabulated the distribution of insect hearing organs across the different insect orders and families, along with details about anatomy and physiology. Busnel (1963) and Sales and Pye (1974) reviewed hearing in moths and their behavior in response to the ultrasonic sounds of bats. Ewing (1989) provides excellent explanations of the physics of sound and vibration production, transmission, reception, behavioral functions, and evolution of sound in insects. Bailey (1991), Fullard and Yack (1993), Hoy and Robert (1996), and Hoy et al. (1998), have reviewed acoustic organs and behavior in various sound-producing insects, and discussed aspects of the evolution of sound. Spangler (1988) has reviewed the role of sound reception and defense behavior in moths.

Energy radiates from a sound source as both high-frequency airborne sound waves and lower-frequency substrate vibrational waves. The two types of energy radiate as different waveforms (Ewing, 1989), and are detected generally by different types of receptors. Thus, although most of the energy from an insect that is calling from a perch on a leaf or stem radiates as airborne sound, some is nearly always transmitted to the substrate as low-frequency vibration (Bailey, 1991). The low-frequency vibrations are not transmitted very far, but insects have receptors capable of transducing both types of energy, and those close to the sound-producer may receive both types.

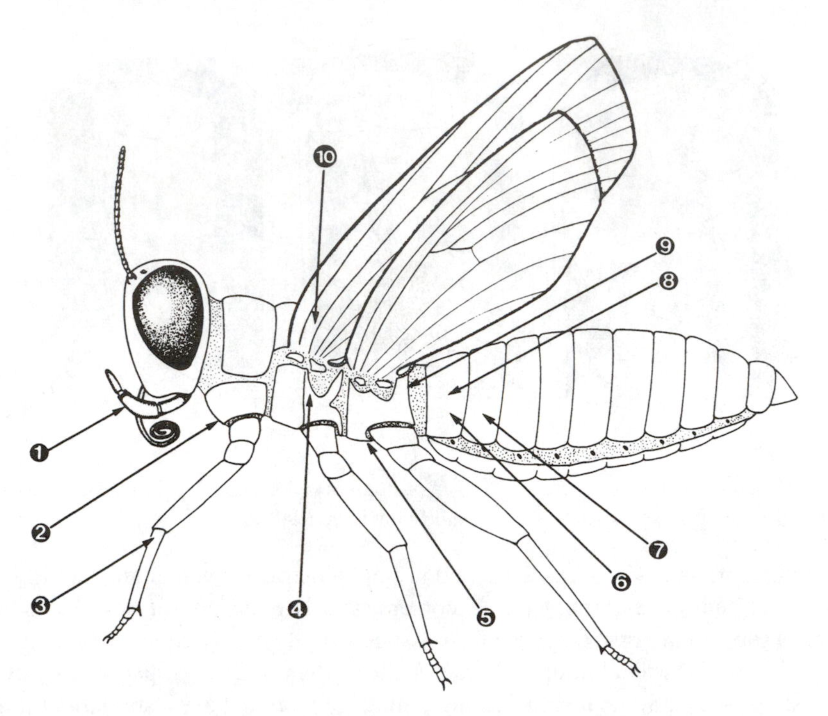

FIGURE 10.4 Diagram to illustrate the multiple places that tympanal organs (functional ears) have been located on insects. 1, Lepidoptera (Sphingoidea); 2, Diptera (Tachinidae); 3, Orthoptera (Ensifera); 4, Hemiptera (Corixidae); 5, Mantodea (Mantidae); 6, Lepidoptera (Geometroidea and Pyraloidea); 7, Hemiptera (Cicadidae); 8, Orthoptera (Acrididae); 9, Lepidoptera (Noctuoidea); 10, Neuroptera (Chrysopidae). (Reproduced with permission from Yack and Fullard, 1993.)

Tympanal organs are specialized for airborne sound pressure waves, and permit sound detection over a relatively long distance. They are sensitive to a wide range of frequencies from 2 kHz up to about 100 kHz (Hoy and Robert, 1996). Typically, in insects, as well as in other animals, tympanal organs are paired. A single pressure receptor is not very efficient at detecting the directionality of the sound source, but two receptors, preferably well separated from each other, can detect directionality by differences in reception at the two locations. Tympanal ears typically have a minimum of three components: (1) a thin cuticular tympanum on the cuticular surface, (2) an air sac or other tracheal structure behind the tympanum, and (3) sensory neurons organized in scolopidia attached to the tympanal membrane or attached near it, so that they vibrate in response to the vibrations of the tympanum (Figure 10.5).

Airborne sound waves cause the tympanum to vibrate, and sensory neurons enclosed in the scolopale cells detect the vibrations and respond first by graded electrical potentials, followed by a burst of spikes in the axon. The air cavity, or tracheal sac, plays an important role as a resonating chamber and in preventing damping of the sound. Some insects have a tympanum that can respond to sound waves striking it from the inside of the air chamber as well as from outside; such tympanal organs are pressure-difference receivers, and they are especially sensitive to directionality of the sound. Some tympanal organs have scolopidia of different lengths, suggesting sensitivity to various frequencies, but function is unproven.

Tympanal organs probably evolved from some early form of mechanoreceptor, probably a stretch-registering proprioceptor, but they evolved independently among the seven orders of insects having tympanal hearing. In addition to tympanal organs, some insects may also hear some sounds with other organs, including Johnston's organ, subgenual organs, scattered simple chordotonal sensilla, and simple hair sensilla.

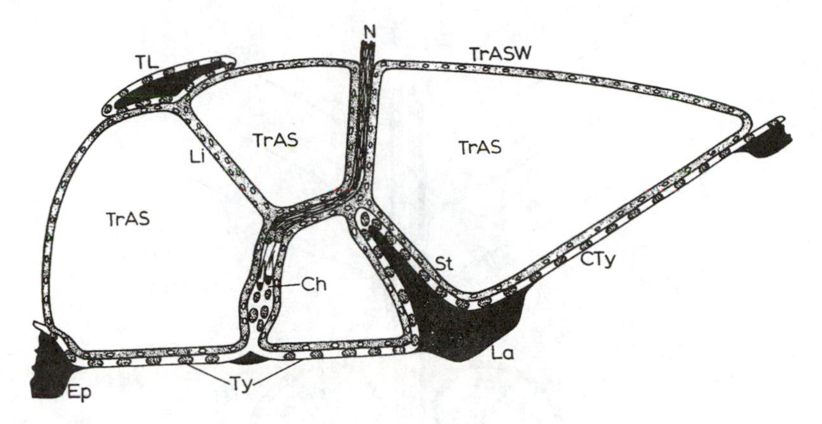

FIGURE 10.5 The left thoracic tympanal organ from a noctuid moth. (From Autrum, 1963.)

10.4.6 Johnston's Organ

Johnston's organ (Figure 10.6) is a large complex chordotonal organ that may consist of several groups or a single grouping of scolopidia located between the second (the pedicel) and third joints of each antenna of most adult insects, although some Apterygota (Collembola and Diplura) do not have a Johnston's organ. A simplified form of the organ occurs in some larvae. Johnston's organ responds to several kinds of stimuli in different insects, including acting as a proprioceptor to indicate movement of the antennae, monitoring wing-beat frequency in relation to flight speed in some Diptera, gravity indicator, detection of ripples at the water surface in gyrinid beetles, and sound reception in mosquitoes and perhaps other insects.

With its location in the second antennal segment, Johnston's organ is positioned to monitor movements of the antennal flagellum, whether due to muscles controlled by the insect or displacements of the antennae by wind and flight. There are variable numbers of scolopidia radially arranged and attached to the wall of the pedicel at one end and to the intersegmental membrane between the pedicel and flagellum. Johnston's organ seems to have reached its apex of development in dipterans in the Families Chironomidae and Culicidae, in which the pedicel is much enlarged and the organ completely fills it. In these small swarming dipterans, the large organ is directionally sensitive and functions in successful swarming and mating. Frequency of sound is detected by the arista, which vibrates in resonance to the sound of the wings of the female in flight. In addition, the males have numerous long hairs on the antennae, and these vibrate in response to the flight sounds produced by the wings of flying females. Their vibration causes the flagellum (the major portion of the length of the antenna) to vibrate. Males of the mosquito *Aedes aegypti* are most sensitive to frequencies from 400 to 650 Hz, closely corresponding to the natural wingbeat frequency of females (Roth, 1948). Johnston's organ functions as a flight speed indicator in adult *Calliphora erythrocephala* (Burkhardt, 1960), and probably also in some other insects such as the housefly, honeybee, and related insects. It is probably an important gravity indicator for most insects, enabling them to have a sense of their body in relation to horizontal and vertical planes because the weight of the antenna excites scolopidia, depending on the pull of gravity relative to the body. Gyrinid water beetles swim at the water surface and avoid colliding with other swimming beetles. They do not "crash" into other beetles or the sides of a small container because Johnston's organ enables them to detect disturbances and ripples in the water created by other beetles, or their own ripples bouncing off the container walls.

10.4.7 Simple Chordotonal Organs

Simple chordotonal structures that have no specific name occur widely over the body of most orders of insects, including adults and larvae. The structures usually consist of only a few scolopidia.

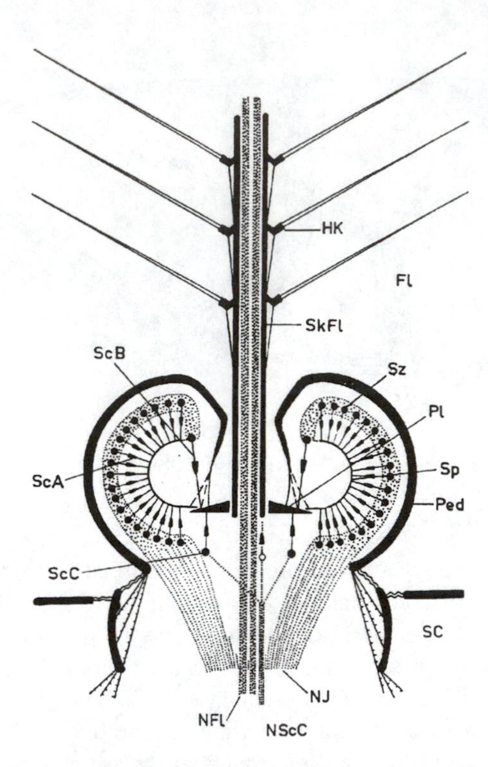

FIGURE 10.6 Diagrammatic illustration of Johnston's organ in the antenna of a culicid mosquito. The circular chitinous plate (Pl) is attached at the base of the first flagellar segment (which is the third antennal segment). The insertions of the scolopidia are the chitinous prongs (Sp). The sensory neurons are arranged in one or two rings (ScA and ScB). The flagellum and Johnston's organ are sound-receiving organs. Fl, flagellum; HK, hair crown of the flagellum; NFl, flagellar nerve; NJ, nerve complex of Johnston's organ; NScC, nerve of the three single scolopidia; Ped, pedicellus; Pl, circular chitin plate; SC, scapus; ScA, outer ring of scolopidia; ScB, inner ring of scolopidia; ScC, single scolopidium; SkFl, inner skeleton of flagellum; Sp, chitin prongs at insertion of scolopidia; Sz, nuclei of sensory cells. (From Autrum, 1963.)

There are about 90 such small chordotonal organs arranged along the length of *Drosophila* larvae (Horridge, 1965). Relatively simple chordotonal organs occur in the legs (on the femur, and sometimes on the tibia); on the wings at the base of the radial and subcostal veins; and within the lumen of the radial vein of many, but not all, insects. Simple chordotonal sensilla occur on the legs in addition to the subgenual organ, and on the antennae, in addition to Johnston's organ, usually at or near the antennal joints, of most insects (Haskell, 1961). Some of the simple chordotonal sensilla respond to certain airborne frequencies, (i.e., are sound receptors), but they are not very sensitive and have a narrow response range. Similar simple chordotonal organs may have been the precursors of tympanal ears in the early evolution of insects (Bailey, 1991).

10.4.8 Thermo- and Hygroreceptors

The literature on insect **thermoreceptors** and **hygroreceptors** has been reviewed by Loftus (1978), Altner and Prillinger (1980), and Altner and Loftus (1985). Experimentally, insects can be shown to respond to warm, moist, and cold air, but conclusive identification of the receptors by which they monitor these environmental changes, and even whether they routinely use such information in their behavioral activities, is tentative or sparse. Receptors tentatively believed, or in a few cases proven, to function as thermo- and hygroreceptors often occur in the same sensillum on the antennae, most frequently as a triad of three neurons, although these types of receptors are not numerous. It has been estimated that the American cockroach *P. americana* has about 1300 sensilla on the

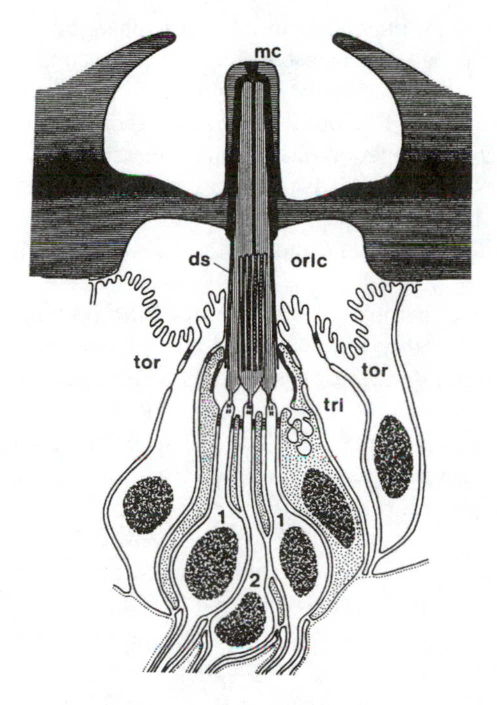

FIGURE 10.7 The triad arrangement of three sensory neurons, and trichogen and tormogen cells in a peg-in-a-pit arrangement of a hygroreceptor. The peg has no pore. The dendritic outer segments of the two type-1 sensory cells (1) enter the lumen of the peg, but the outer segment of the type-2 cell (2) branches into lamellae and ends below the peg. There are three sheath cells, the thecogen (stippled), the trichogen (tri), and the tormogen (tor); ds, dendritic sheath; mc, molting channel of peg; orlc, outer receptor lymph cavity. (Reproduced with permission from Altner and Loftus, 1985.)

antenna that house a thermoreceptor neuron, but this represents only about 0.4% of the receptors on the antennae of a male cockroach (Altner et al., 1983). The most common triad arrangement is one in which one neuron is sensitive to cold air, one to moist air, and one to dry air. The cold receptor responds to sharply falling temperature by a rapid rise in firing rate, the moist air receptor fires more frequently when the humidity rises, and the dry air receptor fires in response to falling humidity. Triads have been found on the antennae of the American cockroach *P. americana*, the migratory locust *Locusta migratoria*, the European walking stick *Carausius morosus*, *Triatoma infestans* (Hemiptera), the honeybee *Apis mellifera*, and the noctuid moth *Mamestra brassicae*. A warm receptor that fires in response to rising temperature has been found in the same sensillum with a cold receptor on the antennae of the mosquito *Aedes aegypti*.

Typically, the cuticular portion of these sensilla has no pore and is set in an inflexible socket; *Sensilla trichoidea*, *S. basiconica*, *S. coeloconica*, and *S. styloconica*, as well as other morphological structures, are known to house receptors believed to be thermo- and hygroreceptors. A few examples of either a thermoreceptor or a hygroreceptor associated with an olfactory receptor in a sensillum with multiple pores have been described. Thus, as in other types of receptors, it is not possible to identify function with certainty based on morphology. In the most common triad arrangement (Figure 10.7, a peg-in-a-pit), typically the dendritic portion of two of the neurons fill the lumen of the peg (or other cuticular arrangement), while the dendrite of the third neuron has short multiple branches, often forming lamellae, and ending beneath the peg. The former cells are called type 1 cells, and the latter is a type 2 cell. In some cases, a type 3 cell has been found in which the outer dendritic segment is slender, like the cilium portion, and ends much before the outer cuticular structure. A few sensilla have been found with four or five sensory neurons. Although experimental evidence is not conclusive, the arrangement and some data suggest that type 1 cells are actually mechanoreceptors that respond to cuticular distortion due to changes of water content in the cuticular portion of the sensillum. If they are mechanosensitive, and can accurately sense cuticular distortions due to water content in the air, they must be well protected from mechanical disturbance, which would create noise in the system. The inflexible socket, short cuticular projection, pit or collar arrangement that is common, and location beneath more massive mechanoreceptors seem likely to protect them from ordinary mechanical disturbance. It should be emphasized, however, that a mechanosensitive functionality is not yet firmly established, and other ways to detect water in the air are possible [for example, by humidity-induced changes in electrolyte concentrations (Loftus, 1976, 1978)].

The type 2 cell may be a thermoreceptor. The dendritic portion of the type 2 cell is more variable among different species than that in type 1 cells. Electrophysiological investigations on a cave dwelling beetle, *Speophyes lucidulus*, showed that a cold and a warm receptor existed in the same sensillum (with a third neuron, possibly also a thermoreceptor). The possibility that the type 2 cell is the thermoreceptor remains quite tentative, but speculation is that the number of lamellae in the distal portion of the dendrite may be correlated with a range of temperatures that can be detected (Corbière-Tichané and Loftus, 1983).

The arrangement of pairs of receptors, such as a warm and cold receptor, or a moist air receptor with a dry air receptor in the same sensillum may improve the ability of the system to discriminate changes in environmental conditions. Each will respond in firing rate to a change in temperature or humidity, but the change will be in the opposite direction. For example, the warm receptor will respond to rising temperature by increasing its rate of firing, while the cold receptor will respond by decreasing its rate of firing. The reverse will occur during cooling. The integration of such information in the central nervous system, if indeed used by the central nervous system, awaits further exploratory research.

10.4.9 Infrared Reception

Many species (about 40; Hart, 1998) of insects have specialized **infrared receptors** and are attracted to forest fires, where they lay their eggs in fire-damaged trees and burned-over debris. Most of the insects are beetles. *Melanophila acuminata*, a beetle in the family Buprestidae, has been most intensively studied and is known to be attracted in large numbers to burning forests (Evans, 1962, 1966a; Apel, 1989), where they mate and begin laying eggs in the burned wood. *M. acuminata* have paired pits adjacent to the mesothoracic coxa on the pleuro-ventral thorax (Figure 10.8). The pits are slightly variable in size, 170 to 320 μm long by 80 to 150 μm wide, by 70 to 100 μm deep. Within each pit are 50 to 100 slightly oblong hollow domes and associated multipore wax glands (Evans, 1966b; Vondran et al., 1995; Schmitz et al., 1997). Each dome is innervated by one bipolar neuron and associated sheath cells (Figure 10.9). The cell body lies just beneath the cuticle at the base of the pit, and its axon passes without synapsing into the metathoracic ganglion. The dendrite of the sensory neuron contains neurotubules and has a **ciliary constriction** near its mid-point. The distal tip of the dendrite is attached at the base of the spherical dome. The dome surface is thin and unsclerotized, possibly allowing the hollow sphere beneath the domed cuticle to change its volume due to absorption of infrared radiation, and thus mechanically stretch the dendritic tip of the neuron. If this is the correct interpretation, then the infrared receptor functions like a modified mechanoreceptor. The beetles respond behaviorally to infrared radiation at 3 μm wavelength (Evans, 1966a), and infrared wavelengths ranging from 2.5 to 4 μm are emitted by intense forest fires. Atmospheric CO_2 and H_2O have narrow bands of strong IR absorption within the same wavelengths, but a window exists at 3.6 to 4.1 μm in which atmospheric components do not absorb strongly. Another insect demonstrated to use heat, and possibly IR, is *Rhodnius prolixus* (Hemiptera: Reduviidae), a blood-feeding insect that is attracted to warm-blooded animals. They are known to orient and approach thermal sources, and recent work with a pure IR source and a cooled IR transmitting

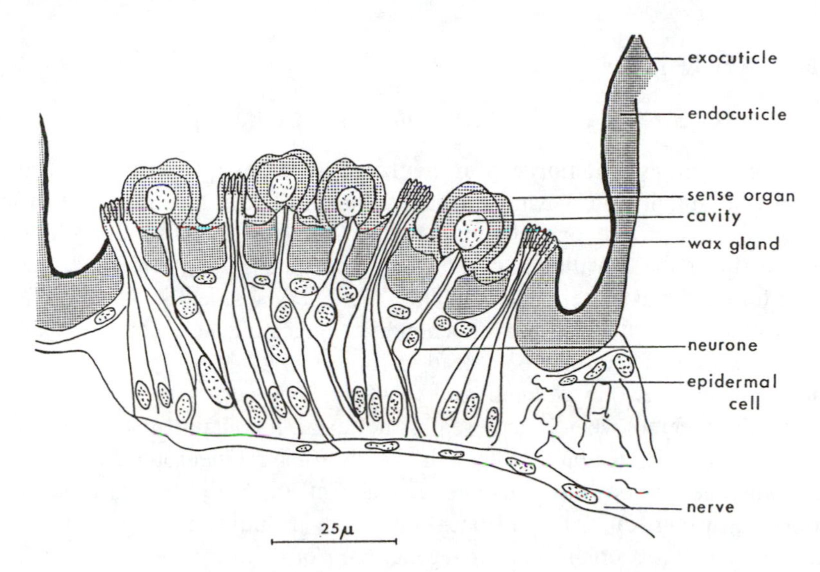

FIGURE 10.8 Diagram of an infrared organ with several receptors shown from the beetle *Melanophila acuminata*. The beetle is attracted to the infrared radiation from forest fires, and thereby locates suitably damaged tree hosts in which to lay its eggs. The receptors probably function as mechanoreceptors, responding to distortions induced by the infrared radiation in the bulbous cavity of each receptor. (Reproduced with permission from Evans, 1966b.)

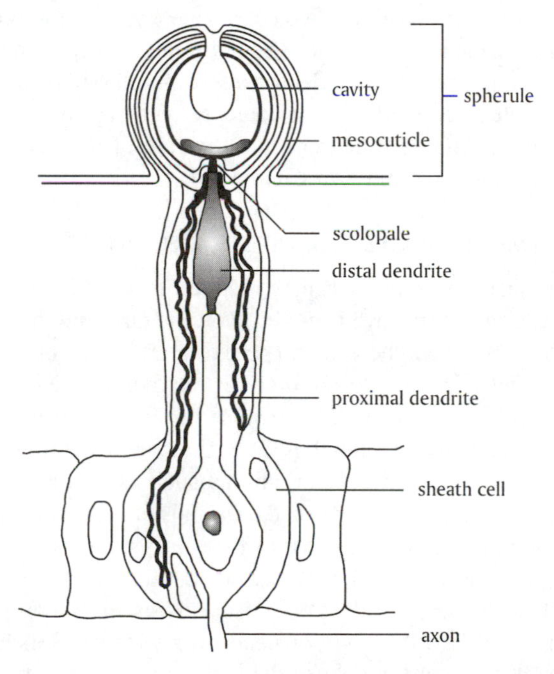

FIGURE 10.9 A schematic diagram of one of the infrared receptors from *Melanophila acuminata*. The spherule is covered by a thin cuticle. The sensory neuron, which has a ciliary constriction, terminates in a scolopale that is attached to the spherule. (Modified from Vondran et al., 1995.)

window between the IR source and the bug strongly indicates that they can detect infrared radiation (Schmitz et al., 2000). It remains unclear whether they might use a mechanoreceptor that is warmed by absorption of IR, as apparently in *M. acuminata* described above, or might possibly have nonthermal IR receptors.

10.5 CHEMORECEPTORS

10.5.1 Olfactory Sensilla: Dendritic Fine Structure

Like other sensory neurons, **chemoreceptor neurons** are bipolar, with the cell bodies located peripherally near the stimulus site. Characteristically, the dendrite of an olfactory neuron comprises a relatively large inner segment connected by a narrow ciliary segment to a smaller outer process extending to the tip of the sensillum. In some sensilla, notably *S. basiconica*, there are many branches of the fine terminal process. Occasionally, the ciliary segment is missing. The large inner segment contains many neurotubules and mitochondria, suggesting a high rate of metabolic activity and utilization of oxygen. There may be a need for high metabolic pump activity in the very fine dendritic endings if there is only a small reserve of the ions necessary for nerve function. The function of the ciliary region, when it is present, is not clear. It contains nine pairs of neurotubules attached to the dendritic membrane. Although the two central neurotubules typical of ciliary structures are sometimes present, they are more often absent. Typically, there is a basal body present. Several theories regarding function have been advanced, including a suggestion that the ciliary organelles might function as organizers for regeneration of dendritic endings following molting. The neurotubules extend into the outer dendritic segment. The cuticular walls of olfactory sensilla contain pores varying from 10 to 100 nm in diameter. Microtubules lead from the inner wall of the pore inward and frequently seem to make direct contact with the dendritic endings. The walls of these microtubules are about 3 nm thick. Chemical molecules enter through the pores, are captured by odorant binding proteins, and are transported across the sensillum liquor (an aqueous medium) to the dendritic endings. Olfactory receptors typically give a phasic-tonic response.

The axons of olfactory sensory neurons are usually small, measuring from 0.1 to 0.2 nm in diameter; they usually pass directly into the deutocerebrum without synapsing, although there are examples of axons from antennal receptors that coalesce or merge by synapsing near the base of the antenna. Greater detail on olfactory neurons and the role of the deutocerebrum in processing input from olfactory neurons can be found in Chapter 14.

10.5.2 Contact Chemoreceptors-Gustatory Receptors

Gustatory receptors are taste receptors and they respond to stimulus molecules in solution. The most carefully studied taste receptors have been those on the tarsi and labellum of blowflies. These receptors are housed in *S. trichoidea*. There are no pore tubules in the cuticle covering the sensillum, the dendrites leading into the sensillum do not branch, and a ciliary region has not been observed. Usually, there is a single pore of about 0.25 to 0.5 nm in diameter near the tip of the hair; some have two pores. The labellar hairs of blowflies contain several neurons in each sensillum. The different neurons are sensitive to sugars, water, anions such as Cl^-, and, in addition, one neuron usually acts as a mechanoreceptor. The unbranched terminals of the dendrites are bathed in a viscous fluid through which stimulating molecules must diffuse. It is convenient to study the response of the labellum sensilla by positioning a capillary electrode containing the stimulating substance (e.g., glucose) over the tip of the hair. This electrode also acts as the recording electrode, while the reference electrode is inserted into the body or head of the insect. Spikes generated in the axon portion of the receptor neurons near the base of the sensillum are conducted back to the tip of the sensillum (by passive transmission) and can be recorded via the capillary electrode on an oscilloscope or other device. The number of spikes per second is used as the indicator of receptor activity.

10.5.3 Specialists vs. Generalists among Chemoreceptors

Among both taste receptors and olfactory receptors, there are some receptors specialized for detection of very specific chemical substances and others capable of responding to a wide variety of chemicals. These are commonly called "specialists" and "generalists," respectively. For example, each antenna of a male *Tela polyphenus* moth bears more than 60,000 sensilla containing about

150,000 sensory neurons. About 60 to 70% of these neurons are specialized for detection of the female produced sex pheromone, about 20% respond to other odors, and the remainder serve a variety of sensory functions. Even the specialists, however, are usually not absolutely specific. A few other chemicals in high concentration may also stimulate them. For example, specialists for 9-oxo-*trans*-2-decenoic acid (the sex pheromone) on the antenna of drone honeybees will also respond to caproic acid if it is presented at 10,000 times greater concentration than the pheromone (10^8 molecules/cc air for the pheromone and 10^{12} molecules/cc air for caproic acid). Dethier (1971) discussed the coding and relaying of stimulus information into the insect brain in relation to specialist and generalist receptors. He described a strict specialist with its axon leading to the brain of an insect as an absolute labeled line. These, he noted, rarely if ever exist in the strictest sense. The information transmitted would be unambiguous but limited to identity of chemical and intensity of stimulus (concentration of chemical). Many absolutely labeled lines, and consequently large sensory nerves (bundles of axons) and central ganglia, would be necessary to accommodate a wide diversity of chemicals. Partially labeled lines, in which each receptor cell is capable of responding to several chemicals, would reduce the number of lines needed into the brain. Still greater capacity for information transmission is displayed by **across-fiber patterning** (Dethier, 1971), in which a receptor responds to several or many chemicals (stimuli) with differing response magnitudes. Thus, the brain could receive information about a specific chemical from a number of receptors, and interpret the profile of responses received. The advantage of across-fiber patterning is that it allows only a few receptors to convey information about a large number of stimuli, because each stimulus will result in a different profile of responses sent to the brain. Dethier (1971) considers this to be the way in which taste and olfactory stimuli are sensed and integrated in *Manduca sexta* caterpillars, which have only 48 taste receptors on the body and about 78 olfactory receptors. Even greater information coding can be achieved in across-fiber patterning by receptors that have different response latencies, rates of adaptation, after effects, and spontaneous activity (a stimulus may increase or inhibit spontaneous activity).

10.5.4 Stimulus-Receptor Excitation Coupling

How is the energy of a chemical stimulus transformed into the electrical energy of the neuron? Many theories (more than 30 according to Amoore et al., 1964) have been advanced. The **stereochemical theory** has gained the most support, and the isolation of pheromone binding proteins and general odorant binding proteins from insects and other organisms has contributed greatly to solidifying the stereochemical theory.

The stereochemical theory is most often associated with the work of J.E. Amoore (for review, see Amoore et al., 1964), although many others have contributed data to support the theory. Amoore's thesis is that the sense of smell (in humans) is based on the geometry of molecules, and he associated the seven primary odors (camphoraceous, musky, floral, pepperminty, ethereal, pungent, and putrid) with molecules having a particular shape. For example, molecules may be round, oblong, kite shaped, have either a positive charge (pungent) or a negative charge (putrid). The receptor site at the dendritic nerve endings should have a complementary shape or charge. Support for the stereochemical theory of odor perception has come from the study of some pheromones that exhibit chirality, and certain highly purified enantiomeric compounds, such as *R*-(–)-carvone and *S*-(+)-carvone. These are the organoleptic compounds in oil of spearmint and oil of caraway, respectively. The two compounds have the same molecular formula and thus are isomers of each other. Their mirror images are not superimposable upon each other, however, and they are called **enantiomers** of each other (see Chapter 14 for more details on enantiomers). The enantiomeric compounds *R*-(+)-limonene and *S*-(–)-limonene have the odors of orange and lemon, respectively, to humans; and *S*-(+)-amphetamine and *R*-(–)-amphetamine smell fecal and musty, respectively. Not all enantiomeric compounds, however, have distinctively different odors to humans.

The stereochemical theory proposed that a receptor site shaped for one of these molecules would not allow the opposite enantiomer to fit. If it smelled differently, it was because that enantiomer fit another receptor site. Further support for the stereochemical theory comes from the actual isolation of some of the receptor molecules (proteins) at the receptor site in insect antennae (reviewed by Breer, 1997; Prestwich and Du, 1997).

10.6 LIGHT AND VISUAL RECEPTORS

Insects often have several types of light sensitive organs. The **compound eyes** allow insects to detect color and they are believed to allow formation of visual images. Compound eyes are especially sensitive motion detectors. **Ocelli** found on both immature and adults of some insects form images but the images are focused below the level of the photosensitive cells. Ocelli probably function primarily to detect the quality and intensity of light, and its presence or absence, but generally very little detail is known about ocellar function. The larval eyes of holometabolous insects are called **stemmata**. Some of them form images, but they likely are poorly resolved. The immatures of hemimetabolous insects have compound eyes similar to the adult compound eyes. The overall structure of compound eyes, ocelli, and stemmata follow the same basic cellular plan, with a light sensitive area, the **rhabdomere**, represented in some form in all. Some insects, particularly dipterous larvae, have cuticular light-sensitive cells without a rhabdomere region located in the cuticle in various parts of the body. Color vision, form discrimination, and detection of plane-polarized light are well developed in some insects and play major roles in their behavior.

10.6.1 COMPOUND EYE STRUCTURE

The compound eyes are composed of multiple functional units called **ommatidia** (singular = **ommatidium**). Each ommatidium is composed of many cells and of functional parts that include the dioptric structures, the photosensitive cells containing the photosensitive pigments, and shielding cells which usually also contain a variety of pigments. The small eyes of Thysanura contain only a few ommatidia (12 in *Lepisma* spp.), while the very large eyes of dragonflies contain as many as 10,000 ommatidia. Adult Collembola, Lepismatidae, Aphaniptera, and Strepsiptera do not have compound eyes, but instead have simple eyes similar to ocelli.

As in most other aspects of insect structure and biology, there is considerable diversity in the structural detail of ommatidia in different insect groups. Two major variations in structure are represented in Figure 10.10, the **photopic** eye of a dipteran; and in Figure 10.11, the **scotopic** eye of a moth. The terms photopic and scotopic are recommended by Goldsmith and Bernard (1974) as replacements for the older usage of apposition and superposition, respectively. Photopic eyes occur in diurnal insects, which are active during the day. The rhabdom extends from the cone to the basement membrane at the proximal limit of the eye. There is little or no movement of the pigment in the shielding cells around the retinular cells. Light can strike the photosensitive pigment in the rhabdom only by entering axially (i.e., directly from above after passing through the cornea and cone of the ommatidium). The pigment-containing cells surrounding the retinular cells effectively shield the rhabdom from any light straying from adjacent ommatidia. An image is focused on the rhabdom or rhabdomeres of retinular cells just below the level of the cone. Scotopic eyes occur in nocturnal and crepuscular insects. The rhabdom is shorter than in photopic eyes and usually extends only about one third the distance from the basement membrane to the cone. The remaining distance to the cone may be filled with thin strands of the retinular cells forming the crystalline tract (Figure 10.11). The crystalline tract is surrounded by pigment cells that allow the pigment to contract into the distal part of the cells (i.e., near the cone) in dim light. This allows light from adjacent ommatidia to pass through and strike the rhabdom below. Pigment disperses throughout the shielding cells upon exposure to bright light or in response to higher environmental temperatures (Nordström and Warrant, 2000). This anatomical type of eye came to be called a superposition eye because it could be demonstrated that it was capable of superimposing images (although likely not in perfect focus) from the visual field of several facets upon a single location. In light-adapted

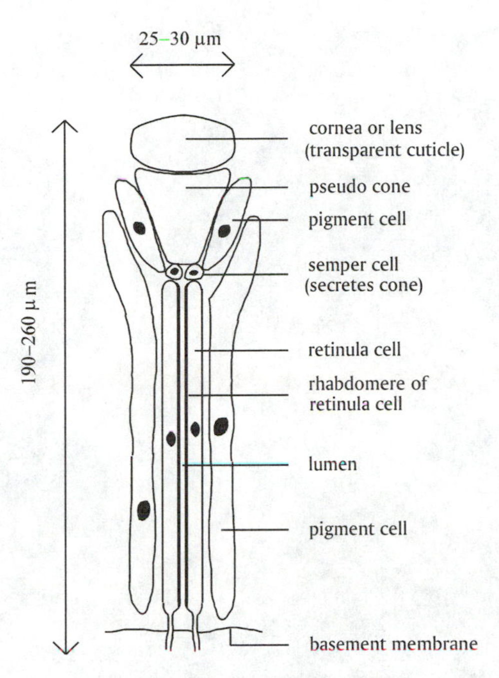

FIGURE 10.10 Diagrammatic representation of the structure of an ommatidium in the photopic compound eye of the tephritid *Anastrepha suspensa.* (Modified from Agee et al., 1977.)

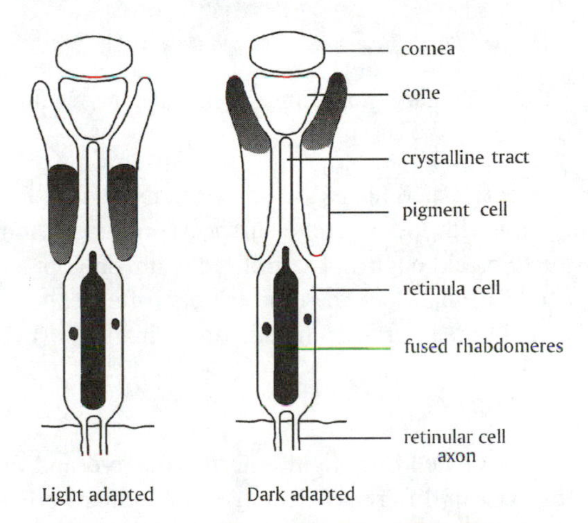

FIGURE 10.11 Diagrammatic illustration of the structure of an ommatidium in the scotopic eye of a nocturnal moth to show shielding pigment distribution in a light-adapted ommatidium and its distribution in a dark-adapted ommatidium. In the light-adapted eye, the dispersed pigment offers protection against light escaping from neighboring ommatidia, while in the dark-adapted state, migration of pigment to the periphery of the ommatidium allows the potential for light to enter from adjacent ommatidia. The latter condition may make for less sharp visual images, but probably allows better visual responses in dim light, which is likely an adaptation for night-flying moths.

scotopic eyes, the shielding pigment migrates throughout the pigment cells and tends to block light from adjacent facets (Figure 10.11). Thus, the superposition eye can then form an apposition image from the light coming into each ommatidium through its own cornea and cone. Migration of the pigment in bright light so as to shield the rhabdomeres may be a protective mechanism to prevent bright light

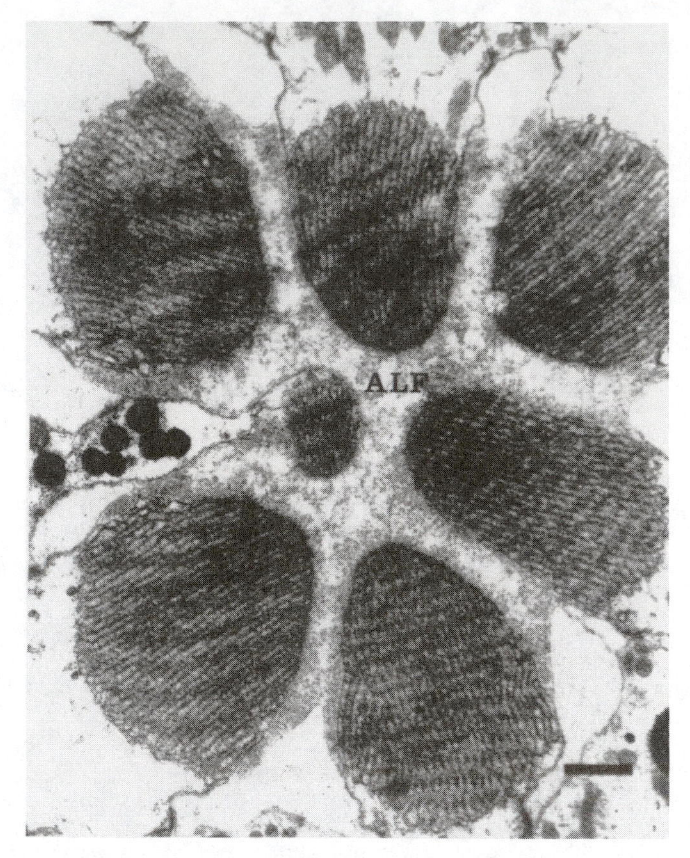

FIGURE 10.12 A transmission electron micrograph of the open rhabdom of a dipteran. (Photograph courtesy of Clay Smith.)

from bleaching the visual pigment. Experiments and observations suggest that the scotopic eye allows increased sensitivity in dim light with loss of some sharpness in image formation. Another morphological variation in eye structure, although not correlated with photopic or scotopic eyes, is that rhabdomeres may be fused into a **rhabdom** in some insects (often called a closed rhabdom, as in honeybee; see Figure 10.21) and remain unfused (called open) in others (Figure 10.12).

10.6.1.1 The Dioptric Structures

The **dioptric structures** refract or bend the light entering the eye and allow it to be focused. A variety of dioptric structures exist in different insect eyes, including cornea, cone, corneal nipples, crystalline tract, and layers of different density in one or more of these structures. All compound eyes and ocelli have a corneal covering that transmits and refracts the light passing through. The cornea is composed of transparent cuticle secreted by corneagenous cells (highly specialized and modified epidermal cells). To allow effective vision, the cornea must transmit a large part of the light striking it (Figure 10.13). About 90% of the light between 400 and 650 nm is transmitted by the cornea of *Manduca sexta.* Wavelengths shorter than 350 nm are very strongly absorbed within the dioptric apparatus of *M. sexta*, and virtually no wavelengths shorter than 300 nm are transmitted.

The cone is another part of the dioptric apparatus in most insects. The light transmission characteristics of the crystalline cone in *M. sexta* are similar to the properties of the cornea. The cone in insects is also very variable in structure and is formed in several ways. In general, there are four cells, sometimes called Semper's cells, that form the cone. In insects without a cone (acone eyes), the cone cells themselves are transparent, but they are not much modified in shape nor do

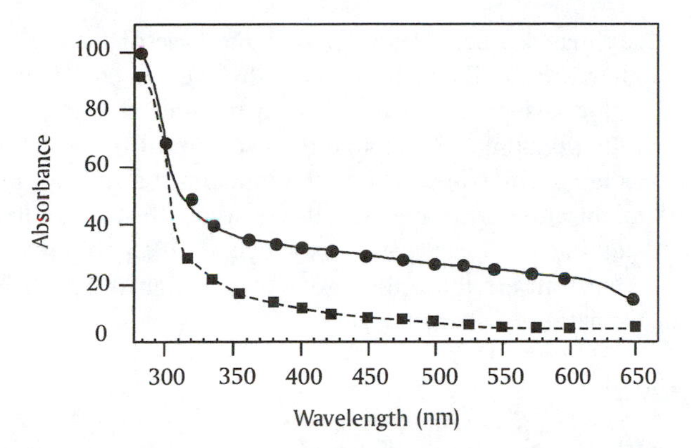

FIGURE 10.13 Variation in the wavelengths of light transmitted by a 20-μm section of the lens (filled squares) and that of a similar thickness of crystalline cone (filled circles) from axial illumination of the compound eye of the moth *Manduca sexta.* (Reproduced with permission from Carlson and Philipson, 1972.)

they secrete crystalline products as in insects with cones in the eye. Acone eyes are considered the most primitive type, and they occur in apterygote insects, and also in some Hemiptera, Coleoptera, and Diptera. Eucone eyes are common and occur in most orders of insects. In these, the cone cells contain a clear, hard intracellular secretion that fills most of the cell volume. The remaining small shell of living cytoplasm in the cone cells becomes pushed to the margin of the cells.

Pseudocone eyes are common in many Diptera, including the Cyclorrhapha (higher Diptera such as the housefly and drosophilid and tephritid fruit flies), and in some Odonata. The pseudocone is a cavity below the cornea filled with a gelatinous or liquid secretion. The cytoplasm of the four cone cells is squeezed into a thin layer beneath the pseudocone.

Exocone eyes occur in some Coleoptera, most notably in fireflies and related beetles. Exocone eyes are not considered homologous to any of the preceding three types. The cone is formed by an inward projection of the cornea, and thus it is cuticular in structure. The four Semper's cells form a short crystalline tract below the exocone, but they secrete no true crystalline core.

The compound eyes of some insects contain additional dioptric components much smaller in dimension than the cone and cornea. These components include (1) a crystalline tract, (2) corneal nipples, (3) corneal layering, and (4) periodic layering of tracheolar structures, commonly referred to as a tapetum (Miller et al., 1968).

Crystalline tracts are found in some butterflies, moths, and in fireflies. The diameter of a tract is one important parameter that determines whether the tract can transmit an image to the rhabdom below. The diameter varies from 2 μm in butterflies to 4 μm in *M. sexta*, to 10 μm in some fireflies. At least in the moths and butterflies, the diameters measured are considered to be too small to allow an image to be transmitted (Miller et al., 1968). Other factors are also important to image transmission and no studies of this nature have been reported from fireflies. Nevertheless, the tracts of moths and butterflies can transmit light to the rhabdom by functioning as a light guide if the refractive index of the tract is greater than that of the surrounding medium so that the light is kept inside the tract and not diffused to the surrounding tissue. The presence of shielding, pigment-filled cells is important is this respect. Finally, to be effective, the light will have to be brought to focus on the rhabdom. On the basis of both theoretical and experimental data, Miller et al. (1968) concluded that the tracts act as light guides in dark-adapted eyes of *M. sexta*, *Elpinor* spp. (a sphingid moth), *Cecropia* and *Polyphemus* (silk moths), and a skipper, *Hobomok* spp.

Corneal nipples are present at the corneal surface of the eyes of many nocturnal Lepidoptera. The nipples probably function as antireflection devices, which in turn should increase the transmission of light into the eye, possibly by as much as 5%. The nipples might also benefit the insect by reducing the reflection of light to predators and parasites, and by reducing or preventing internal reflection of light from the tapetum below the eye.

The tapetum is a thick mat of tracheoles at or near the base of the eye. The shiny surface of the tracheoles reflects light back into the upper parts of the eye and creates eye shine in moths and some other insects. The glow disappears in moths with light-adapted eyes because the shielding pigments move down in the pigment cells enough to absorb most of the light. Very little detailed work has been done on the specific functions of the tapetum, but it may increase sensitivity of dark-adapted eyes and might allow increased sensitivity to contrast patterns (Miller et al., 1968). The reflection of light back into the distal parts of the eye are likely to create blurred images, but insects that have a large tapetum are those that are active at night when the light is dim anyway and not best suited for formation of sharp images.

10.6.1.2 Corneal Layering

The eyes of some dipterans, especially the Tabanidae, show patterns of bright and dark stripes, frequently in colors, in the cornea of the compound eyes. The pattern and colors are caused by the reflection of light from dense (refractive index 1.74) and thin layers (refractive index 1.40) in the corneal cuticle. Deerflies have as many as 20 layers in the cornea. Miller et al. (1968) suggested that the layers might serve as color contrast filters, enabling fast-flying flies to locate more easily a host against the background.

10.6.1.3 Retinular Cells

The receptor neurons in the eyes are called **retinular cells**. They are primary receptors (in contrast to secondary ones) because they send their axons directly into the optic lobe of the brain without synapsing. Eight is the most common number of retinular cells in each ommatidium, arranged in a circle. The number is variable in different insects, however, and ten or eleven have been found in some Lepidoptera, nine in honeybees, four in the dorsal part of the eye of adult dragonflies, and six in the ventral part, and eight in dragonfly nymphs. In the most common situation, the eighth cell, or in honeybees the ninth cell, is short and can be found only in the more proximal region of an ommatidium. The retinular cells are long slender cells extending over the greater part of the length of an ommatidium. They are 60 to 100 μm long in *D. melanogaster* (Wolken, 1957), about 100 μm long in *P. americana* (Wolken and Gupta, 1961), and from 150 to >200 μm long in *Anastrepha suspensa* (Agee et al., 1977).

10.6.1.4 Rhabdomeres

Each retinular cell contains a specialized region called the **rhabdomere** (Figures 10.10, 10.11, and 10.21) which is the site of the light-sensitive pigment. Perpendicular to the long axis of the retinular cell, the cell membrane is extended into thousands of microvilli, typically about 40 to 120 nm in diameter along all or part of the length of the cell. These microvilli make up the rhabdomere. In *D. melanogaster* there are about 60,000 microvilli per cell. Wolken (1968) estimated that there are about 80,000 tubules in each rhabdomere of *P. americana.* The microvilli are the site of some 100,000,000 molecules of rhodopsin per cell (Zuker, 1996). A rhabdomere extends over most of the length of a retinular cell, and is usually oriented toward the circle formed by the group of retinular cells in an ommatidium. The diameter of the rhabdomere is about 1.2 μm in *D. melanogaster* (Wolken, 1957) and about 2 μm in *P. americana* (Wolken and Gupta, 1961). Rhabdomeres fuse at the center of the circle to form a closed or fused rhabdom in honeybees, Lepidoptera, and in many other insects. Alternatively, the rhabdomeres may remain free and face a central hollow cylinder as in some Diptera and Hemiptera. Electron microscope studies reveal that extracellular spaces exist within even the tightest, closed rhabdoms. In compound eyes, the microvilli typically are oriented perpendicular to the long axis of the retinular cell; but in simpler eyes of some arthropods, the direction of microvilli is in line with the long axis of the retinular cell (Phillis and Cromroy, 1977).

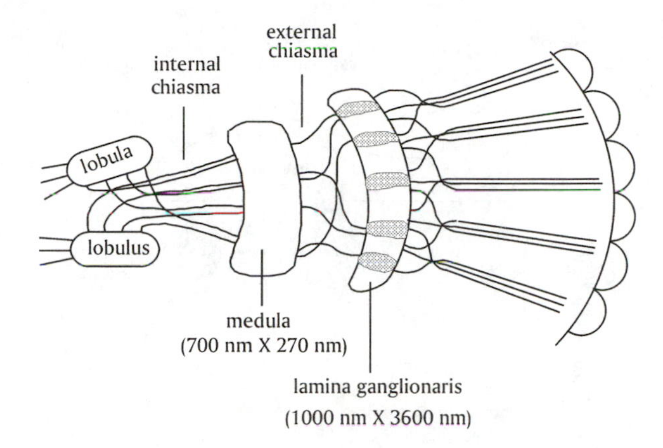

FIGURE 10.14 Basic structure of the optic lobe, illustrating the four major optic neuropils and neural cartridges in the most distal neuropil, the lamina ganglionaris.

10.6.1.5 Electrical Activity of Retinular Cells

Resting potentials ranging from 25 to 70 mv have been recorded across the membranes of retinular cells. The inside of the cell is negative relative to the outside during resting conditions in the dark. In contrast to the polarizing effect of light on vertebrate eye receptors, light acts as a depolarizing stimulus for retinular cells in the compound eye. An electroretinogram (ERG) of electrical activity in response to stimuli can be recorded by placing one electrode on or into the eye and the reference electrode somewhere else in the head. An ERG is a summation of potentials from many retinular cells, and possibly of electrical activity within the optic lobe. Agee (1977) described techniques and equipment for ERG measurements.

Illumination of the eye causes a slow or graded potential that increases with the intensity of the stimulus. The slow potential is the receptor potential and shows the typical characteristics of a graded potential. At higher intensities of illumination, there appears a transient "on" component that has a sharp threshold and a refractory period. Moreover, the "on" component can be abolished by high external K^+ ion concentration. Thus, the "on" component has the characteristics of a spike (action potential). Fuortes (1963) showed that both the transient response and the receptor potential (measured 900 ms after initiation of stimulus) are linear functions of the logarithm of light intensity. The transient "on" potential is a compound action potential of many individual axon responses. Increased illumination brings more and more axons into play and the "on" transient becomes larger.

10.6.2 Neural Connections in the Optic Lobe

The optic lobe of the brain contains three large neuropils (Figure 10.14). The most distal one, the **lamina ganglionaris**, is the site of the first synaptic junctions of retinular cell axons. The lamina is composed of groups of large monopolar neurons, association neurons, and glial cells. In dipterans such as *Musca* and *Calliphora*, which have received detailed **lamina** study, groups of neurons form **neural cartridges** or **neuroommatidia** in the lamina ganglionaris. There are 3000 cartridges in each lobe corresponding to the 3000 ommatidia in the eye sending axons into the cartridges for synaptic connections (Braitenberg, 1972). Within each cartridge of *Musca* and *Calliphora,* there are four monopolar neurons, six branched α- and β-neurons (centrifugal neurons coming from the medulla to the lamina), and six axons coming from retinular cells in six different ommatidia (Figure 10.14). The axons from the retinular cells decussate (cross) before entering the lamina. The retinular neurons synapse with monopolar neurons. Some retinular axons pass through the lamina without synapsing (Boschek, 1971).

Fibers leave the lamina, decussate, and enter the second optic ganglion, the **medulla**. The network of fibers between the lamina and medulla is called the **external chiasma**. Histological evidence

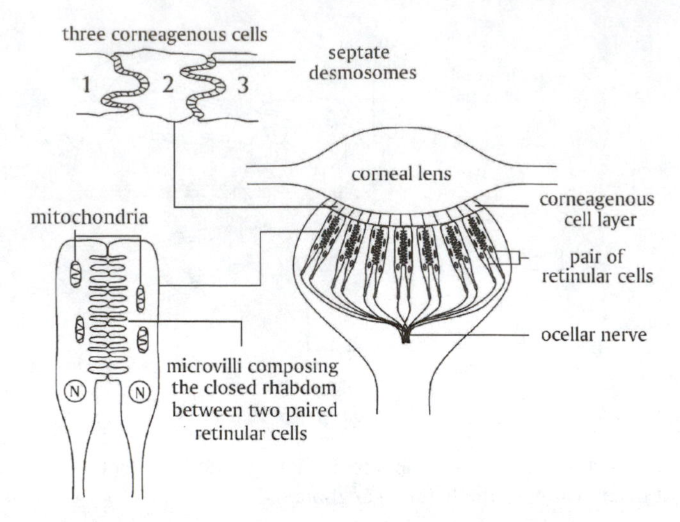

FIGURE 10.15 Diagram of the structure of an ocellus from a worker honeybee. (Modified from Toh and Kuwabara, 1974.)

indicates that within the medulla of *Musca* and *Calliphora* there are regular arrays of repeating units, receiving the input from the lamina (Braitenberg, 1972), but the anatomy of these units has not been studied in detail. Fibers leave the medulla, again decussate, and enter the **lobuli**, the third ganglionic mass. The lobuli consist of several, sometimes separated, masses of cell bodies and neuropil. In *Musca,* there are two masses known as the **lobula** and **lobulus**. Regular repeating units have not been found in the lobuli, and little is known about synaptic connections or input:output interactions.

The large amount of crossing of fibers in the optic lobe of insects is not peculiar to insects, but it is a general feature of visual systems. Its function is not clearly understood, but it would appear to provide a great deal of backup security if only some parts of either the external eye or brain receiving the input suffered damage.

10.6.3 Ocelli

Ocelli have some of the same anatomical features as compound eyes, including a corneal lens, corneagenous cells that secrete the lens, retinular cells, and an ocellar nerve that leads into the protocerebrum (Figure 10.15). The retinular cells typically contain a rhabdomere region with microvilli. The corneal cuticle is probably transparent to various wavelengths of light, but few measurements have been made. Goldsmith and Ruck (1958) showed that the cornea from ocelli of *Blaberus craniifer* (a cockroach) was transparent to light from about 350 to 700 nm.

The cornea of an ocellus covers a large visual field and appears to form an image, but the image is focused beneath the layer of retinular cells and is not transmitted to the brain. The behavior of insects that have the compound eyes covered, but with functional ocelli, further indicates that ocelli do not convey an image to the brain. The retinular axons from ocelli are spontaneously active in the dark. Rather than depolarizing the ocellar retinular cells, light leads to a more stable or increased membrane potential, and hence to a decrease or even cessation in spike activity sent into the brain. The ocelli can signal light on/off information, intensity of illumination, and possibly in some insects may indicate the quality (wavelength) of light.

Ocelli in the cockroach *P. americana* contain a photosensitive pigment that has maximum sensitivity to light of 500-nm wavelength, while ocelli from honeybees contain receptors that show maximum sensitivity to light of 340 and 490 nm, probably indicative of two photosensitive pigments (Goldsmith and Ruck, 1958). Rhabdomere structure of the ocellar retinular cells is similar to that in compound eyes. Honeybees have three ocelli and each contains about 800 retinular cells whose short axons converge on about eight second-order neurons (Toh and Kuwabara, 1974). Thus, the convergence ratio

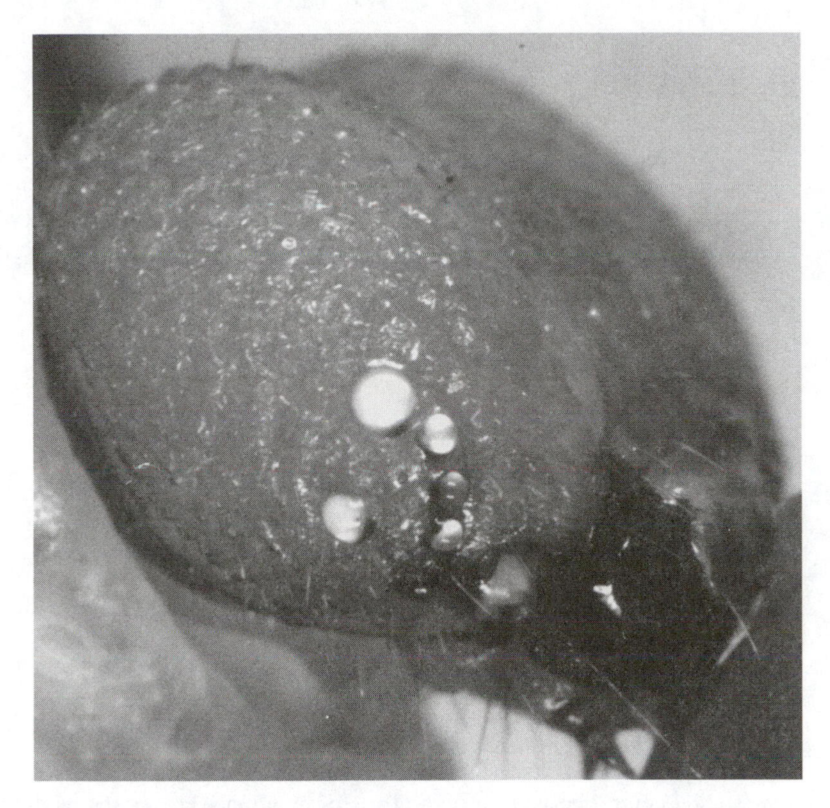

FIGURE 10.16 A photograph of the six stemmata (the large white circle and smaller ones nearby) on the head of the caterpillar of *Achlyodes mithridates*, a butterfly in the family Hesperiidae. (Photograph courtesy of Andrei Sourakov.)

is about 100 retinular neurons to 1 second-order neuron. The second-order neurons from the three ocelli converge and form the ocellar nerve projecting to the posterior of the protocerebrum.

10.6.4 Larval Eyes: Stemmata

The eyes of larvae of holometabolous insects are usually called stemmata. The number of stemmata varies from one in some insects to as many as six on each side of the head in Lepidoptera (Figure 10.16). Their general organization varies considerably among insect groups, but the most well-developed stemmata have an overlying transparent cuticle, a crystalline lens, and a few retinular cells with rhabdomere regions. Some stemmata do not have the crystalline lens. In some insect larvae, the stemmata have two separate rhabdoms: a distal one nearest the overlying cornea and a proximal one below. Each of these rhabdoms is formed by microvillar extensions from only a few cells, and although the dioptric apparatus forms an image that falls on the rhabdomere surfaces, it seems likely that resolution of any image is poor.

Caterpillars frequently move the head from side to side, which may be a behavior that aids them in obtaining a wider field of view with small, multiple stemmata. Some sawfly larvae and tortricid caterpillars have been reported to sense plane-polarized light, but it is difficult to see how this could be adaptive to these larval insects. This, if true, needs to be verified and studied in more detail.

10.6.5 Dermal Light Sense

There are small pockets of photosensitive cells in the cuticle of larvae of Diptera, which have no obvious eyes on the surface of the body. These cup-shaped pockets contain a few cells whose axons

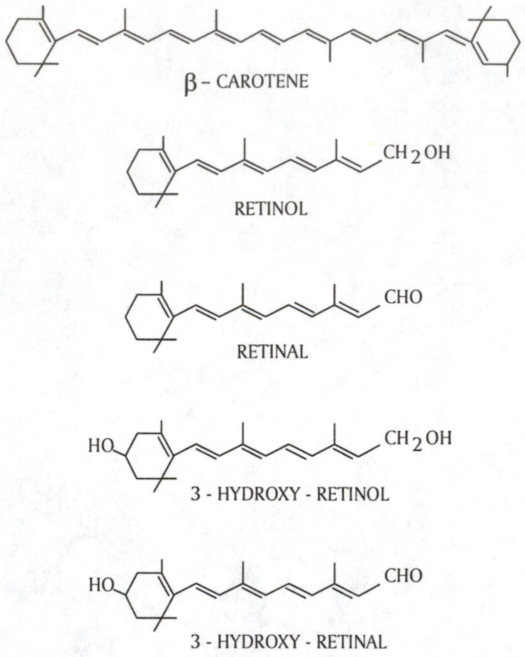

FIGURE 10.17 The chemical structure of the chromophores retinal and 3-hydroxy-retinal found in insect eyes, and β-carotene from which the compounds are derived.

run into the short, stubby central nervous system. Tests of avoidance responses made with monochromatic light indicate that 540-nm light is most effective in stimulating the photosensitive cells. The larvae try to avoid the light, and attempt to crawl into the food or under other debris or cover. Many insect larvae can respond to light when the eyes are blackened or otherwise occluded, but very little information is available on the possible receptors involved.

10.6.6 Chemistry of Insect Vision

The first event leading to a visual image is a photochemical reaction. Light quanta are absorbed by the pigment **rhodopsin** in rhabdomere microvilli. This leads to a chemical change in rhodopsin, which in turn leads to the opening of Na^+ and Ca^{2+} channels, depolarization, and electrical activity in the axon of the retinular cell. Thus, light energy is transduced into chemical energy by the pigment, and then chemical energy is transduced into electrical energy by the neuron. Retinular cells are sensitive to one photon of light.

Rhodopsin is the name for the visual pigment in light-sensitive cells of all animals. It is composed of a protein, **opsin**, and a chromophore, **retinal**, or a closely related molecule. In insect eyes the chromophore (Figure 10.17) can be **retinal** or **3-hydroxy retinal** (Vogt, 1983; Vogt and Kirschfeld, 1984; Smith and Goldsmith, 1990). Opsin is composed of varying numbers and sequences of amino acids. Species specificity and wavelength of light to which rhodopsin is maximally sensitive are dependent on both opsin and chromophore structure (Zhukovsky and Oprian, 1989). Rhodopsin is usually characterized on the basis of its maximum spectral sensitivity, for example, as a UV-sensitive rhodopsin, or a blue-green sensitive rhodopsin, etc. Rhodopsin is one of a large family of **G protein-coupled receptors** involved with ligand binding and message transduction at the cell membrane. **G proteins** are intracellular proteins that are part of the second messenger signaling process. A characteristic of G protein-coupled receptors is that they are transmembrane proteins with seven α-helical transmembrane domains (Figure 10.18), and they bind intracellular G proteins when activated by a signal. Rhodopsin binds its ligand, *cis*-11-retinal,

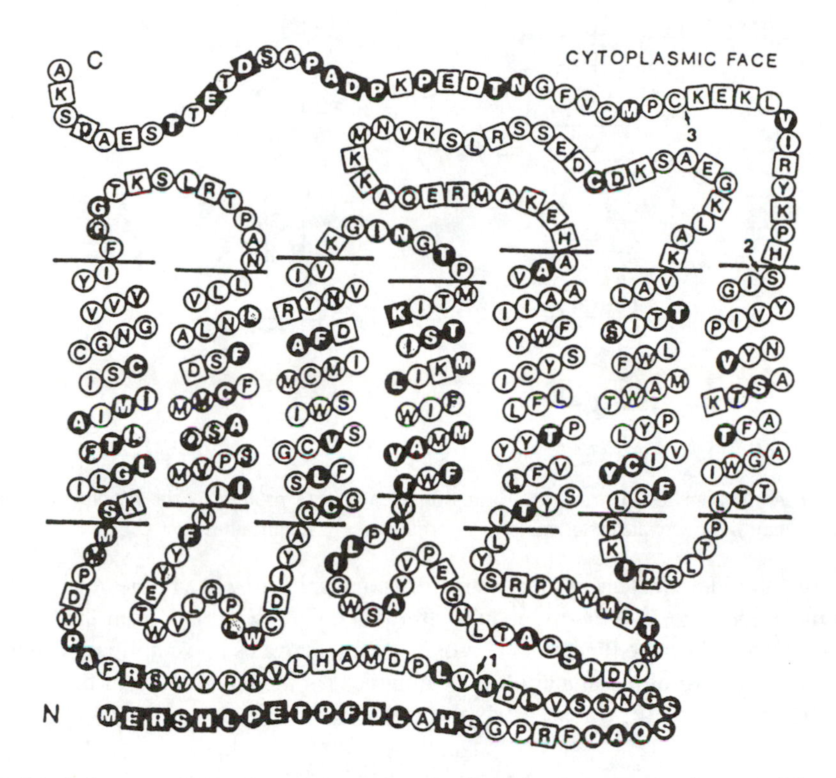

FIGURE 10.18 A diagram of the transmembrane nature of rhodopsin in the membranes of the rhabdomeres of retinular cells. (Reproduced with permission from Mizunami, 1994.)

as a protonated Schiff base, with the ligand forming a covalent bond between the aldehyde group of retinal and the ε-amino group of a lysine residue (Lys-296) in transmembrane-seven of the protein (Stryer, 1975; Dratz and Hargrave, 1983; Zhukovsky et al., 1991). The proton of the Schiff base is counterbalanced by an induced negative charge on a glutamate residue in transmembrane-three (Zhukovsky et al., 1991). The chromophore is bound to the portion of opsin that lies within the plasma membrane of the retinular cell. Amino acid substitutions in opsin, especially in transmembrane-three and transmembrane-six domains, shift the distribution of electrons about the retinal chromophore and within the opsin portion of rhodopsin, and alter the wavelength of maximum light absorption. Both extracellular and intracellular portions of the rhodopsin molecule are important to function. Disulfide bonds in some of the loops of the extracellular portion of the molecule are important in holding the molecule in its characteristic shape and orientation. Portions of the intracellular part of rhodopsin are the sites for G protein recognition and binding. When activated by light, rhodopsin couples mainly with the G protein designated G_t (Smith et al., 1991), although additional G proteins are known with which activated rhodopsin may couple.

10.6.6.1 The Visual Cascade

When a photon of light is absorbed by rhodopsin, **11-*cis*-retinal** is isomerized to **11-*trans*-retinal**. This initiates a conformational change in the protein portion of rhodopsin, and it becomes activated as **metarhodopsin**, a catalyst that activates a cascade of reactions. G protein (also called **transducin**) binds to a loop of the metarhodopsin at the inside surface of the plasma membrane (Figure 10.19). Transducin is composed of three subunits: Tα, Tβ, and Tγ. **Guanosine diphosphate (GDP)** is bound to the Tα subunit as a normal part of its structure. The G protein-metarhodopsin complex is in an activated state, enabling GTP (Guanosine triphosphate) to replace GDP at the Tα

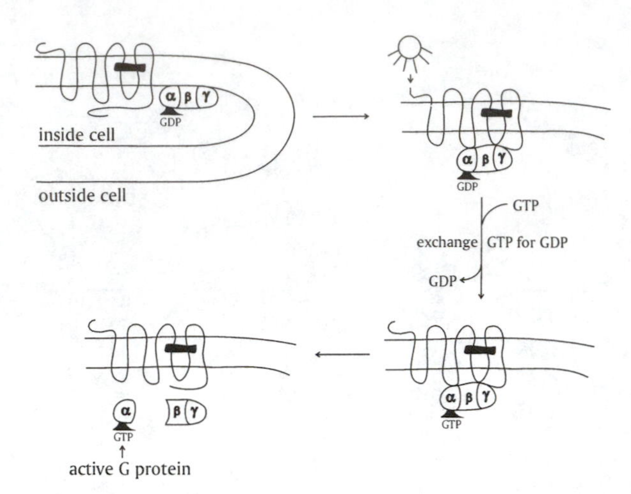

FIGURE 10.19 A diagram to illustrate the formation of active G protein as the second messenger in the visual cascade when light activates rhodopsin in the rhabdomere membranes.

subunit without any additional energy requirement. Transducin, now in a higher energy state, breaks away from metarhodopsin, and the α-subunit with bound GTP separates from the β-γ subunits.

The Tα-GTP protein is the functional G protein, and it activates phospholipase C which cleaves PIP_2 (**phosphatidyl inositol bisphosphate**), a component in the membranes of the microvilli, into DAG (**diacylglycerol**) and **IP_3 (inositol trisphosphate)**:

$$PIP_2 \xrightarrow{\text{Active phospholipase C}} DAG + IP_3$$

Inositol trisphosphate (IP_3) promotes the release of calcium bound to smooth endoplasmic reticulum near the base of the microvilli in retinular cells (Spiegel et al., 1994). Free ionic calcium (Ca^{2+}), opens sodium channels that also admit Ca^{2+} in the base of the microvilli. The inward movement of a few sodium ions generates a receptor (graded) potential (Ranganathan et al., 1991; Hardie, 1991; Hardie and Minke, 1994), and entry of more Ca^{2+} acts as an amplifying mechanism, resulting in more sodium entry. If the receptor potential is strong enough, it may give rise to spikes in the retinular axons. Liberated DAG activates protein kinase C, but its role in the visual process is not clear. It may be involved in the adaptation process in the eye. The β-γ subunits may also have some role in signal transduction, but that role, if any, also remains unclear.

This signaling pathway is extremely fast, and can be turned on and off, and repeat, many times per second. The cascade in *D. melanogaster* is the fastest G protein cascade that has been measured thus far, and it takes only several tens of milliseconds to proceed from light activation of rhodopsin to the generation of a receptor potential (Zuker, 1996). The enzymatic character of the visual cascade at several steps results in amplification of the signal by as much as 10^2 to 10^3 times. Each metarhodopsin can catalyze GTP binding to as many as several hundred Tα subunits before it is deactivated (Yarfitz and Hurley, 1994), and similar amplification occurs in production of IP_3 and in release of Ca^{2+}. Thus, one photon light signal is turned into a barrage of neuronal impulses going into the lamina ganglionaris of the optic lobe.

10.6.6.2 Regulation of the Visual Cascade

How is the visual cascade turned off? One way is that insect metarhodopsin can absorb another photon of light and the metarhodopsin/11-*trans*-retinal complex is reconverted to rhodopsin with 11-*cis*-retinal, a process called **photoisomerization** (Smith and Goldsmith, 1991). This biochemical change terminates the visual response and allows the retinular cell to recover for the next event. In contrast, the metarhodopsin of vertebrate eyes loses its retinal chromophore upon absorption of a photon of light, and a new rhodopsin molecule must be synthesized.

A second regulatory mechanism involves the binding of an inhibitory protein, called **arrestin** (**ARR**), to the metarhodopsin (MET). The metarhodopsin-arrestin complex (MET-ARR) will no longer bind to G proteins, so the amplification cascade is prevented or stopped upon binding of arrestin. **Rhodopsin kinase** phosphorylates the MET-ARR complex (at the expense of ATP) to give a MET-PO_4-ARR complex. Absorption of a photon of light by the MET-ARR-PO_4 complex causes release of ARR, and a phosphatase removes the phosphate to regenerate metarhodopsin. There are also less well-known regulatory processes within the G protein cascade, and in activity of phospholipase C.

Although constant resynthesis of 11-*cis*-retinal and rhodopsin after each visual stimulus is not necessary in insect compound eyes because of photoisomerization of metarhodopsin to rhodopsin, visual pigment molecules do wear out and new ones must be synthesized by insects from time to time. Some insects also shed the rhabdomere portions of the retinular cells from time to time and thus have to synthesize new microvilli and rhodopsin. To synthesize new rhodopsin, 11-*trans*-retinol obtained from β-carotene must be enzymatically converted to 11-*trans*-retinal. A photoisomerase enzyme in eye tissues of honeybees (Smith and Goldsmith, 1990) can be activated by absorbing a photon of light, and it then catalyzes the isomerization of 11-*trans*-retinal to 11-*cis*-retinal. In the dark part of the eye, 11-*cis*-retinal is enzymatically combined with opsin to produce rhodopsin.

10.6.6.3 Color Vision

Color vision, the ability to discriminate between two wavelengths of light, has been reviewed by Briscoe and Chittka (2001). In insects, the test for color vision requires some kind of behavioral action that demonstrates that the insects have indeed discriminated between the two colors. Such careful and thorough work has been done with only a few insects, including honeybees, a few dipterans, an aphid, and possibly a few other insects. Nevertheless, many insects in many different orders have been tested for spectral sensitivity, with the result that receptors sensitive to one or more colors exist in a wide array of insects (Briscoe and Chittka, 2001).

Many, and perhaps most, insects show color sensitivity, which means that they have two or more receptors in the eye that show absorption of various wavelengths of light. Sensitivity to, for example, blue and green wavelengths does not mean that the brain receiving this information orders behavior based on the information received. Autrum and Zwehl (1962) used recorded receptor potentials from single visual cells of honeybee eyes exposed to monochromatic light between 318 and 650 nm to obtain spectral sensitivity curves suggesting four receptor pigments in honeybee eyes, each showing a different absorption maximum. The relative sensitivity of honeybee eyes to different wavelengths is UV > blue-violet > green > yellow. Carlson and Philipson (1972) suggested the presence of four pigments in the eyes of *M. sexta* adults, based on difference spectra obtained from bleaching experiments. The four maxima for light absorption are at approximately 350 nm (UV), 450 nm (blue-green), 490 nm (green), and 530 nm (yellow). After the eyes were bleached (exposed to strong light) at 21°C, pH 7.4 to 8.5, each of the last three pigments absorbed light at 370 nm, indicative of a mixture of bound and free 11-*cis*-retinal (i.e., the rhodopsin had been split into the protein portion and the aldehyde). A small amount of a pigment absorbing maximally at 325 to 330 nm, and possibly indicative of retinol (vitamin A), was also detected. The UV-absorbing visual pigment (wavelength maximum at 350 nm) bleached to an unknown product with an absorption maximum at 290 to 300 nm. Eisner et al. (1969) demonstrated that many flowers have UV reflecting patterns that may be important signals to insects, many of which are sensitive to UV wavelengths. Some butterflies may be among the few insects that have red-sensitive rhodopsins in their eyes (Swihart, 1967; Bernard, 1979).

Little is known of the location of the visual pigments within the rhabdomeres, or whether more than one pigment might exist in the same ommatidium. Mote and Goldsmith (1970) believe on the basis of intracellular recordings and dye-marked sites they recorded electrical activity from both a UV-sensitive retinular cell and a green-sensitive retinular cell within the same ommatidium of *P. americana*. The UV receptor was maximally sensitive at 365 nm while, the green receptor was maximally sensitive at 507 nm. Such experiments are technically difficult because of the small size of the retinular cells and because the location of the recording electrode can be determined only by histological determination of dye location after the recording has been made and the experiment is finished.

Menzel (1975) studied the spectral properties of eyes of *Formica polyctena*, the red wood ant in Europe, by taking advantage of the discovery that pigment migration in the retinular cells was light sensitive. In a fully dark-adapted eye (12 h darkness), the pigment was dispersed away from the rhabdom, while light adaptation (Xenon light at 40,000 lux) caused the pigments to migrate toward the rhabdom (possibly adaptive as a shield against excessive light striking the rhabdom). Evaluations of pigment movements were made from cross section through ommatidia viewed by electron microscopy. Each ommatidium in the central area of the eye contained two cells sensitive to UV (cells 1 and 5), and six cells sensitive to yellow light (cells 2, 3, 4, 6, 7, and 8). Although the UV-detecting cells were smaller, the sensitivity to UV was 20 times that of yellow light.

10.6.7 Nutritional Need for Carotenoids for Good Vision in Insects

All animals, including some insects, that have been critically studied require **β-carotene** or **vitamin A**, or closely related carotenoids, in the diet for normal vision. These carotenoids are readily synthesized by plants, but not by animals. There was a loss in visual sensitivity of 2.4 to 3.0 log units in electrophysiological tests of *Musca domestica* (Goldsmith et al., 1964) and *D. melanogaster* (Zimmerman and Goldsmith, 1971) reared on β-carotene-free diets. Because the insects required so little carotene to satisfy their needs, the diet had to be very highly purified with respect to carotenoid pigments. In some cases, more than one generation had to be reared on the purified diet to exhaust the carotenoid pigments, which were passed through the eggs, in the body of first or second generations. Histological changes in retinular cell structure and in underlying nervous tissue were demonstrated in *M. sexta* (Carlson et al., 1967) and in *Aedes aegypti* (Brammer and White, 1969) reared on carotenoid-free diets.

10.6.8 Detection of Plane-Polarized Light by Insects

Many invertebrates are able to detect **plane-polarized** light in the sky. Light may be thought of in terms of particles or waves. In considering polarized light, it is best to mentally form a picture of light as a sine wave. Most of the light from the sun is not polarized, and the waves vibrate in every conceivable direction. A small percentage of the light becomes polarized by various molecules and small particles encountered in the atmosphere. These polarized waves all vibrate in a specific plane. Both plane of vibration and degree of polarization (as observed by an instrument or animal on earth) vary with the position of the sun above the horizon (Figure 10.20) (Wehner, 1976). The direction of polarization is parallel to the horizon (only) along the path the sun takes toward the zenith and its path as it sinks in the West. At other positions above the horizon, the direction of polarization varies through all possible angles. The directions of polarization are opposite (e.g., +20° and –20°) at points separated by 180°. To further complicate the matter, the angle and degree of polarization vary at each elevation above the horizon. Clearly, using plane-polarized light as a compass is complicated.

In the 1950s, Von Frisch demonstrated that honeybees used plane-polarized light for flight navigation. Houseflies, *Photurus pennsylvanicus* fireflies, Japanese beetles, and several species of ants orient to plane-polarized light under conditions that prevent them from using background reflections as orientation cues. Wehner (1976) concluded that an ant from the North African desert (*Cataglyphis bicolor*) utilized plane-polarized light to travel a direct path to its nest in the ground after having wandered, with many turns, up to 100 meters from its nest in search of food. The cells sensitive to the plane of polarization are the UV-sensitive cells. Wehner determined this by holding

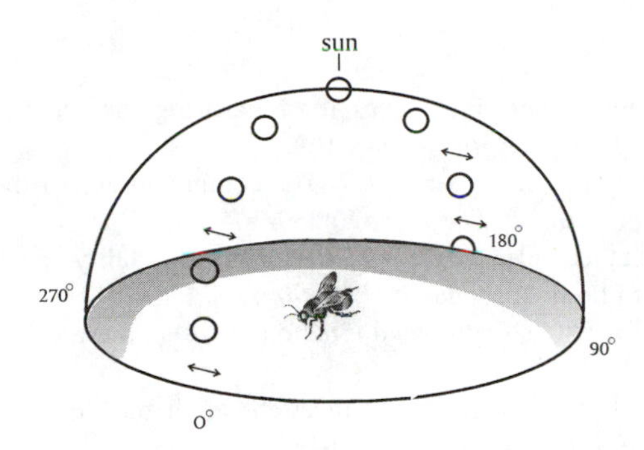

FIGURE 10.20 An illustration of the use of plane-polarized light by honeybees and other insects. (Modified from Wehner, 1976.)

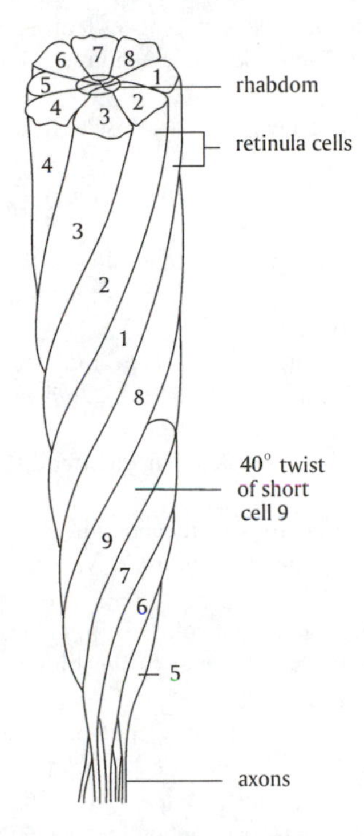

FIGURE 10.21 Ommatidial structure in the eye of the honeybee. About 5500 ommatidia occur in each compound eye. Eight of the retinular cells are elongated, and the ninth is short and confined to the base of the eye. The twisting of the cells may be involved in reception of polarized light; half of the ommatidia are twisted clockwise and half are twisted counterclockwise. The rhabdom is a closed or fused rhabdom in which the rhabdomeres from retinular cells touch each other. (Modified from Wehner, 1976.)

a UV-absorbing shield over ants in the field, thereby causing them to wander aimlessly, unable to locate their underground nest. Menzel (1975) showed that *Formica polyctena*, the red wood ant of northern Europe, uses polarized light as a compass.

The precise way in which these insects determine the plane of polarization and how they measure time lapse (necessary because the plane of polarization changes as the sun moves across the sky) is not known with certainty. Tentative explanations involve the twisting of retinular cells (see Figure 10.21) and the orientation of rhodopsin molecules in the rhabdomere microvilli with respect to the plane of polarization (Wehner et al., 1975; Menzel, 1975; Wehner, 1976).

REFERENCES

Agee, H.R. 1977. Instrumentation and Techniques for Measuring the Quality of Insect Vision with the Electroretinogram. U.S. Dept. Agric. ARS-S-162.

Agee, H.R., W.A. Phillips, and D.L. Chambers. 1977. The compound eye of the Caribbean fruit fly and the apple maggot fly. *Ann. Entomol. Soc. Am.,* 70:359-364.

Altner, H., R. Loftus, L. Schaller-Selzer, and H. Tichy. 1983. Modality-specificity in insect sensilla and multimodal input from body appendages. *Fortschr. Zool.,* 28:17-31.

Altner, H., and R. Loftus. 1985. Ultrastructure and function of insect thermo-and hygroreceptors. *Annu. Rev. Entomol.,* 30:273-296.

Altner, H., and L. Prillinger. 1980. Ultrastructure of invertebrate chemo-, thermo- and hygroreceptors and its functional significance. *Int. Rev. Cytol.,* 67:69-139.

Amoore, J.E., J.W. Johnston, Jr., and M. Rubin. 1964. The stereochemical theory of odor. *Sci. Am.,* 210:42-49.

Apel, K.-H. 1989. Zur Verbreitung von *Melanophila acuminata* DEG (Col., Buprestidae). *Entomol. Nach. Ber.,* 33:278-280.

Autrum, H. 1963. Anatomy and physiology of sound receptors in invertebrates, pp. 412-433, in R.-G. Busnel (Ed.), *Acoustic Behavior of Animals,* Elsevier, Amsterdam, 933 pp.

Autrum, H., and W. Schneider. 1948. Vergleichende Untersuchungen über den Erschütterungssinn der Insekten. *Z. Vergl. Physiol.,* 31:77-88.

Autrum, H., and V. von Zwehl. 1962. Die Sehzellen der Insekten als Analysatoren für polarisiertes Licht. *Z. Vergl. Physiol.,* 46:1-7.

Bailey, W.J. 1991. *Acoustic Behaviour of Insects: An Evolutionary Perspective*, Chapman & Hall, London, 225 pp.

Bernard, G.D. 1979. Red-absorbing visual pigment of butterflies. *Science,* 203:1125-1127.

Boschek, C.B. 1971. On the fine structure of the peripheral retina and lamina ganglionaris of the fly *Musca domestica. Z. Zellforsch.,* 118:369-409.

Braitenberg, V. 1972. I. Anatomy of the Visual System. 1. Periodic structures and structural gradients in the visual ganglia of the fly, pp. 3-15, in R. Wehner (Ed.), *Information Processing in the Visual Systems of Arthropods*, Springer-Verlag, New York, 334 pp.

Brammer, J.D., and R.H. White. 1969. Vitamin A deficiency: Effect on mosquito eye ultrastructure. *Science,* 163:821-823.

Breer, H. 1997. Molecular mechanisms of pheromone reception in insect antennae, pp. 115-130, in R.T. Cardé and A.K. Minks (Eds.), *Insect Pheromone Research: New Directions*, Chapman & Hall, New York, 684 pp.

Briscoe, A.D., and L. Chittka. 2001. The evolution of color vision in insects. *Annu. Rev. Entomol.,* 46:471-510.

Burkhardt, D. 1960. Action potentials in the antennae of the blowfly (*Calliphora erythrocephala*) during mechanical stimulation. *J. Insect Physiol.,* 4:138-145.

Busnel, R.-G. 1963. *Acoustic Behavior of Animals,* Elsevier Publishing Co., Amsterdam, 933 pp.

Carlson, S.D., H.R. Stevens III, J.S. Vandeberg, and W.E. Robbins. 1967. Vitamin A deficiency: Effect on retinal structure of the moth *Manduca sexta. Science,* 158:268-270.

Carlson, S.D., and B. Philipson. 1972. Microspectrophotometry of the dioptric apparatus and compound rhabdom of the moth (*Manduca sexta*) eye. *J. Insect Physiol.,* 18:1721-1731.

Corbière-Tichané, G., and R. Loftus. 1983. Antennal thermal receptors of the cave beetle *Speophyes lucidulus* Delar. *J. Comp. Physiol.,* 153:343-351.

Dethier, V.G. 1955. The physiology and histology of the contact chemoreceptors of the blowfly. *Quart. Rev. Biol.,* 30:348-371.

Dethier, V.G. 1963. *The Physiology of Insect Senses,* Methuen, London, 266 pp.

Dethier, V.G. 1971. A surfeit of stimuli: A paucity of receptors. *Am. Sci.,* 59:706-715.

Dethier, V.G., and A. Gelperin. 1967. Hyperphagia in the blowfly. *J. Exp. Biol.,* 47:191-200.

Dratz, E.A., and P.A. Hargrave. 1983. The structure of rhodopsin and the rod outer segment disk membrane. *Trends Biochem. Sci.,* 8:128-131.

Eisner, T., R.E. Silberglied, D. Aneshansley, J.E. Carrel, and H.C. Howland. 1969. Ultraviolet video-viewing: The television camera as an insect eye. *Science,* 166:1172-1174.

Ewing, A.W. 1989. *Arthropod Bioacoustics: Neurobiology and Behaviour*, Comstock Publishing Associates, Cornell University Press, Ithaca, NY, 260 pp.

Evans, W.G. 1962. Notes on the biology and dispersal of *Melanophila* (Coleoptera: Buprestidae). *Pan-Pac. Entomol.,* 38:59-62.

Evans, W.G. 1966a. Perception of infrared radiation from forest fires by *Melanophila acuminata* De geer (Buprestidae, Coleoptera). *Ecology,* 47:1061-1065.

Evans, W.G. 1966b. Morphology of the infrared sense organ of *Melanophila acuminata* (Buprestidae, Coleoptera). *Ann. Entomol. Soc. Am.,* 59:873-877.

Field, L.H., and T. Matheson. 1998. Chordotonal organs of insects. *Adv. Insect Physiol.,* 27:1-228.

Frazer, J.L. 1985. Nervous system: Sensory system, pp. 287-356, in M.S. Blum (Ed.), *Fundamentals of Insect Physiology,* John Wiley & Sons, New York, 598 pp.

Fullard, J.H., and J.E. Yack. 1993. The evolutionary biology of insect hearing. *Trends Ecol. Evol.,* 8:248-252.

Fuortes, M.G.F. 1963. Visual responses in the eye of the dragon fly. *Science,* 142:69-70.

Gelperin, A. 1967. Stretch receptors in the foregut of a blowfly. *Science,* 157:208-210.

Gilbert, C. 1994. Form and function of stemmata in larvae of holometabolous insects. *Annu. Rev. Entomol.,* 39:323-349.

Goldsmith, T.H., R.J. Barker, and C.F. Cohen. 1964. Sensitivity of visual receptors of carotenoid-depleted flies: A vitamin A deficiency in an invertebrate. *Science,* 146:65-67.

Goldsmith, T.H., and C.D. Bernard. 1974. The visual system of insects, pp. 165-272, in M. Rockstein (Ed.), *The Physiology of Insecta,* Academic Press, New York.

Goldsmith, T.H., and P.R. Ruck. 1958. The spectral sensitivities of the dorsal ocelli of cockroaches and honeybees. *J. Gen. Physiol.,* 41:1171-1185.

Goodman, J.L. 1970. The structure and function of the insect dorsal ocellus. *Adv. Insect Physiol.,* 7:97-195.

Hardie, R. 1991. Whole-cell recordings of the light induced current in dissociated *Drosophila* photoreceptors: evidence for feedback by calcium permeating the light-sensitive channels. *Proc. R. Soc. London B,* 245:203-210.

Hardie, R.C., and B. Minke. 1994. Spontaneous activation of light-sensitive channels in *Drosophila* photoreceptors. *J. Gen. Physiol.,* 103:389-407.

Hart, S. 1998. Beetle mania: An attraction to fire. *Bioscience,* 48:3-5.

Haskell, P.T. 1961. *Insect Sounds,* Quadrangle Books Chicago, copyright by H.F. & G. Witherby Ltd. 1961, Northumberland Press Limited, Gateshead on Tyne, U.K., 189 pp.

Hodgson, E.S. 1956. Electrophysiological studies of arthropod chemoreception. I. General properties of the labelar chemoreceptors of Diptera. *J. Cell. Comp. Physiol.,* 48:51-76.

Hodgson, E.S. 1958. Chemoreception in arthropods. *Annu. Rev. Entomol.,* 3:19-36.

Horridge, G.A. 1965. The Arthropoda: III Insecta, pp. 1030-1055, in T.H. Bullock and G.A. Horridge (Eds.), *Structure and Function in the Nervous System of Invertebrates,* W.H. Freeman & Co., San Francisco.

Hoy, R.R., and D. Robert. 1996. Tympanal hearing in insects. *Annu. Rev. Entomol.,* 41:433-450.

Hoy, R.R., A.N. Popper, and R.R. Fay (Eds.). 1998. *Comparative Hearing: Insects,* Springer-Verlag, New York, 341 pp.

Loftus, R. 1976. Temperature-dependent dry receptor on antenna of *Periplaneta.* Tonic response. *J. Comp. Physiol.,* 111:153-170.

Loftus, R. 1978. Peripheral thermal receptors, pp. 439-466, in M.A. Ali (Ed.), *Sensory Ecology Review and Perspectives,* Plenum Press, New York.

Mazokhin-Porshmyakov. G.A. 1969. *Insect Vision,* translated by R. and L. Masironi, T.H. Goldsmith, Translation editor, Plenum Press, New York, 306 pp.

Menzel, R. 1975. Polarization sensitivity in insect eyes with fused rhabdomes, pp. 372-387, in A.W. Snyder and R. Menzel (Eds.), *Photoreceptor Optics,* Springer-Verlag, Berlin.

Miller, W.H., G.D. Bernard, and J.L. Allen. 1968. The optics of insect compound eyes. *Science,* 162:760-767.

Mizunami, M. 1994. The diversity of ocellar systems. *Adv. Insect Physiol.,* 25:151-265.

Mote, M.I., and T.H. Goldsmith. 1970. Compound eyes: Localization of two color receptors in the same ommatidium. *Science,* 171:1254-1255.

Nordström, P., and E.J. Warrant. 2000. Temperature-induced pupil movements in insect superposition eyes. *J. Exp. Biol.,* 203:685-692.

Phillis, W.A. III, and H.L. Cromroy. 1977. The microanatomy of the eye of *Amblyomma americanum* (Acari: Ixodidae) and resultant implication of its structure. *J. Med. Entomol.,* 13:685-698.

Prestwich, G.D., and G. Du. 1997. Pheromone-binding proteins, pheromone recognition, and signal transduction in moth olfaction, pp. 131-143, in R.T. Cardé and A.K. Minks (Eds.), *Insect Pheromone Research, New Directions*, Chapman & Hall, New York, 684 pp.

Ranganathan, R., G.L. Harris, C.F. Stevens, and C.S. Zuker. 1991. A *Drosophila* mutant defective in extracellular calcium-dependent photoreceptor deactivation and rapid desensitization. *Nature (London),* 354:230-232.

Rees, C.J.C. 1968. The effect of aqueous solution of some 1:1 electrolytes on the electrical response of the type 1 ("salt") chemoreceptor cell in the labela of *Phormia. J. Insect Physiol.,* 14:1331-1364.

Roth, L.M. 1948. A study of mosquito behavior. *Am. Midl. Nat.,* 40:265-352.

Sales, G., and D. Pye. 1974. Countermeasures by insects, pp. 71-97, in *Ultrasonic Communication by Animals*, Chapman & Hall, London, 281 pp.

Schmitz, H., H. Bleckmann, and M. Mürtz. 1997. Infrared detection in a beetle. *Nature,* 386:773-774.

Schmitz, H., S. Trenner, M.H. Hofmann, and H. Bleckmann. 2000. The ability of *Rhodnius prolixus* (Hemiptera: Reduviidae) to approach a thermal source solely by its infrared radiation. *J. Insect Physiol.,* 46:745-751.

Smith, D.P., R. Ranganathan, R.W. Hardy, J. Marx, T. Tsuchida, and C.S. Zuker. 1991. Photoreceptor deactivation and retinal degeneration mediated by a photoreceptor-specific protein kinase C. *Science,* 254:1478-1484.

Smith, W.C., and T.H. Goldsmith. 1990. Phyletic aspects of the distribution of 3-hydroxyretinal in the class Insecta. *J. Mol. Evol.,* 30:72-84.

Smith, W.C., and T.H. Goldsmith. 1991. The role of retinal photoisomerase in the visual cycle of the honeybee. *J. Gen. Physiol.,* 97:143-165.

Spangler, H.G. 1988. Moth hearing, defense, and communication. *Annu. Rev. Entomol.,* 33:59-81.

Spiegel, A.M., L. Teresa, Z. Jones, W.F. Simonds, and L.S. Weinstein. 1994. *G Proteins*, R.G. Landes Company, Austin, TX, 144 pp.

Stryer, L. 1975. *Biochemistry,* W.H. Freeman & Co., San Francisco, 877 pp.

Swihart, S.L. 1967. Maturation of the visual mechanisms in the neotropical butterfly, *Heliconius sarae*. *J. Insect Physiol.,* 13:1679-1688.

Toh, Y., and M. Kuwabara. 1974. Fine structure of the dorsal ocellus of the worker honeybee. *J. Morphol.,* 143:285-306.

Vogt, K. 1983. Is the fly visual pigment a rhodopsin? *Z. Naturforsch.,* 38C:329-333.

Vogt, K., and K. Kirschfeld. 1984. Chemical identity of the chromophore of fly visual pigment. *Naturwissenschaften,* 71:211-213.

Vondran, T., K.-H. Apel, and H. Schmitz. 1995. The infrared receptor of *Melanophila acuminata* De Geer (Coleoptera: Buprestidae): Ultrastructural study of a unique insect thermoreceptor and its possible descent from a hair mechanoreceptor. *Tissue Cell,* 27:645-658.

Wehner, R. 1976. Polarized-light navigation by insects. *Sci. Am.,* 235:106-115.

Wehner, R., G.D. Bernard, and E. Geiger. 1975. Twisted and non-twisted rhabdomes and their significance for polarization detection in the bee. *J. Comp. Physiol.,* 104:225-245.

Wolken, J.J. 1957. A comparative study of photoreceptors. *Trans. N.Y. Acad. Sci.,* 19:315-327.

Wolken, J.J. 1968. The photoreceptors of arthropod eyes, pp. 113-133, in J.D. Carty and G.E. Newell (Eds.), *Invertebrate Receptors*, Symposium Zool. Soc. Lond. No. 23, Academic Press, New York.

Wolken, J.J., and P.D. Gupta. 1961. Photoreceptor structures of the retinal cells of the cockroach eye. IV. *Periplaneta americana* and *Blaberus giganteus*. *J. Biophys. Biochem. Cytol.,* 9:720-724.

Yack, J.E., and J.H. Fullard. 1993. What is an insect ear? *Ann. Entomol. Soc. Am.,* 86:677-682.

Yarfitz, S., and J.B. Hurley. 1994. Minireview. Transduction mechanisms of vertebrate and invertebrate photoreceptors. *J. Biol. Chem.,* 269:14329-14332.

Zacharuk, R.T. 1980. Ultrastructure and function of insect chemosensilla. *Annu. Rev. Entomol.,* 25:27-47.

Zhukovsky, E.A., and D.D. Oprian. 1989. Effect of carboxylic acid side chains on the absorption maximum of visual pigments. *Science,* 246:928-930.

Zhukovsky, E.A., P.R. Robinson, and D.D. Oprian. 1991. Transducin activation by rhodopsin without a covalent bond to the 11-*cis*-retinal chromophore. *Science,* 251:558-560.

Zimmerman, W.F., and T.H. Goldsmith. 1971. Photosensitivity of the circadian rhythm and of visual receptors in carotenoid-depleted *Drosophila*. *Science,* 171:1167-1169.

Zuker, C.S. 1996. The biology of vision in *Drosophila*. *Proc. Natl. Acad. Sci. U.S.A.,* 93:571-576.

11 Circulatory System

CONTENTS

PREVIEW

The principal organ of whole-body circulation in insects is a tubular vessel lying just beneath the dorsal body wall that generally runs the length of the insect. The "blood," usually called hemolymph, typically enters the abdominal portion, the heart, through paired ostial openings, and is pumped anterior through the thoracic portion, the aorta. The aorta sends out branches in a few insects, but generally it is a simple tube with an open end in the head. It is not uncommon for the heartbeat to reverse in some insects, with the contraction wave beginning at the anterior and passing toward the posterior. The heartbeat is myogenic, but modified by nervous input and by neurosecretions. Accessory pulsatile organs or hearts often occur at the base of appendages, such as antennae, legs, and wings, and promote circulation into the appendages. Hemolymph does not transport oxygen, but it is important in the transport of nutrients and hormones and the removal of waste products. It aids locomotion in caterpillars, serves as a hydrostatic force that aids in eversion of various

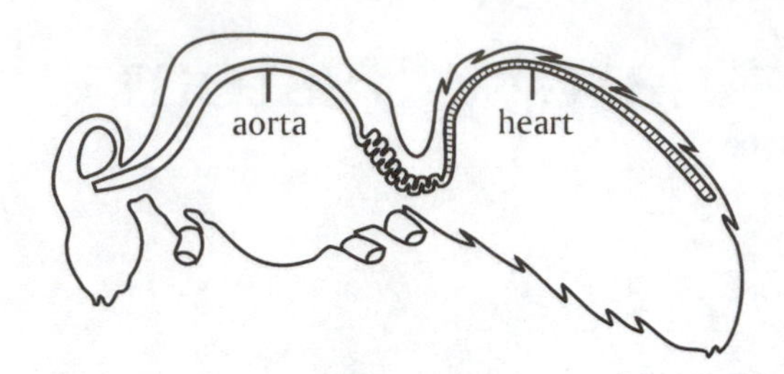

FIGURE 11.1 The dorsal vessel in a worker honeybee. The abdominal portion is called the heart and the thoracic portion is called the aorta. The distinction between heart and aorta is somewhat arbitrary, but the aorta has been defined by some authors as that part of the dorsal vessel that does not have incurrent ostia, but may have excurrent ostia. By this definition, the heart sometimes extends into the thorax. (Drawing modified from *The Hive and the Honey Bee*, Dadant & Sons, editors and publisher, 1975. Permission granted with acknowledgment.)

organs and glandular tissues, and has several other important functions. Hemocytes, free cells in the hemolymph, show variability in structure and function in different insects. Coagulocytes initiate clotting of hemolymph at a wound or when hemolymph is withdrawn from the body. Hemolymph does not clot in some insects. Hemolymph typically contains relatively high concentrations of free amino acids, the disaccharide trehalose, and numerous other chemical substances that are often in transport from the site of synthesis to a site of utilization. The pH of the hemolymph usually is slightly acid, but varies over a small range in different insects. Based on limited studies, it appears that insects regulate hemolymph pH by secretion of acid or base equivalents into the gut, excretion via the renal system and hindgut in some cases, and by increasing respiratory ventilation to control CO_2 content of the hemolymph.

11.1 INTRODUCTION: EMBRYONIC DEVELOPMENT OF THE CIRCULATORY SYSTEM AND HEMOCYTES

The heart is derived from mesodermal tissue in the developing embryo. The abdominal portion of the dorsal vessel is derived from cardioblasts, cells that fuse with each other, modify their shape, and form the dorsal tubular vessel. Eventually, fusion with the cephalic portion of the vessel occurs. In some insects, the heart begins to beat only after dorsal closure. In other insects, the heartbeat may begin before the posterior part of the heart fuses with the anterior portion in the head. Embryonic hemocytes are present and circulate within the cavity of the embryonic heart as they travel anteriorly and are discharged from the cephalic open end in the developing embryo of the water strider *Gerris paludum insularis* (Mori, 1996).

11.2 THE DORSAL VESSEL: HEART AND AORTA

The dorsal vessel (Figure 11.1) consists of two parts, the heart and the aorta. Sometimes, the entire vessel is referred to in the literature as simply the heart. Although heart and aorta are terms borrowed from vertebrate anatomy, they have no real physiological meaning in describing the dorsal vessel of insects; the entire length of the dorsal vessel carries a wave of contraction. The **heart** is the abdominal portion of the vessel, but it may extend into the thorax in some insects. The major criteria for deciding when the heart ends and the aorta begins are the presence of alary muscles and incurrent ostia in the heart portion. The **aorta** does not have alary muscles and lacks incurrent ostia, but it may have excurrent ostia. Both alary muscles and incurrent ostia occur in the thoracic portion of the vessel in many Orthoptera, and consequently, this thoracic portion is still heart, and only a short aorta leads into the head.

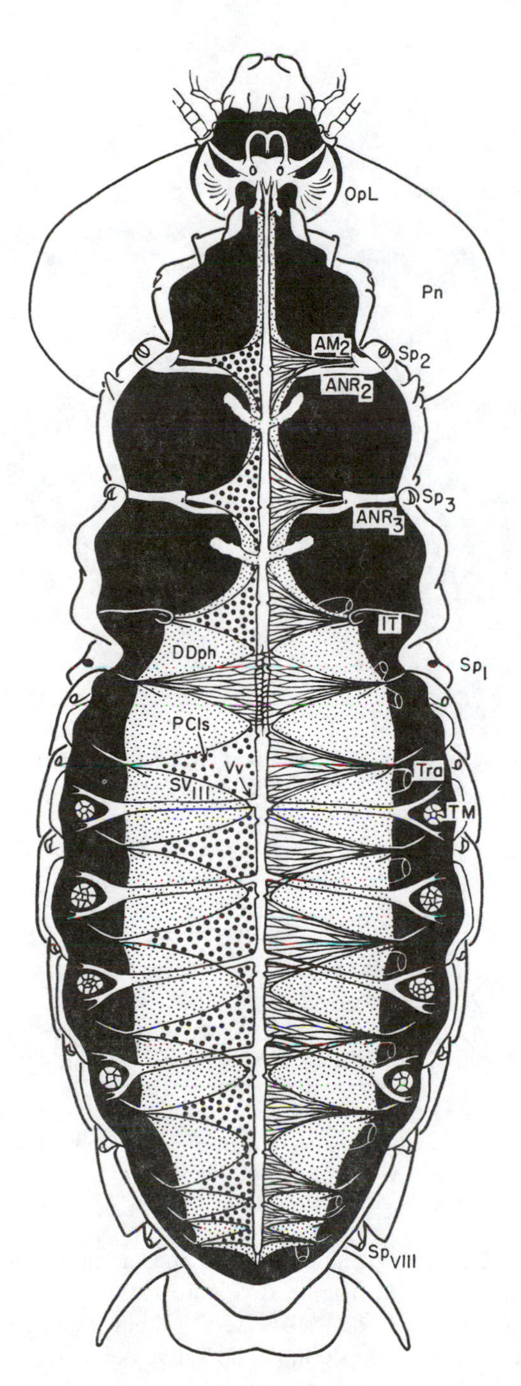

FIGURE 11.2 A diagram of the dorsal vessel, alary muscles, and branches of the dorsal vessel that extend to the pleural region of the abdomen in *Blaberus* sp. (Reproduced with permission from Nutting, 1951.)

Usually, the entire dorsal vessel is a simple tube, but in Orthoptera and Dictyoptera, the heart has several pairs (four pairs in *Blaberus* sp.) of long diverticula (Figure 11.2) passing laterally to tergosternal muscles and fat body tissue. Although the posterior end of the heart is usually closed, it is open in immatures of craneflies (Tipulidae). In the Ephemeroptera (mayflies), the posterior end gives rise to three branches, each passing into one of the three caudal filaments. In some insects,

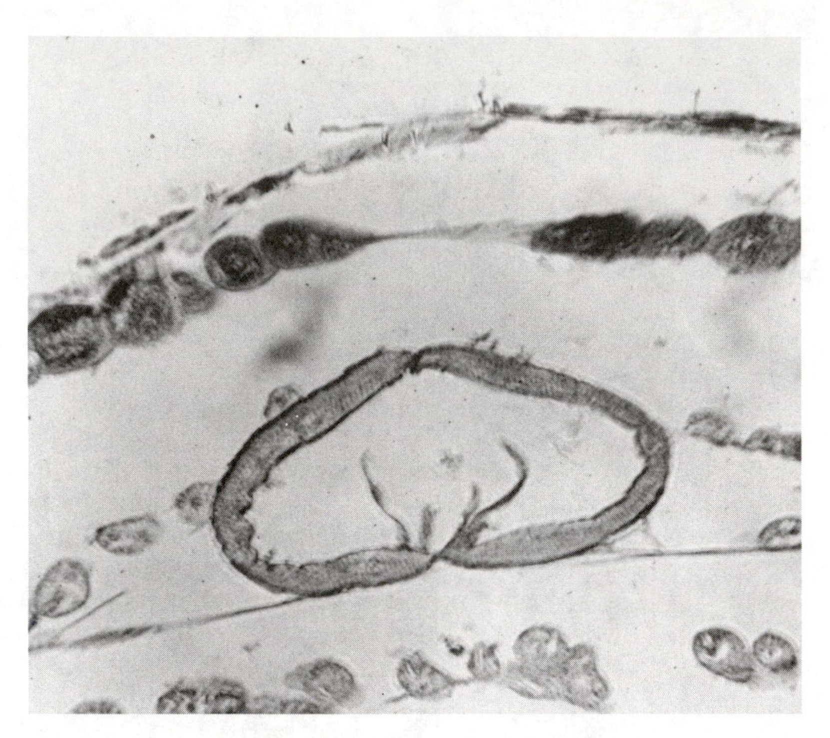

FIGURE 11.3 A cross section of the heart (abdominal portion) of a honeybee. The dorsal diaphragm can be seen passing below the heart and extending laterally, and some pericardial cells are shown. (Photograph by the author.)

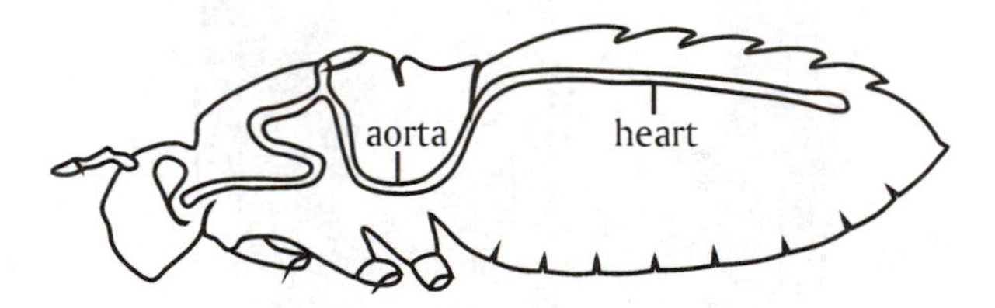

FIGURE 11.4 Loops of the dorsal aorta (the thoracic portion of the dorsal vessel) occur in the thorax of some insects. The drawing of the dorsal vessel in a sphingid moth shows an upward deflection of the dorsal vessel as it passes into the thorax and connects with an accessory pulsatile organ that helps pump hemolymph into the wings. (Modified from Snodgrass, 1935; Heinrich, 1996.)

a chambered effect is created by invaginations of ostia and the attachment of alary muscles, but division into chambers is not complete. The heart (Figure 11.3) contains both circular and longitudinal muscle fibers in most insects. The transverse (T) tubules and longitudinal tubules of the sarcoplasmic reticulum are poorly developed in the dorsal vessel, and T tubules may not penetrate at the Z bands, as is typically the case in most muscle.

Although the abdominal heart lies just beneath the dorsal cuticle, the aorta in the thorax often meanders between the large masses of thoracic flight muscles. In Lepidoptera and Coleoptera, the dorsal vessel passes into the ventral region of the thorax, but immediately rises towards the dorsal body wall (Figure 11.4). Near the dorsal wall it makes a sharp turn and begins to descend in close contact with the ascending loop. At the apex of the dorsal loop, the aorta is joined with a mesothoracic pulsatile organ containing a pair of ostia. This accessory pump aids the flow of hemolymph into the wings. Sections through the thorax in some insects show multiple images of the aorta because of the meandering path it takes.

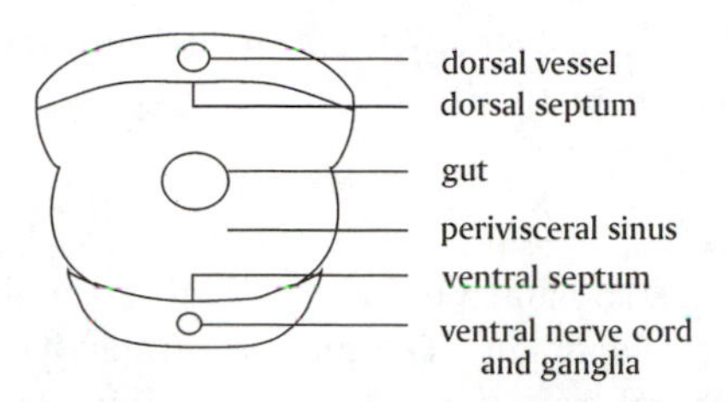

FIGURE 11.5 Diagram showing the partitioning of the body into three regions, the pericardial sinus, the perivisceral sinus, and the perineural sinus, by the dorsal and ventral diaphragms.

Hemolymph pressure is low. Hemolymph enters the dorsal vessel through incurrent openings called **ostia**, which may be guarded by valvular flaps in the abdominal portion of the vessel. Ostia usually exist in pairs. Incurrent ostia are generally confined to the abdomen, but sometimes there are incurrent ostia in the thoracic portion. Occasionally, there are excurrent ostia in the abdomen and thorax, but most of the hemolymph exits from the open anterior end and flows posteriorly through the body cavity, called the hemocoel. There is no distinction between lymph and blood, as in vertebrates. The hemolymph is sometimes referred to as the extracellular fluid. The brain and head glands (corpora allata, and corpora cardiaca) are bathed exceptionally well by the hemolymph. The continuous outflow at the head pushes hemolymph toward the posterior of the insect, where it flows around tissues and organs. It is aspirated into posterior incurrent ostia by relaxation (diastole) of the heart.

In most, but not all, insects, the abdomen is divided by the dorsal and the ventral diaphragm into three partitioned regions: the pericardial sinus, the perivisceral cavity, and the perineural sinus (Figure 11.5). The dorsal vessel lies on the upper surface of the dorsal diaphragm and is partially supported by it. The dorsal diaphragm consists of multiple layers of thin sheets of connective tissue, with small dorsal transverse muscle fibers (the alary muscles) enclosed within the sheets. The diaphragm is developed in grasshoppers as a nearly continuous sheet in the dorsal portion of the abdomen, with hemolymph flow between the pericardial and perivisceral sinuses limited mostly to the extreme posterior region of the abdomen where the diaphragm is not complete. In most other insects, the membrane is both fenestrated and incomplete laterally, so that hemolymph readily passes between the pericardial sinus and the perivisceral sinus. The dorsal diaphragm is present in most insects that have been examined, but in Hemiptera it is greatly reduced and a series of muscles near the posterior end of the heart, along with tracheal connections, support the heart. The dorsal diaphragm usually does not extend into the thorax, although it extends into the thorax to a limited extent in some insects (e.g., some orthopterans). Loose clusters and strings of cells, called **pericardial cells**, often occur on the external surface of the heart, and they also are attached to the dorsal diaphragm at various places in the pericardial sinus. They phagocytize injected particles such as India ink, some dyes, and other small particles, and are presumed to serve a protective phagocytic function.

The alimentary canal, reproductive organs, and some of the fat body lie in the central or perivisceral sinus below the dorsal diaphragm. There is often a ventral diaphragm separating the perineural sinus containing the ventral nerve cord and ganglia from organs in the perivisceral sinus, but the ventral diaphragm usually does not extend into the thorax. It is not even present in the abdomen of all insects. In general, it is present in larvae and adults of Odonata, Orthoptera, Hymenoptera, Ephemeroptera, Lepidoptera, and Neuroptera. It is present only in adults of Mecoptera and lower Diptera, and not present in the higher Diptera (the Cyclorrhapha). The ventral diaphragm is fenestrated, allowing hemolymph to circulate.

Undulations of both the dorsal and ventral diaphragms due to intrinsic muscle fibers, alary muscle action, gut movements, ventilatory movements, and general body muscle action aid circulation of hemolymph and keep it mixed and moving within the body, especially in active insects. Studies on a wide variety of insects and developmental stages (Sláma, 1999, and references therein) suggest that quiescent insects generate micropulses of pressure (coelopulses) in the circulatory system that aid circulation and breathing.

11.2.1 Alary Muscles

The **alary muscles** (Figure 11.2), so named because of their general wing or delta shape in many insects, form part of the dorsal diaphragm. The muscles probably provide support for the heart. The muscle fibers typically fan out from a small point of origin on the lateral wall of the dorsum to a broad insertion on the heart in many insects, presenting the typical delta appearance. In some insects (e.g., many grasshoppers), the origin and insertion are broad and the delta shape is not particularly evident. Some alary muscle fibers pass beneath the heart and extend from lateral side to side. In places, the fibers may also run parallel to the long axis of the heart for a short distance. The pairs of alary muscles tend to agree with the number of pairs of ostia. In addition to support, the alary muscle may assist in the expansion (diastole) of the heart after a contraction wave, and thus aid in pulling hemolymph into the incurrent ostia. They are not necessary for diastole, however, as evidenced by severing them with little or no apparent effect on the heart beat.

11.2.2 Ostia

Ostia are small, slit-like, paired openings in the dorsal vessel that allow hemolymph to enter or leave the vessel. Incurrent ostia allow hemolymph to enter during diastole and excurrent ones permit hemolymph to exit. Some Orthoptera have twelve pairs of incurrent ostia, nine in the abdomen and three in the thorax, but most insects have fewer, with two, three, or five pairs of ostia being common. Ostia more commonly occur in the heart, but may also occur in the aorta. Pairs of ostia are usually located laterally, with one on each side of the heart, but some are ventrally and dorsally located.

Ostial openings tend to occur at the base of shallow pockets or at deeper, funnel-shaped invaginations in the wall of the dorsal vessel, which often give the heart a chambered appearance. Incurrent and excurrent ostia may be difficult to distinguish (Jones, 1977). Excurrent ostia more often occur in the thoracic portion of the vessel, but do occur in the abdomen of some insects. Some ostia do not have developed valves that control hemolymph flow. Most incurrent ostia open with diastole, allowing hemolymph to be forced into the dorsal vessel by general body pressure and/or perhaps slight negative pressure inside the dorsal vessel. Some insects have valve flaps in incurrent ostia that open inward; hemolymph readily passes the valves into the lumen of the heart. During contraction, the valve flaps are forced together, preventing backflow, which may be their primary function.

11.2.3 The Heartbeat

The heartbeat is a wave of contraction (systole), generally originating at the posterior end of the heart and traveling anteriorly. The rate of contractions or beats is highly variable in different insects, and varies with physiological conditions, temperature, species, stage of development, nervous activity, and neurosecretions. The rate may be as slow as 15 beats per minute in the larva of *Lucanus cervus* (Coleoptera), to rates near or higher than 100 beats per minute in several insects (see Beard, 1953; Jones, 1977, for a large number of values). The beat is considered to be **myogenic**, originating in the muscle itself, although the heart beat of many other invertebrates is neurogenic. The heart of the American cockroach *Periplaneta americana*, has one of the most complex systems of innervation; but after careful removal of the lateral cardiac nerve cords, spontaneously active cardiac neurons, and lateral nerves, the heart still beats (Miller and Metcalf, 1968). When the heart is cut into numerous pieces after stripping all neurons away, each piece continues to express a beat, indicating that various parts of the dorsal vessel are capable of acting as a pacemaker. The pacemaker that usually dominates, however, is located at the posterior of the heart, and the contraction wave (systole) usually originates near the posterior of the heart and travels anteriorly. The contraction wave may move very slowly, only 1 or 2 mm/s, so that two or three contraction waves can be seen following each other. In other insects, the propagation rate may be so rapid that it appears as if the entire vessel is contracting simultaneously. In the mealworm *Tenebrio molitor*, the conduction velocity for the contraction wave in the heart was measured at 14 mm/s, but only 1 mm/s in the aorta (Markou and Theophilidis, 2000).

One of the unusual and little-understood features of insect hearts is that the beat can reverse, and originate at or near the anterior and travel posteriorly. Malpighi (about 1669, quoted by Gerould, 1930) observed periodic reversals of the contraction wave, with systole beginning at the anterior of the heart and traveling posteriorly. Beat reversal has been observed in numerous insects in a number of orders (Gerould, 1933), and even prior to hatching (Davis, 1961). Contractions may begin occasionally at both ends and meet near the middle. Reversals are unpredictable, but usually the back-directed beats are slower, and duration of the reversal is usually brief. Certain experimental treatments, such as amputations of parts of the heart, lateral pressure on the dorsal vessel, and ligation near the middle of the heart may alter conduction rates and cause reversal of contraction waves in some insects. Smits et al. (2000) found that the contraction wave in larvae of the tobacco hormworm *Manduca sexta* consistently originates at the posterior end and passes anteriorly, with the dorsal vessel beating at about 34.8 beats per minute. The heart rate is slower in pupae (21.5 beats/min), and irregular in rate, amplitude, and direction, with periods of cardiac arrest from a few seconds to as long as 20 min. Dorsal vessel contraction rates and direction of contraction wave are variable in adult moths, with fast forward heart rates of 47.6 beats/min, slow forward rates of 32.8 beats/min, and reversal of contraction wave (anterior origin) of 32.2 beats/min. The larval heart rate shows no increase in response to activity (induced by prodding), but the adult heart rate rises from about 50 beats/min to as much as 223 beats/min in response to 1 min of prodding (Smits et al., 2000).

11.2.4 Ionic Influences on Heartbeat

Ion influences on heartbeat are variable in different species, and not well studied or understood. The resting potential of the heart in the American cockroach *P. americana* measures 40 to 70 mV, depending on technique, insect variability, and possibly other factors. Although potassium controls the resting potential of the American cockroach heart, the evidence suggests that potassium is not the controlling ion in other insects; and in some, more than one ion may be responsible for the resting potential. The ion or ions responsible for depolarization of the heart are not known for most insects. An action potential can be developed in isolated hearts of *P. americana* and the cecropia moth *Hyalophora cecropia* with sodium-free salines, but the identity of the ion(s) carrying the current is not known in these insects. Consistent with these observations, the heart of the American cockroach is insensitive to tetrodotoxin, a poison that completely blocks sodium channels in nerves, but does not stop the myogenic heart beat. Calcium has been implicated in the depolarization of the heart in some insects, but not others. During depolarization, overshoot potentials up to 20 mV have been recorded for some insects. Saline solutions containing high concentrations of magnesium stop the heart of *P. americana*, a typical response of animal hearts to high magnesium ion concentrations. The heart of *H. cecropia,* however, cannot beat without magnesium in the saline, and this seems consistent with the fact that it is a lepidopteran, which typically contains high potassium and magnesium levels in its hemolymph as a phytophagous feeder.

11.2.5 Nerve Supply to the Heart

The heart receives innervation through lateral neurons from ventral ganglia in the abdomen (or from fused thoracic and abdominal ganglia in some advanced insects) and from a chain of cardiac neurons that lies alongside the heart in some insects, notably Odonata, Orthoptera, Dictyoptera, and Hemiptera. The American cockroach has spontaneously active cardiac ganglion (nerve) cells

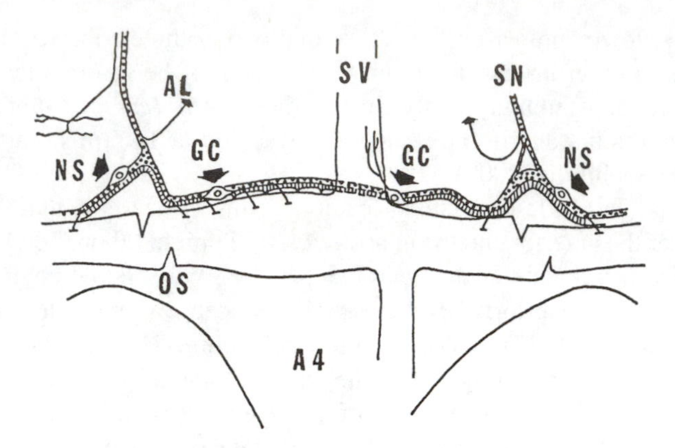

FIGURE 11.6 Diagram of one of the paired chains of cardiac neurons (consisting of cardiac motoneurons and neurosecretory neurons) that run along the heart in the abdomen of the cockroach *Periplaneta americana*. A4, fourth abdominal chamber; AL, alary muscles; GC, ganglion cell or motoneuron; NS, neurosecretory neuron; OS, incurrent ostial valve; SN, segmental nerves; SV, lateral segmental hemolymph vessel. (Reproduced with permission from Miller, 1968.)

(Figure 11.6) that innervate the heart and maintain contact throughout the lateral cardiac chain running parallel with the dorsal vessel (Miller and Thomson, 1968; Miller, 1968). Spontaneously active neurosecretory cells also are located in the cardiac chain, especially near junctions with the lateral segmental nerves. The cardiac neurons in the cockroach are stretch-sensitive motor neurons. They fire at increasing rates as the myocardium stretches in diastole, and cease firing at each systolic contraction. They reinforce the simultaneous contraction of the dorsal vessel. Insects in the orders Diptera and Lepidoptera, and probably other orders, do not have cardiac neurons and a nerve chain paralleling the heart. The heart may still receive a nerve supply from the segmental ganglion.

11.2.6 Cardioactive Secretions

Various secretions and neurosecretory peptides act on the heart to change the rate and amplitude of the heartbeat (Nässel, 1993). **Proctolin**, a neuropeptide produced in motoneurons, interneurons, and neurosecretory cells located at various places in different insects — typically in the brain, but sometimes in other ganglia — stimulates heart rate. Whether this is one of its functions in the normal course of insect life, however, is not clear. **Crustacean cardioactive peptide (CCAP)**, a cardioactive peptide composed of nine amino acids that was first isolated from a crab, has been isolated from *Locusta migratoria*, the migratory locust. It stimulates the heart, but it also is known to have physiological actions unrelated to the heart, so its role in heart action is unclear. One of several peptides isolated from the ventral ganglia of the tobacco hornworm, *M. sexta*, has the same amino acid sequence as CCAP, but is usually designated as CAP_{2a} (Nässel, 1993). CCAP-immunoreactive neurons have been detected in the brain of the mealworm *Tenebrio molitor*. Another cardioactive peptide called **corazonin** has been isolated from brain of the cockroach *P. americana* and neurosecretory cells in the brain of a blowfly are positive to a corazonin-antiserum. Numerous bioactive peptides have been isolated from insects (Nässel, 1993) and shown to have a variety of physiological effects, depending on the *in vitro* bioassays used, but their main or physiological role *in vivo* often remains unclear.

5-Hydroxytryptamine (serotonin) is a neurotransmitter in the nervous system and causes a vigorous increase in heartbeat rate at very low concentrations in isolated or semi-isolated hearts of some insects but not in others. Even in those in which it increases the heart rate, it seems to have little effect when injected into living insects, but this could be because it is quickly deactivated. Its status as a neurotransmitter at the cardiac neurons innervating the heart is at present uncertain. A number of other drugs and potential neurotransmitters, such as octopamine, dopamine, and tyramine

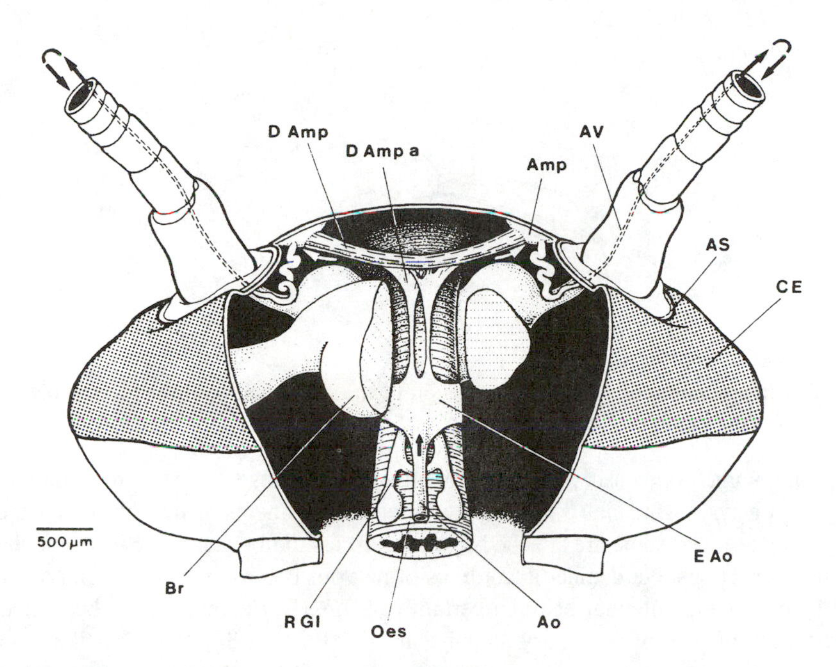

FIGURE 11.7 A. The accessory heart at the base of each antenna in *Periplaneta americana*, the American cockroach. Amp, ampulla of accessory heart; Av, antennal vessel; Ao, aorta; As, antennal sclerite; Br, brain; Ce, compound eye; D Amp, Ampa, dilator muscles of the ampulla; E Ao, enlarged anterior end of the aorta; Oes, esophagus; R Gl, corpora cardiaca and corpora allata, respectively. (Reproduced with permission from Pass, 1985.)

stimulate the semi-isolated hearts of some insects, but their role, if any, in heart function under normal physiological conditions is not clear.

11.3 ACCESSORY PULSATILE HEARTS

Accessory hearts have several shapes and variable morphology, but all are simple pulsatile, sac-like structures. They occur at a number of places in the body, but most commonly at the base of antennae and wings, and within the leg, usually near the femuro-tibial joint, and in the dorsal region of the meso- and metathorax. They assist circulation of hemolymph into and through the appendages. Based on location, they are usually referred to as leg, antennal, or wing hearts. In most cases, they have no connections to the dorsal vessel. They merely aspirate hemolymph into a sinus cavity through ostial openings and pump it out. Less structured ones consist of little more than a pulsating muscle that aids movement of hemolymph. A mesothoracic accessory heart of *Bombyx mori* aids in pumping hemolymph into the wings. In *B. mori,* the metathoracic pulsatile heart is not directly connected to the aorta, although in some other insects it is connected to the aorta. The importance of wing expansion and wing circulation has already been mentioned.

Antennal hearts at the base of the antennae are common in many insects. Good circulation into the antennae is likely to be critical in supplying adequate nutrients to support the large number of sensory structures associated with the antennae. An accessory heart at the base of each antenna in the American cockroach *P. americana* is illustrated in Figure 11.7. A complex array of head accessory hearts occurs in *B. mori*, in which the aorta expands into a large sac on the anterior surface of the supraesophageal ganglion, the "brain." A short transverse tube arises from the sac and terminates in a pair of lateral ampullae, on each side of the head. Each functions as an accessory heart to pump hemolymph into the antenna via an antennal vessel running into the antenna, and to the optic lobe of the brain through a vessel that passes dorsally over the brain to the optic lobe,

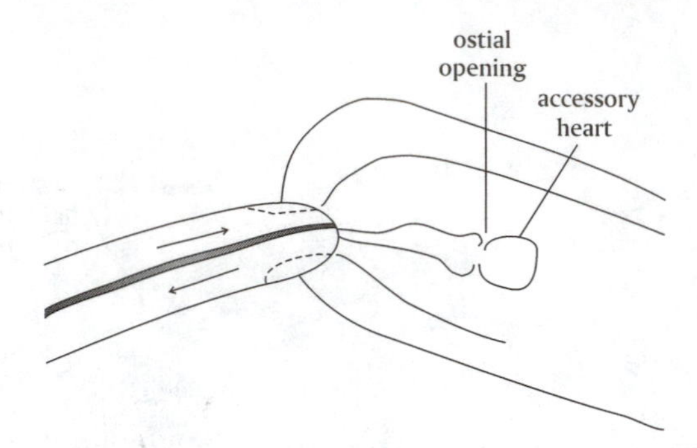

FIGURE 11.8 The accessory leg heart at the junction of the femur with the tibia in the beetle *Notonecta.* (Redrawn from Weber, 1930.)

and ends as an enlarged open sac. Antennal ampullae in some insects do not have direct connection with the aorta. In general, antennal ampullae have simple attachments by tonofibrillae to the epidermis to keep them in place, and some are attached by tiny muscles to the pharynx. Some antennal ampullae have nervous connections, but detailed recordings of nervous control have not been conducted.

The structure of the antennal heart in certain earwigs (Dermaptera) may be indicative of the early evolutionary history of accessory antennal hearts (Pass, 1988). The sac-like ampulla at the base of each antenna in earwigs is connected to an antennal blood vessel that runs to the apex of the antenna. The ampulla does not have a muscular wall, but it is compressed by a small, independent muscle running across it like a belt. A valve-like structure near the origin of the antennal blood vessel prevents hemolymph from flowing back toward the ampulla. When the compression muscle relaxes, the natural elasticity of the ampulla, assisted by the pull of elastic fibers attaching the ampulla to the wall of the head, promote diastole and filling with hemolymph via a ventral ostium.

Leg hearts (Figure 11.8) are common in some Odonata, Hemiptera, Homoptera, and higher Diptera. There may be a membranous septum between the ventral and dorsal regions of the tibia, and hemolymph is pumped into one channel and returns through the other, aided by muscle contractions in the leg that pump the septum.

11.4 HEMOCYTES

Hemocytes are blood cells. They change their appearance and shape (Figure 11.9) from time to time even in the same insect, and they can be distorted in shape by fixation, staining, and other procedures used in collecting and processing hemolymph. Procedures for examining and classifying hemocytes have not been standardized or agreed upon, and various classifications and morphological types have been published. Fixation, spreading, and drying of insect hemocytes on a glass slide, as is commonly done in vertebrate blood analysis prior to staining, tends to result in many bizarre and variable types due to distortion of cell shapes, probably by the drying process.

The electron microscope has been useful in hemocyte classification. The seven most common types of hemocytes found in insects are **prohemocytes**, **plasmatocytes**, **granulocytes**, **spherulocytes**, **adipohemocytes**, **oenocytoids**, and **coagulocytes** (Gupta, 1979a, 1979b). Most insects that have been surveyed have prohemocytes, plasmatocytes, and granulocytes, but the presence of the other types is variable. Only five types (prohemocytes, plasmatocytes, granulocytes, spherule cells or spherulocytes, and oenocytoids) were described from the pink bollworm *Pectinophora gossypiella* (Lepidoptera: Gelechiidae) (Raina, 1976). Eight classes of hemocytes (prohemocytes, plasmatocytes, granular hemocytes, coagulocytes, crystal cells, spherule cells, oenocytoids, and thrombocytoids) were established by Lackie (1988), based on the classification scheme of Rowley and Ratcliffe (1981). Most insects will not have all eight classes, if indeed any have all eight. Application

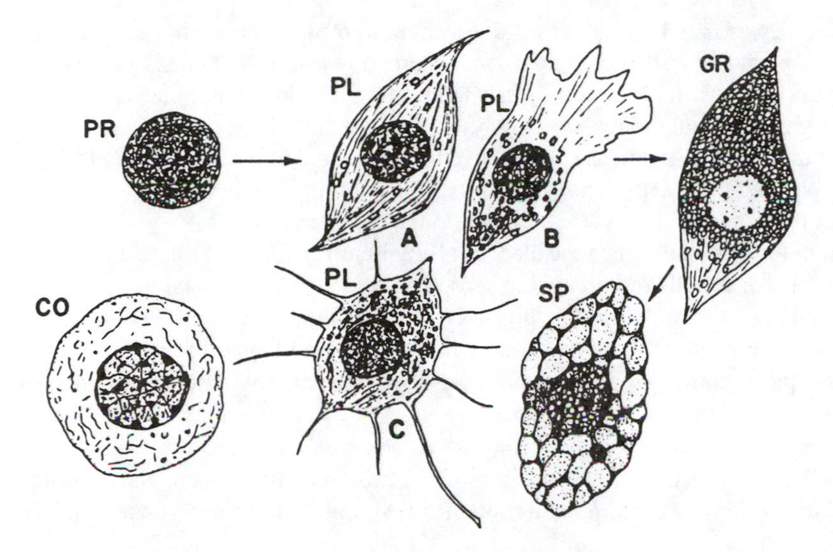

FIGURE 11.9 An illustration of the most common types of hemocytes from insect hemolymph. PR, prohemocyte; PL, plasmatocyte; GR, granulocyte; SP, spherulocyte; CO, coagulocyte (= hyaline hemocyte). Several different shapes of plasmatocytes are shown in A, B, and C. The arrows indicate transformations of cells that are believed to occur. (Reproduced with permission from Woodring, 1985.)

of monoclonal antibodies (MAbs) to the identification and classification of hemocytes has become a useful tool (Gillespie et al., 1997).

Prohemocytes are the smallest hemocytes, and may be stem cells from which some other hemocytes may develop. Prohemocytes are known to divide, and they may differentiate into plasmocytes, which, in turn, may give rise to granulocytes, and these may differentiate into sperulocytes. Although there is some evidence for this pattern of transformation, the evidence that they are the primary source of hemocytes is not conclusive (Gupta, 1979a). Thus, prohemocytes may be one source of new hemocytes. The origin of other hemocytes is uncertain. Prohemocytes are typically round, 6 to 13 μm in diameter, with a relatively large nucleus (70 to 80% of cell volume). The cells stain heavily with Wright's stain, and the nucleus may not be easily discerned. They contain ribosomes and mitochondria, but little endoplasmic reticulum and Golgi membranes. They are not mobile and do not participate in phagocytosis. The primary function of hemocytes may be to divide and give rise to new hemocytes.

Plasmatocytes are small to large polymorphic cells, up to 40 to 50 μm in size, granular or agranular, and round to spindle shaped in wet suspensions, although they lose this shape when dried on a slide. They may be binucleate. "Young" plasmatocytes can be confused with prohemocytes (Gupta, 1979a). They contain lysosomal enzymes and are usually the most numerous of circulating cells (Lackie, 1988). They are phagocytic and participate in encapsulation, nodule formation, and wound-healing (Ratcliffe and Götz, 1990).

Granulocytes are variable in size, spherical or oval, and up to 45 μm in size. The nucleus is usually small, and the cytoplasm is granular. On the basis of histochemical tests, the granules are thought to be glycoproteins and mucopolysaccharides. Granulocytes may arise from plasmatocytes. The precise function of granulocytes is unproven, but some researchers have suggested that they serve storage and possibly secretory functions. They may be involved in cellular defensive functions in various insects, and may be phagocytic in some insects, but in others neither of these functions is established.

Spherulocytes are ovoid to round cells up to about 25 μm in length. They may contain few to many small spherical inclusions that stain for acid mucopolysaccharides (Ashhurst, 1982). Their function is unknown but they may participate in phagocytosis.

Adipohemocytes may be small or large, spherical to oval, and contain lipid droplets. They might be plasmatocytes that are filled with lipids under certain physiological conditions (Gupta, 1979a).

Oenocytoids are variable in size, often large, may be binucleate, and lyse easily, but do not cause hemolymph coagulation when they lyse. They are nonphagocytic. Some evidence indicates that they contain prophenoloxidase, an inactive form of phenoloxidase (Lackie, 1988). Oenocytoids should not be confused with oenocytes, cells found among fat body cells and scattered among epidermal cells in many insects. Oenocytes are not blood cells.

Coagulocytes have also been called **hyaline hemocytes** (Grégoire, 1951) and cystocytes (Lackie, 1988). These cells rupture within seconds after injury, or after taking a hemolymph sample from an insect, and initiate the clotting process (Grégoire and Goffinet, 1979). The cells may contain granules. In phase contrast, they are nearly transparent, and hence the name hyaline hemocytes. The hemolymph of some insects does not coagulate; for example, the hemolymph of larval honeybees, *Apis mellifera*, does not coagulate.

Crystal cells have been described from larval *Drosophila melanogaster*. They contain crystals of prophenoloxidase (Rizke and Rizke, 1980). Thrombocytoids were described from blowfly larvae *Calliphora erythrocephala* as important in wound-healing and in encapsulation (Zachary et al., 1975).

Good light microscope photographs of prohemocytes, plasmatocytes, granulocytes, spherulocytes, and oenocytoids were presented by Arnold and Sohi (1974) from fresh hemolymph of the forest tent caterpillar *Malacosoma disstria*. These authors found that only prohemocytes, plasmatocytes, and granulocytes could be maintained in cell cultures. Cultured prohemocytes and plasmatocytes divided by mitosis. Granulocytes divided but the mechanism was not determined. The size of cells in culture was generally larger than those taken from fresh hemolymph. Hemocytes have been relatively easy to get into tissue or cell culture (Hink, 1976), and numerous cell lines have been developed.

11.4.1 Functions of Hemocytes

Gillespie et al. (1997) have reviewed hemocyte functions, with particular emphasis on their role in immunity. Hemocytes participate in wound-healing by aggregating at the wound site, where some cells phagocytize cellular debris or foreign organisms such as bacteria. Coagulocytes, and possibly granulocytes, in some insects participate in the coagulation of plasma to help plug a wound. Entrapment of hemocytes in the coagulum helps plug the wound. Some of the trapped cells may also play an active role in defense at the wound site.

Some hemocytes contain enzymes that aid in detoxication, including detoxication of some insecticides. Hemocytes participate in nodule formation and encapsulation of foreign objects (Gillespie et al., 1997). Invasions of bacteria may be attacked by nodule formation in which the bacteria are aggregated into nodules containing hemocytes and coagulum. Larger objects or invading parasites may be encapsulated. Plasmatocytes and granular hemocytes aggregate to form thin sheets of cells and plaster themselves around nodules, internal parasites, or other foreign objects. Sometimes, encapsulated objects also become melanized by the action of the phenoloxidase cascade of enzymatic action on tyrosine or other phenolic compounds. *Pseudoplusia includens*, a noctuid moth, attaches granular hemocytes directly to the target object, followed by multiple layers of plasmatocytes adhering to the inner layer of granular hemocytes, and finally a monolayer of granular hemocytes on the periphery of the capsule. The peripheral layer of granular hemocytes dies (apoptosis appears to be induced by substances released by the underlying plasmatocytes), leaving a basal lamina-like layer around the capsule that is important in successful capsule formation (Pech and Strand, 2000). Encapsulated objects often become attached to various tissues or organs in the body. The ability to encapsulate parasitoids appears to be dependent on an evolutionary race between the parasitoid and its host. A parasitoid that is not well adapted to a particular host is likely to have a high percentage of its eggs encapsulated, while a parasitoid well adapted to its host may avoid

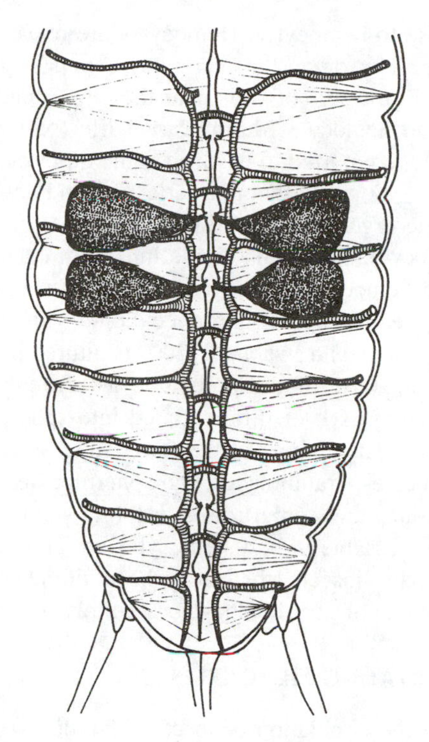

FIGURE 11.10 The hemopoietic organ in the cricket *Gryllus bimaculatus.* (From Hoffmann, 1970.)

significant encapsulation. Similar to nodule formation, the encapsulated object is surrounded by many layers of plasmatocytes that form sheets of cells around the object. Often, but not always, the entire mass becomes melanized, a process that also may be toxic to the offending organism and assist in killing it.

Bacterial invasion of some insects results in synthesis and secretion into the hemolymph of combinations of several antibiotic peptides and proteins. These include lysozyme (14 kDa), the cecropins (4 kDa), the attacin/sarcotoxin II family of proteins (20 to 28 kDa), and the defensins (29 to 34 amino acids), a family of proline-rich antibacterial peptides (18 to 34 amino acids long) with a variety of names depending on source (Gillespie et al., 1997). Fat body tissue is the usual site of synthesis of these proteins, but hemocytes and a variety of other tissues also contribute to the synthesis. Although combinations of the above are commonly secreted, not all are found in the same insect.

One final function of some hemocytes may be to form the basement membrane of some cells (Lackie, 1988), but this is a controversial issue that has not been conclusively resolved.

11.4.2 Hemocytopoietic Tissues and Origin of Hemocytes

During embryological development, hemocytes develop from mesodermal tissue. In the larvae and adults of some insects, circulating hemocytes are known to divide, and some of these may differentiate into other hemocytes. The rate of division of existing hemocytes in larvae of the wax moth, *Galleria mellonella*, seems to be sufficient to account for the numbers of circulating hemocytes (Jones and Liu, 1968). There is evidence for **hemocytopoietic tissues** in some insects, for example, in the cricket *Gryllus bimaculatus* shown in Figure 11.10. (Hoffmann, 1970; Hoffmann et al., 1979). Nutting (1951) described and illustrated similar structures in additional orthopteroid and related insects; but based on dye injection experiments, he concluded that they were phagocytic tissues. Reinvestigation of these structures in light of Hoffmann's work seems worthwhile. Larvae of Lepidoptera have masses of loosely connected cells in a capsule located near the prothoracic spiracles that seem to give rise to hemocytes. Hemocytes are released from the capsules through gaps in the covering surface. The organs disintegrate in pupae and release large numbers of hemocytes into the body. Small clusters of cells near the wing imaginal discs in the commercial silkworm *B. mori*, give rise to hemocytes during larval life (Nittono, 1964). Similar structures without the capsule enclosure are reported to occur in some dipterous larvae, such as the housefly, *Musca domestica* (Arvy, 1954). Some Orthoptera (crickets) and some cockroaches (Dictyoptera) have complex hemocytopoietic organs composed of hollow sacs at the anterior end of the heart. Cells inside the sac are phagocytic, and some divide into stem cells that can further differentiate into several different types of hemocytes. Small groups of cells between pericardial cells located near the dorsal diaphragm are believed to produce hemocytes in *Locusta* (Orthoptera) and *Melolontha* (Coleoptera). Yamashita and Iwabuchi (2001) cultured prohemocytes from larvae of *B. mori* and found that more than 60% of the prohemocytes differentiated into plasmatocytes or granulocytes, and some granulocytes later differentiated into spherulocytes. Some prohemocytes divided into new prohemocytes. The authors of the study suggest that prohemocytes are the stem cells that give rise to plasmatocytes, granulocytes, and spherulocytes, but no oenocytoid cells were produced in culture. Oenocytoids may come from a different stem cell line.

Probably, insects are quite variable in how hemocytes are produced and in whether they have hemocytopoietic tissues. In many insects, there may be multiple origins of hemocytes (Arnold, 1974). In most insects, the source of new hemocytes is simply not known.

11.4.3 Number of Circulating Hemocytes

The number of hemocytes in the circulation of insects is quite variable from species to species, and even within the same individual at different times, depending on its physiological state. Sex, age, stage of development, and activity are known to influence the observed number of cells in some insects. Also, some hemocytes are sessile, attached to various tissue, at least for much of the time, but may be released into the circulation under certain circumstances. Measurements of hemocyte counts per microliter (counts/μl) of hemolymph over time can be influenced by fluctuations in blood volume, which is not nearly so constant as in vertebrates (see topic below).

Counting hemocytes is usually done in a manner similar to the counting of red or white blood cells from a vertebrate with a standard hemocytometer counting chamber. A measured volume of hemolymph is withdrawn and diluted to a known volume with a suitable diluent that does not lyse the cells, and the counting chamber is filled. One frequently used procedure in estimating cell counts is to heat-kill the insect in water at 60°C for 2 min, and then take the hemolymph sample; this is believed to put sessile cells into the circulation and fixes the hemocytes in their normal shape (Woodring, 1985). No really effective general anticoagulants for insect hemolymph are known, but rapid dilution with a saline solution tends to minimize coagulation, as does the heat treatment.

Hemocyte counts for a very large number of insects have been published (Jones, 1977). Different species contain widely differing numbers. For example, the American cockroach, *P. americana* can have as many as 70,000 to 120,000 cells/μl hemolymph; the tobacco hornworm *M. sexta* has been reported to have 8200 cells/μl; wax moth larvae *G. mellonella* have 35,000 cells/μl and the blood-sucking bug *Rhodnius prolixus* can have from 300 to 5000 cells/μl hemolymph. The stage in the development of insects may have a dramatic effect on the number of hemocytes in the hemolymph. For example, 1st and 2nd instars of the Caribbean fruit fly *Anastrepha suspensa* and the housefly *M. domestica* have very few circulating hemocytes. Even large 3rd instars have only a few thousand cells per microliter when they are only 1- and 2-day-old 3rd instars; but as they approach larval maturity and prepare for pupariation, cells rapidly increase in circulation, culminating in up to 30,000 cells/μl hemolymph (Figure 11.11).

The absolute number and kind of hemocytes, and possibly the temporal sequence of their appearance and increase in concentration, may be important to the ability of an insect to defend itself from foreign invaders. For example, eggs of a braconid parasitoid, *Asobara tabida*, more

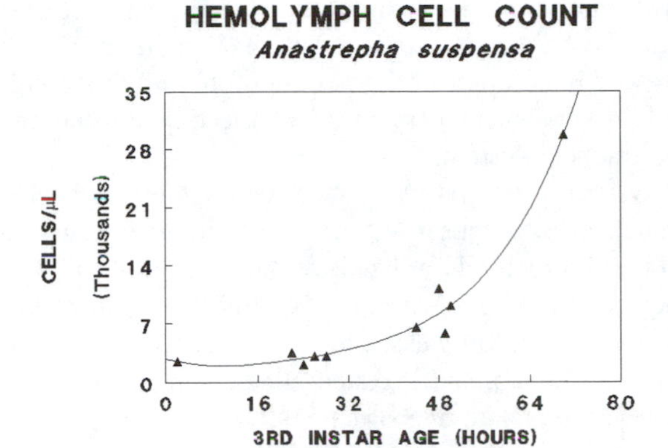

FIGURE 11.11 Age- and time-related increase in numbers of hemocytes in a tephritid fruit fly, *Anastrepha suspensa.* (Data from the author.)

often survive, hatch, and larvae complete development in *D. melanogaster* than in a sister species, *D. simulans*, which usually encapsulates the parasitoid egg before it hatches. Early instars of *D. simulans* that successfully encapsulated the parasitoid eggs had several times more hemocytes than *D. melanogaster* larvae, which usually did not encapsulate the eggs. Individuals of *D. simulans* that successfully encapsulated eggs had greater numbers of hemocytes than those that did not encapsulate eggs (Eslin and Prevost, 1996). The authors of the study suggested that successful encapsulation and defense against parasitism may involve a physiological race between ability of the host to get hemocytes into circulation and the ability of the parasitoid to locate and lay an egg in a very young host before many hemocytes are in circulation.

Some parasitoids may have evolved additional mechanisms to avoid encapsulation, such as mechanisms to destroy host hemocytes and/or inactivate other host defense mechanisms. For example, oviposition by the endoparasitoid *Tranosema rostrale* (Hymenoptera: Ichneumonidae) into its host, the larvae of the spruce budworm *Choristoneura fumiferana* (Lepidoptera: Tortricidae), results in up to 50% reduction in total hemocyte counts and some reduction of phenoloxidase in the host after 3 days. Hemocyte and phenoloxidase reduction is caused by fluid from the calyx tissue of the ovaries that is injected at the same time as the egg. The mechanism(s) by which these actions are accomplished is (are) not known (Doucet and Cusson, 1996). The ability to alter host total hemocytes and phenoloxidase may be mechanisms that have evolved in successful parasitoids enabling parasitization of the host and avoidance of encapsulation or of otherwise being killed.

11.5 THE HEMOLYMPH

11.5.1 FUNCTIONS OF HEMOLYMPH AND CIRCULATION

Circulation of hemolymph through the body of an insect serves a number of functions. The following are listed in no particular order.

1. Hemolymph is important as a sink for carbon dioxide (CO_2). CO_2 is soluble in hemolymph as the bicarbonate ion (HCO_3^-) and substantial amounts may be held in solution in the hemolymph of some insects. In diapausing *H. cecropia* pupae, for example, CO_2 in the gas phase builds up only slowly because most of the CO_2 goes into solution in the hemolymph. This allows the spiracles to remain closed for many minutes until CO_2 in the gas phase reaches a critical level and the spiracles open. Keeping the spiracles closed as much as possible prevents excessive water loss, a critical factor for the closed-system pupa. Similar mechanisms have been demonstrated

in other insects. Hemolymph does not have a role in transport of oxygen, and there is no pigment carrier for oxygen except in a very few species of insects. A limited amount of oxygen is present in solution in the hemolymph, and this oxygen is probably used by cells, but it is a very small percentage of the oxygen that cells need. Oxygen is delivered by the tracheal system.

2. The circulatory system transports nutrients to cells and tissues. Hemolymph is a major storehouse of trehalose, the major insect sugar used for energy, and in particular for flight energy. The main store of lipids utilized by moths in flight is the fat body, and rapid circulation is required to bring released lipids to the flight muscles located in the thorax.
3. The circulatory system delivers waste products, excess water absorbed from food such as a blood meal or plant sap, and ingested allelochemicals or metabolites to excretory organs, or in some cases, to storage structures.
4. The circulatory system is a reservoir of fluid, nutrients, and enzymes. Some of the latter, such as lysozymes and phenoloxidases, act as protective agents that can chemically modify potential toxicants, bacteria, parasitoid eggs and larvae, and other foreign invaders. Some large proteins, such as cecropins, with antibacterial and antifungal activity are induced and transported in the hemolymph after an insect is exposed to bacteria and fungi, or some of their chemical by-products. Trehalose, some storage proteins, and amino acids are typical nutrients present in large quantity in hemolymph.
5. Hemolymph transports hormones from neurohemal organs to target tissues. Many neuropeptides, which have good water solubility, are transported in the hemolymph from their release points in the nervous system to target systems such as Malpighian tubules (diuretic hormone), prothoracic glands (PTTH), pheromone glands (PBAN), and others.
6. Hemolymph is a lubricant and hydraulic support that assists in maintaining body shape and movement, especially of soft-bodied forms such as caterpillars. Pharate adults of dipteran Cyclorrhapha (houseflies, tephritid fruit flies, and related flies) utilize the hydraulic mechanism to force hemolymph into the ptilinum, a balloon-like structure on the front of the head; and as it expands, it breaks open the old puparium. The new adult slowly wiggles out of the puparial case, again utilizing muscles, legs, and hydraulic action. When the fly is out, the ptilinum deflates, collapses inward, and its site is slowly sclerotized into a suture. Some insect glands are extruded by the hydraulic pressure of hemolymph. An example is the pair of osmeterial glands located behind the head on swallowtail butterfly caterpillars that are everted when the caterpillars are handled, probed, or attacked by predators. In late instars, an unpleasant odor (to humans) containing volatile derivatives of butyric acid and other compounds is emitted from the everted glands. The Caribbean fruit fly and some other tephritid fruit flies (Nation, 1989) contract muscles and force hemolymph toward the posterior of the body. The pressure balloons thin, lightly sclerotized lateral pouches at the sides of the abdomen and evert an anal pouch at the tip of the abdomen as part of their pheromone release behavior. Many other insects probably rely, in part, on hydraulic pressure to expose pheromone glands and promote the release of pheromone.
7. Pumping of hemolymph into the wings, and in some cases the secretion of plasticization factors, are critical to proper expansion of the wings in newly emerged adults. Insects such as moth and butterflies typically rest quietly while the wings expand. The mesotergal and metatergal accessory hearts are important in directing hemolymph into the wings, but some insects have other accessory hearts within the wings. Even in mature, older adults of many insects, hemolymph flows through most of the veins of the wings, basically following a pattern of flowing into the wings at the anterior region of the wing and returning through the posterior wing veins. The flow may, however, reverse on occasion in some insects. Wings that are experimentally deprived of circulation become dry and brittle, and tracheae collapse and retract, leaving gas bubbles behind in the wing veins.

When parts of the wings are cut off, hemolymph does not ordinarily hemorrhage out of the wings, but alternative pathways of circulation through cross veins are established and some wing circulation continues.

8. Hemolymph and the hemocytes provide protection from invading bacteria, eggs of parasitoids, and other foreign substances by biochemical "immune-type" reactions, phagocytosis and encapsulation by hemocytes, and wound-healing by coagulation in some insects. Hemolymph does not coagulate in some insects.
9. The circulatory system is important in some insects as a means of heat transfer to prevent excessively high temperatures in the thorax from flight muscle activity, or to hold heat in the thorax and allow thoracic temperature to warm above ambient. Control of thoracic temperature is probably important to many insects in flight when the thoracic wing muscles generate much heat. Many insects do not initiate flight until the temperature in the thorax reaches some critical temperature above ambient. They warm the thorax by "wing-whirring" before attempting to take flight. The tobacco hornworm moth, *Manduca sexta* (Lepidoptera: Sphingidae) (Heinrich, 1970, 1996) needs a thoracic temperature of about 38°C to begin flight. After 2 min of free flight at ambient temperatures ranging from 20 to 30°C, its thoracic temperature is as high as 41 to 42°C, a temperature approaching the maximum the moth can tolerate in continuous flight. *M. sexta* moths will not fly continuously for 2 min at an ambient temperature of 35°C or higher, and those prodded into flight experience thoracic temperatures up to 43.3°C, near the lethal point for many cells, especially neurons. The high temperatures in the thorax are the result of the intense muscular work by the flight muscles. To cool the thorax and prevent the temperature from going too high, the moths circulate hemolymph from the hot thorax into the abdomen where little muscle activity is occurring during flight. Moreover, the abdomen is covered with a much thinner layer of scales (0.5 mm thick layer on abdomen as opposed to 2 mm thick on the thorax), and heat can more readily escape by convection to the atmosphere. When moths were prevented from circulating hemolymph into the abdomen by an experimentally placed ligature in the first segment of the abdomen, the thoracic temperature in flying moths averaged about 23°C above ambient, and increased directly with ambient temperature from 15 to 23°C. When the heart was ligatured, the thorax soon overheated and moths ceased flying at temperatures above 23°C. The advantage of the physiological mechanisms that allow heat in the thorax to either accumulate or to be dissipated gives the moth the opportunity to fly at a wide range of ambient temperatures. Further examples and details of the role of circulation in thermoregulation can be found in Heinrich (1996).

11.5.2 Hemolymph Volume

Hemolymph volume in insects is not constant. Dehydration, physiological stage of development, and other factors can cause large fluctuations in hemolymph volume. Volume may even fluctuate daily with food and water availability (Chapman, 1958). Several methods have been devised for measuring hemolymph volume. One of the simplest is blotting on paper as much fluid as can be removed from the insect and weighing the paper (Figure 11.12). Dye dilution methods following injection of a known volume of dye and dilution of ^{14}C-labeled inulin have been used (Levenbook, 1958; Wharton et al., 1965; Wheeler, 1962; Wheeler, 1963; Levenbook, 1979). Injection of any agent intended for use in dilution calculations may suffer from binding of the agent to tissues. Evans blue binds to the tissues in 3rd instars of the blowfly *Calliphora vicina*, and calculation then overestimates hemolymph volume compared to the use of ^{14}C-labeled inulin (Levenbook, 1979). A micro method has been devised for measurement of hemolymph volume in small insects (housefly) that have very little hemolymph in the body (Shatoury, 1966).

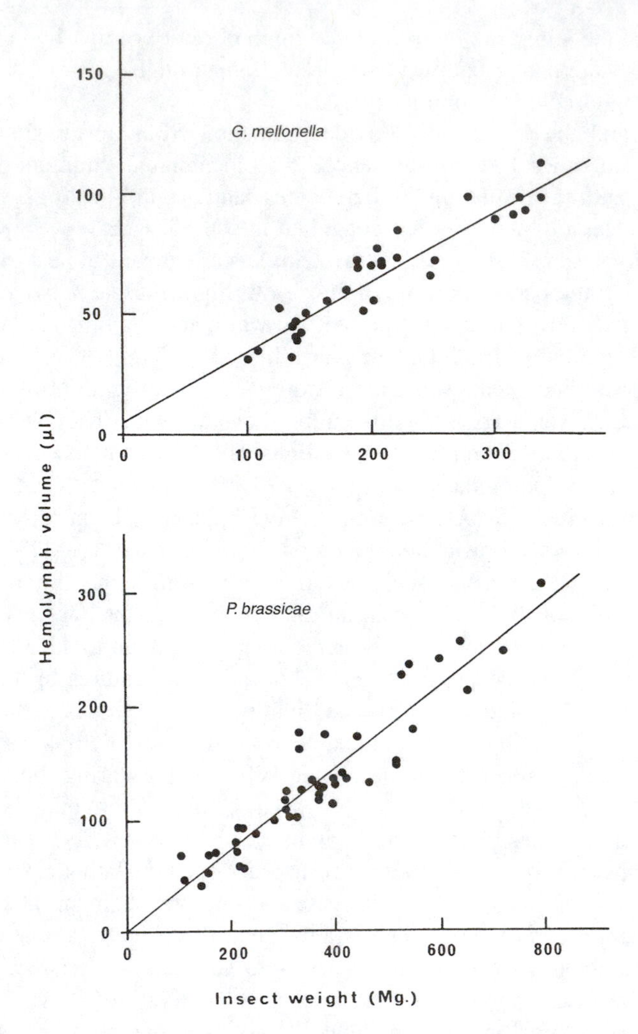

FIGURE 11.12 The graphs show regression lines of hemolymph volume on insect body weight for two lepidopterous caterpillars: wax moth larvae *Galleria mellonella* and larvae of the cabbage butterfly *Pieris brassicae*. Each point represents the blood volume and body weight of one larva. (Reproduced with permission from Gagen and Ratcliffe, 1976.)

Fluctuation in hemolymph volume can distort measurements of the number of cells per microliter and the total concentration of any chemical component in the hemolymph. With the amaranth dye dilution method, Wheeler (1963) found that there is no change in the absolute number of hemocytes/insect in *Periplaneta americana* just prior to a molt, but the number of cells per microliter hemolymph increases due to a decrease in hemolymph volume. During ecdysis, the total number of cells does not change, but the number per microliter decreases because of an increase in hemolymph volume. About 24 h after ecdysis, there is a decrease in the total number of cells.

11.5.3 Coagulation of Hemolymph

The hemolymph of many insects coagulates rapidly at the site of wounds, or when withdrawn from the insect by capillary pipette. However, in some insects, for example, in larvae of the honeybee *A. mellifera*, the hemolymph does not coagulate. Hemolymph fails to coagulate in some Coleoptera, Hemiptera, some adult Lepidoptera, and many Diptera.

Grégoire (1951) and Grégoire and Goffinet (1979) described coagulation in a large number of insect species by observing, with phase microscopy, events occurring immediately after taking a hemolymph sample. Coagulation tends to be a continuous process initiated in all cases by rupture of a single type of hemocyte, the hyaline hemocytes, also called coagulocytes. Varying degrees of plasma clotting occur, from general clotting to a limited reaction. A number of agents, including oxalate, citrate, magnesium sulfate, 2% methylene blue (a reducing agent), cocaine hydrochloride, sodium bisulfite, sodium thiosulfate, ethylenediamine tetraacetic acid (EDTA), and sodium hydrosulfite, can prevent or reduce the clotting reaction in some insects. These reagents are not universally effective, however, and the mechanism by which they interfere with clotting has not been determined. Rapid dilution with a balanced saline solution is probably the easiest and most effective procedure to reduce or prevent clotting in most insects. Heparin, bee venom, and a number of other substances useful in vertebrate blood clotting are usually not effective in insects.

Three types of coagulation occur. Type I coagulation is initiated by immediate rupture of the hyaline hemocytes. Coagulation islands form around each ruptured hyaline hemocyte, and these gradually grow in area until many of them coalesce. This is the predominant type of coagulation in Orthoptera, Dermaptera, some Hemiptera, some Coleoptera, and some Hymenoptera, some Homoptera, Neuroptera, Mecoptera, Trichoptera, and some Lepidoptera.

In Type II coagulation, there is an absence of coagulation islands; instead, a pseudopodial meshwork develops from ruptured hyaline hemocytes. The meshwork gradually expands and traps other hemocytes within the net. This type of coagulation is found in some ground beetles (Carabidae), dragonfly (Odonata) nymphs, several Lepidoptera, and some Coleoptera (Scarabaeidae). Type III coagulation is a combination of events taking place in Types I and II, and is common in Homoptera, many Coleoptera, and Hymenoptera.

11.5.4 Hemolymph pH and Hemolymph Buffers

The great majority of insects have hemolymph that is slightly acidic, but a few have hemolymph that is as alkaline as pH 7.5 or slightly greater; hemolymph pH typically ranges between 6 and 7.5 (Buck, 1953). Measurement within the same species may vary by up to about 0.7 pH units. There is no strong correlation with sex, diet, stage of development, or taxonomic position. The greatest changes occur with metamorphosis; during the pupal stage, or sometimes just prior to pupation, there is a slight increase in acidity, but usually no more than about 0.3 pH units.

Hemolymph pH remains reasonably stable during the life of an individual, which suggests that insects regulate hemolymph pH. Locusts and grasshoppers (and by extrapolation, possibly other insects) regulate hemolymph pH in at least two different ways: (1) by transfer of acid equivalents to the gut when acidosis is caused by a titratable acid (a natural situation might be accumulation of an organic acid from metabolism); and/or (2) by increased ventilation of the tracheal system when acidosis is caused by an increase of CO_2 in the hemolymph, as in vigorous muscular activity. Harrison et al. (1992). demonstrated that (fasting) desert locusts *Schistocerca gregaria* regulate extracellular pH after an experimental injection of HCl into the hemolymph. Recovery is mainly by transfer of about 75% of the acid equivalents from the hemolymph into the alimentary canal, with no evidence that the respiratory system aided the recovery (by eliminating CO_2). The experimentally injected locusts eventually eliminate ammonium urate, which may be a mechanism of compensation for extracellular acidosis (Harrison and Phillips, 1992).

A rise in body temperature results in a biphasic pattern of hemolymph pH regulation in two orthopterans, the two-striped grasshopper *Melanoplus bivittatus* and the locust *Schistocerca nitens* (Harrison, 1988, 1989). Hemolymph pH remains constant at temperatures up to 25°C, and then decreases about 0.017 pH units per °C above 25°C in both species. The two-striped grasshopper accumulates CO_2 in the hemolymph when forced to hop repeatedly over a 5-min period, and hemolymph pH decreases (Harrison et al., 1991). During repeated hopping and in recovery afterward, rates of gas exchange of O_2 and CO_2 increase, with a greater increase in the rate of O_2 transfer than CO_2 during hopping, and a higher rate of CO_2 than O_2 exchange during recovery. The higher rate of O_2 exchange during hopping suggests that the jumping-leg muscles work aerobically, like flight muscles, and that the oxygen functions as the ultimate acceptor for protons released during muscle metabolism. During a 2-min recovery period, the grasshoppers ventilate the tracheal system, rapidly flush excess CO_2 from the body, restore depleted O_2, and return hemolymph to normal resting values.

The western lubber grasshopper *Taeniopoda eques* shows flexible ability to shift between excretion of excess acid or excess base equivalents, depending on physiological condition, in its homeostasis of hemolymph pH (Harrison and Kennedy, 1994). Such flexible ability seems adaptive in a highly polyphagous insect, such as *T. eques*, which may experience acidosis or alkalosis, depending on diet, allelochemicals in the food, type of proteins metabolized (proteins yielding basic, acid, or neutral amino acids, for example), and other metabolic conditions (Harrison and Kennedy, 1994).

Buffering in the hemolymph of *Schistocerca gregaria* at 21°C, pH 7.31, when CO_2 is held constant is provided by (in milliequivalents per liter per pH unit) bicarbonate 20, protein 10, inorganic phosphate 1.6, organic phosphate 1.5, citrate 0.4, and the amino acid histidine 0.1 (Harrison et al., 1990). Thus, nearly 90% of the buffering capacity in hemolymph of these locusts is due to bicarbonate and proteins, with a small additional contribution by inorganic and organic phosphates. Amino acids, although in high concentrations in insect hemolymph, have dissociation constants well removed from typical hemolymph pH. The pK_1 is <3 for ionization of the carboxyl proton, and $pK_2 > 9$ for ionization of the amino group proton for nearly all 20 amino acids that occur in insect hemolymph; only histidine has an ionizable proton from its ring structure with pK of 6.04 in the range of insect hemolymph.

11.5.5 Chemical Composition of Hemolymph

Hemolymph contains many dissolved inorganic and organic substances, colloidally suspended proteins, and lipoproteins. It is about 90% water and 10% solids. Probably, most components contribute to the osmotic pressure and specific gravity of hemolymph. The specific gravity is usually slightly greater than 1. Hemolymph generally has an osmotic pressure, expressed in freezing point depression, of about 0.7 to slightly over 1°C. This is less than some Crustacea [*Cancer* (crab) and *Homarus* (lobster)], and slightly higher on average than that in humans. Osmotic pressure in a number of species, and in different developmental stages of the same species, is not correlated with age, sex, developmental stage, diet, or systematic position among orders (Buck, 1953).

In general, saline solutions that are equivalent to 0.9 to 1.6% sodium chloride (NaCl) will encompass the range of osmotic pressures in insect hemolymph, but such a solution of NaCl will not be balanced ionically with respect to the composition of hemolymph. An isotonic saline developed for tissue perfusion of the adult blowfly *Phormia regina* (Diptera: Calliphoridae) contains 119 m*M* Na, 5.6 m*M* K, 2.4 m*M* Ca, and 1 m*M* Mg, 97 m*M* glutamic acid, 44 m*M* glutamine, 97 m*M* proline, 48 m*M* alanine, and 26 m*M* glycine to give an osmotic pressure of 480 mOs/kg. Some of the sodium is provided by NaOH and some by NaCl; the final pH is 7 (Chen and Friedman, 1975). A general-purpose saline suitable for many insects can be made by dissolving 7.7 g NaCl, 0.36 g KCl, and 0.24 g $CaCl_2 \cdot 2H_2O$ in water to make a liter (Jones, 1977). If the saline is to be used for Lepidoptera and Coleoptera, the NaCl content should be reduced to 0.117 g and KCl increased to 7.46 g. A fixative and staining solution useful for staining and observing hemocytes from the American cockroach and other insects contains, in a final volume of 100 ml water, 0.5 g crystal violet, 1.09 g NaCl, 0.157 g KCl, 0.085 g $CaCl_2$, and 0.017 g $MgCl_2$, and sufficient glacial acetic acid to make the pH 2.9 (Sarkaria et al., 1951). The acetic acid fixes the hemocytes in their natural shape and crystal violet lightly stains them for easier visualization. Numerous salines that have been recommended for use with Lepidoptera failed to adequately support normal heartbeat and neuromuscular transmission in several lepidopterans, but a new saline composed of 12 to 28 m*M* NaCl, 32 to 16 m*M* KCl ($[Na^+] + [K^+] = 44$ m*M*), 9 m*M* $CaCl_2$, 1.5 m*M* NaH_2PO_4, 1.5 m*M* Na_2HPO_4, 18 m*M* $MgCl_2$, 175 m*M* sucrose, pH 6.5, is satisfactory for *Bombyx mori* and several other lepidopterans (Ai et al., 1995, and references therein).

11.5.5.1 Inorganic Ions

The ionic composition of hemolymph plasma is highly variable in different insects. Sodium, potassium, calcium, and magnesium are typical cations, and chloride, phosphate, amino acids, and sometimes bicarbonate are present as anions. Chloride and phosphate are the major anions in the hemolymph of honeybee larvae, but these are not usually the major anions balancing the cations in hemolymph of most insects. Chloride accounts for approximately 7% of the anions in larvae of the horse bot *Gastrophilus intestinalis* (Diptera: Tendipendidae), 12% in larvae of the silkmoth *B. mori*, and up to 39% of anions in larvae of the southern armyworm *Spodoptera eridania* (Lepidoptera: Noctuidae). Amino acids and organic acids account for a substantial part of the anions in some groups, particularly Neuroptera and Lepidoptera. Most insects have rather high levels of amino acids in the hemolymph, and some amino acids can contribute to the cation load.

Sodium and potassium concentrations, and the Na:K ratio in hemolymph are variable. Plasma from Odonata, Diptera, carnivorous Coleoptera, Dictyoptera, and Orthoptera tends to have a relatively high Na:K ratio (19.6 for *P. americana* cockroaches, 21.4 for *Acheta domesticus* crickets, 9.8 for *Schistocerca* locusts), and a Ca:Mg ratio equal to about 1. In general, this has been considered to approximate the early evolutionary condition in insects (Florkin and Jeuneaux, 1974).

Lepidoptera, phytophagous Coleoptera, Hemiptera, Homoptera, and Hymenoptera have a Na:K ratio that is only a few multiples of 1, or even less than 1 (some as low as 0.3 to 0.1, or lower). Some researchers have speculated that the evolution of a low Na:K ratio is related to the evolution of phytophagous food habits, but the correlation is not strong; and if it did evolve this way, it is not strictly diet related in present-day insects. Carnivorous insects often maintain a Na:K ratio that is different from that of the phytophagous insects upon which they feed. Phytophagous insects within some orders do not show Na:K ratios as low as those in Lepidoptera, but neither are they as high as in the Orthoptera or Dictyoptera. Force feeding of large doses of potassium chloride to *P. americana* depresses the Na:K ratio, but never to the level in Lepidoptera, indicating that at least some insects have the ability to regulate ion composition (Buck, 1953).

The very low Na:K hemolymph values in some insects would have consequences for nerve and muscle function were it not for barriers that protect cells. All cells (except hemocytes) are protected from direct contact with the hemolymph by a basement membrane on the hemolymph side, although few experimental data are available to assess how much it actually stops ion movements. The central nervous system and larger nerves especially are protected from direct hemolymph contact by the neurolemma (= perilemma) and perineurium, noncellular and cellular layers, respectively, that surround ganglia and large nerves. Individual neurons are protected by glial sheath cells.

Ion binding to macromolecules in hemolymph occurs in some insects (Weidler and Sieck, 1977). Macromolecules bind more than 20% of the sodium and magnesium, about 16% of the calcium, and about 10% of the chloride in whole hemolymph of the American cockroach, *P. americana*. Potassium is not bound. Similar binding data have been reported for other insects. Differential binding of ions could explain why hemolymph sampled from different body sites does not always have the same composition (Pichon, 1970).

11.5.5.2 Free Amino Acids

One of the interesting features of insect hemolymph is that it contains very large amounts of free amino acids, much more so than the body fluids of other animals. These amino acids contribute to the osmotic value of hemolymph, and account for a substantial portion of the cations and anions of hemolymph. A good physiological explanation for such large quantities of amino acids in hemolymph is not available, but the hemolymph is probably a reservoir of amino acids for protein synthesis, much of which occurs in fat body cells.

11.5.5.3 Proteins

Hemolymph contains a wide variety of different proteins (reviewed by Kanost et al., 1990). Many protein bands can be detected by electrophoresis, and these may change with physiological state, age, sex, and other factors. Some insects (larvae of some Diptera and Lepidoptera have been most intensively studied) synthesize large quantities of one or more storage proteins during late larval life. Calliphorin, named for its original source in *Calliphora erythrocephala* (Diptera: Calliphoridae), has a molecular weight of 540,000 and is composed of subunits of about 85,000. It accounts for up to 60% of the protein in mature blowfly larvae. It is synthesized by the fat body, released into the hemolymph, and finally reabsorbed back into the fat body and stored for use during pupation and adult development. Similar proteins have been described from other insects. Hemolymph proteins include many different enzymes, such as lysozyme, antibacterial and antifungal proteins, and transport proteins that carry hydrophobic substances such as juvenile hormone, cholesterol, diglycerides, hydrocarbons, and other lipoidal substances through the aqueous medium of hemolymph.

One hemolymph enzyme, **phenoloxidase (PO)**, plays an important role in sclerotization of the cuticle and in protecting the insect from foreign invaders (Ashida and Yamazaki, 1990; Nappi et al., 1991; Söderhäll and Aspán, 1993). It exists in the hemolymph and in hemocytes as a proenzyme that can be converted to active phenoloxidase. PO converts a variety of phenols to quinones. The enzyme and some of its products are important in tanning the cuticle. PO activity often causes the hemolymph to melanize (darken) and eventually become dark brown or black at wound sites. When the hemolymph is withdrawn from many insects, it darkens. Active PO develops within minutes in tissue homogenates or drawn hemolymph in some insects, for example, in dipterous larvae. Other insects (lepidopterous caterpillars are an example) activate the prophenoloxidase more slowly over many minutes. PO may not be demonstrable in all stages of an insect; in a number of dipterans it is only present in high titer in the last larval instar (Nation et al., 1995).

11.5.5.4 Other Organic Constituents

Many organic compounds have been found in the hemolymph of insects. One of the main functions of circulating hemolymph is that of a major transport medium to move metabolic compounds synthesized in fat body cells, or hormones synthesized in a variety of neurohemal organs, to sites where they are needed. For example, triacylglycerols stored in fat body cells are released as diacylglycerols and these are transported by lipoproteins (lipophorin) to thoracic flight muscles in Lepidoptera and some other insects.

Insect hemolymph is notable for the large concentration of trehalose in solution, from 0.5 g to as much as 2.5 g/100 ml in some cases (Woodring, 1985). Trehalose is the principal hemolymph sugar in most insects and is rapidly converted to glucose for immediate metabolism in glycolysis and the Krebs cycle. Ecdysteroids, juvenile hormone, PTTH, PBAN, AKH, HTH, diuretic hormone, and a host of other neuropeptides and hormones are transported from the site where they are released into the hemolymph to their targets. The titer of these in the hemolymph varies continuously with the physiological state of the insect.

Hemolymph is a major transporter of metabolic waste products for excretion or storage. Uric acid is transported by hemolymph from the main site of synthesis (the fat body) to the Malpighian tubules for excretion in many insects, or for storage in various sites by some insects. Uric acid is poorly soluble in aqueous solutions, such as hemolymph, and its concentration in hemolymph is low (11.47 ± 0.99 mg/100 ml, mean ± SE, in last instars of the wax moth, *G. mellonella*) (Nation and Thomas, 1965). Hemolymph uric acid levels are variable in tobacco hornworm larvae, *M. sexta*, during development, and the concentration peaks near the middle of the last instar at slightly more than 33 mg/100 ml hemolymph (Buckner and Caldwell, 1980). Possibly, potassium and sodium urate account for much of the uric acid transported in hemolymph because the salts are much more soluble in an aqueous medium than the free acid. A possibility that has not been studied in detail is whether transport may be aided by binding of urates to proteins in the hemolymph.

11.6 THE RATE OF CIRCULATION

Not many measurements of the rate of circulation have been made. Dye injected into the posterior part of the abdomen of the cockroach, *P. americana* can be detected in the head in only 30 s, but takes up to 8 min to reach the tarsus of the mesothoracic leg (Woodring, 1985). Similar results have been found with the large locusts. Some experiments with injected radioisotopes indicate that it takes 15 to 30 min for the isotope to be uniformly distributed in all parts of the body (Craig and Olson, 1951). Slow rates of complete mixing have consequences for determining blood volume by dye or isotope dilution, or for sampling any hemolymph component after some experimental procedure that is expected to influence concentration or distribution. For the purposes of estimating blood volume, it has been recommended that at least 1 h be allowed for complete mixing of ^{14}C-inulin, and longer may be necessary for some insects (Woodring, 1985). Several successive samplings over time would be the best procedure to detect complete mixing. It is likely that in some insects the circulation rate is even slower than the above data indicate. On the other hand, based on the role of circulation in transporting nutrients as a source of flight energy, it is reasonable to assume that the rate of circulation is fairly rapid and efficient. Otherwise, flight could not continue for hours, as it does in some long-distance fliers (mostly the lipid burners). The action of the muscles themselves and the slight flexing action of the thorax during the wing cycle aid the circulation and move hemolymph more rapidly around the body than occurs in an insect at rest.

11.7 HEMOGLOBIN IN A FEW INSECTS

Some species of chironomid larvae, *Chironomus tentans* and others (Diptera: Chironomidae); horse bot larvae (*G. intestinalis* (Diptera: Tendipendidae); and three bugs, *Buocnoa margaritacea*, *Anisops producta*, and *Macrocorixa geoffrey* (Hemiptera) have hemoglobin colloidally suspended in the plasma of the hemolymph. The hemoglobins of the chironomids consist of as many as 12 monomers with molecular weights of about 15,900 each; each monomer may be coded by its own gene. In *C. tentans,* the hemoglobins account for up to 40% of the total proteins in hemolymph.

Insect hemoglobins have strong affinity for oxygen and load to capacity at only a few millimeters Hg partial pressure of oxygen; consequently, insect hemoglobins remain fully saturated and do not release the oxygen for cell use unless the oxygen partial pressure is extremely low. At 7 mmHg, *Chironomus* blood is still completely saturated with oxygen, and half-saturated at 4 mmHg. Thus, how functional the hemoglobin is in supplying oxygen to tissues in the normal ecology of these insects is uncertain.

Whether the *Chironomus* hemoglobins exhibit a Bohr effect (unloading more readily in the presence of high tissue CO_2 concentration) is dependent on pH (they do at pH 7.4 to 7.5), and analysis is complicated by the fact that more than one type of hemoglobin exists in the hemolymph. The large quantity in the hemolymph may have some effect as a pH buffer and help to provide a favorable pH for its unloading to tissues (Agosin, 1978).

Vertebrate hemoglobin exhibits a strong Bohr effect, and in the presence of high CO_2, such as in rapidly respiring tissues, it unloads oxygen more readily to the tissues, functionally a highly adaptive characteristic. If this effect does not occur in the *Chironomus* blood samples at physiological pH, oxygen tension must fall extremely low for the hemoglobin to be of much value to the insect.

Nevertheless, experiments with *Chironomus thummi* and *C. plumosus* suggest that the hemoglobin is beneficial under certain conditions in promoting quicker recovery from enforced anoxia and longer survival under anoxia, and enables filter-feeding behavior at oxygen partial pressures of 12 to 14 mmHg. Species with hemoglobin are more active under conditions of experimentally falling partial pressure of oxygen than species of chironomids that do not have hemoglobin. Species with more hemoglobin in the hemolymph have a tendency to live in lakes with lower oxygen content than species that have little or no hemoglobin (Buck, 1953). Overall, chironomid larvae receive their oxygen supply most of the time by cutaneous respiration even if they do have hemoglobin, but it may improve their ability to survive short periods of very low oxygen tensions.

REFERENCES

Agosin, M. 1978. Functional role of proteins, pp. 93-203, in M. Rockstein (Ed.), *Biochemistry of Insects*, Academic Press, New York.

Ai, H., K. Kuwasawa, T. Yazawa, M. Kurokawa, M. Shimoda, and K. Kiguchi. 1995. A physiological saline for lepidopterous insects: Effects of ionic composition on heart beat and neuromuscular transmission. *J. Insect Physiol.,* 41:571-580.

Arnold, J.W. 1974. The hemocytes of insects, pp. 201-254, in M. Rockstein (Ed.), *Physiology of Insects*, Vol. 5, Academic Press, New York.

Arnold, J.W., and S.S. Sohi. 1974. Hemocytes of *Malacosoma disstria* Hübner (Lepidoptera: Lasiocampidae): morphology of the cells in fresh blood and after cultivation *in vitro*. *Can. J. Zool.,* 52:481-485.

Arvy, L. 1954. Données sur la leucopoièse chez *Musca domestica* L. *Proc. R. Ent. Soc. London A,* 29:39-41.

Ashhurst, D.E. 1982. Histochemical properties of the spherulocytes of *Galleria mellonella* L. (Lepidoptera: Pyralidae). *Int. J. Insect Morph. Embryol.,* 11:285-292.

Ashida, M., and H.I. Yamazaki. 1990. Biochemistry of the phenoloxidase system in insects: With special respect to its activation, pp. 239-256, in E. Ohnishi and H. Ishizaki (Eds.), *Molting and Metamorphosis*, Japan Scientific Societies Press, Tokyo; Springer-Verlag, Berlin.

Beard, R.L. 1953. Circulation, pp. 232-272, in K.D. Roeder (Ed.), *Insect Physiology*, John Wiley & Sons, New York, 1100 pp.

Buck, J.B. 1953. Physical properties and chemical composition of insect blood, pp. 147-190, in K.D. Roeder (Ed.), *Insect Physiology*, John Wiley & Sons, New York, 1100 pp.

Buckner, J.S., and J.M. Caldwell. 1980. Uric acid levels during last larval instar of *Manduca sexta*, an abrupt transition from excretion to storage in fat body. *J. Insect Physiol.,* 26:27-32.

Chapman, R.F. 1958. A field study of the potassium concentration in the blood of the red locust, *Nomadacris septemfasciata* (Serv.), in relation to its activity. *Anim. Behav.,* 6:60-67.

Chen, A.C., and S. Friedman. 1975. An isotonic saline for the adult blowfly, *Phormia regina* and its application to perfusion experiments. *J. Insect Physiol.,* 21:529-536.

Craig, R., and N.A. Olson. 1951. Rate of circulation of the body fluid in adult *Tenebrio molitor* Linnaeus, *Anasa tristis* (de Geer), and *Murgantia histrionica* (Hahn). *Science,* 113:648-650.

Davis, C.C. 1961. Periodic reversal of heart beat in the prolarva of a gyrinid. *J. Insect Physiol.,* 7:1-4.

Doucet, D., and M. Cusson. 1996. Role of calyx fluid in alterations of immunity in *Choristoneura fumiferana* larvae parasitized by *Tranosema rostrale*. *Comp. Biochem. Physiol.,* 114A:311-217.

Eslin, P., and G. Prevost. 1996. Variation in *Drosophila* concentration of haemocytes associated with different ability to encapsulate *Asobara tabida* larval parasitoid. *J. Insect Physiol.,* 42:549-555.

Florkin, M., and C. Jeuneaux. 1974. Hemolymph: Composition, pp. 255-307, in M. Rockstein (Ed.), *The Physiology of Insecta*, Academic Press, New York.

Gagen, S.J., and N.A. Ratcliffe. 1976. Studies on the *in vivo* cellular reactions and fate of injected bacteria in *Galleria mellonella* and *Pieris brassicae* larvae. *J. Invert. Pathol.,* 28:17-24.

Gerould, J.H. 1930. History of the discovery of periodic reversal of heart-beat in insects. *Science,* 71:264-265.

Gerould, J.H. 1933. Orders with heart-beat reversal. *Biol. Bull. Woods Hole,* 64:424-431.

Gillespie, J.P., M.R. Kanost, and T. Trenczek. 1997. Biological mediators of insect immunity. *Annu. Rev. Entomol.,* 42:611-643.

Grégoire, C.H. 1951. Blood coagulation in arthropods. II. Phase contrast microscopic observations on hemolymph coagulation in sixty-one species of insects. *Blood,* 6:1173-1198.

Grégoire, C., and G. Goffinet. 1979. Controversies about the coagulocyte, pp. 189-230, in A.P. Gupta (Ed.), *Insect Hemocytes*, Cambridge University Press, Cambridge.

Gupta, A.P. 1979a. Arthropod hemocytes and phylogeny, pp. 669-735, in A.P. Gupta (Ed.), *Arthropod Phylogeny*, Van Nostrand Reinhold, New York, 762 pp.

Gupta, A.P. 1979b. Hemocyte types: their structures, synonymies, interrelationships and taxonomic significance, pp. 86-127, in A.P. Gupta (Ed.), *Insect Hemocytes*, Cambridge University Press, Cambridge.

Harrison, J.F., C.J.H. Wong, and J.E. Phillips. 1990. Haemolymph buffering in the locust *Schistocerca gregaria*. *J. Exp. Biol.*, 154:573-579.

Harrison, J.F., J.E. Phillips, and T.T. Gleeson. 1991. Activity physiology of the two-striped grasshopper, *Melanoplus bivittatus*: Gas exchange, hemolymph acid-base status, lactate production, and the effect of temperature. *Physiol. Zool.*, 64:451-472.

Harrison, J.F., and J.E. Phillips. 1992. Recovery from acute haemolymph acidosis in unfed locusts. II. Role of ammonium and titratable acid excretion. *J. Exp. Biol.*, 165:97-110.

Harrison, J.F., C.J.H. Wong, and J.E. Phillips. 1992. Recovery from acute haemolymph acidosis in unfed locusts. I. Acid transfer to the alimentary lumen is the dominant mechanism. *J. Exp. Biol.*, 165:85-96.

Harrison, J.F., and M.J. Kennedy. 1994. *In vivo* studies of the acid-base physiology of grasshoppers: The effect of feeding state on acid-base and nitrogen excretion. *Physiol. Zool.*, 67:120-141.

Harrison, J.M. 1988. Temperature effects on haemolymph acid-base status *in vivo* and *in vitro* in the two-striped grasshopper *Melanoplus bivittatus*. *J. Exp. Biol.*, 140:421-435.

Harrison, J.M. 1989. Temperature effects on intra-and extracellular acid-base status in the American locust, *Schistocerca nitens*. *J. Comp. Physiol. B*, 158:763-770.

Heinrich, B. 1970. Thoracic temperature stabilization by blood circulation in a free-flying moth. *Science*, 168:580-582.

Heinrich, B. 1996. *The Thermal Warriors. Strategies of Insect Survival*, Harvard University Press, Cambridge, MA, 221 pp.

Hink, W.F. 1976. A compilation of invertebrate cell lines and culture media, pp. 319-369, in K. Maramorosch (Ed.), *Invertebrate Tissue Culture*, Academic Press, New York.

Hoffmann, J.A. 1970. Les organes hématopoïétiques de deux orthoptères: *Locusta migratoria* et *Gryllus bimaculatus*. *Z. Zellforsch.*, 106:451-472.

Hoffmann, J.A., D. Zachary, D. Hoffmann, and M. Brehélin. 1979. Postembryonic development and differentiation: hemopoietic tissues and their functions in some insects, pp. 29-82, in A.P. Gupta (Ed.), *Insect Hemocytes*, Cambridge University Press, Cambridge.

Jones, J.C. 1977. *The Circulatory System of Insects*, Charles C Thomas, Springfield, IL.

Jones, J.C., and D.P. Liu. 1968. A quantitative study of mitotic divisions in haemocytes of *Galleria mellonella* larvae. *J. Insect Physiol.*, 14:1055-1061.

Kanost, M.R., J.K. Kawooya, J.H. Law, R.O. Ryan, M.C. Van Huesden, and R. Ziegler. 1990. Insect Hemolymph Proteins. *Adv. Insect Physiol.*, 22:299-396.

Lackie, A.M. 1988. Haemocyte behaviour. *Adv. Insect Physiol.*, 21:85-178.

Levenbook, L. 1958. Intracellular water of larval tissues of the southern armyworm as determined by the use of C^{14}-inulin. *J. Cell. Comp. Physiol.*, 52:329-340.

Levenbook, L. 1979. Hemolymph volume during growth of *Calliphora vicina* larvae. *Ann. Entomol. Soc. Am.*, 72:454-455.

Markou, T., and G. Theophilidis. 2000. The pacemaker activity generating the intrinsic myogenic contraction of the dorsal vessel of *Tenebrio molitor* (Coleoptera). *J. Exp. Biol.*, 203:3471-3483.

Miller, T. 1968. Role of cardiac neurons in the cockroach heartbeat. *J. Insect Physiol.*, 14:1265-1275.

Miller, T., and R.L. Metcalf. 1968. Site of action of pharmacologically active compounds on the heart of *Periplaneta americana* L. *J. Insect Physiol.*, 14:383-394.

Miller, T., and W.W. Thomson. 1968. Ultrastructure of cockroach cardiac innervation. *J. Insect Physiol.*, 14:1099-1104.

Mori, H. 1996. Onset of embryonic heart movement in the water strider *Gerris paludum insularis* (Hemiptera: Gerridae). *Ann. Entomol. Soc. Am.*, 89:391-397.

Nappi, A.J., Y. Carton, and F. Frey. 1991. Parasite induced enhancement of haemolymph tyrosinase in a selected immune reactive strain of *Drosophila melanogaster*. *Arch. Insect Biochem. Physiol.*, 18:159-168.

Nässel, D.R. 1993. Neuropeptides in the insect brain: a review. *Cell Tissue Res.*, 273:1-29.

Nation, J.L. 1989. The role of pheromones in the mating system of *Anastrepha* fruit flies, pp. 189-205, in A.S. Robinson and G. Hooper (Eds.), *Fruit Flies, Their Biology, Natural Enemies and Control*, Vol. 3A, Elsevier, Amsterdam.

Nation, J.L., and K.K. Thomas. 1965. Quantitative studies on purine excretion in the greater wax moth, *Galleria mellonella*. *Ann. Entomol. Soc. Am.*, 58:883-885.

Nation, J.L., B. Smittle, and K. Milne. 1995. Radiation-induced changes in melanization and phenoloxidase in Caribbean fruit fly larvae (Diptera: Tephritidae) as the basis for a simple test of irradiation. *Ann. Entomol. Soc. Am.*, 88:201-205.

Nittono, Y. 1964. Formation of haemocytes near the imaginal wing disc in the silkworm, *Bombyx mori* L. *J. Seric. Sci. Jpn.*, 33:43-45.

Nutting, W.L. 1951. A comparative anatomical study of the heart and accessory structures of the orthopteroid insects. *J. Morphol.*, 89:501-597.

Pass, G. 1985. Gross and fine structure of the antennal circulatory organ in cockroaches (Blattodea, Insecta). *J. Morphol.*, 185:255-268.

Pass, G. 1988. Functional morphology and evolutionary aspects of unusual antennal circulatory organs in *Labidura riparia pallas* (Labiduridae, *Forficula auricularia* L. and *Chelidurella acanthopygia* Géné (Forficulidae) (Insecta: Dermaptera). *Int. J. Insect Morphol. Embryol.*, 17:103-112.

Pech, L.L., and M.R. Strand. 2000. Plasmatocytes from the moth *Pseudoplusia includens* induce apoptosis of granular cells. *J. Insect Physiol.*, 46:1565-1573.

Pichon, Y. 1970. Ionic content of haemolymph in the cockroach, *Periplaneta americana*. A critical analysis. *J. Exp. Biol.*, 53:195-209.

Raina, A.K. 1976. Ultrastructure of the larval hemocytes of the pink bollworm, *Pectinophora gossypiella* (Saunders) (Lepidoptera: Gelechiidae). *Int. J. Insect Morphol. Embryol.*, 5:187-195.

Ratcliffe, N.A., and P. Götz. 1990. Functional studies on insect haemocytes, including non-self recognition. *Res. Immunol.*, 141:919-922.

Rizki, R.M., and T.M. Rizke. 1980. Hemocyte responses to implanted tissues in *Drosophila melanogaster* larvae. *Roux' Arch. Dev. Biol.*, 189:207-213.

Rowley, A.F., and N.A. Ratcliffe. 1981. Insects, pp. 471-490, in N.A. Ratcliffe and A.F. Rowley (Eds.), *Invertebrate Blood Cells*, Vol. 2, Academic Press, New York.

Sarkaria, D.S., S. Bettini, and R.L. Patton. 1951. A rapid staining method for clinical study of cockroach blood cells. *Can. Entomol.*, 83:329-332.

Shatoury, H.H. 1966. Micro-determination of insect blood volume. *Nature,* 211:317-318.

Sláma, K. 1999. Active regulation of insect respiration. *Ann. Entomol. Soc. Am.*, 92:916-929.

Smits, A.W., W.W. Burggren, and D. Oliveras. 2000. Developmental changes in *in vivo* cardiac performance in the moth *Manduca sexta*. *J. Exp. Biol.*, 203:369-378.

Snodgrass, R.E. 1935. *Principles of Insect Morphology*, McGraw-Hill, New York.

Söderhäll, K., and A. Aspán. 1993. Prophenoloxidase activating system and its role in cellular communication, pp. 113-129, in J.P.N. Pathak (Ed.), *Insect Immunity*, Oxford & IBH Publishing Co., New Delhi.

Weber, H. 1930. *Biologie der Hemipteren*, Julius Springer, Berlin.

Weidler, D.J., and G.C. Sieck. 1977. A study of ion binding in the hemolymph of *Periplaneta americana*. *Comp. Biochem. Physiol.*, 56A:11-14.

Wharton, R.A., M.L. Wharton, and J. Lola. 1965. Blood volume and water content of the male American cockroach *Periplaneta americana* L. Methods and influence of age and starvation. *J. Insect Physiol.*, 11:391-404.

Wheeler, R.E. 1962. Changes in hemolymph volume during the moulting cycle of *Periplaneta americana*. *Fed. Proc.*, 21:123 (abstract).

Wheeler, R.E. 1963. Studies on the total haemocyte count and haemolymph volume in *Periplaneta americana* (L.) with special reference to the last molting cycle. *J. Insect Physiol.*, 9:223-235.

Woodring, J.P. 1985. Circulatory systems, pp. 5-57, in M.S. Blum (Ed.), *Fundamentals of Insect Physiology,* John Wiley & Sons, New York.

Yamashita, M., and K. Iwabuchi. 2001. *Bombyx mori* prohemocyte division and differentiation in individual microcultures. *J. Insect Physiol.*, 47:325-331.

Zachary, D., M. Brehélin, and J.A. Hoffmann. 1975. Role of the "thrombocytoids" in capsule formation in the dipteran *Calliphora erythrocephala*. *Cell Tissue Res.*, 162:343-348.

12 Respiration

CONTENTS

PREVIEW

Respiration, used in the sense that it means breathing and gas exchange, is a function of the tracheal system, a tubular network that originates at spiracular openings on the body surface and radiates to all parts of the insect body. Spiracular valves at the body surface often may be closed to reduce water loss from the system. Large longitudinal and transverse tracheae can be up to 0.2 mm in diameter, and branches from these become smaller in diameter as they penetrate between cells and tissues and finally terminate as tracheoles. Tracheoles are blind tubules less than 1 μm in diameter. Often, especially in active tissues that demand rapid gas exchange, the tracheoles push against and indent cell membranes, like pushing a finger into a soft balloon, until they terminate within a few micrometers of mitochondria. The tracheal system is very efficient for insects and delivers oxygen to flight muscles in sufficient amounts so that they conduct aerobic metabolism even during flight. In some very small insects, simple diffusion of gases through the system of tracheal tubules may suffice, but most insects actively ventilate the system by muscular pumping motions. Many insects demonstrate discontinuous ventilation, in which the spiracles are tightly closed or flutter almost imperceptibly for varying periods interspersed with bursts of open spiracles. The tracheal lining of tracheae are molted at each molt, but the lining of tracheoles is not molted. Aquatic insects have a tracheal system that is essentially the same in structure as that of terrestrial insects. Many adaptations to an aquatic life occur, however, in the tracheal system, including compressible gas bubbles (gas gills), incompressible gas films (plastrons), and thin cuticular flaps (gills) that take oxygen from water by cutaneous diffusion. Insect eggs often have plastrons as a part of the egg shell that aid gas exchange for the developing embryo. In addition to gas exchange, the tracheal system has been adapted in some insects to serve nonrespiratory functions, such an attachment site for endocrine cells, participation in sound reception, sound production such as a source of hissing sounds, and delivery of a distasteful froth. In all insects, the vast network of tracheae and tracheoles ties cells and tissues together.

12.1 INTRODUCTION

Insects breathe by delivering air through a network of small tubes to within a few micrometers of mitochondria in cells. These tubes, the **tracheae**, arise at openings on the sides of the body, the **spiracles**. In most insects, there are interconnecting longitudinal and transverse tracheal trunks, and sometimes large air sacs that tie together the entire system (Figure 12.1). Larger tracheal tubes send off branches that become smaller in diameter as they ramify to all tissues, with the smallest diameter **tracheoles** (tubes less than 1 μm diameter) (Figure 12.2) touching most cells and even indenting some cells. The air flow may be tidal or directed, depending on the insect and its physiological state. The interconnections make directed air flow possible, in which air enters through one or more anterior spiracles, gets pumped through the body by muscular ventilatory movements, and is expelled through one or more posterior spiracles. Such a directed flow is more efficient than tidal inflow and outflow from the same spiracles because the system is constantly flushed and incoming air is not mixed with used air. In some very simple tracheal systems, the tracheae arising from each spiracle are independent of other tracheae and spiracles, and only tidal flow is possible.

Whitten (1972) has provided a review of comparative anatomy of the tracheal system among insect groups, and various aspects of the physiology of gas exchange and breathing have been reviewed by Miller (1966a), Sláma (1994), Hadley (1994), Snyder et al. (1995), and Lighton (1996). Dyby (1998) has developed a novel method for infiltrating the tracheal system of newly hatched insects for effective visualization.

12.2 STRUCTURE OF THE TRACHEAL SYSTEM

12.2.1 TRACHEAE AND TRACHEOLE STRUCTURE

The tracheal system develops from embryonic ectodermal tissue. Tracheae and tracheoles have an epicuticular lining, comprised primarily of a cuticulin layer that is continuous with the external cuticle. Larger tracheal trunks have an endocuticle layer that gives more strength to the tubular structure. A hydrophobic substance is secreted on the lumen surface of tracheae. The hydrophobic secretion helps prevent water from entering the tracheae and reduces evaporative water loss from the humid, extensive internal tracheal surfaces.

The major distinction between tracheae and tracheoles is one of size. Tubes down to about 1 μm in size are called **tracheae** (singular: **trachea**) while those smaller than 1 μm are **tracheoles**. The lower limit for effective tracheolar diameter is limited by the mean free path of diffusing oxygen molecules, which is about 0.072 μm at 300K (27°C) and 1 atm according to Pickard (1974). The smallest tracheoles that have been observed are slightly smaller than 0.2 μm, or are thus 2 to 3 times the limiting diameter. Tracheoles in the lantern of some fireflies are very specialized in

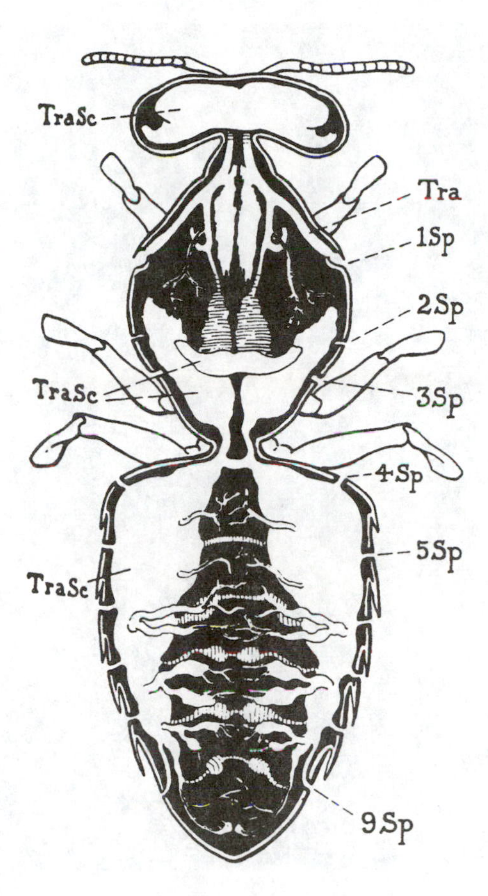

FIGURE 12.1 Large longitudinal tracheal sacs in the body of a honeybee. Sp, spiracle openings in the pleural region of the body; Tra, trachea; TraSc, tracheal sac. (From *The Hive and the Honey Bee*, Dadant & Sons, editors and publisher, 1975. Permission granted with acknowledgment.)

structure, with stiff, reinforcing material in the tracheole to help it resist folding or collapse under what appear to be conditions in which there are rapid and strong osmotic changes across the tracheolar cell membrane when nerve impulses signal the change in permeability (Ghiradella, 1978).

Thickened, tight spirals of the cuticular intima, the **taenidia**, strengthen tracheae and tracheoles, provide elasticity, and help the tubes resist compression and collapse (Figures 12.3 and 12.4). Even tracheoles have taenidial reinforcements, contrary to older reports published prior to availability of the electron microscope. In larger tracheae, the taenidia are up to 450 nm in width and are spaced about 300 nm apart but in tracheoles the taenidia are smaller (50 to 80 nm width) and are spaced further apart than their width. Within the taenidial folds there is a component, probably similar to procuticle, that adds strength to the taenidia (Bordereau, 1975). The cuticular intima is as thick as 200 nm at the taenidial spirals, and as thin as 10 to 40 nm between spirals. The micelles of the cuticulin layer are oriented so that their long axis is parallel to the long axis of a trachea or tracheole in intertaenidial areas, but perpendicular to the long axis (i.e., circular in orientation) within the taenidial thickenings. The orientation of the cuticulin micelles lends strength to the tubes.

When tracheae first form, they have smooth walls but the taenidia soon appear. Locke (1958a) proposed that taenidial formation is the result of expansion and buckling of the tracheal wall. With a mathematical model, he accounted for the frequency of taenidia, tube-wall thickness, and orientation of the cuticulin micelles within the taenidia and intertaenidial regions. The buckling hypothesis may explain taenidial formation, but to date experimental proof for it is lacking, and the buckling theory is not universally accepted.

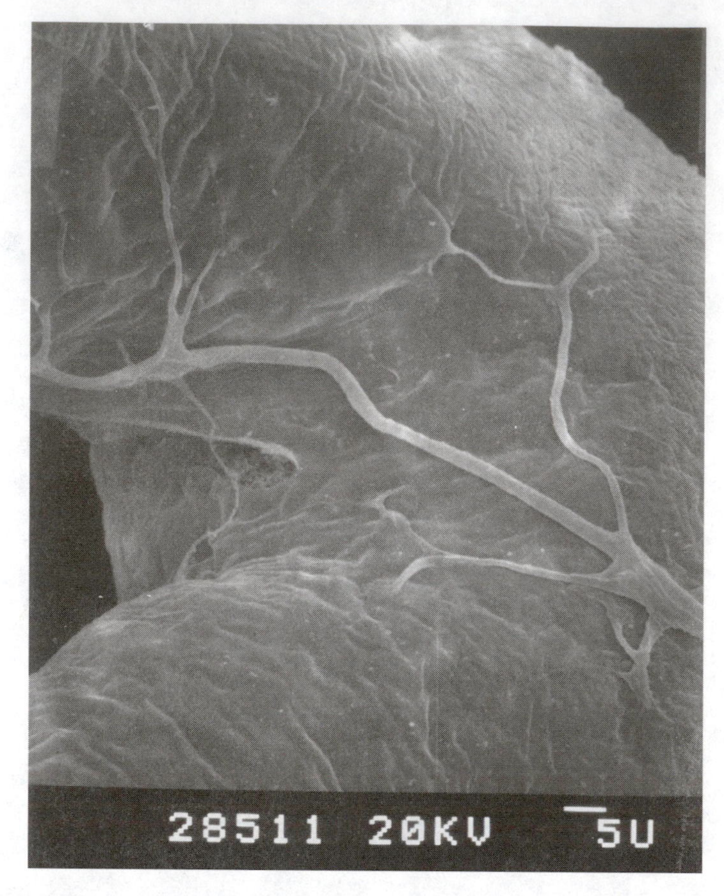

FIGURE 12.2 An SEM view of a tracheal tube branching into many fine tracheoles that disappear beneath the surface of cells in the salivary gland of a male Caribbean fruit fly *Anastrepha suspensa.* (Photograph by the author.)

12.2.2 Spiracle Structure and Function

The openings of the tracheal system at the body surface are called **spiracles**. Spiracles usually occur on the pleural surfaces of the body, typically one on each side of each segment, but numerous variations have evolved. Most terrestrial insects can close the spiracles as an adaptation for water conservation. The simplest closing mechanism consists of folds of the integument that can be pulled together over the spiracular opening by a closer muscle. Many insects have valves or flaps of cuticle (Figures 12.5 and 12.6) that rotate and close over the spiracle. Sometimes, an opener muscle is present, but in most cases opening occurs because of elasticity of the cuticle when the closer muscle relaxes. The spiracle on the prothorax of Orthoptera has both closer and opener muscles; and although the spiracle opens partly by natural elasticity after relaxation of the closer muscle, the opener muscle can cause the spiracle to open more widely for increased ventilation (Miller, 1960a).

The openings can be simple unguarded pores, but frequently there are additional structural details associated with the openings. Typically, a ring of sclerotized cuticle, the peritreme, around the spiracular opening provides reinforcement. In many insects, there is a slightly enlarged chamber or atrium just inside the spiracular opening from which tracheae branch in a variety of directions. The atrium may contain various structural adaptations to filter the air, such as dust-catching setae or hairs in the atrium space (Figure 12.7). Many terrestrial Diptera, Coleoptera, Lepidoptera, and some aquatic insects have thin, perforated partitions, called sieve plates, in the atrium that act as a filter to keep dust particles out of the tracheae. In aquatic insects, it aids in excluding water. A "felt chamber," a dense mat of very fine hairs and setae, occurs just inside the spiracle of some dipterous insects.

FIGURE 12.3 An SEM cut-away view of the inside of a large trachea of a mole cricket *Scapteriscus acletus*, showing the origin of two smaller tracheae and taenidial windings. (Photograph by the author.)

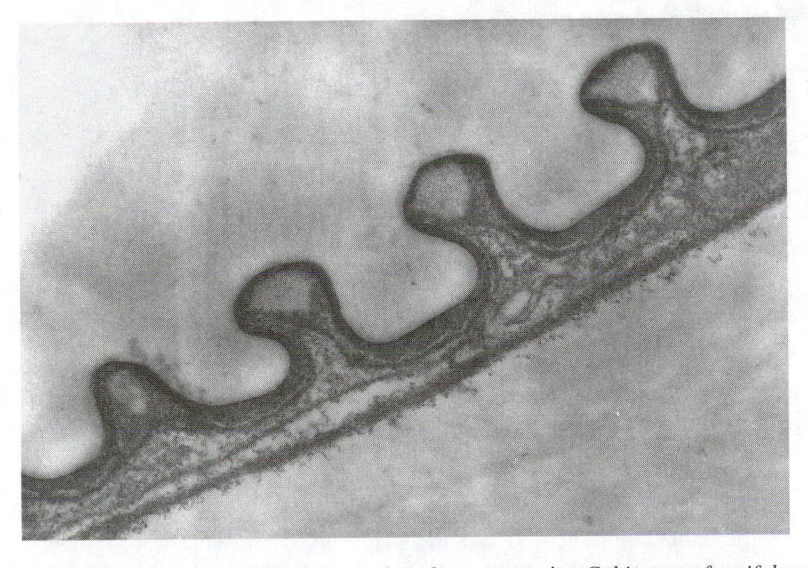

FIGURE 12.4 A lateral view of taenidia in a trachea from a termite *Cubitermes fungifaber.* (Photograph courtesy of Bordereau, 1975.)

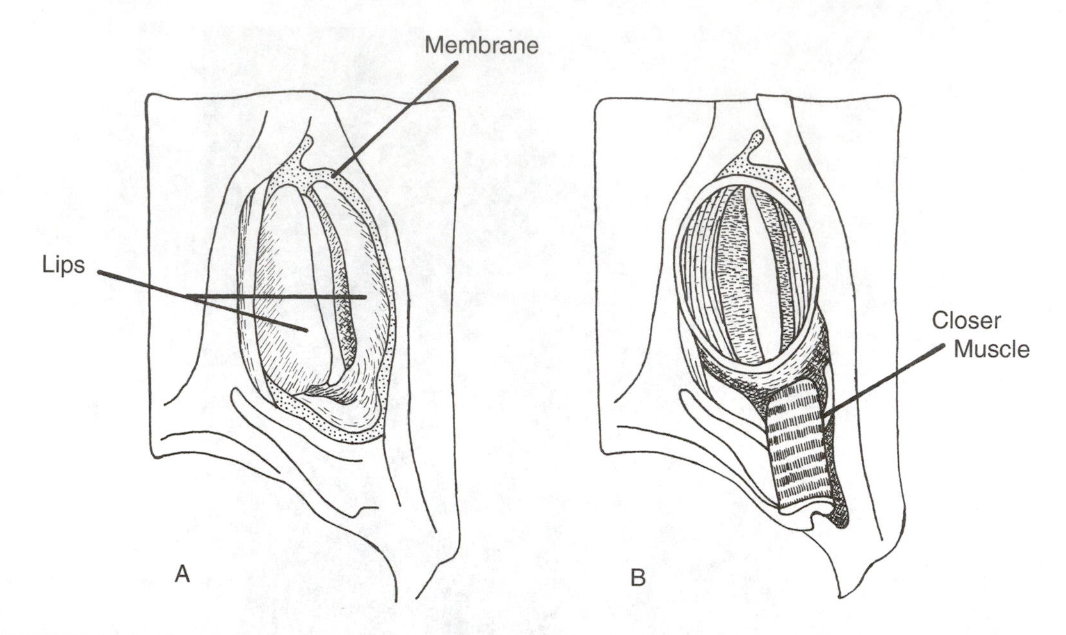

FIGURE 12.5 The second thoracic spiracle of a grasshopper illustrating outer (A) and inner (B) views of spiracular valves. (Modified from Snodgrass, 1935.)

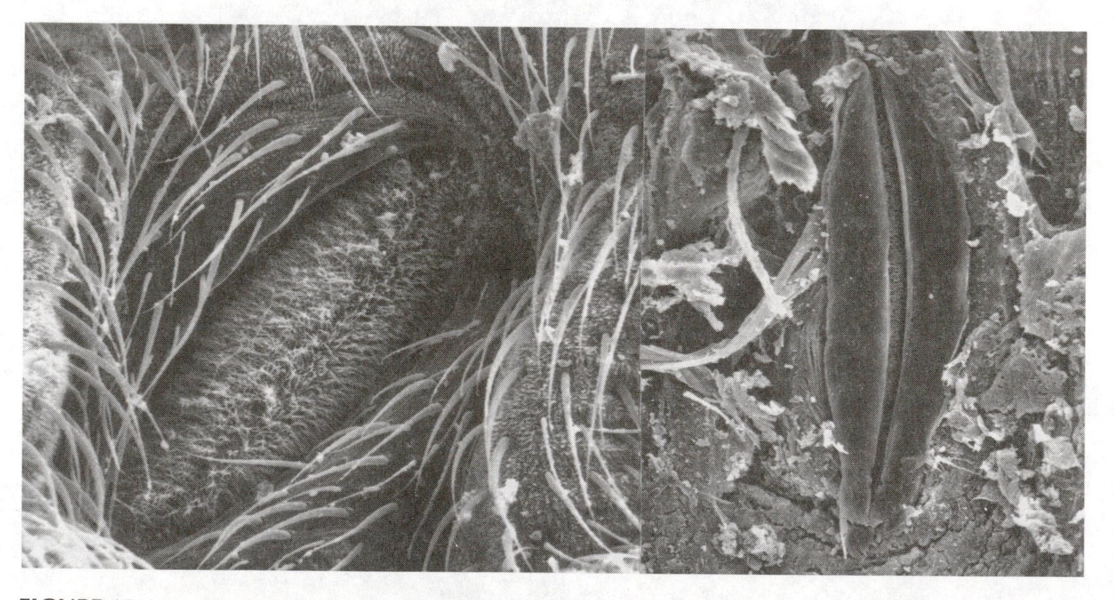

FIGURE 12.6 A scanning electron micrograph of the external (left) and internal (right) views of the valves that close the spiracular opening of a mole cricket *Scapteriscus vicinus*. The many fine setae on the valve and around the spiracle surface may trap a film of air and act as a plastron or compressible gill when the cricket gets flooded in its underground tunnel. (Photographs by the author.)

The muscles associated with the spiracles are innervated by a branch of the median nerve from the ganglion in the same segment or from the ganglion in the anterior segment (Case, 1957). Repetitive action potentials from the median nerve cause contracture of the closer muscle and close

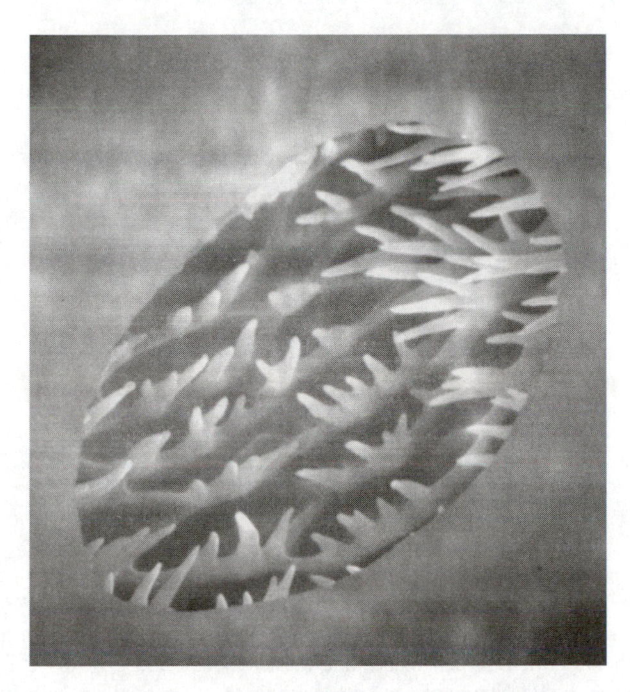

FIGURE 12.7 Setae inside the external opening of a spiracle of the black turpentine beetle *Dendroctonus terebrans* that probably act as a dust and particle filter to protect the tracheal system. (SEM photograph by the author.)

the spiracle. The closer muscle of *Hyalophora cecropia* has a myogenic rhythm leading to slow (graded) pacemaker potentials that give rise to spike discharges (Van der Kloot, 1963). Experimental increments of hyperpolarization of the muscle membrane potential slows the rate by which the pacemaker potentials depolarize the membrane, and strong hyperpolarization stops the ability of the graded potentials to exceed the threshold for action potentials. Prolonged exposure of the closer muscle to CO_2 causes sufficient hyperpolarization that the pacemaker potentials are ineffective and the muscle relaxes and the spiracle opens. Anesthetization of insects with CO_2 is a common practice in laboratory experiments, and in most insects this treatment causes the spiracles to open and results in significant loss of water if the insects are kept anesthetized very long and/or not given access to water after recovery.

A bilateral pair of spiracles occurred on each thoracic segment, and one pair on each abdominal segment in the evolutionary primitive arrangement. Apparently, no openings evolved on the head. Some existing Diplura have 12 spiracles, but most insects have fewer (Edwards, 1953). Some insects have only one or two functional spiracles, and a few, particularly some aquatic ones, have no functional spiracles at all, and tracheae do not open to the outside. Gas exchange between tracheae and the environment in this latter type of system occurs by diffusion through a thin cuticle.

12.2.3 The Tracheal Epithelium

The tracheal system is not an acellular system of tubes and tubules; every part of the system contains living cells. **Tracheal epithelial cells**, derived like cuticular epidermal cells from embryonic ectoderm, surround tracheal tubes and tracheoles throughout all parts of the system. The epithelium cells are flattened, like pieces of a ribbon that might be cut and pasted around a tube. These epithelial cells secrete the cuticulin lining and the hydrophobic compounds at the lumen surface of the tubules. In larger tracheae, the cells secrete a new cuticle lining on their apical side (i.e., toward the old cuticle) prior to each molt, and the old lining of tracheae is shed at each molt.

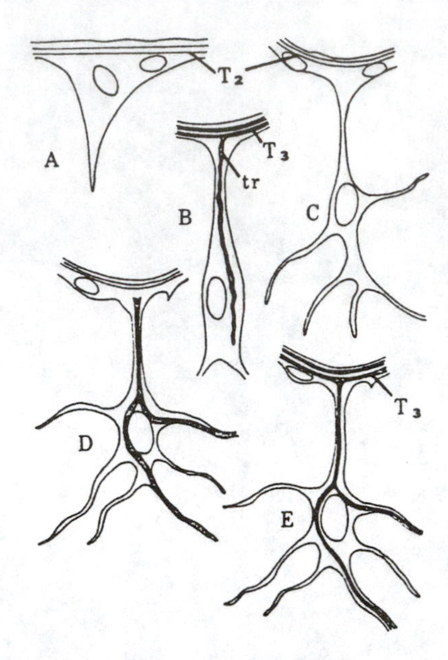

FIGURE 12.8 Stages in the growth of new tracheoles in *Sciara coprophila*, a fungus fly (Diptera). A: Enlargement of a tracheal epithelium cell to become a tracheoblast and growth of a pseudopod toward the area needing more oxygen. B: Beginning of a tracheole within the elongated epithelium cell. C: A tracheoblast that has developed several cytoplasmic filaments before formation of a tracheole has begun. D: Tracheole that has formed in a tracheoblast with multiple cytoplasmic extensions. E: The tracheole will only receive an open contact with the preexisting tracheal trunk at the next molt. (Modified from drawings of Keister, 1948.)

12.2.4 Development of New Tracheoles

The formation of new tracheae and tracheoles varies with respiratory demand of tissues (Locke, 2001). Active larvae of the tenebrionid mealworm *Tenebrio molitor* show a large increase in number and diameter of tracheae during exposure to a 10% oxygen atmosphere, and occluding the spiracles induces growth of new tracheae and tracheoles, particularly on muscles and gut that have a high demand for gas exchange (Locke, 1958b).The process by which new tracheoles form is primarily based on the work of Margaret Keister (1948), who patiently observed events occurring in the cells of *Sciara coprophila* Lintner, a mycetophilid fly whose larvae are so transparent that the internal organs can be observed with transmitted light. The initial event in development of a new tracheole begins when a tracheal epithelial cell begins to grow out, usually in a triangular shape, toward a group of cells that (apparently) need better gas exchange (Figure 12.8). It may be that metabolites from tissue cells induce the tracheal epithelial cell to grow a process out toward the tissue. A tracheal epithelial cell that initiates such growth has been called, by various authors, a tracheoblast, stellate cell, transition cell, and tracheal end cell. Such a cell may be at the end of an existing small trachea, but not necessarily so. As it grows, the tracheoblast tends to become spindle-shaped, and a single unbranched tracheole may form within it at this stage. More often, however, the cell develops multiple finger-like projections (hence the name stellate cell) before a tracheole forms, and then tracheole branches may form in the several fingers. A fluid-filled linear channel begins to appear in the cytoplasm of the tracheoblast and gradually becomes longer as the tracheole grows. The tracheole does not grow from the larger trachea, but grows toward it, the tracheoblast never completely having detached from its home base on the surface of a larger trachea. The new tracheole only becomes air-filled and functional (in the *Sciara* model), however, at the next molt.

Wigglesworth (1954) observed essentially the same processes occurring in the cuticular epithelium of the hemipteran *Rhodnius prolixus*. Columns of epithelial cells grew out from an existing tracheal tube, and new tracheae and tracheoles developed within the outgrowing cells. New tracheae and

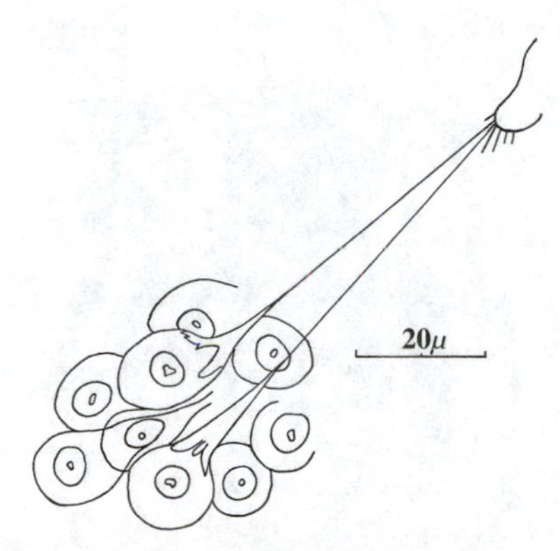

FIGURE 12.9 Contractile strands from epidermal cells pulling a tracheole into an oxygen-deficient region in the epidermis of *Rhodnius prolixus*. (Reproduced with permission from Wigglesworth, 1959.)

tracheoles rapidly grew toward regions made artificially deficient in gas exchange by cutting existing small tracheae to a section of cuticular epithelium, and into transplanted organs. Wigglesworth (1959, 1981) observed that tracheoles were pulled by cytoplasmic strands radiating from the cells needing oxygen (Figure 12.9). In some cases, the tips of tracheoles migrated as much as a millimeter in distance.

Tracheoles also move in cultured *Galleria mellonella* wing discs and in organ transplants in Diptera. Tracheole migration is ecdysone dependent in *Galleria* and may be related to the presence of large numbers of microtubules in cultured wing discs. Microtubule formation is disrupted and tracheole migration is prevented by 10^{-8} *M* vinblastin or 10^{-5} *M* colchicine in the culture medium (Oberlander, 1980).

12.2.5 Air Sacs

Dilations of both primary and secondary tracheae occur in many insects and are called **air sacs** (Figures 12.1 and 12.10). They are variable in size but frequently are large in flying insects such as honeybees, cicadas, many adult Diptera, and some scarab and buprestid beetles. The intima of air sacs may contain typical taenidia, but these may be reduced or irregular. With *Drosophila* as a model, Weis-Fogh (1964a) showed that the air sacs provide a large surface in contact with flight muscles for exchange of gases. The rhythmical squeezing action of working flight muscles pumps the air sacs like a bellows and increases the flow of air through the system. Air sacs sometimes collapse as growing tissues fill the body space; and by collapsing, they make room for the new tissue or organ, with little change in the general body shape. Air sacs serve a hydrostatic function in some aquatic insects and allow more freedom in vibration of the tympanic membrane in some sound-producing insects. They may increase the hemolymph concentration of solutes without necessarily increasing total solute, and may reduce the hemolymph volume by restricting space that the volume of circulating hemolymph must serve.

12.2.6 Molting of Tracheae

Prior to a molt, tracheal epithelial cells secrete a new cuticulin layer on their apical surface (the surface toward the lumen of the trachea). Often, a fluid layer then separates the two cuticulin layers, the old and the new. The ecdysis of the external cuticle pulls the old, air-filled tubes out, leaving behind the newly formed, fluid-filled ones (Figure 12.11). The new tracheal tube rapidly fills with air, and although details are scarce, it seems likely that the fluid in the new system is actively reabsorbed into the surrounding tissue, thus making way for the air. Wigglesworth (1959) found that tracheoles, which are never shed in *Rhodnius*, become cemented to the new tracheae at each molt by a ring of adhesive material.

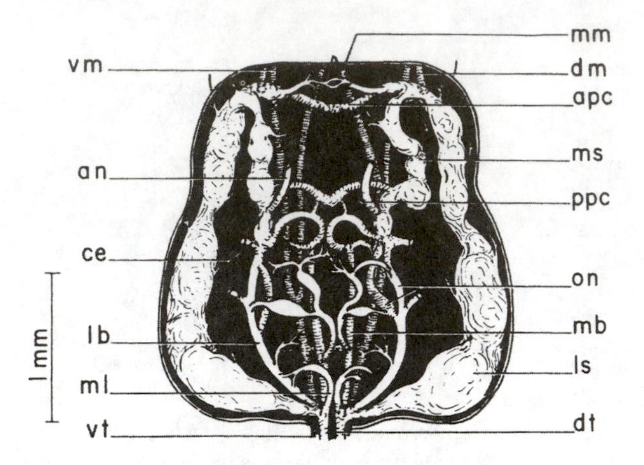

FIGURE 12.10 Tracheal supply and large air sacs in the head of ant *Camponotus pennsylvanicus*: an, branch to antenna; apc, anterior commissure ventral to pharynx; ce, branches to compound eye; dm, dorsal trachea to mandible; dt, dorsal trachea from thorax; lb, trachea lateral to brain; ls, large lateral air sac; mb, trachea medial to brain; ml, loop around muscle; mm, trachea to mouthparts; ms, small median air sac; on, trachea encircling optic nerve; ppc, posterior commissure ventral to pharynx; vm, ventral trachea to mandible; vt, ventral trachea from thorax. (Reproduced with permission from Keister, 1963.)

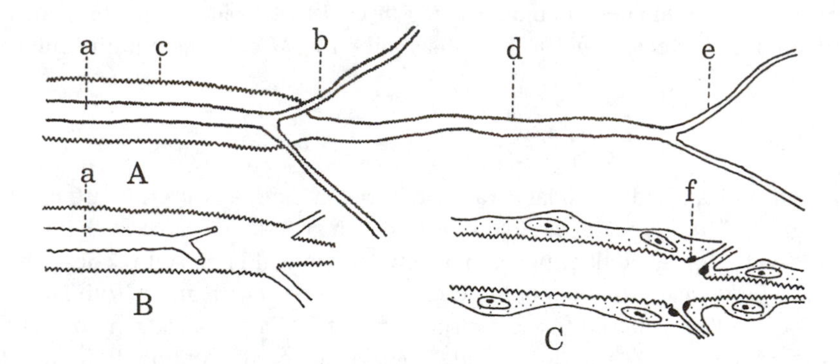

FIGURE 12.11 Molting of tracheae. A: trachea and tracheoles just before a molt of *Rhodnius prolixus*: a, previously existing air-filled trachea; b, previously existing tracheoles; c, new and still fluid-filled trachea; d, e, new trachea and tracheole, respectively, but not yet air-filled and functional. B: Old trachea with short segments of old tracheoles being pulled from the new trachea that is now air-filled. C: Rings of cement reattach the tracheoles to the new trachea. (Modified from Wigglesworth, 1959.)

12.3 TRACHEAL SUPPLY TO TISSUES AND ORGANS

The heart and dorsal musculature are usually aerated by tracheal branches in each segment from the dorsal longitudinal trunks. The visceral and internal reproductive organs receive tracheal branches from the lateral trunks. The ventral nerve cord and ventral musculature are supplied by branches from the ventral trunks or by branches off ventral transverse connectives (Figure 12.12). The legs and wings are supplied by tracheae from the thoracic spiracles, while the head and associated structures are generally supplied with branches from the first spiracle and the dorsal longitudinal trunk.

The only cells or tissues not having direct connections with tracheoles are the hemocytes, the cells that circulate in the hemolymph. Locke (1998) has shown that anoxia causes structural changes in hemocytes and causes sessile hemocytes to be released from the surface of various tissues where they may be attached. Under anoxic conditions, hemocytes accumulate on very thin-walled tufts

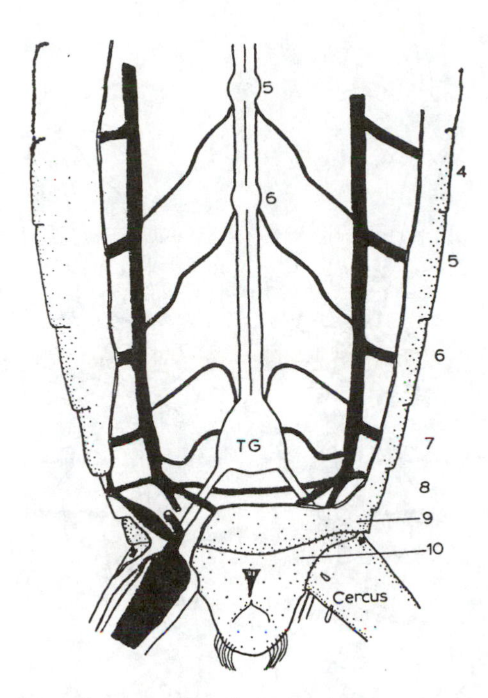

FIGURE 12.12 Tracheae supplying abdominal ganglia in the abdomen of the house cricket *Acheta domesticus*. (Reproduced with permission from Longley and Edwards, 1979.)

of tracheae near the last pair of abdominal spiracles and in the **tokus** compartment at the tip of the abdomen (Figure 12.13). Thus, the highly branched tracheal system of the last segment and the tokus appear to serve the function of a lung for aeration of hemocytes.

12.3.1 Adaptations of Tracheae to Supply Flight Muscles

Working muscles, especially flight muscles, have a very high demand for oxygen (O_2), and there is an extensive tracheal supply to flight muscles enabling them to avoid an oxygen debt even with prolonged flight. For example, an adult blowfly, *Lucilia sericata,* requires about 33 to 50 μl O_2/min/g body weight at rest; but within seconds of taking flight, it increases its O_2 consumption to as much as 1625 μl/min/g body weight. Most of this O_2 is used by flight muscles. This rate of O_2 use is from 30 to 50 times the maximum rate of O_2 used by active vertebrate leg or heart muscle per unit volume. Flight muscles typically have an extensive tracheal supply

Primary tracheae that originate at a thoracic spiracle may (1) pass through the core of a muscle, giving rise to numerous branches (centro-radial system); (2) run along its surface, with branches penetrating deep into the interior of the muscle (latero-radial system); or (3) the tracheae may expand into the air sacs that lie on top of the muscle (latero-linear system) (Figure 12.14) (Weis-Fogh, 1964a). The latter arrangement is common in small insects. Some insects, such as the locust *Schistocerca gregaria*, have each of the three systems in some wing muscles. Secondary tracheae branch from the primary tracheae and radiate into the spaces between muscle fibers, and tertiary tracheae branch from the secondaries. In the tergosternal muscle (a flight muscle) of *Aeshna* spp. dragonflies, tertiary tracheae branch from the secondary tracheae at about 20-μm intervals, and the tertiary tracheae eventually narrow and each branches into 20 to 30 tracheoles. In metabolically active tissues, and especially in flight muscles, tracheoblasts push against the sarcolemma of a muscle fiber, and indent the fiber like a finger pushed into a balloon to become **intracellular** tracheoles. Despite the term "intracellular," the tracheoles are not inside the cells they penetrate, because they still maintain their own cellular epithelium, and the host-cell membrane, although

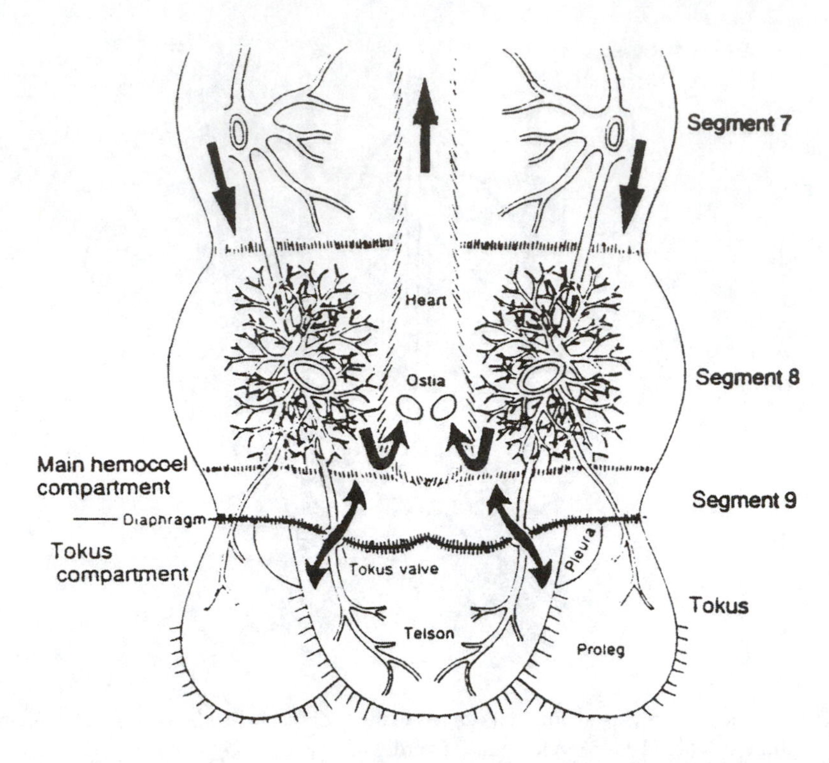

FIGURE 12.13 The tracheal cluster at the 8th abdominal spiracle in lepidopterous larvae and tracheal extensions into the tokus at the tip of the abdomen, sites where hemocytes accumulate and become aerated. (Reproduced with permission from Locke, 1998.)

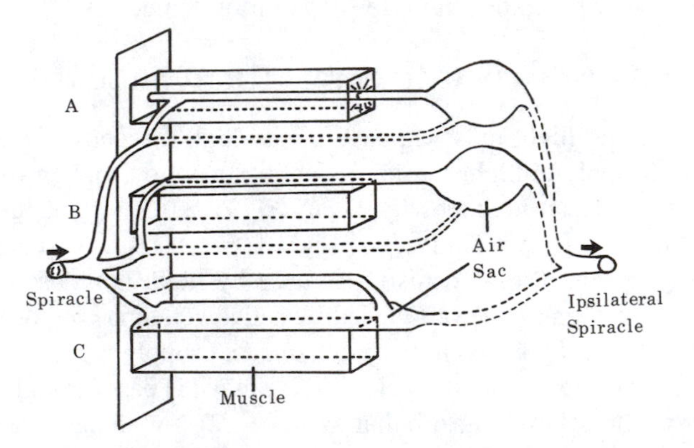

FIGURE 12.14 Three basic types of primary tracheal supply to wing muscles: A: centroradial. B: lateroradial. and C: laterolinear. Arrows indicate the direction of air flow. A muscle shunt, typical of many muscles, is shown by a broken line. (Reproduced with permission from Weis-Fogh, 1964a.)

indented, remains intact. Functionally, intracellular tracheole's bring gas exchange capability within a few tenths of a micrometer of mitochondria.

The tracheal system is very efficient, delivering O_2 through an air path branching into smaller and smaller tracheoles that often end within a few micrometers of mitochondria. The tissues of insects typically maintain constant rates of O_2 consumption when atmospheric O_2 levels are as low as a few kPa (1 Torr = 1 mmHg = 133.32 Pa; 7.5006 mmHg = 1 kPa). This resting safety margin may have evolved in response to several selective pressures, including (1) a relatively low resting metabolic rate that can undergo up to a 50-fold increase during intense activity, such as flight;

(2) muscular pumping mechanisms that ventilate the tracheal system when demand increases; and (3) regulation of tracheal conductance (quantity of gas transferred divided by the partial pressure gradient) (Greenlee and Harrison, 1998). Although the efficiency of the tracheal system enables working flight muscle to perform aerobically at normal atmospheric O_2 levels, the safety margin is reduced so that hypoxic gas mixtures of 5 to 10 kPa O_2 decrease the flight metabolic rate in the dragonfly *Erythemis (Mesothemis) simplicollis* (Harrison and Lighton, 1998).

12.4 VENTILATION AND DIFFUSION OF GASES WITHIN THE SYSTEM

Movement of O_2 and CO_2 through tracheal tubes is promoted by active ventilation in most insects. Simple diffusion may suffice in very small insects; but in all insects, body movements and muscle and gut movements serve to pump the tracheal system and move gases within it, and aid gas exchange within the system.

12.4.1 THE CASE FOR SIMPLE DIFFUSION

The idea that simple diffusion may suffice has a long history. In the early part of the 20th century, August Krogh concluded that the process of diffusion alone was sufficient to supply the O_2 demands of small insects, but some of his assumptions were questioned, and Weis-Fogh re-examined the problem. In brief, Weis-Fogh (1964b) confirmed Krogh's earlier conclusions that diffusion is, indeed, adequate in very small insects, but not in larger ones. Even in small insects, there probably is pumping ventilation by muscle action and body movements. Weis-Fogh (1964b) characterized the area in a microscopic cross section of tissue that is occupied by tracheal tubes as the "hole fraction" value, and he found values ranging from 10^{-1} to 10^{-2} for the secondary tracheal supply to wing muscles, and 10^{-2} to 10^{-3} for the tertiary tracheae and tracheoles to wing muscles. Mathematical expressions (19 different ones in all) were derived to describe diffusion in different tracheal branching systems. The simplest formula for calculating the pressure necessary to cause oxygen to diffuse from the spiracles to the site of use applies to those systems in which the cross-sectional area of all tracheae is constant from the last point of branching, as in *Cossus* (Lepidoptera) and *Rhodnius* (Hemiptera) systems, and can be calculated as

$$\Delta p = mL^2/2aP \qquad (12.1)$$

where Δp is the pressure necessary to cause oxygen to diffuse over the distance L (cm) to satisfy the metabolic rate m, in ml O_2/g tissue/min that is used. P is a permeability constant for O_2 diffusion in an air path (about 12 ml/min/cm^2/atm/cm). The hole fraction is represented by a. Different patterns of branching called for more complex treatments, for example, in systems such as those of *Schistocerca* and *Aeshna* in which the tracheal cross-sectional area decreases with branching.

Weis-Fogh (1964b) concluded that if an insect uses about 25% of the O_2 entering the spiracles, there would be enough pressure drop from the spiracle to the tracheole in small insects to allow O_2 to diffuse at a rate that could supply a few milliliters O_2 per gram tissue per hour. He concluded, however, that diffusion would not be sufficient to supply larger amounts of O_2 or to supply it over distances involved in larger insects, and that active ventilation of the tracheal system was necessary. The fact that as much as 10% of the muscle tissue consists of tracheae, and that they are ventilated, facilitates aerobic metabolism.

12.4.2 ACTIVE VENTILATION OF TRACHEAE

Insects as large as, or larger than, *Drosophila melanogaster* actively ventilate the tracheal system by muscle and body movements (Weis-Fogh, 1967; Lewis et al., 1973; Sláma, 1988, 1994, 1999;

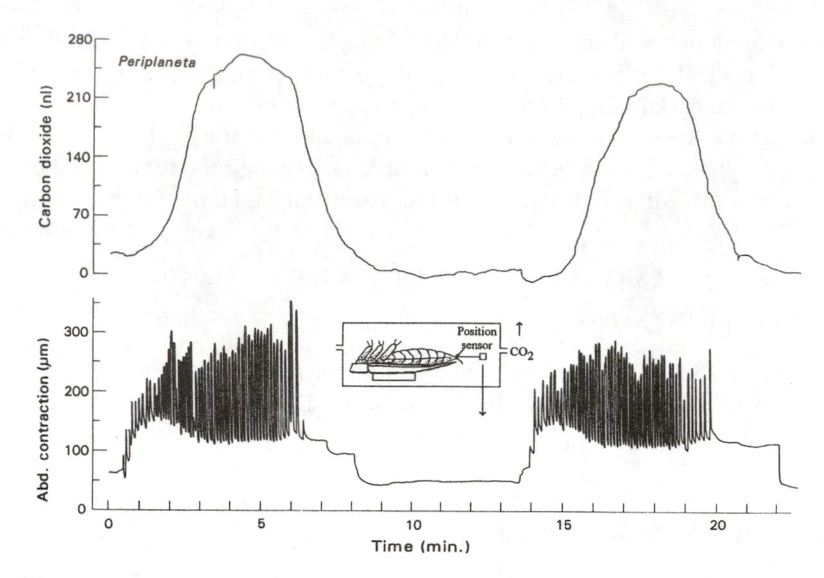

FIGURE 12.15 A recording of the CO_2 release (top) and extracardiac hemocoelic pulsations (coelopulses) of the abdomen (bottom) in an immobilized American cockroach *Periplaneta americana* at 25°C. (Reproduced with permission from Sláma, 1999.)

Lighton and Wehner, 1993; Lighton, 1996). Abdominal pumping may occur by dorsoventral compression of the abdomen or by longitudinal telescoping of abdominal segments. Abdominal pumping is common in large insects at rest and continues during flight. In some fast-flying insects, such as the wasp *Vespa crabro* and possibly in other hymenopterans, abdominal pumping alone is sufficient to supply the wing muscles during flight. Sláma (1988, 1994, 1999, 2000) has recorded periodically repeated **extracardiac miniature pulsations** in hemocoelic pressure mediated from a center in the mesothoracic ganglion (Figures 12.15 and 12.16). Such pulsations, called **coelopulses** by Sláma, appear to be common in many, and perhaps all, insects, including larvae, pupae, and adults. Coelopulses enable insects to breathe through selected spiracles and promote a unidirectional ventilatory stream of air through the body. In pupae of *Tenebrio molitor,* the extracardiac coelopulses produce 30- to 90-µm movements of the abdomen, preventing hemolymph stagnation, promoting flow of hemolymph around organs, and probably pumping small tracheoles that aid air movements (Sláma, 2000).

In dragonflies and locusts, thoracic muscular pumping is important for ventilation of wing muscles, and can be experimentally demonstrated to be sufficient in the absence of abdominal mechanisms. Although abdominal pumping in the locust is vigorous, abdominal action is not necessary to support flight, as demonstrated by selective injury to the nervous system that blocks or greatly reduces abdominal pumping without altering tethered flight. Thoracic pumping alone can adequately ventilate the muscles and body as a result of thoracic movements and wing action during flight.

Thoracic pumping tends to create a tidal flow that moves air in and out of the same spiracles, while abdominal pumping promotes a directed flow in through thoracic spiracles and out through abdominal spiracles. Abdominal pumping can also create a tidal flow in some insects. At rest, abdominal pumping by the locust *Schistocerca gregaria* can move 40 liters of air per kilogram body weight/h in through the thoracic spiracles and out through the abdominal ones. As expected, the volume of air ventilated during flight goes up (180 1/kg body weight/h for the first 5 min of flight, falling to 150 1/kg/h). As much as 16 to 17% of the air flow goes to the head, ventral thorax, and ventral abdomen, where it mainly serves nervous system demand; and about 39% goes to the pterothorax to support flight muscle demand. The remaining 44% goes to other systems in the body (Miller, 1960a, 1960b).

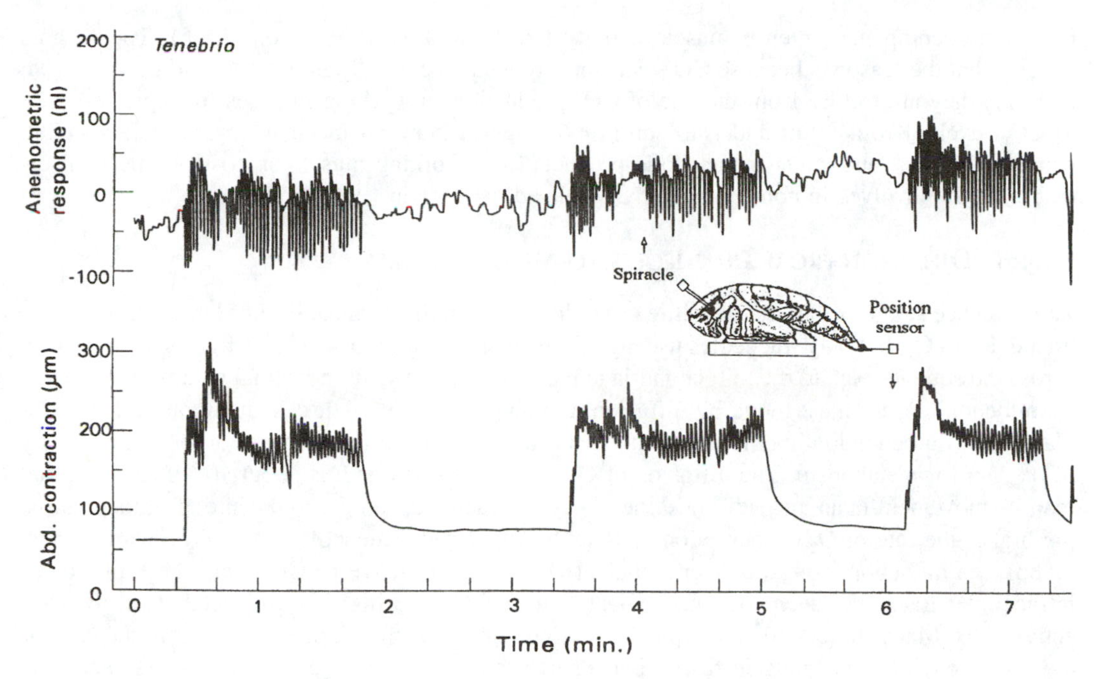

FIGURE 12.16 A recording of the outbursts of tracheal gas (top) synchronized with coelopulses in hemocoelic pressure (bottom, monitored as longitudinal contractions of abdominal segments) of a pupa of the mealworm *Tenebrio molitor* at 23°C. (Reproduced with permission from Sláma, 1999.)

Each abdominal ganglion of *Schistocerca* is capable of initiating ventilatory movements for its segment, but the rhythm that synchronizes the overall movements originates in the metathoracic ganglion (Bustami and Hustert, 2000). The isolated metathoracic ganglion shows efferent discharges consistent with continuous ventilation during activity of the locusts and discontinuous ventilation during quiescent periods. The rhythm can be altered by CO_2 concentration, and to a lesser extent by hypoxia. Local perfusion of the head or thoracic ganglia with gas mixtures indicates that each region can alter the rhythm of ventilation. A non-flying locust is capable of additional ventilatory mechanisms involving longitudinal telescoping of the abdomen, protraction and retraction of the head ("neck ventilation"), and the protraction and retraction of prothorax ("prothoracic ventilation") (Miller, 1960b). These mechanisms do not come into play under normal resting conditions, but do operate for short intervals following periods of great activity, such as after flight.

Dragonflies (*Aeshna* spp.) and wasps (*Vespa crabro*) also vigorously ventilate the tracheal system during flight. In sustained horizontal flight, dragonflies and wasps use about 20 1 O_2/kg body weight/h (Weis-Fogh, 1967). Dragonflies supply the wing muscles entirely from thoracic pumping. Wasps, however, have a very hard, rigid pterothorax that allows little thoracic pumping, but abdominal ventilatory movements increase in amplitude and frequency (to 180/min) during flight.

The way (or ways) in which insects regulate ventilation is not well understood, and diverse mechanisms may be involved, depending on the activity state of the insect, as is the case in grasshoppers (reviewed by Harrison, 1997). *Romalea guttata* and *Schistocerca americana*, two large grasshoppers, regulate the O_2 level (at about 18 kPa) and the CO_2 level (2 kPa) in the large longitudinal tracheal trunks by adjustments in ventilation rate. Experimentally elevating tracheal P_{O_2} above normal or decreasing tracheal P_{CO_2} below normal values decreases ventilatory rate (Gulinson and Harrison, 1996). The ventilation rate is not altered by experimentally changing hemolymph pH, but elevating hemolymph HCO_3^- increases ventilation. In contrast to the influence of tracheal gas tension on resting ventilation, the post-exercise increase in ventilatory rate in *S. americana* and *Melanoplus differentialis* (a grasshopper) is not due to the normal rise in internal

P_{CO_2} that accompanies intense muscle activity (Krolikowski and Harrison, 1996). The authors suggest that the reason is because CO_2 receptors are located centrally in thoracic and head ganglia, and thus they are too far from the site of CO_2 production in working muscles for rapid response to CO_2 levels. Krolikowski and Harrison (1996) suggest humoral mechanisms, nerve activation, neuromodulators, and ionic or metabolic products from working muscle as possible mechanisms that could be involved in controlling activity-related increase in ventilation rate.

12.4.3 Diffusion from Tracheoles to Mitochondria

The evidence suggests that there is little or no difference in the permeability of larger tracheae and tracheoles to O_2, but only tracheoles usually will be close enough to mitochondria for O_2 to diffuse across the aqueous path to the mitochondria in the quantities needed. In its final path from tracheole to mitochondria, O_2 must move by diffusion, crossing tracheole wall, cell membranes, cell cytoplasm, and mitochondrial membranes. This is the slowest part of the pathway for the final delivery of oxygen for metabolism. The diffusion of O_2 in a water-filled path is about 10^6 times less rapid than its movement in an air path; thus, the closer the tracheole can get to the site of mitochondria, the higher the rate of O_2 consumption that can be supported. Tracheoles need to be within about 10 μm of a mitochondrion to deliver sufficient O_2 to support active metabolism. The evolution of intracellular tracheoles seems to be an adaptation to support a high rate of metabolism in large, active cells. Many insect cells are typically 30 to 60 μm in diameter, or even larger in the case of the most active fibrillar muscle fibers; but tracheoles often indent these muscle fibers, and may even touch and indent mitochondria (Afzelius and Gonnert, 1972).

12.5 DISCONTINUOUS GAS EXCHANGE CYCLE

Some insects are able to keep the spiracles tightly closed and/or "apparently closed" for a high percentage of the time. Gas exchange occurs in three periods — the **open**, **flutter**, and **closed** periods (often designated as O, F, and C, respectively) — because of the action of spiracles over the duration of a cycle.

This functional pattern, variously known as **discontinuous release of CO_2**, **passive suction ventilation**, **discontinuous ventilation cycle**, or the **discontinuous gas exchange cycle** (**DGC**) (Lighton and Garrigan, 1995), has been known to occur in some insects for more than half a century (reviewed by Sláma, 1988, 1994; Lighton, 1996). The earliest studies were in diapausing pupae of Lepidoptera, and Schneiderman and Williams (1955) cannulated the spiracles of large cecropia pupae and made the first analyses of intratracheal air (Figure 12.17). Initially, DGC was thought to be limited to quiescent insects in a depressed state of metabolism, but DGC patterns have been observed in a number of different insects in various states of activity, including ants, the cockroach *Periplaneta americana*, a number of adult tenebrionid beetles, the locust *Schistocerca gregaria*, the lubber grasshopper *Romalea guttata*, and adults of additional species (Punt, 1950; Lighton, 1988, 1990, 1991, 1994, 1996; Lighton and Wehner, 1993; Lighton et al., 1993a; Hadley, 1994; Sláma, 1999, 2000; Duncan and Newton, 2000; Vogt and Appel, 2000; Chown and Holter, 2000). DGC behavior also occurs in some arthropods other than insects (Lighton et al., 1993b; Lighton and Duncan, 1995). Data from *Camponotus maculatus* ants showing metabolic rate and various aspects of the discontinuous gas exhange cycle are shown in Table 12.1 from Duncan and Newton (2000).

During DGC, accumulated CO_2 is periodically discharged in episodic bursts during brief intervals when spiracles are open. After a burst, the spiracles are closed for some period of time, which varies from species to species. During the closed interval, O_2 is used by tissues and intratracheal oxygen tension falls. Most insects that exhibit discontinuous gas exchange (DGC) allow the spiracular valve to flutter with an amplitude often imperceptible to the unaided eye during a portion of a cycle. The fluttering (F) phase allows small amounts of O_2 to be sucked into the tracheal

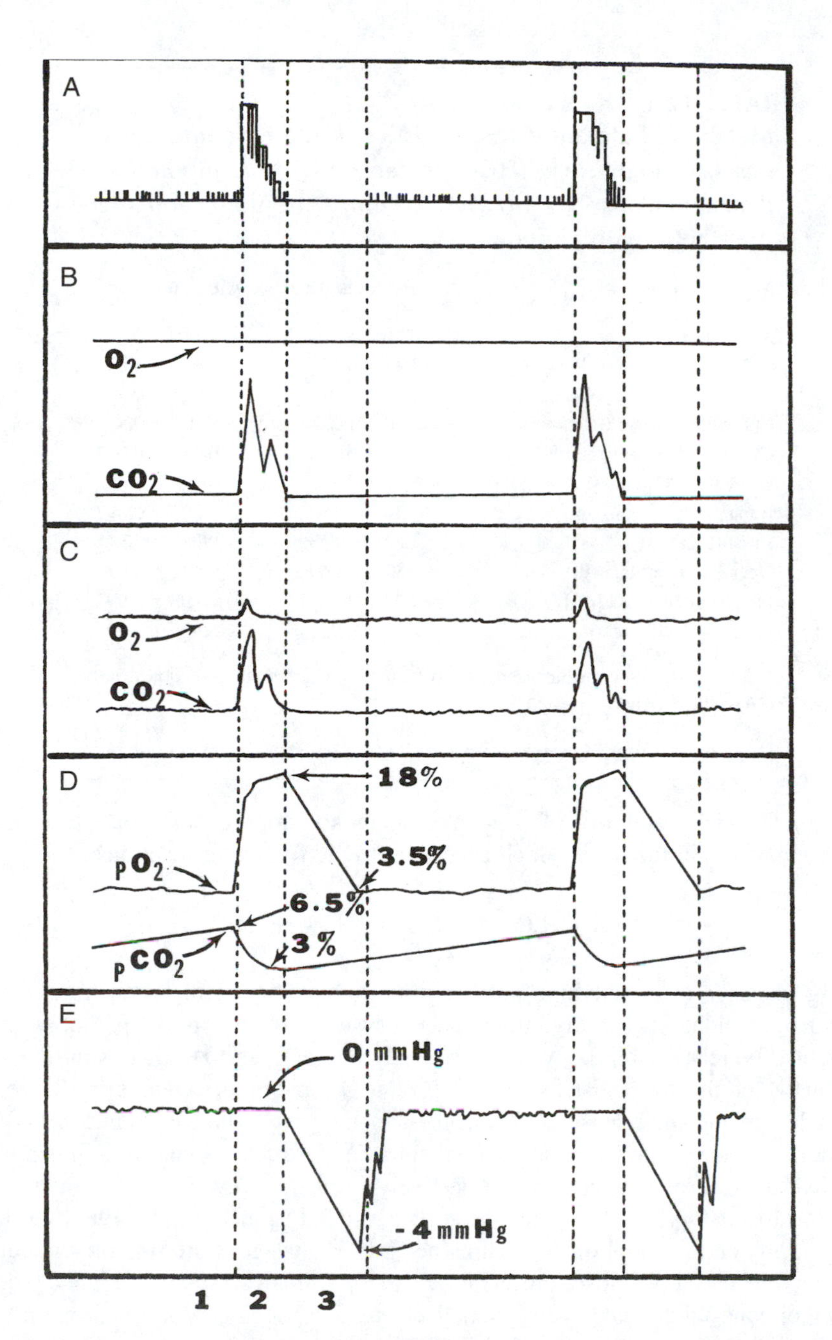

FIGURE 12.17 Correlation of events occurring during the respiratory cycle in a cecropia pupa. A cycle is divided into major phases in which the spiracle is fluttering (1), open (2), or tightly closed (3). A: spiracular opening (rise above the base line) and closing (return to base line). B: Gas exchange measured by manometric methods. C: Gas exchange measured by diaferometric method. D: Tracheal gas composition; E: Intratracheal pressure. (Reproduced with permission from Levy and Schneiderman, 1966b.)

system by the slight negative pressure arising from O_2 consumption by the tissues. The F phase is usually considered to involve convective transfer of O_2. This has given rise to the name "passive suction ventilation" sometimes applied to the process. The slight internal vacuum retards the outward loss of water vapor and CO_2 during the fluttering phase, and the low influx of O_2 lengthens the time to full opening or "burst" of spiracles. CO_2 is produced by tissues even when the spiracles

TABLE 12.1
Metabolic Rate and Characteristics of the Discontinuous Gas Exchange Cycle (DGC) (mean ± SD) in *Camponotus maculatus* Ants (N = 14) at 25°C before Anesthetic and after Anesthetic with Enflurane

Measurement	Before Anesthetic	After Anesthetic
Mass (mg)	10.10 ± 3.75	10.10 ± 3.75
Std. metabolic rate (μW)	12.49 ± 6.11	14.34 ± 8.64
Std. metabolic rate (W/kg)	1.52 ± 1.01	1.26 ± 0.53
CO_2 emission rate (μl/h)	1.96 ± 1.2	1.7 ± 0.8
CO_2 emission rate (ml/g/h)	0.207 ± 0.14	0.172 ± 0.073
Burst rate CO_2 emission (μl/g)	6.51 ± 2.7	8.23 ± 3.6
Burst rate CO_2 emission (ml/g'h)	0.314 ± 0.188	0.293 ± 0.097
Burst length (s)[a]	91.54 ± 18.53	98.43 ± 20.01
Open phase volume (μl)	0.083 ± 0.065	0.071 ± 0.036
Frequency of DGC (mHz)	7.24 ± 2.26	6.93 ± 1.99
Period of DGC (min)	2.54 ± 0.93	2.65 ± 0.99

[a] The mean values for burst length before and after anesthetic are significantly different (paired *t*-test, $p < 0.05$).

Data selected from Duncan and Newton (2000). Reproduced with permission.

are closed, but the high solubility of CO_2 in aqueous solutions enables insects to accumulate bicarbonate ion, HCO_3^-, in the hemolymph according to the following reactions:

$$CO_2 + H_2O \rightarrow H_2CO_3 \rightarrow H^+ + HCO_3^- \qquad (12.2)$$

The buffering capacity of hemolymph aids in solubilizing CO_2 as bicarbonate, and this keeps gaseous CO_2 from building up rapidly in the tracheal system. At some point, probably different for different insects, the relationship between gaseous O_2 and CO_2, and HCO_3^- in solution reaches an equilibrium at which the tracheal tension of CO_2 and O_2, and/or pH change in the hemolymph, triggers spiracle opening and release of CO_2 from the hemolymph as a gas. Sláma (1999) identified specially modified tracheal sacs near abdominal spiracles 2 to 5 in pupae of *Galleria mellonella,* which he calls carboniferous tracheae, that selectively extract dissolved CO_2 from the hemolymph and release it as respiratory gas. Low O_2 tension (Burkett and Schneiderman, 1967, 1974; Lighton, 1996) acts centrally on the ganglion controlling a spiracle to trigger fluttering of spiracular valves, while CO_2 acts directly on the closer muscle of *H. cecropia*. Sláma (1994) presented evidence that the neural center controlling DGC cycles in the lacewing *Chrysopa carnea* (Neuroptera) is located in thoracic ganglia. During quiescent periods at 15°C, the western lubber grasshopper *Taeniopoda eques* can go up to 40 min between bursts, during which time the partial pressure of CO_2 builds to 2.26 kPa with little acidification of the hemolymph (Harrison et al., 1995). These data suggest that when the CO_2 tension in tracheae reaches a threshold between 2 and 2.9 kPa, opening of the spiracle is triggered.

Diapausing *H. crecropia* pupae can tolerate very low O_2 tensions in the tracheae, as low as 5% O_2, before fluttering is triggered. Continued fluttering of the spiracular valve keeps O_2 in the intratracheal air of a cecropia pupa at about 3.5% until the next burst (Figure 12.17). This low level of O_2 is sufficient to support the slow metabolic processes of the diapausing pupa, but it would not likely support an active insect. Intervals between bursts of CO_2, or spiracle opening, increase as

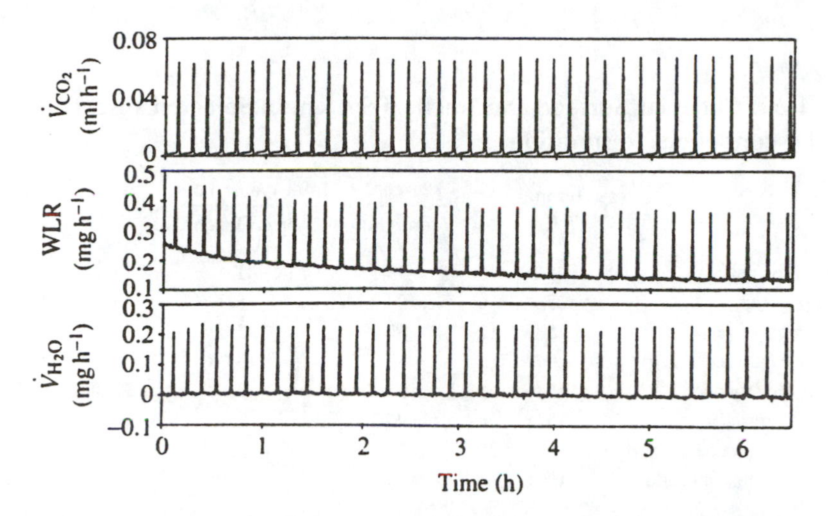

FIGURE 12.18 Discontinuous gas exchange (DGC) measurements recorded with flow-through respirometry techniques in a 30-mg worker ant, *Camponotus vicinus,* under normoxic conditions. Upper trace, rate of CO_2 release; middle trace, total water loss rate; bottom trace, respiratory water loss rate. Mean DGC frequency equals 1.59 mHz and mean rate of CO_2 release equals 6.64 µl/h. (Reproduced with permission from Lighton and Garrigan, 1995.)

the diapausing pupa sinks deeper into the diapause state and requires less O_2 (Schneiderman and Williams, 1955; Levy and Schneiderman, 1966a, 1966b). Discontinuous respiration continues with longer and longer intervals between bursts down to –5°C (Burkett and Schneiderman, 1974); but the tracheal valves freeze closed at lower temperatures, and any additional gas exchange must occur through the cuticle. Thus, discontinuous respiration is highly functional and adaptive for cecropia pupae, which spend the winter buried beneath litter and soil.

It has often been assumed, but not proven, that the functional benefit to an insect of discontinuous ventilation is conservation of water. In diapausing pupae that must pass a long winter under the soil, leaf litter, or other pupation site, water conservation seems quite necessary and a reasonable driving mechanism for the evolution of discontinuous ventilation because the pupa is a closed (to food/water intake) system. Most adult insects that exhibit discontinuous ventilation do so only intermittently, and the remainder of the time they ventilate the system continuously. Surprisingly, some insects do not discontinuously ventilate under conditions that might be expected (based on assumptions) to promote the behavior. For example, the lubber grasshopper *Romalea guttata*, which discontinuously ventilates at times, tends to ventilate continuously when dehydrated, a physiological condition when it presumably has a great need to conserve water (Hadley and Quinlan, 1993; Quinlan and Hadley, 1993). No good explanation has been provided for such behavior. The ant *Camponotus vicinus*, which exhibits discontinuous gas exchange (Figure 12.18), actually loses more water than CO_2 during the period when the spiracles are open (Lighton and Garrigan, 1995). The harvester ant *Pogonomyrmex rugosus* (Table 12.2) and similar desert ants have a relatively high percentage (up to 13%) of body water loss through the tracheal system (Lighton et al., 1993a), although they exhibit discontinuous ventilation at times. Whether the somewhat higher rate of water loss from the tracheal system in these ants is a significant stress for them may depend on how much of the time the ants exhibit discontinuous gas exchange, as well as how much access they have to food with high water content and exposure to environmental extremes. Although it seems somewhat intuitive that water loss might be high from the tracheal system, actual measurements in some insects indicate that more than 90% of total body water loss occurs through the cuticle, with only 2 to 5% typically lost from the tracheal system (Table 12.2). Cyclic release of CO_2 occurs in drywood termites *Incisitermes minor* (Hagen), Formosan subterranean termites *Coptotermes*

TABLE 12.2
Water Loss from Cuticular Surface and Tracheal System in Selected Insects That Have Discontinuous Respiration

Species	Temp. (°C)	N	Percent of Total Water Loss: From Cuticle	From Tracheae	Ref.
Camponotus vicinus (ant)	25	6	98.1	1.9	Lighton (1988)
Cataglyphis bicolor (ant)	25	6	92.0	8.0	Lighton (1988)
Pogonomyrmex rugosus (ant)	25	12	87.0	13.0	Leighton et al. (1993)
Periplaneta americana (cockroach)	20	9	87.0	13.0	Machin et al. (1991)
Romalea guttata (grasshopper)	15	5	97.9	2.1	Hadley and Quinlan (1993)
Taeniopoda eques (grasshopper)	15	6	96.2	3.8	Hadley and Quinlan (1993)
Incisitermes minor (drywood termite)	25	22	93.5	4.7	Shelton and Appel (2000)
Aphodius fossor (scarabaeid beetle)	15	—	95	5	Chown and Holter (2000)

Part of table reproduced with permission from Hadley (1994).

formosanus Shiraki, and Eastern subterranean termites *Reticulitermes flavipes* (Kollar), but the termites do not exhibit a strict DGC pattern, and water loss through the respiratory system is less than 10% of the total daily water loss (Shelton and Appel, 2000, 2001). Populations of the grasshopper *Melanoplus sanguinipes* collected in California at several geographic and altitudinal sites show behavior leading to discontinuous gas exchange, but DGC decreases at elevated temperatures when it might be expected to be more pronounced if respiratory water conservation were important to the grasshoppers. The quantity and melting point of cuticular lipids, however, show strong correlation with lower rates of water loss (Rourke, 2000).

An alternative to the water conservation theory was put forth by Lighton and Berrigan (1995) and amplified by Lighton (1998), who suggested that DGC may occur primarily in insects that experience **hypoxic** (low O_2) and **hypercapnic** (high CO_2) conditions (such as ants living underground), and that it is not necessarily essential to reducing water loss through the tracheal system. Chappell and Rogowitz (2000) also concluded that the hypoxic and hypercapnic environment in which eucalyptus-boring cerambycid beetles live may have had more influence on evolution of discontinuous ventilation than water conservation. Chown and Holter (2000) propose still another explanation for DGC. They suggest that the periodic nature of DGC may be the result of two feedback systems (i.e., sensors detecting low O_2 and those detecting high CO_2) interacting during times of minimal demand, and cite Kauffman (1993), who has described varying effects of interacting feedback systems, ranging from a single steady state to cyclic behavior. In summary, a definitive selection mechanism for evolution of discontinuous gas exchange cycling is not evident, but the mechanism is fairly widespread in many insects, including larvae, pupae, and large and small adults. Water conservation, ecological niche occupied by insects, and interactions of sensory components may be important factors.

12.6 WATER BALANCE DURING FLIGHT

A flying insect can lose large amounts of water due to evaporation from the tracheal surfaces as large volumes of air are ventilated through the system. Physiological and behavioral adaptations, particularly in long-distance flyers such as those that migrate, help prevent desiccation. One adaptation is the use of fat as a flight fuel. Lepidoptera, Orthoptera, and some other groups mobilize triacylglycerols from the fat body, transport the molecules to flight muscles, and oxidize them for energy during flight. In *Schistocerca gregaria,* about 7 g fat/kg body weight/h are burned during flight, resulting in the production of 8.1 g H_2O/kg body weight/h (Weis-Fogh, 1967). This gain of metabolic water helps offset the loss from evaporation. Water loss is related also to relative humidity and temperature of the air. At moderate to high relative humidity and at temperatures between 25 and 30°C, a flying *S. gregaria* locust can stay in water balance during sustained flight; but at very low relative humidity, water will be lost and this is intensified at higher temperatures. Lehmann et al. (2000) suggest that small dipterans, such as the smaller species of *Drosophila*, face high risk of desiccation during flight. They determined that during hovering flight, the average water loss in four species of *Drosophila* is 67.3 ± 36.9 μl/g/h. If the flies metabolize glycogen during flight, they can produce 0.56 mg H_2O/mg glycogen metabolized (Schmidt-Nielsen, 1997). Based on the amount of CO_2 produced during hovering flight (32.4 ± 5.1 ml/g/h) and a factor of 1.19 mg glycogen/ml CO_2 produced, the authors estimate that, on average, the four species produce 21.6 ± 3.4 μl metabolic water/g body mass/h, or about 41.7% of the total water loss during hovering flight.

Behavioral mechanisms such as flight at night when the temperature is usually lower and the relative humidity usually higher, and flight at higher altitudes where temperatures are lower and prevailing winds can blow insects along, can aid insects in reducing water loss. Some insects may make other behavioral adjustments, including being quiescent, seeking shade, and refusing to take flight. For example, tobacco hornworm adult moths will not fly at air temperatures of about 43°C, and this protects them from overheating and dehydration (Heinrich, 1970, 1996), either of which would especially be detrimental to the nervous system and probably other organ systems.

12.7 GAS EXCHANGE IN AQUATIC INSECTS

The tracheal system of most aquatic insects is structurally the same as that of terrestrial insects, that is, open spiracles and an extensive network of tracheae and tracheoles. These aquatic insects breathe air by frequently coming to the surface. Water is prevented from entering the system by the hydrophobic surface of the tracheae and, in some cases, by closed spiracles. Many aquatic larvae have a metaneustic system, with only one functional pair of spiracles near the tip of the abdomen (Keilin, 1944). When the insect comes to the surface, it does so with the posterior end upper most and only the tip of the abdomen bearing the spiracles is held above the surface. Some have a siphon at the tip of the abdomen that is pushed above the surface of the water for gas exchange. During submergence, the spiracles are kept closed. Probably, a small vacuum is created within the tracheal system due to O_2 use while submerged, as in discontinuous release of CO_2, and this aids the intake of O_2 when the larva or pupa comes to the surface. Larvae of *Glossina* spp. (tsetse flies, Diptera: Glossinidae) develop inside the uterus of the mother, an environment in which they get their O_2 through a pair of posterior spiracles after the air enters the vulva of the mother (Zdáreket al., 1996).

Some aquatic insects living under water must be able to break the surface film of water to get air from the surface. Typically, they have structural adaptations of the body (**hydrofuges**) that are difficult to wet, and these facilitate surfacing and hanging at the surface. The hydrofuge repels water, is not easily wetted, and tends to buoy the insect at the surface of the water. Tufts of long hairs surround the siphon in mosquito larvae; and when the mosquito larva comes to the surface, the ring of hydrofuge hairs spread out in a circle on top of the water, keeping the tip of the siphon just above the water level. When the larva submerges, the hydrofuge hairs tend to collapse inward on each other and form a small dome over the tip of the siphon. Oily secretions from cuticular glands coat the hydrofuge regions or hairs and aid in maintaining their hydrofuge nature. Larvae of *Drosophila melanogaster* often live in a wet environment, and three unicellular glands associated with each of the two posterior spiracles (Jarial and Engstrom, 1995) appear to secrete oily substances on to the cuticle around the spiracles (Figure 12.19).

12.7.1 COMPRESSIBLE GAS GILLS

A large number of aquatic insects submerge with a bubble or film of air enclosing one or more spiracles. These gas bubbles or films of air have been called gas gills, and they may be compressible

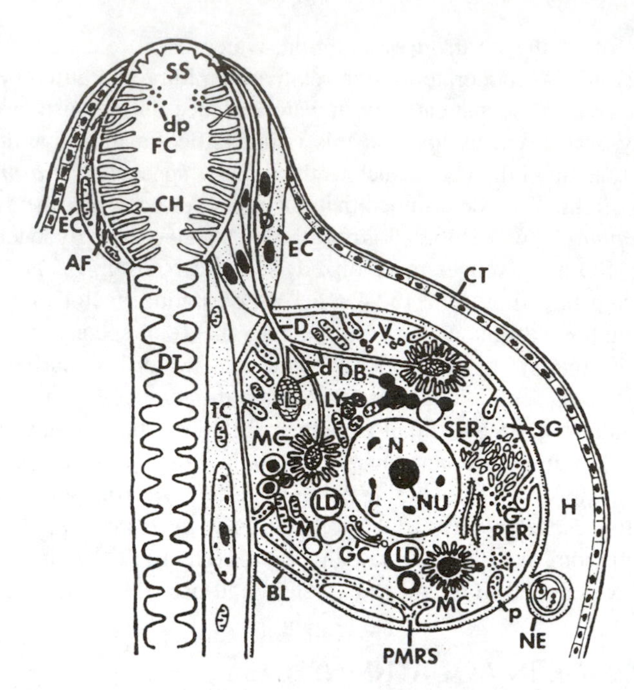

FIGURE 12.19 Basic ultrastructural features of a spiracular gland associated with the dorsal tracheal trunk (DT) and felt chamber (FC) at spiracle on the tip of the abdomen of a *Drosophila* larva. BL, basal lamina; C, chromosome; CT, cuticle; Ch, cuticular hair; d, cuticular ductule; D, main duct; dp, dust particles; EC, epidermal cell; G, glycogen granule; GC, Golgi complex; H, hemocoel; L, lumen of ductule; LD, lipid droplets; LY, lysosome; M, mitochondria; MC, microvillous channel; N, nucleus; Nu, nucleolus; p, pores of PMRS; PMRS, plasma membrane reticular system; r, free ribosomes; RER, rough endoplasmic reticulum; SER, smooth endoplasmic reticulum; SS, spiracular slit; TC, tracheolar epithelial cell; and V, vesicle. (Reproduced with permission from Jarial and Engstrom, 1995.)

and require replenishing periodically, or they may be incompressible and enable the insect to stay submerged indefinitely. **Compressible gas gills**, a bubble or film of air somewhere on the body, are widespread among aquatic members of the Coleoptera (Dryopoidea and Hydrophilidae), Hemiptera (the genera *Gerris* and *Velia*), and Lepidoptera (some arctiids and pyralids). Compressible gills slowly collapse as the O_2 is used, although additional oxygen is usually extracted from the water before the bubble must be renewed at the surface. Air stores in compressible gas gills may be carried in a fine network of hydrofuge hairs. Sometimes, there is a fine, dense set of hairs nearest the body surface from which the gas volume is very slowly used, and a set of longer, larger hairs over these, from which the gas is rapidly used. Alternatively, a gas gill may be carried beneath the elytra as in dytiscid diving beetles, or as a gas bubble on the posterior part of the abdomen. In addition to serving a respiratory function, gas gills also provide buoyancy and a hydrostatic function. When the insect comes to the surface to renew the air store, the buoyancy of the remaining bubble allows the part of the body carrying the bubble to come to the top first. Thus, the air in the bubble is restored with a minimum of exposure of the insect at the surface. Immediately after an insect has surfaced, any air bubble on its body should contain approximately 21% O_2 and 79% N_2, in equilibrium with the atmosphere. As the insect submerges, O_2 will be used for metabolic processes and O_2 tension in the air bubble will fall, while the partial pressure of the nitrogen (p_{N_2}) will rise. In well-aerated water, the gas composition is approximately 33% O_2, 64% N_2, and 3% CO_2. Because O_2 is ordinarily in higher concentration in the water than in the air bubble, O_2 from the water will diffuse into the gas bubble and N_2 in the bubble will begin to diffuse out into the water. Equilibrium pressure in the bubble cannot be attained because O_2 is continually being used by the insect, but this dynamic exchange allows the insect to gain up to 13 times the quantity of O_2 originally carried within the bubble. Eventually, the insect must surface and renew the bubble. The N_2 in the bubble plays a crucial role because its relatively low solubility in water causes it to move out of the bubble more slowly than O_2 enters the bubble. The N_2 remaining in the bubble prevents it from collapsing, giving O_2 from the water a chance to diffuse in. Beetles allowed to fill the bubble at the surface with pure O_2 cannot stay submerged as long as normal because the bubble shrinks rapidly as O_2 is used by the insect, and there is little or no N_2 to keep the bubble expanded. The time that an insect can stay submerged with a compressible gas gill depends on the size of the air bubble, activity of the insect, and O_2 tension in the water. *Dytiscus* diving beetles were able to stay submerged from 3 to 5 min with a bubble that initially contained 19.5% O_2, but then dropped to 2%, requiring them to surface for replenishment (Wigglesworth, 1972).

12.7.2 Incompressible Gas Gills: A Plastron

Incompressible gills do not collapse and O_2 can continue to be extracted indefinitely from the water into the gill (if the water is well aerated), allowing the insect to live underwater. Incompressible gas gills are also called **plastrons**. A plastron consists of any extensive physical meshwork, either of fine hairs or setae, or a meshwork of small pores and channels in the cuticular surface of some insects and eggs, that can hold a volume of air and can present a large water/air interface. When the meshwork is sufficiently extensive, a constant film of air can be held that takes O_2 from the surrounding water, provided that the water is well aerated. Plastrons are common adaptations of insects living in aquatic arrangements and can take many physical forms. A plastron contains a film of air so tightly guarded by a dense network of non-wettable hairs or meshwork of pore that, although the gas equilibrium may change due to O_2 use, water cannot invade the air space. The minimum amount of water/air interface in a meshwork necessary to enable it to function as a plastron has not been determined, but Hinton (1964) suggested that a water/air interface to weight ratio of 15,000 μm^2/mg weight was sufficient to qualify as a plastron. He based this conclusion upon the water/air interface/mg ratio found in the pupa of the fly *Eutanyderus wilsoni*. This fly has the poorest ratio known for an insect obviously adapted for living in water. Most insects with a plastron have a water-air interface of from 10^5 to 10^6 μm^2/mg weight. The thickness of the hair pile will obviously help determine the efficiency of the plastron. *Aphelocheirus aestivalis* (Hemiptera), for example, has from 2×10^8 to 2.5×10^8 hairs/cm^2 forming the plastron. Insects that have 10^6 to 10^8 hairs/cm^2 generally have a very efficient plastron and can usually stay submerged for months. Some Coleoptera and Lepidoptera also have very efficient plastrons, and often have eight or nine spiracles that open into the plastron air space. Agents that lower the surface tension of the water (e.g., soap and alcohols) will cause wetting of the plastron and failure to retain the air space. High pressure, if applied over sufficiently long periods of time, will cause wetting of the plastron, but these high pressures are not likely to occur in the natural habitat. A plastron can work in reverse and extract O_2 from the insect tissue and pass it into water that is very low in O_2 content, as might occur in cases of severe pollution. Insects utilizing a plastron usually live in well-aerated water of streams, lake edges, and intertidal zones.

12.7.3 Use of Aquatic Plants as Air Source

Some insects with a hydrofuge are able to capture and utilize the gas bubbles released by aquatic plants, either by inserting a part of the body bearing a spiracle into the plant tissue (Figures 12.20 and 12.21), or by biting into the air spaces of the plant. Some species of Diptera, Coleoptera, and Lepidoptera independently evolved modifications for piercing aquatic plants for air. Dipterous larvae in the family Ephydridae mine leaves of aquatic or semi-aquatic plants (Deonier, 1993). One species, *Hydrellia pakistanae* Deonier, the Indian hydrilla leaf miner, is a small fly that lays its eggs on leaves of the aquatic weed hydrilla (Buckingham, 1994). The larvae mine the leaves, which usually are underwater much or all of the time, and the larvae get their air either from the plant or

FIGURE 12.20 Mosquito larva, *Mansonia* sp., with its post-abdominal spiracles inserted in a branch of water lettuce, *Pistia stratiotes.* (Photograph by Tom Loyless, courtesy of D.O. Deonier.)

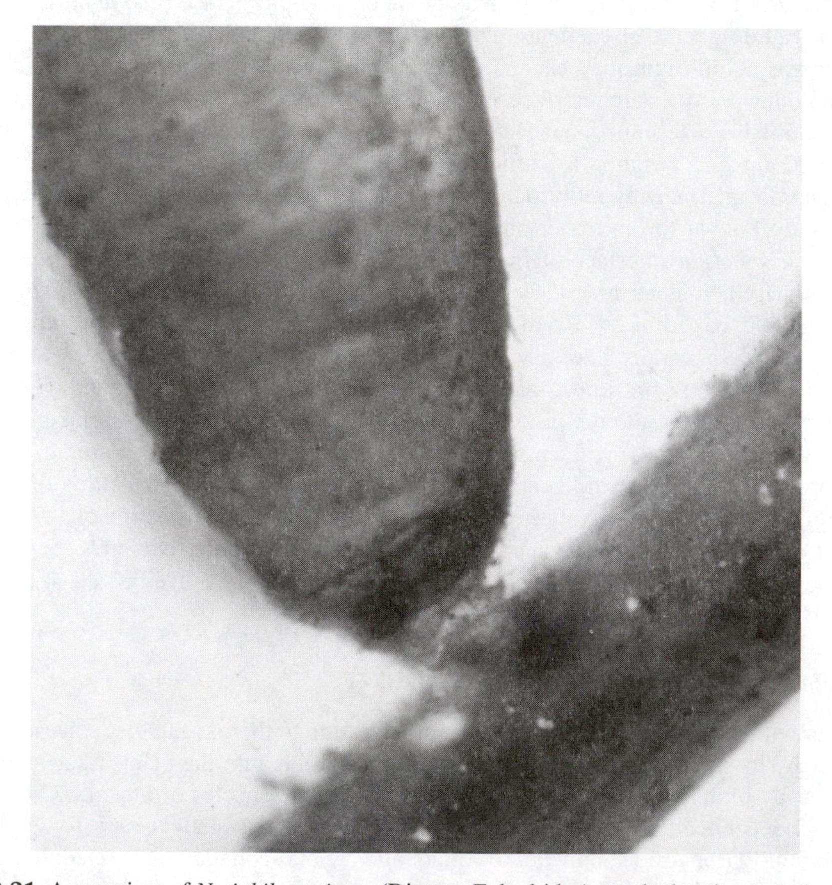

FIGURE 12.21 A puparium of *Notiphila carinata* (Diptera: Ephydridae) attached to the root of a water weed. The larva and pupa live underwater, and obtain their air from the host plant. The larva inserts a pointed, root-piercing spiracle into the plant just before it pupariates, and the pupa obtains its air supply from the plant. (Photograph courtesy of D.O. Deonier.)

by cutaneous diffusion through the cuticle. Before pupariation, a larva leaves the leaf and anchors itself to the stem of the weed by inserting its paired anal spines into the stem, and the pupa gets its air from the plant through the terminal spiracles. Pupae do not survive if mechanically removed from the plant (personal communication, James Kuda). Remarkably, small hymenopteran wasps *(Trichopria columbiana)* hunt and lay their eggs on the pupa under the water. The small wasps enter the water, crawl downward along a hydrilla stem searching for pupae, and upon finding one it inserts an egg through the cuticle of the pupa. The larva of the wasp develops in the pupa and usually has its posterior end oriented with the posterior of the pupa that is inserted into the plant, suggesting that the wasp larva also gets its O_2 from the plant. The adult wasp can stay submerged for several hours, and perhaps much longer in its hunt for a pupa (Deonier, 1971), but its mechanism for obtaining air while submerged in search of a host pupa is not known. It may be able to trap a film of air on the body when it enters the water. Another dipteran, *Notiphila riparia*, similarly pierces submerged plants as a larva and a pupa to obtain an air supply.

Certain coleopterans (the Donaciinae) live in the mud around the roots of aquatic plants, and larvae penetrate the plant roots with a pointed, posterior siphon. The thick mud gives them a resistant surface to push against in inserting the siphon. A spiracle at the end of the siphon allows gas entry. Larvae also bite into the root before pupation and construct a pupal case over the lesion. Air probably continues to be released from the lesion into the pupal case to support the pupa. Some lepidopterous larvae in the genus *Hyrocampa* also insert a respiratory siphon into the air spaces of aquatic plants, while some other lepidopterans bite into the plant.

12.7.4 Cutaneous Respiration: Closed Tracheal System in Some Aquatic Insects

A closed tracheal system without functional spiracles is present in some aquatic insects. The lack of functional spiracles eliminates any chance that water will enter the system, but O_2 must also diffuse through the cuticle, a breathing mechanism called **cutaneous respiration**. Internally, these insects still have an extensive tracheal system like terrestrial insects. Often in larger, more active insects, there are tracheal gills that greatly increase the cuticular surface for gas exchange with the water. Relatively large larvae of Trichoptera, Plecoptera, Odonata, and some Lepidoptera utilize cutaneous respiration that is facilitated by extensive elaborations of thin hair-like or flap-like cuticular extensions from the body surface called **tracheal gills.**

Cuticular extensions from the body surface called tracheal gills (Figure 12.22) occur in several orders of insects and are highly variable in structure and in location on the body. The cuticle of insects with tracheal gills is very thin, and usually large numbers of small tracheae and tracheoles lie just beneath the cuticle. Larvae of Trichoptera, Odonata, and Coleoptera display a highly structured arrangement of tracheoles that are uniform in size and spacing in the gills as an adaptation to utilize optimal functional efficiency with a minimum of tracheoles (Wichard and Komnick, 1974). The tracheoles run parallel to the gill surface and are just beneath the thin cuticle. These characteristics appear to be highly adaptive for trapping O_2 that diffuses through the cuticle. Three caudal gills are characteristic of larvae of the Zygoptera, while tufts of thin gill filaments are located on the head, thorax, abdomen, and coxae of some Plecoptera. In Plecoptera, the gills have a tube-like shape with the center cavity filled with hemolymph that is continuous with the hemocoel of the body (Wichard and Komnick, 1974). The cuticle is thin (0.2 to 1.2 μm thick). The tracheoles have a diameter less than 1 μm, and they indent the epidermal cells so that they lie immediately beneath the body surface (Figure 12.23). The tracheoles are not uniform in size, but vary in diameter from 0.2 to 1.0 μm, and spacing and distribution are less uniform than in Trichoptera. Filamentous gills occur on the abdomen of some Tricoptera, Diptera, and Lepidoptera, and on the thorax and abdomen of a few Coleoptera. Tracheal gills generally are larval structures, but some Trichoptera pupae have them, and they persist as atrophied, probably nonfunctional, structures in some adults of Trichoptera and Plecoptera.

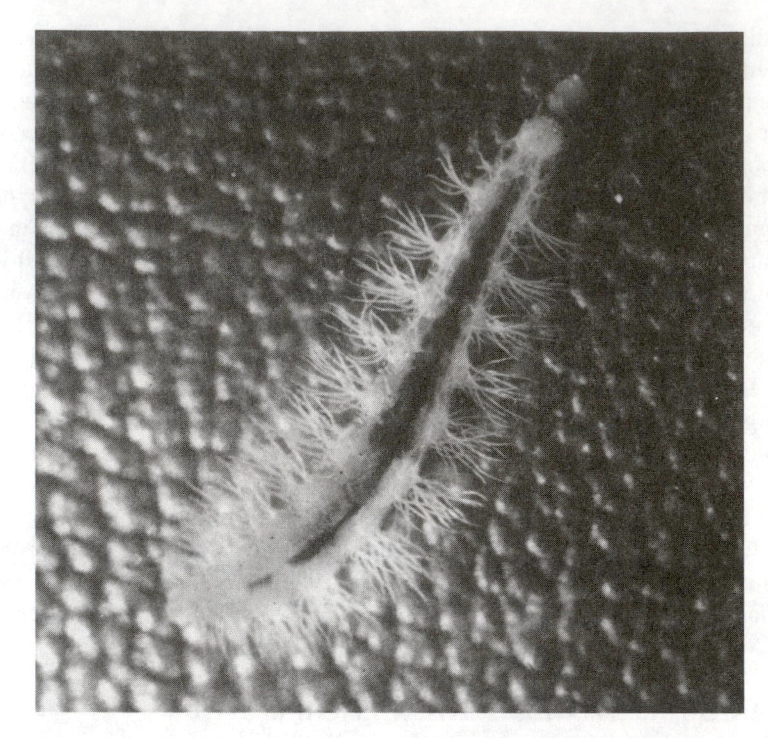

FIGURE 12.22 Filamentous tracheal gills on the larva of *Parapoynx seminealis* (Pyralidae: Lepidoptera) that lives its larval life underwater and obtains air by cutaneous respiration primarily through the tracheal gills. (Photograph courtesy of Dale Habeck.)

Pupae of the simuliids (dipterans) develop plastrons on the gill surface at the last larval molt. The plastron fills with air shortly before the larval-pupal ecdysis; and after ecdysis, the plastron expands into its functional appearance before the cuticle hardens. The entire structure of the gill, except a small area at the base, bears a plastron in gills of *Simulium ornatum* (Miller, 1966b).

After O_2 diffuses across the thin cuticle of the typical gill surface and enters the tracheae, it is probably distributed to different body tissues much as in terrestrial insects by diffusion and by body and muscular movements. Some researchers have raised questions about the respiratory function of tracheal gills because some insects from which the gills were removed did not show much, if any, change in behavior or in O_2 consumption. Probably in these insects there is a high level of general cutaneous respiration that is assisted by the gills.

Movement of water over the gill and body surface of aquatic insects is important in maintaining a fresh supply of oxygenated water in contact with the body, and most use undulations of the body and/or movements of the gills themselves to create ventilatory currents of water. Larvae of some dragonflies (Anisoptera) draw water into the rectum by elastic expansion of the body as dorsoventral compressor muscles relax. Typically, there are six main gill folds as extensions of the cuticular intima in the anterior part of the rectum that extract O_2 from the water. The water is pumped out by dorsoventral compression of the abdomen. The rate of ventilation varies with several factors, including the O_2 content of the water. About 85% of the water in the rectum is renewed during each pumping cycle, and 25 to 50 cycles/min have been recorded. Larvae will also come to the surface and ventilate the rectum with air when the O_2 content of the water is very low.

12.8 RESPIRATION IN ENDOPARASITIC INSECTS

Many hymenopterans and dipterans are parasitic on other insects, and have been minimally studied with respect to respiration, perhaps for the obvious reasons that they are usually small

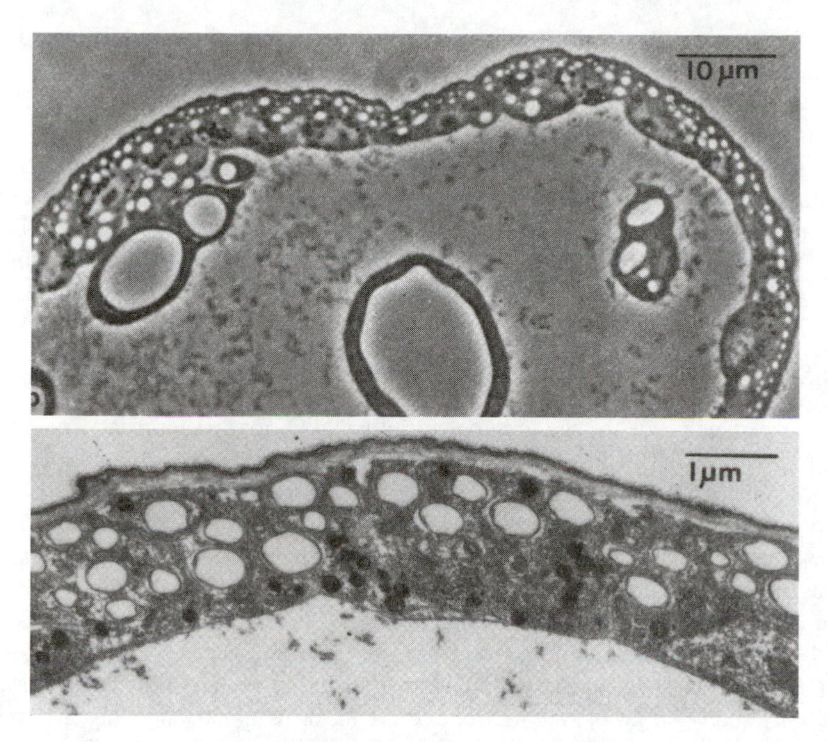

FIGURE 12.23 Top: Cross section through a gill filament of a *Perla* species (Perlidae: Plecoptera) showing many small tracheoles and a few larger tracheae just beneath the thin cuticle of the gill filament. Bottom: Higher magnification to show more clearly the tracheoles (tracheal tubes less than 1 μm in diameter) in a cross section of a gill filament. (Reproduced with permission from Wichard and Komnick, 1974.)

and hidden in the body of the host. Cutaneous respiration is probably very important. The 1st instar often has a fluid-filled tracheal system, necessitating cutaneous respiration. Although air displaces the fluid at the molt into the 2nd instar, the mechanism for clearing the fluid from the system is not clear. Possibly, the fluid is reabsorbed into the tissues of the larva. The spiracles become functional only just before the larva is ready to leave the host to pupate. Many chalcid wasps and tachinid flies that hatch from eggs laid on the surface of the host insect eat their way into the host. They have a metapneustic system and orient the posterior pair of spiracles at the body surface of the host so that they breathe air directly. Chalcid wasps remain in contact with the hollow pedicel of the egg from which they hatched at the host's integumental surface. Some tachinids stimulate the host's integument to become invaginated into a sheath around the parasite, leaving it with an air opening at the host's surface. The larvae of bot flies, *Hypoderma* spp., migrate to the skin of the vertebrate host where they bore a tiny opening to the surface through which gas exchange occurs. In earlier instars that are migrating, respiration is presumably cutaneous.

12.9 RESPIRATORY PIGMENTS

Only a few insects have **respiratory pigments** in the hemolymph or cells. *Chironomus* spp. larvae have a small **hemoglobin** composed of two chains with a MW of 31,400. The hemoglobin is present in the hemolymph but not in the hemocytes. It has a high affinity for O_2 and is 50% saturated at a p_{O_2} of 0.6 mmHg at 17°C. Its loading curve is not shifted by CO_2 tension or temperature, as in the case of vertebrate hemoglobin. The implication of such strong affinity of the hemoglobin for O_2 is that it will unload O_2 to the tissues only at a site of extremely low O_2 deficit. Moreover, the expected high CO_2 at such a site will not promote unloading as in the case of vertebrate hemoglobin.

Its principal function may be to aid recovery from anaerobic conditions and to provide limited O_2 to some critical tissues, such as the nervous system.

Hemoglobin occurs within certain cells of *Gastrophilus* spp. (horse bots) and in larvae and adults of some beetles (Notonectidae, *Anisops* and *Buenoa* spp.) (Mill, 1974). In contrast to the situation in *Chironomus* larvae, the hemoglobin of the beetle *Anisops pellucens* is only 50% saturation at a p_{O_2} of 28 mmHg at 24°C, giving it much more functional potential. The beetle may get up to 75% of the O_2 consumed during a normal dive from its hemoglobin (Miller, 1966b). Although CO_2 tension causes little tendency to unload, increased temperature does shift the curve to the right and increases unloading of O_2 in actively working tissues such as muscles.

12.10 RESPIRATION IN EGGS AND DEVELOPING EMBRYOS

The embryo developing inside the egg must obtain sufficient O_2 for development. Contradictory to an intuitive approach, the majority of aquatic and semi-aquatic insects lay eggs with no special respiratory structures incorporated into the shell, while eggs of a majority of terrestrial insects contain special structures for respiration, including an extensive, inner chorionic meshwork that can function as a plastron when the egg is submerged in well-aerated water (Hinton, 1969). Gas exchange in eggs with no special respiratory structure occurs by simple diffusion through interstices in the egg shell (Hinton, 1969). Even in eggs with special respiratory structures, the eggs do not have a large enough water/air interface per unit weight to enable effective function of the air space as a plastron for very long. Plastrons on eggs laid in a terrestrial environment may help prevent an O_2 deficiency if the eggs are subjected to only short periods of wetting from dew, rain, or temporary flooding. A 4-mm diameter raindrop falling on an egg can exert up to about 30 cmHg pressure, but for only a fraction of a second; thus, plastrons of eggs exposed directly to rainfall usually do not become wet.

The eggs of many terrestrial insects are frequently laid in wet environments, including decaying organic matter, animal manure, fruits, leaves, and stems of plants. To be effective, the plastron must resist the action of naturally occurring surfactants that are often present in animal dung and decaying flesh. Eggs laid in sites containing natural surfactants resist wetting better than those eggs laid in sites where surface-active agents are less likely to be encountered. Clearly this greater resistance to wetting in the presence of surface active agents is an adaptation to the egg environment. The egg shell, the **chorion**, may contain many small, twisting tubules called **aeropyles** that connect the inner part of the egg shell with the outside air. The aeropyles present little open surface to the air interface (e.g., 536 μm^2 in eggs of the hemipteran *Rhodnius prolixus*) and water loss from the egg is not greatly increased.

Some eggs have the plastron elevated on a **respiratory horn** (Hinton, 1961) that may be up to 10 to 13 mm long, with the plastron covering most of the surface of the horn. When the egg is completely submerged in water, the length of the longer horns make the gradient for gas exchange favorable only in the proximal part of the horn. It may be that long horns are useful as a conduit to atmospheric air when the egg is not too deep in water.

12.11 NONRESPIRATORY FUNCTIONS OF THE TRACHEAL SYSTEM

Tracheae serve the functions of connective tissue, typically tying cells and tissues together. Organs such as Malpighian tubules frequently are tied to each other, and to other structures such as the gut by tracheae. Tracheae are important as a structural base for at least two important endocrine tissues: (1) the diffuse cluster of cells making up the prothoracic glands that are attached to the prothoracic tracheae near the prothoracic spiracle in larvae of Lepidoptera, and (2) the epitracheal glands attached to the ventral surface of the major ventrolateral tracheal trunk connection to each spiracle in lepidopterous larvae (Žitňan, et al., 1996).

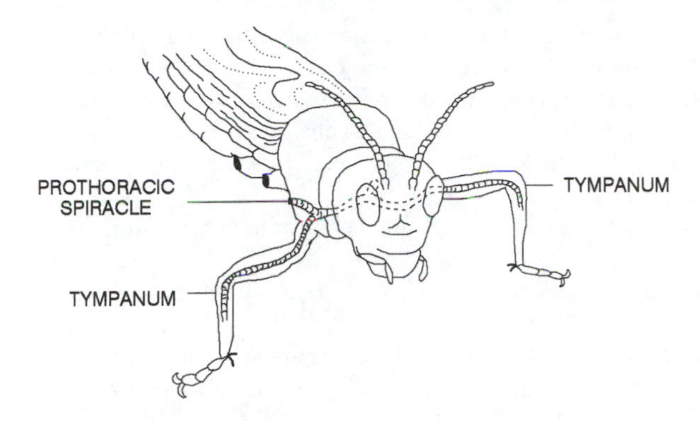

FIGURE 12.24 Drawing of the interconnected prothoracic leg tracheae forming a pathway between the two tympanic organs located just below the tibia of the prothoracic legs of the cricket *Teleogryllus commodus*. The large tracheal pathway acts as a sounding board to increase sensitivity, and allows increased directional sensitivity to sound waves impinging on the two tympanal membranes. (Redrawn from Hill and Boyan, 1976.)

Air sacs that back sound-producing organs in insects, particularly in cicadas (Homoptera) that produce very loud sounds, are important to loudness or sound production and its modulation (Claridge, 1985). A large air sac backs up to the tymbal located on each side of the first abdominal segment of male cicadas. The size of the air sac varies with different species, and its size and tuning are partly responsible for the species-specific quality of the sound produced by male cicadas. In many cicada species, the air sacs are broadly tuned to resonate over a range of frequencies, spanning the natural vibration frequency of the tymbal (Pringle, 1954). Some cicada species are able to damp the air sac resonance and produce complex pulses of sound. Most cicada calls generally have a characteristic sound frequency varying from 4 to 7 kHz (Pringle, 1954), but an Australian species, *Cystosoma saundersii*, is able to call and resonate at about 1 kHz because of a very large air sac. The interconnected prothoracic tracheae extending across the prothorax and into the prothoracic legs of crickets act as a resonator of sounds. They aid the cricket in discriminating direction of sounds reaching the tympanal membrane located just below the joint of the femur with the tibia (i.e., the knee joint) on each prothoracic leg (Figure 12.24). Hissing cockroaches expel air forcefully from certain spiracles and produce a hissing sound when they are disturbed. The sound is apparently a warning signal intended to scare away a potential predator or parasite. Lubber grasshoppers *R. guttata* release a yellow-colored foam containing quinones from spiracle 4 on the side of the attack when an ant attacks or when stimulated by probing in the laboratory (Roth and Stay, 1958).

REFERENCES

Afzelius, B.A., and N. Gonnert. 1972. Intramitochondrial tracheoles in flight muscle from the hornet *Vespa crabro*. *J. Submicrosc. OSC Cytol.*, 4:1-6.

Bordereau, C. 1975. Croissance des Trachées au cours de L'evolution de la physogastrie chez la reine des termites superieurs (Isoptera: Termitidae). *Int. J. Insect Morphol. Embryol.*, 4:431-465.

Buckingham, G.R. 1994. Biological control of aquatic weeds, pp. 413-480, in D. Rosen, F.D. Bennett, and J.L. Capinera (Eds.), *Pest Management in the Subtropics*, Vol. 1, Intercept Limited, Andover, U.K.

Burkett, B.N., and H.A. Schneiderman. 1967. Control of spiracles in silk moths by oxygen and carbon dioxide. *Science,* 156:1604-1606.

Burkett, B.N., and H.A. Schneiderman. 1974. Discontinuous respiration in insects at low temperatures: Intratracheal changes and spiracular valve behavior. *Biol. Bull.,* 147:294-310.

Bustami, H.P., and R. Hustert. 2000. Typical ventilatory pattern of the intact locust is produced by the isolated CNS. *J. Insect Physiol.,* 46:1285-1293.

Case, J.F. 1957. The median nerves and cockroach spiracular function. *J. Insect Physiol.,* 1:85-94.

Chappell, M.A., and G.L. Rogowitz. 2000. Mass, temperature and metabolic effects on discontinuous gas exhange cycles in eucalyptus-boring beetles (Coleoptera: Cerambycidae). *J. Exp. Biol.,* 203:3809-3820.

Chown, S.L., and P. Holter. 2000. Discontinuous gas exchange cycles in *Aphodius fossor* (Scarabaeidae): A test of hypotheses concerning origins and mechanisms. *J. Exp. Biol.,* 203:397-403.

Claridge, M.F. 1985. Acoustic signals in the Homoptera: Behavior, taxonomy, and evolution. *Annu. Rev. Entomol.,* 30:297-317.

Deonier, D.L. 1971. A systematic and ecological study of Nearctic *Hydrellia* (Diptera: Ephydridae). *Smithson. Contrib. Zool.,* 68:1-147.

Deonier, D.L. 1993. A critical taxonomic analysis of the *Hydrellia pakistanae* species group (Diptera: Ephydridae). *Insecta Mundi,* 7:133-158.

Duncan, F.D., and R.D. Newton. 2000. The use of the anaesthetic, enflurane, for determination of metabolic rates and respiratory parameters in insects, using the ant, *Camponotus maculatus* (Fabricius) as the model. *J. Insect Physiol.,* 46:1529-1532.

Dyby, S.D. 1998. Method for visualizing the tracheal system of newly hatched insects. *Ann. Entomol. Soc. Am.,* 91:350-352.

Edwards, G.A. 1953. Respiratory systems, pp. 55-95, in K.D. Roeder (Ed.), *Insect Physiology*, John Wiley & Sons, New York, 1200 pp.

Ghiradella, H. 1978. Reinforced tracheoles in three firefly lanterns: Further reflections on specialized tracheoles. *J. Morphol.,* 157:281-300.

Greenlee, K.J., and J.F. Harrison. 1998. Acid-base and respiratory responses hypoxia in the grasshopper *Schistocerca americana. J. Exp. Biol.,* 201:2843-2855.

Gulinson, S.L., and J.F. Harrison. 1996. Control of resting ventilation rate in grasshoppers. *J. Exp. Biol.,* 199:379-389.

Hadley, N.F. 1994. Ventilatory patterns and respiratory transpirations in adult terrestrial insects. *Physiol. Zool.,* 67:175-184.

Hadley, N.F., and M.C. Quinlan. 1993. Discontinuous CO_2 release in the eastern lubber grasshopper *Romalea guttata* and its effect on respiratory transpiration. *J. Exp. Biol.,* 177:169-180.

Harrison, J.F. 1997. Ventilatory mechanism and control in grasshoppers. *Am. Zool.,* 37:73-81.

Harrison, J.F., N.F. Hadley, and M.C. Quinlan. 1995. Acid-base status and spiracular control during discontinuous ventilation in grasshoppers. *J. Exp. Biol.,* 198:1755-1763.

Harrison, J.F., and J.R.B. Lighton. 1998. Oxygen-sensitive flight metabolism in the dragonfly *Erythemis simplicicollis. J. Exp. Biol.,* 201:1739-1744.

Heinrich, B. 1970. Thoracic termperature stabilization by blood circulation in a free-flying moth. *Science,* 168:580-582.

Heinrich, B. 1996. *The Thermal Warriors. Strategies of Insect Survival*, Harvard University Press, Cambridge, MA, 221 pp.

Hill, K.G., and G.S. Boyan. 1976. Directional hearing in crickets. *Nature (London),* 262:390-391.

Hinton, H.E. 1961. The structure and function of the egg-shell in the Nepidae (Hemiptera). *J. Insect Physiol.,* 7:224-257.

Hinton, H.E. 1964. The respiratory efficiency of the spiracular gill of *Simulium. J. Insect Physiol.,* 10:73-80.

Hinton, H.E. 1969. Respiratory systems of insect egg shells. *Annu. Rev. Entomol.,* 14:343-368.

Houlihan, D.F. 1969. The structure and behaviour of *Notiphila riparia* and *Erioptera squalida*, two root-piercing insects. *J. Zool.,* 159:249-267.

Jarial, M.S., and L. Engstrom. 1995. Fine structure of the spiracular glands in larval *Drosophila melanogaster* (Meig.) (Diptera: Drosophilidae). *Int. J. Insect Morphol. Embryol.,* 24:1-12.

Kauffman, S.A. 1993. *The Origins of Order, Self Organization and Selection in Evolution*, Oxford University Press, Oxford.

Keilin, D. 1944. Respiratory system and respiratory adaptations in larvae and pupae of Diptera. *Parasitology,* 36:1-66.

Keister, M. 1948. The morphogenesis of the tracheal system of *Sciara. J. Morphol.,* 83:373-423.

Keister, M. 1963. The anatomy of the tracheal system of *Camponotus pennsylvanicus* (Hymenoptera: Formicidae). *Ann. Entomol. Soc. Am.,* 56:336-340.

Krolikowsi, K., and J.F. Harrison. 1996. Haemolymph acid-base status, tracheal gas levels and the control of post-exercise ventilation rate in grasshoppers. *J. Exp. Biol.,* 199:391-399.

Lehmann, F.-O., M.H. Dickinson, and J. Staunton. 2000. The scaling of carbon dioxide release and respiratory water loss in flying fruit flies (*Drosophila* spp.). *J. Exp. Biol.,* 203:1613-1624.

Levy, R.I., and H. Schneiderman. 1966a. Discontinuous respiration in insects. II. The direct measurement and significance of changes in tracheal gas composition during the respiratory cycle of silkworm pupae. *J. Insect Physiol.,* 12:83-104.

Levy, R.I., and H. Schneiderman. 1966b. Discontinuous respiration in insects. IV. Changes in intratracheal pressure during the respiratory cycle of silkworm pupae. *J. Insect Physiol.,* 12:465-492.

Lewis, G.W., P.L. Miller, and P.S. Mills. 1973. Neuro-muscular mechanisms of abdominal pumping in the locust. *J. Exp. Biol.,* 59:149-168.

Lighton, J.R.B. 1988. Simultaneous measurement of oxygen uptake and carbon dioxide emission during discontinuous ventilation in the tok-tok beetle, *Psammodes striatus*. *J. Insect Physiol.,* 34:361-367.

Lighton, J.R.B. 1990. Slow discontinuous ventilation in the Namib dune-sea ant, *Camponotus detritus* (Hymentoptera, Formidicidae). *J. Exp. Biol.,* 151:71-82.

Lighton, J.R.B. 1991. Ventilation in Namib Desert tenebrionid beetles: Mass scaling, and evidence of a novel quantitized flutter phase. *J. Exp. Biol.,* 159:249-268.

Lighton, J.R.B. 1994. Discontinuous ventilation in terrestrial insects. *Physiol. Zool.,* 67:142-162.

Lighton, J.R.B. 1996. Discontinuous gas exchange in insects. *Annu. Rev. Entomol.,* 41:309-324.

Lighton, J.R.B. 1998. Notes from underground: Towards ultimate hypotheses of cyclic, discontinuous gas-exchange in tracheate arthropods. *Am. Zool.,* 38:483-491.

Lighton, J.R.B., and D. Berrigan. 1995. Questioning paradigms: Caste-specific ventilation in harvester ants, *Messor pergandei* and *M. julianus* (Hymenoptera: Formicidae). *J. Exp. Biol.,* 198:521-530.

Lighton, J.R.B., and F.D. Duncan. 1995. Standard and exercise metabolism and the dynamics of gas exchange in the giant red velvet mite, *Dinothrombium magnificum*. *J. Insect Physiol.,* 41:877-884.

Lighton, J.R.B., and D. Garrigan. 1995. Ant breathing: Testing regulation and mechanism hypotheses with hypoxia. *J. Exp. Biol.,* 198:1613-1620.

Lighton, J.R.B., and R. Wehner. 1993. Ventilation and respiratory metabolism in the thermophilic desert ant, *Cataglyphis bicolor* (Hymenoptera, Formicidae). *J. Comp. Physiol.,* 163:12-17.

Lighton, J.R.B., D. Garrigan, F.D. Duncan, and R.A. Johnson. 1993a. Respiratory water loss during discontinuous ventilation in female alates of the harvester ant, *Pogonomyrmex rugosus*. *J. Exp. Biol.,* 179:233-244.

Lighton, J.R.B., L. Fielden, and Y. Rechav. 1993b. Characterization of discontinuous ventilation in a non-insect, the tick *Amblyomma marmoreum* (Acari: Ixodidae). *J. Exp. Biol.,* 180:229-245.

Locke, M. 1958a. The formation of tracheae and tracheoles in *Rhodnius prolixus*. *Quart. J. Microscop. Sci.,* 99:29-46.

Locke, M. 1958b. The coordination of growth in the tracheal system of insects. *Quart. J. Microscop. Sci.,* 99:373-391.

Locke, M. 1998. Caterpillars have evolved lungs for hemocyte gas exchange. *J. Insect Physiol.,* 44:1-20.

Locke, M. 2001. The Wigglesworth Lecture: Insects for studying fundamental problems in biology. *J. Insect Physiol.,* 47:495-507.

Longley, A., and J.S. Edwards. 1979. Tracheation of abdominal ganglia and cerci in the house cricket *Acheta domesticus* (Orthoptera, Gryllidae). *J. Morphol.,* 159:233-244.

Machin, J., P. Kestler, and G.J. Lampert. 1991. Simultaneous measurements of spiracular and cuticular water loss in *Periplaneta americana*: Implications for whole-animal mass loss studies. *J. Exp. Biol.,* 161:439-453.

Mill, P.J. 1974. Respiration: Aquatic insects, pp. 403-467, in M. Rockstein (Ed.), *The Physiology of Insecta,* Vol. VI, 2nd ed., Academic Press, New York.

Miller, P.L. 1960a. Respiration in the desert locust. I. The control of ventilation, *J. Exp. Biol.,* 37:224-236.

Miller, P.L. 1960b. Respiration in the desert locust. III. Ventilation and the spiracles during flight. *J. Exp. Biol.,* 37:264-278.

Miller, P.L. 1966a. The regulation of breathing in insects. *Adv. Insect Physiol.,* 3:279-354.

Miller, P.L. 1966b. The function of haemoglobin in relation to the maintenance of neutral buoyancy in *Anisops pellucens* (Notonectidae, Hemiptera). *J. Exp. Biol.,* 44:529-543.

Oberlander, H. 1980. Morphogenesis in tissue culture: Control by ecdysteroids, pp. 423-438, in M. Locke and D.S. Smith (Eds.), *Insect Biology in the Future,* Academic Press, New York.

Pickard, W.F. 1974. Transition regime diffusion and the structure of the insect tracheolar system. *J. Insect Physiol.,* 20:947-956.

Pringle, J.W.S. 1954. A physiological analysis of cicada song. *J. Exp. Biol.,* 31:525-556.
Punt, A. 1950. The respiration in insects. *Physiol. Comp.,* 2:59-74.
Quinlan, M.C., and N.F. Hadley. 1993. Gas exchange, ventilatory patterns, and water loss in two lubber grasshoppers: quantifying cuticular and respiratory transpiration. *Physiol. Zool.,* 66:628-642.
Richards, A.G. 1953. Structure and development of the integument, pp. 1-22, in K.G. Roeder (Ed.), *Insect Physiology,* John Wiley & Sons, New York, 1200 pp.
Roth, L.M., and B. Stay. 1958. The occurrence of para-quinones in some arthropods, with emphasis on the quinone-secreting tracheal glands of *Diploptera punctata* (Blattaria). *J. Insect Physiol.,* 1:305-318.
Rourke, B.C. 2000. Geographic and altitudinal variation in water balance and metabolic rate in a California grasshopper, *Melanoplus sanguinipes. J. Exp. Biol.,* 203:2699-2712.
Schmidt-Nielsen, K. 1997. *Animal Physiology,* Cambridge University Press, Cambridge, 607 pp.
Schneiderman, H.A., and C.M. Williams. 1955. An experimental analysis of the discontinuous respiration of the Cecropia silkworm. *Biol. Bull.,* 109:123-143.
Shelton, T.G., and A.G. Appel. 2000. Cyclic CO_2 release and water loss in the western drywood termite (Isoptera: Kalotermitidae). *Ann. Entomol. Soc. Am.,* 93:1300-1307.
Shelton, T.G., and A.G. Appel. 2001. Carbon dioxide release in *Coptotermes formosanus* Shiraki and *Reticulitermes flavipes* (Kollar): Effects of caste, mass, and movement. *J. Insect Physiol.,* 47:213-224.
Sláma, K. 1988. A new look at insect respiration. *Biol. Bull. Woods Hole,* 175:289-300.
Sláma, K. 1994. Regulation of respiratory acidemia by the autonomic nervous system (coelopulse) in insects and ticks. *Physiol. Zool.,* 67:163-174.
Sláma, K. 1999. Active regulation of insect respiration. *Ann. Entomol. Soc. Am.,* 92:916-929.
Sláma, K. 2000. Extracardiac vs. cardiac haemocoelic pulsations in pupae of the mealworm (*Tenebrio molitor* L.). *J. Insect Physiol.,* 46:977-992.
Smith, D.S. 1964. The structure and development of flightless Coleoptera: A light and electron microscopic study of the wings, thoracic exoskeleton and rudimentary flight musculature. *J. Morphol.,* 124:107-183.
Snodgrass, R.E. 1935. *Principles of Insect Morphology,* McGraw-Hill, New York.
Snyder, G.K., B. Sheafor, D. Scholnick, and C. Farrelly. 1995. Gas exchange in the insect tracheal system. *J. Theor. Biol.,* 172:199-207.
Van der Kloot, W.G. 1963. The electrophysiology and the nervous control of the spiracular muscle of pupae of the giant silkmoths. *Comp. Biochem. Physiol.,* 9:317-333.
Vogt, J.T., and A.G. Appel. 2000. Discontinuous gas exchange in the fire ant, *Solenopsis invicta* Buren: Caste differences and temperature effects. *J. Insect Physiol.,* 46:403-416.
Weis-Fogh, T. 1964a. Functional design of the tracheal system of flying insects as compared with the avian lung. *J. Exp. Biol.,* 41:207-226.
Weis-Fogh, T. 1964b. Diffusion in insect wing muscle, the most active tissue known. *J. Exp. Biol.,* 41:229-256.
Weis-Fogh, T. 1967. Respiration and tracheal ventilation in locusts and other flying insects. *J. Exp. Biol.,* 47:561-587.
Whitten, J.M. 1972. Comparative anatomy of the tracheal system. *Annu. Rev. Entomol.,* 17:373-402.
Wichard, W., and H. Komnick. 1974. Structure and function of the respiratory epithelium in the tracheal gills of stonefly larvae. *J. Insect Physiol.,* 20:2397-2406.
Wigglesworth, V.B. 1954. Growth and regeneration in the tracheal system of an insect, *Rhodnius prolixus* (Hemiptera). *Quart. J. Microscop. Sci.,* 95:125-137.
Wigglesworth, V.B. 1959. The role of the epidermal cells in the "migration" of tracheoles in *Rhodnius prolixus* (Hemiptera). *J. Exp. Biol.,* 36:632-640.
Wigglesworth, V.B. 1972. *The Principles of Insect Physiology,* 7th ed., Chapman & Hall, London.
Wigglesworth, V.B. 1981. The natural history of insect tracheoles. *Physiol. Entomol.,* 6:121-128.
Zdárek, J., F. Weyda, M.M.B. Chimtawi, and D.L. Denlinger. 1996. Functional morphology and anatomy of the polypneustic lobes of the last larval instar of tsetse flies, *Glossina* spp. (Diptera: Glossinidae). *Int. J. Insect Morphol. Embryol.,* 25:235-248.
Žitňan, D., T.G. Kingan, J.L. Hermesman, and M.E. Adams. 1996. Identification of ecdysis-triggering hormone from an epitracheal endocrine system. *Science,* 271:88-91.

13 Excretion

CONTENTS

PREVIEW

The excretory system consists of two organ systems working together: the Malpighian tubules and the hindgut. The Malpighian tubules typically arise at the junction of the mid- and hindgut. Protons secreted into the Malpighian tubule lumen by a membrane proton pump provide the driving force for urine formation. Potassium ions (K^+) enter Malpighian tubule cells from the hemolymph side through potassium channels, and are then secreted into the tubule lumen by a membrane antiporter mechanism that exchanges H^+ for K^+. Fluid from the hemolymph follows the osmotic gradient created by K^+ movement across tubule cells and carries dissolved substances into the tubule lumen. The tubules transfer the accumulated urine to the hindgut, where selective reabsorption by the ileum and rectum retains necessary substances within the body, while allowing waste products and excesses of useful substances to be voided with the fecal wastes. Specializations in hindgut epithelial cells facilitate reabsorption processes. The excretory system plays a major role in homeostasis of hemolymph, cells, and tissues by helping control levels of electrolytes, water, acid-base equivalents, and nitrogen metabolites. Homeostasis is challenged by food habits, habitat, and metabolic state of the insect. Some excretory products may be stored in the body or cuticle where they may offer protection from predation and parasitism. Insects have a high surface-to-volume ratio and challenge their excretory system to conserve water rather than excrete it. Other insects ingest large liquid meals, such as vertebrate blood, plant phloem sap, or xylem sap, each food consisting of more water than needed, so they rapidly excrete the water and concentrate the nutrients. Some insects (mainly Lepidoptera and Coleoptera) have a cryptonephridial system of Malpighian tubules in which the distal ends of the tubules are held closely to the surface of the rectum in many loops and folds. The loops are more extensive in insects that live in very dry environments, apparently aiding in water reabsorption

13.1 INTRODUCTION

Excretion can be broadly defined as any process that eliminates the interaction of harmful substances with cells and tissues. Even useful substances, such as glucose, amino acids, and certain ions, can be harmful if present in excess amounts. Nitrogenous metabolites, ions, water, and ingested chemicals are substances that an insect may need to excrete. Allelochemicals from plant tissues eaten may be excreted from the body or stored in some inert location in the body. For example, *Manduca sexta*, the tobacco hornworm, efficiently excretes nicotine ingested with its host food, the leaves of tobacco plants (Baldwin, 1991), while *Danaus plexippus*, the monarch butterfly, stores cardenolides from milkweed in the cuticle of the larva and adult, where they act as chemical protection from predators (Brower, 1969).

Maddrell (1971) defined **storage excretion** so as to include materials with future potential use, such as glycogen as a storage form of glucose, or amino acids stored as proteins. He viewed **deposit excretion** as waste material of no further use that needed to be removed from harmful interaction with the tissues. The distinction between stored and deposit excretion, however, is very subtle, and not agreed upon by everyone.

In insects, both the **Malpighian tubules** and the **hindgut** function together as excretory organs. The Malpighian tubules collect a filtrate from the hemolymph and pass this primary urine to the hindgut. Additional components are secreted into the excreta by the hindgut, and some substances are reabsorbed into the hemolymph (Figure 13.1). The term "**excreta**" describes the material actually eliminated from the anus by insects because it is a mixture of undigested materials passing through the gut, substances acted upon and possibly modified by bacterial action in the gut, and urinary materials from the Malpighian tubules. Insect excretion has been reviewed frequently and extensively. Formation of the primary tubule urine was reviewed by Maddrell (1977, 1980), Phillips (1981), Bradley (1985), Spring (1990), Nicolson (1993), and Pannabecker (1995). Selective reabsorption in the hindgut has been reviewed by Phillips et al. (1986). Bursell (1967) and Cochran (1975, 1985a) presented particularly thorough reviews of nitrogen excretion, and Cochran (1985b) summarized the overall function of the excretory system in insects.

13.2 THE MALPIGHIAN TUBULES

Malpighian tubules, the first of the two systems involved in excretion, are long, tubular structures, usually arising at the junction of the mid- and hindgut and terminating blindly in the hemocoel. Some variations in the gross morphology of Malpighian tubule systems are shown in Figure 13.2. The tubules vary in number from 2 to more than 100 in various insect species. Collembola, Aphidae, and some Thysanura lack Malpighian tubules altogether, and other cells and glands take over the functions of excretion. In some members of the Lepidoptera and Coleoptera, the distal ends of the tubules are embedded in the wall of the rectum (see later section). This arrangement, called **cryptosolenic** or **cryptonephridial** tubules, appears to be a modification that aids water conservation. Tracheal connections to Malpighian tubules are numerous, and are indicative of a high metabolic demand for oxygen. A small spiral muscle (Figure 13.3) frequently runs along the surface of a tubule, promoting coiling movements that assist proximal flow of fluid and increase hemolymph in contact with the tubule. Several structural types of tubules may occur in the same insect (Figure 13.4).

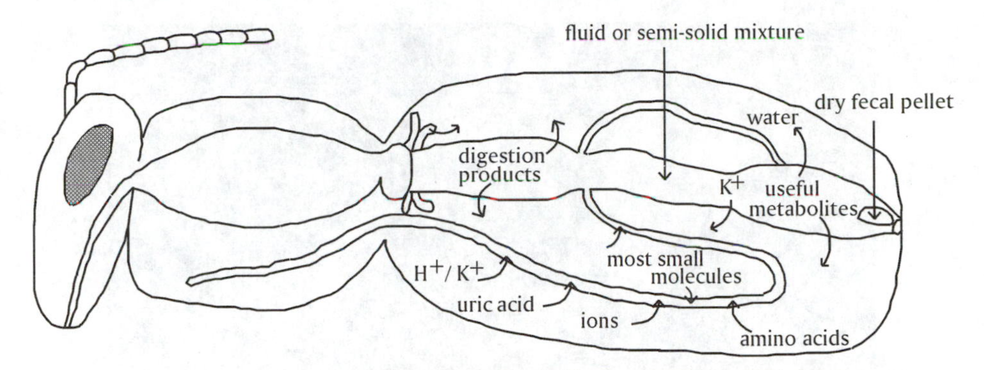

FIGURE 13.1 A generalized scheme of excretion showing the collection of fluid in the Malpighian tubules (promoted by active secretion of protons followed by an antiporter exchange of K^+ for H^+ in the tubule lumen), and extensive reabsorption of water, K^+, and useful substances from the hindgut, primarily the rectum.

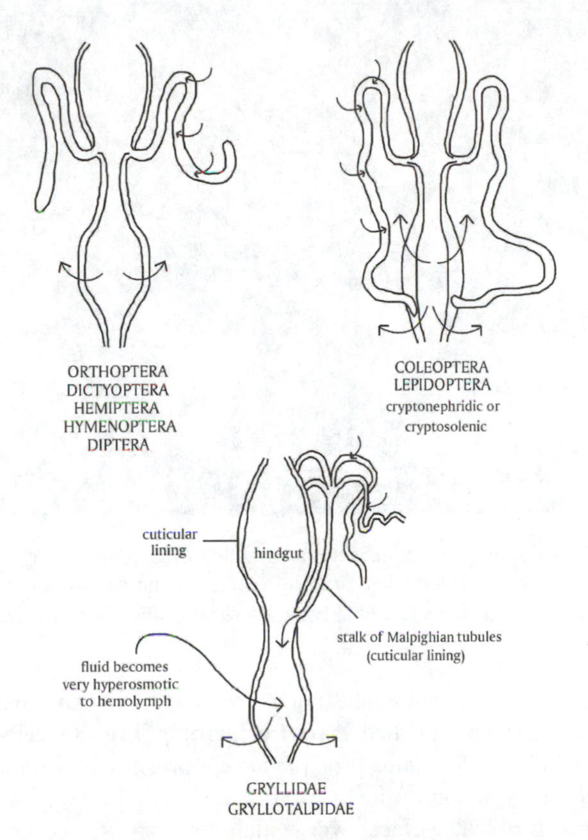

FIGURE 13.2 Variations in the Malpighian tubule systems of insects. The majority of insects have Malpighian tubules that originate at the junction of the mid- and hindgut, and terminate as blind tubules in various regions of the hemocoel. Cryptosolenic tubules in which the distal ends of the tubules lie on top of the terminal part of the hindgut, occur in most Lepidoptera and Coleoptera. The Malpighian tubules arise from a short, cuticle-lined stalk in gryllid crickets and mole crickets.

13.3 ULTRASTRUCTURE OF MALPIGHIAN TUBULE CELLS

A single layer of epithelial cells (usually comprising two to five cells) surrounds the lumen of a tubule. Several different cell types have been identified, but their specific functions have not been

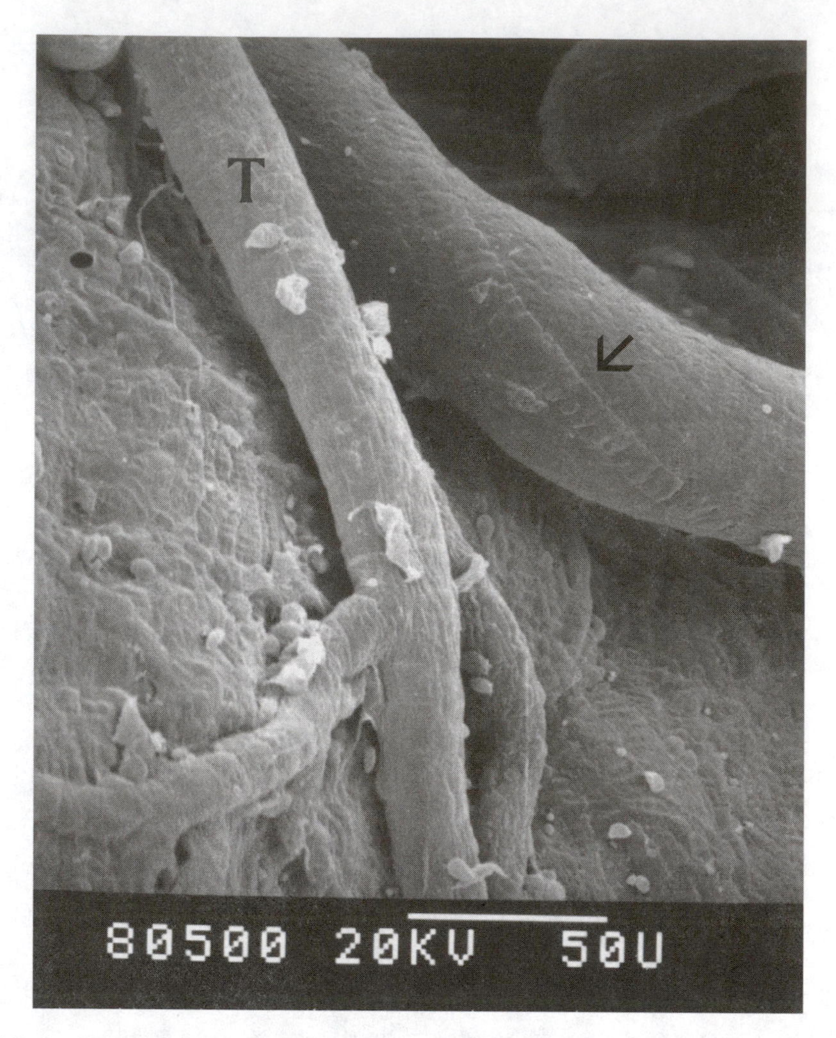

FIGURE 13.3 An SEM photo of the small muscle (arrow) that often spirals along the length of a Malpighian tubule of some insects, in this case, the cricket *Gryllus assimilis*. Contraction of the muscle throws the tubules into tight coils in some insects. The muscle probably serves to keep the tubule moving through the hemolymph. (Photograph by the author.)

elucidated in many cases. Some, but not all, tubule cells have a **brush border** of **microvilli** on the apical surface, and these have been called **Type 1** or **principal tubule cells** (Figure 13.5). Cells in the distal half of the tubules in *Rhodnius prolixus* have a brush border on the apical surface of the cells and are involved in formation of the primary urine. Cells in the proximal half of the tubules have a relatively smooth apical surface over which reabsorption occurs, probably by energy-requiring mechanisms (Wigglesworth, 1931).

Tubule cells are thin, sheet-like cells that wrap around the tubule lumen. Water and hemolymph components entering at the basal side of the cells have only a short distance to traverse to reach the apical surface where they may be secreted, or diffuse, into the tubule lumen. Malpighian tubule cells are characterized by extensive infolding of the basal membrane (the membrane on the hemolymph side), creating many twisting channels that reach 5 to 10 μm or more into the cell (O'Donnell et al., 1985) (Figure 13.6). Potassium ions in the hemolymph enter Malpighian tubule cells through potassium ion gates in these infolded channels in the basal membrane (Nicolson and Isaacson, 1990), and water and dissolved substances in the aqueous medium of the hemolymph follow the osmotic gradient (Figure 13.7). Extensive membrane surfaces presented by both the

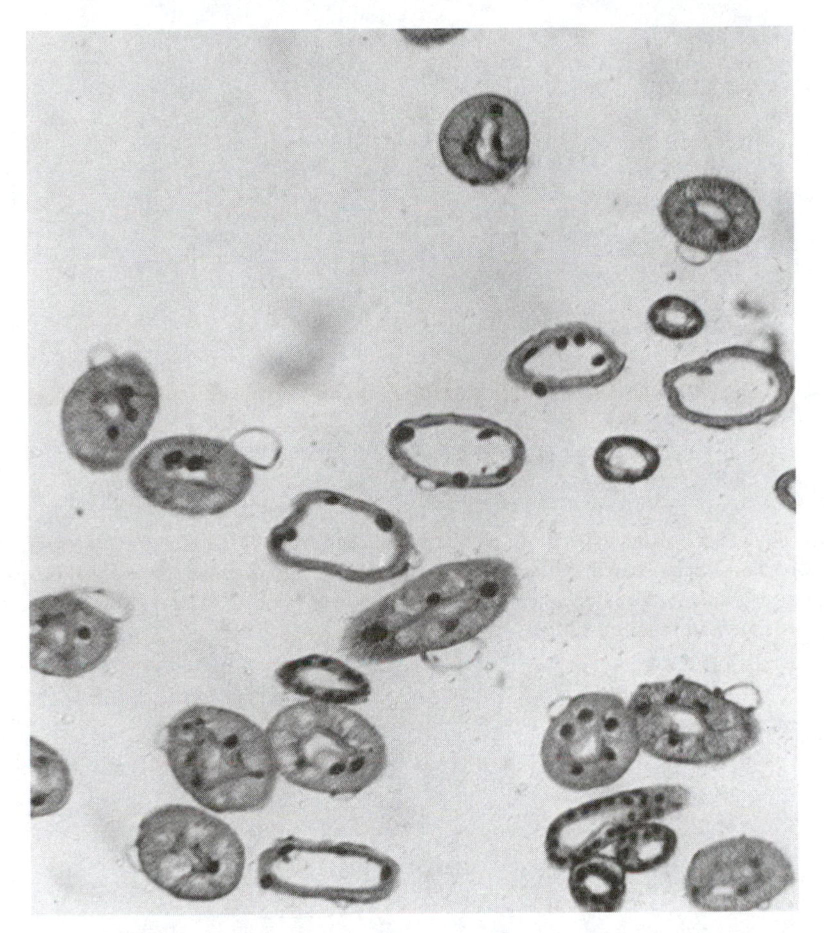

FIGURE 13.4 A photograph from a histological preparation showing the three types of Malpighian tubules in a mole cricket *Scapteriscus vicinus*.

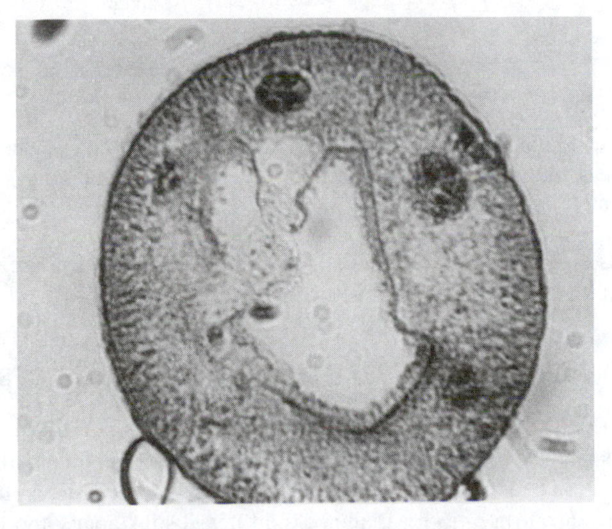

FIGURE 13.5 A cross section through the primary type of Malpighian tubule in *Scapteriscus vicinus,* showing microvilli on the lumen surface of the cell.

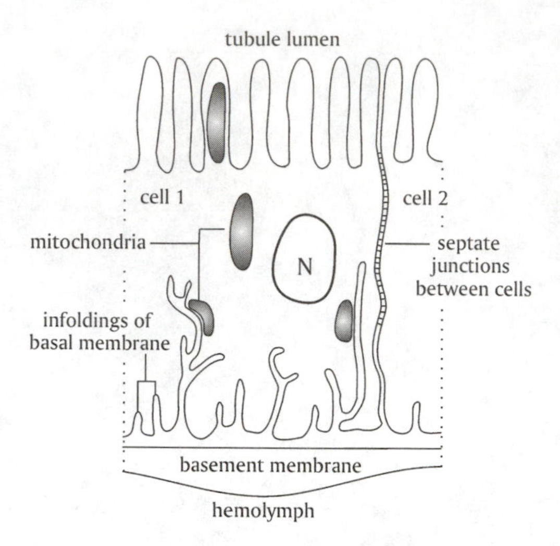

FIGURE 13.6 The general structure of a Malpighian tubule cell from the proximal tubule segment of the last instar of *Drosophila melanogaster* that illustrates extensive basal infoldings, a relatively short path across the narrow cell, and long microvilli on the apical (lumenal) surface of the cells. The microvilli often contain large mitochondria, as shown in this illustration.

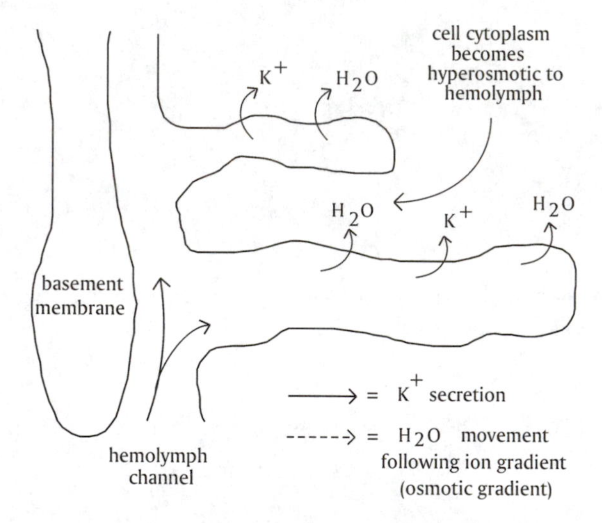

FIGURE 13.7 A diagram to illustrate the influx of K^+ through potassium ion channels in the basal infoldings of Malpighian tubule cells, and the passive movement of water and dissolved solutes following the microosmotic gradients set up by K^+ movement.

basal and apical faces of Malpighian tubule cells and large mitochondria are indicative of specialization for active transport, as well as passive diffusion.

13.4 FORMATION OF PRIMARY URINE IN MALPIGHIAN TUBULES

The primary urine formed in the lumen of the Malpighian tubules is a **filtrate of the hemolymph** (Ramsay, 1953, 1955a, 1955b, 1956, 1958), and it contains most of the small ions and molecules (sugars, amino acids, ions, as well as other components) that occur in the hemolymph. The urine:hemolymph concentration ratio for many of the filtered substances approaches unity, indicating passive movement across the tubule cell membranes, but some components are actively secreted and their urine:hemolymph ratio is always greater than one.

13.5 A PROTON PUMP IS THE DRIVING MECHANISM FOR URINE FORMATION

Urine formation in Malpighian tubules relies on a **proton pump** (see Figures 2.9 and 2.10 in Chapter 2 for more details on the proton pump) in the apical membrane (the side facing the lumen) of Malpighian tubule cells that actively secretes protons (H^+) into the tubule lumen against an electrochemical gradient (Wieczorek et al., 2000; Hopkin et al., 2001). The pump consists of a V_1 complex of proteins in the cytoplasm of the "principal" cells (the primary cells in Malpighian tubules, but occasional other cells occur, such as stellate cells) of the Malpighian tubules, and an ion channel formed by the transmembrane Vo complex embedded in the lipid bilayer of the apical membrane. The pump causes the tubule lumen to become positive (as much as +30 mV or more in some insects) to the hemolymph, and creates highly variable gradients in pH across the apical membrane of principal cells. The proton gradient provides the energy for an **antiporter mechanism** that exchanges K^+ for H^+ across the apical membrane (Forgac, 1989; Weltens et al., 1992; Maddrell and O'Donnell, 1992; Zhang et al., 1994; Wieczorek et al., 2000). The net result is that K^+ is secreted into the tubule lumen and concentrated against an electrochemical gradient. In some insects that take blood meals rich in Na^+ (e.g., mosquito adults, *R. prolixus*, and tsetse fly adults), Na^+ is actively transported by the pump mechanism. The pump is probably regulated by dissociation/reassociation (Kane and Parra, 2000), and is likely under genetic regulation at the transcriptional and post-transcriptional levels (Wieczorek et al., 1999). Cations such as K^+ and/or Na^+, and probably anions, must enter the Malpighian tubule cells (at the basolateral surface) from the hemolymph in order for secretion into the lumen at the apical face of cells to continue. Entry of ions at the basolateral membrane surface, however, is not well understood. Secretion of cations (H^+, Na^+, and K^+) across the apical membrane appears to be electrically coupled with Cl^- transport (Beyenbach, 1995; Dijkstra et al., 1995) in the basolateral membrane of tubule cells, providing for balance and steady-state conditions between entry of cations from the hemolymph across the basolateral membrane and secretion across the apical membrane (Beyenbach et al., 2000b). Chloride ions may be transported by a paracellular pathway (i.e., between adjacent cells) in some insects, such as *Aedes aegypti* (Beyenbach et al., 2000a, 2000b) or by a transcellular pathway in others [O'Donnell et al. (1998), in tubules of *Drosophila melanogaster*]. There is evidence that the route for Cl^- transport may vary within the same organism in response to variable physiological conditions (Dijkstra et al., 1995) and/or hormonal stimulation (Yu and Beyenbach, 2001). Based on measurements of the transepithelial potentials across Malpighian tubule cells and lumen, Ianowski and O'Donnell (2001) suggest a stoichiometry of Na^+:K^+:2Cl^- co-transport across the basolateral membrane of tubule cells in *R. prolixus*.

Although the secretion of K^+ to the tubule lumen has been known for half a century (Ramsay, 1953, 1955a, 1955b), a cellular/molecular explanation was not known until discovery of the proton pump. The formation of urine volume is highly dependent on the K^+ concentration in the bathing hemolymph or saline (Figure 13.8), but fluid formation stops even when the K^+ concentration in the bathing saline is high if the H^+ pump is inhibited (Bertram et al., 1991; Weltens et al., 1992). Molecules such as sugars, amino acids, and some allelochemicals in the hemolymph follow the osmotic gradient created by the transport of K^+ and other ions. In addition to secreting K^+, Malpighian tubule cells in some insects also secrete Na^+ (Figure 13.8), some other ions, and organic molecules. One organic molecule, the amino acid proline, is actively secreted into the urine by tubules of the desert locust *Schistocerca gregaria*; and after passage into the hindgut, it is used as an energy source for ATP production to fuel ion pumps in the hindgut epithelium (Phillips et al., 1994).

The process driven by the proton pump has been called a **standing gradient process** (Berridge and Oschman, 1969). Although it probably accounts for most of the urine formed, there may be additional processes by which substances enter the tubule lumen. Wessing and Eichelberg (1975) suggested that there might be a number of mechanisms operating in various insects to account for some components in the urine, and they presented electron micrograph evidence that they interpreted

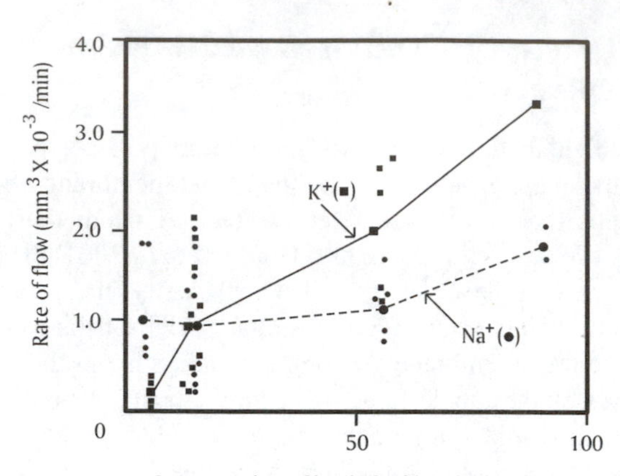

FIGURE 13.8 The rate of urine formation by an isolated Malpighian tubule from the stick insect *Carausius morosus* as a function of K^+ or Na^+ concentration in the bathing saline formation. Although the tubules of *C. morosus* do not strongly secrete Na^+, the tubules of some other insects secrete Na^+ and K^+, and sometimes other ions. (Reproduced with permission from Ramsay, 1955b.)

as indicative of more than one process operating in tubule cells of *Drosophila melanogaster* (Wessing and Eichelberg, 1978). Additional processes might include transport of substances enclosed in vesicles (Riegel, 1966; Linton and O'Donnell, 2000), free movement of substances through the cell cytoplasm, and passage of substances between adjacent cells by movement through the intercellular space. None of these seems likely to be mutually exclusive of others, and several mechanisms might operate in the same insect.

The rates of urine formation and ion secretion are controlled by diuretic peptide hormones and certain non-peptide compounds, such as 5-hydroxytryptamine (5-HT or serotonin) (reviewed by O'Donnell and Spring, 2000). The peptide compounds fall into two major classes: those similar to vertebrate corticotropin-releasing factor (called CFC-related peptides) and smaller kinins. The CFC-type peptides range in size from 30 to 46 amino acids, while the kinins are smaller, comprised of 6 to 15 amino acids. Although both types stimulate urine formation, they act through different mechanisms. CFC-peptides (and 5-HT) stimulate adenylate cyclase and raise the level of cAMP, while the kinins thus far studied activate the Ca^{2+}-signaling pathway. The Malpighian tubule secretion rate is typically controlled by the interaction of several of these compounds. Diuretic factors may exhibit synergism (i.e., greater activity in combination than the additive effects of each alone) or, alternatively, effects of multiple compounds may only be additive, but cation and anion pathways are controlled separately by different second messengers. In one case, an inhibitory interaction is known. In the blood-feeding hemipteran *Rhodnius prolixus*, two factors, 5-HT and diuretic hormone (DH), act synergistically to stimulate urine formation in the tubules; but a third peptide [cardioacceleratory peptide 2b (CAP_{2b})] acts as an antidiuretic hormone when part of the mixture. Its effects are mediated through stimulation of cGMP, which then inhibits the action of 5-HT. The synergistic action of these hormones probably benefits an insect by reducing quantities of hormones released and ensuring that all tubules respond rapidly. In insects that take a large fluid meal, such as mosquitoes, some hemipterans, and phloem and xylem feeders, the synergistic action of hormones may compensate for hemolymph dilution as water from the large meal is absorbed into the hemolymph.

The primary urine formed by the Malpighian tubules is **isosmotic** or sometimes slightly **hyposmotic** to the hemolymph. Malpighian tubules are not capable of producing primary urine that is appreciably **hyperosmotic** to the hemolymph. The proximal tubules may modify urine by reabsorption of some substances (e.g., in *R. prolixus*), but many insects transfer the tubule fluid to the hindgut with few or no changes in its chemical composition or volume. The hindgut then proceeds to concentrate waste products by reabsorbing water and useful substances.

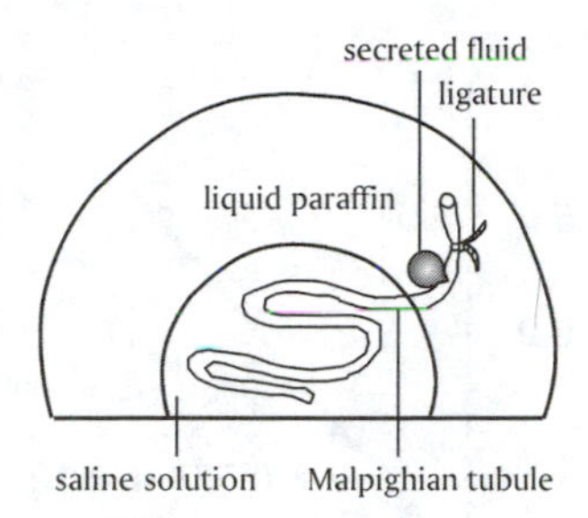

FIGURE 13.9 An illustration of the isolated Malpighian tubule technique for studying urine formation in response to hormones, ions, inhibitors, or other agents dissolved in the bathing saline. (Modified from Ramsay, 1954.)

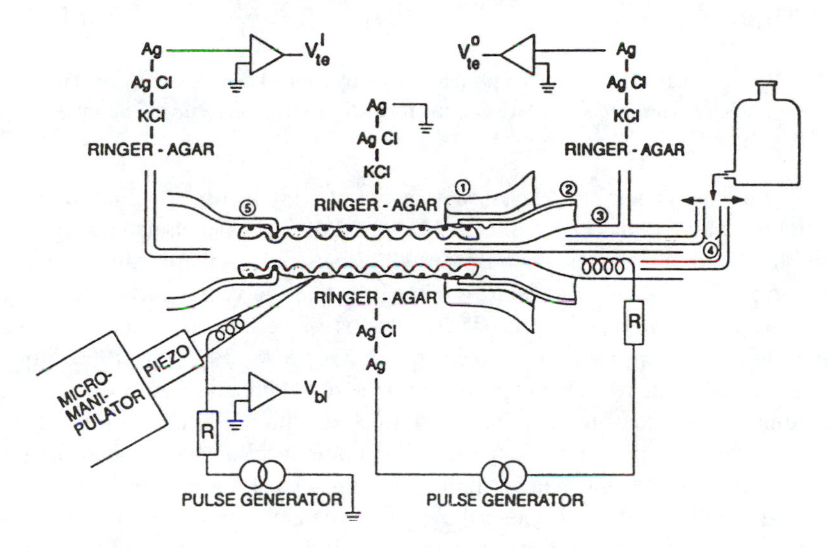

FIGURE 13.10 Arrangement for experimental perfusion of an isolated tubule. (Reproduced with permission from Leyssens et al., 1992.)

The use of isolated tubules (Figures 13.9 and 13.10), a technique devised by Ramsay (1954), continues to be an important research technique for elucidating physiology of the tubules (Nicolson and Hanrahan, 1986; Isaacson et al., 1989; Hegarty et al., 1991; Leyssens et al., 1992; Leyssens et al., 1993). Isolated tubules secrete a droplet of urine that can be measured volumetrically by assuming that the droplet has the dimensions of a sphere. Typical secretion rates, measured in nanoliters per minute (nl/ml), over a period of several hours, are shown in Figure 13.11. With current ultrasensitive techniques, sufficient fluid can be recovered for microchemical analyses (Beyenbach, 1995, and references therein).

13.6 SELECTIVE REABSORPTION IN THE HINDGUT

13.6.1 ANATOMICAL SPECIALIZATION OF HINDGUT EPITHELIAL CELLS

The **hindgut** is the second system that completes the excretion process by selectively reabsorbing some substances into the hemolymph, leaving others in the lumen, and actively secreting some substances into the hindgut lumen. The rectal cuticular lining has greater permeability than the cuticular lining on foregut cells, and the epithelial cells of the hindgut are specialized for both active secretion and active reabsorption. Phillips and Dockrill (1968), who removed and tested the permeability of the cuticular lining from the hindgut of *S. gregaria*, found that molecules with a molecular weight of 300 to 500 crossed the membrane slowly, and molecules having a radius larger

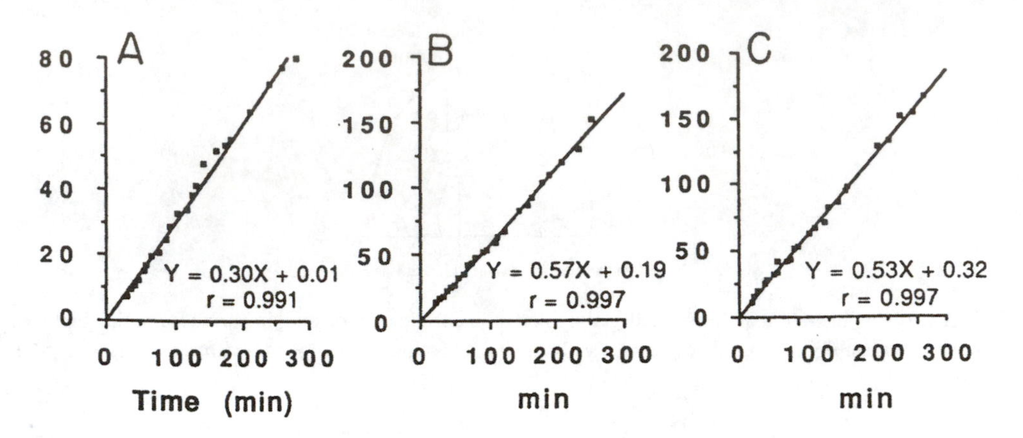

FIGURE 13.11 The cumulative formation of primary urine by an isolated tubule. A, B, and C show results from three separate tubules. The *y*-axis units are nl urine formation. (Reproduced with permission from Hegarty et al., 1991.)

than about 0.5 to 0.6 nm penetrated very slowly or not at all. Glucose (0.42 nm radius, MW 180) penetrated readily, while trehalose (radius 0.52 nm, MW 342) penetrated much more slowly. Although this should not be taken as the norm for all insects, it is probably similar in other insects.

In the rectum, small groups of cells are variously called the **rectal cells**, **rectal pad cells**, or **rectal papillae cells** in different insects. These groups of cells have special modifications for reabsorption. In Diptera, four to six finger-like papillae (Gupta and Berridge, 1966; Hopkins, 1967) are attached to the wall of the rectum and project into the rectal lumen (Figure 13.12). The chitinous lining on the lumenal surface of the papillae is continuous with the lining on the inner wall of the rectum. The cells of a rectal papilla are large, usually cuboidal cells that surround a central channel in the papilla that opens into the hemolymph space through a valve (Figure 13.13). Fluid that crosses the rectal papillae cells and enters the central channel is returned to the hemolymph. A small tracheal trunk and a nerve pass into the central cavity, and the tracheal trunk branches into many finer tracheae and tracheoles, suggesting a high demand for oxygen for performance of metabolic and secretory work. In rectal papillae cells of the mosquito *Aedes aegypti,* the lateral cell membranes are elongated into extensive inward-directed folds of membranes lying very close to each other and projecting nearly to the basal and apical surfaces of the cells (Hopkins, 1967). These elaborate membrane folds create many membrane-bound channels and spaces within cells in papillae. Cell nuclei are large and prominent. The apical cell membrane (facing the lumen of the rectum) of rectal papillae cells is also greatly infolded, and large mitochondria are usually associated closely with the intercellular channels created by the extensive membrane folding.

Rectal pad cells are common in many insects, and typically are enlarged, columnar to cuboidal cells arranged in six clusters separated by smaller, squamous cells between the "pads" of absorptive cells. The rectal pad cells in the cockroach *Periplaneta americana* (Phillips, 1981) have highly folded cell membranes at the surface of the rectal lumen that present 10 to 20 times the surface area of smooth membranes. Mitochondria are located near and within the apical folds, and often occur in compact stacks in conjunction with the highly infolded lateral membranes. The extensive infoldings of membranes create many intracellular channels that collect fluid (with dissolved solutes) from the rectum and direct it toward channels between adjacent cells (intercellular channels). The intercellular channels lead to the basal membrane of pad cells where water and useful solutes reenter the hemolymph. The membranes of the intercellular channels are straight and smooth; thus, they present relatively little surface for back-diffusion into the rectal pad cells. Using a micropuncture technique to withdraw minute amounts of fluid from various regions of the rectal pad cells of the American cockroach *P. americana*, Wall and Oschman (1970) found that fluid from

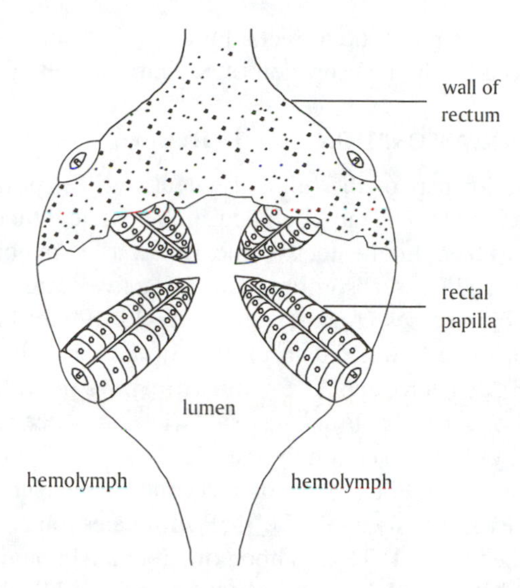

FIGURE 13.12 A diagram of the rectal papillae in the rectum of adult dipterans and adult siphonapterans (fleas).

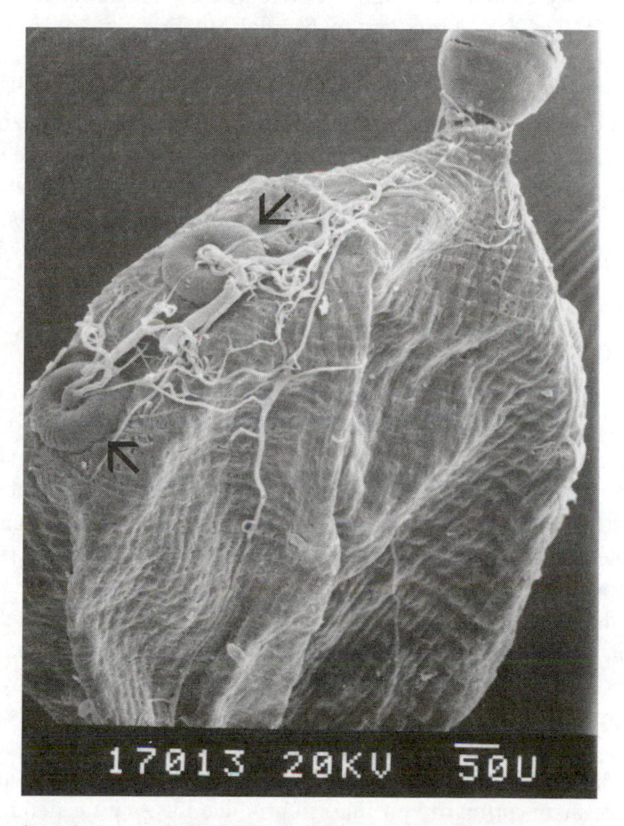

FIGURE 13.13 A scanning electron micrograph of the hemolymph side of the rectum of the tephritid fruit fly *Anastrepha suspensa,* showing the exit to the hemolymph of two of the four rectal papillae. Note the extensive tracheal network and the large trachea entering each papilla.

the basal subepithelial sinus is hyposmotic to rectal lumen fluid. This was interpreted to mean that its ion load had been reduced by the reabsorption of K^+ into the cells for recycling.

13.6.2 Secretion and Reabsorption in the Ileum

The ileum is the most anterior part of the hindgut, occurring just posterior to the origin of the Malpighian tubules in most insects. The most detailed studies of ileal function have been conducted in the desert locust *S. gregaria.* In the locust, the **ileum** is a major site for isosmotic fluid reabsorption, for active Na^+ and Cl^- reabsorption, and for active secretion of proline as an energy source to support metabolic processes (Audsley et al., 1992a, 1992b; Phillips et al., 1994, 1988). The driving mechanism for ion and water reabsorption in the ileum is an electrogenic Cl^- pump (Phillips et al., 1986, 1988). A neuropeptide, the **ion transport peptide** (**ITP**) isolated from the fused corpus cardiacum of *S. gregaria* stimulates Na^+, Cl^-, and water reabsorption, and promotes passive reabsorption of K^+ by electrical coupling (Audsley et al., 1992a, 1992b; Phillips et al., 1994; Harrison, 1995; Meredith et al., 1996). ITP needs a second messenger, which is probably cAMP based on observations that exogenously applied cAMP stimulates ion and fluid reabsorption. Some uncertainties still exist in the way that ITP acts upon the ileum. Although ITP inhibits H^+ secretion (i.e., inhibits formation of NH_4^+) in the ileum, cAMP stimulates NH_4^+ formation (Audsley et al., 1992b). The gene encoding ITP has been investigated by Meredith et al. (1996), who prepared a cDNA encoding a peptide with 130 residues that may be a propeptide of ITP, but the mechanism by which the active ITP is derived from this prohormone was not elucidated. The net result from movement of the excretory contents through the ileum of the desert locust is that the volume of fluid is reduced as Na^+, Cl^-, K^+, and fluid are reabsorbed (the ions by active mechanisms and fluid following the osmotic gradients).

The ileum plays a major role in **acid-base balance** (see Section 13.7.3) by secretion of H^+ into the lumen, formation of NH_4^+, and reabsorption of HCO_3^-. Metabolism of reabsorbed nonessential amino acids (alanine, asparagine, glutamine, serine, and proline) present in the urine releases energy for ATP synthesis, an essential power source needed to drive the active reabsorption of Na^+ and Cl^-. About 80% of the ammonia produced in the epithelial cells from metabolism of these amino acids is transported (mechanism not elucidated) into the lumen (Phillips et al., 1994) where it is excreted as NH_4^+.

13.6.3 Reabsorption in the Rectum

The **rectum** is the final and major site for reabsorption of ions, water, and nutrients, and it is capable of reabsorbing fluid against strong osmotic gradients, ultimately producing a very concentrated **hyperosmotic** excreta in many insects. The driving mechanism for cation and water reabsorption, as in the ileum, is an electrogenic Cl^- pump under the influence of a neuropeptide hormone, **chloride transport stimulating hormone** (**CTSH**), from the corpora cardiaca. It acts on the rectal epithelium to promote active Cl^- absorption (Phillips, 1964; Phillips et al., 1986, 1994; Harrison, 1995). The pump provides the energy for K^+ reabsorption. The exact route followed by K^+ as it crosses the rectal epithelial cells to reenter the hemolymph or Malpighian tubules varies in different insects, depending on the anatomy of the rectum and the cells involved in the process. Water from the rectal lumen and dissolved solutes follows the osmotic gradient created by ion absorption. The result is that water is reabsorbed into the epithelial cells against an increasingly strong concentration gradient in the rectal lumen. The excreta in the rectum becomes very pasty or even dry in many insects as water is reabsorbed. The rectal epithelial cells actively reabsorb amino acids from the lumen, and metabolize them (primarily proline) to produce the ATP needed to energize the pump. Proline is metabolized within mitochondria by the proline dehydrogenase pathway (Chamberlin and Phillips, 1982,1983). Thus, active ion secretion and eletrogenic pumps play a major role in the formation of primary urine in the Malpighian tubules and then in reclaiming water from the hindgut. Several different hormones are responsible for regulating the different functions.

13.7 THE ROLE OF THE EXCRETORY SYSTEM IN MAINTAINING HOMEOSTASIS

Dynamic changes in salt, water, acid-base, and nitrogen amounts occur from time to time in all organisms as a result of food ingested, environmental conditions, and metabolism. Regulatory mechanisms that respond rapidly to these changes are necessary to preserve the integrity of cells and tissues. For example, herbivores ingest relatively large amounts of potassium and little sodium with their plant-based diet, while blood-feeders such as some hemipterans and mosquitoes ingest relatively large amounts of sodium (mostly as sodium chloride) and little potassium with their food. Ingestion of plant phloem or xylem sap results in an excessive intake of water, and usually more sugar and some amino acids than needed. Nitrogen metabolites from proteins, amino acids, and purines must be disposed of by all cells. Maintenance of the constancy of the internal environment of cells, tissues, and organisms is the process of **homeostasis**, and the excretory system plays a major role.

13.7.1 ELECTROLYTE HOMEOSTASIS

Beyenbach (1995) reviewed mechanisms for maintaining **electrolyte homeostasis**, with special emphasis on the blood-feeding mosquito *A. aegypti* as a model insect. Larval and adult *A. aegypti* live in different habitats, have different food habits, and control Malpighian tubule function by different hormones. Adult female mosquitoes need a blood meal in order to mature each batch of eggs, but with the blood comes a large salt (NaCl) load that must be excreted. Sodium excretion is an active process and occurs in Malpighian tubule cells of the adult mosquito in response to stimulation from the **mosquito natriuretic peptide** (**MNP**) released from the corpora cardiaca (Wheelock et al., 1988; Beyenbach, 1995). A proton pump coupled with a H^+/Na^+ exchange mechanism secretes sodium into the tubule lumen. The pump appears to work as previously described for Malpighian tubules in general, except that Na^+, rather than K^+, is the principal ion exchanged for protons pumped into the tubule lumen.

Prior to a blood meal, urine forms slowly in isolated tubules from *A. aegypti* at about 0.4 nl/min, and the measured **transcellular resistance**, **Rc**, across the tubule cells is high, keeping cation and water movement low. Feeding on a blood meal stimulates the release of MNP, and cAMP is produced as a second messenger at the inner surface of the basolateral membrane of tubule cells. cAMP acts selectively to open Na^+ channels in the basolateral membrane. As Na^+ enters the tubule cell from the hemolymph, the Rc falls to about 40% of its prefeeding value (Wheelock et al., 1988). Movement of water into tubule cells follows the osmotic gradient. Urine flow rates as high as 2.8 nl/min in hormone stimulated tubules are promoted by the apical membrane H^+-pump coupled with H^+/Na^+ exchange. The ion flux generated by MNP and cAMP is specifically an increase in secretion of sodium. Potassium movement is not influenced. The voltage in the lumen of the Malpighian tubules increases from about +52 mV in unfed mosquitoes to about +70 mV in fed mosquitoes (lumen positive to hemolymph in both cases).

The large chloride load from the blood meal must also be excreted, and chloride (Cl^-) moves from hemolymph to tubule lumen in a passive transport pathway between the cells (called the **paracellular pathway**) (Pannabecker et al., 1993). The permeability of the paracellular pathway is increased by leucokinin-VIII, a neuropeptide (Wang et al., 1996), although it is not known whether this or a similar hormone is secreted by the mosquito.

Larval *A. aegypti* live in fresh water, and in response to an increase in salinity they secrete **5-hydroxytryptamine** (**serotonin**) into the hemolymph, leading to an increase in cAMP formation in the Malpighian tubules (Clark and Bradley, 1993). Serotonin and cAMP stimulate fluid and ion (Na^+ and K^+) secretion rates in isolated larval tubules, but the urine is not concentrated with respect to the ions (Clark and Bradley, 1996). The blood-feeding hemipteran *R. prolixus* also secretes Na^+ (and K^+) into the lumen of its Malpighian tubules. Hematophagous behavior may have driven the evolution of Na^+ secretion by Malpighian tubule cells, thus enabling blood-feeders to regulate ion homeostasis after a large, salty meal.

Beyenbach (1995) has reviewed three potential physiological processes through which *A. aegypti* may regulate (control) rates of ion and fluid excretion. These processes are (1) the proton pump that supplies energy for Na^+ and K^+ secretion to the tubule lumen; (2) the resistance Rc across the tubule cells that control ion channels in the basolateral membrane; and (3) the resistance of the passive transport pathway for chloride movement. Regulation of the proton pump has not been demonstrated in mosquito Malpighian tubules (Beyenbach, 1995). However, Rc and basolateral ion channels are regulated by the natriuretic peptide and cAMP in mosquitoes. The passive transport pathway between tubule cells may be a function of extracellular secretion of a leucokinin-type peptide in adult mosquitoes.

13.7.2 Water Homeostasis

Water homeostasis is very important to insects because they have a high surface-to-volume ratio and their food often has variable water content. Although many physiological systems and behavior are related to water conservation, ridding the body of excess water is primarily a function of the excretory system. Water excretion and retention are regulated by hormones. **Diuretic hormones** promote fluid formation and rapid excretion by the Malpighian tubules, while the currently known **antidiuretic hormones** act on the hindgut (with one exception) and promote water reabsorption. The exception is an antidiuretic hormone demonstrated from water-deprived, dehydrated house crickets *Acheta domesticus* that inhibits fluid formation by the Malpighian tubules without action upon the hindgut (Spring et al., 1988).

13.7.2.1 Diuretic Hormones

Currently, more than 20 insect diuretic hormones are known. All are neuropeptides. They increase the formation of fluid by Malpighian tubules in insects in the orders Orthoptera, Lepidoptera, Diptera, Dictyoptera, and Coleoptera (Wheeler and Coast, 1990). 5-Hydroxytryptamine (5-HT, serotonin) also stimulates urine formation (Barrett and Orchard, 1990; Maddrell et al., 1991).

Rhodnius, the first insect studied in detail with respect to a neuropeptide diuretic hormone (Maddrell, 1963, 1964a, 1964b, 1966), takes one large blood meal in each instar, and within hours rapidly excretes a large volume of urine. This action leaves concentrated proteins from the blood meal in the midgut and rids the body of excess water and Na^+. Within minutes after feeding starts, the large volume of blood ingested greatly distends the abdomen and activates stretch receptors located in the tergosternal muscles near the lateral edge of abdominal segments 2 to 7. Rapid circulation of hemolymph is induced, caused at least in part by vigorous peristaltic movements of the alimentary canal. Diuresis starts within 3 minutes after feeding, initiated by secretion of 5-HT and a neuropeptide diuretic hormone synthesized in large neurosecretory cells in the fused mesothoracic ganglion (Figure 13.14). The neuropeptide is released from a series of enlarged axonal endings of abdominal nerves originating from the mesothoracic ganglion (Maddrell, 1966). The hormone is transported to the Malpighian tubules, the target tissue, by rapid circulation of hemolymph. By the time diuresis is complete, *Rhodnius* has lost 40% of its freshly fed weight. The osmotic concentration of its hemolymph first falls due to the dilution effect of absorbing so much water from the midgut; but after diuresis, the osmotic concentration is about the same as before feeding. Malpighian tubules and possibly other tissues rapidly destroy the diuretic hormone, and sustained release of the hormone is necessary to maintain the excretion of a large volume of fluid.

Diuresis is also under similar hormonal control in the cotton stainer *Dysdercus fasciatus* (Hemiptera), which takes plant sap. Medial neurosecretory cells of the brain are the principal source of the hormone in *D. fasciatus*, but some activity is present also in corpora cardiaca and in the mesothoracic ganglion. Extracts from median NSC accelerated urine flow rate from the normal value of 3.1×10^{-3} mm^3 to 9.87×10^{-3} mm^3 per minute (Berridge, 1966).

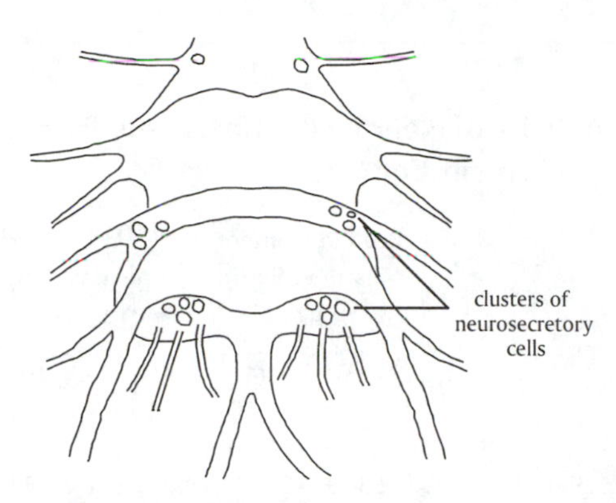

FIGURE 13.14 The large thoracic ganglion containing neurosecretory cells (NSC) that are the synthesis site of the diuretic hormone in *Rhodnius prolixus*. The neuropeptide is released from swollen axons arising from the posterior of the ganglion. The complex ganglion contains fused abdominal ganglia. (Modified from Maddrell, 1963.)

A diuretic neuropeptide (**MAS-DH**) with 41 amino acid residues has been isolated from *M. sexta* (Kataoka et al., 1989). It has some sequence similarity to **corticotropin releasing factor** (**CRF**) and **urotensin I**, two vertebrate neuropeptides with hormonal activity, and to a toxin, **sauvagine**, from the skin of a South American tree frog. A receptor for MAS-DH has been characterized from the Malpighian tubules of 5th instars of tobacco hornworms (Reagan et al., 1993). MAS-DH rapidly binds to the receptor and in a reversible manner (Lehmberg et al., 1991). A diuretic neuropeptide isolated from *Locusta migratoria* (Mordue and Morgan, 1985; Proux et al., 1987; Lehmberg et al., 1991, 1993) consists of two antiparallel 9-amino acid peptides joined by two disulfide bonds, as follows:

Cys-Leu-Ile-Thr-Asn-Cys-Pro-Arg-Gly-NH_2
| |
NH_2-Gly-Arg-Pro-Cys-Asn-Thr-Ile-Leu-Cys

Malpighian tubules from the cricket *A. domesticus* are stimulated to secrete fluid by tissue extracts (Spring and Hazelton, 1987; Coast, 1988) (Table 13.1) of corpora cardiaca (Figures 13.15 and 13.16), corpora allata (Coast, 1989), and some other parts of the central nervous system (Coast and Wheeler, 1990). Tubules in this cricket have three physiologically and morphologically different segments (Kim and Spring, 1992), and different extracts may act on different segments by different mechanisms.

The beetle *Tenebrio molitor* lives in dry grain and grain products, and needs no water other than that already present in the food and that derived from metabolism. Surprisingly, it produces a diuretic hormone (Nicolson, 1991). Why would an insect living in a water-impoverished environment evolve a diuretic hormone? Nicolson (1991) suggests that its function is to act as a "**clearance hormone**" to flush the Malpighian tubules and hindgut. The tubule fluid may be passed into the hindgut to help move the very dry food residue through the gut, and may help promote a countercurrent flow in the midgut. Fluid reaching the rectal region is nearly all reabsorbed by the cryptonephridic tubules and returned to the hemolymph.

Another tenebrionid beetle, *Onymacris plana*, living in the desert also has a diuretic hormone, and it too may have a flushing function. Nicolson and Hanrahan (1986) found that an isolated Malpighian tubule from *O. plana* typically produces about 3 nl/min/tubule; but with stimulation by a diuretic hormone from the corpora cardiaca, the tubule forms 40 to 60 nl, or sometimes up to 100 nl/min/tubule. They speculated that a flushing function might be beneficial to the insects by aiding removal of plant allelochemicals eaten with the food. The rectum reabsorbs the water and conserves it for reuse.

TABLE 13.1
The Diuretic Activity of Neuroendocrine Tissue Extracts Expressed as a Function of Protein Content

Tissue	Protein Content Mean ± SE (n) μg	Increased Rate of Secretion pl/mm/min Mean ± SE (n)	pl/mm/min μg protein
Brain	35.53 ± 2.54(4)	195 ± 9.3 (10)	5.49
Corpora cardiaca	4.57 ± 0.83 (4)	256 ± 26.1 (8)	56.02
Corpora allata	3.26 ± 0.31 (4)	189 ± 10.3 (9)	57.98
Subesophageal ganglion	22.34 ± 0.73 (4)	168 ± 16.2 (11)	7.52
Thoracic ganglion	39.65 ± 3.63 (4)	138 ± 21.5 (10)	3.48
Abdominal ganglia	10.19 ± 0.67 (4)	181 ± 19.7 (7)	17.76
Terminal abdominal ganglion	27.05 ± 1.61 (4)	150 ± 17.0 (10)	5.55

Reproduced with permission from Coast, 1989.

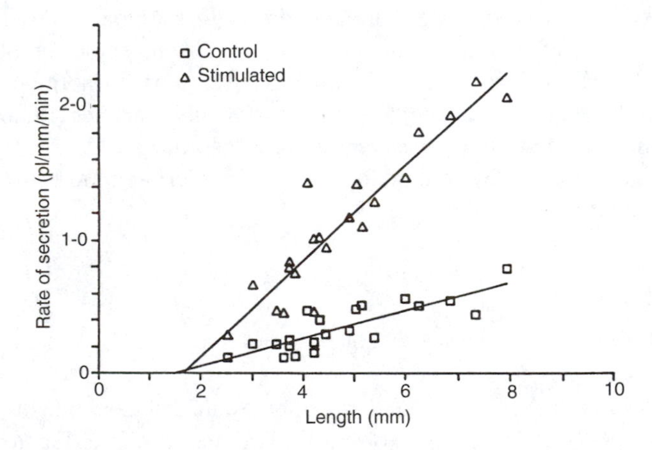

FIGURE 13.15 Fluid formation in an isolated tubule of the cricket *Acheta domesticus* is a function of the length of the tubule immersed in the bathing saline and stimulation by an extract of corpora cardiaca. (Reproduced with permission from Coast, 1988.)

13.7.2.2 Antidiuretic Hormones

In general, insects that feed on dry solid food probably need to conserve water rather than excrete it; thus, evolution of an **antidiuretic hormone** might be expected. Mills (1967) and Goldbard et al. (1970) found evidence for both an antidiuretic and a diuretic hormone in *P. americana*. The antidiuretic hormone promotes water conservation and is ordinarily the predominant hormone. A diuretic hormone was demonstrated in *P. americana* only by depriving male cockroaches of water for 3 days, after which the thirsty insects drank enough water at one time to stimulate the release of the diuretic hormone, thus temporarily increasing water excretion. Ligation experiments indicated that the diuretic hormone was released from the posterior part of the abdomen, and extracts of various tissues indicated that the terminal abdominal ganglion was the source. Fluid reabsorption across the cryptonephric complex of larval *M. sexta* has been demonstrated with an antidiuretic factor extracted from the brain/CC/CA complex of the caterpillars by Liao et al. (2000), who suggest that the factor involves cAMP increase and activation of a Cl^- pump in the cryptonephric system.

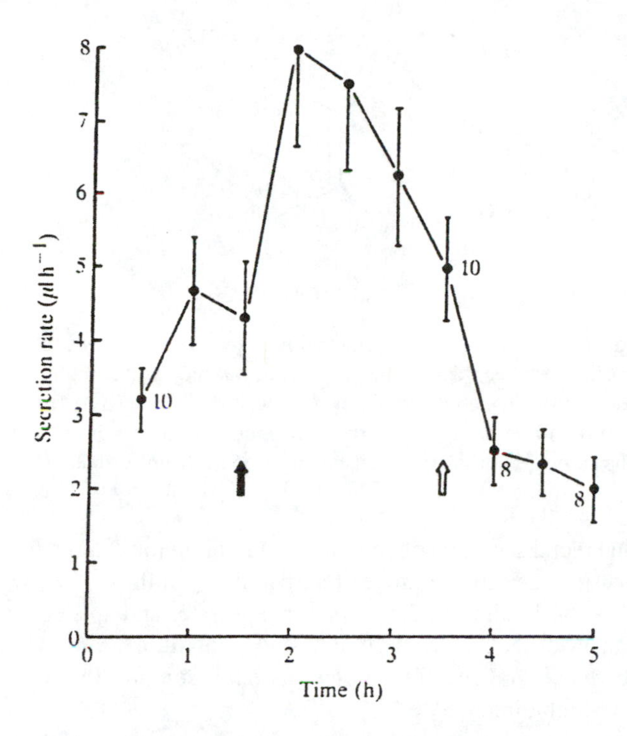

FIGURE 13.16 The stimulating effect of adding one homogenized pair of corpora cardiaca to the bathing saline surrounding an isolated tubule of *Acheta domesticus*, as a homogenate added at the filled arrow. The bath was changed with three rinses of fresh saline at the open arrow. The numbers represent the number of replicate preparations. (Reproduced with permission from Spring and Hazelton, 1987.)

13.7.3 Acid-Base Homeostasis

The excretory system is important in maintaining the **acid-base balance** of body fluids and tissues (Harrison, 2001). Acidosis or alkalosis may be experienced by an insect, depending on various foods, the presence of certain types of chemical compounds in plants eaten, the type of proteins metabolized (whether proteins yield a high proportion of acidic, basic, or neutral amino acids), and metabolic conditions such as exercise (e.g., flight) that produce acids in the tissues (Harrison and Kennedy, 1994). The western lubber grasshopper *Taeniopoda eques* exhibits flexibility in shifting between excretion of excess acid or excess base equivalents (Harrison and Kennedy, 1994), depending on need. This flexibility and regulatory ability in a highly polyphagous insect seems adaptive, and many other insects may show similar ability (Harrison and Kennedy, 1994).

Acid-base regulation has been most thoroughly studied in the desert locust *S. gregaria* (see Phillips et al., 1994; Harrison and Kennedy, 1994, and references therein). Locusts experimentally injected with HCl excrete most of the acid equivalents by secretion of protons by the hindgut epithelium. The Malpighian tubules participate only marginally. Secretion of H^+ and formation of ammonium ions (NH_4^+) in the ileum is a principal mechanism for excreting excess acid equivalents (Harrison, 1994; Harrison and Kennedy, 1994; Phillips et al., 1994). The ileum is a major site of **ammoniagenesis**, the formation of ammonia from precursors, in locusts in which hindgut cells specifically metabolize amino acids and glucose for energy (Peach and Phillips, 1991). The excess nitrogen from amino groups is incorporated into the formation of ammonia. These metabolites are present in the lumen of the ileum, having come with urine formed in the Malpighian tubules. Ammonia and urate are about equal in concentration in the fluid within the Malpighian tubules;

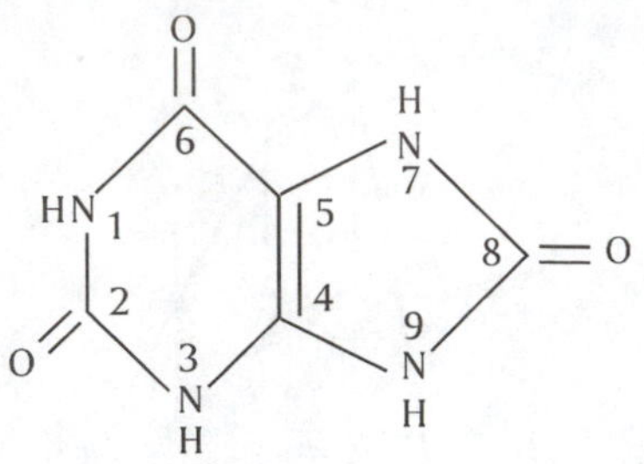

URIC ACID

FIGURE 13.17 The structure of uric acid, the principal nitrogenous excretory product for most insects. The source of nitrogen atoms has not been determined in insects, but they almost certainly come from the amino nitrogen of amino acids that are metabolized. Carbon at position-2 is derived from the carbon of formate, carbon-4 from the carboxyl carbon of glycine, carbon-5 from the α-carbon of glycine, carbon-6 from carbon dioxide derived from the carboxyl carbon of glycine, and carbon-8 from formate. (Modified from Barrett and Friend, 1970.)

but because of ileal and rectal secretion of ammonia/ammonium ions, about half of the total nitrogen excreted by desert locusts is ammonia nitrogen (Harrison and Phillips, 1992). Some of the ammonia reacts with uric acid in the hindgut to form ammonium urate, and spares sodium and potassium, which also react with uric acid to form sodium and potassium urate. Thus, excretion of total ammonia nitrogen serves several functions in locusts (Harrison and Phillips, 1992; Phillips et al., 1994; Harrison, 1995), including:

1. Ammonium urate allows the insects to conserve Na^+, an ion that is not high in the food of locusts.
2. Conversion of ammonia (NH_3) to ammonium (NH_4^+) in the ileal cells is equivalent to removal of protons (H^+), and excretion of ammonia is more than sufficient to explain the recovery of hemolymph pH after a load of HCl is injected into the hemocoel.
3. Excretion of ammonia by locusts conserves water (because of precipitation of not-very-soluble ammonium urate salt).
4. Increases nitrogen excretion by 25% more than excretion of only sodium or potassium urate.

13.7.4 Nitrogen Homeostasis

Harrison (1995) has addressed the fact that nitrogen, although known to be a growth-limiting nutrient for some insects, is nevertheless excreted in several forms by insects. Excess nitrogen that must be excreted may come from ingestion of proteins, leading to an imbalance of amino acids; they save essential amino acids and may metabolize the nonessential ones as energy sources. Insects also ingest nitrogen in the nucleic acids of their food. Excess protein nitrogen is excreted as **uric acid** (a purine) (Figure 13.17), as purines related to uric acid, as **ammonia** or **ammonium salts**, and in several other (usually minor) forms. Nitrogen from nucleic acids is also excreted as uric acid (or as related purine metabolites or metabolites of uric acid). Typically, only trace or small amounts of urea are excreted. The complete sequence of enzymes in the ureotelic pathway for synthesis of urea has not been found in insects, but arginase, a primary enzyme in the pathway, is active in fat body of the abdomen and thorax throughout the life cycle of *Aedes aegypti* mosquitoes. Although uric acid is the primary excretory product, small amounts of urea are excreted (Dungern and Briegel, 2001). Bursell (1967) and Cochran (1975) provided thorough reviews of early literature on nitrogen excretory products in insects.

13.7.4.1 Ammonia Excretion

Ammonia is a product of protein and amino acid metabolism. Free ammonia cannot be stored in tissues or cells because it is a very strong base influencing pH, and in its free form it is very toxic to all cells. It must be rapidly excreted or transformed into a less toxic compound. If water is readily available for dilution, ammonia can be excreted as the free base or as an ammonium salt. Animals that excrete ammonia as their primary nitrogenous waste product are described as **ammoneotelic**.

Ammonia is a major excretory product for larval stages of some Diptera that live in very wet environments. Larvae of *Calliphora erythrocephala*, the common blowfly; *Wohlfahrtia vigil*, a sarcophagid fly (Brown, 1936), *Phormia regina*, a blowfly; and *Lucilia cuprina*, the sheep ked (Hitchcock and Haub, 1941) excrete ammonia into wet surroundings that dilute it to nontoxic levels. *Lucilia sericata*, another blowfly, excretes up to 15-fold more ammonia than uric acid (Brown, 1938). Although uric acid is synthesized, most of it is stored in the tissues (storage excretion). Allantoin, a breakdown product of uric acid, is also excreted by larvae of *L. sericata*. In these dipterans, ammonia excretion ceases at pupariation, and the adults excrete uric acid and, in a few cases, some allantoin.

Staddon (1955, 1959) found that most of the nitrogen excreted by the aquatic larva of the neuropteran *Sialis lutaria* and the odonate *Aeshna cyanea* is ammonia. Although some aquatic insects excrete substantial amounts of ammonia, many synthesize and excrete uric acid (or a further metabolic derivative of uric acid).

Some terrestrial insects excrete the majority of their excretory nitrogen as ammonia or ammonium salts. The American cockroach *P. americana* excretes ammonia as a major excretory product (Mullins and Cochran, 1972), but the precise mechanism(s) of its excretion has not been elucidated. Ammonia and ammonium nitrogen can account for from 10 to 46% of the total nitrogen excreted by the desert locust, *S. gregaria* according to Harrison (1995), who cautions that previous studies on the distribution of nitrogen in excreta of terrestrial insects may have missed more labile ammonia and ammonium nitrogen by the methods and techniques employed. Ammonia nitrogen is rapidly lost from fecal pellets, especially if they are dried prior to analysis. In some cases, ammonium urate may be lost because it is poorly soluble unless excretory material is extracted with a large volume of aqueous solvent (Harrison, 1995). Fecal pellets should be collected within minutes after being excreted, deposited in acid solution, and kept frozen until analysis (Harrison, 1995).

Most animals, including insects, synthesize ammonia into less toxic compounds such as urea (mammals) or uric acid (birds, reptiles, insects, Dalmatian dog). Enzymes involved in amino acid metabolism and ammonia production include amino transaminases (or transferases), glutamic acid and alanine dehydrogenases, L- and D-amino acid oxidases, adenosine deaminase, and monoamine oxidase. All of these enzymes have been detected in a number of insects (Cochran, 1975).

The **amino transaminases** are widely distributed in insect tissues and enable the amino group of one amino acid to be transferred to a ketoacid, thereby forming a new amino acid. For example,

$$\text{L-Aspartic acid} + \alpha\text{-Ketoglutaric acid} \rightleftharpoons \text{L-Glutamic acid} + \text{Oxaloacetate}$$

Although ammonia is not directly released in transaminase reactions, the reactions provide a way to interconvert nonessential amino acids and to make amino acids available to enzymes that deaminate with release of ammonia, such as glutamic dehydrogenase and alanine dehydrogenase in the following reactions:

$$\text{L-Glutamic acid} + NAD^+ + H_2O \rightarrow \alpha\text{-Ketoglutarate} + NADH + NH_3$$

$$\text{Alanine} + NAD^+ + H_2O \rightarrow \text{Pyruvate} + NADH + NH_3$$

The ammonia formed in these reactions is rapidly excreted or converted into less toxic compounds. The α-ketoglutaric acid and pyruvate formed can be metabolized through the Krebs cycle as an energy source (Cochran, 1975).

Ammonia production may also come from turnover and replacement of an insect's own nucleic acids, as well as from the metabolism of nucleic acids ingested with food. The enzymes adenosine deaminase, guanine deaminase, and adenine deaminase, all of which give rise to ammonia from metabolism of nucleic acids or their derived products, have been reported from various insects.

13.7.4.2 Uric Acid Synthesis and Excretion

Uric acid is synthesized in insects from protein nitrogen as well as from nucleic acid nitrogen. The major portion is synthesized from protein nitrogen simply because insects ingest relatively much more protein nitrogen (or amino acid nitrogen) in their diet than nucleic acid nitrogen (Cochran, 1975). In birds (and presumably in insects), synthesis of 1 mole of uric acid from NH_3 requires the expenditure of 8 moles ATP (Cochran, 1975). The 8 ATP/uric acid formed ignores other costs that may be incurred, such as transport across cell membranes and maintenance of enzymatic machinery. Thus, excretion of uric acid as the main product of protein metabolism is energetically costly. Its advantages are that it rids the body of four nitrogen atoms, which make up 33.3% of the molecular weight of uric acid (MW of uric acid = 168.11; 4 N = 56), and it is very insoluble in water. It often reaches concentrations that promote precipitation from solution in the Malpighian tubules and hindgut, and as a precipitate it does not contribute to osmotic values across cells lining the tubules or hindgut.

The fat body is the primary site for uric acid synthesis. Barrett and Friend (1970) found that glycine contributes a carboxyl carbon to position-4 and an α-carbon to position-5 during synthesis of uric acid in *R. prolixus*. Formate contributed carbons to positions 2 and 8. The labeled carboxyl carbon of glycine was converted rapidly in *Rhodnius* to $^{14}CO_2$, and much of the $^{14}CO_2$ formed gave rise to labeled carbon-6 in uric acid. The α-carbon of glycine can also contribute to the synthesis of formate, so that ultimately some or even much of carbon-8 of uric acid might also come from glycine. Although the origin of the nitrogen atoms in uric acid from *Rhodnius* was not determined, it seems reasonably certain that they are derived from NH_3 resulting from metabolism of proteins, amino acids, and nucleic acids (Barrett and Friend, 1970).

The final steps in the synthesis of uric acid involve the conversion of hypoxanthine and xanthine to uric acid (Figure 13.18). The enzyme that catalyzes the two-step conversion is **xanthine dehydrogenase** (Irzykiewicz, 1955). Its counterpart in vertebrates (birds and reptiles) is xanthine oxidase, a true oxidase because molecular oxygen can accept the protons removed from hypoxanthine and xanthine. The insect enzyme does not work without NAD^+ or FAD^+ (or a synthetic acceptor such as methylene blue) as the proton acceptor, so it is a true dehydrogenase. Transfer of electrons from reduced cofactors through the electron transport system could yield 6 (NADH) or 4 ($FADH_2$) ATP/uric acid molecules. Evolution of xanthine dehydrogenase in insects, instead of xanthine oxidase, may have occurred as a potential way to recover some of the costs of urate synthesis (Cochran, 1975).

Uric acid and related uricotelic compounds are admirably adaptive excretory products for animals that live with water stress and that need to conserve water. These compounds have limited solubility; and when they crystallize from solution, they reduce the osmotic work required to reabsorb water from the rectum. Uric acid is the least soluble of the compounds (6 mg/100 ml of water) followed by allantoin (60 mg/100 ml), hypoxanthine (70 mg/100 ml), and xanthine (260 mg/100 ml) (Bursell, 1967). Uric acid may crystallize as free uric acid and/or as a salt. Data presented by Harrison (1995) shows that ammonium urate is much less soluble than either sodium or potassium urate, and it has the advantage that the NH_4^+ rids the body of additional nitrogen and acid equivalents.

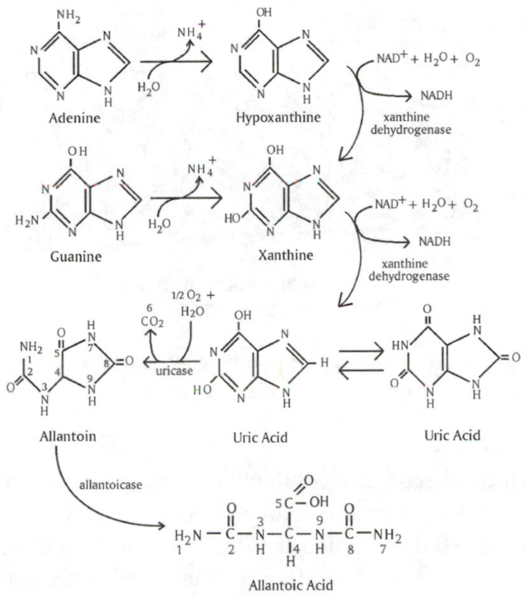

FIGURE 13.18 The final stages in the synthesis of uric acid, and various metabolic breakdown products of uric acid that some insects excrete.

Storage excretion of uric acid, or deposition in various parts of the body, is common in cockroaches and some other insects. Crystals of uric acid may occur in the hemolymph, and in other tissues, especially in fat body. Cochran (1973) found high levels of crystalline uric acid in the fat bodies of 14 species of cockroaches. Male cockroaches deposit uric acid in accessory glands associated with their reproductive tract, and deposit urates from the glands on the outside of the sperm packet, or spermatophore, that they produce and insert into the female at mating. Mullins and Keil (1980) found that labeled uric acid of male *Blattella germanica* cockroaches could be recovered in female German cockroaches, and in their oothecae after mating. They suggested that the urates represented a nitrogen source for the female and a **paternal investment** by the male in her progeny.

Paternal investment behavior is also shown by the tropical cockroach *Xestoblatta hamata*, in which females feed on urates deposited in the genital chamber by males during mating. Females transfer the urates to their terminal oocytes and the ootheca (Schal and Bell, 1982). Under both field and laboratory conditions, male *X. hamata* choose high protein foods, which are known to result in greater production of urates in cockroaches (Haydak, 1953), and they feed opportunistically on substances containing uric acid, such as bird droppings.

The Malpighian tubules of the American cockroach, *P. americana*, do not clear uric acid from the hemolymph, and the cockroaches do not excrete uric acid with the fecal wastes (except in small amounts under crowded conditions, which may be because they have cannibalized another cockroach and ingested uric acid that simply passes through the gut) (Mullins and Cochran, 1972). Most of the nitrogen excreted by American cockroaches is ammonia nitrogen (Mullins and Cochran, 1972).

Razet (1966) found that many insects excrete small to large percentages of their excretory nitrogen as **allantoin**, a breakdown product of uric acid catalyzed by the enzyme **uricase** (Figure 13.18). Uricase is widespread in insects and in their tissues, but no particular advantage is known for the conversion of uric acid to allantoin, nor for it as an excretory product. A few insects excrete some **allantoic acid**, an oxidation product of allantoin, but the enzyme **allantoicase** is not widespread in insects, and, if present at all, allantoic acid is a very minor excretory product.

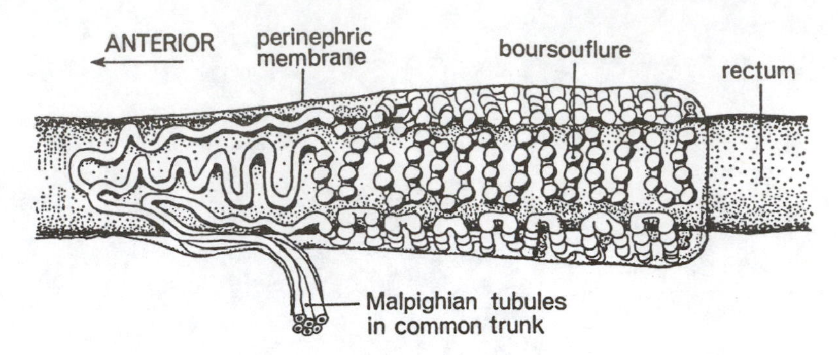

FIGURE 13.19 The cryptosolenic or cryptonephritic tubule system characteristic of most Lepidoptera and Coleoptera. (Reproduced with permission from Grimstone et al., 1968.)

Animals that excrete most or all of their excretory nitrogen as uric acid are described as **uricotelic**. Occasionally, it has been questioned whether insects really fit the definition of uricotelism (discussed in Cochran, 1975) because some insects do not excrete most of their nitrogen as uric acid. Bursell (1970) proposed extending the definition of uricotelism to include excretion of allantoin and allantoic acid because both are derived from further metabolism of uric acid. Cochran (1975) concurred with the broader definition, and extended it further to include excretion of the uric acid precursors hypoxanthine, xanthine, and guanine excreted by some insects (Morita, 1958; Mitchell et al., 1959; Nation and Patton, 1961; Nation, 1963; Mitlin and Vickers, 1964; Nation and Thomas, 1965). Thus, Bursell (1970) and Cochran (1975) concluded that, as a group, insects should still be considered uricotelic in excretion although some excrete a variety of nitrogenous compounds and a few excrete relatively little or no uric acid.

13.8 CRYPTONEPHRIDIAL SYSTEMS

Many families of Coleoptera, Lepidoptera, and some saw-fly larvae (Hymenoptera) have an arrangement of Malpighian tubules in which the distal ends of the tubules are enveloped within a membrane and held close to the surface of the rectum (Figure 13.19). This arrangement is known as a **cryptosolenic** or **cryptonephridic** tubule system. It appears to be an arrangement that enables very efficient conservation of water. Insects living in the driest habitats and eating very dry food have the most extensive development and network of cryptonephridial tubules.

Cryptonephridic tubules do not penetrate the lumen of the rectum, but lie on the outer surface of the rectum, encased within a **perinephric chamber** bounded by the **perinephric membrane** (Figure 13.20). The perinephric membrane is composed of thin, elongated cells that seal the tubules from the hemocoel and hemolymph at the initial point of contact with the gut (Saini, 1964). The tubules do not terminate immediately after contacting the rectum, but typically are thrown into many loops and convolutions, with segments running radially around the rectum as well as looping anteriorly and posteriorally along the length of the rectum (Saini, 1964). The perinephric membrane follows the various convolutions and turns, always enclosing the tubules like a blanket. Several layers of tubules may lie on the rectum in those insects that live in the driest environments (Saini, 1964). A small **perirectal space** occurs between the epithelial cell layer of the rectum and the innermost layer of tubules within the perinephric chamber.

The food of insects, particularly that of Coleoptera, is variable in water content. Some Coleoptera are phytophagous, eating food with relatively high water content; while others feed on dry stored grain products or similar dry food materials. In phytophagous Coleoptera, the posterior rectal region containing the fecal pellet just prior to its being expelled is smaller and has fewer convolutions of the tubules than in Coleoptera that feed on dry food (Saini, 1964). Coleoptera living

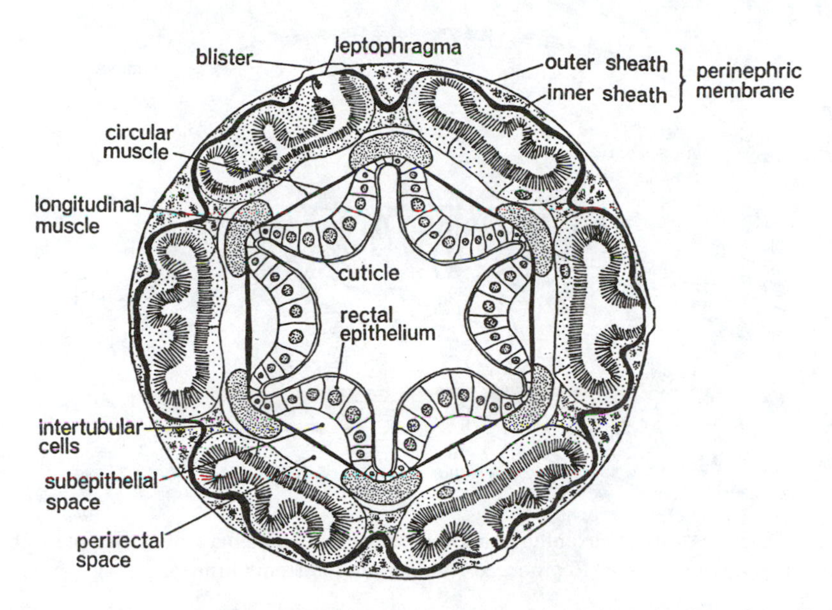

FIGURE 13.20 Cross-sectional view of the rectum with cryptonephridial tubules from the yellow mealworm *Tenebrio molitor.* (Reproduced with permission from Grimstone et al., 1968.)

in very dry environments, such as *Tenebrio molitor* and other grain infesting beetles and weevils, have a large cryptonephridial system enabling them to extract more water from the fecal pellet.

At frequent points in most of the Coleoptera, a cryptonephridic tubule in the outermost layer makes contact with the outer perinephric membrane through a single, highly modified cell of the tubule wall called the **leptophragma cell**. These points of contact when a tubule is separated from the hemolymph only by the thin cell membrane of the leptophragma cell and the very thin (at this particular point) perinephric membrane are called **leptophragmata**. Only two families of beetles, Ptinidae and Anobiidae, do not have leptophragmata, but none of the Lepidoptera studied by Saini (1964) have them.

In *T. molitor* and some other beetles, the thin perinephric membrane is expanded into a ***boursouflure***, a French word meaning a blister (Figure 13.21), above each leptophragma cell. The exact function of the boursouflure is uncertain. Ramsay (1964) and Grimstone et al. (1968) proposed that it may be a site of active secretion of ions from the hemolymph into the perinephric tubules, thus creating high osmotic values in the cryptonephridic tubules to aid passive movement of water down the osmotic gradient from rectal lumen to tubule lumen. There is some evidence (Maddrell, 1971) that a high-molecular-weight compound (probably proteinaceous) is secreted into the perirectal space of *T. molitor*, and that it first absorbs water from the rectum, with the water passed on to the tubules.

Cryptonephridic tubules of Lepidoptera have neither leptophragmata nor leptophragma cells. Thus, regardless of the function of leptophragma cells in Coleoptera, they are not essential to the system in Lepidoptera. As in the Coleoptera, the extent of layering of cryptonephridial loops on the rectum is correlated with the habitat and food eaten by larvae (Saini, 1964). In larvae feeding on green plant matter, the innermost layer of tubules extends only about one third the length of the anterior portion of the rectum, and there is only a single layer of tubules not packed very close together in the posterior half of the anterior rectum. In those larvae living in very dry conditions and eating dry food (e.g., Galleriidae, Phycitidae, and Tineidae), the convoluted tubules are packed closely together in both the inner and outer layers and extend the entire length of the anterior rectum. There is no cryptonephridic system of tubules in the aquatic Lepidoptera, *Paraponyx* (= *Nymphula*) *stratiotata* and *Cataclysta lemnata* (Pyraustidae). In those insects, the distal ends of

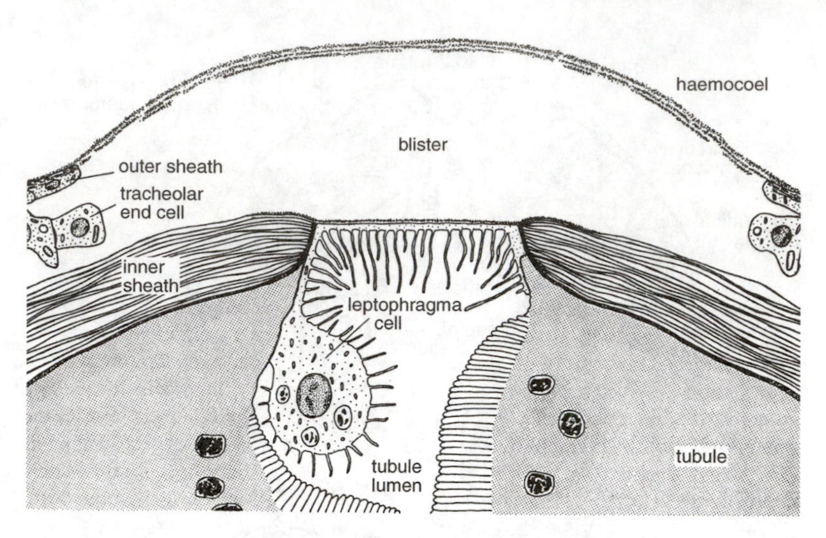

FIGURE 13.21 Diagram of the boursouflure, and underlying leptophragma cell and cryptonephritic tubule cells in a larva of *Tenebrio molitor.* (Reproduced with permission from Grimstone et al., 1968.)

the Malpighian tubules merely lie on the rectum in association with some fat body tissue and tracheae. Although the crytptonephric complex has been well studied anatomically, physiological studies are sparse.

REFERENCES

Audsley, N., C. McIntosh, and J.E. Phillips. 1992a. Isolation of a neuropeptide from locust corpus cardiacum which influences ileal transport. *J. Exp. Biol.,* 173:261-274.

Audsley, N., C. McIntosh, and J.E. Phillips. 1992b. Actions of ion-transport peptide from locust corpus cardiacum on several hindgut transport processes. *J. Exp. Biol.,* 173:275-288.

Baldwin, I.T. 1991. Damage-induced alkaloids in wild tobacco, pp. 47-69, in D.W. Tallamy and M.J. Raupp (Eds.), *Phytochemical Induction by Herbivores*, Wiley Interscience, New York.

Barrett, F.M., and Friend, W.G. 1970. Uric acid synthesis in *Rhodnius prolixus*. *J. Insect Physiol.,* 16:121-129.

Barrett, M., and I. Orchard. 1990. Serotonin-induced elevation of cyclic AMP levels in the epidermis of the blood-sucking bug *Rhodnius prolixus*. *J. Insect Physiol.,* 36:625-634.

Berridge, M.J. 1966. The physiology of excretion in the cotton stainer, *Dysdercus fasciatus* Signoret. IV. Hormonal control of excretion. *J. Exp. Biol.,* 44:553-566.

Berridge, M.J., and J.L. Oschman. 1969. A structural basis for fluid secretion by Malpighian tubules. *Tissue Cell,* 1:247-272.

Bertram, G., Schleithoff, L. Zimmermann, P., and Wessing, A. 1991. Bafilomycin A_1 is a potent inhibitor of urine formation by Malpighian tubules of *Drosophila hydei*: Is a vacuolar-type ATPase involved in ion and fluid secretion? *J. Insect Physiol.,* 37:201-209.

Beyenbach, K.W. 1995. Mechanism and regulation of electrolyte transport in Malpighian tubules. *J. Insect Physiol.,* 41:197-207.

Beyenbach, K.W., D.J. Aneshansley, T.L. Pannabecker, R. Masia, D. Gray, and M.-J. Yu. 2000a. Oscillations of voltage and resistance in Malpighian tubules of *Aedes aegypti*. *J. Insect Physiol.,* 46:321-333.

Beyenbach, K.W., T.L. Pannabecker, and W. Nagel. 2000b. Central role of the apical membrane H^+-ATPase in electrogenesis and epithelial transport in Malpighian tubules. *J. Exp. Biol.,* 203:1459-1468.

Bradley, T.J. 1985. The excretory system: Structure and physiology, pp. 421-465, in G.A. Kerkut and L.I. Gilbert (Eds.), *Comparative Insect Physiology, Biochemistry, and Pharmacology,* Vol. 4, Pergamon Press, New York.

Brower, L.P. 1969. Ecological chemistry. *Sci. Am.,* 220:22-29.

Brown, A.W.A. 1936. The excretion of ammonia and uric acid during the larval life of certain muscoid flies. *J. Exp. Biol.,* 13:131-139.

Brown, A.W.A. 1938. The nitrogen metabolism of an insect (*Lucilia sericata* Meig). I. Uric acid, allantoin, and uricase. *Biochem. J.,* 32:895-902.

Bursell, E. 1967. The excretion of nitrogen in insects. *Adv. Insect Physiol.,* 4:33-67.

Bursell, E. 1970. *An Introduction to Insect Physiology*, Academic Press, London.

Chamberlin, M.E., and J.E. Philips. 1982. Regulation of hemolymph amino acid levels and active secretion of proline by Malpighian tubules of locusts. *Can. J. Zool.,* 60:2745-2752.

Chamberlin, M.E., and J.E. Philips. 1983. Oxidative metabolism in the locust rectum. *J. Comp. Physiol. B,* 151:191-198.

Clark, T.M., and T.J. Bradley. 1993. Short term changes in hemolymph properties of larval *Aedes aegypti* in response to physiological challenges affect Malpighian tubule secretion rates *in vitro*. *Am. Zool.,* 33:43A.

Clark, T.M., and T.J. Bradley. 1996. Stimulation of Malpighian tubules from larval *Aedes aegypti* by secretagogues. *J. Insect Physiol.,* 42:593-602.

Coast, G.M. 1988. Fluid secretion by single isolated Malpighian tubules of the house cricket, *Acheta domesticus*, and their response to diuretic hormone. *Physiol. Entomol.,* 13:381-391.

Coast, G.M. 1989. Stimulation of fluid secretion by single isolated Malpighian tubules of the house cricket, *Acheta domesticus*. *Physiol. Entomol.,* 14:21-30.

Coast, G.M., and C.H. Wheeler. 1990. The distribution and relative potency of diuretic peptides in the house cricket, *Acheta domesticus*. *Physiol. Entomol.,* 15:13-21.

Cochran, D.G. 1973. Comparative analysis of excreta from twenty cockroach species. *Comp. Biochem. Physiol.,* 46A:409-419.

Cochran, D.G. 1975. Excretion in insects, pp. 177-281, in D.J. Candy and B.A. Kilby (Eds.), *Insect Biochemistry and Function*, Chapman & Hall, London.

Cochran, D.G. 1985a. Nitrogen excretion in cockroaches. *Annu. Rev. Entomol.,* 30:29-49.

Cochran, D.G. 1985b. Nitrogen excretion, pp. 467-506, in G.A. Kerket and L.I. Gilbert (Eds.), *Comprehensive Insect Physiology, Biochemistry and Pharmacology*, Vol 4., Pergamon, New York

Dijkstra, S., A. Leyssens, E.van Kerkhove, W. Zeiske, and P. Steels. 1995. A cellular pathway for Cl^- during fluid secretion in ant Malpighian tubules: Evidence from ion-sensitive microelectrode studies? *J. Insect Physiol.,* 41:695-703.

Dungern, P. von, and H. Briegel. 2001. Protein catabolism in mosquitoes: Ureotely and uricotely in larval and imaginal *Aedes aegypti*. *J. Insect Physiol.,* 47:131-141.

Forgac, M. 1989. Structure and function of vacuolar class of ATP-driven proton pumps. *Physiol. Rev.,* 69:765-796.

Goldbard, G.A., J.R. Sauer, and R.R. Mills. 1970. Hormonal control of excretion in the American cockroach. II. Preliminary purification of a diuretic and an antidiuretic hormone. *Comp. Gen. Pharmacol.,* 1:82-86.

Grimstone, A.V., A.M. Mullinger, and J.A. Ramsay. 1968. Further studies on the rectal complex of the mealworm, *Tenebrio molitor* L. (Coleoptera, Tenebrionidae). *Phil. Trans. R. Soc. Ser. B,* 253:343-382.

Gupta, B.J., and M.J. Berridge. 1966. Fine structural organization of the rectum in the blowfly, *Calliphora erythrocephala* (Meig.) with special reference to connective tissue, tracheae and neurosecretory innervation in the rectal papillae. *J. Morphol.,* 120:23-82.

Harrison, J.F. 1994. Respiratory and ionic aspects of acid-base regulation in insects: An introduction. *Physiol. Zool.,* 67:1-6.

Harrison, J.F. 1995. Nitrogen metabolism and excretion in locusts, pp. 119-131, in P.J. Walsh and R. Wright (Eds.), *Nitrogen Metabolism and Excretion,* CRC Press, Boca Raton, FL.

Harrison, J.F. 2001. Insect acid-base physiology. *Annu. Rev. Entomol.,* 46:221-250.

Harrison, J.F., and M.J. Kennedy. 1994. *In vivo* studies of the acid-base physiology of grasshoppers: The effect of feeding state on acid-base and nitrogen excretion. *Physiol. Zool.,* 67:120-141.

Harrison, J.F., and J.E. Phillips. 1992. Recovery from acute haemolymph acidosis in unfed locusts II. Role of ammonium and titratable acid excretion. *J. Exp. Biol.,* 165:97-110.

Haydak, M.H. 1953. Influence of the protein level of the diet on the longevity of cockroaches. *Ann. Entomol. Soc. Am.,* 46:547-560.

Hegarty, J.L., B. Zhang, T.L. Pannabecker, D.H. Petzel, M.D. Baustian, and K.W. Beyenbach. 1991. Dibutyryl cAMP activates bumetanide-sensitive electrolyte transport in Malpighian tubules. *Am. J. Physiol.*, 261: C521-C529.

Hitchcock, F.A., and J.G. Haub. 1941. The interconversion of foodstuffs in the blowfly (*Phormia regina*) during metamorphosis. I. Respiratory metabolism and nitrogen excretion. *Ann. Entomol. Soc. Am.*, 34:17-25.

Hopkin, R., J.H. Anstee, and K. Bowler. 2001. An investigation into the effects of inhibitors of fluid production by *Locusta* Malpighian tubule Type I cells on their secretion and elemental composition. *J. Insect Physiol.*, 47:359-367.

Hopkins, C.R. 1967. The fine-structural changes observed in the rectal papillae of the mosquito *Aedes aegypti*, L. and their relation to the epithelial transport of water and inorganic ions. *J. R. Micros. Soc.*, 86:235-252.

Ianowski, J.P., and M.J. O'Donnell. 2001. Transepithelial potential in Malpighian tubules of *Rhodnius prolixus*: Lumen-negative voltages and the triphasic response to serotonin. *J. Insect Physiol.*, 47:411-421.

Irzykiewicz, H. 1955. Xanthine oxidase of the clothes moth, *Tineola bisselliella*, and some other insects. *Aust. J. Biol. Sci.*, 8:369-377.

Isaacson, L.C., S.W. Nicolson, and D.W. Fisher. 1989. Electrophysiological and cable parameters of perfused beetle Malpighian tubules. *Am. J. Physiol.*, 257:R1190-R1198.

Kane, P.M., and K.J. Parra. 2000. Assembly and regulation of the yeast vacuolar H^+-ATPase. *J. Exp. Biol.*, 203:81-87.

Kataoka, H., R.G. Troetschler, J.P. Li, S.J. Kramer, R.L. Carney, and D.A. Schooley. 1989. Isolation and identification of a diuretic hormone from the tobacco hornworm, *Manduca sexta*. *Proc. Natl. Acad. Sci. U.S.A.*, 86:2976-2980.

Kim, I.S., and J.H. Spring. 1992. Excretion in the house cricket: Relative contribution of distal and mid-tubule to diuresis. *J. Insect Physiol.*, 38:373-381.

Lehmberg, E., R.B. Ota, K. Furuya, D.S. King, S.W. Applebaum, H.-J. Ferenz, and D.A. Schooley. 1991. Identification of a diuretic hormone of *Locusta migratoria*. *Biochem. Biophys. Res. Commun.*, 179:1036-1041.

Lehmberg, E., D.A. Schooley, H.-J. Ferenz, and S.W. Applebaum. 1993. Characteristics of *Locusta migratoria* diuretic hormone. *Arch. Insect Biochem. Physiol.*, 22:133-140.

Leyssens, A., P. Steels, E. Lohrmann, R. Weltens, and E. Van Kerkhove. 1992. Intrinsic regulation of K+ transport in Malpighian tubules (*Formica*): Electrophysiological evidence. *J. Insect Physiol.*, 38:431-446.

Leyssens, A., S.-L. Zhang, E. Van Kerkhove, and P. Steels. 1993. Both dinitrophenol and Ba^{2+} reduce KCl and fluid secretion in Malpighian tubules of *Formica polyctena*: The role of the apical H^+ and K^+ concentration gradient. *J. Insect Physiol.*, 39:1061-1073.

Liao, A., N. Audsley, and D.A. Schooley. 2000. Antidiuretic effects of a factor in brain/corpora cardiaca/corpora allata extract on fluid reabsorption across the cryptonephric complex of *Manduca sexta*. *J. Exp. Biol.*, 203:605-615.

Linton, S.M., and M.J. O'Donnell. 2000. Novel aspects of the transport of organic anions by the Malpighian tubules of *Drosophila melanogaster*. *J. Exp. Biol.*, 203:3575-3584.

Maddrell, S.H.P. 1963. Excretion in the blood-sucking bug, *Rhodnius prolixus* Stål. I. The control of diuresis. *J. Exp. Biol.*, 40:247-256.

Maddrell, S.H.P. 1964a. Excretion in the blood-sucking bug, *Rhodnius prolixus* Stål. II. The normal course of diuresis and the effect of temperature. *J. Exp. Biol.*, 41:163-176.

Maddrell, S.H.P. 1964b. Excretion in the blood-sucking bug, *Rhodnius prolixus* Stål. III. The control of the release of the diuretic hormone. *J. Exp. Biol.*, 41:459-472.

Maddrell, S.H.P. 1966. The site of release of the diuretic hormone in *Rhodnius* — A new neurohaemal system in insects. *J. Exp. Biol.*, 45:499-508.

Maddrell, S.H.P. 1971. The mechanisms of insect excretory systems. *Adv. Insect Physiol.*, 8:199-331.

Maddrell, S.H.P. 1977. Insect Malpighian tubules, pp. 541-569, in B.L. Gupta, R.B. Moreton, J.L. Oschman, and B.J. Wall (Eds.), *Transport of Ions and Water in Animals*, Academic Press, London.

Maddrell, S.H.P. 1980. Characteristics of epithelial transport in insect Malpighian tubules, pp. 427-463, in F. Bonner and A. Kleinzeller (Eds.), *Current Topics in Membranes and Transport*, Vol. 14, Academic Press, New York.

Maddrell, S.H.P., and M.J. O'Donnell. 1992. Insect Malpighian tubules: V-ATPase action in ion and fluid transport. *J. Exp. Biol.,* 172:417-430.

Maddrell, S.H.P., W.S. Herman, R.L. Mooney, and J.A. Overton. 1991. 5-Hydroxytryptamine: A second diuretic hormone in *Rhodnius prolixus*. *J. Exp. Biol.,* 156:557-566.

Meredith, J., M. Ring, A. Macins, J. Marschall, N.N. Cheng, D. Theilmann, H.W. Brock, and J.E. Phillips. 1996. Locust ion transport peptide (ITP): Primary structure cDNA and expression in a baculovirus system. *J. Exp. Biol.,* 199:1053-1061.

Mills, R.R. 1967. Hormonal control of excretion in the American cockroach. I. Release of a diuretic hormone from the terminal abdominal ganglion. *J. Exp. Biol.,* 46:35-41.

Mitchell, H.K., E. Glassman, and E. Hadorn. 1959. Hypoxanthine in *rosy2* and *maroon-like* mutants of *Drosophila melanogaster*. *Science,* 129:268-269.

Mitlin, N., and D.H. Vickers. 1964. Guanine in the excreta of the boll weevil. *Nature,* 203:1403-1404.

Mordue, W., and P.J. Morgan. 1985. Chemistry of peptide hormones, pp. 153-183, in G.A. Kerkut and L.I. Gilbert (Eds.). *Comprehensive Insect Physiology, Biochemistry and Pharmacology,* Vol. 7, Pergamon Press, New York.

Morita, T. 1958. Purine catabolism in *Drosophila melanogaster*. *Science,* 128:1135.

Mullins, D.E., and D.G. Cochran. 1972. Nitrogen excretion in cockroaches: uric acid is not a major product. *Science,* 177:699-701.

Mullins, D.E., and C.B. Keil. 1980. Paternal investment of urates in cockroaches. *Nature,* 283:567-569.

Nation, J.L. 1963. Identification of xanthine in excreta of the greater wax moth, *Galleria mellonella* (L). *J. Insect Physiol.,* 9:195-200.

Nation, J.L., and R.L. Patton. 1961. A study of nitrogen excretion in insects. *J. Insect Physiol.,* 6:299-308.

Nation, J.L., and K.K. Thomas. 1965. Quantitative studies on purine excretion in the greater wax moth, *Galleria mellonella*. *Ann. Entomol. Soc. Am.,* 58:983-885.

Nicolson, S.W. 1991. Diuresis or clearance: Is there a physiological role for the "diuretic hormone" of the desert beetle *Onymacris*. *J. Insect Physiol.,* 37:447-452.

Nicolson, S.W. 1993. The ionic basis of fluid secretion in insect Malpighian tubules: Advances in the last ten years. *J. Insect Physiol.,* 39:451-458.

Nicolson, S.W., and L. Isaacson. 1990. Patch clamp of the basal membrane of beetle Malpighian tubules: Direct demonstration of potassium channels. *J. Insect Physiol.,* 36:877-884.

Nicolson, S.W., and S.A. Hanrahan. 1986. Diuresis in a desert beetle? Hormonal control of the Malpighian tubules of *Onymacris plana* (Coleoptera: Tenebrionidae). *J. Comp. Physiol. B,* 156:407-413.

O'Donnell, M.J., and J. Machin. 1991. Ion activities and electrochemical gradients in the mealworm rectal complex. *J. Exp. Biol.,* 155:375-402.

O'Donnell, M.J., and J.H. Spring. 2000. Modes of control of insect Malpighian tubules: synergism, antagonism, cooperation and autonomous regulation. *J. Insect Physiol.,* 46:107-117.

O'Donnell, M.J., S.H.P. Maddrell, H. le B. Skaer, and J.B. Harrison. 1985. Elaborations of the basal surface of the cells of the Malpighian tubules of an insect. *Tissue Cell,* 17:865-881.

O'Donnell, M.J., M.R. Rheault, S.A. Davies, P. Rosay, B.J. Harvey, S.H.P. Maddrell, K. Kaiser, and J.A.T. Dow. 1998. Hormonally-controlled chloride movement across *Drosophila* tubules is via ion channels in stellate cells. *Am. J. Physiol.,* 43:R1039-R1049.

Pannabecker, T. 1995. Physiology of the Malpighian tubule. *Annu. Rev. Entomol.,* 40:493-510.

Pannabecker, T.L, T.K. Hayes, and K.W. Beyenbach. 1993. Regulation of epithelial shunt conductance by the peptide leucokinin. *J. Membr. Biol.,* 132:63-76.

Peach, J.L., and J.E. Phillips. 1991. Metabolic support of chloride-dependent short-circuit current across the locust (*Schistocerca gregaria*) ileum. *J. Insect Physiol.,* 37:255-260.

Phillips, J.E. 1964. Rectal absorption in the desert locust, *Schistocerca gregaria* Forskål. II. The nature of the excretory process. *J. Exp. Biol.,* 41:68-80.

Phillips, J.E. 1981. Comparative physiology of insect renal function. *Am. J. Physiol.,* 241: R241-R257.

Phillips, J.E., and A.A Dockrill. 1968. Molecular sieving of hydrophilic molecules by the rectal intima of the desert locust (*Schistocerca gregaria*). *J. Exp. Biol.,* 48:521-532.

Phillips, J.E., J. Hanrahan, M. Chamberlin, and B. Thomson. 1986. Mechanisms and control of reabsorption in insect hindgut. *Adv. Insect Physiol.,* 19:329-422.

Phillips, J.E., N. Audsley, R. Lechleitner, B. Thomson, J. Meredith, and M. Chamberlin. 1988. Some major transport mechanisms of insect absorptive epithelia. *Comp. Biochem. Physiol.,* 90A:643-650.

Phillips, J.E., R.E. Thomson, N. Audsley, J.L. Peach, and A.P. Stagg. 1994. Mechanisms of acid-base transport and control in locust excretory system. *Physiol. Zool.,* 67:95-119.

Proux, J.P., C.A. Miller, J.P. Li, R.L. Carney, A. Girardie, M. Delaage, and D.A. Schooley. 1987. Identification of an arginine vasopressin-like diuretic hormone from *Locusta migratoria. Biochem. Biophys. Res. Comm.,* 149:180-186.

Ramsay, J.A. 1953. Active transport of potassium by the Malpighian tubules of insects. *J. Exp. Biol.,* 30:358-369.

Ramsay, J.A. 1954. Active transport of water by the Malpighian tubules of the stick insect, *Dixippus morosus* (Orthoptera, Phasmidae). *J. Exp. Biol.,* 31:104-113.

Ramsay, J.A. 1955a. The excretory system of the stick insect *Dixipus morosus* (Orthoptera, Phasmidae). *J. Exp. Biol.,* 32:183-199.

Ramsay, J.A. 1955b. The excretion of sodium, potassium and water by the Malpighian tubules of the stick insect, *Dixipus morosus* (Orthoptera, Phasmidae). *J. Exp. Biol.,* 32:200-216.

Ramsay, J.A. 1956. Excretion by the Malpighian tubules of the stick insect, *Dixipus morosus* (Orthoptera, Phasmidae): calcium, magnesium, chloride, phosphate and hydrogen ions. *J. Exp. Biol.,* 33:697-709.

Ramsay, J.A. 1958. Excretion by the Malpighian tubules of the stick insect, *Dixippus morosus* (Orthoptera, Phasmidae): amino acids, sugars and urea. *J. Exp. Biol.,* 35:871-891.

Ramsay, J.A. 1964. The rectal complex of the mealworm *Tenebrio molitor* L. (Coleoptera, Tenebrionidae). *Phil. Trans. R. Soc. Ser. B,* 248:279-314.

Razet, P. 1966. Les Éléments terminaux du catabolisme azoté chez les insectes. *Annee Biol.,* 5:43-73.

Reagan, J.D., J.P. Li, R.L. Carney, and S.J. Kramer. 1993. Characterization of a diuretic hormone receptor from the tobacco hornworm, *Manduca sexta. Arch. Insect Biochem. Physiol.,* 23:135-145.

Riegel, J.A. 1966. Micropuncture studies of formed body secretion by the excretory organs of the crayfish, frog, and stick insect. *J. Exp. Biol.,* 44:379-385.

Saini, R.S. 1964. Histology and physiology of the cryptonephridial system of insects. *Trans. R. Entomol. Soc., London,* 116:347-392.

Schal, C., and W.J. Bell. 1982. Ecological correlates of paternal investment of urates in a tropical cockroach. *Science,* 218:170-172.

Spring, J.H. 1990. Endocrine regulation of diuresis in insects. *J. Insect Physiol.,* 36:13-22.

Spring, J.H., and S.R. Hazelton. 1987. Excretion in the house cricket (*Acheta domesticus*): Stimulation of diuresis by tissue homogenates. *J. Exp. Biol.,* 129:63-81.

Spring, J.H., A.M. Morgan, and S.R. Hazelton. 1988. A novel target for antidiuretic hormone in insects. *Science,* 241:1096-1098.

Staddon, B.W. 1955. The excretion and storage of ammonia by the aquatic larva of *Sialis lutaria* (Neuroptera). *J. Exp. Biol.,* 32:84-94.

Staddon, B.W. 1959. Nitrogen excretion in nymphs of *Aeshna cyanea* (Müll.) (Odonata, Anisoptera). *J. Exp. Biol.,* 36:566-574.

Wall, B.J., and J.L. Ochsman. 1970. Water and solute uptake by rectal pads of *Periplaneta americana. Am. J. Physiol.,* 218:1208-1215.

Wang, S., A.B. Rubenfeld, T.K. Hayes, and K.W. Beyenbach. 1996. Leucokinin increases paracellular permeability in insect Malpighian tubules. *J. Exp. Biol.,* 199:2537-2542.

Weltens, R., A. Leyssens, S.L. Zhang, E. Lohrmann, P. Steels, and E. Van Kerkhove. 1992. Unmasking of the apical electrogenic H pump in isolated Malpighian tubules (*Formica polyctena*) by use of barium. *Cell. Physiol. Biochem.,* 2:101-116.

Wessing, A., and D. Eichelberg. 1975. Ultrastructural aspects of transport and accumulation of substances in the Malpighian tubules. *Fortschr. Zool.,* 23:148-172.

Wessing, A., and D. Eichelberg. 1978. Malpighian tubules, rectal papillae and excretion, pp. 1-42, in M. Ashburner and T.R.F. Wright (Eds.), *The Genetics and Biology of Drosophila*, Vol. 2c, Academic Press, London.

Wheeler, C.H., and G.M. Coast. 1990. Assay and characterization of diuretic factors in insects. *J. Insect Physiol.,* 36:23-34.

Wheelock, G.D., D.H. Petzel, J.D. Gillet, K.W. Beyenbach, and H.H. Hagedorn. 1988. Evidence for hormonal control of diuresis after a blood meal in the mosquito *Aedes aegypti. Arch. Insect Biochem. Physiol.,* 7:75-89.

Wieczorek, H., G. Grüber, W.R. Harvey, M. Huss, and H. Merzendorfer. 1999. The plasma membrane H^+ V-ATPase from tobacco hornworm midgut. *J. Bioenerg. Biomembr.,* 31:67-74.

Wieczorek, H., G. Grüber, W.R. Harvey, M. Huss, H. Merzendorfer, and W. Zeiske. 2000. Structure and regulation of insect plasma membrane H^+ V-ATPase. *J. Exp. Biol.,* 203:127-135.

Wigglesworth, V.B. 1931. The physiology of excretion in a blood-sucking insect, *Rhodnius prolixus* (Hemiptera, Reduviidae). III. The mechanism of uric acid excretion. *J. Exp. Biol.,* 8:443-451.

Yu, M.-J., and K.W. Beyenbach. 2001. Leucokinin and the modulation of the shunt pathway in Malpighian tubules. *J. Insect Physiol.,* 47:263-276.

Zhang, S.-L., A. Leyssens, E. Van Kerkhove, R. Weltens, W. Van Driessche, and P. Steels. 1994. Electrophysiological evidence for the presence of an apical H-ATPase in Malpighian tubules of *Formica polyctena*: Intracellular and luminal pH measurements. *Pfluegers Arch. Eur. J. Physiol.,* 426:288-295.

14 Pheromones

CONTENTS

PREVIEW

Semiochemicals are chemicals produced and released for communication functions. The communication may be intraspecific, and the semiochemicals are then called pheromones. Semiochemicals also function interspecifically, between species, and these are sometimes called allelochemicals. Pheromones are often characterized by the type of behavior elicited in the receiving organism, such as sex attraction, alarm, trail-following, and numerous other categories. Interspecific semiochemical classification is usually based on the nature of who benefits from the chemical message. Interspecific semiochemicals include allomones benefiting the sender, kairomones benefiting the receiver, and synomones benefiting both sender and receiver. The same chemical substance may serve more than one function, such as a sex pheromone that attracts a potential mate and predator or parasitoid, and thus functions as a pheromone and a kairomone. Social insects have evolved a further elaboration of semiochemical function in which one chemical may elicit several different bahaviors within the colony, depending on social context and/or when the chemical is released. For example, queen substance in honeybees controls colony unity and behavior, suppresses ovary development in genetically female worker bees, and serves in the proper context as a sex pheromone to attract male bees to mate with the queen. This multiple functionality is called pheromone parsimony. Several thousand semiochemicals have been identified, and they comprise a large variety of chemical structures, functional groups, and molecular variations, including geometric and positional isomers and chirality. Research continues to focus on chemical identification of new semiochemicals, and on understanding receptor physiology and the nervous processes involved in responding to a semiochemical. Sex pheromones are usually blends of several chemicals. Sometimes, the opposite sex will respond to the major chemical, or to a partial blend, but in some species only the full blend in the correct proportions will attract a potential mate. Knowledge of what components and blend proportions will attract becomes particularly important in attempting to formulate pheromones for practical application in traps for population monitoring. Information processing of odor plumes is an active area of research. Odor plumes tend to be discontinuous pulses of chemical in the air, and their discontinuities appear to be important to the detection process by the receiving insect. Research shows that the male tobacco hornworm moth detects (with antennal receptors) pheromone quality, quantity, and pulse rate of the pheromone plume several times per second. Sex pheromones often function as species isolating mechanisms, and population and geographic differences suggest evolution in progress in pheromone blends based on what is currently believed to be a single species in several well-studied cases. Pheromone production in insects is regulated by hormones, with the pheromone biosynthesis activating neuropeptide (or PBAN) as one important hormone. Juvenile hormone (JH) may regulate pheromone production in some insects. Much of the research on pheromones derives from the desire to use pheromones in the control of insect populations. Currently, the most effective uses are as monitoring tools for population appearance or increase, and in mating disruption techniques.

14.1 INTRODUCTION

Communication through exchange of chemical signals is undoubtedly the oldest language known. It probably evolved with life itself, first as a means of intracellular communication, later as a means of intercellular exchange of signals, and finally as a vehicle for communication between organisms. **Semiochemicals** are signaling chemicals produced by an organism to send a message. Semiochemicals elicit changes in the behavior or physiology of the receiving organism. Many examples of chemical communication can be found in living organisms, ranging from the simplest unicellular plants, to microorganisms, to animals. The secretion of hormones and "second messengers" are examples of internal secretions that influence internal physiology, biochemistry, and behavior of the producing organism, but these are not called semiochemicals.

14.2 CLASSES OF SEMIOCHEMICALS

Semiochemicals have been defined as **intraspecific** agents influencing the physiology or behavior of members of the same species as that of the producer (**pheromones**), or as **interspecific** agents influencing a different species than that of the producer (**allomones** and **kairomones**) (Regnier, 1971). Some workers have suggested use of the term "**infochemical**" as a subcategory of semiochemicals, with redefinition of the types of infochemicals based on a context-specific rather than chemical-specific basis (Dicke and Sabelis, 1988; Vet and Dicke, 1992). The term "pheromone" was coined by Karlson and Luscher (1959) from two Greek words, *pherein*, meaning to carry, and *hormon*, meaning to excite. In the early 1960s when only a few pheromones had been identified, it was thought that a pheromone would be a single specific chemical. The sensitivity and sophistication of the work at that time generally resulted in identification of the major component in what later turned out to be blends in all the insects from which the first few pheromones were identified. Nearly all sex pheromones, and many other types of pheromones, are blends of components, and the total blend generally is considered to be the pheromone. The blend or ratio of components allows more information to be encoded, and allows greater discrimination among closely related species, which often use the same, or some of the same, components. Pheromones often serve multiple functions. For example, the male-produced sex pheromone hydroxydanaidal is derived from pyrrolizidine alkaloids by male *Utetheisa ornatrix* moths (family Arctiidae). The pheromone and plant-derived alkaloids are used by the female moth to evaluate the fitness of male moths, resulting in a model for sexual selection (Eisner and Meinwald, 1995).

Pheromones have been identified from more than 1600 species of insects in 90+ families from 9 orders (Roelofs, 1995). Lists of Lepidoptera from which compounds have been identified as pheromonal components are available electronically (Arn et al., 1992, 1999). A pheromone that has an effect on the physiology or biochemistry of an animal, such as suppression of ovary development in bees or stimulation of maturation in the desert locust, is called a **primer** pheromone because it "primes the pump" by requiring a finite period of time for its action to be effective. It causes relatively slow changes in the physiology, and usually behavior, of the animal. In contrast, a sex pheromone that attracts a potential mate is called a **releaser** pheromone because the pheromone almost instantly releases some behavior, such as upwind searching behavior or attempted mating.

Allomones, kairomones, and **synomones** influence the behavior or physiology of members of a different species than that of the producer. They are interspecific semiochemicals, and sometimes are also called allelochemicals because they act between species. Whittaker and Feeny (1971) used the term "allelochemical" to describe "chemicals significant to organisms of a species different from their source, for reasons other than food ...".

Defensive secretions are **allomones**. Allomones benefit the producer. An example of an allomone is the defensive spray of a bombardier beetle directed against attacking ants. A recent review (Berenbaum, 1995) includes a discussion of the theory of defensive secretions in both plants and animals. On the other hand, a **kairomone** often works to the detriment of the producer, and to the benefit of the receiver. An example is the odor of a prey, whether rabbit or lepidopterous caterpillar, that leads a parasite or predator to seek and find it. The host odors that attract a phytophagous insect to its food plant are often called kairomones. **Synomones** are mutually beneficial to producer and receiver, and operate in cases of mutualism or commensalism. Semiochemicals are used by insects in a variety of ways. Some well-known cases have been described in which a chemical can play more than one of the roles noted above. The sex pheromone of *Dendroctonus frontalis*, a bark beetle, not only attracts a potential mate, but may attract the clerid predator *Thanasimus dubius*. Thus, it functions as a pheromone to the beetles and as a kairomone to the predator. Many parasitoids and predators are evidently in an evolutionary race with their hosts to home-in on semiochemical components, while the hosts tend to modify the chemicals or blend they release (reviewed by Vet and Dicke, 1992).

14.3 IMPORTANCE OF THE OLFACTORY SENSE IN INSECTS

Examples of nearly incredible sensitivity to pheromonal chemicals occur in some insects, and especially among representatives of moths (Lepidoptera). Most moths mate in low light intensity during some part of the evening or night hours. Although the eyes of moths are adapted for low light intensity, they probably rely less on vision than on olfaction, and certainly olfaction is important for long-distance orientation. Perhaps because of feeding and mating activity in dim light, their ability to detect chemicals in the air has evolved to a very high degree of sensitivity. The antennae of many male Lepidoptera are plumose, with thousands of small hairs containing pheromone-sensitive sensory neurons (Figure 14.1). Although female Lepidoptera usually produce a sex pheromone that attracts males, many male Lepidoptera also release pheromonal components from hair

FIGURE 14.1 The plumose antenna of a male gypsy moth. There are thousands of pheromone receptors on the tiny hairs of the antennae. (Photograph courtesy of USDA.)

pencils that can be extruded from pouches at the tip of the abdomen. Figures 14.2 and 14.3 show hair pencils of the male moth *Heliothis virescens* and those of the male danaine butterfly *Idea leuconoe*, respectively.

The male commercial silk moth, *Bombyx mori*, is a good example of an insect with very high sensitivity to its sex pheromone, and was the first insect from which a sex pheromone component was chemically identified (Butenandt et al., 1959). The male has about 17,000 olfactory receptors on its antennae, and about 50% of these are tuned to detect the sex pheromone (Schneider, 1974). The advantage of having a very large number of receptors tuned to the pheromone is that sensitivity to low concentrations in the air is greatly increased. These male pheromone receptors fire spontaneously, with each antenna sending about 1600 impulses per second into the brain even when not stimulated by pheromone. According to established information theory in biological systems, for the pheromone stimulus to be perceived in the brain, the firing rate must increase by about 3 times the square root of the noise (noise in this case is the spontaneous rate). Thus, 3 times the square

FIGURE 14.2 Hair pencils everted from the tip of the abdomen by a male *Heliothis virescens* in response to a pheromone source. The hair pencils release male pheromone components that act on the female. (Photograph courtesy of Dr. Peter Teal, USDA.)

root of the noise is equal to about 120 firings per second, and an antenna must increase its firing rate to about 1720 impulses per second. Electrophysiological recordings from the antenna indicate that when a *single* pheromone molecule strikes a receptor, there is a receptor response. When a total of about 200 molecules of the sex pheromone simultaneously strike the antenna of the male moth (i.e., 200 receptor neurons are activated), the male moth responds by searching upwind for the source of the chemical. The male of the commercial variety of silkworm, however, has lost the ability to fly because it is too heavy. When it detects the sex pheromone of the female, motor output to the wing muscles causes them to start vibrating, but it merely walks upwind in a zig-zag path toward the source of the pheromone.

14.4 THE ACTIVE SPACE CONCEPT

The **active space** is the physical space in which the concentration of a pheromone is sufficiently high to cause a behavioral effect in the receiving individual, and it has been mathematically modeled (Bossert and Wilson, 1963; Wilson and Bossert, 1963). The active space is influenced by the

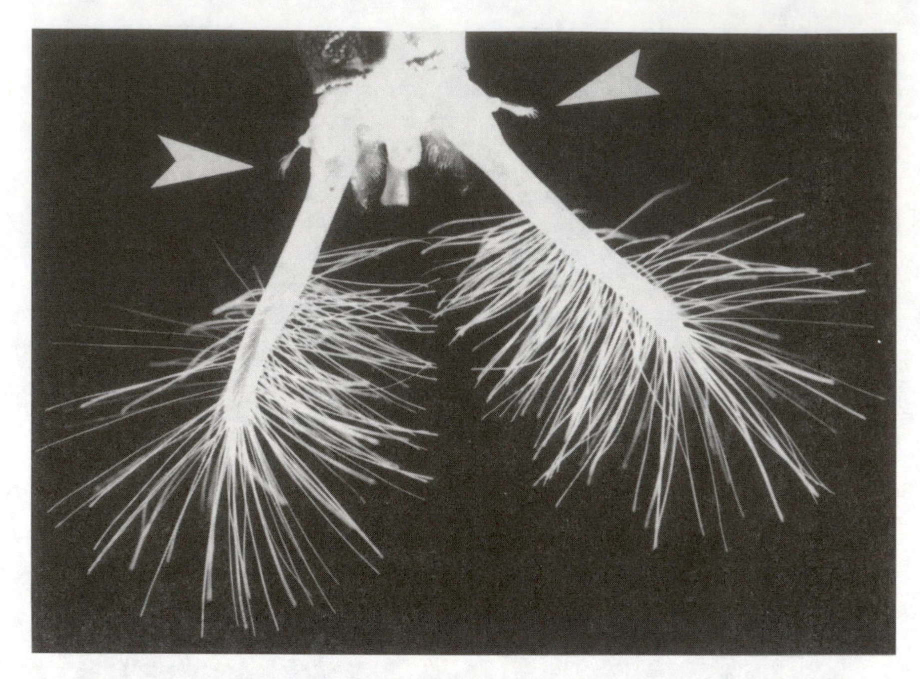

FIGURE 14.3 Hair pencils of the male giant danaine butterfly, *Idea leuconoe*. The arrows point to a small pair of hair pencils on the sheath of the larger pair. (Reproduced with permission from Nishida et al., 1996. Photograph courtesy of Ritsuo Nishida and colleagues.)

sensitivity of the receiver, the quantity of chemical produced and released by the sender per unit time, the volatility of the chemical(s) involved, and environmental factors such as wind velocity and temperature. The female silkworm moth contains only about 0.01 μg pheromone in her body; but if the total were released instantly and uniformly distributed through air moving at about 1.6 km/h, the active space for the male response system theoretically would extend for 4560 m (2.8 miles) downwind in a swath 215 m wide and 108 m high. Any male within that active space, or wandering into its periphery, should be excited to fly upwind. If uniformly distributed, the female can release sufficient chemical in that space to potentially attract about 1 billion moths.

The possibility of a male detecting a single female up to 4.5 km distant stretches the imagination. Almost certainly, such uniform distribution of pheromone never exists in nature because of changes in wind current and direction, and the presence of buildings, trees, and other objects between the female and the male that disrupt the air flow and create turbulence. In addition, pheromone is absorbed by many plants and objects, reducing the amount in the air. Nevertheless, the success of small insects in locating each other under semi-dark conditions, as most moths do, is impressive.

14.5 PHEROMONES CLASSIFIED ACCORDING TO BEHAVIOR ELICITED

Pheromones are often described or classified by the behavior elicited from the receiver (Shorey, 1973). There are sex pheromones, aggregation pheromones, alarm pheromones, egg-laying pheromones, brood-tending pheromones, recruitment pheromones, trail-following pheromones, and territory-marking pheromones, to name a few. Either sex, or both, can produce one or several pheromones.

Trail-following pheromones have been widely studied. Ants, for example, have evolved highly sensitive mechanisms for trail-following (Wilson, 1962; Wilson and Bossert, 1963; Kern et al., 1997; Janssen et al., 1997). They live underground where it always dark and, in general, their eyes are often small and their visual sense is not acute. However, their olfactory sense is very keen, and they use the sense of smell to follow a chemical trail. Each ant reinforces the trail as it follows it. The trail chemicals come from a gland, the Dufour's gland, near the tip of the abdomen. If the trail gets old — no ants pass over it and the trail is not reinforced with new chemical deposition — then the trail vanishes. The trail is easy to demonstrate by allowing a column of ants to establish a trail over a glass slide or strip of paper laid on the ground. When the trail is well established, the slide or paper can be turned 90°, whereupon the trail is completely disrupted. Ants arriving at the point where the trail is interrupted wander around searching for the trail. If the trail is reestablished after a few minutes by turning the slide again to its original position, the ants pick up the trail and continue. This simple experiment has been the basis for some of the behavioral tests to isolate and chemically identify the trail pheromone components. Other insects that use trail pheromones include termites (Matsumura et al., 1968; Grace et al., 1988; Reinhard and Kaib, 1995), some caterpillars (Capinera, 1980; Fitzgerald and Costa, 1986; Fitzgerald, 1993; Fitzgerald and Underwood, 1998), and bumblebees (Svensson and Bergström, 1977; Bergman and Bergström, 1997).

14.6 PHEROMONE PARSIMONY

Pheromone parsimony refers to the fact that the same pheromonal compound, sometimes synergized by additional compounds, can serve multiple functions, depending on ecological and behavioral contexts (Blum, 1996). The phenomenon is prevalent in social insects. Examples of pheromone parsimony can be found in the alarm pheromones that often also serve, in the proper context, as defensive allomones, attractants, trail pheromones, antimicrobial agents, and as releasers of several additional behavioral actions. The pheromones produced and released by the queen of a colony of social insects frequently have several functions, depending on the social context. For example, the principal queen substance from the queen honeybee *Apis mellifera* is (*E*)-9-oxo-2-decenoic acid (9-ODA). It releases behavior in worker bees that makes them attempt to form a cluster around the queen (called a queen retinue) and lick 9-ODA and other pheromonal components from her body. The workers subsequently spread the components throughout the colony by communal feeding. Distributed in this way, 9-ODA acts in conjunction with several other queen-produced components to suppress ovary development of workers (a primer function), and inhibits construction of queen cells in which new queens might be reared. 9-ODA also releases mating behavior in male bees (drones), but only when the drones are flying and when the queen substance is released from several meters up in the air, the normal site for mating by honeybees (Gary, 1962). Drones will attempt to mate with a variety of objects, including inanimate queen mimics and dead queens if 9-ODA is released from them and if they are displayed in the air, for example, on a flag pole. In the colony, drones exhibit no mating behavior.

14.7 CHEMICAL CHARACTERISTICS OF PHEROMONES

To have sufficient volatility, airborne pheromones must be relatively small molecules. In general, the molecular weight must be less than 200, or not much over 200, to get the volatility needed. Not all pheromones are airborne; some insect pheromones are contact pheromones and some crustaceans have waterborne pheromones. No particular chemical structure is used exclusively by insects as a pheromone, but low-molecular-weight acids, esters, alcohols, aldehydes, ketones, epoxides, lactones, hydrocarbons, terpenes, and sesquiterpenes are common components in pheromones. A few large molecules serve as pheromones, but they are predominantly contact pheromones. For example, (*Z*)-9-tricosene (a hydrocarbon composed of 23 carbons) is the sex pheromone of the housefly *Musca domestica*. It has very low volatility. Male houseflies are stimulated to make contact with any small dark object, such as a knot in a black shoelace, which was used as one of the bioassay devices for the pheromone. They detect the pheromone, if it is present, after landing upon the object and then attempt to mate with it. Even larger cuticular hydrocarbons on the surface

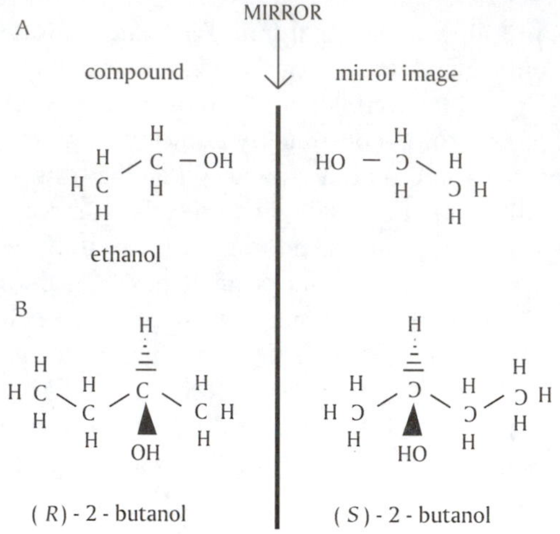

A

MIRROR

compound

mirror image

ethanol

B

(*R*) - 2 - butanol

(*S*) - 2 - butanol

FIGURE 14.4 The concept of chirality. A: The mirror images of ethanol can be superimposed upon each other because ethanol does not have a stereo center. B: The mirror images of 2-butanol are not superimposable, and the carbon bearing the –OH group is a stereo center. On the left is the *R* enantiomer, and on the right is the *S* enantiomer.

of female tsetse flies serve as contact pheromones. In general, a male tsetse fly cannot distinguish a female from a male until it contacts the body surface of the fly. The chemoreceptors may be located on the tarsi of the male.

Pheromones often serve as species isolating mechanisms, and the receptors on the antennae of the opposite sex must be tuned to the pheromone of its species. There are several ways that species specificity is coded in pheromones. One of the most common adaptations is the use of a blend of two or more chemical components in the pheromone, as well as variations in different blend ratios when the same components are incorporated into the pheromone. Many of the sex pheromone components that have been identified from moths are acetates and alcohols with a backbone of 14 to 16 carbons with one or more double bonds in the molecule. The position of the double bond offers many variations on each backbone. Still further specificity can be encoded in (*E*) or (*Z*) configuration of the groups or atoms on the carbons at the double bond. Finally, **chirality** in molecules provides another way that insects can specify their pheromone signal (Mori, 1984; Silverstein, 1988). The word "**chiral**" is derived from the Greek word for hand. Chiral compounds have a three-dimensional shape analogous to a person's hands. Although the right hand is a mirror image of the left hand, the two hands are not perfectly superimposable upon each other with respect to matching shape when both palms are down or both up. All objects, including chemical molecules, have a mirror image; but like human hands, not all chemical molecules can be superimposed upon each other. Ethanol, for example, can be superimposed upon itself, but 2-butanol can exist in two three-dimensional shapes that cannot be superimposed upon each other (Figure 14.4). When chemical compounds are mirror images of each other, but the two images cannot be superimposed, the compounds are called **enantiomers** of each other, and they possess stereo centers. One feature that makes for chirality is when a carbon atom is attached to four different groups; such an arrangement produces a stereo center. A molecule with one stereo center is a chiral molecule. Many of the most common biochemical compounds in living organisms are chiral, such as amino acids and glucose. Most enzymes that utilize amino acids to synthesize proteins accept only L-amino acids, and D-glucose is the form synthesized into glycogen, trehalose, and chitin.

A molecule may have more than one stereo center, but this gets more complicated, and such a molecule may or may not be chiral. It is not uncommon, however, for pheromone molecules to have more than one stereo center. Other atoms in addition to carbon may also be a source of chirality in a molecule, a situation not often relevant to pheromones. The designation of the two

TABLE 14.1
Priority in the Cahn-Ingold-Prelog System for Determining the Absolute Configuration of a Molecule Bearing Some of the More Common Substituents Found in Insect Pheromones

Atom or Group	Priority
Hydrogen	1
Methyl	2
Ethyl	3
n-Propyl	4
n-Butyl	5
n-Pentyl	6
n-Hexyl	7
Isopentyl	8
Isobutyl	9
Allyl	10
Isopropyl	14
Vinyl	15
Cyclohexyl	17
Acetyl	36
Carboxy	38
Amino	43
Hydroxy	57
Fluoro	68
Chloro	74

Data from Cahn (1964a, 1964b).

enantiomers of a chiral compound as *R* or *S* is based on the Cahn-Ingold-Prelog system (simple sugars and amino acids have been excepted from the system because of the long-term usage of D- and L-designations). Chemists assign a priority value to chemical groups, such as –OH, $-CH_3$, $-CH_2CH_3$, etc., based on certain priority rules (Table 14.1) (Cahn, 1964a, 1964b). The molecule is viewed by looking at it so that the group of lowest priority is behind the molecule, or farthest from the eye (Figure 14.5). Then, the decreasing priority order of the remaining groups is noted, and if the order decreases clockwise around the chiral carbon, the designation is *R*, from the Latin word *rectus* for right. If the order decreases counterclockwise, the designation is *S*, from Latin for sinister, meaning left.

Enantiomers are optically active; each rotates the plane of polarized light, but in different directions, either clockwise, designated dextrorotatory (+), or counterclockwise, designated levorotatory (–). The direction of rotation of plane-polarized light is completely independent of the *R* or *S* designation. Thus, a particular molecule might be *R*-(+)-, or it might be *R*-(–)-. Whatever it turns out to be, its enantiomer will have the opposite designation.

For chirality in a pheromone to provide specificity, a receptor has to recognize the difference in the two enantiomers. The opposite enantiomer of a chiral pheromone component may have no effect at the sensory neuron, may stimulate, or may inhibit the response to the natural enantiomer. A good example of enantiomeric differentiation at the receptor site occurs in the Japanese beetle, *Popillia japonica* Newman (family Scarabaeidae). Tumlinson et al. (1977) identified (*R,Z*)-5-(1-decenyl) dihydro-2(3H)-furanone as the pheromone produced by female beetles, and showed inhibition by the *S,Z* enantiomer. Mirror images of the enantiomers are shown in Figure 14.6. A **racemic mixture** (i.e., an equal mixture of the *R* and *S* configurations) of the

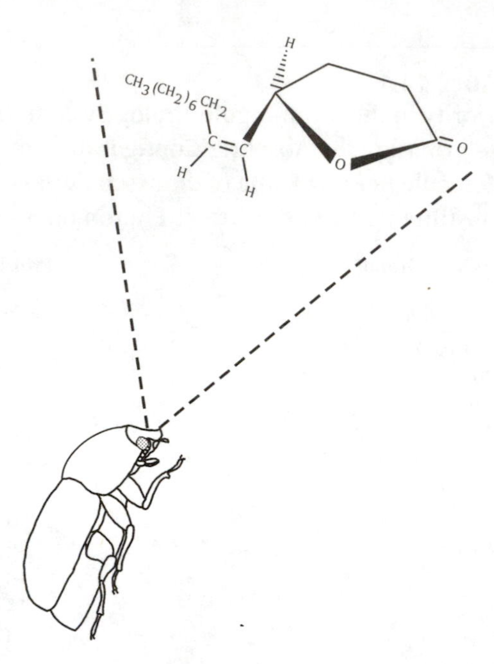

FIGURE 14.5 An illustration to show how one determines whether a molecule has the *R* or *S* configuration. In the drawing, a male Japanese beetle *Popillia japonica* views the chiral pheromone produced by a female as a chemist would view it, with the group of lowest priority behind the molecule. The natural enantiomer of the Japanese beetle pheromone has the *R* configuration because the priority of the groups on the chiral carbon decrease in priority in a clockwise direction.

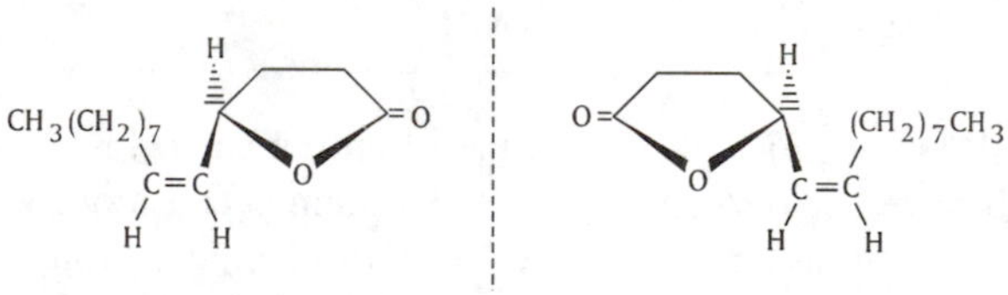

H
$CH_3(CH_2)_7$
C=C
H H
O
=O

(*R,Z*) - 5 - (1-decenyl)
dihydro - 2(3H)
furanone

H
O=
O
$(CH_2)_7CH_3$
C=C
H H

(*S,Z*) - 5 - (1-decenyl)
dihydro - 2(3H)
furanone

FIGURE 14.6 The *R* and *S* enantiomers of the Japanese beetle pheromone depicted as mirror images of each other. The natural enantiomer that the males respond to is (*R,Z*)-(1-decenyl)-dihydro-2(3H)-furanone.

synthetic (*Z*) isomer is completely inactive in field tests, and as little as 1% of the unnatural synthetic (*S,Z*) enantiomer mixed with the natural (*R,Z*) enantiomer substantially reduces male response. When a trap contains 5% of the (*S,Z*) enantiomer, capture of beetles is reduced to the level of an empty trap. One can only guess at what happens at the receptor level in this case because the receptor has not been isolated. One possibility is that the (*S,Z*) enantiomer binds rapidly, and perhaps preferentially, to the receptor and cannot be removed. Thus, it might block binding of the natural enantiomer. Other explanations, however, are possible, including physiological actions within the brain of the insect.

Grain beetles in the family Cucujidae illustrate the probable utilization of pheromone chirality in evolution and isolation of species (Table 14.2) (Oehlschalager et al., 1987). The pheromonal compounds produced by males during feeding act as aggregation pheromones; they attract both sexes. Starved males are not good producers. One possible evolutionary factor operating in the selection of these pheromones is that they may indicate a food source, and both sexes could benefit

TABLE 14.2
Similarities and Differences in the Pheromone Blend Components and Enantiomeric Composition of Blend Components in Closely Related Grain Beetles in the Family Cucujidae

Species	Pheromone
Cryptolestes ferrugineus	(*S*,*Z*)-3-dodecen-11-olide
Oryzaephilus mercator	(*R*,*Z*)-3-dodecen-11-olide
	(*R*,*Z*,*Z*)-3,6-dodecadien-11-olide
O. surinamensis	(*R*,*Z*,*Z*)-3,6-dodecadien-11-olide
	(*R*,*Z*,*Z*)-5,8-tetradecadien-13-olide
	(*Z*,*Z*)-3,6-dodecadienolide
C. turcicus	(*R*,*Z*,*Z*)-5,8-tetradecadien-13-olide, 85%
	(*S*,*Z*,*Z*)-5,8-tetradecadien-13-olide, 15%
	(*R*,*Z*)-5-tetradecen-13-olide, 33%
	(*S*,*Z*)-5-tetradecen-13-olide, 67%
C. pusillus	(*S*,*Z*)-5-tetradecen-13-olide

Data from Oehlschlager et al. (1987).

by responding to the pheromone. Mixed sexes of each species are attracted best to blends of compounds produced by feeding males of their own species.

14.8 INSECT RECEPTORS AND THE DETECTION PROCESS

Pheromones are generally detected by olfactory receptors located primarily on the antennae. Many male Lepidoptera have pheromonal receptors specialized for reception of the sex pheromone; these receptors are relatively insensitive to other chemicals. A sex pheromone receptor on the antenna typically consists of one or two nerve cells housed within a seta or fine "hair" on the antenna. The entire structure is called the sensillum (Figure 14.7). There often are many thousands of such sensilla on each antenna of a male moth. Each seta has microscopic pores along its length, through which the airborne molecules enter the sensillum and make contact with the sensory neuron.

Kaissling (1987) postulated that the following six steps occur as a part of the process of semiochemical detection.

1. **Adsorption** of an odor molecule by sensory hairs (setae) on the antennae
2. **Penetration** of the molecule through pores in the setal wall
3. **Receptor binding** of the molecule and transport to the sensory nerve endings (It is believed that molecules adsorb to the cuticular surface and then move into a pore, although an occasional molecule may hit a pore directly. Processes that may promote movement of adsorbed molecules into a pore are unknown.)
4. **Membrane alteration**, probably opening of sodium channels
5. **Receptor potential generation**, a graded potential, followed by spikes in the axon hillock region
6. **Inactivation** of the odor molecule and removal (Inactivation clears the receptor so that it can respond again.)

14.8.1 PHEROMONE BINDING PROTEINS

After a molecule enters a pore, it must cross the **sensillum liquor** (lymph, extracellular fluid) to reach the dendritic nerve endings. Most pheromonal molecules are lipid soluble, and the sensillum

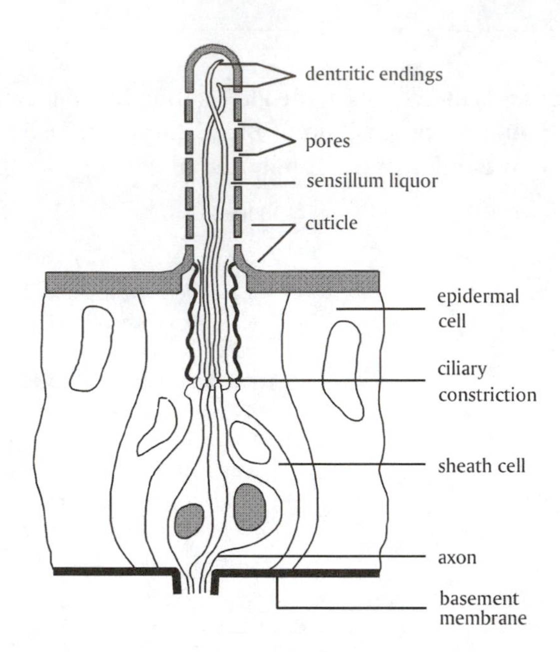

FIGURE 14.7 Basic structure of an olfactory receptor with pores and dendritic nerve endings in the cuticular hair.

liquor is an aqueous medium that does not readily dissolve lipids. Thus, the molecule must usually bind with a specific binding protein, which then transports it across or through the sensillum liquor. Pheromones bind to **pheromone binding proteins** (**PBPs**) (Vogt and Riddiford, 1981; Vogt, 1987; Klein, 1987), a subset of a larger group of **odorant binding proteins** (**OBPs**) known from both vertebrates and invertebrates. A number of PBPs and GOBPs have been isolated and sequenced (Prestwich and Du, 1997).

General OBPs (**GOBPs**) of Lepidoptera bind a variety of odorant molecules associated with food, habitat, and oviposition substrates (Breer et al., 1990a; Vogt et al., 1991a, 1991b). GOBPs may occur in other insects, perhaps in males in some cases as well as in females; but currently, little research has been done in groups other than Lepidoptera.

PBPs are synthesized in male Lepidoptera just prior to adult emergence and are localized in the extracellular fluid (the sensillum liquor) of pheromone-responsive sensilla on the male antennae (Vogt et al., 1989). Auxiliary cells associated with the receptor cell appear to be responsible for synthesis (Steinbrecht et al., 1992). By binding the (usually) hydrophobic pheromone molecule, the PBP aids in transport of the pheromone through the aqueous sensillum lymph so that it makes contact with specific receptor proteins in the dendritic nerve endings (Vogt and Riddiford, 1981; Prestwich, 1993b; Breer, 1997; Prestwich and Du, 1997). One current hypothesis is that the PBP with bound protein may attach directly to the dendritic membrane where subsequent activation of a G protein coupled cascade of events results in a receptor potential in the dendrite (Prestwich and Du, 1997). Specific binding proteins may serve as filters that protect the dendritic endings from many other chemical molecules of the air that also enter sensilla pores (Prestwich and Du, 1997). Some evidence suggests that PBPs may also be involved in destruction of the pheromone after signal transduction (Vogt et al., 1985; Prestwich, 1993a).

Prestwich and Du (1997) determined the active site for pheromone binding in a binding protein isolated from *Antheraea polyphemus* using photoaffinity labeling. The major component of the *A. polyphemus* sex pheromone, 6*E*,11*Z*-hexadecadienyl acetate, has only one binding site; but a second pheromonal component, 4*E*,9*Z*-tetradecadienyl acetate can bind in two slightly different ways (Du et al., 1994).

14.8.2 Signal Transduction and Receptor Response

At the dendritic ending, the pheromone probably combines with a receptor protein in the dendritic membrane, although a specific protein receptor has not been identified in an insect as yet. G proteins, cAMP, and inositol triphosphate (IP_3) are involved in the odor transduction process in several vertebrates (Buck and Axel, 1991; Ressler et al., 1993; Ngai et al., 1993), and some or all of these may mediate and amplify the pheromone signal at the dendritic ending in insects. The concentration of cAMP, however, is low in antennae, and its concentration is not stimulated by pheromone (Breer et al., 1990a; Ziegelberger et al., 1990), so it seems an unlikely participant. High doses of pheromone elevate cGMP in the antennae, but only slowly and after a long delay (Boekhoff et al., 1993); thus, it also seems unlikely to mediate the fast responses needed.

Some evidence suggests that IP_3 might be a participant in pheromone response. Phospholipase C, the enzyme that hydrolyses phosphatidyl inositol bisphosphate (PIP_2) to inositol trisphosphate (IP_3) and diacylglycerol (DAG), has high activity in the antenna (Breer et al., 1990a; Boekhoff et al., 1990a). Moreover, IP_3 shows a rapid phasic increase followed by a tonic decline in male antennae after sex pheromone application (Boekhoff et al., 1990b, 1993). Kinetic measurements following pheromone application indicate that IP_3 reaches a stimulus-dependent maximum in about 50 ms, and declines to the basal level within a few hundred milliseconds (Breer et al., 1990b). Such a time course is consistent with the observed ability of some insects to resolve several pheromone pulses per second (Marion-Poll and Tobin, 1992). The slow increase and sustained elevation of cGMP, and experiments indicating that elevated cGMP modify the response to pheromone may mean that it is involved in the adaptation of pheromone receptors exposed to high and sustained pheromone levels (Breer, 1997).

Electrophysiological recordings from single cells and from the whole antenna have been used to study the response to pheromone components, and in bioassay of potential pheromone components during pheromone identification. The technically more easily accomplished recordings from the antenna have been used more frequently. The procedure requires the mounting of an antenna between two electrodes. The antenna can be left attached to the head of the insect, but frequently it is severed from the head and mounted. The response is called the EAG (Figure 14.8), the electroantennogram, and it is a summed potential from many receptors responding simultaneously or in rapid sequence (Schneider, 1957). The technique has been used effectively in pheromone identification (Roelofs, 1984), and in combination with a gas chromatographic-mass spectrometer (GC-MS) technique by splitting the column effluent, sending part to a flame ionization detector or ion trap, and part to the antenna (Arn et al., 1975; Cossé et al., 1995). In such a situation, the antenna is called an EAD, an electroantennogram detector (Figures 14.9 and 14.10). Portable EAG devices have been designed and built to measure pheromone concentrations in the field (Baker and Haynes, 1989; Sauer et al., 1992; Rumbo et al., 1995; Karg and Sauer, 1995; Leal et al., 1997). A design scheme for a portable EAG device is shown in Figure 14.11.

Single-cell recordings have usually been made extracellularly by carefully placing a glass capillary electrode filled with saline over a single hair on the antenna. After a relatively long latency (15 ms or longer), slow graded potentials with superimposed spikes can be obtained when the pheromone is pulsed over the antenna (Kaissling, 1986). Single-sensillum recording also has been adapted for use as a GC-MS detector (Wadhams, 1982).

Axons from sex pheromone receptor neurons located peripherally project into the antennal lobe of the deutocerebrum as labeled-lines (pheromone-specific neurons), while general food, host odors, and environmental odors are transmitted by across-fiber patterning (Masson and Mustaparta, 1990). The concept of labeled-lines and across-fiber patterning lines was described by Dethier (1972) from studies with taste receptors. It currently appears that an odor is defined in the deutocerebrum by an across-glomeruli pattern based on input that the glomeruli receive from peripheral receptors (Todd and Baker, 1997).

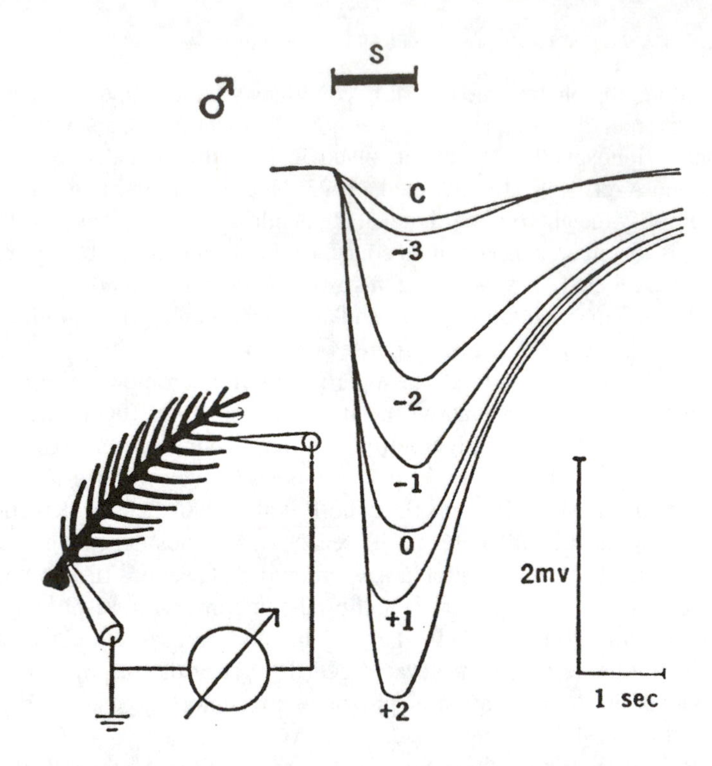

FIGURE 14.8 A typical series of electroantennogram (EAG) recordings from an antenna of a male silkmoth *Bombyx mori* in response to a control puff of air (C) or a puff containing pheromone (numbers indicate log μg of bombykol on odor source). Response to 0.001μg to 100 μg bombykol on the odor source is indicated by the increasingly larger EAG responses. Each stimulus lasts for 1 s. An indication of the magnitude of the response is given by the scale line of 2 mV. The responses are characteristic of graded or slow potentials, that is, slow rise time, slow decay, and increasing magnitude in response to increasing stimulus. (From Boeckh et al., 1965.)

Pheromone receptors on the antennae of some insects — and perhaps on most insects — are spontaneously active and display a constant low level of firing. Contact with a pheromone component to which a receptor is sensitive usually results in an increase in firing. In some cases, a receptor is differentially sensitive to a particular component of a pheromone blend, and it typically responds to low concentrations of that component by a large increase in rate of firing. Exposure of a specialist receptor to large amounts of other blend components may cause an increase in firing to nonspecific components, but usually at a much lower rate (i.e., it is less sensitive to the other components) (Almaas and Mustaparta, 1990; Berg and Mustaparta, 1995; Berg et al., 1995).

Most insects with pheromone receptors appear to have more than one type of receptor, each being sensitive to one or more of the blend components of the pheromone. It may be that when a responding insect flies upwind in a plume of pheromone composed of several components, each of its several receptor types responds to one component while ignoring other components in the blend. Thus, the nerve activity going into the deutocerebrum and other parts of the brain is via labeled-lines, and the brain presumably has to integrate the input as the relative activities of different receptor neurons (Mustaparta, 1997). Few neurophysiological details are available to substantiate or refute this concept.

14.8.3 Pheromone Inactivation and Clearing of the Receptor

When pheromone molecules have made contact with the dendritic endings, it is important to destroy or inactivate them in order to clear the receptor active site and allow the receptor to be sensitive

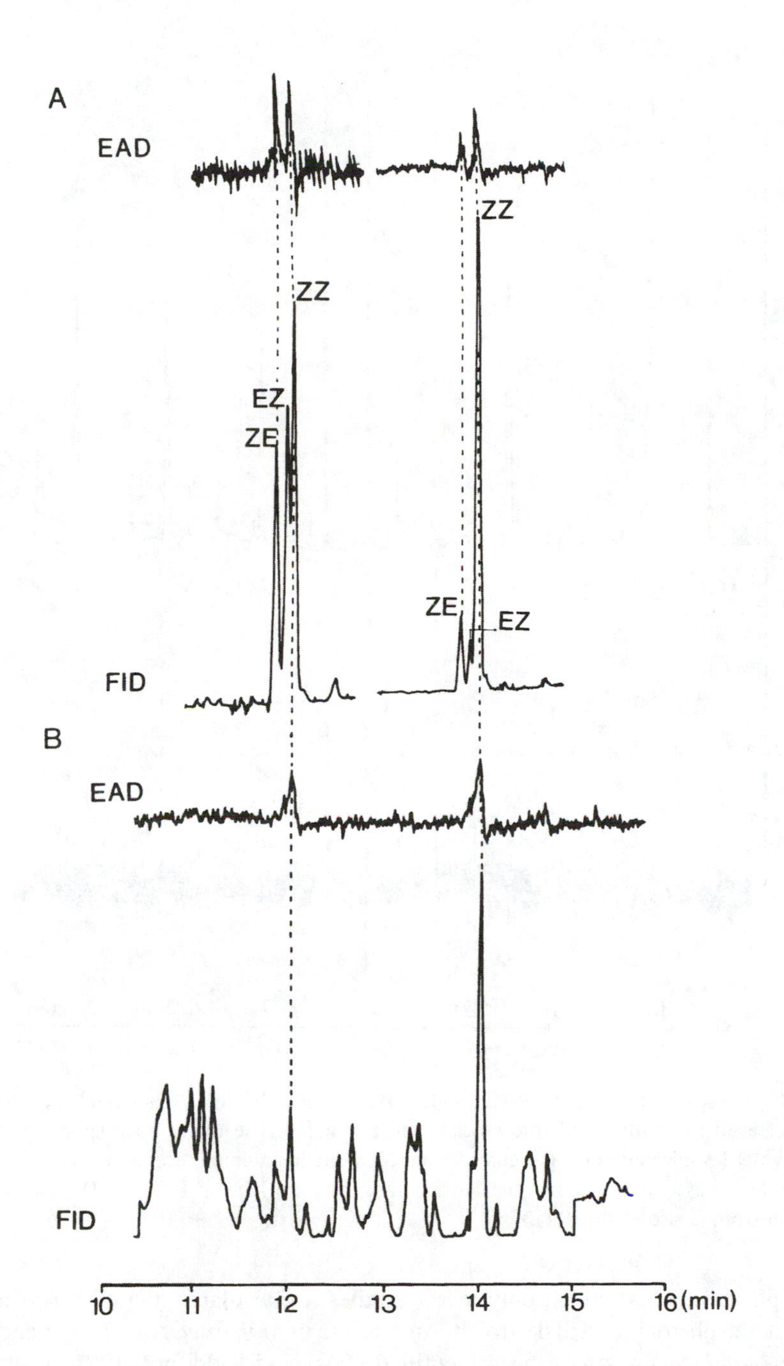

FIGURE 14.9 Use of the EAG response as an electroantennogram detector (EAD) in conjunction with a flame ionization detector (FID) in a gas chromatograph to recognize behaviorally active peaks in a gas chromatogram. A: The EAD responses were made by the antenna of a male *Idea aversata* geometrid moth to the effluent from a gas chromatograph (GC). The GC-FID response is shown to be a mixture of *Z*7,*E*9-, *E*7,*Z*9, and *Z*7,*Z*9-dodecadienyl acetate and a mixture of *Z*9,*E*11-, *E*9,*Z*11-, and *Z*9,*Z*11-tetradecadienyl acetate. B: Response of the male antenna to female gland extract is shown as the EAD with the FID response to the gland extract. (Reproduced with permission from Zhu et al., 1996.)

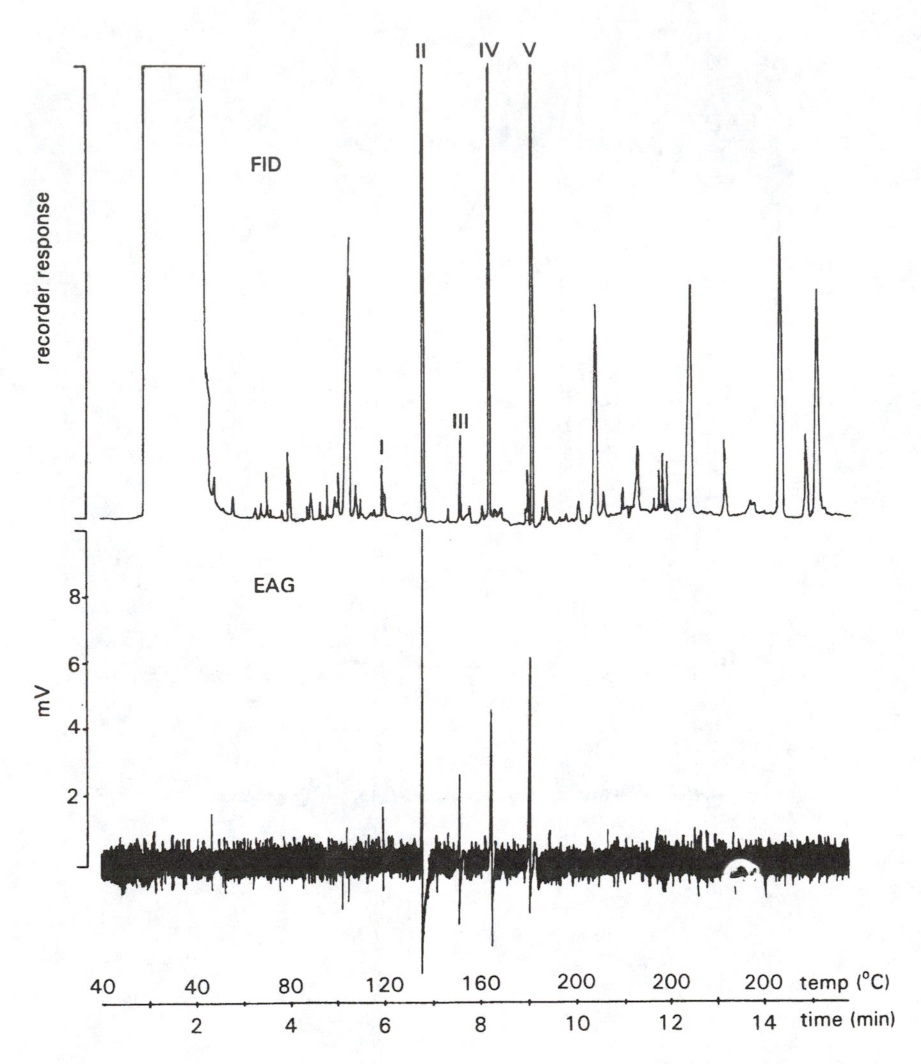

FIGURE 14.10 The lower trace shows EAG (or EAD) responses by female antennal receptors to some of the volatiles released by "calling" Mediterranean fruit fly males. The upper trace shows FID responses to male-released volatiles injected into a gas chromatograph. Female antennal receptors respond strongly to only some of the male-released volatiles, and the EAG-active ones are labeled I, II, III, IV, and V. (Reproduced with permission from Cossé et al., 1995.)

to incoming pheromone. Although only a few studies are available, the evidence indicates that enzymes attack the pheromone and destroy it. An esterase in *Antheraea polyphemus* and an aldehyde oxidase from *Manduca sexta* have been identified (Vogt and Riddiford, 1981; Vogt et al., 1985; Klein, 1987; Rybczynski et al., 1989). In less than 0.5 s the esterase in *M. sexta* antennae can destroy a million pheromone molecules, and this seems consistent with the rapid changes in upwind or casting behavior that males make in a changing pheromone plume (Vogt, 1987). The pheromone-destroying enzymes may also aid an insect by destroying small amounts of pheromone slowly leaking into the pores after adsorption on the antennal surface (Breer, 1997). Such a slow, persistent pheromone leak might create high background noise in the system.

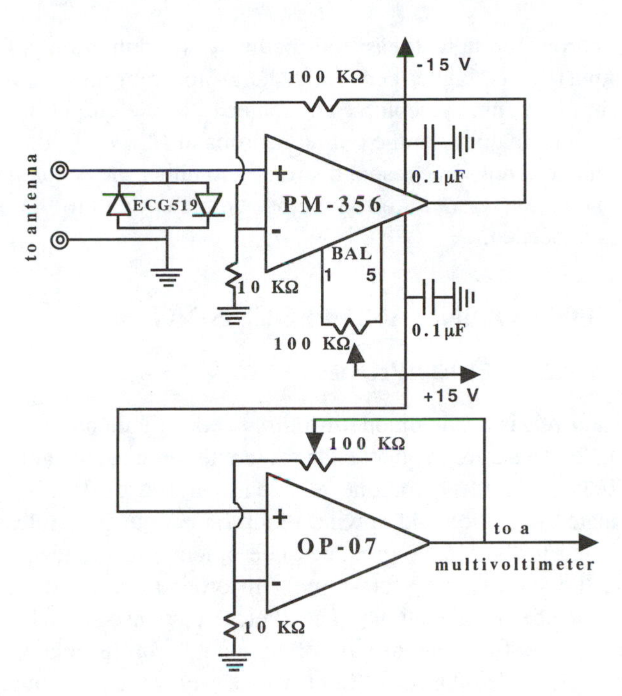

FIGURE 14.11 A schematic circuit diagram for constructing a portable or field EAG measuring unit. (Reproduced with permission from Leal et al., 1997.)

14.8.4 Do Insects Smell the Blend or Just the Major Component(s)?

Pheromone specialists have been divided over the issue of whether the responding insects smell and respond to only one or perhaps two major components in the pheromone blend, or whether the entire blend is necessary for response. Actually, the behavior of male moths in this respect is quite variable, depending on species. It appears that different species utilize different strategies to locate a mate. In some, the major component may be sufficient for the complete behavioral response and mating; while in others, a more complex or complete blend is necessary. Males of the red-banded leafroller moth *Argyrotaenia velutinana* and Oriental fruit moth *Grapholita molesta* are typical of insects that need the blend. Very few males of the red-banded leafroller moth fly upwind toward a pheromone source containing only the major pheromonal component. Significantly more males of the Oriental fruit moth give behavioral displays (wing fanning) up to 60 meters away from a pheromone source when the source contains the correct proportions of a three-component pheromone blend, as opposed to when the source contains only the major component (Linn et al., 1987). Alternatively, a high percentage of the males of a number of other insects (cabbage looper *Tricoplusia ni*; cotton bollworms *Helicoverpa armigera*, *H. zea*, and *Heliothis virescens*) take flight, fly the typical zig-zag pattern in a pheromone plume, and may contact the source and attempt mating when exposed to only the major component of the blend (Mayer and McLaughlin, 1992; Kehat and Dunkelblum, 1990; Vickers et al., 1991).

A model (reviewed by Christensen, 1997) based on *Manduca sexta* diversity in neuronal input and output in the MGC provides insight into the possible neuronal explanations of the blend vs. major component response. Males usually do not fly upwind when exposed to only one component; they need the correct blend. Nevertheless, in *M. sexta,* some olfactory receptors on the antennae of males respond best when exposed to the major component of the blend (component A), and others respond best to a second component (B). Similar responses are obtained from some receptors on the antennae of male *H. zea*. The input neurons synapse in different glomeruli in *M. sexta*, and in the same glomeruli in *H. zea*. The outputs from the glomeruli go to higher brain centers through local and projection neurons through at least four pathways (A output only, B output only, A or B output, and blend output). Thus, higher centers in the protocerebrum receive a variety of inputs, depending on exposure of antennal receptors. The majority of the output interneurons in *M. sexta* respond best to a blend. The majority of the output neurons in *H. zea* respond strongly to the major component (A, in the model), but some respond strongly to either the A component or to the blend. *H. zea* males also respond behaviorally to the major component and to the blend. Detailed study of many more species is needed.

14.9 INFORMATION CODING AND PROCESSING

14.9.1 THE STRUCTURE OF ODOR PLUMES

A female moth typically releases pheromone in pulses and a filamentous plume snakes out from the female (Figure 14.12). There are frequent changes in pheromone concentration within the plume (Murlis and Jones, 1981). Pulsed pheromone released at a rate of 3 pulses per second is more effective in causing male orientation and upwind flight than continuous release (Kaissling, 1986). A male insect responding to the female-produced pheromone must detect, process, initiate flight commands, and clear the sensory receptors rapidly in order to respond effectively to the rapid changes that occur in a pheromone plume. The typical response of a flying male insect to a pheromone plume is **optomotor anemotaxis**, or upwind flight (Kennedy, 1940; Kennedy and Marsh, 1974; Marsh et al., 1978). Once in flight, its own movement through the air prevents an insect from determining wind direction except by its visual displacement over the ground. Flight directly into the wind direction causes the image received by the eyes to move in line with the body axis, while a cross wind causes the insect to experience lateral drift. Usually, a male flies in a zigzag pattern upwind in response to the sex pheromone. Its behavior changes from upwind flight to casting from side to side in as little as 0.5 s when it loses the plume (Kennedy et al., 1980), indicating that its sensory and nervous system can detect and respond rapidly.

Male silkworm moths *B. mori* are too heavy to fly, and they walk upwind in a zigzag pattern while vibrating their wings at 40 to 50 Hz (Kanzaki and Shibuya, 1986; Kanzaki, 1997). Loudon and Koehl (2000) determined that the wings are flapped through a stroke angle of 90 to 110° at about 40 Hz, directing an unsteady flow of air (average speed, 0.3 to 0.4 m/s) toward the antennae. Air flow over the antennae is about 15 times faster that than produced by walking and 560 times faster through the spaces between the sensory hairs; they fail to move upwind in a steady plume of pheromone, and only respond when the stimulus is pulsed. Many species of flying moths exhibit similar behavior, and resort to casting from side to side without forward movement, or come to rest on some convenient substrate in a steady plume of pheromone (Kramer, 1997). Experiments have demonstrated that discontinuous pulsing of the pheromone in the plume is much more important to upwind flight than the concentration of pheromone in the plume (Kramer, 1986).

Attempts to explain the mechanism underlying zigzag and casting behavior have led to many experiments and numerous arguments. One idea has been that the male moth initiates flight upwind, but cannot steer a perfect course; so, sooner or later, it comes to the edge of the active space. It then makes a turn back into the active space. Experiments show, however, that males of some species make turns in clean air with no pheromone, and within a (presumably) uniform cloud of pheromone (Kennedy, 1983). Such observations led to the idea that perhaps zigzag flight is not an essential component of optomotor anemotaxis, and an internal turn generator that operates independently of the odor plume has been proposed (Wright, 1958; Kennedy and Marsh, 1974; Baker et al., 1984).

A second explanation is the **flight imprecision model** (Mafra-Neto and Cardé, 1995), based on experimental data from gypsy moths and a computer simulation program (Preiss and Kramer, 1986a, 1986b). Data from these experiments and flight simulations lend support to the idea that turns are course corrections caused by the inability of the moth to head straight upwind. Presumably,

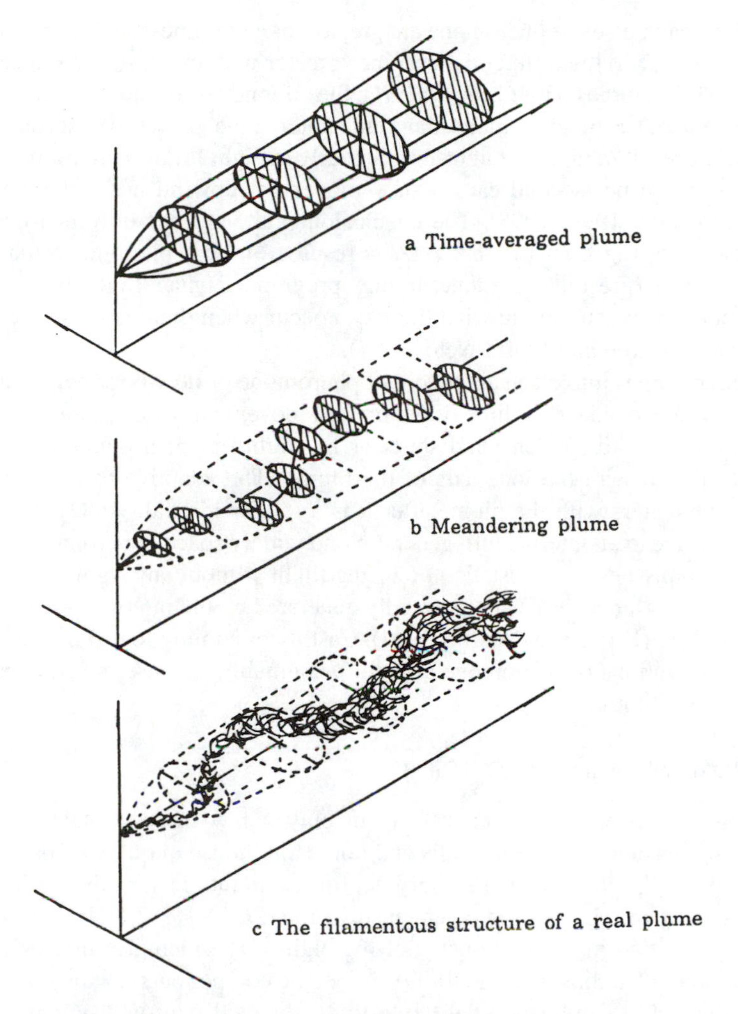

FIGURE 14.12 A schematic diagram to illustrate the structure of a pheromone plume in the air: a time-averaged approach (a), a more realistic meandering plume (b), and the discontinuous, filamentous structure of a real plume (c). (Reproduced with permission from Murlis et al., 1992.)

it would fly directly upwind if it could, and when its movement relative to the ground (substrate) indicates lateral drift, it corrects course.

A third possibility is that zigzag flight is caused by blend quality that does not perfectly mimic the natural blend released by a calling female (Witzgall and Arn, 1990, 1991; Witzgall, 1997). For example, several different investigators have observed that males of *Lobesia botrana*, *Grapholita molesta*, and *Eupoecilia ambiguella* are able to fly nearly straight upwind without the zigzag movement in response to a calling female. They fly the zigzag pattern, however, when exposed to synthetic pheromone, which is presumed, at best, to be slightly different from the blend released by the female.

Regardless of the arguments about mechanisms operating during upwind orientation, the microstructure of the pheromone plume controls the optomotor anemotaxis response. Pheromone does not disperse in the natural environment of insects as a uniform cloud; it travels as a plume of discrete filaments or eddies ranging in size from less than a centimeter to many meters (Murlis et al., 1992; Murlis, 1997) interspersed with pheromone-free air gaps. The gaps become larger and the pheromone filaments smaller at greater distances from the source. Thus, the signal gets broken into discrete stimuli lasting many milliseconds and reoccurring several times per second as a moth flies upwind. The peak pulses of pheromone are greater than the time-averaged mean concentration by up to a factor of 10, and the differential becomes greater with increasing distance from the point source (Murlis, 1997). Studies with *Cadra cautella*, the almond moth, show that males surge upwind with each turbulent pulse of pheromone they encounter. Five pulses per second of pheromone released into the air result in rapid, straight-line or nearly straight flight with few or shallow zigzags, but less than one pulse per second causes slow movement upwind and wide zigzag excursions (Mafra-Neto and Cardé, 1994, 1995). The conclusion (although not universally accepted as the only explanation of zigzag flight) is that zigzags result from low pheromone filament encounter rate combined with a (presumed) counterturning program. Zigzag flight is not a necessity of optomotor anemotaxis and straight upwind flight can occur when there is a high rate of pheromone filament encounters (Cardé and Mafra-Neto, 1997).

When a moth emerges into clean air devoid of pheromone or host-odor plume, it begins casting (zigzagging) from side to side, with little or no forward movement. Under natural conditions, males may lose the plume due to sudden wind shifts or by flight out of a plume because the upwind direction is not aligned with the lone axis of the plume. The adaptive value of casting is that it may maximize encounters with the plume after it is lost (David et al., 1983). Those who believe that zigzag flight is due to an internal turn generator and widely spaced pheromone filaments believe that casting may simply be a manifestation of zigzag flight without any significant encounters with a pulse of pheromone. Hence, turns are repeatedly generated by the internal generator in the central nervous system (CNS) (Kuenen and Cardé, 1994). Casting or coming to rest on the substrate occurs in very high concentrations of pheromone as well, presumably because the antennal receptors are not receiving pulsed stimuli.

14.9.2 Pheromone Signal Processing

Signal processing has been studied intensively in only a few insects. Males of *M. sexta* have approximately 10^5 sensilla with single walls and pores that house about 3×10^5 receptor neurons. Each sensillum typically contains two sensory neurons, one that is sensitive to (*E,Z*)-10,12-hexadecadienal (EZ-10,12-16:AL) and the other sensitive to (*E,E,Z*)-10,12,14-hexadecatrienal (E,E,Z-10,12,14-16:AL), the two main components of the eight C16 aldehydes in the *M. sexta* female-produced pheromone (Tumlinson et al., 1989). All eight components are important and give the best results in the field (Tumlinson et al., 1994), but detailed neurophysiological data are only available for the two main components. The antennal receptor neurons respond to pheromone aldehydes by opening Na^+, K^+, and Ca^{2+} channels, which are not ligand (pheromone) gated, but mobilized through the second messenger system of G-proteins (Stengl et al., 1992). Axons from olfactory receptors on the antennae pass through the antennal nerve and enter the large antennal lobe (AL) of the deutocerebrum. In moths, the antennal nerve breaks into two branches as it enters the AL (Hansson, 1997). One branch carries axons from non-pheromone olfactory receptors to **glomeruli** in a mechano- and taste-sensitive region of the AL, while the second branch carries axons from pheromone receptors to glomeruli in the **macroglomerular complex** (**MGC**). In the MGC, incoming axons synapse with antennal lobe interneurons that interconnect parts of the AL (local interneurons) or pass to other parts of the brain, including the protocerebrum (projection interneurons). Similarly, organized olfactory glomeruli occur in a wide variety of organisms in which olfaction is very important, including vertebrates and other (non-insect) invertebrates (Ache, 1991; Hildebrand, 1995, 1996; Christensen, 1997).

In all Lepidoptera that have been investigated, the MGC is a sex-specific region found only in the AL of males. In male *M. sexta* (Camazine and Hildebrand, 1979; Rospars and Hildebrand, 1992; Christensen et al., 1993), the large globular glomerulus near the point at which the antennal nerve enters the AL is called the **cumulus** (because of its resemblance to the cumulus cloud shape), and the ring or donut-shaped area beneath it is called the **toroid** (Hansson, 1997). Similar glomeruli varying in shape and size are also known in a number of other male moths, including *Bombyx* *mori, Antheraea polyphemus, Agrotis segetum, Tricoplusia ni, Spodoptera littoralis, Helicoverpa zea*, and *Heliothis virescens*. All have one large glomerulus at the entrance of the antennal nerve into the AL and several smaller satellite glomeruli beneath the large one.

Females in some other insect groups that receive information about male-produced sex pheromones may have a similar structure in the deutocerebrum (Anton and Hansson, 1994), but studies are limited. Glial cells invest the glomeruli and provide protection. In addition to glomeruli, the AL contains lateral and medial groups of cell bodies of the interneuron associated with the glomeruli. The cell bodies of the primary olfactory neurons are located peripherally in the antennae, as is characteristic of sensory neurons in insects.

M. sexta males can detect the pheromone quality, quantity of pheromone, and the frequency of pulses in pheromone plumes. Some pheromone receptor neurons in the male act as pheromone generalists that respond to either of the two aldehyde components or to the total pheromone blend, while other pheromone specialists discriminate between the two aldehydes and respond differently to them (Christensen and Hildebrand, 1990). Thus, the response to either aldehyde or to a blend of both is sent to the MGC as information about the blend (i.e., its quality).

A subset of pheromone-specialist neurons provides further discrimination by responding in opposite ways to the two aldehydes. For example, some of the neurons are stimulated by (*E,Z*)-10,12-hexadecadienal, resulting in excitation; but exposure to (*E,E,Z*)-10,12,14-hexadecatrienal inhibits these same neurons. The opposite scenario can also occur, that is, (*E,E,Z*)-10,12,14-hexadecatrienal may stimulate some neurons while (*E,Z*)-10,12-hexadecadienal inhibits them. A blend of aldehyde components results in a unique mixture of inhibitory and excitatory responses, depending on the mixing of these two input channels. These pheromone specialists respond and recover rapidly enough to detect the natural intermittent pheromone release by the female at frequencies of about ten pheromone plumes or pulsed releases per second.

In some moth species (*A. segetum*, *S. littoralis*, and *M. sexta*) receptor neurons responding to different pheromone components synapse in different glomeruli in the MGC. For example, receptor neurons of male *M. sexta* that respond when one of the sex pheromone components [(*E,Z*)-10,12-hexadecadienal] is blown onto the antenna have terminal arborizations in the toroid, while receptors responding to (*E,E,Z*)-10,12,14-hexadecatrienal, a second component, terminate in the cumulus (Hansson et al., 1991). In some other species (*H. virescens* and *A. polyphemus*) receptor neurons tuned to different pheromone components synapse in the same MGC glomeruli, with some neurons also possibly projecting to a second glomerulus (Hansson, 1997).

Juvenile hormone (JH) has been shown to play a role in nervous system regulation of certain behaviors and nervous system structure in a few insects (Gadenne and Anton, 2000, and references therein). In allatectomized mature male *Agrotis ipsilon*, the proportion of low-threshold AL interneurons sensitive to the female sex pheromone is lower than in intact males. Injection of JH restores (in allatectomized males) or induces (in intact males) a larger proportion of low-threshold interneurons, and increases the specificity of *A. ipsilon* males for its own female-produced blend compared to the very similar blend from a closely related species (Gadenne and Anton, 2000).

A number of examples are known of male moths that have receptor neurons on the antennae sensitive to one or more components that inhibit pheromone response. These inhibitory components may serve as isolating mechanisms for closely related species that share blend components. In all cases known, the receptor neurons that detect inhibitory compounds synapse in a glomerulus that does not receive input from pheromone components (Hansson, 1997).

In the antennal lobe, labeled lines and across-fiber patterning occur together because most receptor neurons make synaptic contacts with many different local interneurons (Christensen et al., 1993). Male and female insects have dimorphic numbers of receptors sensitive to sex pheromone on the antennae. Moths and cockroaches, in particular, have large numbers of receptors on the male antennae that respond mainly to the sex pheromone produced by females. Sex-specific sex pheromone receptors do not occur (or do not occur in large numbers) on the antennae of female moths and cockroaches.

14.10 GEOGRAPHIC AND POPULATION DIFFERENCES AND EVOLUTION OF PHEROMONE BLENDS

Ips pini, the pine engraver, is a major tree-killer in the Great Lakes area of the United States, and in California and Idaho. Males locate a new host tree, bore into the phloem layer, feed, and release the compound 2-methyl-6-methylene-2,7-octadien-4-ol, also known as ipsdienol, as the principal component of its pheromone blend. California and Idaho populations are attracted to *R*-(–)-ipsdienol, and attraction is reduced if *S*-(+)-ipsdienol is added. On the other hand, New York populations are most attracted to a 50:50 blend of *R*-(–)-ipsdienol and *S*-(+)-ipsdienol. Beetles from Wisconsin respond preferentially to a 75:25 mixture of the *S*-(+):*R*(–). The beetles also show regional differences in response to a minor pheromonal component, lanierone (Miller et al., 1997). What could be promoting such population changes?

One possible answer is predation pressure. Two major predators of the pine engraver are the adult beetles *Thanasimus dubius* (family Cleridae) and *Cylistix cylindrica* (family Histeridae). These beetles use the pine engraver pheromone as a kairomone, leading them to their prey, which they aggressively attack and eat. Both predators show strong preference for attraction to a mixture of 25% *S*-(+):75% *R*-(–)-ipsdienol. Populations may be evolving aggregation pheromone blends that are less effective in attracting their predators.

If sex pheromone blends evolve, then there must be concomitant evolution in both sexes. The gland of the producer must produce a different blend, and the receptors of the receiver must respond to the changing blend. The turnip moth, *Agrotis segetum*, appears to provide a good example of co-evolution between chemical pheromone and receptor. There are three populations of the moth: one in France, one in Sweden, and one in the general area of Armenia/Bulgaria. The pheromone is multicomponent, but female moths in the French population produce a large amount of (*Z*)-5-decenyl acetate, and males have many receptors on the antenna that respond electrophysiologically to this component. Moths in the Swedish population produce less (*Z*)-5-decenyl acetate, and males have a smaller population of receptors on their antennae that respond to the compound. Finally, female moths in the Armenian/Bulgarian population produce very little (*Z*)-5-decenyl acetate, and males have very few receptors for it. Clearly, the evolution of changes in female production of (*Z*)-5-decenyl acetate has been correlated with changes in receptor specificity in males. The mechanism driving the changes has not been elucidated (Hansson et al., 1990).

The specificity in pheromone structure and in the receptors that detect the pheromone have been described in a series of closely related scarab beetles (Leal, 1997, 1999) and an illustration of some of the compounds involved is shown in Figure 14.13. Similar evolutionary specificity has occurred in many insect groups.

14.11 HORMONAL CONTROL OF PHEROMONE SYNTHESIS AND RELEASE

Pheromone production in some insects, and perhaps in most if not all Lepidoptera, is under hormonal control (see Cardé and Minks, 1997; Holman et al., 1990, for reviews). Early evidence for possible hormonal control of pheromone production and/or release came from studies on the cockroaches *Byrsotria fumigata* and *Pycnoscelus surinamensis* (Barth, 1964, 1965). Allatectomized cockroaches do not produce sex pheromone, leading to the hypothesis that JH from the corpora allata is probably a pheromonatropic hormone in cockroaches.

A polypeptide hormone that controls the synthesis of the sex pheromone in moths, called **PBAN (pheromone biosynthesis activating neuropeptide)** (Raina and Klun, 1984), was first isolated from the subesophageal ganglion of the moth *Helicoverpa* (formerly *Heliothis*) *zea*, and determined to consist of 33 amino acids (MW = 3900) (Raina et al., 1989). PBAN is now known from several different sources, and that from *H. zea* is currently called **Hez-PBAN**. PBAN from *Bombyx mori* is called **Bom-PBAN-I** (Kitamura et al., 1989; Raina and Gäde, 1988), and a second PBAN isolated

A

B

C

D

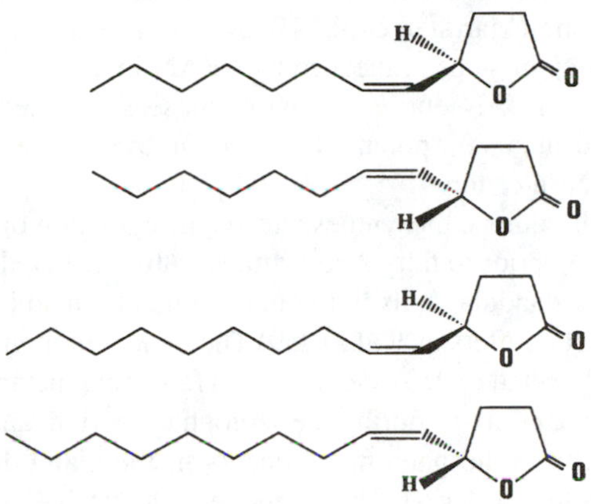

FIGURE 14.13 Illustration of the specificity of sex pheromones and of the receptors that detect the pheromones. Compounds A and C are pheromonal components of the female-produced sex pheromone of a scarab beetle, *Anomala octiescostata*; and C is the pheromone of the female Japanese beetle, *Popillia japonica*. Males of the Japanese beetle can detect and are inhibited by D. Males of *A. octiescostata* cannot detect the unnatural enantiomers (B and D) and are neither stimulated nor inhibited by them in experimental studies. A, (*R*)-buibuilactone; B, (*S*)-buibuilactone; C, (*R*)-japonilure; D, (*S*)-japonilure. Compounds A and C are pheromonal components of the scarab beetle. (Reproduced with permission from Leal, 1999.)

from *B. mori* is called **Bom-PBAN-II** (Kitamura et al., 1990). PBAN isolated from the gypsy moth *Lymantria dispar*, is called **Lyd-PBAN** (Masler et al., 1994). The PBANs belong to a class of peptides called **pyrokinins** and those isolated thus far have 33 or 34 amino acid residues. All PBANs share, with other pyrokinins, a common C-terminal sequence of five amino acids — Phe-X-Pro-Arg-Leu-NH_2 (X can be Gly, Ser, Thr, or Val) — that is required for biological activity. Substitution in the X position is extremely critical to pheromonotropic activity; putting glycine in the X position causes loss of activity, whereas a molecule containing threonine in position X is active (Abernathy et al., 1995).

Pyrokinin-type peptides have several different functions in insects. For example, one has **myotropic** activity in an *in vitro* cockroach hindgut assay (Nachman et al., 1986) and another (Bom-DH) functions as an **egg diapause hormone** in *B. mori*, (Imai et al., 1991). Some PBAN peptides have myotropic activity *in vitro* in the cockroach hindgut assay (Nachman and Holman, 1991). Some degree of cross-reactivity in the different bioassays is common, and is probably due to the characteristic C-terminal peptide sequence. Some of the cockroach and locust pyrokinins also stimulate sex pheromone synthesis in *B. mori* females (Fonagy et al., 1992; Kuniyoshi et al., 1992).

Teal et al. (1996) suggested that there may be a number of neuropeptides involved with pheromone production, and that some of the neuropeptides may have other physiological functions as well. They cite as evidence the fact that genes encoding for PBAN in *H. zea* and in *B. mori* encode for propheromones. In *B. mori* (Kawano et al., 1992), the propheromones give rise to Bom-PBAN, Bom-DH, a peptide with homology to Pss-Pt (a pheromonotropic peptide from the army worm *Pseudaletia separata*), and other peptides.

14.11.1 Mode of Action of PBAN

In the redbanded leafroller, *Argyrotaenia velutinana*, PBAN regulates pheromone biosynthesis by increasing the supply of octadecanoyl and hexadecanoyl fatty acids needed for pheromone biosynthesis, although the exact mechanism by which it does this is not clear (Tang et al., 1989). PBAN regulates the Δ11 desaturase (Roelofs and Jurenka, 1997), an enzyme widely distributed in Lepidoptera. The enzyme introduces a double bond into the pheromone precursor in *Mamestra brassicae* (Bestmann et al., 1989) and *Chrysodeixis chalcites* (Alstein et al., 1989). In two lepidopterans,

B. mori and *Spodoptera littoralis*, PBAN influences the reduction of the fatty acid to the alcohol precursor of the pheromone (Martinez et al., 1990; Arima et al. 1991). Fang et al. (1992) were unable to determine the step or steps influenced by PBAN in *Manduca sexta* females, but they did determine that injection of PBAN during the photophase stimulated pheromone production and that putative fatty acid precursors for potential conversion to the aldehyde pheromone were present but not changed by PBAN injection.

PBAN may control pheromone biosynthesis through regulation of fatty acid biosynthesis in *H. zea,* or it may control a step prior to fatty acid synthesis (Jurenka et al., 1991). The sex pheromone of *H. zea* is secreted from glandular cells in the intersegmental membrane located between the 8th and 9th abdominal segments (Jefferson et al., 1969) The major component (~92%) of the pheromone is (*Z*)-11-hexadecenal (Klun et al., 1979), derived from fatty acid metabolism. The females normally synthesize and release the pheromone during the scotophase period, and only then can it be detected in the glandular tissue; little or no pheromone occurs in the gland during the photophase (Raina et al., 1986; Teal and Tumlinson, 1989). The hormone is detectable in the hemolymph only during the time when the pheromone is being produced (Raina and Klun, 1984). Injection of exogenous PBAN into a female moth can cause pheromone synthesis independent of the photoperiod. As little as 0.06 pmol of Hez-PBAN injected into a female moth stimulated pheromone biosynthesis, and 2 pmol stimulated maximum pheromone synthesis; higher doses produced no additional increase in synthesis.

PBAN is produced in the **subesophageal ganglion** (**SEG**) of *H. zea* and *B. mori*. In studies with *H. zea*, Raina et al. (1989) found that the PBAN is released from the SEG and transported by the hemolymph to the pheromone gland, its target. Teal et al. (1989) have evidence that PBAN may be transported through the ventral nerve cord to the terminal abdominal ganglion (TAG), where it causes the release of some second messenger (unidentified as yet) that acts on the pheromone gland cells. They propose that the target for PBAN is the **terminal abdominal ganglion** (**TAG**), and that nerves from the TAG must be intact to the pheromone gland for extracts of brain-SEG, applied to the TAG, to ultimately elicit pheromone synthesis. Ma and Roelofs (1995) showed in the female European corn borer, *Ostrinia nubilalis*, that PBAN was synthesized in three sets of NSC (neurosecretory cells) in the SEG and released from the corpora cardiaca. Although PBAN immunoreactivity was present throughout the ventral nerve cord, complete removal of the nerve cord did not alter female response to exogenous PBAN. In the gypsy moth, *Lymantria dispar*, PBAN-immunoreactive material can be detected in the SEG and cells in each segmental ganglion (Golubeva et al., 1997), and transection of the ventral nerve cord disrupts pheromone production in females. It thus appears that the origin, transport, and release mechanism of PBAN vary with species. A similar hormone, a sort of generic PBAN, probably functions in most, if not all, Lepidoptera.

14.12 BIOSYNTHESIS OF PHEROMONES

Most insects that have been studied synthesize their pheromonal components from small metabolic pool precursor molecules. Some insects modify precursors obtained in the food. For example, scolytid bark beetles use one or more of the terpenes in the host tree they feed on as a pheromone precursor (Figure 14.14). Males of *Bactrocera dorsalis* fruit flies are strongly attracted to a Hawaiian lei flower, feed on phenylpropanoid compounds on the petals (Figure 14.15), and use the compounds to make a pheromone that attracts female flies. Many moths use mono-unsaturated alcohols, acetates, or aldehydes with 12 to 16 carbons as pheromonal components. The starting material for synthesis of these pheromonal components is a saturated fatty acid synthesized from the acetate pool. Radiolabeled tracer studies show that common fatty acid starting materials are stearic (C18:COOH), palmitic (C16:COOH), myristic (C14:COOH), and lauric (C12:COOH) acids. Depending on the number of carbons in the pheromone component, one or more of these fatty acids is chain-shortened in the pheromone gland by β-oxidation. A Δ11 desaturase enzyme

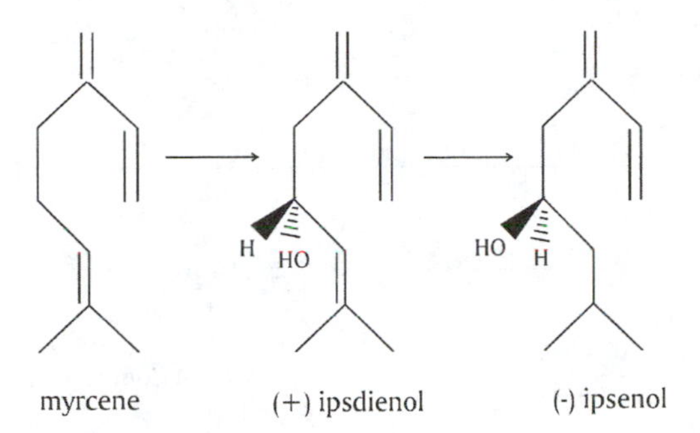

FIGURE 14.14 Some bark beetles use a naturally occurring compound in their host tree phloem resin as a precursor to synthesize pheromonal components. For example, *Ips confusus* uses naturally occurring myrcene to synthesize ipsdienol and ipsenol.

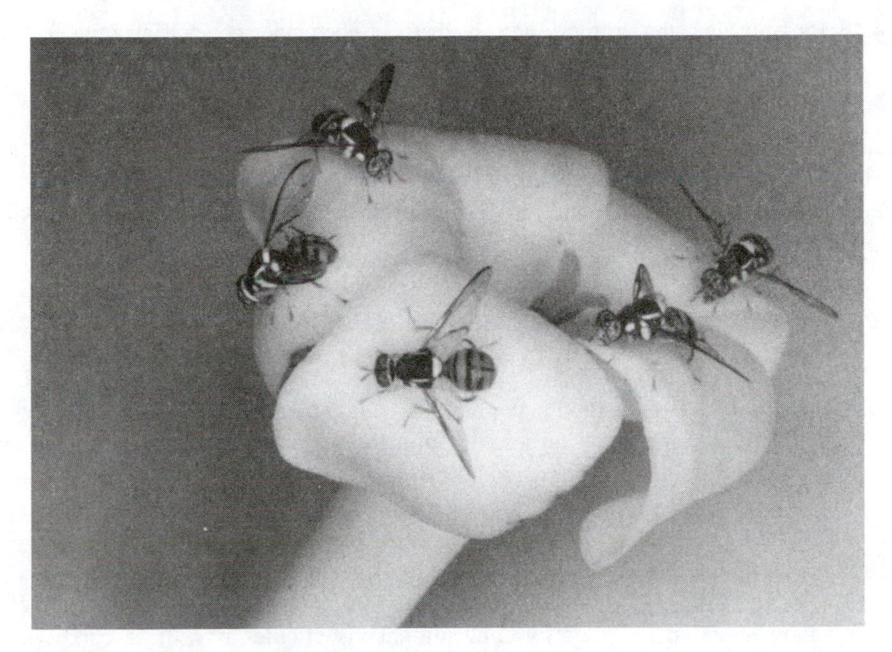

FIGURE 14.15 Males of the Oriental fruit fly, *Bactrocera dorsalis*, collect naturally occurring phenylpropanoid compounds from the petals of the Hawaiian lei flower, *Fagraea berteriana*, and use the compounds to make a male-produced pheromone, *trans*-coniferyl alcohol, that attracts female flies. (Reproduced with permission from Nishida et al., 1997. Photograph courtesy of Ritsuo Nishida and colleagues.)

(Figure 14.16) introduces a double bond, and *E* and *Z* isomers can be produced. The moths reduce mono-unsaturated fatty acid intermediates to alcohols, acetates, or aldehydes to produce their pheromonal components (Roelofs and Bjostad, 1984; Teal and Tumlinson, 1986; Morse and Meighen, 1986; Bjostad et al., 1987; Roelofs and Wolf, 1988).

Some moths use di-unsaturated pheromonal components, and two double bonds can be introduced by the Δ11 desaturase acting before and after chain-shortening. Deuterium-labeled palmitic and myristic acids were used to demonstrate production of (*E*)-11-14 acetate and (*E,E*)-9,11 acetate, major components of the pheromone of *Epiphyas postvittana* (Bellas et al., 1983).

In addition to the Δ11 desaturase, a Δ10 desaturase has been identified in the New Zealand leafroller *Planotortrix excessana*, (Foster and Roelofs, 1988) and a Δ9 desaturase in the brown-headed leafroller *Ctenopseutis obliquana*. Roelofs and Wolf (1988) speculate that in the early evolutionary stages of these Tortricidae, the β-oxidation step was used to shorten oleic and/or

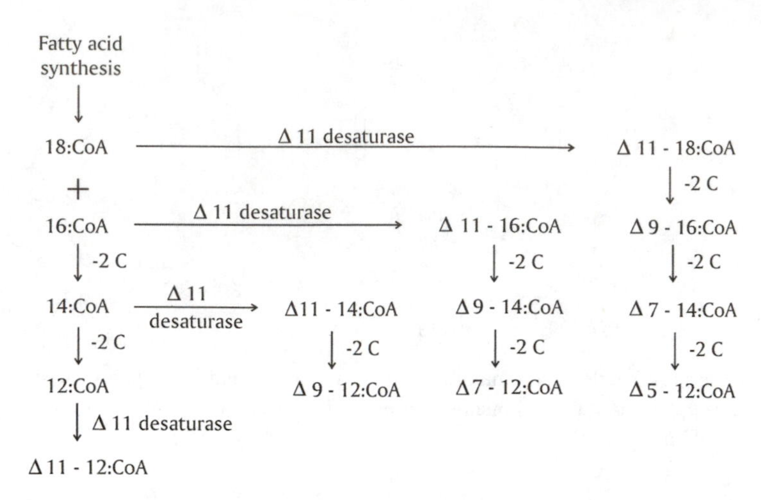

FIGURE 4.16 Biosynthetic pathways for *de novo* synthesis of pheromone components from general precursors in the general metabolic pool by female moths. (Reproduced with permission from Roelofs and Wolf, 1988.)

palmitoleic acids from which shorter-chain pheromone components were synthesized. Later, evolution of the Δ10 and Δ11 desaturases made it possible for the moths to biosynthesize a wide range of unique pheromonal components and may have paved the way for evolution of many different species-specific blends.

14.13 PRACTICAL APPLICATIONS OF PHEROMONES

Much of the stimulus for identification of pheromones has come from the expectation that pheromones would have practical application in population management of insects. The most widespread and successful use of pheromones has been in **monitoring** insect emergence (in the spring or summer) and population build-up. Deployed in traps, pheromones can indicate the presence of pest insects, timing of emergence, flights, and movement into a crop. When large numbers are caught, the decision may be made to apply control measures such as a pesticide.

Direct control with pheromones is also possible in some cases. Direct control includes (1) **mass trapping**, (2) **lure and kill**, and (3) **mating disruption**. Mass trapping and luring large aggregations to trap trees where they can be killed by conventional insecticides have been tried with limited success in population control of bark beetles (Borden, 1997).

Disruption of mating has been one of the successful applications of pheromone to direct insect control of a number of lepidopteran species, although it, like other control procedures, sometimes fails. Current successes and some reasons for failure in Lepidoptera have been reviewed by Cardé and Minks (1995), Sanders (1997), Arn and Louis (1997), Staten et al. (1997), and Suckling and Karg (1997). Mating disruption is currently more costly than conventional insecticide treatments if environmental issues are not considered. Disruption of mating in tortricid moths in Switzerland and Germany costs 2 to 4 times as much as conventional insecticides. Reducing the cost of pheromone, more effective delivery systems (Figure 14.17) that release pheromone in controlled, pulsed amounts and only when the target insect is flying (Shorey and Gerber, 1996; Shorey et al., 1996; Baker et al., 1997), and application of the least amount of pheromone that will work are ways to reduce costs (Arn and Louis, 1997). Disruption of bark beetle aggregations with inhibitors of the aggregation response has enjoyed some success and has a promising future (Borden, 1997).

Successful mating under field conditions involves several behavioral actions, such as long-range orientation (upwind flight in response to pheromone) followed by close-range courtship (possibly also involving pheromone, vision, and other sensory modes such as mechanoreception). Disruption

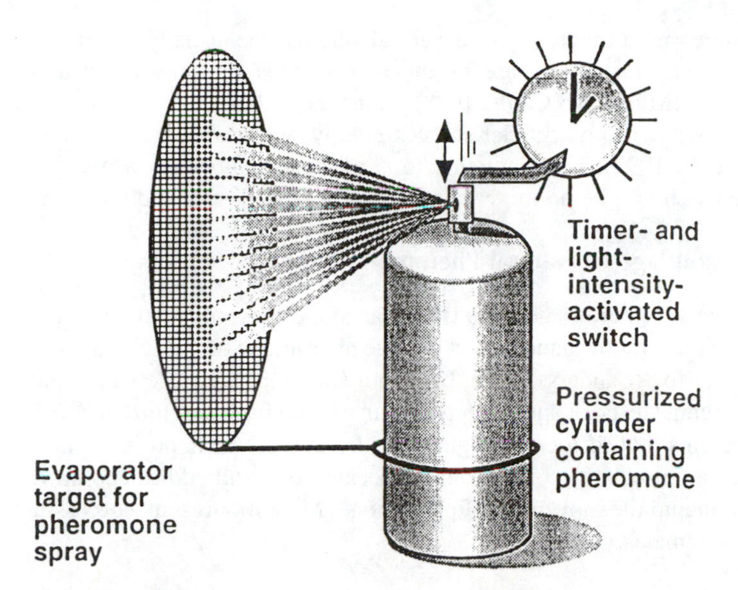

FIGURE 14.17 One design for a puffer device that disperses pheromone from a pressurized canister at timed intervals. (Reproduced with permission from Shorey et al., 1996.)

tactics may target any or combinations of these behaviors. Upwind flight, in theory and practice, has been more amenable to disruption (Sanders, 1997) than close-range behaviors.

14.13.1 Mechanisms Operating in Mating Disruption

How mating disruption occurs is poorly understood and may occur by one or more of the following mechanisms:

1. Sensory fatigue, which can be divided into the component parts of adaptation of receptors at the periphery of the insect (the antenna) and/or habituation in the central nervous system
2. Competition between natural and synthetic sources (false trail-following)
3. Camouflage of the natural pheromone trail
4. Use of blend imbalance and antagonists that stop the response to pheromone

14.13.1.1 Sensory Fatigue

Males of many moth species show failure to successfully find and mate with a female in wind tunnel tests when exposed to high concentrations of pheromone. Males also preconditioned by periodic exposure to pheromone exhibit reduced response and/or fail to find and attempt mating with a female when exposed to pheromone in subsequent wind tunnel tests. Preexposure to pulses of pheromone are more effective than exposure to a constant concentration of pheromone in creating subsequent failure or low response (Kuenen and Baker, 1981).

14.13.1.2 False Trail-Following

Pheromone that is widely dispersed in the environment may cause male moths to follow the synthetic pheromone plume as opposed to the plume from a female. For this mechanism to work, the synthetic pheromone sources must be at least as attractive as the natural female, and there must be sufficient artificial sources to make it improbable that a male will find a female by chance. Presumably, males will spend virtually all their energy and time in following false trails. To make the artificial pheromone sources as attractive as the natural pheromone usually requires that the complete pheromone blend and rate of release by the female must be known and used in the synthetic pheromone sources (Minks and Cardé, 1988). In theory and practice, the method works best at low population density when the artificial sources greatly outnumber the feral females (Webb et al., 1990; Howell et al., 1992). For disruption to work against *Lobesia botrana* in grape vineyards, population density should be no more than four pairs per 10 m^2 (Feldhege et al., 1995).

14.13.1.3 Camouflage of Natural Pheromone Plume

Camouflage of natural pheromone trails is similar to the previous method, but is predicated on the assumption that a male moth cannot detect the true pheromone trail if it is surrounded by pheromone or in a pheromone fog (Sanders, 1997). In such a situation, the male would either remain at rest, as in some wind tunnel tests, or spend all its time in casting back and forth in search of discontinuous pheromone filaments that, if located, signal it to fly upwind. It is probably not possible in a field situation to have a uniform fog of pheromone because of wind eddies and air disturbances due to vegetation breaking up the movement of pheromone. Male moths may successfully find significant numbers of feral females.

14.13.1.4 Pheromone Antagonists and Imbalanced Blends

In a few cases, the use of pheromone antagonists and incomplete blends show promise in field tests. For example, mating disruption of some lepidopterans has been achieved with incomplete blends. Disruption of mating in the navel orangeworm was based on the presence of inhibitors of the natural pheromone (Curtis et al., 1987), and the pea moth, *Cydia nigricana*, is inhibited by a pheromone blend containing attraction inhibitors (Bengtsson et al., 1994). The female tortricid moth *Eupoecilia ambiguella* produces (*Z*)-9-dodecenyl acetate and males are inhibited if a synthetic blend contains more than about 0.1% of the *E* isomer. A technical grade of Z9-12:AC that has been used successfully in mating disruption contains a small percentage of the *E* isomer (Arn and Louis, 1997). Disruption of mating in *E. ambiguella* is the most important method of control of this moth in grape vineyards in Switzerland and parts of Germany (Arn and Louis, 1997). Aggregations of some species of bark beetles can be reduced by the application of several available inhibitors of the aggregation response, and these inhibitors may have good potential for protection of valuable specimen and ornamental trees (Borden, 1997).

Any of the methods in which pheromones are used to control a population have the potential to select for survival and reproduction of those (possibly few) individuals that respond to variable blend ratios, different release rates, or that somehow compensate for possible inhibitors.

REFERENCES

Abernathy, R.L., R.J. Nachman, P.E.A. Teal, O. Yamashita, and J.H. Tumlinson. 1995. Pheromonotropic activity of naturally occurring pyrokinin insect neuropeptides (FXPRLamine) in *Helicoverpa zea. Peptides,* 16:215-219.

Ache, B.W. 1991. Phylogeny of smell and taste, pp. 3-18, in T.V. Getchell (Ed.), *Smell and Taste in Health and Disease*, Raven Press, New York.

Almaas, T.J., and H. Mustaparta. 1990. Pheromone reception in tobacco budworm moth, *Heliothis virescens. J. Chem. Ecol.,* 16:1331-1347.

Altstein, M., M. Harel, and E. Dunkelblum. 1989. Effect of a neuroendocrine factor on sex pheromone biosynthesis in the tomato looper, *Chrysodeixis chalcites* (Lepidoptera: Noctuidae). *Insect Biochem.,* 19:645-649.

Anton, S., and B.S. Hansson. 1994. Central processing of sex pheromone, host odour, and oviposition deterrent information by interneurons in the antennal lobe of the female *Spodoptera littoralis* (Lepidoptera: Noctuidae). *J,. Comp. Neurol.,* 350:199-214.

Arima, R., K. Takahara, T. Kadoshima, F. Numazaki, T. Ando, M. Uchiyama, H. Nagasawa, A. Kitamura, and A. Suzuki. 1991. Hormonal regulation of pheromone biosynthesis in the silkworm moth, *Bombyx mori* (Lepidoptera: Bombycidae). *Appl. Entomol. Zool.,* 26:137-147.

Arn, H., E. Stadler, and S. Rauscher. 1975. The electroantennographic detector — A selective and sensitive tool in the gas chromatographic analysis of insect pheromones. *Z. Naturforsch.,* 30c:722-725.

Arn, H., M. Tóth, and E. Preisner. 1992. *List of Sex Pheromones of Lepidoptera and Related Attractants*, 2nd ed., International Organization for Biological Control, Montfavet, France.

Arn, H., M. Tóth, and E. Preisner. 1999. The Pherolist. Internet edition. http://www.nysaes.cornell.edu/pheronet

Arn, H., and F. Louis. 1997. Mating disruption in European vineyards, pp. 377-382, in R.T. Cardé and A.K. Minks (Eds.), *Insect Pheromone Research. New Directions*, Chapman & Hall, New York, 684 pp.

Baker, T.C., M.A. Willis, and P.L. Phelan. 1984. Optometer anemotaxis polarizes self-steered zigzagging in flying moths. *Physiol. Entomol.,* 9:365-376.

Baker, T.C., and K.F. Haynes. 1989. Field and laboratory electroantennographic measurements of pheromone plume structure correlated with Oriental fruit moth behavior. *Physiol. Entomol.,* 14:1-12.

Baker, T.C., A. Mafra-Neto, T. Dittl, and M.E. Rice. 1997. A novel controlled-release device for disrupting sex pheromone communication in moths, pp. 141-149, in P. Witzgall and H. Arn (Eds.), *Technology Transfer in Mating Disruption*, IOBC wprs Bulletin, Vol. 20(1), Avignon, France.

Barth, R.H., Jr. 1964. The mating behavior of *Brysotria fumigata*. *Behavior,* 23:1-30.

Barth, R.H., Jr. 1965. Endocrine control of a chemical communication system. *Science,* 149:882-883.

Bellas, T.E., R.J. Bartell, and A. Hill. 1983. Identification of two components of the sex pheromone of the moth, *Epiphyas postvittana* (Lepidoptera, Tortricidae). *J. Chem. Ecol.,* 9:503-512.

Bengtsson, M., G. Karg, P.A. Kirsch, J. Löfqvist, A. Sauer, and P. Witzgall. 1994. Mating disruption of pea moth *Cydia nigricana* F. (Lepidoptera: Tortricidae) by a repellent blend of sex pheromone and attraction inhibitors. *J. Chem. Ecol.,* 20:871-887.

Berenbaum, M.R. 1995. The chemistry of defense: theory and practice. *Proc. Natl. Acad. Sci. U.S.A.,* 92:2-8.

Berg, B.G., and H. Mustaparta. 1995. The significance of major pheromone components and interspecific signals as expressed by receptor neurons in the oriental tobacco budworm moth, *Helicoverpa assulta. J. Comp. Physiol. A,* 177:683-694

Berg, B.G., J.H. Tumlinson, and H. Mustaparta. 1995. Chemical communication in heliothine moths. IV. Receptor neuron responses to pheromone compounds and formate analogues in the tobacco budworm moth *Heliothis virescens*. *J. Comp. Physiol. A,* 177:527-534.

Bergman, P., and G. Bergström. 1997. Scent marking, scent origin, and species specificity in male premating behavior of two Scandinavian bumblebees. *J. Chem. Ecol.,* 23:1235-1251.

Bestmann, H.J., M. Herrig, A.B. Attygalle, and M. Hupe. 1989. Regulatory steps in sex pheromone biosynthesis in *Mamestra brassicae* L. (Lepidoptera: Noctuidae). *Experientia,* 45:778-781.

Birch, M.C., G.M. Poppy, and T.C. Baker. 1990. Scents and eversible scent structures of male moths. *Annu. Rev. Entomol.,* 35:25-58.

Blum, M.S. 1996. Semiochemical parsimony in the Arthropoda. *Annu. Rev. Entomol.,* 41:353-374.

Bjostad, L.B., W.A. Wolf, and W.L. Roelofs. 1987. Pheromone biosynthesis in lepidopterans: Desaturation and chain shortening, pp. 77-120, in G.D. Prestwich and G.J. Blomquist (Eds.), *Pheromone Biochemistry,* Academic Press, New York.

Boeckh, J., K.E. Kaissling, and D. Schneider. 1965. Insect olfactory receptors. *Cold Spring Harbor Symposium Quant. Biol.,* 30:263-280.

Boekhoff, I., K. Raming, and H. Breer. 1990a. Pheromone-induced stimulation of inositoltrisphosphate formation in insect antennae is mediated by G-proteins. *J. Comp. Physiol.,* 160:99-103

Boekhoff, I., J. Strotmann, K. Raming, E. Tareilus, and H. Breer. 1990b. Odorant-sensitive phospholipase C in insect antennae. *Cell. Signal.,* 2:49-56.

Boekhoff, I., E. Seifert, S. Goggerle, M. Lindemann, B.W. Kruger, and H. Breer. 1993. Pheromone-induced second-messenger signaling in insect antennae. *Insect Biochem. Mol. Biol.,* 23:757-762.

Borden, J.H. 1997. Disruption of semiochemical-mediated aggregation in bark beetles, pp. 421-438, in R.T. Cardé and A.K. Minks (Eds.), *Insect Pheromone Research. New Directions*, Chapman & Hall, New York, 684 pp.

Bossert, W.H., and E.O. Wilson. 1963. The analysis of olfactory communication among animals. *J. Theoret. Biol.,* 5:443-469.

Breer, H., I. Boekhoff, J. Strotmann, K. Raming, and E. Tareilus. 1990a. Molecular elements of olfactory signal transduction in insect antennae, pp. 77-86, in D. Schild (Ed.), *Information Processing of Chemical Sensory Stimuli in Physiological and Artificial Systems*, Springer-Verlag, Berlin.

Breer, H., I. Boekhoff, and E. Tarelius. 1990b. Rapid kinetics of second messenger formation in olfactory transduction. *Nature,* 345:65-68.

Breer, H. 1997. Molecular mechanisms of pheromone reception in insect antennae, pp. 115-130, in R.T. Cardé and A.K. Minks, *Insect Pheromone Research. New Directions*, Chapman & Hall, New York, 684 pp.

Buck, L., and R. Axel. 1991. A novel multigene family may encode odorant receptors: a molecular basis for odor recognition. *Cell,* 65:175-187.

Butenandt, A., R. Beckmann, D. Stamm, and E. Hecker. 1959. Über den Sexual-Lockstoff des Seidenspinners *Bombyx mori*. Reindarstellung und Konstitutionsermittlung. *Z. Naturforsch.,* 14b:283-284.

Cahn, R.S. 1964a. An introduction to the sequence rule. *J. Chem. Ed.,* 41:116-125.

Cahn, R.S. 1964b. Errata. *J. Chem. Ed.,* 41:508.

Camazine, S.M., and J.G. Hildebrand. 1979. Central projections of antennal sensory neurons in mature and developing *Manduca sexta*. *Soc. Neurosci. Abstr.,* 5:155.

Capinera, J.L. 1980. A trail pheromone from the silk produced by larvae of the range caterpillar *Hemileuca olivae* (Lepidoptera: Saturnidae) and observations on aggregation behavior. *J. Chem. Ecol.,* 3:644-655.

Cardé, R.T., and A.K. Minks. 1995. Control of moth pests by mating disruption: Successes and constraints. *Annu. Rev. Entomol.,* 40:559-585

Cardé, R.T., and A.K. Minks (Eds.). 1997. *Insect Pheromone Research. New Directions*, Chapman & Hall, New York, 684 pp.

Cardé, R., and A. Mafra-Neto. 1997. Mechanisms of flight of male moths to pheromone, pp. 275-290, in R.T. Cardé and A.K. Minks (Eds.), *Insect Pheromone Research. New Directions*, Chapman & Hall, New York, 684 pp.

Christensen, T.A., and J.G. Hildebrand. 1990. Representation of sex-pheromonal information in the insect brain, pp. 142-150, in K.B. Døving (Ed.), *Proceedings X International Symposium on Olfaction and Taste*, Graphic Communication Systems, Oslo.

Christensen, T.A. 1997. Anatomical and physiological diversity in the central processing of sex-pheromone information in different moth species, pp. 184-193, in R.T. Cardé and A.K. Minks (Eds.), *Insect Pheromone Research. New Directions*, Chapman & Hall, New York, 684 pp.

Christensen, T.A., B.R. Waldrop, I.D. Harrow, and J.G. Hildebrand. 1993. Local interneurons and information processing in the olfactory glomeruli of the moth *Manduca sexta*. *J. Comp. Physiol. A,* 173:385-399.

Cossé, A.A., J.L. Todd, J.G. Millar, L.A. Martínez, and T.C. Baker. 1995. Electroantennographic and coupled gas chromatographic-electroantennographic responses of the Mediterranean fruit fly, *Ceratitis capitata*, to male-produced volatiles and mango odor. *J. Chem. Ecol.,* 21:1823-1836.

Curtis, C.F., J.D. Clark, D.A. Carlson, and J.A. Coffelt. 1987. A pheromone mimic: disruption of mating communication in the navel orangeworm, *Amyelois transitella*, with *Z,Z*-1,12,14-heptadecatriene. *Ent. Exp. Appl.,* 44:249-255.

David, C.T., J.S. Kennedy, and A.R. Ludlow. 1983. Finding a sex pheromone source by gypsy moths, *Lymantria dispar*, released in the field. *Nature,* 303:804-806.

Dethier, V.G. 1972. A surfeit of stimuli: A paucity of receptors. *Am. Sci.,* 59:706-715.

Dicke, M., and M.W. Sabelis. 1988. Infochemical terminology: Based on cost-benefit analysis rather than on origin of compounds? *Funct. Ecol.,* 2:131-139.

Du, G., C.-S. Ng, and G.D. Prestwich. 1994. Odorant binding by a pheromone binding protein: Active site mapping by photoaffinity labeling. *Biochemistry,* 33:4812-4819.

Eisner, T., and J. Meinwald. 1995. The chemistry of sexual selection. *Proc. Natl. Acad. Sci. U.S.A.,* 92:50-55.

Fang, N., J.H. Tumlinson, P.E.A. Teal, and H. Oberlander. 1992. Fatty acyl pheromone precursors in the sex pheromone gland of female hornworm moths, *Manduca sexta* (L.). *Insect Biochem. Mol. Biol.,* 22:621-631.

Feldhege, M., F. Louis, and H. Schmutterer. 1995. Untersuchungen über Falterabundanzen des Bekreuzten Traubenwicklers *Lobesia botrana* Schiff. im Weinbau. *Anz. Schädlingsk. Pflanzenschutz Umweltschutz,* 68:85-91.

Fitzgerald, T.D., and J.T. Costa. 1986. Trail-based communication and foraging behavior of young colonies of the forest tent caterpillar *Malacosoma disstria* Hubn. (Lepidoptera: Lasiocampidae). *Ann. Entomol. Soc. Am.,* 79:999-1007.

Fitzgerald, T.D. 1993. Trail and arena marking by caterpillars of *Archips cerasivoranus* (Lepidoptera: Tortricidae). *J. Chem. Ecol.,* 19:1479-1489.

Fitzgerald, T.D., and D.L.A. Underwood. 1998. Communal foraging behavior and recruitment communication in *Gloveria* sp. *J. Chem. Ecol.,* 24:1381-1396.

Fónagy, A., L. Schoofs, S. Matsumoto, A. De Loof, and T. Mitsui. 1992. Functional cross-reactivities of some locustamyotropins and *Bombyx* pheromone biosynthesis activating neuropeptide. *J. Insect Physiol.,* 38:651-657.

Foster, S.P., and W.L. Roelofs. 1988. Sex pheromone biosynthesis in the leafroller moth *Planotortrix excessana* by Δ10 desaturation. *Arch. Insect Biochem. Physiol.,* 8:1-9.

Gadenne, C., and S. Anton. 2000. Central processing of sex pheromone stimuli is differentially regulated by juvenile hormone in a male moth. *J. Insect Physiol.,* 46:1195-1206.

Gary, N. 1962. Chemical mating attractants in the queen honeybee. *Science,* 136:773-774.

Golubeva, E., T.G. Kingan, M.B. Blackburn, E.P. Masler, and A.K. Raina. 1997. The distribution of PBAN (pheromone biosynthesis activating neuropeptide)-like immunoreactivity in the nervous system of the gypsy moth, *Lymantria dispar. Arch. Insect Biochem. Physiol.,* 34:391-408.

Grace, J.K., D.L. Wood, and G.W. Frankie. 1988. Trail-following behavior of *Reticulitermes hesperus* Banks (Isoptera: Rhinotermitidae). *J. Chem. Ecol.,* 14:653-667.

Hansson, B.S., M. Toth, C. Löfstedt, G. Szos, M. Subchev, and J. Lofqvist. 1990. Pheromone variation among eastern European and a western population of the turnip moth *Agrotis segetum. J. Chem. Ecol.,* 16:1611-1622.

Hansson, B.S., T.A. Christensen, and J.G. Hildebrand. 1991. Functionally distinct subdivisions of the macroglomerular complex in the antennal lobe of the male sphinx moth *Manduca sexta. J. Comp. Neurol.,* 312:264-278.

Hansson, B.S. 1997. Antennal lobe projection patterns of pheromone-specific olfactory receptor neurons in moths, pp. 164-183, in R.T. Cardé and A.K. Minks (Eds.), *Insect Pheromone Research. New Directions*, Chapman & Hall, New York, 684 pp.

Hildebrand, J.G. 1995. Analysis of chemical signals by nervous systems. *Proc. Natl. Acad. Sci. U.S.A.,* 92:67-74.

Hildebrand, J.G. 1996. Olfactory control of behavior in moths: Central processing of odor information and the functional significance of olfactory glomeruli. *J. Comp. Physiol. A,* 178:5-19.

Holman, G.M., R.J. Nachman, and M.S. Wright. 1990. Insect neuropeptides. *Annu. Rev. Entomol.,* 35:201-217.

Howell, J.F., A.L. Knight, T.R. Unruh, D.F. Brown, J.L. Krysan, C.R. Sell, and P.A. Kirsch. 1992. Control of codling moth in apple and pear with sex pheromone-mediated mating disruption. *J. Econ. Entomol.,* 85:918-925.

Imai, K., T. Konna, Y. Nakazawa, T. Komiya, M. Isobe, K. Koga, T. Goto, T. Yaginuma, K. Sakakibara, K. Hasegawa, and O. Yamashita. 1991. Isolation and structure of diapause hormone of the silkworm moth, *Bombyx mori. Proc. Jpn. Acad.,* 67:98-101.

Janssen, E., B. Hölldobler, F. Kern, H.-J. Bestmann, and K. Tsuji. 1997. Trail pheromone of myrmicine ant *Pristomyrmex pungens. J. Chem. Ecol.,* 23:1025-1034.

Jefferson, R.N., H.H. Shorey, and R.E. Rubin. 1969. Sex pheromones of noctuid moths. XVI. The morphology of the female sex pheromone glands of eight species. *Ann. Entomol. Soc. Am.,* 61:861-865.

Jurenka, R.A., E. Jacquin, and W.L. Roelofs. 1991. Control of the pheromone biosynthetic pathway in *Helicoverpa zea* by the pheromone biosynthesis activating neuropeptide. *Arch. Insect Biochem. Physiol.,* 17:81-91.

Kaissling, K.-E. 1986. Chemo-electrical transduction in insect olfactory receptors. *Annu. Rev. Neurosci.,* 9:121-145.

Kaissling, K.-E. 1987. Transduction processes in olfactory receptors of moths, pp. 33-43, in J.H. Law (Ed.), *Molecular Entomology,* Alan R. Liss, New York.

Kanzaki, R. 1997. Pheromone processing in the lateral accessory lobes of the moth brain: Flip-flopping signals related to zigzagging upwind walking, pp. 291-303, in R.T. Cardé and A.K. Minks (Eds.), *Insect Pheromone Research. New Directions*, Chapman & Hall, New York, 684 pp.

Kanzaki, R., and T. Shibuya. 1986. Descending protocerebral neurons related to the mating dance of the male silkworm moth. *Brain Res.,* 377:378-382.

Karg, G., and A.E. Sauer. 1995. Spatial distribution of pheromone in vineyards treated for mating disruption of the grape vine moth *Lobesia botrana* measured with electroantennograms. *J. Chem. Ecol.,* 21:1299-1314.

Karlson, P., and M. Luscher. 1959. Pheromones: A new term for a class of biologically active substances. *Nature,* 183:55.

Kawano, T., H. Kataoka, H. Nagasawa, A. Isogai, and A. Suzuki. 1992. cDNA cloning and sequence determination of the pheromone biosynthesis activating neuropeptide of the silkworm, *Bombyx mori. Biochem. Biophys. Res. Commun.,* 189:221-226.

Kehat, M., and E. Dunkelblum. 1990. Behavioral responses of male *Heliothis armigera* (Lepidoptera: Noctuidae) moths in a flight tunnel to combinations of components identified from female sex pheromone glands. *J. Insect Behav.,* 3:75-83.

Kennedy, J.S. 1940. The visual responses of flying mosquitoes. *Proc. Zool. Soc. London A,* 109:221-242.

Kennedy, J.S., and D. Marsh. 1974. Pheromone-regulated anemotaxis in flying moths. *Science,* 184:999-1001.

Kennedy, J.S., A.R. Ludlow, and C.J. Sanders. 1980. Guidance system used in moth sex attraction. *Nature,* 288:475-477.

Kennedy, J.S. 1983. Zigzagging and casting as a programmed response to wind-borne odour: A review. *Physiol. Entomol.,* 8:109-120.

Kern, F., R.W. Klein, E. Janssen, H.-J. Bestmann, A.B. Attygalle, D. Schäfer, and U. Maschwitz. 1997. Mullein, a trail pheromone component of the ant *Lasius fuliginosus. J. Chem. Ecol.,* 23:779-792.

Kitamura, A., H. Nagasawa, H. Kataoka, T. Inoue, S. Matsumoto, T. Ando, and A. Suzuki. 1989. Amino acid sequence of pheromone-biosynthesis-activating neuropeptide (PBAN) of the silkworm, *Bombyx mori. Biochem. Biophys. Res. Commun.,* 163:520-526.

Kitamura, A., H. Nagasawa, H. Kataoka, T. Ando, and A. Suzuki. 1990. Amino acid sequence of pheromone-biosynthesis-activating neuropeptide-II (PBAN-II) of the silkworm, *Bombyx mori. Agric. Biol. Chem.,* 54:2495-2497.

Klein, U. 1987. Sensillum-lymph proteins from antennal olfactory hairs of the moth *Antheraea polyphemus* (Saturniidae). *Insect Biochem.,* 17:1193-1204.

Klun, J.A., J.R. Plimmer, B.A. Bierl-Leonhardt, A.N. Sparks, and O.L. Chapman. 1979. Trace chemicals: The essence of sexual communication systems in *Heliothis* species. *Science,* 204:1328-1330.

Kramer, E. 1986. Turbulent diffusion and pheromone-triggered anemotaxis, pp. 59-67, in T.L Payne, M.C. Birch, and C.E.J. Kennedy (Eds.), *Mechanisms in Insect Olfaction*, Clarendon Press, Oxford.

Kramer, E. 1997. A tentative intercausal nexus and its computer model on insect orientation in windborne pheromone plumes, pp. 232-247, in R.T. Cardé and A.K. Minks (Eds.), *Insect Pheromone Research. New Directions*, Chapman & Hall, New York, 684 pp.

Kuenen, L.P.S., and T.C. Baker. 1981. Habituation vs. sensory adaptation as the cause of reduced attraction following pulsed and constant sex pheromone preexposure by *Trichoplusia ni. J. Insect Physiol.,* 27:721-726.

Kuenen, L.P.S., and R.T. Cardé. 1994. Strategies for recontacting a lost pheromone plume: casting and upwind flight in the male gypsy moth. *Physiol. Entomol.,* 19:15-29.

Kuniyoshi, H.R.A. Nagasawa, A. Suzuki, R.J. Nachman, and G.M. Holman. 1992. Cross-reactivity between pheromone biosynthesis activating neuropeptide (PBAN) and nyotropic pyrokinin insect peptides. *Biosci. Biotech. Biochem.,* 56:167-168.

Leal, W.S., J.I.L. Moura, J.M.S. Bento, E.F. Vilela, and P.B. Pereira. 1997. Electrophysiological and behavioral evidence for a sex pheromone in the wasp *Bephratelloides pomorum* congeneric to a parthenogenetic species. *J. Chem. Ecol.,* 23:1281-1289.

Leal, W.S. 1997. Evolution of sex pheromone communication in plant-feeding scarab beetles, pp. 505-513, in R.T. Cardé and A.K. Minks (Eds.), *Insect Pheromone Research. New Directions*, Chapman & Hall, New York, 684 pp.

Leal, W.S. 1999. Enantiomeric anosmia in scarab beetles. *J. Chem. Ecol.,* 25:1055-1066.

Linn, C.E., Jr., M.G. Campbell, and W.L. Roelofs. 1987. Pheromone components and active spaces: What do moths smell and where do they smell it? *Science,* 237:650-652.

Loudon, C., and M.A.R. Koehl. 2000. Sniffing by a silkworm moth: Wing fanning enhances air penetration through the pheromone interception by antennae. *J. Exp. Biol.,* 203:2977-2990.

Ma, P.W.K., and W.L. Roelofs. 1995. Sites of synthesis and release of PBAN-like factor in the female European corn borer, *Ostrinia nubilalis. J. Insect Physiol.,* 41:339-350.

Mafra-Neto, A., and R.T. Cardé. 1994. Fine-scale structure of pheromone plumes modulates upwind orientation of flying moths. *Nature,* 369:142.

Mafra-Neto, A., and R.T. Cardé. 1995. Effect of the fine-scale structure of pheromone plumes: Pulse frequency modulates activation and upwind flight of almond moth males. *Physiol. Entomol.,* 20:229-242.

Marion-Poll, F., and T.R. Tobin. 1992. Temporal coding of pheromone pulses and trains in *Manduca sexta. J. Comp. Physiol.,* 171:505-512.

Marsh, D., J.S. Kennedy, and A.R. Ludlow. 1978. An analysis of anemotactic zigzagging flight in male moths stimulated by pheromone. *Physiol. Entomol.,* 3:221-240.

Martinez, T., G. Fabrias, and F. Camps. 1990. Sex pheromone biosynthetic pathway in *Spodoptera littoralis* and its activation by a neurohormone. *J. Biol. Chem.,* 265:1381-1387.

Masler, E.P., A.K. Raina, R.M. Wagner, and J.P. Kochansky. 1994. Isolation and identification of a pheromonotropic neuropeptide from the brain-subesophageal ganglion complex of *Lymantria dispar*: A new member of the PBAN family. *Insect Biochem. Mol. Biol.,* 24:829-836.

Masson, C., and H. Mustaparta. 1990. Chemical information processing in the olfactory system of insects. *Physiol. Rev.,* 70:199-245.

Matsumura, F., H.C. Coppel, and A. Tai. 1968. Isolation and identification of termite trail-following pheromone. *Nature,* 219:963-964.

Mayer, M.S., and J.R. McLaughlin. 1992. Discrimination of female sex pheromone by male *Trichoplusia ni* (Hubner). *Chem. Senses,* 16:699-710.

Miller, D.R., K.E. Gibson, K.F. Raffa, S.J. Seybold, S.A. Teale, and D.L. Wood. 1997. Geographic variation in response of pine engraver, *Ips pini*, and associated species to pheromone, Lanierone. *J. Chem. Ecol.,* 23:2013-2031.

Minks, A.K., and R.T. Cardé. 1988. Disruption of pheromone communication in moths: Is the natural blend really more efficacious? *Ent. Exp. Appl.,* 49:25-36.

Mori, K. 1984. The significance of chirality: Methods for determining absolute configuration and optical purity of pheromones and related compounds, pp. 323-370, in H.E. Hummel and T.A. Miller (Eds.), *Techniques in Pheromone Research*, Springer-Verlag, New York.

Morse, D., and E.A. Meighen. 1986. Pheromone biosynthesis and role of functional groups in pheromone specificity. *J. Chem. Ecol.,* 12:335-351.

Murlis, J. 1997. Odor plumes and the signal they provide, pp. 221-231, in R.T. Cardé and A.K. Minks (Eds.), *Insect Pheromone Research. New Directions*, Chapman & Hall, New York, 684 pp.

Murlis, J., and C.D. Jones. 1981. Fine scale structure of odour plumes in relation to insect orientation to distant pheromone and other attractant sources. *Physiol. Entomol.,* 6:71-86.

Murlis, J., J.S. Elkinton, and R.T. Cardé. 1992. Odor plumes and how insects use them. *Annu. Rev. Entomol.,* 37:505-532.

Mustaparta, H. 1997. Olfactory coding mechanisms for pheromone and interspecific signal information in related moth species, pp. 144-163, in R.T. Cardé and A.K. Minks (Eds.), *Insect Pheromone Research. New Directions*, Chapman & Hall, New York, 684 pp.

Nachman, R.J., G.M. Holman, and B.J. Cook. 1986. Active fragments and analogs of the insect neuropeptide leucopyrokinin: Structure-activity studies. *Biochem. Biophys. Res. Commun.,* 137:936-942.

Nachman, R.J., and G.M. Holman. 1991. Myotropic insect neuropeptide families from the cockroach *Leucophaea maderae*: Structure-activity relationships, pp. 194-214, in J.J. Menn, T.J. Kelly, and E.P. Masler (Eds.), *Insect Neuropeptides: Chemistry, Biology and Action*, American Chemical Society, Washington, D.C.

Ngai, J., A. Chess, M.M. Dowling, N. Necles, E.R. Macgno, and R. Axel. 1993. Coding of olfactory information: topography of odorant expression in the catfish olfactory epithelium. *Cell,* 72:667-680.

Nishida, R., S. Schulz, C.S. Kim, H. Fukami, Y. Kuwahara, K. Honda, and N. Hayashi. 1996. Male sex pheromone of a giant danaine butterfly, *Idea leuconoe. J. Chem. Ecol.,* 22:949-972.

Nishida, R., T.E. Shelly, and K.Y. Kaneshiro. 1997. Acquisition of female-attracting fragrance by males of Oriental fruit fly from a Hawaiian lei flower, *Fagraea berteriana. J. Chem. Ecol.,* 23:2275-2285.

Oehlschalager, A.C., G.G.S. King, H.D. Pierce, Jr., A.M. Pierce, K.N. Slessor, J.G. Millar, and J.H. Borden. 1987. Chirality of macrolide pheromones of grain beetles in the genera *Oryzaephilus* and *Cryptolestes* and its implications for species specificity. *J. Chem. Ecol.,* 13:1543-1554.

Preiss, R., and E. Kramer. 1986a. Pheromone-induced anemotaxis in simulated free flight, pp. 69-79, in T.L. Payne, M.C. Birch, and C.E.J. Kennedy (Eds.), *Mechanisms in Insect Olfaction*, Clarendon Press, Oxford.

Preiss, R., and E. Kramer. 1986b. Mechanism of pheromone orientation in flying moths. *Naturwissenschaften,* 73:555-557.

Prestwich, G.D. 1993a. Bacterial expression and photoaffinity labeling of a pheromone binding protein. *Protein Sci.,* 2:420-428.

Prestwich, G.D. 1993b. Chemical studies of pheromone receptors in insects. *Arch. Insect Biochem. Physiol.,* 22:75-83.

Prestwich, G.D., and G. Du. 1997. Pheromone-binding proteins, pheromone recognition, and signal transduction in moth olfaction, pp. 131-143, in R.T. Cardé and A.K. Minks (Eds.), *Insect Pheromone Research. New Directions*, Chapman & Hall, New York, 684 pp.

Raina, A.K., and J.A. Klun. 1984. Brain factor control of sex pheromone production in the female corn earworm moth. *Science,* 225:531-533.

Raina, A.K., J.A. Klun, and E.A. Stadelbacker. 1986. Diel periodicity and effect of age and mating on female sex pheromone titer in Heliothis zea (Lepidoptera: Noctuidae). *Ann. Entomol. Soc. Am.,* 79:128-131.

Raina, A.K., H. Jaffe, T.G. Kempe, P. Keim, R.W. Blacher, H.M. Fales, C.T. Riley, J.A. Klun, R.L. Ridgway, and D.K. Hayes. 1989. Identification of a neuropeptide hormone that regulates sex pheromone production in female moths. *Science,* 244:796-798.

Raina, A.K., and G. Gäde. 1988. Insect peptide nomenclature. *Insect Biochem.,* 18:785-787.

Reinhard, J., and M. Kaib. 1995. Interaction of pheromones during food exploitation by the termite *Schedorhinotermes lamanianus*. *Physiol. Entomol.,* 20:266-272.

Regnier, F.E. 1971. Semiochemicals — structure and function. *Biology of Reproduction,* 4(3):309-326.

Ressler, K.J., S.L. Sullivan, and L.B. Buck. 1993. A zonal organization of odorant receptor gene expression in the olfactory epithelium. *Cell,* 73:597-609.

Roelofs, W.L. 1984. Electroantennogram assays: rapid and convenient screening procedures for pheromones, pp. 131-159, in H.E. Hummel and T.A. Miller (Eds.), *Techniques in Pheromone Research*, Springer-Verlag, New York.

Roelofs, W.L. 1995. Chemistry of sex attraction. *Proc. Natl. Acad. Sci. U.S.A.,* 92:44-49.

Roelofs, W.L., and L.B. Bjostad. 1984. Biosynthesis of lepidopteran pheromones. *Bioorg. Chem.,* 12:279-298.

Roelofs, W.L., and W.A. Wolf. 1988. Pheromone biosynthesis in Lepidoptera. *J. Chem. Ecol.,* 14:2019-2031.

Roelofs, W.L., and R.A. Jurenka. 1997. Interaction of PBAN with biosynthetic enzymes, pp. 42-45, in R.T. Cardé and A.K. Minks, *Insect Pheromone Research. New Directions*, Chapman & Hall, New York, 684 pp.

Rospars, J.P., and J.G. Hildebrand. 1992. Anatomical identification of glomeruli in the antennal lobes of the male sphinx moth *Manduca sexta*. *Cell Tissue Res.,* 270:205-227.

Rumbo, E.R., D.M. Suckling, and G. Karg. 1995. Measurement of airborne pheromone concentrations using electroantennograms: Interactions between environmental volatiles and pheromone. *J. Insect Physiol.,* 41:465-471.

Rybczynski, R., J. Reagan, and M. Lerner. 1989. A pheromone-degrading aldehyde oxidase in the antennae of the moth *Manduca sexta*. *J. Neurosci.,* 9:1341-1353.

Sanders, C.J. 1997. Mechanisms of mating disruption in moths, pp. 333-346, in R.T. Cardé and A.K. Minks (Eds.), *Insect Pheromone Research. New Directions*, Chapman & Hall, New York.

Sauer, A.E., G. Karg, U.T. Koch, J.J. De Kramer, and R. Milli. 1992. A portable EAG system for the measurement of pheromone concentrations in the field. *Chem. Senses,* 17:543-553.

Schneider, D. 1957. Electrophysiologische Untersuchungen von Chemo- und Mechanorezeptroren der Antenne des Seidenspinners *Bombyx mori* L. *Z. Vergl. Physiol.,* 40:8-41.

Schneider, D. 1974. The sex-attractant receptor of moths. *Sci. Am.*, July 1974, pp. 28-35.

Shorey, H.H. 1973. Behavioral responses to insect pheromones. *Annu. Rev. Entomol.,* 18:349-380.

Shorey, H.H., and R.G. Gerber. 1996. Use of puffers for disruption of pheromone communication of codling moths (Lepidoptera: Tortricidae) in walnut orchards. *Environ. Entomol.,* 25:1401-1405.

Shorey, H.H., C.B. Sisk, and R.G. Gerber. 1996. Widely separated release sites for disruption of sex pheromone communication in two species of Lepidoptera. *Environ. Entomol.,* 25:446-451.

Silverstein, R.M. 1988. Chirality in insect communication. *J. Chem. Ecol,* 14:1981-2004.

Staten, R.R., O. El-Lissy, and L. Antilla. 1997. Successful area-wide program to control pink bollworm by mating disruption, pp. 383-396, in R.T. Cardé and A.K. Minks (Eds.), *Insect Pheromone Research. New Directions*, Chapman & Hall, New York.

Steinbrecht, R.A., M. Ozaki, and G. Ziegelberger. 1992. Immunocytochemical localization of pheromone-binding protein in moth antennae. *Cell Tissue Res.*, 270:287-302.

Stengl, M., H. Hatt, and H. Breer. 1992. Peripheral processes in insect olfaction. *Annu. Rev. Physiol.*, 54:665-681.

Suckling, D.M., and G. Karg. 1997. Mating disruption of the light brown apple moth: Portable electroantennogram equipment and other aspects, pp. 411-420, in R.T. Cardé and A.K. Minks (Eds.), *Insect Pheromone Research. New Directions*, Chapman & Hall, New York.

Svensson, B.G., and G. Bergström. 1977. Volatile marking secretions from the labial gland of north European *Pyrobombus* D.T. males (Hymenoptera, Apidae). *Insectes Soc.*, 24:213-224.

Tang, J.D., R.E. Carlton, R.A. Jurenka, W.A. Wolf, P.L. Phelan, L. Sreng, and W.L. Roelofs. 1989. Regulation of pheromone biosynthesis by a brain hormone in two moth species. *Proc. Natl. Acad. Sci. U.S.A.*, 86:1806-1810.

Teal, P.E.A., and J.H. Tumlinson. 1986. Terminal steps in pheromone biosynthesis by *Heliothis virescens* and *H. zea*. *J. Chem. Ecol.*, 12:353-366.

Teal, P.E.A., and J.H. Tumlinson. 1989. Neuronal induction of pheromone biosynthesis by *Heliothis zea* (Broddie) during the photophase. *Can. Entomol.*, 121:43-46.

Teal, P.E.A., J.H. Tumlinson, and H. Oberlander. 1989. Neural regulation of sex pheromone biosynthesis in *Heliothis* moths. *Proc. Natl. Acad. Sci. U.S.A.*, 86:2488-2492.

Teal, P.E.A., R.L. Abernathy, R.J. Nachman, N. Fang, J.A. Meredith, and J.H. Tumlinson. 1996. Pheromone biosynthesis activating neuropeptides: functions and chemistry. *Peptides*, 17:337-344.

Todd, J.L., and T.C. Baker. 1997. The cutting edge of insect olfaction. *Am. Entomol.*, Fall 1997:174-182.

Tumlinson, J.H., M.G. Glein, R.E. Doolittle, T.L. Ladd, and A.T. Proveaux. 1977. Identification of the female Japanese beetle sex pheromone: Inhibition of male response by an enantiomer. *Science*, 197:789-792.

Tumlinson, J.H., M.M. Brennan, R.E. Doolittle, E.R. Mitchell, A. Brabham, B.E. Mazomenos, A.H. Baumhover, and D.M. Jackson. 1989. *Arch. Insect Biochem. Physiol.*, 10:255-271.

Tumlinson, J.H., E.R. Mitchell, R.E. Doolittle, and D.M. Jackson. 1994. Field tests of synthetic *Manduca sexta* sex pheromone. *J. Chem. Ecol.*, 20:579-591.

Vet, L.E.M., and M. Dicke. 1992. Ecology of infochemical use by natural enemies in a tritrophic context. *Annu. Rev. Entomol.*, 37:141-172.

Vickers, N.J., T.A. Christensen, H. Mustaparta, and T.C. Baker. 1991. Chemical communication in heliothine moths. III. Flight behavior of male *Helicoverpa zea* and *Heliothis virescens* in response to varying ratios of intra- and interspecific sex pheromone components. *J. Comp. Physiol. A*, 169:275-280.

Vogt, R.G. 1987. The molecular basis of pheromone reception: its influence on behavior, pp. 385-431, in G.D. Prestwich and G.L. Blomquist (Eds.), *Pheromone Biochemistry*, Academic Press, Orlando, FL.

Vogt, R.G., and L.M. Riddiford. 1981. Pheromone binding and activation by moth antennae. *Nature*, 293:161-163.

Vogt, R.G., L.M. Riddiford, and G.D. Prestwich. 1985. Kinetic properties of a pheromone degrading enzyme: The sensillar esterase of *Antheraea polyphemus*. *Proc. Natl. Acad. Sci. U.S.A.*, 82:8827-8831.

Vogt, R.G., A.C. Kohne, J.T. Dubnau, and G.D. Prestwich. 1989. Expression of pheromone binding proteins during antennal development in the gypsy moth *Lymantria dispar*. *J. Neurosci.*, 9:3332-3346.

Vogt, R.G., J.D. Prestwich, and M.R. Lerner. 1991a. Odorant-binding-protein subfamilies associate with distinct classes of olfactory receptor neurons in insects. *J. Neurobiol.*, 22:74-84.

Vogt, R.G., R. Rybcynski, and M.R. Lerner. 1991b. Molecular cloning and sequencing of general odorant-binding proteins GOBP1 and GOPB2 from the tobacco hawk moth *Manduca sexta*: comparison with other insect OBPs and their signal peptides. *J. Neurosci.*, 11:2972-2984.

Wadhams, L.J. 1982. Coupled gas chromatography-single cell recording: A new technique for use in the analysis of insect pheromone. *Z. Naturforsch.*, 37c:947-952.

Webb, R.E., B.A. Leonhardt, J.R. Plimmer, K.M. Tatman, V.K. Boyd, D.L. Cohen, C.P. Schwalbe, and L.W. Douglass. 1990. Effect of racemic disparlure released from grids of plastic ropes on mating success of gypsy moth (Lepidoptera: Lymantriidae) as influenced by dose and population density. *J. Econ. Entomol.*, 83:910-916.

Whittaker, R.H., and P.P. Feeny. 1971. Allelochemics: Chemical interactions between species. *Science*, 171:757-770.

Wilson, E.O. 1962. Chemical communication among workers of the fire ant *Solenopsis saevissima* (Fr. Smith). 1. The organization of mass-foraging. *Anim. Behav.*, 10:134-147.

Wilson, E.O., and W.H. Bossert. 1963. Chemical communication among animals. *Recent Progr. Hormone Res.,* 19:673-716.

Witzgall, P. 1997. Modulation of pheromone-mediated flight in male moths, pp. 265-290, in R.T. Cardé and A.K. Minks (Eds.), *Insect Pheromone Research. New Directions*, Chapman & Hall, New York, 684 pp.

Witzgall, P., and H. Arn. 1990. Direct measurement of the flight behavior of male moths to calling females and synthetic sex pheromones. *Z. Naturforsch.,* 45c:1067-1069.

Witzgall, P., and H. Arn. 1991. Recording flight tracks of *Lobesia botrana* in the wind tunnel, pp. 187-193, in I. Hrdy (Ed.), *Insect Chemical Ecology,* SPB Academic Publishers, The Hague.

Wright, R.H. 1958. The olfactory guidance of flying insects. *Can. Entomologist,* 90:81-89.

Zhu, J., N. Ryrholm, H. Ljungberg, B.S. Hansson, D. Hall, D. Reed, and C. Lofstedt. 1996. Olefinic acetates, Δ-9,11-14:OAc and Δ-7,9-12:OAc used as sex pheromone components in three geometrid moths, *Idaea aversata, I. straminata,* and *I. biselata* (Geometridae, Lepidoptera). *J. Chem. Ecol.,* 22:1505-1526.

Ziegelberger, G., M.J. Van den Berg, K.-E. Kaissling, S. Klump, and J.E. Schultz. 1990. Cyclic GMP levels and guanylate cyclase activity in pheromone-sensitive antennae of the silkmoths *Antheraea polyphemus* and *Bombyx mori. J. Neurosci.,* 10:1217-1225.

15 Reproduction

CONTENTS

PREVIEW

Insects display great diversity in modes of reproduction. Most insects reproduce in the adult stage by laying eggs, but a few produce gametes and reproduce during an immature stage, a process known as paedogenesis. Some insects (e.g., aphids, some flies) give birth to live young. Although sexual reproduction by union of male and female gametes is typical, certain insects reproduce some or all of the time by laying unfertilized eggs (parthenogenesis). Oocytes accumulate yolk and cytoplasm and develop in egg chambers or follicles in the ovarioles within the ovary. Ovarioles may contain nurse cells (meroistic ovarioles) in several different configurations relative to the developing oocyte, or there may be no special nurse cells (panoistic ovarioles). Nurse cells provide nutrients and gene products for the developing oocyte. Presumably, similar components are provided by other cells in panoistic ovarioles. Developing oocytes in most females incorporate into the yolk large glycolipoproteins called vitellogenins that are synthesized in fat body cells and transported to the ovaries by hemolymph. Higher Diptera evolved small protein-lipid complexes (called yolk proteins) for incorporation in the yolk. Yolk proteins are synthesized in both fat body and follicular epithelial cells under the influence of a different set of genes than the vitellogenins of other insects. In most insects, several hormones regulate oogenesis, synthesis of yolk proteins, and additional aspects of reproduction such as pheromone production, mating, and oviposition. The hormones controlling these processes are not the same in all groups of insects. When maturation of the oocyte is nearly complete, a vitelline membrane is secreted, followed by secretion of the egg shell, or chorion. Eggs are fertilized after the chorion is put on as the egg passes down the median oviduct and past the opening to the spermatheca where sperm are stored. Sperm enter through the micropyle, a twisting channel through the chorion. Male insects produce sperm in the testes. Males typically transfer sperm to the female tract by insertion of the aedeagus into the reproductive tract of the female, or by incorporating the sperm into a spermatophore, a protein sac, that may be inserted into the opening of the female's reproductive tract or formed there as mating occurs. A spermatophore is usually viewed as an investment of protein nutrition by the male to the next generation. There are at least three chromosomal systems for gender determination in insects; in some insects, the male is heterogametic, while in other species the female is heterogametic. Hymenoptera and some coccids (Homoptera) have haploid males and diploid females. The ratio of sex chromosome to autosomes and/or the presence of sex-determining genes are two mechanisms known to determine gender in most insects.

15.1 INTRODUCTION

With the diversity in insect life histories and ecology, it should not be surprising to find great diversity in the details of mating behavior, endocrine regulation of egg development and pheromone production, and the physical structures associated with reproduction. This chapter primarily deals with internal reproductive structures, and with the physiology of gamete production and yolk deposition in the developing eggs. Sex pheromones involved with mate attraction have been dealt with in Chapter 14.

15.2 FEMALE REPRODUCTIVE SYSTEM

15.2.1 Structure of Ovaries

An excellent review of the functional and comparative anatomy of the ovarian system has been presented by Bonhag (1958), the principal source for the information presented here and for details on model types of ovarioles below. The general anatomy of the female internal organs, consisting of the **paired ovaries**, **lateral oviducts**, **common oviduct**, **spermatheca**, and **accessory glands**, is illustrated in Figure 15.1. The internal reproductive structures of both sexes are located dorsally to the alimentary tract. Each ovary consists of one to many **ovarioles**, with each ovariole containing a string of "egg-shaped" chambers called **follicles**. A follicle is separated from the preceding one by a constriction and a bit of interfollicular tissue. Each ovariole is typically enclosed in an epithelial sheath of variable structure in different insects. Striated muscle fibers are often associated with this outer epithelial sheath.

The number of ovarioles per ovary varies in different species, and even to some extent within a species. Viviparous Diptera have one (tsetse fly *Glossina* spp.) or two (*Melophagus* spp. and *Hippoboscа* spp.); the American cockroach *Periplaneta americana* has eight; *Drosophila melanogaster* has from 10 to 30; the blowfly *Calliphora erythrocephala* has about 100; and termite queens (Isoptera) may have up to 2000. Just one ovary containing one ovariole occurs in some aphids, and Collembola have sac-like ovaries that do not contain ovarioles.

Insect ovaries and the ovarioles are classified into two main types, depending on whether there are nurse cells associated with the developing egg (**meroistic ovaries**) or no nurse cells (**panoistic ovaries**) (Figure 15.2). Meroistic ovaries can also be divided into two types: **polytrophic** and **telotrophic**. Panoistic ovaries are considered to be the earliest type to evolve, and occur in present-day Thysanura, Odonata, Plecoptera, Dictyoptera, and Isoptera. They evolved secondarily in

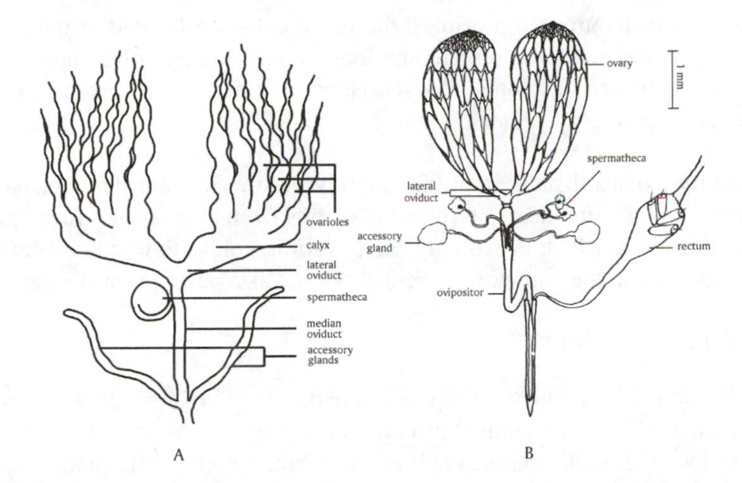

FIGURE 15.1 A: The internal reproductive structures of a female milkweed bug *Oncopeltus fasciatus*. B: Internal structures of a female Caribbean fruit fly *Anastrepha suspensa*. There are three spermathecae in most of the tephritid fruit flies. (A, by the author; B, modified from Dodson, 1978.)

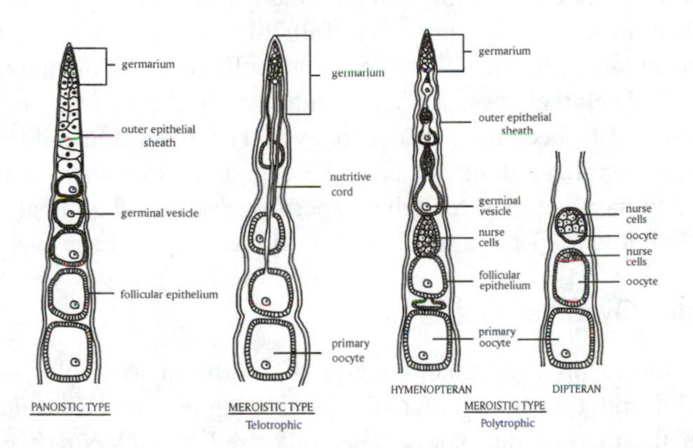

FIGURE 15.2 Major types of ovary structure in insects. The panoistic ovary is typical of Orthoptera and Dictyoptera with no nurse cells. Meroistic telotrophic ovaries have nurse cells in the germarial region and cytoplasmic strands extend to the developing oocytes. Coleoptera and Hemiptera have telotrophic ovaries. Meriostic ovaries may be polytrophic, as for example in Hymenoptera and higher Diptera. In polytrophic ovaries, the nurse cells occur in an adjacent follicle (Hymenoptera) or in the follicle with the developing oocyte (higher Diptera). In all cases, the nurse cells pass nutrients and gene products (mRNAs) to the developing oocyte.

Ephemeroptera, Orthoptera, and Siphonaptera (Bonhag, 1958). Meroistic ovaries occur in most of the Holometabola (except Siphonaptera, noted above), and in Hemiptera, Dermaptera, Psocoptera, Anoplura, and Mallophaga among the Hemimetabola. Meroistic ovaries contain nurse cells arranged in one of two ways:

1. In polytrophic ovarioles, each oocyte is closely associated with nurse cells in its follicle. Most of the Holometabola, some Coleoptera (Adephaga), Dermaptera, Psocoptera, Anoplura, and Mallophaga have polytrophic ovarioles. The nurse cells may be present in the follicle containing the developing oocyte as in higher Diptera, or they may occupy a separate follicle adjacent to that of the developing oocyte, as in the honeybee *Apis mellifera* and other Hymenoptera.

2. In acrotrophic or telotrophic ovarioles, the nurse cells are located at the distal apex of the ovariole, in the germarial region, and long connecting nutritive chords extend from the nurse cells to each developing oocyte. Hemiptera and some Coleoptera (polyphaga) have the telotrophic arrangement (Bonhag, 1955, 1958).

Each ovariole terminates distally in a thin, slightly elastic filament that attaches to the dorsal diaphragm or dorsal cuticle. Frequently, terminal filaments from all ovarioles fuse into a suspensory ligament that is likewise attached. Proximally each ovariole connects with the lateral oviduct, and the two lateral oviducts join the common or medial oviduct as a passage for the eggs to the outside.

15.2.1.1 Panoistic Ovarioles

Panoistic ovarioles do not have nurse cells. The panoistic ovary in *Thermobia domestica* (Packard) the firebrat (Thysanura), is a well-studied example of a panoistic ovary. Each ovary is composed of only five ovarioles, with each ovariole enclosed in a thin, largely membranous, epithelial sheath containing occasional nuclei. The distal germarial region contains oogonia that can undergo mitosis to produce additional oogonia. Proximal to the germarium, and in the area nearest the germarium, are young oocytes not yet arranged in single file, although more proximally they become arranged into a single string of developing oocytes. Many small prefollicular nuclei are present in a common cytoplasm, but later these nuclei acquire cell boundaries and arrange themselves around each developing oocyte as a follicular epithelium. Interfollicular tissue separates each follicle from the next above it. The follicular epithelial cells secrete a noncellular, thick membranous tunica between themselves and the outside epithelial sheath. The terminal oocyte, the most proximal one, sequesters yolk and increases in size to become a mature oocyte. The follicular epithelial cells secrete the chorion (the egg shell). Fertilization occurs after the egg shell is in place as the egg passes down the median oviduct and past spermatheca where sperm are stored. The final maturation divisions of the egg nucleus do not take place in most eggs until the egg has been ovulated.

15.2.1.2 Telotrophic Ovarioles

There are nurse cells located in the germarial region of telotrophic ovarioles, and long cytoplasmic cords or strands extend from the germarium to the developing oocytes. Nutrients and maternal gene products pass down the cytoplasmic cords. The milkweed bug *Oncopeltus fasciatus* has been intensively studied as a model of an insect with a telotrophic ovary (Bonhag, 1955, 1958). The germarial region at the distal end of each of the seven ovarioles in each ovary is primarily occupied by the apical trophic tissue. These cells send long cytoplasmic strands to each developing oocyte through which nutrients and genetic messages pass to the oocytes. Proximal to the apical trophic tissue are the young oocytes, all of which are produced during the nymphal stage. No additional oocytes are produced in the milkweed bug during the adult stage (Wick and Bonhag, 1955).

Each ovariole is covered by an inner envelope composed of a single layer of cells and an outer, thin syncytial epithelial sheath. As the terminal oocyte matures, these two outer layers become stretched very thin and generally are evident only in the region around the interfollicular tissue dividing each follicle from the preceding one. A single layer of follicular epithelium is arranged around each oocyte, and as the oocyte enlarges and grows toward maturity, the follicular epithelial cells first become large, binucleate rounded cells, and finally become squamous. Prior to the egg moving down the oviduct, the **chorion** is secreted by the follicular epithelium.

Detailed studies of the telotrophic ovary in the yellow mealworm *Tenebrio molitor* have been made as a model for the polyphaga group of Coleoptera. Many of the basic features are the same as in the milkweed bug, but there are some differences. The publication by Schlottman and Bonhag (1956) should be consulted for more details.

15.2.1.3 Polytrophic Ovarioles

In polytrophic ovarioles, the nurse cells are located in the egg chambers, either in the same egg chamber with the oocyte or in an adjacent chamber. The earwig *Anisolabis maritima* has five ovarioles in each ovary, with each ovariole enclosed in a syncytial outer epithelial sheath. Each ovariole consists of a terminal filament, a germarium, and string of egg chambers. Each follicle contains one developing oocyte and one **trophocyte** or nurse cell. Initially, the trophocyte is the larger of the two cells in a follicle, but this changes as the oocyte matures. The cells of the follicular epithelium divide by mitosis to accommodate the need for more cells to surround the growing oocyte during the previtellogenesis growth; but later during vitellogenesis, the follicular epithelial cells mainly grow and change shape, becoming more thin and squamous as they stretch to cover the enlarging oocyte.

The number of nurse cells or trophocytes varies in different insects. Earwigs are somewhat unusual in having just one trophocyte for each oocyte. Many insects have multiple trophocytes per oocyte: Lepidoptera typically have 5 nurse cells per oocyte, the diving beetle *Dytiscus marginalis* has 15 trophocytes per oocyte, the gyrinid beetle *Dinuetes nigrior* has 7 trophocytes per oocyte, *Drosophila* and higher Diptera have 15 nurse cells per oocyte, and the honeybee *Apis mellifera* has 48 nurse cells (occupying the follicle preceding the oocyte). In advanced polytrophic ovarioles in which the trophocytes occupy a separate follicle adjacent to the oocyte, an oocyte process typically extends into the nutritive follicle through which nutrients and maternal gene products are passed to the oocyte.

Multiple trophocytes and a cell destined to become the oocyte are produced by mitotic division of an oogonium to produce two cells, with successive mitotic division of these to produce four, etc., to give a group of sister cells, all diploid in chromosome number. For example, in *Drosophila* eight divisions produce 16 cells, and 15 become trophocytes and 1 becomes the oocyte. In *Drosophila* (and probably other insects), interconnecting cytoplasmic strands, often called ring canals, allow nutrients and gene products to pass from nurse cells to the developing oocyte (Figure 1.8, Chapter 1).

The nurse cells of some insects amplify rRNA genes so that they produce a large complement of ribosomes for the egg (Kafatos et al., 1985), but only a few copies are put into the oocytes of other insects (Schäfer and Kunz, 1987). In the polytrophic ovary of dipterans, including *D. melanogaster*, rRNA is not amplified, but ribosomes are supplied by the highly polyploid nurse cells. The oocyte also usually receives some nutrients from the layer of follicle cells surrounding it.

Typically, several oocytes in various stages of development occur in each ovariole. For example, in the housefly, oogonial division begins during early pupation and the first egg chamber is formed before emergence of the adult. At the beginning of oviposition several days after emergence, there is one mature egg in each ovariole and secondary oocytes already in various stages of development.

15.2.2 Nutrients for Oogenesis

The availability of nutrients during oogenesis is a major limiting factor in the ability of an insect to successfully reproduce (Wheeler, 1996). In addition, mating (Gillott and Friedel, 1977) and physical activity such as flight influence the physiological availability of nutrients in many insects. Mating is a stimulus that induces mobilization of reserves in some females. The male may make nutrient contributions to the female during courtship and mating (Boggs, 1990) by offering nuptial gifts of food, and nutrients may be obtained from seminal fluid and a spermatophore.

Oogenesis, the formation of eggs, requires incorporation of relatively large amounts of protein and lipids, and thus is an energy-intensive activity in most insects. Insects that live only short lives as adults typically accumulate nutrients for oogenesis during their larval stage. Adults with longer lives often have a period of preoviposition development of the ovaries and usually require nitrogenous foods for maximum growth of ovaries and egg production. Among Diptera, the terms "**autogenous**," the ability to develop a first set of eggs without an exogenous nitrogen source, and "**anautogenous**," the need for a protein or nitrogen source as an adult to develop eggs, are used to describe the requirements for a blood meal to provide nutrients for egg development. Spielman (1971) suggests that some individuals in all populations are likely to show autogeny. Autogenous individuals have been found in populations of two anautogenous higher dipterans (*Sarcophaga bullata* and *Musca domestica*) (Baxter et al., 1973; Robbins and Shortino, 1962). Some species of mosquitoes are autogenous while others are anautogenous. The autogenous species can mature one set of eggs without a nitrogen source as an adult, but subsequent egg development depends on taking a blood meal. Anautogenous species of mosquitoes need a blood meal to develop the first and each subsequent set of eggs. Some parasitic insects produce small (20 to 200 μm), almost yolkless eggs that are deposited in a host (another insect) where the developing embryo absorbs nutrients from the host through a thin chorion shell (Fisher, 1971; Flanders, 1942; Wheeler, 1996).

Physical activity, and especially flight, which demands so much energy, can compete with the ovary for nutrients. Oogenesis is generally inhibited during migratory flights, or when lengthy flights must be taken to locate a new host or suitable oviposition site (Wheeler, 1996). A new brood of scolytid bark beetles (e.g., the genera *Ips* and *Dendroctonus*) emerging from the log or tree in which they have developed often must fly some distance in seeking a new suitable tree to colonize. When a suitable one has been selected, the wing muscles of some females degenerate, making nutrients available quickly for oogenesis. After the mating flight, queen ants break off their wings (Fletcher and Blum, 1981), and the wing muscles degenerate to make nutrients available for the first oogenesis cycle.

15.2.3 Hormonal Regulation of Ovary Development and Synthesis of Egg Proteins

Hormones control ovary growth, synthesis of **vitellogenin (Vg)** by fat body cells, and uptake of Vg by the developing oocytes (Hagedorn, 1985; Adams and Filipi, 1988). In some groups, only **juvenile hormone (JH)** produced by the corpora allata appears to be involved in regulating reproductive biology, while in others both JH and **ecdysone** (produced by the ovaries) (Hagedorn et al., 1975) are important; and in still other groups, JH, **ecdysteroids**, and **additional hormones** are known.

1. *Dictyoptera* (*cockroaches*). **JH** has numerous pleiotropic actions on the development and reproduction of cockroaches, including maturation of gonads, production of attractant and courtship pheromones, "calling" behavior and pheromone release, and sexual receptivity (Schal et al., 1997). It appears to be the only hormone that is involved in controlling fat body synthesis of vitellogenins, ovary growth, and uptake of the Vgs (Figure 15.3) in the German cockroach *Blatella germanica* and *Leucophaea maderae*. Mating, high-quality nutrition, social interactions, and the presence of vitellogenic ovaries influence JH synthesis by the corpora allata (Schal et al., 1997). One or more additional factors may be involved in the decline of vitellogenin synthesis in *Blattella germanica* late in the gonatropic cycle because JH levels remain high while vitellogenin production is declining (Martín et al., 1995).
2. *Orthoptera*. **JH** is the principal hormone that stimulates fat body cells and the ovary, and the **adipokinetic hormone** (from the corpora cardiaca) inhibits Vg mRNA translation at the end of an egg production cycle in *Locusta migratoria* (Bownes, 1986). The evidence suggests that JH controls the Vg gene because JH analogs promote transcription of a gene coding for vitellogenin in the fat body (Glinka and Wyatt, 1996). JH also regulates uptake of the Vgs by the developing oocytes, and the mature ovary appears to have feedback to the brain and/or corpora, which regulates down the production of JH until the primary set of eggs is laid. JH III is the major JH in several gryllid crickets (*Acheta*

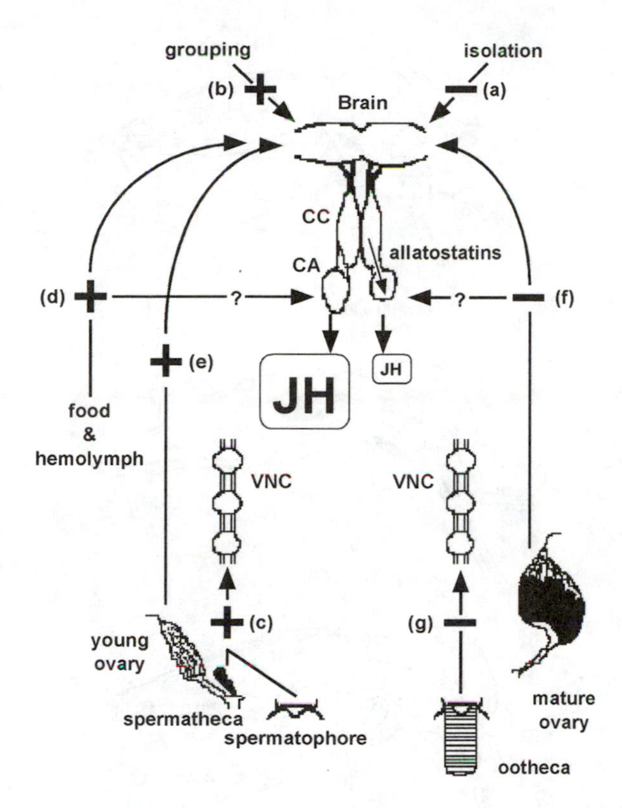

FIGURE 15.3 A model to suggest external and internal regulation of JH synthesis in female German cockroaches *Blattella germanica*. Inhibitory regulators are indicated by a minus sign and positive regulators by a plus sign. JH synthesis is inhibited by social isolation, a mature ovary, or an ootheca (an egg case carried for some time by a female and finally deposited before eggs hatch). Social interaction, mating, food availability, and a young ovary stimulate JH synthesis. (Reproduced with permission from Schal et al., 1997; figure courtesy of Coby Schal.)

domesticus, *Teleogryllus commodus*, and *Gryllus bimaculatus*) (Strambi et al., 1997). The ovaries of *A. domesticus* synthesize ecdysone and conjugate most of it with fatty acids to form ecdysone 22-fatty acyl esters (Whiting et al., 1997). The precise role of the ecdysone or fatty acyl esters is not known.

3. *Diptera*. Diptera exhibit complex control of reproduction involving multiple hormones. Hormonal control of reproduction in mosquitoes has been reviewed by Klowden (1997) and in the higher Diptera (Cyclorrhapa = Muscamorpha) by Yin and Stoffolano (1997). Female mosquitoes, and some other dipterans, are hematophagous, that is, foragers on blood. Some mosquitoes, such as *Aedes aegypti* and *Culex pipiens pipiens,* are anautogenous and must have a blood meal to mature the first set of eggs. Other mosquitoes, including *Ae. taeniorhynchus*, *Ae. atropaplus*, and *Culex pipiens molestus*, are autogenous. They can mature one set of eggs using stored reserves from their larval life, but require a blood meal for subsequent egg development. Hormonal controls appear to be similar in both kinds of mosquitoes (Klowden, 1997). Autogeny and anautogeny also occur in the cyclorrhaphous Diptera.

In anautogenous mosquitoes, hormonal control of ovary and egg development is conveniently divided into **previtellogenic**, **vitellogenic**, and **postvitellogenic** stages (Figure 15.4). Soon after an adult female mosquito emerges, the **previtellogenic stage** begins with JH secretion from the corpora allata (CA). In *Ae. aegypti,* JH III is the only JH known (Baker et al., 1983). The stimulus for JH

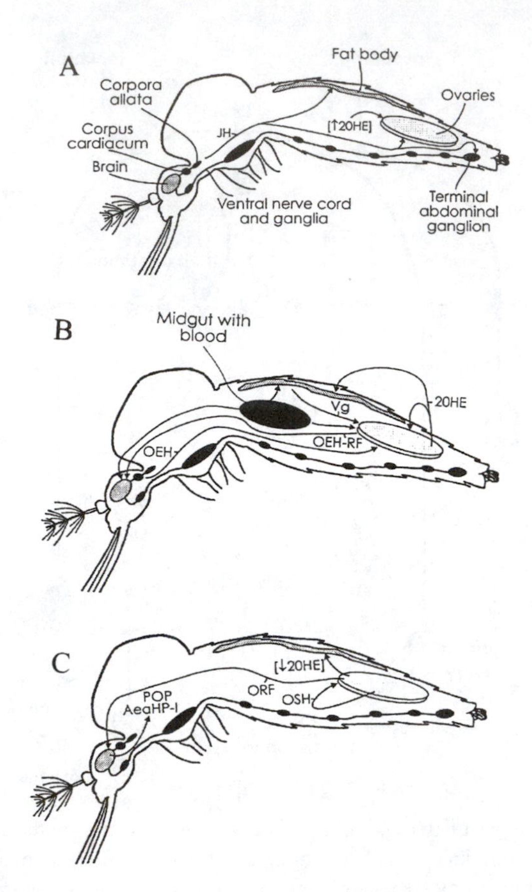

FIGURE 15.4 A: Hormonal control of previtellogenesis in mosquitoes. Paired CA produce JH prior to a blood meal, and JH acts on fat body and ovaries to make them competent to respond to later hormones and secrete vitellogenin. JH also acts on the terminal abdominal ganglion and mediates mating acceptance. Circulating 20-hydroxyecdysone left from the pupal-adult molt initiates follicle formation in the ovary. B: Hormonal action during vitellogenesis. The blood meal is digested and nutrients are incorporated into the fat body. Ovarian ecdysteroidogenic hormone-releasing factor (OEH-RF) is released from the ovaries. OEH-RF causes the corpora cardiaca to release OEH, which stimulates competent ovaries to synthesize ecdysone. Ecdysone is converted to 20-hydroxyecdysone (20-HE) by the fat body and other tissues, and it stimulates the fat body to synthesize and release vitellogenin (Vg). 20HE also stimulates the separation of additional follicles from the germarium. Vg is taken up by oocytes. C. Postvitellogenic hormonal controls. The fat body ceases to produce Vg in response to falling titers of 20HE. An oostatic hormone (OSH) is produced by growing primary follicles and inhibits development of secondary follicles. Hormones that inhibit host-seeking behavior and promote pre-oviposition behavior are released by ovarian releasing factor (ORF). (Reproduced with permission from Klowden, 1997.)

secretion in not known, but it is not the taking of the blood meal itself, because secretion occurs prior to the meal. JH has at least three actions:

1. It makes females receptive to mating.
2. It stimulates previtellogenic growth of ovaries.
3. It prepares the fat body so that it is competent for responding to later hormones and secretion of vitellogenins.

Follicles begin to separate from the growing previtellogenic ovary, possibly under influence of residual 20-hydroxyecdysone remaining from the pupal to adult transformation (Whisenton et al.,

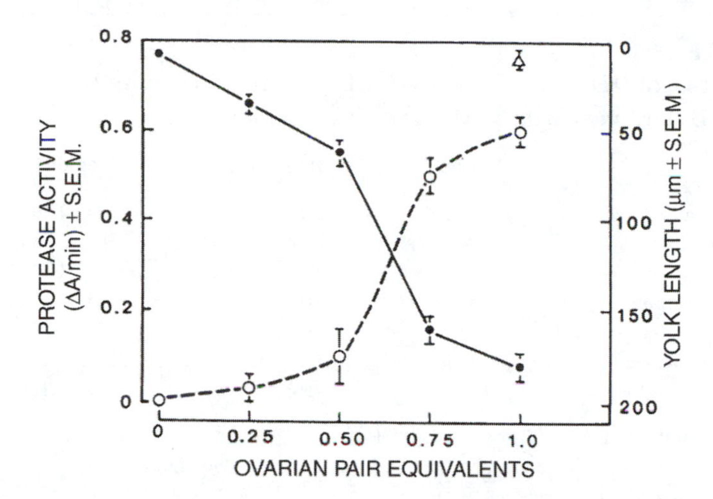

FIGURE 15.5 The inhibition of yolk deposition in oocytes (measured as yolk length, broken line) and decrease in protease activity (solid line) in the midgut of *Aedes aegypti* females 24 hours after injection of different concentrations of oostatic hormone (oostatic hormone expressed as ovarian pair equivalents). Each point is the mean ± SE of 15 determinations. (Reproduced with permission from Borovsky, 1988.)

1989). Soon after emerging, females seek a blood meal, which provides proteins, other nutrients, and initiates the **vitellogenic stage**. With availability of nutrients, the ovary, stimulated by JH, releases **corpora cardiaca stimulating factor (CCSF)**, a neuropeptide probably produced somewhere in the nervous system (before the blood meal) and stored in the young ovary. The target cells for CCSF are in the corpora cardiaca, which releases **egg development neurohormone (EDNH)** (Hagedorn et al., 1979), recently renamed with the more descriptive name as the **ovarian ecdysteroidogenic hormone I (OEH)**. OEH (MW = 8803) is a polypeptide comprising 86 amino acids. OEH is synthesized in brain medial neurosecretory cells and stored in the CC. There is evidence that both humoral and nervous stimuli are important in causing the release of OEH from the CC (Klowden, 1987). Follicular epithelium cells in the ovary respond to OEH by producing and releasing **ecdysone** into the circulating hemolymph. Ecdysone is rapidly converted to **20-hydroxyecdysone** by many types of cells throughout the body, including the target fat body cells. Fat body cells respond to 20-hydroxyecdysone by synthesizing vitellogenins (but only if the fat body has previously been exposed to JH; see function 3 for JH stated above). A receptor for 20-hydroxyecdysone is expressed in fat body and ovary (Cho et al., 1995). JH also stimulates Vg synthesis (Wyatt et al., 1987; Bradfield et al., 1989; Wu et al., 1987; Hagedorn, 1989).

The postvitellogenic stage terminates Vg production in the fat body when the complement of primary oocytes are approaching or have reached maturity (one mature egg per ovariole). The ovaries (the exact site is not established) release an **oostatic hormone**, called the **trypsin modulating oostatic factor (TMOF)** in *Aedes aegypti* by Borovsky (1982, 1988) and simply **oostatic hormone (OSH)** by Klowden (1997). The oostatic hormone stops the uptake of yolk by secondary oocytes, thus stopping their growth until the primary set of eggs is laid. The exact mechanism by which oostatic hormone works is somewhat contentious. Borovsky (1988) and Borovsky et al. (1990, 1994) present evidence that the function of TMOF is to stop the synthesis of late trypsin enzyme (see Chapter 2 for role of early and late trypsin in digestion) in midgut cells, interrupting blood meal digestion and denying secondary oocytes nutrients (Figure 15.5). The overall function of the oostatic hormone is to keep the ovary and abdomen from becoming overly distended by too many eggs growing to maturity at the same time. Females now seek an appropriate site for oviposition. Hormones inhibiting host-seeking behavior and stimulating pre-oviposition searching

TABLE 15.1
Identity of JHs in Cyclorrhaphous Dipterans (Biosynthesized *in vitro*) and Determined with Physiochemical Methods

Species	Insect Stage	JH Identity[a]	Method[b]
Calliphora vicina	Larvae	JHB_3	TLC, HPLC
C. vomitoria	Adults	JHB_3, trace JH III	HPLC
Drosophila melanogaster	Larvae, adults	JHB_3, JH III	TLC, HPLC, GC-MS
Lucilia curpina	Larvae, adults	JHB_3	TLC, GC-MS
Musca domestica	Larvae	JHB_3	TLC, HPLC
Phormia regina	Adults	JHB_3, JH III, MF	TLC, HPLC, GC-MS
Neobellieria bullata	Larvae	JHB_3	TLC, HPLC

[a] JHB_3, juvenile hormone III bisepoxide; JH III, juvenile hormone III; MF, methyl farnesoate.
[b] TLC, thin layer chromatography; HPLC, high-pressure liquid chromatography; GC-MS, gas chromatography/mass spectrometry.

Modified from Yin and Stoffolano (1997).

behavior are activated by an ovarian releasing factor (Figure 15.4C). Exactly how the inhibition of trypsin synthesis is released when eggs are laid is not clear.

Alternative mechanisms for the action of oostatic hormones may exist. Oostatic hormone activity has been demonstrated in *Ae. atropalpus*, an autogenous mosquito that does not need to feed on blood for the first set of eggs nor seems likely to depend on trypsin enzyme activity in the midgut to mature the first set of eggs (Kelly et al., 1984, 1986). Both an oostatic hormone (Adams et al., 1968) and an ecdysteroidogenin hormone (Adams et al., 1997) have been demonstrated in the housefly *Musca domestica.* In response to purified extracts of the ecdysteroidogenin, housefly ovaries *in vitro* produced approximately 500 pg each of ecdysone, 20-hydroxyecdysone, and makisterone A.

JH plays a major role in reproduction of **cyclorrhaphous Diptera**, but Yin and Stoffolano (1997) suggest that dipterans are too diverse a group (with diversity even within the two major divisions, the Cyclorrhapha and the Nematocera) to expect JH (and by extrapolation other hormones) to play a common role. Adult life history and nutrition determine, in large part, the way in which JH and the neuroendocrine system exert their control over reproduction (Yin and Stoffolano, 1997). Juvenile hormone III bisepoxide (**JHB_3**) as well as **JH III** and **methyl farnesoate** have been identified in a number of cyclorrhaphous larval and adult dipterans (Table 15.1). In the blowfly *Phormia regina* (Meigen), JH influences mating behavior in both sexes, fat body development, ovary growth and development, and pinocytotic uptake of vitellogenin by oocytes (Yin et al., 1989), but JH is synthesized by the corpora allata only at low levels until a protein meal is taken by the flies (Yin et al., 1995). Biosynthesis of vitellogenin is primarily controlled by **ecdysteroids** released from the ovaries (Yin et al., 1990). In contrast to mosquitoes, there presently is no published evidence for an ovarian ecdysteroidogenic hormone (OEH) to stimulate ecdysteroid synthesis nor for the necessity of the ovaries to be exposed first to JH to make them competent for further function (Yin and Stoffolano, 1997). Liver-fed flies produce approximately normal levels of vitellogenin but oocytes do not sequester it when JH production is suppressed by precocene II treatment (Yin et al., 1989); however precocene-treated flies sequester vitellogenin and mature eggs if they are rescued by treating them with methoprene, a JH mimic (Stoffolano et al., 1992).

Apterygota. Thermobia domestica, the firebrat, represents the present-day success of a very early evolutionary group of insects. These insects continue to secrete **ecdysteroids** and **molt as adults**. **JH** is implicated in control of ovary development and oogenesis as indicated by allatectomy or treatment with precocenes, either of which prevents egg development. Both procedures also prevent the secretion of JH, and it likely that JH is the main hormone controlling vitellogenesis. Fat body and ovaries are involved in producing the two large Vg molecules that go into the eggs (Rousset and Bitsch, 1993), but details relating to precise endocrine controls on synthesis are not available. JH III is the major JH of *Euborellia annulipes* (Lucas) (Dermaptera) another group that evolved early with strong sister group relationships with Dictyoptera, Isoptera, and Mantoidea but methyl farnesoate is also present in the CA and the medium in which glands are incubated (Rankin et al., 1997).

Coleoptera. Engelmann (1983) has shown that JH is responsible for vitellogenin synthesis in the Colorado potato beetle *Leptinotarsa decemlineata*, the yellow mealworm *Tenebrio molitor*, and in some other beetles.

Hemiptera. JH promotes vitellogenin synthesis and uptake by the oocytes of *Rhodnius prolixus* (Davey, 1993, 1997), and the milkweed bug *Oncopeltus fasciatus*, *Pyrrhocoris apterus*, and *Triatoma protracta* (Engelmann, 1983).

Lepidoptera. Ramaswamy et al. (1997) and Bellés (1998) suggest evolution of flexibility in the hormonal control of vitellogenesis in Lepidoptera. **Ecdysteroids** control vitellogenesis in those species that start vitellogenesis in the larval or early pupal stages, with progression to a combination of ecdysteroids and JH in those that start vitellogenesis prior to emergence in the pharate adult stage, and finally only JH controls egg protein synthesis in those species that begin vitellogenesis after adult emergence.

The cecropia moth *Hyalophora cecropia* initiates Vg synthesis early in the prepupal stage, and the silkmoth *Bombyx mori* begins Vg synthesis in the early pupal stage. In these two moths JH does not seem to have a role, and 20-hydroxyecdysone stimulates Vg synthesis (Tsuchida et al., 1987). 20-Hydroxyecdysone stimulates Vg synthesis in the gypsy moth, in which Vg synthesis begins late in the last instar, and experimental treatment with JH inhibits Vg synthesis (Fescemyer et al., 1992). Some moths, including some pyralid moths, use falling ecdysteroid concentrations to induce Vg synthesis, and eggs mature before eclosion of adults. In the fall armyworm *Spodoptera frugiperda*, both JH and ecdysteroids promote Vg synthesis, but sequestering of Vg by the oocytes is under JH control (Sorge et al., 2000). Only JH induces vitellogenin synthesis in the monarch butterfly *Danaus plexippus* (Pan and Wyatt, 1971), the moth *Heliothis virescens* (Zeng et al., 1997), and in a number of other lepidopterans that begin vitellogenesis after emergence of the adult (Cusson et al., 1994). Males transfer JH to female *H. virescens* during mating (Park et al., 1998; Ramaswamy et al., 2000), and mating itself stimulates the CA in females to synthesize JH II (and small amounts of JH I and III) and causes inhibition of JH esterase that could potentially destroy JH transferred or synthesized.

15.3 VITELLOGENINS AND YOLK PROTEINS

15.3.1 BIOCHEMICAL CHARACTERISTICS OF VITELLOGENINS AND YOLK PROTEINS

The egg yolk is rich in proteins and lipids. Sex-limited proteins present in the hemolymph that are incorporated into developing eggs were discovered initially in *Hyalophora cecropia*, the cecropia silkmoth (Telfer, 1954), and since have been found in many different groups of insects. Some early work suggested the proteins were sex specific and found only in females; later work has shown varying, but small, amounts of the same proteins in males of some species, including some lepidopterans, hemipterans, orthopterans, and honeybees. In some males, egg proteins can be induced with hormone treatments (Shirk et al., 1983; Wyatt, 1991).

Although synthesis of the egg proteins occurs in both fat body (Keely, 1985) and the follicular cells of the ovary in some insects, the major source of the proteins is the fat body in most insects.

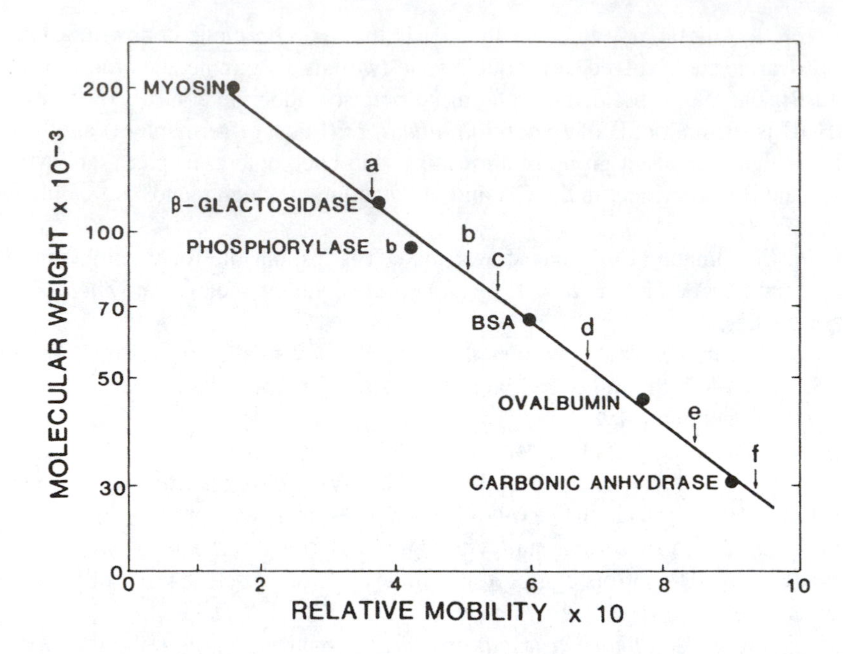

FIGURE 15.6 Molecular weight determination of *Aedes aegypti* vitellin subunits determined by electrophoresis in the presence of sodium dodecyl sulfate (SDS). Arrows indicate the mobility of purified vitellin (50 μg) subunits. Molecular weights are as follows: a, 116,000; b, 83,000; c, 75,000; d, 54,000; e, 36,000; f, 29,000. (Reproduced with permission from Borovsky and Whitney, 1987.)

The proteins are called **vitellogenins** (**Vg**) while they are being produced by the fat body and during transport to the ovaries by the hemolymph (Pan et al., 1969), but after incorporation into developing eggs, the proteins are known as **vitellins**.

Insect vitellogenins are typically large **glycolipoproteins**, from 400 to 600 kDa, that are composed of small (40 to 60 kDa) and large (120 to 200 kDa) subunits (Figure 15.6) (Kunkel and Nordin, 1985; Shirk, 1987; Borovsky and Whitney, 1987; Yano et al., 1994a, 1994b). Large insect Vgs contain 7 to 15% lipid consisting primarily of phospholipids and diacylglycerol (Raikhel and Dhadialla, 1992), and apoproteins (Zeng et al., 1997). Vgs usually exist as **dimer**s, but **monomers** are known from the cockroach *Nauphoeta cinereae* (Imboden et al., 1987). Genes controlling synthesis of Vgs have been identified and cloned from a number of insects, including *Locusta migratoria*, *Aedes aegypti*, *Anopheles gambiae*, and *Drosophila melanogaster* (Bownes, 1986; Wyatt, 1991, and references therein). The apoproteins of insect and vertebrate vitellogenins diverged from a superfamily of proteins controlled by genes with an ancient heritage (Speith et al., 1985; Blumenthal and Zucker-Aprison, 1997; Wahli, 1988).

Egg proteins of Diptera fall into two classes that split along the lines of the two suborders of Diptera. Lower Diptera in the suborder Nematocera, including mosquitoes and some other dipterans, have Vgs similar in structure to those of other insects, that is, large glycolipoproteins composed of small and large subunits. A major evolutionary shift in gene control of yolk proteins occurred in the higher Diptera (suborder Cyclorrhapha), and the proteins that go into the yolk are not homologous with the Vgs of other insects (Romans et al., 1995). Consequently, these proteins are not called Vgs, and instead are called **yolk proteins** (**YPs**). As noted in Section 15.2.3, a cascade of hormones is often involved in controlling synthesis of Vgs and YPs, with JH and ecdysteroids playing important roles. In nonfeeding moths synthesis of Vgs and their uptake by oocytes appear to be controlled by ecdysteroids during prepupal, pupal, or pharate adult development (depending on species), and experimental addition of exogenous JH inhibits Vg production (Satanarayana et al., 1994, and references therein).

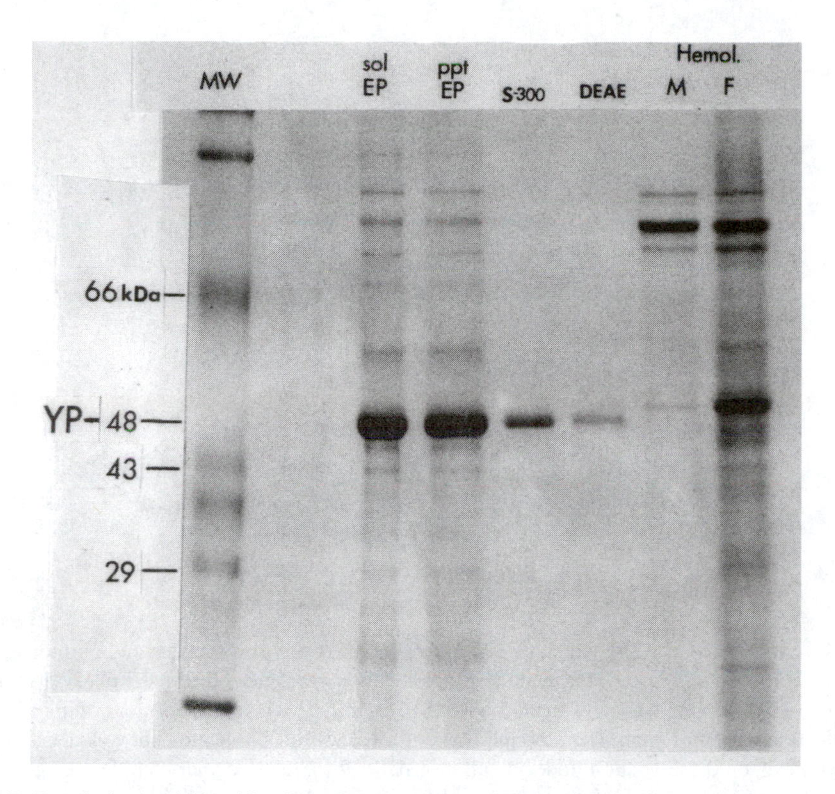

FIGURE 15.7 Resolution and identification of yolk protein (YP) from eggs of the Caribbean fruit fly *Anastrepha suspensa* on 10% SDS-PAGE stained for protein with Coomassie Blue. Lane a, molecular mass standards; lane b, soluble proteins from oviposited eggs; lane c, ammonium sulfate precipitated egg proteins; lane d, combined YP fractions from S-300 separation; lane e, combined YP fractions from DEAE separation; lane f, 5 μl hemolymph from 3- to 4-day-old males; lane g, 5 μl hemolymph from 3- to 4-day-old females. (Photograph courtesy of Al Handler and Paul Shirk.)

15.3.2 Yolk Proteins of Higher Diptera

Known **YPs** in higher Diptera are **small polypeptides** (Figure 15.7) composed of small subunits. For example, in *D. melanogaster* and some other higher dipterans, the yolk protein is composed of three subunits, YP1, YP2, and YP3 (Figure 15.8) (46, 45, and 44 kDa, respectively), and each is coded by single-copy genes on the X chromosome (Bownes et al., 1993). The numbers of YPs differ in several different *Drosophila* spp., but in all investigated, the YPs are small polypeptides. Small YPs have been found in a number of other higher Diptera, including blowflies, fleshflies, houseflies, and several tephritid fruit flies (Huybrechts and DeLoof, 1982; DeBianchi et al., 1985; Handler and Shirk, 1988; Rina and Savakis, 1991; Martínez and Bownes, 1994).

The genetic controls of YPs and Vgs are different. The YPs of *D. melanogaster* are under control of a family of genes different from that controlling the more widespread Vgs (Bownes et al., 1993). The *Drosophila* genes have greater sequence similarity to genes controlling mammalian triacylglycerol lipase than to Vg genes of other insects (Bownes et al., 1988; Tepstra and AB, 1988; Baker, 1988), and the YP genes and vertebrate lipase genes may have evolved from ancestral progenitors (Kirchgessner et al., 1989). YPs may **sequester ecdysteroids** and make them available to the developing embryo as the proteins are digested during embryogenesis. Bownes et al. (1988) found that degradation of the YPs from *Drosophila* releases ecdysteroid in proportion to the degree of enzyme attack by protease and esterase. Ecdysteroids may have multiple functions in the embryo, but one function that seems likely is to promote the secretion of a cuticle. Some embryos secrete and molt more than one cuticle during embryogenesis (Bownes et al., 1988).

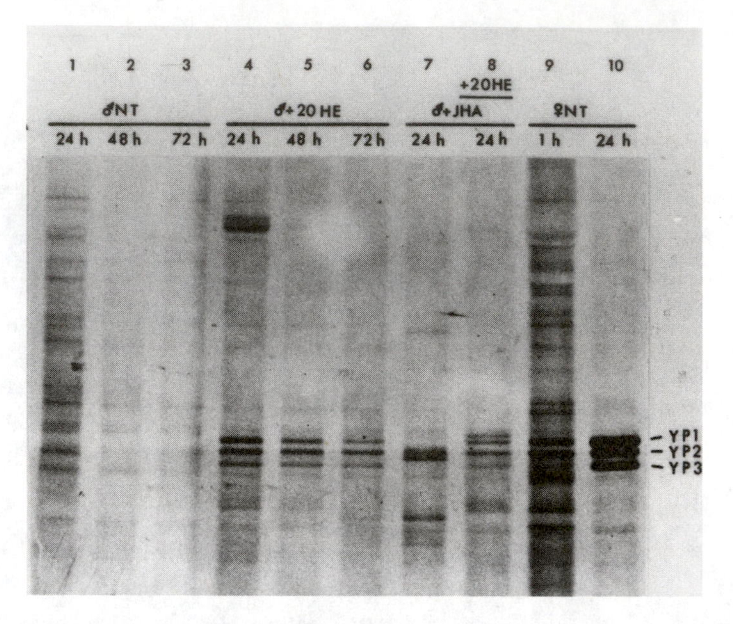

FIGURE 15.8 Yolk proteins (YP) 1, 2, and 3 in normal adult males and females and hormonally stimulated males of *Drosophila melanogaster*. Hormonally stimulated flies were injected with 0.3 µl of 0.1 m*M* 20-hydroxyecdysone (20 HE) or were topically treated with 0.16 µg ZR-515 juvenile hormone mimic in acetone 8 h prior to collection of hemolymph. Hemolymph was collected at times indicated above lanes and subjected to gel electrophoresis on 0.1% sodium dodecyl sulfate (SDS), 9 to 12% polyacrylamide slab gels. Lanes 1, 2, and 3 from left represent males with no treatment; lanes 4, 5, and 6 represent males treated with 20-HE; lane 7 shows hemolymph from ZR-515-treated male; lane 8 represents ZR-515 + 20 HE-treated males; and lanes 9 and 10 represent normal untreated females. Bands in lane 7 that nearly match YPs also appeared in untreated 24-h-old males (lane 1). These polypeptides do not migrate exactly with the YPs and are not immunoprecipitable and probably are not YPs. They may be synthesized by remaining larval fat body cells because they are not present in 2- and 3-day-old males (lanes 2 and 3) when larval fat body has disappeared. (Photograph courtesy of Paul Shirk.)

15.4 SEQUESTERING OF VITELLOGENINS AND YOLK PROTEINS BY OOCYTES

15.4.1 PATENCY OF FOLLICULAR CELLS

Oocytes take up proteins through channels between the follicular cells (Figure 15.9) (Telfer, 1961, 1965; Davey, 1981; Raikhel and Dhadialla, 1992). JH acts with a membrane receptor to promote Vg uptake by promoting widening of the intercellualr spaces in the follicular epithelium. A Na^+/K^+-ATPase is activated and the cells shrink. **Phosphatidylinositol** and **protein kinase C** are involved (Ilenchuk and Davey, 1987; Davey, 1993). The opening of spaces between follicular cells is called **patency**. Egg proteins and experimentally added dyes readily pass through the spaces between follicle cells when patency has occurred. Patency is inhibited by ouabain and metabolic poisons that stop or reduce Na^+/K^+ ATPase activity, and by colchicine and cytochalasin B that inhibit cytoskeletal elements such as microtubules (Davey, 1981). At the end of vitellogenesis, new junctions between follicular cells seal the interfollicle cell channels and protein uptake rapidly falls (Rubenstein, 1979; Koller et al., 1989).

The Vgs and YPs bind to specific receptors at the surface of the oocyte plasma membrane between and at the base of microvilli. The receptor-protein complex tends to sink inward at the oocyte surface, forming a pit, with a **clathrin protein coat** on the cytoplasmic side (Raikhel, 1984, 1987; Raikhel and Dhadialla, 1992). The pits continue to invaginate, close up, and become pinched

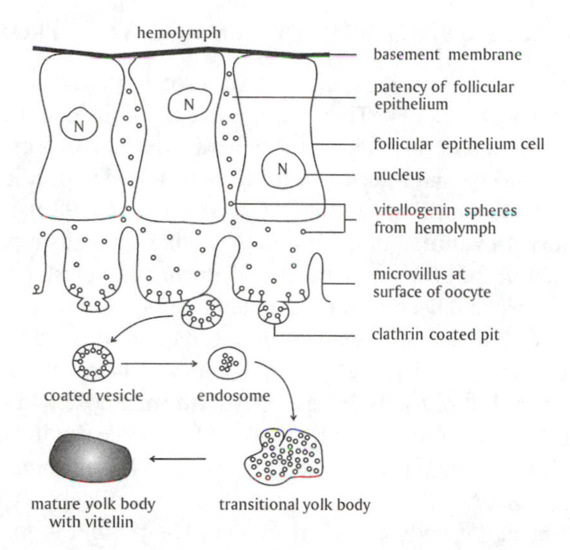

FIGURE 15.9 A schematic diagram to illustrate the uptake of vitellogenin by a developing oocyte. The proteinaceous vitellogenin (Vg) is primarily synthesized in fat body cells. It is transported by the hemolymph and passes between spaces in follicular epithelial cells that have shrunk (shrinking of the cells is hormonally induced and is called patency). Vg is bound to receptors on the surface of the developing oocyte. The membrane with bound vitellogenin invaginates and pinches off as small vesicles. Within the vesicles, the vitellogenin is released from the receptors, and the receptor molecules probably return to the membrane where they bind more vitellogenin. Vesicles filled with vitellogenin are called endosomes, and they coalesce as transitional yolk bodies that finally becomes mature yolk bodies containing vitellin, as the protein is called after it is stored in the yolk bodies. (Modified from Raikhel and Dhadialla, 1992.)

off as small coated vesicles (150 to 190 nm in diameter) inside the oocyte. Coated vesicles have been isolated from the ovaries of *Locusta migratoria* locusts (Röhrkasten and Ferenz, 1987) and the clathrin heavy chains have a molecular weight of 180,000.

The clathrin coat soon dissociates from the vesicles containing the receptor-protein molecules, and the vesicles are then called **endosomes**. Bound egg proteins dissociate from the receptor within the endosome, probably as a result of ATP-dependent acidification within the endosome (Stynen et al., 1988). The clathrin molecules and receptors probably recycle to the oocyte surface for reuse.

Endosomes coalesce into a larger transitional **yolk body**, and the egg proteins, now called vitellins (Vns), begin to crystallize. Additional Vns are added to the transitional yolk body until it becomes a mature yolk body (Figure 15.9). In general, Vgs and Vns have the same immunological properties, chemical composition, and physical properties. Some Vns derived from YPs are sulfated, but it is not known whether this is a general property of Vns (Baeuerle and Huttner, 1985; Dhadialla and Raikhel, 1990).

15.4.2 Egg Proteins Produced by Follicular Cells

Proteins produced in the ovary, usually by follicular cells lining the follicle, do not enter the hemolymph, but are passed directly into the developing oocytes. In several *Drosophila* species that have been studied, the proportion of the several YPs synthesized in the fat body and in the follicular cells varies with species, but follicular cells are a major source of some YPs. **Paravitellogenin** is a 70-kDa protein produced in the follicular cells of *H. crecropia* and taken into the oocyte (Bast and Telfer, 1976; Telfer et al., 1981). A similar, perhaps homologous, protein comprising up to 25% of the egg proteins is produced in the follicular cells of *B. mori* (Sato et al., 1990). Proteins produced in the follicular cells of Indian meal moths, *Plodia interpunctella*, and in several mosquitoes are incorporated into eggs (Bean et al., 1988; Borovsky and Van Handel, 1980).

15.4.3 Proteins in Addition to Vitellogenin and Yolk Proteins in the Egg

Developing eggs often contain varying amounts of proteins that are sequestered from the hemolymph in addition to Vgs and YPs. They are not usually considered to be Vgs because they are present in small amounts and/or do not have the general structure of Vgs. **Lipophorin**, a general lipid transport protein of insects, and **microvitellin**, a small female-specific protein, are present in eggs of *H. cecropia* (Telfer and Pan, 1988) and *Manduca sexta* (Law, 1989), respectively. In addition, *M. sexta* eggs contain **insecticyanin**, a blue biliprotein giving the eggs a pale blue color. A group of storage proteins of about 30 kDa are synthesized during larval stages of both sexes of the commercial silkmoth *B. mori*, and become the dominant hemolymph proteins during larval life and persist into the pupal stage (Izumi et al., 1981). Substantial quantities are taken into the developing eggs, but they are not usually considered to be vitellogenins. One of these is a glycoprotein with a molecular weight of 55,000. It contains 2% carbohydrate (mannose and traces of amino sugars) and 4% lipid. It is the second major protein in the yolk and is used early in embryogenesis, although some of the original vitellins are still present at hatching (Irie and Yamashita, 1983). The hemolymph of the female migratory locust *L. migratoria*, has a low concentration of a 21-kDa monomer whose synthesis is stimulated in the fat body by JH treatment. The protein is female specific and taken into developing oocytes (Zhang and Wyatt, 1990). In most cases, little is known about the fate of these proteins once they enter the oocyte.

15.5 FORMATION OF THE VITELLINE MEMBRANE

Near the termination of vitellogenesis, a thin protein sheet, the **vitelline membrane**, is secreted at the inner surface of the follicle cells (Raikhel and Dhadialla, 1992). The vitelline membrane in *D. melanogaster* is composed of numerous proteins ranging from 14 to 130 kDa (Fargnoli and Waring, 1982) that are encoded by a family of genes (Wyatt, 1991).

15.6 THE CHORION

The egg shell, the **chorion**, is composed of a number of sclerotized proteins. It contains no chitin. With very few exceptions, it is not mineralized like the egg shell of birds. The chorion is placed on the egg while it is still in the ovary, and before fertilization. Sperm, which are released from the spermatheca as the egg passes down the common oviduct, have to enter the egg through a small, usually twisted channel, the **micropyle**, that passes through the various layers of the chorion. More than one micropyle channel is not uncommon; although most Diptera have only one, *Locusta* have 35 to 43 openings. Follicular epithelial cells secrete chorionic proteins on the outer surface of the vitelline membrane, thus enclosing it on the inside of the chorion. The follicular epithelial cells lay down proteins sequentially, indicative of the sequential expression of a superfamily of genes, to produce a laminar structure. In wild silkmoths, *Antheraea polyphemus*, chorion formation requires about 2 days, and more than 100 low-molecular-weight proteins are secreted (Lecanidou et al., 1986). A large gene family controls the secretion of chorion proteins in *A. polyphemus* without gene amplification (Kafatos et al., 1987). By contrast, in *D. melanogaster* only about 20 proteins are secreted under control of a small family of genes (Waring and Mahowald, 1979). The single-copy genes are amplified 20- to 80-fold in the follicle cells about 15 hours prior to transcription (Spradling and Mahowald, 1980). These multiple gene copies (after transcription) allow the follicle cells to secrete a large amount of chorionic proteins in a short time, and the chorion is completed in about 5 hours (Hammond and Laird, 1985). The proteins become sclerotized to produce a tough, water-impermeable covering for the egg and developing embryo. When secretion of the chorion is complete, the old follicular epithelial cells are sloughed off as the egg passes into the median oviduct. The chorion does not contain chitin and, except in a few dipterans, no significant quantity of minerals. Intricate surface sculpturing is characteristic of many insect eggs. Although hormonal control of choriogenesis in *D. melanogaster* has not been demonstrated, recent work has shown that the DNA site to which a chorion gene transcription factor, CF1, binds has part of the sequence of the ecdysone response element. This suggests the possibility of hormonal control (Shea et al., 1990; Wyatt, 1991).

15.7 GAS EXCHANGE IN EGGS

Many eggs have a porous, gas-filled meshwork near the inner (yolk) side. In some cases, this is a **plastron,** a surface that is not easy to wet. Several channels, called aeropyles, connect the meshwork or plastron to the external surface of the egg. The function of such hard-to-wet structures is to supply oxygen to the developing embryo if the egg becomes submerged under water for some time, or when the natural site for laying the eggs is in wet decaying organic matter, animal manure, fruits, or similar plant tissues. When eggs have a plastron surface, tests have shown that the plastron surface resists the wetting action of raindrops, which can exert up to about 30 cmHg pressure for about a millsecond. When there is an egg plastron, it is usually part of the chorion itself, but some eggs have the plastron surface on **respiratory horns** or filaments protruding from the egg. These might give a submerged egg the opportunity to have the plastron surface above the fluid medium if it were not deep.

15.8 MALE REPRODUCTIVE SYSTEM

The internal organs of the male reproductive system are the **paired testes**, **vas deferens**, **accessory glands**, and **ejaculatory duct** (Figure 15.10). All parts of the system may produce secretions that aid the transfer of sperm to the female (Happ, 1992). Each testis generally consists of a number of tubes or **follicles** in which **spermatozoa** are matured (Figure 15.11). Follicles can vary in number from 1 to greater than 100 and can be incompletely separated from each other, such as lepidopteran testis, or the testes may consist of several lobes, each with several follicles. In Diptera, the testes consist of a simple, elongated, and undivided sac (French and Hoopingarner, 1965). **Zones of maturation** stages of sperm exist along the length of a typical follicle. The distal part of a testicular tubule is a **germarium** in which repeated mitotic divisions give rise to undifferentiated, diploid **spermatogonia**. In a **growth zone** (Zone I), the spermatogonia divide by mitosis into many diploid **spermatocytes** enclosed within a cyst or capsule of somatic cells. All spermatocytes within a sac or cyst generally arise from the same spermatogonial cell and their development is synchronized. The spermatocytes may undergo more mitotic divisions; there are five to eight divisions in Acrididae and seven in *Melanoplus*; but eventually in the "zone of maturation" (Zone II), **meiosis** and haploid **spermatids** are produced. A spermatid has completed its meiotic divisions but is an immature sperm. According to Jones (1978), meiosis in *Schistocerca gregaria* males depends on the presence of ecdysteroids. Normally, four sperm are produced from each spermatocyte. In Zone III, the **region of transformation**, the mature sperm develop. Insect spermatozoa tend to be very long (300 μm in *Rhodinus prolixus*) and have a slender head region, probably as an evolutionary adaptation to the necessity to navigate the micropyle. Usually, the mature sperm remain bundled together in Zone III.

Many insects contain mature sperm in the late pupal stage, while others may require several days as an adult to mature sperm. In *R. prolixus* and grasshoppers, the accessory glands in males are influenced in their development by secretions from the corpora allata. In contrast to the situation in most vertebrates, sperm survival within the genital tract of female insects may be prolonged for weeks, months, or even years. Honeybee queens have been known to lay fertilized eggs after several years (8 to 9 years in one reported case), and queen ants were reported to contain viable sperm after 15 years. Spermatozoa survive in female *R. prolixus* for about a month, and for about 10 weeks in *Schistocerca gregaria*.

For a very long period of time it was thought that spermatogenesis in insects was not under hormonal control; but since the 1970s, it has become clear that ecdysteroids are synthesized in the

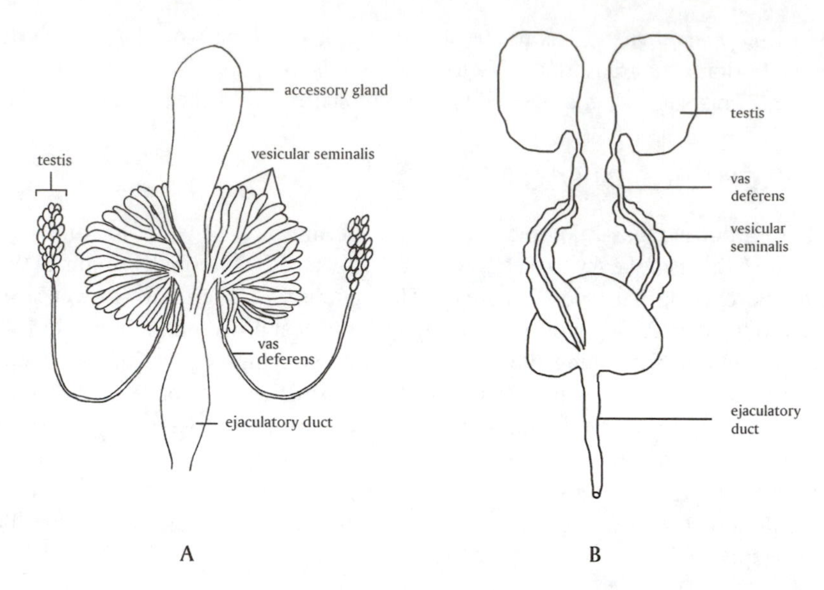

FIGURE 15.10 Drawing of (A) the male reproductive system from the American cockroach *Periplaneta americana* and (B) from the milkweed bug *Oncopeltus fasciatus*.

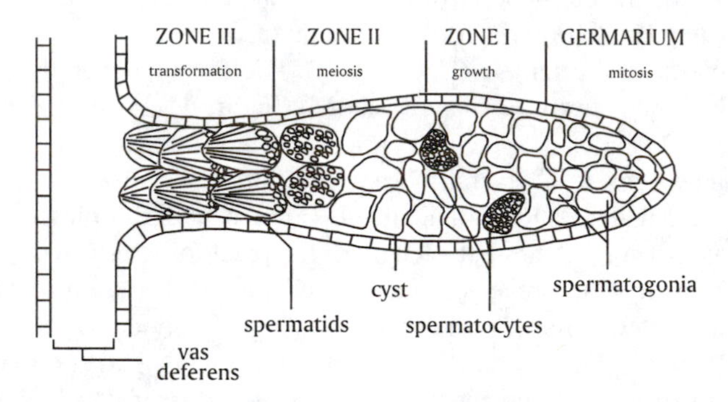

FIGURE 15.11 Zones of maturation of spermatozoa that can be observed in the testes of some insects.

testes of flies, crickets, mosquitoes, and various lepidopterans, and that ecdysteroids stimulate spermatogenesis in several insects (Wagner et al., 1997, and numerous references therein). These observations bring spermatogenesis in insects more in line with known hormonal controls of spermatogenesis in other animals. Wagner et al. (1997) has described an ecdysiotropic peptide from the brain of gypsy moths that stimulates synthesis of ecdysteroids (at reasonable hormonal levels of about 10^{-13} to 10^{-15} M) in testes of larval and pupal gypsy moth males.

15.8.1 Apyrene and Eupyrene Sperm of Lepidoptera

Most Lepidoptera produce two types of sperm: **apyrene** sperm without a nucleus and nucleated **eupyrene** sperm. Only the latter type can fertilize an egg. The two types of sperm are produced by major differences in the meiotic process (Garvey et al., 2000). Sperm dichotomy appeared early in the evolution of Lepidoptera, but apyrene sperm are not present in the Micropterigidae, one of the most primitive families of Lepidoptera (Sonnenschein and Hauser, 1990; Hamon and Chauvin, 1992). Eupyrene sperm are usually packaged into bundles, while apyrene sperm are dissociated as single, but immobile cells. Both types of sperm are incorporated into the **spermatophore** that is formed in the female bursa copulatrix at the time of mating by secretions from the male. Maturation division (to produce the haploid number of chromosomes) of the eupyrene sperm and development of motility in both types of sperm occur in the spermatophore of some insects (Osanai et al., 1987, 1989). He et al. (1995) demonstrated a correlation between decrease of apyrene sperm in the spermatheca of an army worm *Pseudaletia separata* and remating patterns, and suggested that the number of apyrene sperm in the spermatheca may influence female remating.

Male *B. mori* have an **endopeptidase**, called **initiatorin**, in secretions of the posterior segment of the ejaculatory duct (Osanai et al., 1989) that is important in activation of both apyrene and eupyrene sperm and in maturation of the eupyrene sperm. Initiatorin is a **serine endoprotease** that is active at pH 9.2. It digests the surface coat of apyrene sperm most easily, and these sperm become motile before the euprene sperm are completely freed from their bundles. Their vigorous movements serve to stir the viscous contents of the spermatophore, aiding the liberation of the eupyrene sperm and facilitating metabolic reactions that promote eupyrene sperm maturation. In addition, initiatorin converts an inactive **procarboxypeptidase** secreted by the ampulla, the region of the ejaculatory duct where the vasa deferentia and ducts from the accessory glands converge and empty, into an active **carboxypeptidase**. The carboxypeptidase digests proteins and liberates arginine and other amino acids (Kasuga et al., 1987). Arginine is subsequently converted to glutamate, which is metabolized by the sperm to support motility (Aigaki et al., 1987). By virtue of its production in the terminal portion of the ejaculatory duct, initiatorin is kept away from the sperm and the procarboxypeptidase until ejaculation at mating. Similar processes likely occur in other Lepidoptera.

15.8.2 Male Accessory Glands

Many males have **accessory glands** associated with the reproductive tract. The accessory glands, which have varied morphology in different insects (Chen, 1984; Davey, 1985), empty into the ejaculatory duct. Their secretions are used to form the **spermatophore** in some insects; or if no spermatophore is formed, the secretions are added to the sperm prior to transfer to the female. Some of the secretory products may stimulate contractions in the reproductive tract of females, thus aiding movement of sperm into the spermathecae of the female.

Accessory glands are present in *Drosophila* species and are called **paragonial glands**. The glands synthesize more than 85 proteins (Coulthart and Singh, 1988; Stumm-Zollinger and Chen, 1985). One of the proteins, sometimes called a "sex peptide," is passed at mating to the female, which is then much less receptive to remating for 6 to 9 days (Chen et al., 1988). The peptide also stimulates oviposition (Chen, 1984). Genes controlling the synthesis of a number of the accessory gland proteins have been identified (reviewed in Happ, 1992). Accessory glands are lacking in some insects; for example, they are not present in some dipterans, including the housefly (*Musca domestica*) and stable fly (*Stomoxys calcitrans*).

15.8.3 Transfer of Sperm

Some insects transfer packets or bundles of sperm to the female reproductive tract by insertion of the **aedeagus** into the reproductive tract of the female. Many insects produce a **spermatophore**, which contains the sperm and is transferred to the female. Many orders and families produce spermatophores. Accessory glands secrete **spermatophorins**, proteins that form the spermatophore. Mealworm adults (*Tenebrio molitor*) contain up to eight cell types in the wall of the accessory glands that secrete different proteins in a sequential manner so that specific layers of the spermatophore are formed (Happ, 1987; Shinbo et al., 1987; Grimnes et al., 1986). Spermatophorin production is stimulated in the mealworm by **20-hydroxyecdysone** (Yaginuma et al., 1988), but JH stimulates production in the hemipteran, *R. prolixus* (Gold and Davey, 1989). In *R. prolixus* the spermatophore consists of a pear-shaped mass of transparent mucoprotein in a sol or gel state, depending on the pH. The semen is contained in a slit inside the jelly mass. The protein jelly is secreted in the accessory glands and is first fluid at the pH (about 7) in the glands. As the fluid moves down the reproductive system, the pH decreases to about 5.5 in the bulbous ejaculatorius and intromittent organ. This is evidently at or near the isoelectric point of the secretion and it gels. Spermatozoa are released from the spermatophore after it is deposited in the bursa copulatrix of the female *Rhodnius*. Mechanical abrasion of the spermatophore or the action of proteolytic enzymes, or both, may play a role in releasing sperm. Although it appears that sperm are moved to the spermathecae to be stored without active participation from the female, contractions in the oviducts of the female induced by secretions from the male probably help to force sperm toward the spermathecae. Formation of the spermatophore results in loss of protein from the male, but the effects of this or consequences for the male usually have not been evaluated. The protein content of the spermatophore has often been viewed as male investment in the next generation. In at least one cricket the loss of protein during spermatophore formation and transfer amounted to 40% of the body weight.

15.9 GENDER DETERMINATION

Insects have at least three chromosomal systems for gender determination, with variations existing within the types [reviewed by Lauge (1985) and Wyatt, (1991)]. In **type 1**, probably the most primitive mechanism, the male is heterogametic, or **maleXY** and femaleXX. Type 1 occurs in many different groups, including *D. melanogaster* and other Diptera. A variation within this type is the loss of the Y chromosome, so that the male is **XO**, as found in Odonata, Orthoptera, and among some groups in some orders. The female is heterogametic (**femaleZW**, **maleZZ**) in **type 2**, found in Lepidoptera and the closely related Trichoptera. **Type 3**, in which **females are diploid** while **males are haploid**, is present in Hymenoptera and in some coccids in the order Homoptera.

Variations in the types above include loss or suppression of chromosomes during the early cleavage stages in embryogenesis, resulting in only the germ cells retaining a complete set of chromosomes; subsequently, the sex of individuals is controlled by differences in the incomplete chromosome sets retained by somatic cells. In a few insects, gender can be determined by prevailing temperature, as in subarctic mosquitoes, and, in some gall midges, by available nutrition (Nöthiger and Steinmann-Zwicky, 1985, 1987)

The genetic mechanisms by which the several chromosomal patterns lead to gender determination is variable and poorly known for all except a few insects. Two broad mechanisms are known: the **ratio of sex chromosomes to autosomes** and the presence of **sex-determining genes**. Gender in *D. melanogaster* is determined by the ratio of sets of sex chromosomes to autosomes, or X:A (A = autosome set). Although the Y chromosome carries genes for factors necessary for the production of motile sperm, it does not carry gender-determining genes. The ratio 2X:2A = 1, as in normal females, (or the ratio of 3X:2A in aneuploids) results in female phenotype. Conversely, a ratio of 0.5, as in normal males (1X:2A), or a smaller ratio (1X:3A), produces male phenotypes. Intermediate ratios are known that result in mosaic intersexes in which the individual contains both male and female cells. Because males have only one X chromosome, genes on it are twice as active in males as in females, a process known as **dosage compensation**, but details of how this is brought about are not clear. Although vast genetic information in *D. melanogaster* has resulted in better understanding of gender determination and early development, especially as controlled by genes, these fruit flies cannot be considered representative of insects in general. Chromosomal ratios are ultimately expressed in specific gene actions, and mechanisms are not well understood in insects. Specific **gender-determining genes** control the phenotype of some insects, including the housefly *M. domestica*, the silkworm moth *B. mori,* and some mosquitoes (Baker and Sakai, 1976).

In *Drosophila*, the expression of the chromosome ratio is related to the expression of a set of genes involved in somatic gender determination. The X:A ratio is irreversibly fixed at the blastoderm stage in the embryo (Sanchez and Nöthiger, 1983) by genes (*sisterless-a* and *sisterless-b*, and possibly others) coding for certain proteins that probably act as transcription factors (Torres and Sanchez, 1989). The genes and/or protein products are involved in determining the X:A ratio. The proteins are crucial for neural tube formation, and may be expressed prior to establishment of dosage compensation, so that they exert a greater effect in a female (2X) than in a male (1X), at a time when gender determination is being established (Torres and Sanchez, 1989; Hodgkin, 1990). Another gene, *daughterless* (*da*), known to be required in females (but not males) acts synergistically in unknown ways with the *sisterless-a* and *-b* genes in determining femaleness, possibly in allowing a female-specific expression of a main regulatory on/off gender determination gene, *Sex-lethal* (*Sxl*). Once expression of *Sex-lethal* begins, its activity is maintained by positive feedback from its own gene products in females, and it sets in motion a cascade of gene actions (Baker and Belote, 1983; and briefly reviewed in Wyatt, 1991) that leads to differentiation into a female. In males, *Sex-lethal* does not seem to be involved in causing maleness, but it does regulate a set of genes controlling dosage compensation of the X chromosome, so that the male, with only 1 X, realizes twofold expression of the X-linked genes (Hazelrigg, 1987; Gergen, 1987). Maintenance of female sexual behavior patterns requires sustained expression of at least one of the cascade genes, transformer (tra+) in the adult female. When genetic females with certain *tra* genotypes are reared at high temperature (29°C) or adults females are transferred from 16 to 29°C, they display male courtship behavior (Belote and Baker, 1987; Wyatt, 1991). Presumably at 29°C, the *tra* gene cannot be expressed properly and female behavior is abnormal.

Nöthiger and Steinmann-Zwicky (1985) have proposed a unifying model for gender determination that depends on a primary signal that is monitored by a key gene whose activity state, off or on, controls gender differentiation genes. When the key gene is "on," a female or male can be determined, depending on a second "male/female" switch mechanism. When the key gene is "off," a female cannot be produced and only male development is possible.

REFERENCES

Adams, T.S., A.M. Hintz, and J.G. Pomonis. 1968. Oostatic hormone production in houseflies, *Musca domestica*, with developing ovaries. *J. Insect Physiol.*, 14:983-993.

Adams, T.S., and P.A. Filipi. 1988. Vitellin and vitellogenin concentrations during oogenesis in the first gonotrophic cycle of the housefly, *Musca domestica. J. Insect Physiol.*, 29:723-733.

Adams, T.S., J.W. Gerst, and E.P. Masler. 1997. Regulation of ovarian ecdysteroid production in the housefly, *Musca domestica. Arch. Insect Biochem. Physiol.*, 35:135-148.

Aigaki, T., H. Kasuga, and M. Osanai. 1987. A specific endopeptidase, BAEE esterase, in the glandula prostatica of the male reproductive system of the silkworm, *Bombyx mori. Insect Biochem.*, 17:323-328.

Baeuerle, P.A., and W.B. Huttner. 1985. Tyrosine sulfation of yolk proteins 1, 2, and 3 in *Drosophila melanogaster. J. Biol. Chem.*, 260:6434-6439.

Baker, B.S., and J.M. Belote. 1983. Sex determination and dosage compensation in *Drosophila melanogaster. Annu. Rev. Genetics,* 17:345-393.

Baker, F.C., H.H. Hagedorn, D.A. Schooley, and G. Wheelock. 1983. Mosquito juvenile hormone: Identification and bioassay activity. *J. Insect Physiol.*, 29:465-470.

Baker, M.E. 1988. Is vitellogenin an ancestor of apolipoprotein B-100 of human LDL and human lipoprotein lipase? *Biochem. J.*, 255:1057-1060.

Baker, R.H., and R.K. Sakai. 1976. Male determining factor on chromosome 3 in the mosquito, *Culex tritaeniorhynchus. J. Hered.*, 67:289-294.

Bast, R.E., and W.H. Telfer. 1976. Follicle cell protein synthesis and its contribution to the yolk of the cecropia moth oocyte. *Develop. Biol.*, 52:83-97.

Baxter, J.A., A.M. Mjeni, and P.E. Morrison. 1973. Expression of autogeny in relation to larval population density of *Sarcophaga bullata* Parker (Diptera: Sarcophagidae). *Can. J. Zool.*, 51:1189-1193.

Bean, D.W., P.D. Shirk, and V.J. Brookes. 1988. Characterization of yolk proteins from the eggs of the Indian meal moth, *Plodia interpunctella. Insect Biochem.*, 18:199-210.

Belote, J.M., and B.S. Baker. 1987. Sexual behavior: Its genetic control during development and adulthood in *Drosophila melanogaster. Proc. Natl. Acad. Sci. U.S.A.*, 84:8026-8030.

Bellés, X. 1998. Endocrine effectors in insect vitellogenesis, pp. 71-90, in G.M. Coast and S.G. Webster (Eds.), *Recent Advances in Arthropod Endocrinology*, Cambridge University Press, Cambridge, U.K., 406 pp.

Blumenthal, T., and E. Zucker-Aprison. 1997. Evolution and regulation of vitellogenin genes, pp. 3-19, in J.D. O'Connor (Ed.), *Molecular Biology of Invertebrate Development*, Alan R. Liss, New York.

Boggs, C.L. 1990. A general model of the role of male-donated nutrients in female insects' reproduction. *Am. Nat.*, 136:598-617.

Bonhag, P.F. 1955. Histochemical studies of the ovarian nurse tissues and oocytes of the milkweed bug, *Oncopeltus fasciatus* (Dallas). *J. Morphol.*, 96:381-439.

Bonhag, P.F. 1958. Ovarian structure and vitellogenesis in insects. *Annu. Rev. Entomol.*, 3:137-160.

Borovsky, D. 1982. Release of egg development neurosecretory hormone in *Aedes aegypti* and *Aedes taeniarhynchus* induced by an ovarian factor. *J. Insect Physiol.*, 28:311-316.

Borovsky, D. 1988. Oostatic hormone inhibits biosynthesis of midgut proteolytic enzymes and egg development in mosquitoes. *Arch. Insect Biochem. Physiol.*, 7:187-210.

Borovsky, D., D.A. Carlson, P.R. Griffin, J. Shabanowitz, and D.F. Hunt. 1990. Mosquito oostatic factor: A novel decapeptide modulating trypsin-like enzyme biosynthesis in the midgut. *FASEB J.*, 4:3015-3020.

Borovsky, D., C.A. Powell, J.K. Nayar, J.E. Blalock, and T.K Hayes. 1994. Characterization and localization of mosquito-gut receptors for trypsin modulating oostatic factor using a complementary peptide and immunocytochemistry. *FASEB J.*, 8:350-355.

Borovsky, D., and P.L. Whitney. 1987. Biosynthesis, purification, and characterization of *Aedes aegypti* vitellin and vitellogenin. *Arch. Insect Biochem. Physiol.*, 4:81-99.

Borovsky, D., and E. van Handel. 1980. Synthesis of ovary-specific proteins in mosquitoes. *Int. J. Invert. Reprod.*, 2:153-163.

Bownes, M. 1986. Expression of the genes coding for vitellogenin (yolk protein). *Annu. Rev. Entomol.*, 32:507-531.

Bownes, M., A. Shirras, M. Blair, J. Collins, and A. Coulson. 1988. Evidence that insect embryogenesis is regulated by ecdysteroids released from yolk proteins. *Proc. Natl. Acad. Sci. U.S.A.*, 84:1554-1557.

Bownes, M., E. Ronaldson, D. Mauchline, and A. Martínez. 1993. Regulation of vitellogenesis in *Drosophila*. *Int. J. Insect Morphol. Embryol.*, 22:349-367.

Bradfield, J.Y., R.L. Berlin, and L.L. Keeley. 1989. Contrasting modulations of gene expression by a juvenile hormone analog. *Insect Biochem.*, 20:105-111.

Chen, P.S. 1984. The functional morphology and biochemistry of insect male accessory glands and their secretions. *Annu. Rev. Entomol.*, 29:233-255.

Chen, P.S., E. Stumm-Zollinger, T. Aigaki, J. Balmer, M. Bienz, and P. Böhlen. 1988. A male accessory gland peptide that regulates reproductive behavior of female *D. melanogaster*. *Cell*, 54:291-298.

Cho, W.-L., M.Z. Kapitskaya, and A.S. Raikhel. 1995. Mosquito ecdysteroid receptor: Analysis of the cDNA and expression during vitellogenesis. *Insect Biochem. Mol. Biol.*, 25:19-27.

Coulthart, M.B., and R.S. Singh. 1988. Differing amounts of genetic polymorphism in testes and male accessory glands of *Drosophila melanogaster* and *Drosophila simulans*. *Biochem. Genet.*, 26:153-164.

Cusson, M., C.G. Yu, K. Carruthers, G.R. Wyatt, S.S. Tobe, and J.N. McNeil. 1994. Regulation of vitellogenin production in armyworm moths, *Pseudaletia unipuncta*. *J. Insect Physiol.*, 40:129-136.

Davey, K.G. 1981. Hormonal control of vitellogenin uptake in *Rhodnius prolixus* Stål. *Am. Zool.*, 21:701-705.

Davey, K.G. 1985. The male reproductive tract, pp. 1-14, in G.A. Kerkut and L.I. Gilbert (Eds.), *Comprehensive Insect Physiology, Biochemistry and Pharmacology*, Vol. 1, Pergamon Press, Oxford.

Davey, K.G. 1993. Hormonal control of egg production in *Rhodnius prolixus*. *Am. Zool.*, 33:397-402.

Davey, K.G. 1997. Hormonal controls on reproduction in female Heteroptera. *Arch. Biochem. Physiol.*, 35:443-453.

DeBianchi, A.G., M. Coutinho, S.D. Pereira, O. Marinotti, and H.J. Targa. 1985. Vitellogenin and vitellin of *Musca domestica*. Quantification and synthesis by fat bodies and ovaries. *Insect Biochem.*, 15:77-84.

Dhadialla, T.S., and A.S. Raikhel. 1990. Biosynthesis of mosquito vitellogenin. *J. Biol. Chem.*, 265:9924-9933.

Dodson, G. 1978. Morphology of the reproductive system in *Anastrepha suspensa* (Loew) and notes on related species. *Florida Entomol.*, 61:231-239.

Engelmann, F. 1983. Vitellogenesis controlled by juvenile hormone, pp. 259-270, in R.G.H. Downer and H. Laufer (Eds.), *Endocrinology of Insects*, Alan R. Liss, New York.

Fargnoli, J., and G.L. Waring. 1982. Identification of vitelline membrane proteins in *Drosophila melanogaster*. *Develop. Biol.*, 92:306-314.

Fecemyer, H.W., E.P. Masler, R.E. Davis, and T.J. Kelly. 1992. Vitellogenin synthesis in female larvae of the gypsy moth, *Lymantria dispar* (L.): Suppression by juvenile hormone. *Comp. Biochem. Physiol.,* 103B:533-542.

Fisher, R.C. 1971. Aspects of the physiology of endoparasitic Hymenoptera. *Biol. Rev.,* 46:243-278.

Flanders, S.E. 1942. Oosorption and ovulation in relation to oviposition in the parasitic Hymenoptera. *Ann. Entomol. Soc. Am.,* 35:251-266.

Fletcher, D.J.C., and M.S. Blum. 1981. Pheromonal control of dealation and oogenesis in virgin queen fire ants. *Science,* 212:73-75.

French, A., and R. Hoopingarner. 1965. Gametogenesis in the housefly, *Musca domestica*. *Ann. Entomol. Soc. Am.,* 58:650-657.

Garvey, L.K., G.M. Gutierrez, and H.M. Krider. 2000. Ultrastructure and morphogenesis of apyrene and eupyrene spermatozoa in the gypsy moth (Lepidoptera: Lymantriidae). *Ann. Entomol. Soc. Am.,* 93:1147-1155.

Gergen, J.P. 1987. Dosage compensation in *Drosophila*: evidence that *daughterless* and *Sex-lethal* control X chromosome activity at the blastoderm stage of embryogenesis. *Genetics,* 117:177-185.

Gillott, C., and T. Friedel. 1977. Fecundity-enhancing and receptivity-inhibiting substances produced by male insects: A review. *Adv. Invertebr. Reprod.,* 1:199-218.

Glinka, A.V., and G.R. Wyatt. 1996. Juvenile hormone activation of gene transcription in locust fat body. *Insect Biochem. Mol. Biol.,* 26:13-18.

Gold, S.M.W., and K.G. Davey. 1989. The effect of juvenile hormone on protein synthesis in the transparent accessory gland of male *Rhodnius prolixus*. *Insect Biochem.,* 19:139-143.

Grimnes, K.A., C.S. Bricker, and G.M. Happ. 1986. Ordered flow of secretion from accessory glands to specific layers of a spermatophore of mealworm beetles: Demonstration with a monoclonal antibody. *J. Exp. Zool.,* 240:275-286.

Hagedorn, H. 1985. The role of ecdysteroids in reproduction, pp. 205-261, in G.A. Kerkut and L.I. Gilbert (Eds.), *Comprehensive Insect Physiology, Biochemistry and Pharmacology,* Vol. 8, Pergamon Press, Oxford, UK.

Hagedorn, H. 1989. Physiological roles of hemolymph ecdysteroids in the adult insect, pp. 279-289, in J. Koolman (Ed.), *Ecdysone, from Chemistry to Mode of Action*, Thieme, Stuttgart, Germany.

Hagedorn, H.H., J.D. O'Connor, M.S. Fuchs, B. Sage, D.A. Schlaeger, and M.K. Bohm. 1975. The ovary as a source of an ecdysone in an adult mosquito. *Proc. Natl. Acad. Sci. U.S.A.,* 72:3255-3259.

Hagedorn, H.H., J.P. Shapiro, and K. Hanaoka. 1979. Ovarian ecdysone secretion is controlled by a brain hormone in an adult mosquito. *Nature,* 282:92-94.

Hammond, M.P., and C.D. Laird 1985. Chromosome structure and DNA replication in nurse and follicle cells of *Drosophila melanogaster*. *Chromosoma,* 91:267-278.

Hamon, C., and G. Chauvin. 1992. Ultrastructural analysis of spermatozoa of *Korscheltellus lupulinus* L. (Lepidoptera: Hepialidae) and *Micropterix calthella* L. (Lepidoptera: Micropterigidae). *Int. J. Insect Morphol. Embryol.,* 21:149-160.

Handler, A.M., and P. Shirk. 1988. Identification and analysis of the major yolk polypeptides from the Caribbean fruit fly, *Anastrepha suspensa* (Loew). *Arch. Insect Biochem. Physiol.,* 9:91-106.

Happ, G.M. 1987. Accessory gland development in mealworm beetles, pp. 433-442, in J.L. Law (Ed.), *Molecular Entomology,* Alan R. Liss, New York.

Happ, G.M. 1992. Maturation of the male reproductive system and its endocrine regulation. *Annu. Rev. Entomol.,* 37:303-320.

Hazelrigg, T. 1987. The *Drosophila white* gene: A molecular update. *Trends Genet.,* 3:43-47.

He, Y., T. Tanaka, and T. Miyata. 1995. Eupyrene and apyrene sprerm and their numerical fluctuations inside the female reproductive tract of the army worm *Pseudaletia separata*. *J. Insect Physiol.,* 41:689-694.

Hodgkin, J. 1990. Sex determination compared in *Drosophila* and *Caenorhabditis*. *Nature (London),* 244:721-728.

Huybrechts, R., and A. DeLoof. 1982. Similarities in vitellogenin and control of vitellogenin synthesis within the genera *Sarcophaga*, *Calliphora*, *Phormia* and *Lucilia* (Diptera). *Comp. Biochem. Physiol.,* 72B:339-344.

Ilenchuk, T.T., and K.G. Davey. 1987. Effects of various compounds on Na/K-ATPase activity, JH I binding capacity and patency response in follicles of *Rhodnius prolixus*. *Insect Biochem.,* 17:1085-1088.

Imboden, H., R. König, P. Ott, A. Lustig, U. Kämpfer, and B. Lanzrein. 1987. Characterization of the native vitellogenin and vitellin of the cockroach, *Nauphoeta cineraea*, and comparison with other species. *Insect Biochem.,* 17:353-365.

Irie, K., and O.Yamashita. 1983. Egg-specific protein in the silkworm, *Bombyx mori*: Purification, properties, localization and titre changes during oogenesis and embryogenesis. *Insect Biochem.,* 13:71-80.

Izumi, S., J. Fujie, S. Yamada, and S. Tomino. 1981. Molecular properties and biosynthesis of major plasma proteins in *Bombyx mori*. *Biochim. Biophys. Acta,* 670:222-229.

Jones, T. 1978. The blood/germ cell barrier in male *Schistocerca gregaria*: The time of its establishment and factors affecting its formation. *J. Cell Sci.,* 31:145-163.

Kafatos, F.C., W. Orr, and C. Delidakis. 1985. Developmentally regulated gene amplification. *Trends Genet.,* 1:301-306.

Kafatos, F.C., N. Spoerel, S.A. Mitsialis, H.T. Nguyen, C. Romano, J.R. Lingappa, B.D. Mariani, G.C. Rodakis, R. Lecanidou, and S.G. Tsitilou. 1987. Developmental control and evolution in the chorion gene families of insects. *Adv. Genet.,* 24:223-242.

Kasuga, H., R. Aigaki, and M. Osanai. 1987. System for supply of free arginine in the spermatophore of *Bombyx mori*. Arginine-liberating activities of contents of male reproductive glands. *Insect Biochem.,* 17:317-322.

Keely, L.L. 1985. Physiology and biochemistry of the fat body, pp. 211-248, in G.A. Kerkut and L.I. Gilbert (Eds.), *Comprehensive Insect Physiology, Biochemistry and Pharmacology*, Vol. 3, Pergamon Press, Oxford, U.K.

Kelly, T.J., M.J. Birnbaum, C.W. Woods, and A.B. Borkovec. 1984. Effects of housefly oostatic hormone on egg development neurosecretory hormone action in *Aedes atropalpus*. *J. Exp. Zool.,* 229:491-496.

Kelly, T.J., E.P. Masler, M.B. Schwartz, and S.B. Haught. 1986. Inhibitory effects of oostatic hormone on ovarian maturation and ecdysteroid production in Diptera. *Insect Biochem.,* 16:273-279.

Kirchgessner, T.G., J.-C. Chuat, C. Heinzmann, J. Etienne, S. Guilhot, K. Svenson, D. Ameis, C. Pilon, L. D'Auriol, A. Andalibi, M.C. Schotz, R. Galibert, and A.J. Lusis. 1989. Organization of the human lipoprotein lipase gene and evolution of the lipase gene family. *Proc. Natl. Acad. Sci. U.S.A.,* 86:9647-9651.

Klowden, M.J. 1997. Endocrine aspects of mosquito reproduction. *Arch. Insect Biochem. Physiol.,* 35:491-512.

Klowden, M.J. 1987. Distention-mediated egg maturation in the mosquito, *Aedes aegypti*. *J. Insect Physiol.,* 33:83-87.

Koller, C.N., T.S. Dhadialla, and A.S. Raikhel. 1989. A study of receptor-mediated endocytosis of vitellogenin in mosquito oocytes. *Insect Biochem.,* 19:693-702.

Kunkel, J.G., and J.H. Nordin, 1985. Yolk proteins, pp. 84-111, in G.A. Kerkut and L.I. Gilbert (Eds.), *Comprehensive Insect Physiology, Biochemistry and Pharmacology*, Vol. 1, Pergamon Press, Oxford.

Lauge, G. 1985. Sex determination: Genetic and epigenetic factors, pp. 295-318, in G.A. Kerkut and L.I. Gilbert (Eds.), *Comprehensive Insect Physiology, Biochemistry and Pharmacology*, Vol. 1, Pergamon Press, Oxford.

Law, J.H. 1989. Egg proteins other than vitellin. *5th Int. Congr. Invert. Reprod.,* Nagoya. (Abstract p. 45).

Lecanidou, R., G.C. Rodakis, T.H. Eickbush, and F.C. Kafatos. 1986. Evolution of the silk moth chorion gene superfamily: Gene families CA and CB. *Proc. Natl. Acad. Sci. U.S.A.,* 83:6514-6518.

Martín, D., M.-D. Piulachs, and X. Bellés. 1995. Patterns of haemolymph vitellogenin and ovarian vitellin in the German cockroach, and the role of juvenile hormone. *Physiol. Entomol.,* 20:59-65.

Martínez, A., and M. Bownes. 1994. The sequence and expression pattern of the *Calliphora erythrocephala* yolk protein A and B genes. *J. Mol. Evolution,* 38:336-351.

Nöthiger, R., and M. Steinmann-Zwicky. 1985. A single principle for sex determination in insects. *Cold Spring Harbor Symp. Quant. Biol.,* 50:615-621.

Nöthiger, R., and M. Steinmann-Zwicky. 1987. Genetics of sex determination in eukaryotes, pp. 271-300, in W. Hennig (Ed.), *Structure and Function of Eukaryotic Chromosomes, Results and Problems in Cell Differentiation*, Vol. 14, Springer-Verlag, Berlin.

Osanai, M., T. Aigaki, and H. Kasuga. 1987. Arginine degradation cascade as an energy-yielding system for sperm maturation in the spermatophore of the silkworm, *Bombyx mori*, pp. 185-195, in H. Mohri (Ed.), *New Horizons in Sperm Research*, Japanese Scientific Societies Press, Tokyo.

Osanai, M., H. Kasuga, and T. Aigaki. 1989. Induction of motility of apyrene spermatozoa and dissociation of eupyrene sperm bundles of the silkworm, *Bombyx mori* by initiatorin and trypsin. *Invertebr. Reprod. Dev.,* 15:97-103.

Pan, M.-L., W.J. Bell, and W.H. Telfer. 1969. Vitellogenic blood protein synthesis by insect fat body. *Science,* 165:393-394.

Pan, M.-L., and G.R. Wyatt. 1971. Juvenile hormone induces vitellogenin synthesis in the monarch butterfly. *Science,* 174:503-505.

Park, Y.I., S. Shu, S.B. Ramaswamy, and A. Srinivasan. 1998. Mating in *Heliothis virescens*: Transfer of juvenile hormone during copulation by male to female and stimulation of biosynthesis of endogenous juvenile hormone. *Arch. Insect Biochem. Physiol.,* 38:100-107.

Raikhel, A.S. 1984. The accumulative pathway of vitellogenin in the mosquito oocyte; a high resolution immuno- and cytochemical study. *J. Ultrastruct. Res.,* 87:285-302.

Raikhel, A.S. 1987. Monoclonal antibodies as probes for processing of yolk protein in the mosquito; a high-resolution immunolocalization of secretory and accumulative pathways. *Tissue Cell,* 19:515-529.

Raikhel, A.S., and T.S. Dhadialla. 1992. Accumulation of yolk proteins in insect oocytes. *Annu. Rev. Entomol.,* 37:217-251.

Ramaswamy, S.B., S. Shu, G.N. Mbata, A. Rachinsky, Y.I. Park, L. Crigler, S. Donald, and A. Srinivasan. 2000. Role of juvenile hormone-esterase in mating-stimulated egg development in the moth *Heliothis virscens*. *Insect Biochem. Mol. Biol.,* 30:785-791.

Ramaswamy, S.B., S. Shu, Y.I. Park, and F. Zeng. 1997. Dynamics of juvenile hormone-mediated gonadotropism in the Lepidoptera. *Arch. Insect Biochem. Physiol.,* 35:539-558.

Rankin, S.M., J. Chambers, and J.P. Edwards. 1997. Juvenile hormone in earwigs: Roles in oogenesis, mating, and maternal behaviors. *Arch. Insect Biochem. Physiol.,* 35:427-442.

Rina, M., and C. Savakis. 1991. A cluster of vitellogenin genes in the Mediterranean fruit fly, *Ceratitis capitata*: Sequence and structural conservation in dipteran yolk proteins and their genes. *Genetics,* 127:769-780.

Robbins, W.E., and T.J. Shortino. 1962. Effect of cholesterol in the larval diet on ovarian development in the adult housefly. *Nature,* 194:502-503.

Röhrkasten, A., and H.-J. Ferenz. 1987. Coated vesicles from locust oocytes: Isolation and characterization. *Int. J. Invertebr. Reprod. Dev.,* 12:341-346.

Romans, P., Z. Tu, Z. Ke, and H. Hagedorn. 1995. Analysis of a vitellogenin gene of the mosquito, *Aedes aegypti*, and comparisons to vitellogenins from other organisms. *Insect Biochem. Mol. Biol.,* 25:939-958.

Rousset, A., and C. Bitsch. 1993. Comparison between endogenous and exogenous yolk proteins along an ovarian cycle in the firebrat *Thermobia domestica* (Insects, Thysanura). *Comp. Biochem. Physiol.,* 104B:33-44.

Rubenstein, E.C. 1979. The role of an epithelial occlusion zone in the termination of vitellogenesis in *Hyalophora cecropia* ovarian follicles. *Dev. Biol.,* 71:115-127.

Sanchez, L., and R. Nöthiger. 1983. Sex determination and dosage compensation in *Drosophila melanogaster*: Production of male clones in XX females. *EMBO J.,* 2:485-491.

Sato, Y., S. Inagaki, and O. Yamashita. 1990. Egg-specific protein in the silkworm, *Bombyx mori*: Gene structure, expression and post-translational modification, pp. 91-95, in M. Hoslin and O. Yamashita (Eds.), *Advances in Invertebrate Reproduction,* Elsevier, Amsterdam.

Satyanarayana, K., J.Y. Bradfield, G. Bhaskaran, and K.H. Dahm. 1994. Stimulation of vitellogenin production by methoprene in prepupae and pupae of *Manduca sexta*. *Arch. Insect Biochem. Physiol.,* 25:21-37.

Schäfer, M., and W. Kunz. 1987. Ribosomal gene amplification does not occur in the oocytes of *Locusta migratoria*. *Dev. Biol.,* 120:43-52.

Schal, C., G.L. Holbrook, J.A.S. Bachmann, and V.L. Sevala. 1997. Reproductive biology of the German cockroach, *Blattella germanica*: Juvenile hormone as a pleiotropic master regulator. *Arch. Insect Biochem. Physiol.,* 35:405-426.

Schlottman, L., and P. Bonhag. 1956. Histology of the ovary of the adult mealworm *Tenebrio molitor* L. (Coleoptera, Tenebrionidae). University of California Publications in Entomology, University of California Press, Berkeley, Vol. 11, No. 6, pp. 351-394.

Shea, M.J., D.L. King, M.J. Conboy, B.D. Mariani, and F.C. Kafatos. 1990. Proteins that bind to *Drosophila* chorion *cis*-regulatory elements: A new C_2H_2 zinc finger protein and a C_2C_2 steroid receptor-like component. *Genes Devel.,* 4:1128-1140.

Shinbo, H., T. Yaginuma, and G.M. Happ. 1987. Purification and characterization of a proline-rich secretory protein on an insect spermatophore. *J. Biol. Chem.,* 262:4794-4799.

Shirk, P.D. 1987. Comparison of yolk production in seven pyralid moth species. *Int. J. Invertebr. Reprod. Dev.,* 11:173-188.

Shirk, P.D., P. Minoo, and J.H. Postlethwait. 1983. 20-Hydroxyecdysone stimulates the accumulation of translatable yolk polypeptide gene transcript in adult male *Drosophila melanogaster*. *Proc. Natl. Acad. Sci. U.S.A.,* 80:186-190.

Sonnenschein, M., and Ch. L. Hauser. 1990. Presence of only eupyrene spermatozoa in adult males of the genus *Microtpterix* Hubner and its phylogenetic significance (Lepidoptera: Zeugloptera, Micropterigidae). *Int. J. Insect Morphol. Embryol.,* 19:269-276.

Sorge, D., R. Nauen, S. Range, and K.H. Hoffmann. 2000. Regulation of vitellogenesis in the fall armyworm, *Spodoptera frugiperda* (Lepidoptera: Noctuidae). *J. Insect Physiol.,* 46:969-976.

Speith, J., K. Denison, S. Kirtland, J. Cand, and T. Blumenthal. 1985. The *C. elegans* vitellogenin genes: short sequence repeats in the promoter regions and homology to the vertebrate genes. *Nucl. Acids Res.,* 13:5283-5295.

Spielman, A. 1971. Bionomics of autogenous mosquitoes. *Annu. Rev. Entomol.,* 16:231-248.

Spradling, A.C., and A.P. Mahowald. 1980. Amplification of genes for chorion proteins during oogenesis in *Drosophila melanogaster. Proc. Natl. Acad. Sci. U.S.A.,* 77:1096-1100.

Stoffolano, Jr., J.G., M.-F. Li, B.-X. Zou, and C.-M. Yin. 1992. Vitellogenin uptake, not synthesis, is dependent on juvenile hormone in adults of *Phormia regina* (Meigen). *J. Insect Physiol.,* 11:839-845.

Strambi, A., C. Strambi, and M. Cayre. 1997. Hormonal control of reproduction and reproductive behavior in crickets. *Arch. Insect Biochem. Physiol.,* 35:393-404.

Stumm-Zollinger, E., and P.S. Chen. 1985. Protein metabolism of *Drosophila melanogaster* male accessory glands. 1. Characterization of secretory proteins. *Insect Biochem.,* 15:375-383.

Stynen, D., R.I. Woodruff, and W.H. Telfer. 1988. Effect of ionophores on vitellogenin uptake by *Hyalophora* oocytes. *Arch. Insect Biochem. Physiol.,* 8:261-276.

Telfer, W.H. 1954. Immunological studies of insect metamorphosis. II. The role of a sex-limited blood protein in egg formation by the cecropia silkworm. *J. Gen. Physiol.,* 37:539-558.

Telfer, W.H. 1961. The route of entry and localization of blood proteins in the oocytes of saturniid moths. *J. Biophys. Biochem. Cytol.,* 9:747-759

Telfer, W.H. 1965. The mechanism and control of yolk formation. *Annu. Rev. Entomol.,* 10:161-184.

Telfer, W.H., E. Rubinstein, and M.-L. Pan. 1981. How the ovary makes yolk in *Hyalophora*, pp. 637-654, in F. Sehnal, A. Zabza, J. Menn, and B. Cymborowski (Eds.), *Regulation of Insect Development and Behavior*, Wroclaw Technical University Press, Wroclaw, Poland.

Telfer, W.H., and M.-L. Pan. 1988. Adsorptive endocytosis of vitellogenin, lipophorin, and microvitellogenin during yolk formation in *Hyalophora. Arch. Insect Biochem. Physiol.,* 9:339-355.

Tepstra, P., and AB, G. 1988. Homology of *Drosophila* yolk proteins and the triacylglycerol lipase family. *J. Mol. Biol.,* 202:663-665.

Torres, M., and L. Sanchez. 1989. The *scute* (T4) gene acts as a numerator element of the X:A signal that determines the state of activity of *Sex-lethal* in *Drosophila. EMBO J.,* 8:3079-3086.

Tsuchida, K., M. Nagata, and A. Suzuki. 1987. Hormonal control of ovarian development in the silkworm, *Bombyx mori. Arch. Insect Biochem. Physiol.,* 5:167-177.

Wagner, R.M., M.J. Loeb, J.P. Kochansky, D.B. Gelman, W.R. Lusby, and R.A. Bell. 1997. Identification and characterization of an ecdysiotropic peptide from brain extracts of the gypsy moth, *Lymantria dispar. Arch. Insect Biochem. Physiol.,* 34:175-189.

Wahli, W. 1988. Evolution and expression of vitellogenin genes. *Trends Genet.,* 4:227-232.

Waring, G.L., and A.P. Mahowald. 1979. Identification and time of synthesis of chorion proteins in *Drosophila melanogaster. Cell,* 16:599-607.

Wheeler, D. 1996. The role of nourishment in oogenesis. *Annu. Rev. Entomol.,* 41:407-431.

Whisenton, L.R., J.T., Warren, M.K. Manning, and W.E. Bollenbacher. 1989. Ecdysteroid titers during pupal-adult development of *Aedes aegypti*: Basis for a sexual dimorphism in the rate of development. *J. Insect Physiol.,* 35:67-73.

Whiting, P., S. Sparks, and L. Dinan. 1997. Endogenous ecdysteroid levels and rates of ecdysone acylation by intact ovaries *in vitro* in relation to ovarian development in adult female house crickets, *Acheta domesticus. Arch. Insect Biochem. Physiol.,* 35:279-299.

Wick, J.R., and P.F. Bonhag. 1955. Postembryonic development of the ovaries of *Oncopeltus fasciatus* (Dallas). *J. Morphol.,* 96:31-60.

Wu, S.-J., J.-Z. Zhang, and M. Ma. 1987. Monitoring the effects of juvenile hormones and 20-hydroxyecdysone on yolk polypeptide production of *Drosophila melanogaster* with enzyme immunoassay. *Physiol. Entomol.,* 12:355-361.

Wyatt, G.R., K.E. Cook, H. Firko, and T.S. Dhadialla. 1987. Juvenile hormone action on locust fat body. *Insect Biochem.,* 17:1071-1074.

Wyatt, G.R. 1991. Gene regulation in insect reproduction. *Invertebr. Reprod. Dev.,* 20:1-35.

Yaginuma, T., H. Kai, and G.M. Happ. 1988. 20-Hydroxyecdysone accelerates the flow of cells into the G_1 phase and the S phase in a male accessory gland of a mealworm pupa. *Dev. Biol.,* 126:173-181.

Yano, K., M.T. Sakurai, S. Izumi, and S. Tomino. 1994a. Vitellogenin gene of the silkworm, *Bombyx mori*: Structure and sex-dependent expression. *FEBS Lett.,* 356:207-211.

Yano, K., M.T. Sakurai, S. Watabe, S. Izumi, and S. Tomino. 1994b. Structure and expression of mRNA for vitellogenin in *Bombyx mori. Biochim. Biophys. Acta,* 1218:1-10.

Yin, C.-M., B.-X. Zou, and J.G. Stoffolano, Jr. 1989. Precocene II treatment inhibits terminal oöcyte development but not vitellogenin synthesis and release in the black blowfly, *Phormia regina* (Meigen). *J. Insect Physiol.,* 36:375-382.

Yin, C.-M., B.-X. Zou, S.-X. Yi, and J.G. Stoffolano, Jr. 1990. Ecdysteroid activity during oögenesis in the black blowfly, *Phormia regina* (Meigen). *J. Insect Physiol.,* 36:375-382.

Yin, C.-M., B.X. Zou, M.G. Jiang, M.F. Li, W.H. Qin, T.L. Potter, and J.G. Stoffolano. 1995. Identification of juvenile hormone III bisepoxide (JHB3), juvenile hormone III and methyl farnesoate secreted by the corpus allatum of *Phormia regina* (Meigen), *in vitro* and function of JHB3 either applied alone or as a part of a juvenoid blend. *J. Insect Physiol.,* 41:473-479.

Yin, C.-M., and J.G. Stoffolano, Jr. 1997. Juvenile hormone regulation of reproduction in the cyclorrhaphous Diptera with emphasis on oogenesis. *Arch. Insect Biochem. Physiol.,* 35:513-537.

Zeng, E., S. Shu, and S.B. Ramaswamy. 1997. Vitellogenin and egg production in the moth, *Heliothis virescens. Arch. Biochem. Physiol.,* 34:287-300.

Zhang, J.-Z., and G.A. Wyatt. 1990. A new member of a low molecular weight hemolymph protein family from *Locusta migratoria*, pp. 385 (abstract), in H.H. Hagedorn, J.C. Hildebrand, Kidwell, M.G., and J.H. Law (Eds.), *Molecular Insect Science,* Plenum Press, New York.

Appendix

The purpose of this appendix is to provide some background information about insects and their near relatives for those who may not be trained as invertebrate biologists or entomologists.

THE ARTHROPODA

Despite their differences, arthropods share a number of characteristic features, including a chitinous exoskeleton that must be molted periodically as the animal grows, jointed legs, a well-developed ventral nervous system, and an open circulatory system with a dorsal vessel or heart. The body of arthropods contains a hemocoel through which the blood flows freely once it leaves the dorsal vessel, although many members of this group show remnants of peripheral vessels that direct the flow of hemolymph. The hemocoel is a cavity derived from the embryonic blastocoel and is not a true coelom, which is defined as a cavity lined by mesoderm. Various arthropods have gills (some aquatic ones), a tracheal system, book lungs, or book lungs and a tracheal system for gas exchange. Some arthropods have a blood pigment that aids in transport of oxygen to the tissues. The phylum Arthropoda is divided by some authorities into five subphyla.

1. **Subphylum Trilobita**. This is an extinct group of marine arthropods that flourished in the Cambrian period some 500 million years ago. They are very common in the fossil record, were probably quite diverse, and probably included a number of classes.
2. **Subphylum Chelicerata**. The Chelicerata are commonly divided into four classes. The class Eurypterida, an extinct group called giant water scorpions, is known from fossils of the Paleozoic era. Some were as large as 2.5 meters (m), and were probably predators. The class Pycnogonida contains the relatively rare and exotic sea spiders, marine arthropods found in the oceans and especially in shallow water near the North and South poles. Nearly all members of the class Merostomata are extinct, but surviving remnants include the horseshoe crabs, living relics of an ancient line of Chelicerata having changed little over 350 million years. Horseshoe crabs are marine bottom-feeders living in shallow water along the coasts of North and South America, China, Japan, and the East Indies. *Limulus polyphemus*, common in North America, has been important in physiological studies, particularly in studies of the compound eyes. The class Arachnida includes about 60,000 species, most of which are carnivores. This class includes spiders, ticks, mites, scorpions, whipscorpions, daddy longlegs, and a few less-common relatives.

   All members of the Chelicerata subphylum typically have the body divided into two regions: a cephalothorax and an abdomen. They do not have antennae. The name for the group comes from the structure of the first pair of mouthparts, called the chelicerae, which may be pincer-like or fang-like, but not mandibulate. The second pair of mouth appendages are the pedipalps, which serve in a variety of functions in different groups of the Chelicerata, including food manipulation, locomotion, defense, and copulation.

   The arachnids are the most diverse of the Chelicerata today, and members exhibit many morphological and physiological adaptations for life in varying terrestrial habitats. The body tends to be divided into a cephalothorax and an abdomen, although the latter is not evident in ticks and daddy longlegs. Arachnids do not have compound eyes or antennae, but some have simple eyes on the cephalothorax. They typically have six pairs of jointed appendages, two pairs of which are mouthparts. In spiders, the chelicerae, the first pair of appendages, are fang-like structures that are used to inject poison into the prey. Some spiders use the second pair (the pedipalps) to manipulate and chew food. In some species, the pedipalps are gustatory sensory organs, and in others they serve in courtship display and in sperm transfer. The remaining four pairs of appendages are used for walking. Some arachnids use tracheal tubes in gas exchange; others use book lungs; and some have both book lungs and tracheal tubes. Book lungs consist of a series of thin plates (similar to pages in a book, and hence the name) paired ventrally at as many as four sites in some arachnids. The plates contain blood vessels and gas exchange occurs as blood flows through the plates. A respiratory pigment, the copper-containing protein hemocyanin, is present in the blood. Air reaches the book lungs through slits in the outer body wall.

3. **Subphylum Crustacea**. Crabs, barnacles, shrimps, brine shrimps, crayfish, lobsters, fairy shrimps, water fleas, sand hoppers, and sow (pill) bugs are crustaceans. Crustaceans are a diverse group and their body morphology is highly variable, but most have two pairs of antennae on the head, mandibulate mouthparts, and compound eyes as adults. The mandibles may be modified for biting and chewing, or piercing and sucking. Most crustaceans also have a second pair of mouthparts called the maxillae, which are used for holding and manipulating food. Additional appendages on the body are specialized for walking, swimming, sperm transfer, carrying eggs and young, or serve as sensory structures. Nearly all crustaceans live in aquatic habitats in marine environments or in fresh water. Aquatic species generally have gills and a respiratory pigment (hemocyanin) for gas exchange. Two large antennal glands that open at the base of each antenna serve as excretory organs. Sow bugs are terrestrial and have a tracheal system for respiration.
4. **Subphylum Uniramia**. This largest group of all living animals contains five classes: the Chilopoda (centipedes); Diplopoda (millipedes); Pauropoda (0.5 to 2-mm arthropods living in leaf litter and soil and resembling centipedes, although not necessarily closely related to them); Symphyla (a small group of small arthropods with mouthparts that resemble those of insects in the view of some, but not all, authorities); and the largest group of all other animals put together, the Insecta. Centipedes, millipedes, pauropodans, and symphylans share many characteristics with each other, including a long trunk with many legs; a five- or six-segmented head; and living in leaf litter, loose soil, rotting wood, and similar moist habitats. Some authorities recommend placing them together in the class Myriapoda, or even raising the group to a subphylum level.

THE CLASS INSECTA

There are more than 750,000 described species of insects, with new ones being described on a continuing basis. Some authorities, such as E.O. Wilson of Harvard University, suggest that there are millions of species yet undescribed. Most authorities agree that insects are the most numerous and diverse of all animals, and most diverse in number of species and individuals on Earth. Authorities disagree on the number of orders of insects. Arnett (2000) lists 30 orders (Table 1). The order with the largest number of insects, the Coleoptera (beetles and weevils), contains more than 300,000 described species. Readers interested in more systematic and taxonomic details can consult one of the several general entomology textbooks listed at the end of this chapter.

Adult insects, and some immature ones, typically have six segmented legs, with one pair attached to each of the three thoracic segments. Some larval insects, such as Hymenoptera larvae and Diptera larvae, are legless. Other larval insects, such as caterpillars (Lepidoptera), have fleshy prolegs that are not jointed and that arise from various segments of the larval body. All insects molt from time to time as they grow to fill the old cuticular exoskeleton; they secrete a new exoskeleton beneath the old one before the old is shed, and then ecdysis of the old cuticle occurs.

Many orders of insects (described as Holometabola by some authorities and in this book) undergo complete metamorphosis, with egg, larva, pupa, and adult forms. Others have a gradual

TABLE 1
A Listing of Insect Orders[a] and Type of Metamorphosis According to Arnett

Ametabolous (no external metamorphosis)	Hemimetabolous (gradual metamorphosis with a naiad)[b]	Paurometabolous (gradual metamorphosis with a nymph)	Holometabolous (complete metamorphosis)
Collembola	Ephemeroptera	Grylloblattodea	Neuroptera
Protura	Odonata	Orthoptera	Coleoptera
Entotrophi	Plecoptera	Phasmatodea	Trichoptera
Microcoryphia		Dictyoptera	Lepidoptera
Thysanura		Isoptera	Mecoptera
Mallophaga		Dermaptera	Hymenoptera
Anoplura		Embioptera	Diptera
		Zoraptera	Siphonaptera
		Psocoptera	
		Hemiptera	
		(Orders with a prepupal and pupal stage)	
		Homoptera	
		Thysanoptera	

[a] Nearly every entomology book has slightly different names for some of the orders, and not infrequently different ways of listing them. The author does not intend by the present listing to sanction a particular kind of systematics. The table is presented as an example and as a terminology aid for readers not specializing in insects.
[b] *Naiad* is a term used in the literature to describe the aquatic, immature form of certain groups of insects.

metamorphosis (the Hemimetabola and Paurometabola) in which the immature insect, sometimes called a naiad or nymph, respectively (Arnett, 2000), looks much like the adult without wings (or with the beginning growth of wings in the later instars). Many authorities now describe all immature insects as larvae. There is no pupal stage in those with gradual metamorphosis, and the last instar molts into the adult.

The term "instar" is sanctioned by the Entomological Society of America as a term to describe an immature insect (larva or nymph) between ecdyses (1st instar, 2nd instar, etc.). One should not say or write "2nd larval instar" because it is redundant. Instar has also been used in the literature to describe the duration of time between ecdyses, but Romoser and Stoffolano (1998) recommend use of the word "stadium" to describe the length of time spent between ecdyses.

The body of insects is divided into segments, and there is typically a clearly defined **head**, **thorax**, and **abdomen**. The head in arthropods has evolved from the fusion of a number of segments — from three to seven, depending on different authorities — but it superficially appears to be all one unit in most insects. Each of the primitive segments of the head probably bore appendages, and these have evolved into the antennae and mouthparts. A pair of antennae occurs on the head of adult insects and some immature ones. The Protura lack antennae. The antennae have evolved into a wide variety of shapes in different groups. They bear a variety of sensory structures, many of which are olfactory. Authorities generally agree that the primitive mouthparts were of the mandibulate type (Figure 1). Other types of mouthparts, such as piercing and sucking, are derived from mandibulate components. The mandibulate type consists of paired ventrolateral **mandibles** and **maxillae**, the (ventral) **labium**, the (dorsal) **labrum**, and the **hypopharynx**. The labium and labrum extend beyond the true mouth and form a pre-oral cavity. The hypopharynx is part of the ventral surface of the head capsule and it extends into the pre-oral cavity between the labium and labrum. The salivary gland duct empties into the pre-oral space between the hypopharynx and

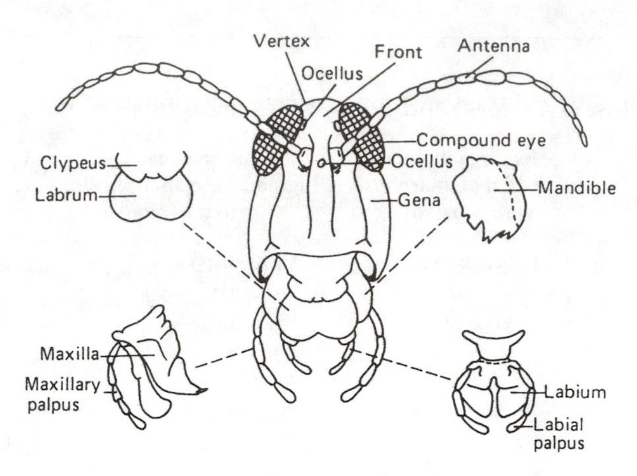

FIGURE 1 The head and mouthparts of a grasshopper, a typical insect with mandibulate mouthparts. (Reproduced with permission from Arnett, 2000.)

labium. The maxillae each bear nonsegmented lobes called the **galea** and **lacinia**, and a segmented appendage, the **maxillary palpus**. The labium is derived from the fusion of two primitive segments, and it also bears several nonsegmented lobes and a pair of segmented **labial palps**. The palps and various lobes of the mouthparts bear tactile and gustatory receptors.

Part of the success of insects can be attributed to the plasticity of their mouthparts and their evolution to support diverse food habits. Chewing mouthparts, often modified for specific trophic functions, are present in many orders of insects. Piercing and sucking mouthparts have evolved in the Hemiptera, adult Siphonaptera, and some Diptera; and sucking mouthparts occur in some Diptera, Hymenoptera, and Lepidoptera. The mouthparts may be reduced or vestigial in non-feeding adults, such as some Lepidoptera, and in some endoparasitic insects.

The **thorax** is divided longitudinally into three segments: the **prothorax**, **mesothorax**, and **metathorax.** In schematic form, each of the thoracic segments is like a box with a slightly rounded top, the **tergum**, that laps over considerably on the sides, two side plates (the paired **pleura**, singular **pleuron**); and a ventral plate, the **sternum**. Each thoracic segment bears a pair of segmented legs attached between the sternal and pleural plates. The mesothoracic and metathoracic segments of many adult insects bear paired wings attached by small wing sclerites (small pieces of cuticle that act like hinges) at the interface between the tergal and pleural plates. Some adult insects in the advanced orders in which wings are typically present are wingless, and this is considered to be a secondarily evolved condition from winged ancestors. Wings probably evolved only once in some early ancestor of winged insects.

Several factors (complex behavior, external and internal morphology, physiology and biochemistry, size, food habits, flight, exoskeleton, and metamorphosis) have contributed to the success of insects in becoming the most diverse and largest group of animals. In fact, just about anything one describes about insects must have contributed to their success. Because this is a book about the physiology and biochemistry of insects, those features are the ones that are stressed, but many morphological, behavioral, and genetic factors could also be cited.

One feature that certainly sets insects apart from other arthropods — and most other animals as well — is the evolution of wings. Flight has been a major factor in the success of insects, allowing rapid, wide dispersal; escape from enemies; and searching for mates, food, and habitats. When and how wings evolved is not clear, and authorities have proposed a number of theories. The evolution of wings is briefly discussed in Chapter 9, but for a more comprehensive discussion of the various theories, the reader can consult Gillott (1995).

Another major evolutionary step was the evolution of a pupal stage. This, too, has led to numerous theories (Gillott, 1995). One advantage of the pupal stage is that it allows the larval and adult forms to have very different food habits and to occupy different habitats, thus reducing intraspecific competition.

The adaptability of insects to changing environmental conditions over the millions of years since their appearance on Earth is a major contributor to their success. Adaptability must reside in genetic diversity and in the complex physiological systems that evolved in insects. Insects have a well-developed nervous system enabling complex individual behavior, and social behavior in ants, termites, some bees and wasps, and to a limited extent in some other groups. Virtually all insects have an extraordinary array of sensory receptors that enable them to gather information about their internal and external environments. An exoskeleton provides protection from the external environment, controls water loss from a body that has a very high surface-to-volume ratio, and provides for skeletal muscle attachments. The tracheal system, a system of air-filled tubes, arborizes like the human capillary system to virtually every cell in the body, and allows air to move through an air path to within a few micrometers of mitochondria. Consequently, insects nearly always respire aerobically, even during periods of prolonged flight. Thus, they get the maximum energy release from the breakdown of carbohydrates, and some groups can metabolize fatty acids for even greater amounts of energy during flight. The use of semiochemicals in communication and location of mates, food plants, and prey is well developed and seems to have reached an apex in moths, which fly at night and depend on olfaction to find food and mates. Sex pheromones play an important role in sexual isolation today, and probably has done so over millions of years. A great radiation of insects occurred during the evolution of flowering plants about 140 million years ago, and insects feed on nearly every type of plant. Their food habits and alimentary canal structure evolved together, so that diversity in food is reflected in the great diversity in gut structure. The excretory system, based on the principle that most small molecules enter the Malpighian tubules and are passed to the hindgut for selective reabsorption of physiologically useful molecules while those not reabsorbed are excreted with the feces, must have been very important in adapting to evolving food plants. Small size has in itself been a major factor in success. Insects live in many diverse microhabitats, and a small body requires relatively less food to grow and to sustain life. They have a very complex endocrine system involving steroid hormones, neurosecretions, neuromodulators, biogenic amines, second messengers, and possibly a unique hormone, the juvenile hormone. Hormones and neurosecretions regulate growth, metabolism, behavior, molting, metamorphosis, excretion, circulation, reproduction, and probably many processes yet to be discovered. Genetic diversity has enabled them to adapt to changing environmental and food conditions over several hundred millions of years.

THE EVOLUTION OF INSECTS

Insects are generally believed to have evolved from a line of ancient annelid-like ancestors, a line that probably also gave rise to the early onychophorans, fossils of which are well preserved in Cambrian marine deposits dating from more than 500 million years ago. Modern onychophorans have some characteristics of annelid worms and arthropods. Extant onychophorans comprise a group of about 65 species of caterpillar-like velvet worms in the genus *Peripatus* living in moist tropical habitats. Onychophorans have a segmented body, a pair of antennae, and from 14 to 43 pairs of short, unsegmented legs. They are similar to annelid worms in having a thin, permeable, flexible cuticle, a pair of nephridia (excretory organs) in each segment, and an annelid-like nervous system. Like insects, they have claws, an open circulatory system, a tracheal system, a hemocoel, insect-like mandibles, and salivary glands. Some authorities in the past considered onychophorans as a class in the Arthropoda, but they are often placed into the separate phylum Onychophora. At one time or another, authorities have proposed a single evolutionary line for the insects (monophyletic origin), diphyletic origins, and polyphyletic origins (Gillott, 1995). Based on the fossil record, the first insects evolved as wingless forms during the Devonian period of the Paleozoic era, about 400 million years ago. The early fossil insects looked much like some thysanurans do today.

Great radiation in the evolution of insects occurred in the Carboniferous period (360 million years ago, in the Paleozoic era) when Earth was dominated by large primitive vascular plants and later in this period by ferns and gymnosperms. A second great expansion of insects occurred during the Cretaceous period about 140 million years ago in the Mesozoic period when flowering plants were expanding and gymnosperms were declining. Co-evolving and adapting with flowering plants led to the success and expansion of several groups of insects, including beetles, bees, and plant-feeding Hemiptera.

CITED AND SELECTED REFERENCES

Arnett, R.H., Jr. 2000. *American Insects, A Handbook of the Insects of America North of Mexico*, 2nd ed., CRC Press, Boca Raton, FL.

Chapman, R.F. 1998. *The Insects, Structure and Function*, 4th ed., Cambridge University Press, New York.

Evans, H.E. 1984. *Insect Biology, A Textbook of Entomology*, Addison-Wesley, Reading, MA.

Gillott, C. 1995. *Entomology*, 2nd ed., Plenum Press, New York.

Gould, J.L., and W.T. Keeton. 1996. *Biological Science*, 6th ed., W.W. Norton & Co., New York, pp. 540, 680-691.

Romoser, W.S., and J.G. Stoffolano, Jr. 1998. *The Science of Entomology*, 4th ed., McGraw-Hill, Boston.

Solomon, E.P., L.R. Berg, D.M. Martin, and C. Villee. 1996. *Biology*, 4th ed., Saunders College Publishing, Orlando, FL, pp. 636-649.

Index

A

C

D

E

F

G

H

I

J

K

L

M

N

O

P

Q

R

S

T

U

V

W

X

Y

Z